Urban Ecology

Science of Cities

How does nature work in our human-created city, suburb, and exurb/periurb? Indeed how is ecology – including its urban water, soil, air, plant, and animal foundations – spatially entwined with this great human enterprise? And how can we improve urban areas for both nature and people?

Urban Ecology: Science of Cities explores the entire urban area: from streets, lawns, and parks to riversides, sewer systems, and industrial sites. The book presents models, patterns, and examples from hundreds of cities worldwide. Numerous illustrations enrich the presentation. Cities are analyzed, not as ecologically bad or good, but as places with concentrated rather than dispersed people. Urban ecology principles, traditionally adapted from natural-area ecology, now increasingly emerge from the distinctive features of cities. Spatial patterns and flows, linking organisms, built structures, and the physical environment highlight a treasure chest of useful principles.

This pioneering interdisciplinary book opens up frontiers of insight, as a valuable source and text for undergraduates, graduates, researchers, professionals and others with a thirst for solutions to growing urban problems.

Richard T. T. Forman is the PAES Professor of Landscape Ecology at Harvard University, where he teaches ecological courses in the Graduate School of Design and in Harvard College. His research and writing include landscape ecology, road ecology, urban ecology, land-use planning and conservation, the netway system, and linking science with spatial pattern to mesh nature and people on the land. His previous title, Urban Regions: Ecology and Planning Beyond the City, was published by Cambridge University Press in 2008.

Urban Ecology

Science of Cities

Richard T. T. Forman
Harvard University, USA

Shaftesbury Road, Cambridge CB2 8EA, United Kingdom

One Liberty Plaza, 20th Floor, New York, NY 10006, USA

477 Williamstown Road, Port Melbourne, VIC 3207, Australia

314–321, 3rd Floor, Plot 3, Splendor Forum, Jasola District Centre, New Delhi – 110025, India

103 Penang Road, #05–06/07, Visioncrest Commercial, Singapore 238467

Cambridge University Press is part of Cambridge University Press & Assessment, a department of the University of Cambridge.

We share the University's mission to contribute to society through the pursuit of education, learning and research at the highest international levels of excellence.

www.cambridge.org
Information on this title: www.cambridge.org/9780521188241

© Cambridge University Press & Assessment 2014

This publication is in copyright. Subject to statutory exception and to the provisions of relevant collective licensing agreements, no reproduction of any part may take place without the written permission of Cambridge University Press & Assessment.

First published 2014

A catalogue record for this publication is available from the British Library

Library of Congress Cataloging-in-Publication data
Forman, Richard T. T.
Urban ecology : science of cities / Richard T. T. Forman.
   pages cm
Includes bibliographical references and index.
ISBN 978-1-107-00700-0 (hardback) – ISBN 978-0-521-18824-1 (paperback)
1. Urban ecology (Biology)   I. Title.
QH541.5.C6F67 2014
577.5'6–dc23
2013032913

ISBN   978-1-107-00700-0   Hardback
ISBN   978-0-521-18824-1   Paperback

Cambridge University Press & Assessment has no responsibility for the persistence or accuracy of URLs for external or third-party internet websites referred to in this publication and does not guarantee that any content on such websites is, or will remain, accurate or appropriate.

Dedicated to Adrian William Forman

Contents

Foreword ix
Preface xii
Acknowledgments xiv

Part I – Framework

1. **Foundations** 1
   Urban ecology concept 2
   Routes to the present 4
   Urban attributes and ecological assays 11
   People and their activities 16
   Using urban ecology for society's solutions 26

2. **Spatial patterns and mosaics** 31
   Scale, human, and nature patterns 32
   Urban–rural gradient as spatial model 38
   Patch–corridor–matrix and other spatial models 44
   Neighborhood mosaics and their linkages 49
   Urban regions, metro areas, cities 58

3. **Flows, movements, change** 65
   The nature of flows and movements 65
   Flows around boundaries and mosaics 69
   Change-over-time principles 73
   Urbanization 76
   Rates and trajectories of ecological change 83

Part II – Ecological features

4. **Urban soil and chemicals** 91
   The essence of urban soil 91
   Key natural and human processes 97
   Soil texture and associated properties 101
   Life in the soil 105
   Urban soil chemicals 113
   The urban underground 119

5. **Urban air** 125
   Microclimate 126
   Energy and radiation 129
   Urban heat 133
   Diverse airflows 139
   Air pollutants and effects 143

6. **Urban water systems** 149
   Urban flows in the water cycle 150
   Groundwater 151
   Clean water supply 158
   Sewage and septic wastewater 160
   Stormwater and pollutants 170

7. **Urban water bodies** 175
   Urban wetlands and ponds 175
   Constructed basins, ponds, wetlands, biofilters 179
   Urban streams 182
   Urban rivers 186
   Flooding by river and stream 193
   Urban coastal zones 197

8. **Urban habitat, vegetation, plants** 205
   Urban vegetation and habitat 205
   Urban plants 208
   Urban plant biology 215
   Trees and shrubs 223
   Plant community structure and dynamics 230
   Plants and urban habitat fragments 238

9. **Urban wildlife** 241
   Species types 242
   Vertical structures, vegetation layers, and animals 255
   Spatial habitat patterns and animals 259
   Wildlife movement 264
   Changing urban wildlife and adaptation 268

Part III – Urban features

10 Human structures   275
  Railways   276
  Roads and associated features   279
  Hard surfaces and cracks   287
  House plots, gardens, lawns   292
  Buildings   306

11 Residential, commercial, industrial areas   314
  City residential areas   315
  Suburban and peri-urban/exurban residential areas   321
  City center   326
  Commercial sites dispersed across the urban area   332
  Industrial areas   336

12 Green spaces, corridors, systems   343
  Urban agriculture   344
  Parks   349
  Diverse large greenspaces   353
  Green corridors and networks   362
  Integrated urban greenspace system   366

Epilogue   372
Appendix A: Positive and negative attributes of an urban region   379
Appendix B: Equations   381
References   386
Index   444

Foreword

The term "urban revolution" was introduced by Gordon Childe in 1936 to highlight the powerful process of transforming agricultural societies to large complex urban centers. His model describes how communities, beginning around 9000 years ago, grew from tens or hundreds to thousands of people. In 3100 BC, Memphis, Egypt was the largest city in the world with over 30 000 residents. Today, the Cairo metropolitan area has over 17 million inhabitants and ranks 15th on the list of the world's largest cities. Mega-cities such as Tokyo, Seoul, Mexico City, and New York have over 20 million urban dwellers and continue to grow. The scale and complexity of the urbanization process originally depicted by Childe has little resemblance to what is happening today.

In the short history of humans on our planet, the number, population size, spatial extent, rate of growth, and degree of environmental impact of cities are unprecedented. Today, cities and towns face a myriad of formidable environmental challenges concerning food production, energy, water, waste management, and pollution, as well as social challenges in regard to jobs, poverty, and human health and wellbeing. I propose that as a result of the current rate and magnitude of urbanization around the globe, we are on the cusp of a new "urban revolution." The goal and rallying call of this revolution is "We Want Healthy, Liveable, Sustainable, and Resilient Cities and Towns."

Modern cities have developed primarily based on the best planning, engineering, architectural, and design standards of the day and have been driven by societal and economic requirements and constraints. This resulted in cities having largely been built and managed as distinct entities where people, buildings, roads, rails, nature, water, energy, and money were studied, planned, and managed separately in professional, academic, and administrative silos.

Over the past 25 years, human settlements have increasingly been regarded and treated as complex ecosystems. Ecosystems can be simply defined as specific places on Earth along with all the organisms that live there and the associated nutrient and energy flows. The ecosystem concept implies a complex system of interacting components with discernible feedbacks between components. Thus, the vegetation in a city park can influence energy use in adjacent buildings and the wellbeing of the residents and workers in the neighborhood. Ecosystem boundaries are not fixed but depend on the questions or problems being addressed. Therefore, an entire city can be viewed as an ecosystem or its smaller components such as lake ecosystems, woodland ecosystems, and residential community ecosystems can be legitimate units of study and management. Ecologists propose that a healthy ecosystem is one that is stable and sustainable while maintaining its organization and autonomy over time and its resilience to stress. Hence, a key tool to achieving the goals of this new "urban revolution" is the incorporation of ecological knowledge and principles into the management and creation of cities in order to develop healthy, liveable, sustainable, and resilient urban ecosystems.

In the 1980s, Richard Forman's ideas and research on landscapes started another revolution, in this case a "landscape ecology revolution" in the way we see, manage, develop, and use our world. His groundbreaking papers and books on landscape and road ecology changed my view of the world and I don't think I am alone. When I ride in a car, bus, or plane, or even when I look out the windows of very tall buildings, I no longer see only static views of vegetation, waterways, buildings, and roads; I now see a dynamic, multidimensional landscape powered by the actions of humans and ubiquitous ecological processes. Richard has provided us with the terminology, tools, and methods to describe and analyze the towns and cities in which we live and work; the farmlands that produce our food; the forest, lake, and mountain regions in which we take our vacations; and the remote regions of the world where humans rarely tread. His pioneering patch–corridor–matrix, and subsequent land mosaic model of landscape structure has passed the test of time and has been adopted throughout the world in order to achieve more positive environmental outcomes. While the use of an ecosystem perspective appropriately represents cities as complex adaptive systems and provides the tools to assess levels of sustainability and resilience, the adoption of Richard's land mosaic model provides the tools to create sustainable and resilient cities and towns.

Richard's seminal book *Land Mosaics* summarizes the ecology of heterogeneous landscapes and includes comprehensive discussions of how landscape structure and composition (i.e., the land mosaic) affect the flows of water, nutrients, animals, wind, and people. A recent search of Google Scholar revealed that this book has been cited in over 4000 publications. Thus, Richard's "landscape ecology revolution" has had far-reaching global effects that have influenced how ecological and social scientists conduct their research, how policy makers and land managers conserve plants and animals, and how planners, designers, and landscape architects create more sustainable human settlements.

The current worldwide interest in creating sustainable and resilient cities has resulted in an increasing call for locally relevant ecological information and principles to guide urban development and management. Unfortunately, there has been a mismatch between the questions that planners, designers, and decision-makers are asking urban ecologists, and the questions that urban ecologists are asking to advance the science of urban ecology. Planners, designers, and managers are asking questions that are relevant to their day-to-day decision-making such as: How much green space is necessary to reduce the impacts of climate change? What design and construction techniques can be put into practice to minimize energy consumption? How much connectivity is required in an urban landscape to support diverse plant and animal communities? How can we design cities to improve human wellbeing? In contrast, most urban ecologists are conducting basic research designed to attain a better understanding of the structure and function of urban ecosystems.

Over the past 25 years, urban ecologists have produced a large body of studies from cities around the world that provide important insights into how urbanization is affecting ecological and social patterns and processes. However, the results of these studies have proven to be somewhat lacking when called upon to address the pressing questions from practitioners. This is because they have primarily been focused on single cities or single organisms and have been primarily funded and designed to advance the basic science of urban ecology, rather than to address the applied research questions being asked by practitioners. To be fair, there are urban ecologists working in cities around the world, especially in Europe, who have been actively addressing applied research questions. Recently, there have been calls within the discipline of urban ecology to bridge the gap between basic and applied urban ecology research by increasing the interactions between scientists and practitioners, by adopting a comparative approach to the study of cities and towns, and by identifying more general principles regarding the effects of urbanization on ecological patterns and processes.

As a result of this current state of affairs, there has been no urban ecology textbook published to date for students, planners, designers, and policy makers interested in the practical aspects of creating healthy, livable, sustainable, and resilient cities and towns. Of course that is until I began reading the book in your hand. I am very impressed with the content and approach of this volume and feel it will no doubt make a significant contribution to the future development of the study and practice of the discipline of urban ecology. As I have written in a recent history of urban ecology, Richard approaches the study and practice of urban ecology from a different perspective than the mainstream academics in the field.

Richard has utilized his extensive ecological knowledge and experience, as well as his two decades of teaching suburban and urban ecology at Harvard University in the Graduate School of Design to bring the full force of this revolutionary landscape scale approach to the study, design, and management of cities and towns. He has carefully chosen the appropriate topics for chapters to provide his readers with both the basic principles of his unique landscape mosaic approach to urban ecology and practical examples of how they can be applied. As with Richard's previous books, the number of references he cites is indicative of the extent of his scholarship. I am certain that both students and professionals alike will find these references a valuable resource for years to come. The breadth of the subject matter and examples presented in the book no doubt came from the many scientists, landscape architects, planners, designers, engineers, and policy makers that Richard has worked with in cities around the world.

In conclusion, I would have to say that Richard Forman has had the unique education, university environment, and international experience to write the first comprehensive urban ecology text book that will guide the new urban revolution. This innovative book provides the foundation and inspiration for creating healthy, livable, sustainable, and resilient cities and towns in the future.

Mark J. McDonnell
Director, Australian Research Centre for Urban Ecology
Associate Professor, University of Melbourne

Preface

Most of us call urb our home. Today's giant urban areas grow upward in population, with a fast-march outward. Now urban footprints, the agricultural and natural lands needed to sustain us, more than cover the globe. This great urban enterprise thoroughly interacts with ecology, reflecting a yet more powerful force. Nature molds our urban world, from "natural disasters" and resource scarcities to treasured plants and wildlife around us. Even pollutant and waste accumulation, green marketing, and natural landscapes pictured on our walls and in our dreams highlight the ecological dimension. The clashing and collaborating of these two giants – urb and ecology – lead to this book.

My lens focuses on spatial pattern, how it molds and responds to flows/movements, and how they all change. The pattern and process of mosaics, now a centerpiece of ecology from which most other components nicely follow, are central. Indeed, spatial pattern emerges as an especially useful handle for planners, engineers, landscape architects, park managers, pollution experts, architects, transportation specialists, hydrologists, and more. All can easily use the principles of urban ecology to build more-promising futures.

Although other sciences underpin cities – chemistry, soil science, meteorology, microbiology, and more – ecological science fills the core. Indeed, ecology incorporates key features of these fields. Ecology as metaphor, marketing, sociology, or motherhood receives bare mention. Ecologists use the central concept of ecology – interactions among organisms and the environment – to study and understand the ecology of forests, lakes, populations, ecosystems, soils, whole landscapes, even regions. Tying this core theme to urban areas produces the highly useful concept of urban ecology used in this book:

> *Interactions of organisms, built structures, and the physical environment, where people are concentrated.*

Plants, animals, and microbes are the organisms, roads and buildings the predominant built structures. Soil, water, and air comprise the physical environment. Cities, towns, and adjacent built areas are the prime human concentrations.

Of course, many other interactions highlighted in sister disciplines, professions, and human activities are important in urban areas. For instance, public health, urban agriculture and bird watching emphasize human–organism interactions. Architecture, engineering, and construction link humans and built structures. Water supply, flood disasters and meteorology link humans and the physical environment. Sociology, employment, and retail shopping highlight human–human interactions. Adding "interactions with built structures" and "where people are concentrated" to the traditional core of ecology pinpoints urban ecology as a basic science, promising a diversity of highly useful applications. This intriguing subject offers challenge, discovery, and societal solutions now.

The perspective is global. Limiting our view to a single nation or region feels like a Mozart symphony with most of the orchestra missing. But using models, data, examples, and figures from a breadth of cities worldwide, we gain understanding, and discover great patterns for our own places. I have lived in urban regions of ten nations, and while writing, visited numerous US cities, five Chinese cities, five Spanish cities, four Brazilian cities, three English cities, two Mexican cities, Calgary, Dublin, Berlin and Paris. To feel my subject, I lived temporarily in a balconied old-narrow-street apartment at the heart of several million people, observing street trees change, birds in a courtyard, successional habitats, scores of industrial sites, soil under streets, plenty of parks, water flows, restaurant wastes, spatial patterns galore, even cracks underfoot. I became a keen city watcher.

For convenience, the general term "urban area" is used for all scales, from megalopolis and city to neighborhood and housing development. The chapters ahead include parks and other greenspaces, but ecologically explore essentially the entire urban area – streets, walls, lawns, industrial sites, sewer systems, artifact-rich soil, aerial components, roofs, commercial centers, parks, dumps, and much more. Also, rather than outlining the newest hypotheses, I emphasize patterns and processes with reasonable evidence and broad application. Together, these approaches open up frontiers of insight and provide dependability for users.

High winds, scorching sun, frigid nights, big floods, insect outbreaks, and the seasonal flower explosion periodically highlight the power of nature. Meanwhile, the day-to-day expression of urban nature – pleasant temperatures, pouring rain, tree shading, venerable trees, birds singing, flies flying, soil growing grass, stormwater running off, microbes decomposing, clouds moving over us – permeates the city. People and nature are thoroughly intertwined in cities.

Most urban residents like the nature around them. In contrast, most ecologists consider urban nature and ecological conditions to be severely degraded, bulging with bad contaminants, invasive weeds, waste sites, sewage overflows, traffic pollutants, pigeons, pests, and pathogens. While I cannot eradicate my own tiltings, cities and ecological conditions are inherently neither good nor bad. Rather than judging urban nature, I attempt to objectively analyze and portray the distinctive ecological dimensions. Occasionally, urban areas are compared with natural or agricultural landscapes mainly to enhance our understanding of urban patterns. The book is urbanocentric rather than natural-land-centric.

Urban ecology has roots in many related fields, and benefits from the obvious goal of improving conditions for people packed together. Two recent salutary trends have been especially important. In one, a few integrated studies, notably in and around Berlin, Baltimore, Phoenix, Melbourne, Seattle, and Sheffield/London, have combined multi-investigator, multi-disciplinary, and relatively long-term study. In the second, several edited books and an occasional authored one on urban ecology have appeared, each containing useful information and insights, and together suggesting a rich promising picture. Publications with strong applied dimensions provide additional perspectives.

The time has arrived to pull the science together in a coherent and comprehensive form, pinpointing synergies where pieces of the picture are juxtaposed. The pages ahead attempt to catalyze urban ecology as accessible and appealing to students who will carry the field to greater heights, to research scholars pushing exciting frontiers, and to professionals improving the built areas around us. Moreover, the informed public can gain "eurekas," while pursuing a better future in ever-more-crowded urban living space. Discovery and delight pop off the pages.

In 1992 I began teaching a Harvard course on urban and suburban ecology. From the outset we attempted to discover or develop an intellectual core of urban ecology. Landscape ecology emerged as a key integrative ingredient, along with fine-scale urban pattern, process and, change. Gradually that core coalesced. Urban ecology principles, traditionally adapted from natural-area ecology, now increasingly emerge from the distinctive features of cities.

Broadening the perspective from city to urban region, another key step, was catalyzed by an analysis-and-planning project for the Barcelona urban region (*Mosaico territorial para la region metropolitana de Barcelona*; Forman 2004b). That challenge convinced me, an ecologist, that urban regions are really important globally for natural systems and their human uses, and especially as cities expand outward in the years just ahead. So, to help jump-start our understanding of urban ecology, I analyzed the spatial ecological and human patterns in urban regions of 38 small-to-large cities worldwide (*Urban Regions: Ecology and Planning Beyond the City*; Forman, 2008). Using scores of spatial analyses, this effort highlighted patterns and principles, plus the importance of the ring-around-the-city to the city, and vice versa.

The book in your hand is the essential complement to the urban region work. In effect, peeling back our familiar human layer reveals the fundamental natural and built patterns of a city, how it works, and how it changes. Lots of lucid patterns and processes appear. The world of eternal flows, especially in urban networks, emerges. Plenty of principles based on these are articulated. Worldwide forms of repeated spatial patterns, such as road network, city center, building plot, even cracks in a surface, are compared ecologically. Lacunae and research frontiers galore are evident. Scores of solutions for human application are mentioned, but left for applied experts and professionals to develop and use.

Today urban regions are the place for most of us, "*Homo sapiens urbanus*," and for many more in the years just ahead. Cultural and natural resources within the region enrich us. It is our annual home range; over years we become familiar with and care about it. Our sense of place is increasingly the urban region. Here ecology is in our heart, and on our lips. The pages ahead move urban ecology to the forefront of our mind.

Acknowledgments

I am immensely grateful to the following colleagues and students for important contributions to this book: Anita Berrizbeitia, Michael W. Binford, Anthony J. Brazel, Stephan Brenneisen, Mark Brenner, Peter Del Tredici, Sarah W. Dickinson, Matthew Girard, Gary R. Hilderbrand, Michael C. Hooper, Stephanie E. Hurley, Jason J. Kolbe, Jonathan Losos, Mary E. Lydecker, Mark J. McDonnell, Steward T. A. Pickett, Peter G. Rowe, Hashim Sarkis, Hilary Swain, John C. Swallow, Jianguo (Jingle) Wu, my treasured colleagues in Barcelona, and the impressive teaching fellows in Harvard's ESPP Ecology and Land-Use Planning course. I salute the remarkable students in my classes who were encouraged to improve the world; we learned together, even shared eurekas.

I thank the Harvard University Graduate School of Design, the Harvard Forest, the Harvard University Center for the Environment, and the Universidad de Alcala-Madrid for milieus of ideas and reflection time. I also thank the Zofnass Program for Sustainable Infrastructure at Harvard University and Spiro Pollalis for funding support. It is a special pleasure to thank the talented editors of Cambridge University Press for this and two previous books, which were of highest quality and well marketed.

Once again it has been a joy to work with Taco I. Matthews, whose thoughtful approach, communication skill, and enormous talent created the superb artwork for this book.

I deeply appreciate Andrew F. Bennett, Jose Vicente de Lucio, Jessica M. Newman, Daniel Sperling, and Michael T. Wilson, who played especially important roles. Lawrence Buell and Barbara L. Forman have been catalytic forces throughout, helping me intellectually transform energized ponderings and paper scribblings into the sequence of ideas and discoveries ahead.

Cover photograph: Sao Paulo aerial view courtesy and with permission of Silvio Soares Macedo.

Part I Framework

Chapter 1

Foundations

In short, then, it takes the whole region to make the city.
Patrick Geddes, Cities in Evolution, *1914*

But for all our buildings and lights and roads, for all our signs and words, that human presence is only a thin film stretched over mystery. Let sunlight flame in a blade of grass, let night come on, let thunder roar and tornado whirl, let the earth quake, let muscles twitch, let mind curl about the least pebble or blossom or bird, and the true wildness of this place, of all places, reveals itself …
Scott Russell Sanders, Staying Put: Making Home in a Restless World, *1993*

Imagine a glorious day in your favorite city being energized at every turn. Sparkling clear air. Hardly any traffic. People alive, interesting. Appealing architecture and gardens (Figure 1.1). Amazing cultural events. Delightful diverse shopping. Food the best. Saw everything … relished it all.

Suddenly a friend appears, an ecologist. Comparing notes, she or he is equally enthusiastic. Luxuriant native street trees with lots of lichens. Clear water in the city pond. Bicycle routes and long walking routes busily used. Songbirds zipping along a shrubby tree strip between parks. No dog droppings. Wind blowing the smokestacks' noxious air out of the city. Green walls and balconies facing each other over streets. Restaurants with rat-proof dumpsters. Elongated grass-and-flower depressions for riverside floodwaters. The intriguing list goes on. Chuckling, together you have seen almost everything, yet seemingly in two different cities.

But what about the invisibles? The what? We saw "everything" but maybe we missed some important things. For instance, in this urban ecology no-one directly sees the sounds and vibrations around us (Figure 1.2). The smells and gases we breathe are invisible. Turbulent and streamline air flows hit us unseen. We do not see what happens in tree canopies over us, on the roofs of buildings, or in the atmosphere with organisms further overhead. Nor is the soil just under us visible, the deeper underground infrastructures, the fish movements in the river, the river bottom, or the far side of the city. Indeed, we cannot even see the multitude of microbes right around, on, and in us. We mainly miss the active organisms and processes at night. And we do not have the time or patience to see the really slow flows and slow changes eternally occurring around us.

Interesting, but are all those things important compared with what we see? Well, consider a few examples. Up in that atmosphere, pollen and seeds and even spiders with tiny parachutes move across the city, ozone smog forms, and sky radiation is generated. The tree canopies contain bird nests and numerous insects, and evapo-transpire water to the air. Traffic noise inhibits successful avian reproduction, while vibrations from traffic and trains compact soil. Soil itself is a cornucopia of roots, microbes, and soil animals, with water and oxygen flowing downward, and carbon dioxide upward. The underground urban infrastructure contains raceways for cockroaches and rats to reach buildings, as well as stormwater and sewage wastewater to enrich water bodies. Underwater fish are feeding, being eaten, even migrating. Harbor and river bottoms boast a rich interacting mix of sediment, worms, pollutants, carbon dioxide, even sometimes oxygen. At nighttime, migrating songbirds are hitting towers and skyscraper windows, cats are roaming, slugs are eating plants, nighthawks are catching insects, and garbage is being ravished. Slow flows and changes are also really hard to see – plants growing, termites chewing, water-table dropping, species diversity changing, plants adapting, pests becoming pesticide resistant, species ranges expanding, pipes rusting, wood foundations decaying, and sea-level rising. While the list of invisible organisms, interactions and processes could go on and on, do these examples seem important ecologically? And for society?

Foundations

Figure 1.1. Glimpsing a garden of nature in a city center. A wide range of planting designs, architectural forms, and urban patterns provides rich experience for people, and microhabitat diversity for species. Sevilla, Spain. From R. Forman photo.

Let's find a little restaurant with something to drink and explore this urban ecology a bit more. Maybe a book on the subject would highlight lots of invisibles and visibles, opening doors to insight and delight all around us. Indeed, these revelations could be foundations for making where we live much better.

Urban ecology concept

We have just become an urban species, *Homo sapiens* "*urbanus.*" Half the human population now lives in urban areas. The proportion grows, and the number of urbanites skyrockets. The next two billion people will all be urban, half joining today's urban poor. These newcomers will squeeze in now within a single generation. How welcoming is our land, our urban space, our planet?

Meanwhile two mammoth changes are engulfing us. First urbanization, the "urban tsunami," easily visible today, sweeps swiftly and powerfully across the land. Seemingly inexorable, yet not. And second, natural systems degrade – freshwater dries up, biodiversity plummets, climate changes, soil thins, and unpolluted places disappear. Two familiar drumbeats. We pick at the problems. Or simply shrug, and consider them too large, too complex to solve.

Addressing such trajectories requires understanding of natural areas, forestry areas, agricultural areas, and dry areas of the globe. Ecologists for over a century have analyzed and educated us about natural systems

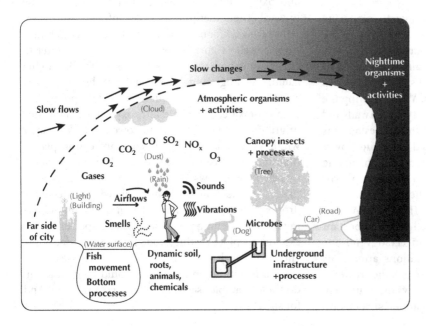

Figure 1.2. The invisibles and visibles for understanding an urban area. Invisible objects in bold type; visible objects in light type and parentheses.

there. Yet the overlooked ecology of built areas has now emerged as of core importance. Urban ecology is the ecology of right where we live.

Envisioning the subject at two spatial scales is a useful way to start. First, cities lie at the center of urban regions (Forman, 2008). In effect, an essentially all-built metropolitan area visible from outer space is surrounded by a ring-around-the-city. The metro-area and its urban-region ring are interdependent, that is, tied together by in-and-out flows and movements. Cities are no longer viable units, no longer make sense, whereas urban regions make good sense. An ecology of urban regions.

Second, urban areas are mosaics. The spatial pattern or arrangement of patches and corridors is extremely diverse and ecologically important (Forman, 1995, 2008; Wu, 2004; Pickett et al., 2009). Indeed, most people and most decisions focus on these finer-scale spots or areas within the urban region. Urban ecology highlights all the spaces, not only parks and other greenspaces, but also the rich variety of built spaces. An ecology of these spatial patterns, especially where most of us live in metropolitan areas, is the topic at hand. An ecology of urban mosaics.

To some, urban and ecology are contrasts (McIntyre et al., 2000), or even an oxymoron. Recent work by urban ecologists should dispel this perspective. The two concepts overlap and are quite compatible. Another familiar ecological perspective is that the urban or human component is "bad," that is, has a negative effect on nature or ecological conditions. No such assumption is made here. People can have both negative and positive effects on nature. Furthermore, nature has both negative and positive effects on people (Forman, 2010a).

Ecologists have focused on understanding "natural" patterns and processes, those minimally affected by humans, and thus have largely avoided urban areas. For example, of 6157 articles published during 1995–2000 in nine leading ecological journals, only 25 (i.e., 0.2%) dealt with cities (Benton-Short and Short, 2008). As seen in the previous section, the core of urban ecology must focus on, and understand, the central patterns and processes of urban areas.

Ecology is the study of organisms interacting with the environment. "Environment" here is overwhelmingly understood by ecologists to refer to the physical environment dominated by air, water, and soil (not the built environment of roads and buildings). With research mainly in relatively natural areas, "organisms" has normally meant plants, animals, and microbes (microorganisms).

Although humans are obviously organisms, ecologists have mainly excluded people in their research, or considered humans as an outside factor causing effects. A humans-as-inside-or-outside-of-an-ecosystem discussion is endless (McDonnell and Pickett, 1993; Alberti *et al.*, 2003; Head and Muir, 2006). Meanwhile lots of major disciplines, including economics, sociology, transportation, engineering, and architecture, all focused on human activities and including interactions with the environment, carry on. One could include humans as a key part of ecology, and then much of the field would be logically subdivided and dispersed into pieces within these other big human-centered disciplines. However, it seems wiser to maintain and further build on the core strength of ecology, with its basic focus on plants, animals, and microbes.

Sister disciplines and professions will welcome and use principles developed by a strong vibrant urban ecology. Tying these conceptual threads together leads nicely to the following urban ecology concept (Figure 1.3):

Urban ecology studies the interactions of organisms, built structures, and the physical environment, where people are concentrated.

Organisms refer to plants, animals, and microbes. Built structures are buildings, roads, and other human

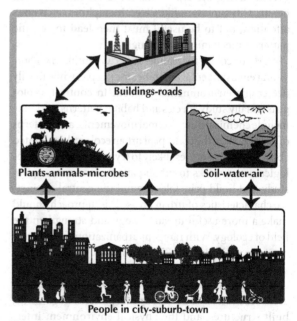

Figure 1.3. Urban ecology concept. Interactions of organisms, built structures, and the physical environment, where people are aggregated. Adapted from Forman (2010a).

constructions. The physical environment refers to air, water, and soil/earth. Where people are concentrated primarily refers to cities, suburbs, and towns.

Built structures are a key to urban ecology. Organism–environment interactions are simple ecology, whereas inserting buildings and roads in the interactions transforms the subject to urban ecology.

Urban ecology is useful for many allied fields focused on different interacting factors. Sociology highlights people–people interactions. Recreation and aesthetics commonly focus on people–organisms interactions. Architecture, housing, and transportation emphasize people–built structure interactions. Engineering and weather reports focus on people–environment interactions. Economics concentrates on people–environment–built structure interactions, while public health highlights people–organisms–built structure interactions. The distinctiveness of urban ecology promises much of use to each of these major human-centered fields.

Appealing metaphors and symbols often enhance understanding, though normally are not conceptual research frameworks (Grove and Burch, 1997). City as powerhouse. City as system of arteries. City teeming like an anthill. City as "second nature." City mimicking the human body. Urban development as natural process. Neighborhood change as ecological succession. City as living system or natural system. Ecosystem health. City functions like a tree. Metaphors catch one's attention, but to be useful must then lead to specific patterns, mechanisms or changes.

Urban ecology for planners typically emphasizes providing environmental amenities for people, while ideally decreasing environmental impacts. In contrast, ecologists usually study species and habitat patterns, and may include chemical flows, animal movements, and patterns of change. With the concept of urban ecology highlighted above, ecologists are less likely to try to fit traditional ecological frameworks to urban patterns, and more apt to study and build principles around the central distinctive characteristics of urban areas. This approach should make a more useful urban ecology and strengthen the field of ecology. With rampant urbanization, future ecology texts are likely to be 25% urban ecology.

Coalescence of the preceding themes leads to the intellectual core of urban ecology. Specifically, urban areas are mosaics of diverse spatial pattern. Organisms, built structures, and the physical environment interact. Flows and movements through the mosaic create a dynamic system. Urban areas markedly change over time. Urban ecology theory and principles lead to applications for society.

Routes to the present

Three intriguing and brief histories bring us up to date: (1) cities; (2) ecology and environment; and (3) urban ecology.

Cities and history

We begin with the population growth of cities. Then the key concepts and terms used to understand urban ecology are spelled out.

Changing city size

The first population centers that might be called cities appeared some 5000–6000 years ago in at least Mesopotamia (today's Iraq), Egypt, and the Indus Valley (today's Pakistan) (Pacione, 2005; Benton-Short and Short, 2008). Early cities also emerged in the Huang Ho Valley (today's China), Greece, Rome, and Maya land (today's Middle America). Damascus might be the oldest continuously inhabited city. Over time, the largest city worldwide has moved around, e.g., Constantinople (now Istanbul) with 700 000 people in 1700, to Peking with 1.1 million people in 1800, to Tokyo today (Berry, 1990). By the end of the 19th century the UK and Australia were largely urban nations.

In 1850 human history had produced 1 billion people on Earth (Platt, 2004), 10% of whom were urban. Two billion and 20% urban were reached 80 years later; 3 billion and 30% arrived 30 years after that. Each new billion people thereafter arrived in only 12–15 years. Today the 6.5+ billion people on Earth are half urban. More than 4% of the Earth's land surface is urban.

The next billion is coming fast. United Nations statistics point to an Earth in 2040 with 8.5+ billion, approaching two-thirds urban. Since the rural population worldwide is expected to remain essentially constant at 3 billion, the next 2 billion people will join the existing 3 billion urban people. Of the new urbanites, half will be middle-income and wealthy, perhaps mainly settling in suburban/exurban/peri-urban areas and near city center. The other half of the arrivals, 1 billion, is expected to double the population of urban poor to 2 billion. The rapidly growing informal- or squatter-settlement component of the urban poor particularly targets and covers urban greenspaces.

Where on Earth are we humans, and where will we soon be? In 1970, Asia had 37% of the world's population, rising to about half in 2005 (UN-Habitat, 2005). For Europe the comparable figures dropped noticeably from 31% to 19%. Northern America (USA and Canada) figures also dropped. Latin America (and the Caribbean) changed little. Africa increased from 6% to 10% of the world's people. Overall these relative growth rates are expected to continue for the upcoming few decades.

Today Europe, Northern America, and Latin America each have about 75% of their population urban. In sharp contrast, Asia and Africa are each about 40% urban. Megacities are commonly highlighted as centers, powerhouses and hubs. However, small cities are much more numerous, remain widely distributed across the land, and provide quite different human benefits and ecological characteristics.

The size of population centers is yet more informative. For instance in the USA, nine cities have >1 million people, 52 have 250K (250 000) to 1 million, 172 have 100K–250K, 363 have 50–100K, 644 have 25–50K, 1435 have 10–25K, and 16 772 population centers have fewer than 10 000 people each (Platt, 2004). Also, a rather constant 15% (±2%) of the total 300 million population live in each community-size category (see equations, Appendix B). Thus, virtually the same number of residents lives in large cities as in tiny communities.

The 22 *megacities* worldwide with >10 million population are currently most abundant in Asian regions (cities listed by population size in each region) (UN Population Division, 2007; Benton-Short and Short, 2008): East Asia area (Tokyo, Shanghai, Osaka, Beijing, Seoul); South Asia (Mumbai, Delhi, Kolkata, Dhaka, Karachi); Southeast Asia (Jakarta, Manila); Latin America (Mexico City, Sao Paulo, Buenos Aires, Rio de Janeiro); North America (New York, Los Angeles); Europe/Russia (London, Moscow); and Africa (Cairo, Lagos). Still, more than half of the world's urban people live in cities of less than 500 000.

Plato in the 4th century BC said that when a city reached 50 000 people, that is enough … a new city should be founded. A few decades ago a leading urban planner suggested 25 000 to 250 000 as the optimum city size (Lynch, 1981). Cities commonly grow for long periods and decrease for short periods (e.g., Leipzig, Detroit, Baltimore, Sao Paulo, Mexico City). Recently some megacities, after rapid population growth, have grown little (Newman and Jennings, 2008). This may be a temporary pause or may reflect some limit to concentrated population growth. Are there limits to city population size?

Some scholars have suggested that travel time might provide an answer. The "Marchetti constant" of an average of approximately one hour of travel per person per day seems to apply in many cities of different types and sizes (Kenworthy and Laube, 2001). Travel time has widely shaped the size and form of cities, so many remain "one hour wide" (Newman and Jennings, 2008). People can get to most places they need by transit or traffic in less than a half hour. Thus, high-density cities can become larger without being "dysfunctional," whereas low-density cities reach the apparent travel-time threshold at a smaller population size. The density of buildings or people in an urban area (Theobold, 2004; Pacione, 2005) is of particular ecological importance, both for the area and its surroundings.

The size of land-use units or districts within a city also affects city size. Indeed, "mixed-use patterns," rather than large separate residential, industrial, and shopping areas, reduce transportation time and cost. Planning that arranges people's primary needs in proximity reduces the travel-time limitation on city population size.

Bioregional limits constrain city size as well (Newman and Jennings, 2008). Thus, local water supply, food, energy, and materials from the ring-around-the-city are cost effective and reduce dependence on imports. The carrying capacity idea of a city in balance with the resources of its urban region is an especially appealing goal (Mumford, 1961; Rees and Wackernagel, 1996, 2008; Forman, 2008). Reducing consumption, waste production, and air and water pollution should also affect city population size.

As the major concentrations of human residential, commercial and industrial activity, cities inherently are the primary users of energy and residential water, and the primary emitters of greenhouse gas. One may ask whether all cities today are in "ecosystem decline." That is, has the human use of environmental resources exceeded the environmental carrying capacity everywhere? Have ecological footprints outstripped the land, so we need more than one Earth's surface to sustain today's human population (UN Population Division, 2007)? Are any cities living effectively within environmental limits? If so, we should learn from them.

Many cities have a published natural history describing especially the key greenspaces, plants, and animals present (Kieran, 1959; Houck and Cody, 2000; Forrest and Konijnendijk, 2005; Wein, 2006). Yet apparently the "history of a city's nature" is rare. Thus, for Boston's four centuries, the dramatic changes in greenspaces, water bodies, wildlife, bird populations, and much more are elucidated (Mitchell, 2008). This provides a much-needed complement to the familiar sequence of military, economic, social and other human changes constituting most histories. Indeed, Boston's natural history also

Foundations

highlights the changing efforts and successes in protecting, even enhancing, natural conditions over time.

Key concepts and terms for urban areas

As for all major subjects, a few key terms are particularly useful in understanding cities and urban areas (Forman, 2008). To sense the problem of choosing terms useful worldwide, consider some common land-use terms in particular countries: tip (UK), biotope (Germany), rodeo (USA), bush (Australia), rink (Canada), fengshui (China), favela (Brazil), shrine (Japan), allotment (South Africa), and polder (The Netherlands). In the UK, a city with a cathedral may be an urban area, and sprawl refers to unregulated (by government) rather than low-density spread of housing.

As much as practical, concepts and terms in this book are used for clarity and applicability worldwide. "Urban" pertains or relates to city. I extend the concept slightly in using the general term, *urban area*, referring to city- or town-related spaces where people and buildings are concentrated (*Webster's College Dictionary*, 1991; Hartshorn, 1992; World Resources Institute, 1996; Hardoy *et al.*, 2004; UN-Habitat, 2006). Thus, urban area applies broadly at different scales to, for instance, megalopolis, urban region, city, suburb, neighborhood, or housing development.

Specifically, *megalopolis* refers to a group of adjoining urban regions (such as Washington-Baltimore-Wilmington-Philadelphia-New York-Hartford-Boston or Amsterdam-Utrecht-The Hague-Rotterdam) (Hanes, 1993). *Urban region* is the area of active interactions between a city and its surroundings (e.g., the 80-km-radius irrigated-rice floodplain encompassing Bangkok, or Philadelphia and its surrounding farmland areas now squeezed by New York, Wilmington, and other regions) (Figure 1.4). *Metro area* (metropolitan area) is the continuous essentially all-built area of a city and its adjacent suburbs (e.g., as seen in a satellite image). [Note that this spatial concept applies in all regions, and avoids the USA sprawl-and-car concept of a "commuter shed" (Hartshorn, 1992); in most nations working in a city also means living in or adjacent to it]. A *city* is a relatively

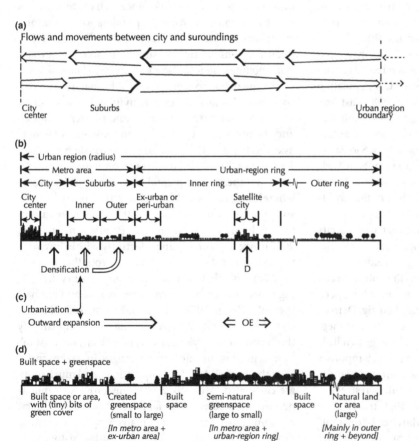

Figure 1.4. Concepts and terms for urban ecology. Metro area extends outward to the edge of the essentially continuous all-built area. Some suburbs extend beyond the metro area, and include some or all of the exurban or peri-urban zone. The urban-region ring also contains separate towns and villages. (a) Width of arrows roughly proportional to amount of flows and movements. (b) Concepts and terminology used in this book. (c) The two components of urbanization in different areas [see (b)] of the urban region. (d) Examples for bits of green cover (in built space) = window box, street trees, back yard space, green roof; examples of created greenspaces = city plaza/square, cemetery, vacant lot, dump, waterworks, golf course. Based on Forman (2008) and other sources.

large or important municipality or population center (*Webster's College Dictionary*, 1991; Hardoy et al., 2004; UN-Habitat, 2006). A *suburb* is a mostly residential municipality or town close to a city, and may be within, partially within, or outside the metro area.

Three terms describe the area outside the all-built metro area (Forman, 2008). The *urban-region ring* refers to the area between the metro-area border and the urban-region boundary. A usually narrow zone adjacent to or close to the metro-area border, typically characterized by some recent housing developments and other evidence of urbanization, is perhaps best called *exurban*. If the narrow zone contains considerable agriculture, it can be called either *peri-urban* or exurban, whereas if natural land predominates, exurban seems to be the more appropriate term (Theobold, 2004; Pacione, 2005; Vince et al., 2005; McGregor et al., 2006; Tacoli, 2006; Maconachie, 2007).

Urbanization is a land-change process of densification and/or outward expansion (Figure 1.4c) (Pacione, 2005; Forman, 2008). *Densification* refers to increasing the density of people or buildings (e.g., by changing low-rise to high-rise housing or the conversion/loss of greenspace to buildings, as in Portland, Oregon, USA). *Outward expansion* refers to city-related development beyond the metro-area border, effectively a suburbanization process. The expansion may occur in many different ways, including development along transportation corridors (e.g., Grenoble, France), by bulges around the metro-area border (history of London), or by dispersed sprawl [e.g., Las Vegas (USA) and Chicago] (Forman, 2008). Consistent with the dictionary concept, *sprawl* is the process of distributing built structures in an unsatisfactory ("awkward") spread-out (rather than compact) pattern. Compact-nucleus expansion, illustrated by concentrated growth on the edges of many European towns, is an urbanization alternative to sprawl.

An urban area is basically covered by two components, built spaces or areas and greenspaces. Both are exceedingly diverse and important. *Greenspaces* are mainly covered by plants and, though publicly or privately owned, are large enough to be public greenspace. *Built spaces* or areas are mainly covered with human-made structures, but commonly contain small areas of plant cover (Figure 1.4d). Thus, greenspaces include golf courses and most abandoned sand/gravel-extraction sites and capped dumps, while built spaces include most housing developments, active dumps, and parking lots.

The familiar general term, "land use," is used as equivalent to the slightly more-technical term, "land cover," which refers to the area where a specific type of greenspace or built space predominates (Breuste, 2009). Thus, a particular *land use* is considered to be a single land cover with one or more uses or roles.

Three types of "environment" may be recognized (Hardoy et al., 2004): natural environment (dominated by organisms, and with little human influence); physical environment (air, soil, or water characteristics predominant, with little role of organisms); and built environment (area dominated by buildings or other human artifacts). *Nature* refers to what humans have not made or strongly altered (*Webster's College Dictionary*, 1991; Forman, 2008). *Natural system* focuses on the structure and functioning of nature (dominated by air, soil, water, plants, animals, and/or microbes), and *ecosystem* highlights a natural system where organisms play central roles (in contrast to groundwater, earthen fill, and atmospheric systems). *Habitat* refers to a relatively distinct area and its environmental conditions, where an organism or group of organisms mainly lives.

Natural habitats or systems are found in four easily recognized forms: built area; created greenspace; semi-natural greenspace; and natural area (Figure 1.5) (Forman, 2008, 2010a). As suggested above, a *built area* contains continuous closely spaced buildings typically with roads and other human structures present, as in various residential and commercial areas. Within the built area, tiny spaces covered with plants are described as *green cover*, as for example a grassy entranceway to a building, street trees, backyard of a house plot, or a tiny unmaintained weedy patch. A *created greenspace* is a small or large area mainly covered by plants that was formed by, or is intensively used or maintained by, people, such as a grass-tree city park, golf course, or farmland. A *semi-natural area* is a large or small space resembling a natural ecosystem but significantly altered or degraded by people, sometimes with created unbuilt spaces intermixed, such as a woodland park (Figure 1.5), or greenway, or wetland for pollutant treatment (Haber, 1990; Millard, 2008; Cilliers and Siebert, 2011). A *natural area* is unplanted and without intensive human management or use, such as a relatively large marsh, forest, or shrub area with little human usage, usually in the outer urban-region ring (Peterken, 1996; van Bohemen, 2005; Kowarik and Korner, 2005). These four categories, from built to natural, represent a broad sequence of ecological degradation by human activities, which significantly alter

7

Foundations

Figure 1.5. Semi-natural oak woods (*Quercus*) in a city park. Rather dense canopy, understory, shrub layer, and herbaceous layer suggest natural conditions, whereas the paths, bench, constructed pond, and overflowing trash bin indicate a significant human impact. London. R. Forman photo courtesy of Jessica Newman.

natural vertical structure, horizontal pattern, and/or flows and movements.

Ecology, environment, and history

Barely a century and a half old, ecology as a discipline catapulted to the front line for society in the 1960s when an "environmental crisis" was suddenly recognized (Carson, 1962; Bartuska and Young, 1994; McNeill, 2000). Ecology was highlighted as a core subject for both understanding and solution. Quickly it became a familiar word in kitchens, drinking places, and diplomats' conferences.

Ecology appeared in the 1860s in Germany, and by the 1890s was a recognized scientific discipline in Europe, tying together animal and plant ecology plus freshwater and marine biology (Worster, 1977; McIntosh, 1985; Forman, 2010a). In the USA the field emerged in the Midwest about 1900, focusing on ecological succession. Professional societies and journals were founded in 1912–15, and modern ecology emerged in the 1940s–50s, highlighting ecosystem, theoretical, evolutionary, community and systems ecology. Many subspecialties have evolved over time, including the recent development of landscape ecology, conservation biology, and urban ecology. These diverse flavors of ecological science naturally have generated variations in defining ecology, e.g., in emphasizing vegetation, population dynamics, ecosystem flows, evolutionary adaptation, or interaction with the physical environment. Fortunately, despite these variations, ecologists of diverse types almost all ascribe to the traditional core concept of ecology, as the "study of interactions of organisms and the environment."

In a two-century history of society's "big ideas" – religion, science/rationalism, nationalism, hard-work-makes-land-productive, communism, and economic growth – the idea of environmentalism barely made a sound (McNeill, 2000). But it hit the headlines and became a household word in the 1960s–70s, associated with a whole set of issues – wetlands, wolves, foaming rivers, and choking air – and in the wake of Rachel Carson's 1962 book, *Silent Spring*. Environmental organizations, political parties, laws, regulations, and some visible successes rapidly followed in developed and certain developing nations. International conferences and treaties further spurred environmentalism into our consciousness.

Then suddenly in the 1990s–2000s, urbanization (especially sprawl) and global climate change further pushed environmentalism to the forefront, as one of the big ideas of history. The primarily scientific component of this, ecology, emerged as a core field for societal solutions. Within this, embryonic urban ecology is growing rapidly.

Not surprisingly, diverse subjects and terms have also appeared in the overlap areas of ecology and other fields. Consider environmental engineering, eco-criticism, social ecology, political ecology, environmental design in architecture, ecological/natural-resource/environmental economics, human ecology, global ecology, eco-city and ecopolis, sustainable development, road ecology, green infrastructure, industrial ecology, deep ecology, even green marketing (Park *et al.*, 1925; Ma, 1985; Costanza *et al.*, 1997a, 1997b; Roseland, 2001; Buell, 2001, 2005; Steiner, 2002; Forman *et al.*, 2003; Babbitt, 2005; van Bohemen, 2005; Allenby, 2006). Together such subjects represent hybrid vigor, the opening of frontiers of discovery and knowledge, and importance to society.

Eight major concept areas today describe the field of ecology (Smith, 1996; Cain *et al.*, 2011; Morin, 2011):

(1) physiological organism–environment ecology; (2) population growth and regulation; (3) competition and predation; (4) community/habitat and succession; (5) ecosystem and biogeochemical cycling; (6) freshwater ecology; (7) marine biology; and (8) landscape, regional, and global ecology. Professional journals, organizations, meetings, academic programs, courses, research grants, and research programs sustain these subfields and their growth. More than 25 000 people today identify themselves primarily as ecologists.

Urban ecology and history

Early phases

The roots and development of urban ecology are highlighted in two recent reviews (Sukopp, 2008; McDonnell, 2011). The term was used in the 1920s by a group of sociologists drawing analogies from the science of ecology (Park et al., 1925). However, urban ecology as a scientific discipline really emerged in the 1970s–80s (Stearns, 1970; Nix, 1972; Duvigneaud, 1974; Stearns and Montag, 1974). Thus, although overlaps exist, it is convenient to briefly consider the pre-1970 period and the post-1970 period.

Floras of urban botanical gardens, cemeteries, tree-planted spaces, and indeed of whole cities [e.g., in German cities, Montpellier (France), and Leningrad] were published from the 1500s on (Sukopp, 2002, 2008). Floras of castles, ruins, and urban areas (Rome, Paris, Palestine, London) appeared from the 1600s on. Urban plant migration studies, especially by a Swiss botanist A. Thellung in the 1910s, were published from the 1700s on (Schroeder, 1969; Pysek, 1995b). Urban bird and mammal studies appear from the 1800s on, though studies of animals of economic importance appear >1000 years ago (Gilbert, 1991; Owen, 1991; Klausnitzer, 1993; Erz and Klausnitzer, 1998; Sukopp, 2008). Urban vegetation studies appeared from the 1950s on [Berlin, Prague, Brno (Czech Republic)] (Murcina, 1990; Pysek, 1993; Sukopp, 2002). Urban environmental conditions relative to ecology are also important in urban ecology, including studies of phenology from the 1700s, and of microclimate, soils, and air pollution from the 1800s (Sukopp, 2008).

Ecological studies of World War II bombed sites and rubble surfaces from the 1940s on highlighted flora, fauna, and vegetation dynamics (Salisbury, 1943; Pfeiffer, 1957; Gilbert, 1992), and represent an important step underlying modern urban ecology. Newer studies of whole cities from the 1940s–50s on emphasized distinctive urban sites, plant communities, and changing species composition (e.g., London, Paris, New York, Vienna) (Sukopp, 2008).

Ecosystem studies of urban areas, in some cases by teams of researchers focusing on flows of nutrients and materials, appeared in the 1970s–80s (e.g., Brussels, Berlin, Hong Kong) (Nix, 1972; Duvigneaud, 1974; Stearns and Montag, 1974; Boyden et al., 1981; Sukopp, 1990). A focus on urban trees also appeared from the 1970s on (Grey and Deneke, 1992; Rowntree, 1986). Vegetation and the ecosystem concept were linked in Tokyo (Numata, 1982). A particularly nice balance and synthesis of urban microclimate, soil, water, plants, vegetation, and animals was published mainly for UK cities (Gilbert, 1991).

Also since at least the 1970s, urban nature has been scientifically linked with human health, welfare, and culture, highlighting a human ecology dimension (e.g., Tokyo, Hong Kong) (Boyden et al., 1981; Numata, 1998). Human ecology as a field linking urban planning and social patterns with ecological science has continued to evolve (Steiner and Nauser, 1993; Steiner, 2002).

The major linkage between the 1970s–80s urban ecology work and the current phase goes through Berlin and Central Europe, especially the work of H. Sukopp, P. Pysek, and later I. Kowarik (Sukopp et al., 1990, 1995; Pysek, 1993; Pysek et al., 2004; Breuste et al., 1998; Kowarik and Korner, 2005; Sukopp, 2008). An active researcher, editor of books, and catalyst for the field, especially in Northern and Central Europe, Sukopp highlighted the changing urban vegetation and flora, but welcomed contributions from diverse researchers, ecological fields and geographies. Vegetation or "biotope" mapping in cities was a foundation of this work (Schulte et al., 1993; Pysek, 1995a; Schulte and Sukopp, 2000). In 1995 Berlin and London were probably the best known major cities ecologically.

The current phase of urban ecology

Throughout both the early phase and the current phase, dispersed perceptive pioneers have contributed, and continue to contribute, an unending sequence of diverse insights and important results to our understanding of urban ecology. These individual scholars or small groups work in large and small cities, different geographic settings, and diverse cultures. Examples are: M. Soule et al. (1988), top predator effect on urban species diversity; A. von Stulpnagel et al. (1990), park size and air cooling; J. Owen (1991), ecology of a house plot or garden; M. Godde et al. (1995), urban habitats and plant/animal diversity; and, yes, R. Forman and D. Sperling (2011), netway system for reconnecting the land.

The current phase of urban ecology perhaps emerged in the late 1990s with the establishment of multidisciplinary, integrative and long-term studies of a few temperate-zone cities (Grimm et al., 2000; Wu, 2008): New York, Baltimore, Phoenix, Seattle, and Melbourne. Research in Sheffield/London is similar in scope. This work added impetus and integrated knowledge. It changed the field from the domain of dispersed individual scholars to the initial coalescence of an embryonic field.

Numerous edited books from the 1980s to 2010s have catalyzed the field and effectively sketched out its current and evolving core (McDonnell, 2011): Sukopp et al. (1990, 1995), Platt et al. (1994), Breuste et al. (1998), Konijnendijk et al. (2005), Kowarik and Korner (2005), Carreiro et al. (2008), Marzluff et al. (2008), McDonnell et al. (2009), Gaston (2010), Muller et al. (2010), Niemela et al. (2011) and Richter and Weiland (2012). Also key books on urban climatology (Gartland, 2008; Erell et al., 2011), soils (Craul. 1992, 1999; Brown et al., 2000), water (Baker, 2009), and geography (Hartshorn, 1992; Pacione, 2005) provide important components for urban ecology.

Five books with the benefits of single authorship offer valuable integration and depth in key areas. O. L. Gilbert (1991), as mentioned, highlights the basic ecological components of urban ecology, especially for UK cities. C. P. Wheater (1999) has a similar content but is less detailed and appeals to audiences beyond ecology. M. Alberti (2008) highlights concepts from ecology through the eyes of a planner, and provides many stimulating ideas. R. T. T. Forman (2008) highlights the urban region, within which a city functions, as a key viable unit for ecological analysis and planning. F. R. Adler and C. J. Tanner (Adler and Tanner, 2013) usefully apply some basic ecological concepts to the built environment. The book in your hand thus delves into urban areas from megalopolis to micro-site, developing ecological principles based on the urban characteristics.

The present book focuses squarely on the science of ecology and urban areas (Grimm et al., 2000; Pickett et al., 2001; Alberti et al., 2003; Niemela et al., 2009). Naturally this science is of considerable use and value to various human disciplines. For example, engineering, planning, and landscape architecture incorporate components into their fields, contribute to the understanding of urban ecology, and may have tailored definitions to their diverse fields (Geddes, 1914, 1925; Spirn, 1984; Deelstra, 1998; Beatley, 2000b; Pickett et al., 2001, 2013; Hough, 2004; Alberti, 2008; Forman, 2008; Nassauer and Opdam, 2008; Musacchio, 2009; Reed and Hilderbrand, 2012). Social science does as well (Pickett et al., 2001; Alberti, 2008; Muller et al., 2010; McDonnell, 2011). At a much earlier time sociology saw promising analogies with the then-emerging science of ecology (Park et al., 1925; Hawley, 1944; Catton and Dunlap, 1978), and subsequent thinking from this approach may have been retained in part in the broad field of human ecology (Steiner and Nauser, 1993). The role of social science, engineering, and other fields in urban ecology will of course remain dynamic. As in the evolution of landscape ecology (Zonneveld and Forman, 1990; Forman, 1995; Farina, 2006), an ecumenical approach without attempting to draw boundaries lets the highest quality and most valuable theory-and-application work simply define the core of a field, in this case urban ecology.

Today's major urban-ecology approaches and centers of research (Sukopp et al., 1990; Nilon and Pais, 1997; Breuste et al., 1998; Jenerette and Wu, 2001; Pickett et al., 2001; Luck and Wu, 2002; Grimm et al., 2003, 2008; van der Ree and McCarthy, 2005; Kowarik and Korner, 2005; Wu, 2008; Alberti, 2008; Forman, 2008; McDonnell et al., 2009; Lepczyk and Warren, 2012; Richter and Weiland, 2012) include: (1) habitat/biotope mapping and related analyses (especially in Berlin and Central Europe); (2) species types and richness (Berlin, Melbourne); (3) city-to-rural gradient (Melbourne, Baltimore); (4) modeling and biogeochemical/material flows (Phoenix, Seattle); (5) coupled biophysical-human systems (Phoenix, Baltimore, Seattle); and (6) urban-region spatial patterns, processes, and changes (worldwide analyses).

The concept of a city-to-rural (urban-to-rural) gradient has been an especially useful concept in catalyzing urban ecology research (McDonnell and Pickett, 1990; McDonnell et al., 1993; McDonnell and Hahs, 2008; McDonnell, 2011). Just as the 19th- and 20th-century, lichenologists and botanists studied lichens and other plants along lines from outside the city to city center (Le Blanc and Rao, 1973; Schmid, 1975), numerous ecological phenomena have now been studied and compared along such gradients worldwide. This approach is familiar and convenient for ecologists and is likely to continue, even as research increasingly turns to the more difficult, but especially valuable, two-dimensional studies of urban mosaics. Differentiating the ecology "in" and "of" cities (typically "in" = single component, small space, within a city; "of" = interdisciplinary and multi-scale) has helped spur broad-scale urban-ecology thinking and research (Grimm *et al.*, 2000; Pickett *et al.*, 2001; Alberti, 2008; Wu, 2008; McDonnell, 2011). Presumably, ongoing research, including multidisciplinary small-space as well as single-component multi-scale studies, will reduce the value of or need for such a separation.

Today the field is developing in two ways. First, dispersed scholars continue to publish research results from large and small cities in diverse regions and cultures worldwide. These studies crack open frontiers on an array of subjects the researchers perceive to be important. Together these provide both specific insights and broaden the field. Second, several research teams carry out relatively integrated research from a few cities on related subjects with a logical focus. These focused research studies deepen understanding in their domains, and together provide comparisons, linkages, and breadth. This dual and synergistic approach represents a strong model for developing urban ecology into the future.

Urban attributes and ecological assays

In the late 19th century, suppose the field of ecology had begun in cities, rather than in woods, shrublands, farmland, and ponds (Worster, 1977; McIntosh, 1985). Surely its central themes would have involved muddy roads and puddles, cobblestone streets, horses and dung, dust, coal-burning smoke, lots of dumps, streams/rivers for waste removal, connected backyards/garden areas, privies and human-wastewater, pigeons and house sparrows, rail facilities, rats, diseases, noise (Slabbekorn *et al.*, 2007), and burned buildings. An ecology of natural ecosystems might have evolved separately, or perhaps hand-in-hand with the ecology of cities. Irrespective, today urban ecology would be much further along.

Solidly rooted in natural areas, often with a limited or modest human imprint, how does an ecologist get started in a city? Buildings and roads are packed together. People and vehicles and an array of human-made objects cover the place. Flows and movements are channeled through networks. Flying animals and bits of green, from sidewalk mosses to whole parks, catch our attention. Where do species live in a city? Urban species, urban habitats, and their spatial arrangement. That is a good way to begin.

Distinctive attributes, hierarchical scales, and gradients

Consider a few widespread unusual or distinctive ecological characteristics. All are familiar to urban ecologists and even to the observant public.

1. *Habitats and species*
   - Usually diverse intermixed greenspaces and built patches cover the area.
   - Small sites tend to have few species, whereas large areas are often species rich.
   - Planted ornamentals, as well as spontaneous colonized species, are widespread.
   - Generalist species survive and predominate in urban conditions.

2. *Patches and areas*
   - Housing developments and house plots emphasize rectilinear repetition.
   - Boundaries are overwhelmingly straight, abrupt, and in high density.
   - Mowed grassy areas range from abundant to essentially absent.
   - Widespread impervious surfaces absorb solar radiation, generate heat, and greatly increase stormwater runoff.
   - Air and water are often heavily polluted.

3. *Corridors and flows*
   - Rectilinear road networks channel hordes of moving vehicles and people.
   - Underground branching conduits permeate and connect the place.
   - Animal movement is often along stepping stones rather than continuous strips.
   - Watercourses are channelized and flood-prone areas common.

4. *Change*
   - Many ecological changes are human-caused, rapid, and drastic.
   - Abundant species from afar endlessly arrive, while both native and non-native species appear and disappear.
   - The city expands directionally over suburbs, and suburbs over rural land.

For a natural or agricultural landscape, these patterns would be bizarre. In urban areas they predominate.

Most of the distinctive patterns are not even mentioned, and none is emphasized, in today's ecology texts.

Such patterns tend to fit together and suggest a novel model. In essence, buildings and roads, as well as people, vehicles, and diverse other artifacts at the core, strongly imprint and determine ecological patterns, processes, and changes of urban areas.

Before examining urban attributes in more detail, we should briefly step back to consider them in broader perspectives. First, a range of *spatial scales* characterizes any subject, including urban areas (Milne, 1988; Wiens, 1989; Pacione, 2005). For example, in urban ecology the following scales seem especially important: (1) megalopolis; (2) urban region; (3) metro area; (4) city; (5) major land-use type (residential, commercial, etc.); (6) neighborhood; (7) block; (8) building; and (9) micro-site (wall, roof, basement, and so forth). At any of these levels one could focus in on a different component, such as suburb instead of city, or tiny park instead of building. Furthermore, pinpointing the scale of a study is important, because a particular object, such as an animal population or soil condition, often differs at each level of scale.

This *hierarchy* of scales highlights another key dimension in understanding urban ecology. Typically the object or pattern of interest at a particular scale is strongly affected by characteristics at three scales (O'Neill et al., 1986; Forman, 2008). First, characteristics at the next broader or higher scale encompass and tend to control the object. Second, characteristics at the next finer or lower scale help control and explain the internal mechanisms or functions of the object. Third, other objects at the same scale interact competitively or collaboratively with the object considered. These three hierarchical interactions effectively mold the form of an object and determine how it functions.

The other broad perspective for urban attributes is spatial pattern, especially gradient and mosaic patterns (see Chapter 2). Measuring ecological attributes along a line or transect across a sequential *gradient* of different land uses or habitats is of prime interest here. Such studies provide insight into spatial arrangement and changing spatial pattern along a slice of the land.

Three gradients would seem to be of particular interest in urban ecology. The most familiar case is city-to-rural. This is effectively a radius from city center through inner suburb, outer suburb, exurban (or peri-urban) area, agricultural land, to natural land. One or more of the types may be absent in certain radii.

A second important gradient is vertical, e.g., extending sequentially from bedrock through the zone of underground infrastructure, surface fill or soil (with roots, microbes, and soil animals; Chapter 4), vegetation layers (herbaceous, shrub, understory, sub canopy and canopy) or building levels, and on upward through the atmosphere layers (Chapter 5). The vertical gradient thus integrates many dimensions of urban ecology.

A third gradient of ecological interest, perhaps as yet unstudied, is circular. A ring, for instance just outside the all-built metro area, might slice through and highlight a rich diversity of large patches and fine-scale heterogeneity, housing developments and farmland, and radial transportation strips of development. This circular gradient would normally portray dynamic rapid land-use changes.

Note that all three gradient types add considerable ecological insight if done in different locations. Radius gradients along different radii, vertical gradients in different land uses, and circular gradients at different locations between city center and urban-region boundary, in each case produce different results. Studying multiple gradients of a type pinpoints the average pattern and the variability present, and facilitates comparisons.

Urban objects of ecological study

The basic concept of urban ecology highlights organisms, built structures, and the physical environment in populated areas, so urban objects in each of these areas are illustrated below. Also we explore urban flows, movements, and changes.

Built structures, physical environment, and organisms

Built structures. Chicago's surface is dominated by tar (tarmac or asphalt) (21%), lawn (20%), buildings (17%), and cement (12%) (Nowak, 1994). A Chicago suburb is predominantly lawn (33%), herbaceous vegetation (20%), tar (12%), and buildings (8%). A study of suburban and urban woods around Wilmington (USA) found the following human impacts in order of abundance (Matlack, 1993): recent dumps, grass clippings, lawn extensions, building rubble, children's huts, and woodpiles. A study of European cities identified hundreds of distinctive urban forms – plazas/squares, intersections, areas around historical and religious structures, tiny parks, and so forth (Kostof, 1992).

A list of built objects only or mainly found in urban areas would be extensive. Consider skyscraper, intercity train station, art museum/concert hall, large city park, major sewage-treatment facility, subway system, group of high-rise residential buildings, major shopping center, dense street network, fine-scale soil heterogeneity, major port, convention center, and so forth.

Even the peri-urban or exurban area teems with built objects related to urban uses (Figure 1.6) (Hersperger et al., 2012). The objects in the figure are also interesting because in a geographic-information-system (GIS) study of land uses, they are largely eliminated by the computer. In the 47 km² area sampled around Zurich (Switzerland), 1012 objects of 80 types (excluding signs and structures along roads) were discovered. These human objects serve major roles for society, including 56% for recreation, 21% infrastructure, and 14% agriculture. All three roles serve the peri-urban people, but perhaps much more important is how the objects serve people of the big city, such as food products being transported inward and recreationists going outward.

Most of the objects are probably of little ecological importance (Figure 1.6). However, 26 of the object types seem to be of ecological importance, slightly over half positive and the rest negative. Clusters of farm buildings would be the prime negative objects, while bins

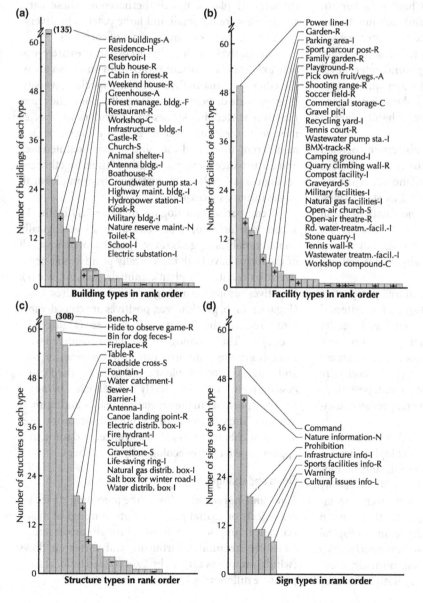

Figure 1.6. Human-made objects and their roles in a peri-urban area. Objects were identified by bicycling and walking in forty-seven 1 km² (0.39 mi²) plots in Canton Aargau, Switzerland (15% urban, 45% agricultural, 37% forest, 3% unproductive land). + = typically more ecological benefits or positives than negatives; − = more negatives than positives. General human uses: I = infrastructure; A = agriculture; H = housing; S = spiritual need; C = commercial; L = cultural; F = forestry; N = nature protection. (a) Buildings; (b) facilities; (c) structures; (d) signs. Adapted from Hersperger et al. (2012).

for dog feces and nature information signs are the most frequent ecologically positive objects. Since numerous objects of the 26 ecologically important types are present, it would be valuable to learn the cumulative impact as well as the spatial arrangement of ecological impacts. Analogous multitudes of human objects exist in suburbs and a city. A research frontier awaits urban ecologists.

Physical environment. A still-broader approach studies *habitats*. These are where organisms and species mainly live, and thus strongly represent the physical environmental conditions of soil, water, and air. Again the list is extensive – soil composition, fill, earth, hill, slope, and valley (see Chapter 4). For air, atmosphere level, composition, and air quality are characteristic (see Chapter 5). And for water, we see wetland, river, stream, pond, harbor, riverside, stream corridor, lakeshore, seacoast, groundwater level (water-table), water quantity, and water quality (see Chapters 6 and 7). The degree of habitat heterogeneity, i.e., how packed together different habitats are, is particularly informative.

Organisms. Organisms and objects associated with them also range widely in urban areas. Obvious organisms are trees, shrubs, birds, and soil animals. Organisms can also be characterized as street trees, ornamentals, lawn grass, pests, non-natives, wall vines, disease carrying, and so forth. Ecologists not only study these as individual organisms, but also as "populations" of individuals, especially their variations in abundance or density.

A natural or "ecological community" perspective often focuses on the species living together at a site and their interactions. Thus, *species richness* (i.e., diversity or number), *composition* (the actual species present), and *dominance* (or relative abundance) of species are major attributes to assay. Different *types of species* are of prime interest, including rare, dominant, pest, non-native, rapidly reproducing, herbivore, predator, dangerous, and keystone species.

Change over time in the ecological community or vegetation is *ecological succession*. Ecologists usually focus on changing vertical, horizontal, and/or biological structure (Ricklefs and Miller, 2000). Typically succession follows major disturbance, even vegetation clearing. Ecological processes, rather than human management or activity, predominate in ecological succession. In this book we refer to *successional habitats* as sites in an early stage of succession, normally characterized by abundant herbaceous vegetation that may contain shrubs plus tree seedlings and saplings (Adams et al., 2006).

Eleven contrasting forms or types of ecological-succession sites seem to characterize urban areas worldwide. These range from microsites, such as a window box or the base of a sign, to long discontinuous corridors along railways and highways (Figure 1.7). The prime causative mechanism differs for each of the successional forms. All 11 cases have little or no human maintenance, and manifest vegetation change in the early phase of ecological succession (before a site is essentially covered by a canopy of woody plants). Also all cases are dominated by "spontaneously" colonized (unplanted) plants, mostly herbaceous. These early successions are on small and large patches in diverse locations, as well as on many types of corridors.

Cumulatively such early succession sites are probably extremely important ecologically in urban areas. Together they maintain high plant species diversity. They provide dispersed sites for pollinators. They may function as stepping stones and corridors for species movement across urban areas. By beginning succession at different times and escaping human maintenance and disturbance for varied lengths of time, together the ecological succession sites sustain many examples of all stages in early succession.

A comparison (analogous to Table 7.1) of the 11 succession forms or types based on 28 ecologically related variables suggests several patterns of interest. All of the forms have both ecologically positive and negative roles (relative to plants, animals, soil, air, water). Negatives strongly outweigh positives in three cases (Figure 1.7a, b, g). However, positives are considerably more frequent than negatives in six cases (Figure 1.7c, d, e, f, j, k). The 28 ecological variables considered were roughly grouped into three categories: general; air, soil, and water; vegetation, plants, and animals. The 11 succession forms are most clearly differentiated by negatives and positives in the vegetation, plants, and animals category. Thus, overall, early ecological succession sites are major characteristics of urban areas, and some sites appear to be far better ecologically than others.

Flows and changes

Flows and movements. Most of the preceding attributes are objects or spatial patterns. Lots of things of major ecological importance move through urban areas. Consider streamline, turbulent, and vortex air flows (winds), as well as various breezes created by local temperature differences.

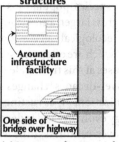
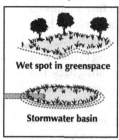

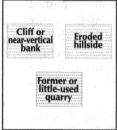
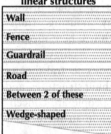
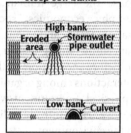
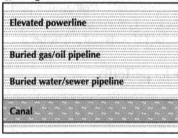
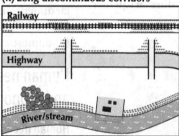

Figure 1.7. Eleven alternative forms of urban successional habitats. Such sites (variously labeled neglected, abandoned, fallow or ruderal) are largely covered with "spontaneous" herbaceous vegetation, often with scattered shrubs, even small trees, present. Based on Gilbert (1991), Godde et al. (1995), Wheater (1999), Boada and Capdevila (2000), Del Tredici (2010), and other sources.

Urban water flows are exceedingly diverse. Rainfall, stream/river flow, and floodwaters are readily seen. Stormwater drains and pipes carry rainwater and snowmelt away. But also a clean-water supply is typically piped into urban areas. Human wastewater may leave buildings in a pipe and flow to a sewage treatment facility or to a local water body. Wastewater may also leave a building and abruptly enter a small adjacent septic system for cleaning and distribution into the soil.

Organisms mainly move by locomotion in the urban environment. Birds and bats and insects fly overhead, while mammals, reptiles, amphibians and many invertebrates run, walk, or crawl about. Fish and some other aquatic organisms swim by. Other aquatic organisms are carried one-way by water flow, analogous to the movement of spores, pollen, and tiny organisms caused by wind. Many of the vertebrate species have several types of movement. Territorial defense, foraging for food in a home range, dispersing to mate and set up a new home range, and even migrating back and forth between locations over time are characteristic.

Change over time. Change occurs over different time scales. These may range from minutes/hours to day/night, week/weekend, seasonally, over years, over decades, over centuries, even over millennia. Some species or ecological systems may be relatively resistant to change, while others are resilient and rapidly bounce back after an alteration. With the intensity of changes in the urban environment, some species or habitats are neither sufficiently resistant nor resilient, so that change leads to a quite different ecological condition or state afterward.

Minutes/hours changes include earthquakes, tsunamis, sinkhole formation, and the flight of flocking birds to a different park. Diurnal changes occur in CO_2 evolution and plant evapo-transpiration, as well as animals either being active at daytime when people are, or at night with few people about. Week/weekend ecological studies barely exist, but the cycles of commuters and of weekend recreationists suggest that animals could readily respond, even adapt. The changing seasons are familiar, with weather, phenology, dormancy, and migration responses of organisms.

The over-years changes seem especially widespread in urban areas. Shops change, informal squatter settlements appear, termites degrade wooden foundations, species populations change, shrubs and small trees grow larger, and vacant lots appear and disappear. Early ecological succession is conspicuous on many sites, which are often transformed to buildings or other land uses at this time scale.

Over-decades changes describe a city's outward urbanization, pipes rusting and bursting, vehicle technology evolving, and indeed sea-level rise (Figure 1.8). The over-centuries time scale has been useful, for instance, to understand shipping and port patterns, vegetation types relative to land-use change in Sydney (Benson and Howell, 1990), as well as the arrival time of plants to cities in Europe (Sukopp, 1990). Cities come and go (Mumford, 1961). Very few cities have persisted over millennia.

People and their activities

With a concentrated human population, nearly everything people do in urban areas has ecological implications. The large human fields of urban history, geography, economics, sociology, housing, transportation, water, and public health will be noticeably enhanced by adding a strong urban-ecology dimension. Before moving on to the book's central themes of interactions among organisms, built structures, and the physical environment where people are concentrated, we briefly introduce diverse human interactions with the environment.

For illustration, we briefly highlight three dissimilar dimensions, each of considerable ecological importance: (1) human needs, social patterns, and economics; (2) domestic animals; and (3) pests and public health.

Human needs, social patterns, and economics

Human needs

The United Nations traditionally has highlighted four *basic human needs*: water, food, health, and shelter. A fifth – energy – is often added. Water means a relatively clean water supply; health means age-related reasonably good health; shelter usually refers to housing; and energy indicates fuel for cooking, and in cool climates, heating. All have major implications, positive and negative, for the ecological conditions of a city or urban area.

For most cities a clean water supply requires a reservoir(s) with surrounding land protection (Chapter 6)

Figure 1.8. Change in shipping over centuries plus big bridge and cars over decades. Philadelphia (foreground) and Camden, New Jersey (background). Historic cargo ship. Industrial and stormwater pollutants wash directly into the river water. R. Forman photo.

(Forman, 2008). Small cities using groundwater or streams typically have insufficient water quantity and poor water quality. Without adequate land protection around a reservoir, water quantity fluctuates and quality may require extensive chemical treatment. The water supply and its surroundings are key sources of biodiversity in an urban region.

Food for urban people may be imported from afar and depend on distant agricultural land plus long-distance transportation. On the other hand, urban agriculture produces food locally, thus avoiding both distant farmland and transportation (Chapter 12). Urban agriculture ranges from window-box tomatoes and herbs, to backyard vegetable gardens and market gardening (truck farming) near the city. All types of urban agriculture may affect biodiversity and species movement, as well as soil, water, and air conditions.

Energy may be locally generated usually in small amounts, such as from geo-thermal, tidal, wave, wind or solar sources, all of which have ecological effects. But the concentrated urban human population today normally depends on a major input of fossil fuel, i.e., oil, gas, and/or coal. Transportation by ship, pipeline or rail involves key urban land uses, as well as leaks and spills of ecological import. Pipeline and rail corridors into a city are doubtless major routes for species movement. Mostly within the metro area, the fossil fuel is combusted in power plants, industries and vehicles, with another range of ecological effects.

Several key urban features are not included as basic human needs. Transportation by train, bus, car, and bicycle is omitted. Mainly people can walk (supplemented by public transit), especially in a city where their needs could be met close by. Waste disposal, from garbage and rubbish to human wastewater, is missing from the list. Is that a basic human need? Certainly the urban ecology implications are large. Recreation, education, religion, aesthetics, private spaces, family, security, and employment, while often desirable, are not included. Quality of life is not listed as a basic human need. Nonetheless, essentially all of these fundamental human dimensions affect ecological conditions in urban areas, and vice versa.

Social patterns

Social patterns focus on groups of people, their interactions, and their organizational and spatial arrangements (Grove and Burch, 1997; Pacione, 2005; Berke *et al.*, 2006; Warren *et al.*, 2010; Redman, 2011). With thousands of people packed together, clearly this is a central component of urban areas. For convenience, we consider social conditions and interactions from three perspectives: (1) interactions with one or a few people; (2) interactions with a group; and (3) spaces for social interactions.

First, family activities occur both within a residence and in outdoor private space, if any, where gardening, lawn mowing, relaxing, and children playing are common. Beyond the residence, people walk, join friends for a meal, recreate, and engage in mate-finding. For all of these a sense of security is important. Where cars are abundant, many people spend considerable time driving, often alone, in traffic. The ecological dimensions of these personal activities are far-reaching and introduced in later chapters.

Second, people interact in groups of all sorts. Neighborhood activities, government or political activities, and meetings of varied social groups illustrate the breadth of group interactions. People join crowds on pilgrimages or attending concerts and sporting events. Shopping normally involves interacting with pedestrians and diverse shop personnel. Using public transport is a group-joining activity. Social interactions may be extensive in an emergency or disaster event.

Third, spaces in urban areas are planned and distributed to enhance different social interactions. Parks are favorite meeting places as well as relaxation and recreation spaces. Neighborhoods, from old wealthy areas to rich or poor residential areas, even informal squatter settlements, often have walkways, ball fields, and other amenities as meeting places. Neighbors may meet at the water fountain or well, the public toilet, or indeed the dump. Creating an effective arrangement of spaces that combines residential, shopping (commercial), and employment (e.g., industry) areas in proximity would tend to sustain a stable community. Stability over time normally increases social interactions and a sense of community.

A few activities illustrate the countless ecological dimensions of these social patterns. Gardening and lawn mowing may improve soil and water conditions, as well as enhance plant species diversity and wildlife movement. Or they may cause the opposite. Walking or using public transport, rather than driving, causes less aerial nitrogen-oxide emission, petroleum-related pollution reaching water bodies, and traffic blockage of wildlife movement routes. Distributing parks so that they are, say, a 10-minute walk from everyone's residence provides stepping stones for wildlife movement across the city. Combining homes, shopping and employment in mixed-use areas reduces transportation impacts. In essence, social patterns and urban ecology dimensions have manifold reciprocal interactions, both positive and negative.

Economics

Before introducing specific economic imprints and activities in an urban area, it is valuable to briefly consider key perspectives linking economics and ecology. These include growth economics, regulatory economics, ecological economics, and environmental economics.

Growth economics, which has especially developed since the 1940s (McNeill, 2000) is usefully highlighted in the phrase, "Let the free market determine it" (Forman, 2008). Three components are central. Human consumers are the main players. Preferences and tastes are the primary driving force. And the resource base is virtually limitless, since substitution or technology can overcome shortages. Success normally means sustained growth, avoiding too-high and too-low periods.

Regulatory economics, on the other hand, depends on government regulations, laws, other limitations, and sometimes planning for the future, to head off economic problems or crises (Jones, 2002; Perman *et al.*, 2003; Box, 2011). Short-term regulations may be reactive or proactive and open up opportunity, but they also intrude on the free market and business, sometimes suppressing innovation. Population growth and economic growth are related in complex ways (Ray, 1998; Rogers *et al.*, 2006). Irrespective, cities worldwide are mushrooming in population, and overall the outward expansion is worse ecologically than internal densification.

Ecological economics seems to have emerged as the combination of resource and environmental dimensions (Perman *et al.*, 2003; National Research Council, 2005b). Thus, natural resources represent input capital for an economy (Costanza *et al.*, 1997b; De Groot, 2006; Kareiva *et al.*, 2011). Meanwhile *environmental economics* focuses on the by-products of production and wastes of consumption, especially air and water pollutants. Many other important human effects on natural systems should also be highlighted.

Compared with the growth and regulatory approaches above, ecological economics has a noticeably different set of core attributes, including (Costanza *et al.*, 1997a; Jones, 2002; Forman, 2008): (1) humans are a part of the overall natural and human system; (2) preference and technology adapt to changing ecological conditions; (3) human intelligence can manage for a goal; (4) resources are finite; (5) both the long-term and short-term future are important; and (6) prudence based on uncertainty addresses resource constraints.

Economic models are traditionally weak in considering spatial arrangement and natural systems, both especially important in urban areas (Daily and Ellison, 2002; Burchell *et al.*, 2005; Forman, 2008). For instance, the core of environmental economics is the degradation by pollution of water bodies, soil, biodiversity, natural communities, wildlife, and even recreation, aesthetics and public health. Consider clean water, which is increasingly in short supply and expensive in urban regions. Surrounding a water source with natural vegetation is the best way to protect it both long-term and short-term. This natural vegetation is one of nature's services or ecosystem services (Daily, 1997; Daily and Ellison, 2002; Millennium Ecosystem Assessment, 2005; National Research Council, 2005b), or landscape services (Termorshuizon and Opdam, 2009), providing a huge value to the population. Wetlands that absorb stormwater and reduce flooding, nearby areas for recreation, vegetated stream corridors reducing erosion and sedimentation, and natural soils that absorb and break down pollutants are nature's services. These provide large, mainly non-market economic values.

Looking specifically within an urban area pinpoints its economic imprint and activity. Manufacturing and industry, a city-center central business district (CBD), office and business centers, centers of employment, and several types of shopping areas directly highlight economic activity. Also one sees transportation, housing of varied types, and construction sites as indirect measures of economic conditions. Wealthy, middle-income, poor, informal-squatter, ethnic, old, and new neighborhoods are quite separated or mixed, and are close to or far from employment. These spatial patterns tell much about both economic and ecological conditions.

Consider a few more ecological linkages with economic patterns. Old and new neighborhoods tend to have different species, densities, and sizes of trees (Schmid, 1975; Whitney, 1985). So do wealthy and poor neighborhoods. Soil erosion is often a problem in informal settlements, partly because of their location often on steep slopes and partly because of the density of people squeezed into one- and two-level housing.

Ecological finiteness is a special concern in urban areas where so much of the land is impervious surface. Many habitat types are rare. Indeed, only one lake, semi-natural woods, stunning natural view, rock outcrop, rare-species population, large wetland, stream corridor, or sequence of key stepping stones may be present. To prevent disappearance and consequent urban impoverishment, such rare ecological features require sustained protection. Economic activity in an urban setting of scarce resources can be appropriately integrated with ecological conditions for the long term.

Domestic animals

Companion animals or pets are packed into urban areas at high density and represent a key linkage between human and ecological conditions. Pets include fish, birds, lizards, turtles, snakes, frogs, invertebrates, hamsters, mice, monkeys and much more. However, dogs and cats are the predominant companion animals, and are most likely to interact with wildlife in urban areas. In addition to companionship and comfort, pets may provide security, education for children, and improved human health (Kellert, 2005).

Biophilia, the inherent human attraction to, linkage with, and dependence on nature, is well illustrated in urban areas (Kellert, 2005; Kellert *et al.*, 2008). Familiar examples are the more rapid recovery from illness, or the increased productivity and satisfaction of workers, when viewing vegetation rather than bare walls. But the biophilic benefits of nature right around us range from lower stress levels to enhanced mental development in children. Plants along hallways, outdoor green walls, tropical fish tanks in schools and corporate office spaces, and viewing cute or active animals all have biophilic roles for people. Yet vegetation also enhances environmental conditions, such as improving interior air quality, reducing airborne particulate pollution above streets, and providing insect food and nest sites for birds.

Farm animals, while sometimes pets in urban areas, are usually kept in small numbers for food. Chickens, rabbits, goats, cows, and even pigs may provide meat, eggs, or milk for families or local shops. Other values of urban farm animals include pelts or hides (e.g., rabbits, cows), garbage recycling (pigs), riding (horses), and spiritual or sacred dimensions (cows).

Domestic animals are concentrated in pet stores, animal shelters, varied farm-animal venues, animal pounds, kennels for temporary care, and veterinary facilities. These facilities scattered through the urban area deal with large amounts of animal food and animal waste. Pests from cockroaches to mice, rats and other animals are permanently attracted to such sites.

Presumably pesticides are widely used. Stormwater runoff and sewage wastewater are heavily dosed with nitrogen and other pollutants from these urban-animal facilities. Water bodies then receive the pollutants.

Finally domestic animals are subject to, and vectors for, numerous diseases. These include viruses, bacterial infections, and parasites (MacPherson *et al.*, 2000). In areas of unvaccinated dogs, wildlife are infected with more canine distemper and parvovirus (Fiorello *et al.*, 2006). In the next two sections we consider dogs and cats in more detail relative to urban ecology.

Dogs

While some dogs (*Canis familiaris*) remain indoors, most spend a portion of their day outdoors, either on a leash or free-ranging. Since urban dogs are typically fed by their owners, generally dog predation on wildlife is uncommon. "Feral" dogs and cats have no owners. Dog packs in urban areas are usually limited to locations with few people and near sources of garbage for food.

Consider the *home ranges*, the area covered in daily movements, of free-ranging dogs in a small city (Figure 1.9) (Berman and Dunbar, 1983). The average home range size is 3.5 hectares (8.6 acres) in this small city, though of course home ranges vary enormously according to the breed or size of dog. Thus, the beagle (0.01 ha) and spaniel (0.02 ha) essentially only use a single city block in their daily movements. In contrast, the home range for collie (5.4 ha) includes five city blocks and for afghan (7.9 ha) seven city blocks. Overall, urban home ranges tend to be small for such dogs because being fed at home means that foraging for food is essentially unnecessary.

Much of these residential home-range areas is composed of buildings and fenced-off yard areas. Thus, the effective or "accessible home range" for the dogs averages only 1.7 ha (4 acres). Accessible home ranges differ little from total home ranges for the smaller sedentary dogs, but differ considerably for the wider-ranging breeds. The dog home ranges are composed of a core area of dense movement, plus a usually larger travel area of occasional movement. Superimposing home ranges of the different dogs emphasizes that a travel area commonly extends into another dog's home range, though usually not into its core area where territorial defense might be an issue. In general these dogs do not gravitate toward the park (Figure 1.9, left side) where more wildlife diversity might be expected.

Foundations

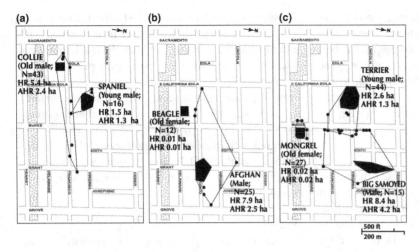

Figure 1.9. Home ranges of dogs in residential area of a small city. Based on 105 samples between 06.00 and 24.00 h from April through October in a 48-hectare (120-acre) middle-income area of Berkeley, California. The free-ranging dogs are without human supervision and have immediate unrestrained access to public property. Black = dense core area for a dog; remaining enclosed area = sparse travel area; speckled area = park; N = number of observations; HR = home range area based on connecting the outlying sightings; AHR = accessible home range, i.e., on streets, sidewalks, parks, and unfenced private property. No aggressive behavior and no dog pack of 2+ individuals observed. (1 ha = 2.5 acres.) Adapted from Berman and Dunbar (1983).

The sight, sound (barking), or smell of a dog may cause wildlife to respond with any of three *avoidance behaviors* (Liddle, 1997): (1) flushing, that is, stress and moving away; (2) changing to a new habitat; and (3) direct damaging contact. Flushing is usually a temporary movement and the wildlife may return after the dog has left. Flushing distances, such as between person and animal, are known for some wildlife such as herons and deer in natural areas, but seem to be poorly known for dogs in urban areas. Changing to a new habitat usually results from repeated flushings of an animal, and tends to reduce populations and species richness in the original habitat. Direct damaging contact, such as a dog catching a rabbit, eliminates only one animal, unless other animals witness the event and thereafter avoid the site.

Dog walking may cause all three wildlife behavioral responses. Thus, for visitors to an urban nature-protection area, dog walkers penetrated the furthest in, and off-leash dogs were considered to have the most intrusive impact on wildlife (Antos *et al.*, 2007). But the smell of dogs also affects wildlife. Canine scent-marking by urine and feces near trails may cause some wildlife to avoid the area (Miller and Hobbs, 2000). In the UK, certain urban tree bases regularly squirted by dog urine as "signal or communication posts" may be covered by a green alga, *Prasiola crispa*, presumably benefitting from the added nitrogen (Gilbert, 1991). Imagine the amount of dog urine and dog feces produced daily in a city or region (Matter and Daniels, 2000).

Direct mortality by dogs is perhaps best illustrated by the iconic yet scarce koala, which feeds on leaves atop tall eucalypt trees in Australia. In suburban areas, koalas must periodically cross the ground between trees, where the slow-moving animals are easily subject to dog-kill (Prevett, 1991).

The effectiveness of requiring dogs to be on leashes (via a dog leash law) in urban parks was evaluated for bird diversity in three major habitat types in Edmonton (Canada) (Forrest and St. Clair, 2006). In none of the habitats was there a significant effect of leashing dogs on avian species diversity, or on the diversity of probable-breeding species. However, dogs, especially unleashed, seem to have the greatest effect on mid-size and large mammals.

The home range of dogs that are occasionally leash-restrained by owners in a Queens, New York City study was reported to be about 2 ha (5 acres) (Rubin and Beck, 1982). Free-ranging dogs without restraints covered five times that area. However, free-ranging dogs at a dump with ample food had a home range of about 2 ha, the same as the occasionally restrained dogs that feed at home (Rubin and Beck, 1982).

Dog packs of typically 2–5 individuals remaining together are frequent where home feeding and leashed dog-walking are less. In a rural New Mexico (USA) community, feral dogs in packs had a home range averaging 0.14 km^2 (14 ha = 0.05 mi^2) when pups were small, expanding to 1.6 km^2 after pups were about 4 months old (Daniels and Bekoff, 1989). In the small city of Fort Collins (Colorado, USA), about 80% of the free-ranging dogs showed evidence of being owned, while 17% were in packs (Rubin and Beck, 1982; Lehner *et al.*, 1983). In the larger cities of Sacramento (California) and Baltimore, with a higher human population density, 33% and 49% of the dogs respectively were reported to be in packs. For Sacramento, in city center 30% of the dogs were in packs, and in the suburbs 21%.

Finally, >95% of the free-ranging dogs in an Isoso (Bolivia) area were carrying both the canine distemper virus and canine parvovirus (Fiorello et al., 2006). Since many opportunities existed for contact between the domestic dogs and wild carnivores, the disease risk for native wildlife is considerable. Water bodies in and downslope of urban areas may also be at risk of disease from dogs. Considerable amounts of dog feces are produced in an urban area where the dog density is 20 to 100+ times the density of native carnivores (Figure 1.10) (Matter and Daniels, 2000). Fecal coliform bacteria (E. coli) from dog waste throughout the city and suburb are stormwater-washed into drains and water bodies. Nitrogen from dog urine follows the same route into water bodies. Also, park areas dedicated for use by off-leash dogs may have a buildup of dog waste, and consequent impacts on nearby water (Lee et al., 2009).

Cats

Since being domesticated in the populated areas of Egypt some 4000 years ago, cats (Felis catus) have been prime human companions. In addition to being pets, cats retain their instinct to hunt and are often maintained to reduce mouse, and even rat, populations. Estimates of cat abundance range from >400 million worldwide to some 100 million in the USA alone (Marks and Duncan, 2009).

(a)

(b)

Figure 1.10. Urban dogs and cats. (a) Meeting place of companion dogs in a park; columns for the 52 Chinese peoples in background. Beijing. R. Forman photo. (b) Free-ranging urban cats by the Coliseum underground, where lions and tigers once prowled before shows. Rome. Photo courtesy and with permission of Brent C. Forman.

Relatively few cats remain indoors and hardly any outdoor cats function well on leashes. Thus, *free-ranging cats* outdoors are of primary ecological interest (Figure 1.10). Three types are recognized: (1) cats maintained or fed by owners; (2) stray cats recently lost or escaped from owners; and (3) feral cats. About two-thirds of the outdoor cats in the USA are considered to be owner-maintained, while the rest are free-ranging strays and feral cats.

The density of cats correlates much more with human density than with prey density (Sims et al., 2008). In an area of dispersed houses the cat density is some 20 to >100 times the density of native predators (Kays and DeWan, 2004). The ratio is doubtless much higher in city areas, although the density of scavengers such as rats and raccoons often far exceeds cat density (Jackson, 1951; Childs, 1986; Courchamp et al., 1999). Cats generally cannot control urban rat populations. Free-ranging cat density is higher in suburbs than in rural areas. Meanwhile, urban cat owners may have only half as many outdoor cats as owners in rural areas (Lepczyk and Warren, 2012). Also in suburbs cat density typically exceeds dog density (Campos et al., 2007).

Four ecological studies provide insight into the actual density of free-ranging cats in urban areas. A 4.2 km² urban area of Bristol (UK) had an average of 229 owner-maintained cats per km² (595/mi²) (Baker et al., 2005). Another UK study found 986 cats associated with 618 households, an average of 1.6 cats/household (Woods et al., 2003). An estimated 800 to 3100 free-ranging cats were found along 120 km (72 mi) of road in a Michigan (USA) suburb (Lepczyk et al., 2003). Finally, in a 33-ha (83-acre) city area in two neighborhoods of Brooklyn (New York), the density of owner-maintained free-ranging cats (pets) was 48/km² (125/mi²). Meanwhile, 134 feral cats per km² (348/mi²) were also present (Haspel and Calhoon, 1991). If the preceding studies are broadly representative, it appears that free-ranging cat density increases from rural to suburban to city area, and that feral cats may predominate outdoors in a large city.

Food is particularly important in determining how cats, as generalists, use urban space. Three types of food are available: (1) owner-provided food (cat food); (2) garbage; and (3) prey animals caught by cats. The total amount of available food in urban areas typically well exceeds what cats consume (Haspel and Calhoon, 1991). Owner-provided food is normally stable and dependable, so cats dependent on this have no need to forage, and thus have rather-small urban home ranges. Garbage supply, as found in house or building garbage containers, restaurants/hotels, grocery stores, recycling dumpster containers, and rubbish dumps (tips), is often also a dependable food source (Sims et al., 2008). Cat density around these sources is doubtless related to the amount of accessible garbage.

The third food source, prey animals, follows from domestic cats' instinct for hunting. Prey animals normally vary markedly in density and location, so hunting and predation mean much larger distances moved. Predation may especially occur along the edges of woods where two habitats adjoin and where vegetation and wildlife may be dense (Marks and Duncan, 2009). Although urban cats may catch prey, it is normally a tiny portion of the diet, because of the ready and widespread availability of garbage and owner-provided food.

In the Brooklyn study, the average home range was 1.8 ha (4.5 acres) for female cats, and 2.6 ha for males. Home-range sizes were essentially the same in two city neighborhoods. Feral cats especially are more active and travel longer distances at night (when fewer people are around in built areas) (Haspel and Calhoon, 1991; Barrat, 1997). Thus, a suburban study of ten cats found the daytime (diurnal) home range to average 2.7 ha (6.8 acres), whereas the nocturnal home range averaged 7.9 ha, and with considerable variation among cats (Barrat, 1997). Movements more than 100–200 m from home only occurred at night.

Feral cats concentrate where food resources and shelter are close together, and where few people are present (McCarthy et al., 2013)(Figure 1.10b). Without an owner's home for shelter, feral cats are vulnerable to daily and seasonal weather variations. In Rome, groups of female feral cats collectively occupy areas with the best shelter, which is used for rearing kittens. The females attack and chase away all males and almost all females (Natoli, 1985).

Predation by domestic cats on native wildlife has been reported to be significant on certain islands (Atkinson, 1989). The relative importance of cat predation remains uncertain in rural areas, where little natural vegetation and few animals of conservation interest survive. There human activities around homes and farm buildings alter the fauna, e.g., by attracting house mice and house sparrows and largely eliminating sensitive wildlife species. Consequently, almost all rural cat predation is of common species benefitting from humans.

In considering the roles of urban and suburban cats following, it is important to note that, although cats as hunters kill prey, little information is available evaluating whether the number killed significantly decreases a species population. Common urban animal species doubtless make babies much faster than cats kill animals, especially with ample garbage or cat food normally available. Nonetheless, the question of the effect of cat predation is important because the uncommon wildlife species in urban areas have small population sizes. Such populations could be significantly reduced by cat predation.

A comparative study of two parks in California is particularly interesting (Hawkins et al., 1999). One park had no cats, and the second park had 25 cats that were fed daily. The first park ended up with twice as many bird species, including two ground-nesting species absent from the second park. Trapping small mammals indicated that the first park had 80% native deer mice and 20% house mice, whereas the second park with cats had 79% house mice and 15% deer mice. Thus, a population of fed cats apparently lowered the abundance of native bird and rodent populations, and radically changed species composition.

Do cats kill more birds or more mammals? The following studies provide real data. In the urban Bristol study mentioned above, with an average of 229 pet cats per km^2, the predation rate based on animals brought home was calculated to be 21 prey killed per cat per year (Baker et al., 2005). Twice as many birds were killed as mammals. Bird predation was highest in spring and summer, probably reflecting the availability of young. For three species [house sparrow (*Passer domesticus*), dunnock (*Prunella modularis*), and robin (*Erithacus rubecula*), all among the most common birds present], the estimated predation rates were high relative to the estimated production of birds. This suggests that the study area could be a "dispersal sink," where birds lost to cat predation are replaced by immigrating juvenile birds, which in turn are at particular risk of predation.

The Michigan study above only studied predation on birds, and concluded that the 800–3100 cats present along 120 km of suburban roads killed 16 000 to 47 000 birds during the breeding season, or 1 bird killed/km/day. On average each cat killed 1 bird per week. Of the 23 bird species killed, two were considered as of "conservation concern."

Other studies conclude that small mammals are more abundant in the cat diet. The 986 cats in 618 UK households brought home (killed) during April through August were 69% mammals, 24% birds, 4% amphibians, 2% fish, 1% reptiles, and some invertebrates (Woods et al., 2003). These included 44+ species of wild birds, 20 wild mammal species, 4 reptiles, and 3 amphibian species. On average a cat brought home one animal every 2 weeks. Free-ranging urban cats in Reading (UK) brought home an average of 18 prey animals/yr (one every 3 weeks), with most of the predation by a few cats (only 20% had 24 prey/yr) (Thomas et al., 2012).

In a suburban and rural area in Hungary, small mammals were 61–82%, and birds 2–7%, of cat prey (Biro et al., 2005). A village study (Bedfordshire, UK) found a similar 64% of the cat prey to be small mammals, especially wood mice, field voles, and common shrews (Churcher and Lawton, 1989). Bird prey was only in the majority in winter when mammals were mostly underground. Young cats also chased frogs and butterflies, while old cats had little predation. House sparrows (*Passer domesticus*) were 16% of the prey, and it is hypothesized that cat predation may represent 1/3 to 1/2 of sparrow mortality. The authors suggest that cats may only bring home half of their prey, and that, despite higher previous estimates, about 100 killed prey per year is the maximum for a free-ranging mid-age cat.

In short, domestic cats as inherent hunters kill small mammals, birds, and other animals. In urban areas of diverse types, little evidence is available that predation rates have a significant effect on population sizes of prey, except at local spots (Sims et al., 2008). Nevertheless, the question needs study because most species of ecological and conservation interest are present in low numbers, and increasing their numbers and movement patterns may be considered an urban conservation goal. In essence, domestic cats overall kill lots of common small mammals and birds, but few animal species of conservation importance.

A few final observations on urban cats broaden the perspective in diverse directions. A suburban/rural study in Southeastern Brazil found that invertebrates were the prime prey of free-ranging cats, and that mammals represented 20%, and birds still less of the prey (Campos et al., 2007). Cats are well-known vectors for many animal and human diseases, including parasites, bacteria, and viruses (MacPherson et al., 2000). Little seems to be known about the effect of concentrated feces from so many outdoor cats in the limited greenspaces of urban areas. With an overabundance of food for cats, overall predation, competition, territorial behavior, and starvation typically remain at low levels in urban areas. Since hunting and hunger are apparently controlled by different areas of the brain, cat predation rates may not be affected by the availability of cat food at home (Warner, 1985). Finally, domestic cats introduced into Australia are generally considered to be significant ecological problems due to their predation on the distinctive native fauna. Of special concern is the edge portion of a natural area adjacent to homes, where many cats and many conservation-important animals may exist.

Pests and public health

Bags split, cockroaches lurk, pipes burst, sewage facilities overflow, and pathogens multiply. Unlike the preceding topic of cats and dogs, numerous research projects and books focus on pests and public health. Thus, the point of this brief section is to illustrate a range of public health issues, and how related ecological conditions affect people, as well as vice versa, in urban areas. But first, garbage trucks break down, house mice lurk, stormwater drains clog, cracks form, termite-filled foundations crumble, toilets leak, vehicles crash, floods happen, clouds of mosquitoes descend, pollen pours in, pollutants accumulate, pigeon flocks poop (defecate), dumpsters overflow, and rats lurk. These are essentially inexorable urban processes, all causing significant effects. All thoroughly link humans to urban ecology, i.e., to organisms, built structures, and the physical environment.

Pests

Pests are organisms that annoy humans. The common response is to apply pesticides, including insecticides, herbicides, fungicides, and many others depending on the type of pest. Because pesticide application is so frequent and has continued for so long, in effect many pest species have genetically adapted and are pesticide resistant (Robinson, 1996). In some rural and suburban areas, "integrated pest management" (IPM) is sometimes recommended or used (Carpenter, 1983; Robinson, 1996; Mizell and Hagen, 2000). This means that a battery of approaches is used to reduce pest numbers, such as by increasing species diversity, accelerating ecological succession, reducing or adding water and nutrients, altering the topsoil, adding competitors, adding predators, and, if appropriate, adding a limited amount of pesticide.

Plant pests tend to defoliate trees, including street trees (Tello et al., 2005). Ornamental plantings of flowers, shrubs, and trees throughout the urban area may suffer from plant pests, though pest resistance has often been selected for in advance for ornamentals (Mizell and Hagan, 2000; Benedikz et al., 2005). Vegetables and fruits are often damaged by plant pests as gardeners well know (Mougeot, 2005; Tallaki, 2005). Among the solutions that may work are growing adjacent plants known to repel pests such as rabbits. Pests of lawn grass include vertebrates, invertebrates, fungi, and bacteria (Mizell and Hagen, 2000).

Not surprisingly, insects are the most familiar and numerous urban pests (Ehler and Frankie, 1978; Hickin, 1985; Robinson, 1996). Cockroaches, termites, carpenter ants, mosquitoes, flies, midges, and the list goes on (Fowler, 1983; Wheater, 1999; Aluko and Husseneder, 2007). Insecticides have traditionally been the solution of choice. Insects, however, may readily adapt to insecticides, heat, and other urban conditions (Ehler and Frankie, 1978; Parmesan, 2006; Angilletta et al., 2007; Evans, 2010). Urban alternatives for limiting insects include planting certain plants, adding bat boxes, using natural substances, improving water drainage, adding fish, and other ecological approaches (Flanders, 1986; Raupp, 2012). Invertebrate pests also include slugs, spiders, and ticks.

Vertebrate pests often make headlines in urban areas (Hickin, 1985; Adams and Leedy, 1991; Bolen, 2000; Adams et al., 2006). For example, the top-ten urban vertebrate pests in the USA (in order based on number of national and regional records during 1994–2003) (Adams et al., 2006) are: (1) raccoons (73 084); (2) coyotes (56 081); (3) skunks (54 198); (4) beavers (45 958); (5) deer (36 943); (6) geese (34 808); (7) squirrels (28 975); (8) opossums (21 871); (9) foxes (20 635); and (10) blackbirds (19 228) (*Procyon; Canis; Mephitis; Castor; Odocoileus; Branta canadensis; Sciurus, Tamiasciurus; Didelphis; Vulpes, Urocyon; Quiscalus, Agelaius, Turdus*, etc.).

In each of nine regions of the nation, on average six of these species are among the top-ten pests, along with four other species (Adams et al., 2006). The Middle Atlantic Region, with only three of the above species in the top ten, diverges by far the most from the national pattern. In addition to the list above, top-ten regional pest species in order are: woodchucks; black bears; hawks/owls; pigeons; bobcats; crows; black rats; coots; ducks; armadillos; vultures; bats; cougars; woodpeckers; gophers; gulls; mourning doves; sparrows; mink; and blue jays. Many other urban vertebrates are regarded as pests, including wild boar (*Sus scrofa*) in Berlin and elephants in African and South Asian cities. Lists of airport wildlife posing a threat to aircraft worldwide would be quite interesting. Reports of pests commonly stimulate people and government to act in eliminating or reducing the perceived pest problem.

The wide range of vertebrates listed above as pests may be annoying to certain people, yet all of the species are also much appreciated by other urban residents. Both the diversity and abundance of such species in urban areas is impressive. Indeed, all of the vertebrates also play important positive ecological roles in specific urban habitats, as well as urban regions as a whole.

Public health

Ecological changes and conditions play central roles in urban public health (McNeill, 2000; Millennium Ecosystem Assessment, 2005). Chemicals in air and water cause many illnesses and deaths but microorganisms (microbes) are the prime source of disease and death. Before introducing the human diseases, it is important to remember that microbes are "everywhere." Indeed, many are central to well-functioning natural systems and human life. For example, consider the bacteria and fungi that decompose dead leaves and garbage, soil-enriching nitrogen-fixing bacteria in legume roots, mycorrhizal fungi upon which many trees depend, productive phytoplankton in the sea, zooplankton as a key stage in aquatic food chains, essential bacteria in the gut of animals and us, and so forth.

The public health issues introduced below are illustrative rather than exhaustive; many others including tuberculosis, typhus, and smallpox could be added. Also they highlight physical health, not mental or emotional health, for instance related to genetics, stress, or disasters (Frumkin et al., 2004; Kaplan and Kaplan, 2008; Tzoulas and Greening, 2011). Lots of indirect issues could be added, such as urbanization or increasing urban heat (Chapter 5), which may make people more susceptible to illness and disease (Knowlton et al., 2004; Gartland 2008). Public health issues have repeatedly altered human history (McNeill, 2000; Ponting, 2007; Benton-Short and Short, 2008). Human population density and associated rate of disease transmission are keys to urban public health. But also ecological change and conditions underlie most illnesses and diseases. Public health issues focus around four urban characteristics: (1) water-related; (2) wildlife-related; (3) air pollution; and (4) home, plants, and sprawl.

Water-related. Stormwater runoff cleans the urban surface by carrying heavy metals, organic toxic substances, and some pathogens to nearby water bodies, adding to water pollution levels. In many cities human wastewater flows in open ditches or channels to nearby water bodies, greatly increasing the pathogenic bacteria levels (Costa-Pierce et al., 2005). In other cities, the wastewater flows in sewer pipes to a sewage treatment facility. Such facilities may completely clean the wastewater, or clean it except during heavy rains that cause combined sewer overflows (CSOs) putting raw sewage directly into water bodies (Benton-Short and Short, 2008). Most urban sewage treatment facilities only partially clean the wastewater (providing primary or secondary treatment), leaving pathogens and/or pollutants to enter water bodies. Urban wastewater is sometimes used in aquaculture to grow food (Costa-Pierce et al., 2005; Mukherjee, 2006), a process warranting continued public-health overview.

According to the United Nations, contaminated water, especially in developing nations, is overwhelmingly the major cause of disease-related deaths. Drinking non-potable (unclean) water leads to gastro-enteritic diseases (e.g., the protozoa *Girardia* and *Cryptosporidium*), diarrhea, hepatitis virus, and much more. Many chemicals in water from industry, transportation, and other sources also produce toxic effects.

Wildlife-related. Animal-transmitted diseases are extremely diverse, each with a specific pattern, and cumulatively represent a major public health focus. Arthropods and rodents especially, but also domestic animals, carry diseases to people. Various intermediate hosts or vectors may be involved, from snails to deer and farm animals.

Mosquitos, favored by standing water where their larvae grow, carry many diseases including the malaria protozoan and the equine encephalitis and West Nile viruses (Carpenter, 1983; Robinson, 1996; Patz et al., 2000). Ticks on deer, rodents, pets, plants, and soil may carry the bacteria for Lyme disease or Rocky Mountain spotted fever (Hayes and Piesman, 2003; Adams et al., 2006). Fleas on rodents, pets, and other vertebrates may carry the plague bacterium, as highlighted in the 14th century Black Deaths of Europe and Asia, and in the 20th century plague outbreaks killing >10 million people in India (Robinson, 1996; Collinge et al., 2005).

Mammals such as bats, mongoose, skunks, raccoons, and foxes may carry the rabies virus directly to a person, or the virus may be transmitted indirectly via a pet dog or cat (Rosatte et al., 1991; Bolen, 2000). Armadillos may carry the leprosy bacteria. Large amounts of bird or bat droppings (guano) may contain the fungus *Histoplasma* causing histoplasmosis, a human disease.

Rodents, especially rats and mice, are particularly important transmitters of disease in urban areas. These rodents carry salmonellosis, trichinosis, tularenia, plague, and more. Mice tend to concentrate around buildings and kitchens, while rats typically aggregate around dumps, markets/supermarkets, restaurant/hotel areas, storage warehouses, and harbors. Stormwater pipe systems connecting streets and building basements are favorite runways for ready rodent movement. Garbage around home containers, dumpster containers, and dumps is a major rat attractor.

Some years ago, in the midst of a swine encephalitis outbreak, I recall feeling like almost the only visitor in Kuala Lumpur and Malacca (Malaysia). Hundreds of people died and thousands of pigs were killed. The virus normally only resided in flying-fox fruit bats, but urbanization pushing into rainforest placed the bats and people's pigs in close proximity (Breed et al., 2006). The virus was transmitted from bat to pig to person. This *zoonotic disease*, which effectively jumps from wild animals to people (humans "suddenly" become susceptible), is of special worldwide public-health concern. Avian flu or influenza spreading from migrating wild birds to poultry to people, another example, is particularly worrisome because migrating birds cover such large distances.

Air pollution. Industry and transportation are major sources of air pollutants of public health importance (Chapter 5) (Frumkin et al., 2004). Particulate matter (PM_{10}, and especially the smaller $PM_{2.5}$ particles) accumulating in lungs causes respiratory disease. Heavy metals such as lead (Pb) are important airborne pollutants degrading health. Concentrated organic compounds in air near certain industries are toxic. Smog, produced in the air by combining nitrogen oxides (NOx) with hydrocarbons (VOCs) in the presence of sunlight, is an important contributor to asthma and emphysema. Moving vehicles are major sources of NOx.

Home, plants, and sprawl. Cooking with wood and other types of biomass coats the lungs with particulates and toxics, causing disease and death. Gardens may use fertilizer with excessive heavy metals and other pollutants (Maconachie, 2007). Residents often over-use pesticides, plus petroleum-based chemicals associated with mowing and vehicles, which reach air and water. A surprising percentage of septic systems function poorly, so pathogenic bacteria such as *E. coli* accumulate in the soil and water. Pollen allergy seems related to the abundance of early successional sites, or a low diversity of trees dominated by species producing voluminous wind-dispersed pollen (Kopecky, 1990; Carinanos and Casares-Porcel, 2011). Finally, low-density homes dependent on vehicle driving are associated with high levels of obesity (Frumkin et al., 2004; Ewing et al., 2008).

A few urban infectious diseases, such as HIV, Chagas disease, Lyme disease, hantavirus, and rabies, seem likely to continue no matter how people attempt to improve or modify ecological conditions (Millennium Ecosystem Assessment, 2005). But the majority of diseases, including malaria, leishmaniasis, schistosomiasis, Japanese encephalitis, cryptosporidiosis, dengue fever, and cholera, could be greatly reduced with ecological solutions.

The varied diseases introduced above can be and are addressed with individual solutions, such as reducing point pollution sources, improving sewage treatment facilities, spraying, and so forth (Hardoy et al., 2004; Tzoulas et al., 2007). Some combined solutions including reducing urban heat, separating stormwater and wastewater flows in different pipe systems, and an ambitious new semi-permanent system for garbage and making rats rare, would provide considerable public health and economic value. However, a big-picture integrated approach, focused on broad-scale ecological patterns and processes and the design of whole urban regions or metro areas, would doubtless offer still-greater cumulative benefits for public health.

Using urban ecology for society's solutions

We first briefly examine the quite-different urban environments worldwide as the product of regional and cultural human perspectives. Then urban ecology in key disciplines and professions dealing with urban areas is introduced. Finally we consider the use of ecology in solutions to big urban problems.

Cultural and regional perspectives on the urban environment

Not surprisingly, the urban environment and urban ecology are perceived differently depending on culture, geography, climate, land use, and history (Gilbert, 1991; Hardoy et al., 2004; McGregor et al., 2006; Forman, 2008). Comparing six major regions of the world, the individual characteristics and perspectives clearly differ (Table 1.1). Adding the ten concepts or characteristics together highlights huge and fundamental differences from region to region in human perception, as well as the formation and use of urban areas. Yet despite these striking differences, people in all regions apparently strongly believe in nature protection and an environment based on good ecological principles. A puzzle to ponder.

"Visual core" differences (Table 1.1) are particularly important because they tend to daily reinforce the image or distinctiveness of a city for its people. The people's image in turn strongly colors their uses and planning for the future. The second through ninth concepts listed in Table 1.1 effectively contrast cities as products of human perception. The ninth concept listed, "time perspective," is also especially informative. Some of the reasons for a short or long time perspective are mentioned. Time perspective says much about people's sense of place, caring or stewardship, and planning ahead. The last concept listed, "solutions to problems," emphasizes how things get done, how the urban area has been and can be molded for the future.

As urban ecology evolves, research questions and interpretations of results can be expected to differ geographically. But cities are no more diverse than, for instance, vegetation types or rural land uses worldwide. Ecologists calculate an average and a variance for data. Core theories and principles, plus models highlighting worldwide and local variability, will likely describe the encompassing urban ecology of the future.

Using urban ecology principles

Spatial planning principles and patterns

Principles combined with creativity are used to accomplish planning and design goals. Stating goals at the outset helps determine solutions and achieve success. Thus, a person wishing to have an ecologically designed house plot may choose among goals such as high species richness; native species; certain favorite species; pest reduction; diverse aquatic habitat conditions; wetland zonation; erosion control; wind control; recycling; movements and flows; ever-changing conditions; or a natural ecosystem (Forman, 1995). The solutions for these different goals are all ecological but produce extremely different house plots.

Urban, city, and regional planning represents a body of theory and practice of major importance to society. With strong roots in both the natural and social sciences, pioneers such as Frederick Law Olmsted left indelible marks on major cities including Boston and New York. In recent decades urban and city planning has shifted its foundations to socioeconomic and policy dimensions. Focusing on housing, jobs, transportation, and economic development, a rich array of planning principles is used (Lynch and Hack, 1996; Pacione, 2005; Berke et al., 2006).

In the current phase of limited environmental planning, several cognate fields have stepped in to contribute valuable principles (Craul, 1992; Forman

Table 1.1. Urban environment perspectives in different regions and cultures. Brief encapsulations attempt to highlight the essence, but of course miss the richness of pattern within and among cities of a region. Perspectives are more characteristic of traditional or small cities than very large "globalized" cities. Mitsch and Jorgensen (2004) and van Bohemen (2005) consider China and the West.

Concept	Tropical Latin Am.	North Africa	East Asia	Australia	North America	Europe
Visual core	Business and plaza central	Seamless white buildings	Recent buildings and signs	British imprint, eucalypts	Central business district	Historical buildings, tourists
People and parks	Dense, and informal settlers	Fairly dense, few parks	Extremely dense, few parks	Not dense, many parks	Fairly dense, parks	Very dense, parks
Energy use	Oil and people	Oil	People, oil, and coal	Oil and coal	Oil, gas, and coal	Oil, coal, and gas
Recycling, re-use	Limited	Limited	Important	Encouraged	Acceptable	Strongly encouraged
Urban agriculture	Some	Tiny courtyard gardens	Much in tiny spaces	Limited	Limited, field crops	Community gardens
Air and water	Widely polluted	Scarce water, polluted	Highly polluted to ±clean	Rather clean	Rather clean	Clean and polluted areas
Nature in city	Diverse and self-sufficient	Mostly in watered courtyards	Treasured tiny spots, meanings	Tidy parks, exotics, and wild bush	City parks and street trees	Tidy parks and lines of plane trees
Surrounding context	Rainforest and pasture	Irrigation and pastoral dry land	High rises and market gardening	Bush and planned suburb	Suburban sprawl and nature areas	Farmland and leisure spaces
Time perspective	Medium (family, mañana)	Long (extended family)	Very short and very long	Rather short	Short (people often move)	Long (history and place)
Solutions to problems	Government	Family and government	Government	Government and self-sufficient	Public–private partners	Government and people

and Hersperger, 1997; Hough, 2004; Marsh, 2010; Erell et al., 2011; Butler and Davies, 2011; Pickett et al., 2013). Most striking, however, is the array of urban ecology principles now available to planners and other urban practitioners (Sukopp, 2008; Alberti, 2008; Marzluff et al., 2008; Grimm et al., 2008; Pickett et al., 2009). An especially useful and usable summary groups the urban land-use planning principles into five categories (Forman, 2008). In essence, each principle is a statement of importance and broad applicability:

1. Patch sizes, edges, and habitats: 30 principles.
2. Natural processes, corridors, and networks: 24 principles.
3. Transportation nodes: 16 principles.
4. Communities and development: 31 principles.
5. Land mosaics and landscape change: 20 principles.

At some point in the planning process, planners draw on the store of spatial patterns known to produce good results. Most such spatial patterns are pieces of the overall puzzle to creatively solve. Patterns known not to work well are avoided. Still, urban-ecology-based spatial patterns for planning have been enumerated into three useful categories (Forman, 2008): good, bad, and interesting patterns. Each category is subdivided into the following groups: (1) metro area, city, and urban region characteristics; (2) nature, forest/woodland, and food production; (3) water; and (4) transportation, development, industry, and pollution. A fifth "hazards" group is added under bad patterns.

Such spatial patterns can be fit together for urban planning that seriously addresses natural environmental dimensions, along with housing, jobs, transportation, and economic development.

Use in key disciplines and professions

I think the fundamental objective for our future urban regions is to "mold the land so nature and people both thrive long term." But why, for example, are we repeatedly stuck with ecologically sterile parks, hot air, and dirty water bodies (Figure 1.11)?

What problems can be solved partially or entirely by using urban ecology principles? Each of us could make a long and distinctive list. Indeed, suppose ten of us gathered around a round table, each representing one of the major disciplines and professions focused on the city. We pause for five minutes, individually listing key problems where urban ecology could provide valuable solutions. Then we read our lists to the group, and a lively discussion follows, frequently punctuated by discoveries and "eurekas," as overlap and convergent interests emerge.

Here is an example of the problems and issues where urban ecology principles may be important (Forman, 1999), but your round table may well do better. The background of each person is given with the primary objective of the field or profession (in parenthesis).

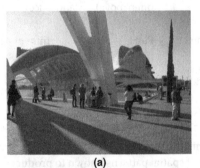

Figure 1.11. Parks illustrating two models of urban people and nature. (a) One tree, 20 people, and architectural inspiration. Valencia, Spain. (b) Lawn, trees, and paved walkways, plus the author in a one-sided conversation on urban ecology. De Hoge Veluwe, Netherlands. R. Forman photos.

1. *Engineering* (*Design/construction of infrastructure and systems*). Impermeable surfaces. Urban heat island. Diverse types of foundations. Stormwater ponds and depressions. Hard structures in streams and rivers. Infrastructure. Piped systems. Vegetated green walls. Dammed reservoirs. Land protection for reservoirs.
2. *Architecture* (*Create well-designed buildings*). Materials and leaching of toxic chemicals. Green roofs. Solar heat. Water use and recycling. Greenhouse-gas emissions. Interior biophilic designs. Vegetated walls. Wind turbulence and accelerated streamline airflows. Natural ventilation. Natural light. Mixed multi-use housing.
3. *Landscape architecture* (*Aesthetics and nature in spaces for people*). Trees for the greatest heat benefit. Connected street-tree corridors along and across roads. Biodiversity hot spots. Connections between parks. Park design for nature. Diversity of native-species plantings. Riversides and stream corridors. Stormwater ponds, wetlands, and biofilter depressions.
4. *Transportation* (*Safe and efficient mobility*). Traffic flows. Urban spread. Quieter road surfaces. Quieter tires. Quieter trucks. Diverse public transit systems. Multi-modal nodes. Netway and pods system. Transit-oriented development (TOD). Air quality. Wildlife underpasses and overpasses. Vegetated noise barriers and soil berms.
5. *Water resources* (*Maintain water quantity and quality*). Sea level rise. Flooding. Stormwater. Wastewater sewer system. Wetlands. Water supply. Land protection. River and floodplain. Streams/pipes. Shorelines. Aquifers. Reservoirs. Groundwater and water table. Agricultural, stormwater, industrial, and wastewater pollutants. Nitrogen and phosphorus. Constructed basins/ponds/wetlands.
6. *Public health* (*Services to protect/improve community health*). Wetlands. Air quality. Water quality. Toxic chemicals. Sewage flows, pathogenic bacteria, and treatment. Land protection. Mosquitoes, midges, malaria habitats.
7. *Economics* (*Production/distribution/consumption of goods and services*). Ecological economics. Environmental economics. Park and natural resource values. Commercial areas. TOD. Non-sprawl residential development. Real estate land

values. Adaptability/flexibility/resilience/ stability. Long-term horizon.
8. *Urban planning (Mold the arrangement and condition of a community).* Urban spread. Informal squatter housing. Sea level rise. Residential development. Commercial development. Industrial development. Transportation. Jobs. Land protection. Protect economic value.
9. *Government-politics (Directing of the public affairs of a community).* Land protection. Park management. Stormwater. Sewage wastewater. Adaptability/flexibility/stability. Many of the urban planning items.
10. *Religion (Universe-related beliefs, often with moral code and superhuman).* Ethics. Help the poor. Informal squatter housing. Ecology of religious properties/buildings. Intermixed affordable housing. Honor nature's beings.

Although we may perceive the land quite differently (Meinig, 1979), one characteristic ties all of the disciplines and professions together. Just as in the previous section on people's perspectives worldwide, all ten of these fields subscribe to nature protection and an environment based on good ecological principles. In fact, that goal is commonly the second stated goal (after the primary one in parentheses) in each field. This coalescence of fields on ecological conditions highlights the synthesizing role and expected value and use of urban ecology.

For many issues facing cities, urban ecology has only been incorporated in selected places or in minimalistic ways or mainly symbolically. A strengthened solid urban ecology should lead to widespread dependable solutions for society. Each discipline will be strengthened as well.

Use of urban ecology for big solutions

Clearly urban ecology principles, either alone or combined with other fields, can play useful roles in solutions to many or most of the specific problems or issues raised above. Yet the principles can also be used, and may be more significant, in addressing big issues involving large areas. Too seldom do we address or solve whole city, whole metro-area, or whole urban-region issues.

Before introducing examples, consider two terms occasionally used, eco-city and ecopolis, focused at the broad scale. "Eco-city" seems to refer to an idealized cultural/ecological condition where urban people and their activities are in balance with natural processes, so that the latter do not degrade (Roseland, 2001; Register, 2006; Wittig, 2008; Suzuki et al., 2010). But the ideal seems to have been used as a framework to justify technical planning solutions for specific areas within a city, effectively economic development that degrades nature. "Ecopolis" seems to refer to a conceptual urban-planning framework for sound urban development, based on four general principles, three main themes (responsible city, living city, participating city), and 18 guiding models (Tjallingii, 1995). Again the term describes an urban-development methodology, using the term environment, but instead focusing on people and social patterns. At present both "eco" terms seem akin to green marketing and nearly devoid of the science of ecology and urban ecology (Sze and Gambirazzio, 2013).

"Urban sustainability," also has a nice ring to it, though many scholars have noted that it is essentially an oxymoron. It is hard to envision a city of thousands or millions attaining and maintaining a viable long-term balance between nature and people (Forman, 2008). Also, everyone seems to define the term to suit one's own interests. Scholars have made some headway in elucidating scientific aspects of sustainability (Kates et al., 2001; Clark and Dickson, 2003; Antrop, 2006; Wu, 2012). Although I essentially avoid the term, four perspectives might be useful (Forman, 1990; Rees, 2002; Newman et al., 2009; Wu, 2010), i.e., sustainability as (1) a never-achieved ideal goal or endpoint; (2) a cumulative product of a multitude of fine-scale solutions; (3) a "black box" or system of unending inputs, outputs, and changing box sizes; and (4) a prime-footprints model that integrates the urban area with dispersed major ecological-footprint areas plus the connecting transportation links (Forman, 2008).

Societal solutions at the broad scale, as suggested by the prime-footprints concept, tend to be scarce. Examples range from the New York parkway-highway system and London's greenbelt to Seoul's former greenbelt, a TVA dam-system project (Southeastern USA), Stockholm's green wedges, Portland's (Oregon, USA) urban growth boundary, the large-woods-and-fields land around Moscow and Berlin, and the cleanup of Lake Washington in the city of Seattle (USA) (Forman, 2008). Such broad-scale solutions can be enormously successful economically, socially, and environmentally, especially in the long term.

Consider some big-picture perspectives for planning and solutions with an urban-ecology foundation. All are at the city-wide, metro-area-wide, or urban-region-wide scale. These could be replicated or tailored to cities worldwide. Pilot projects, as well as full solutions, could turn around current major degradation trends to form wonderful urban areas ahead.

- *Public health.* Addressing all or most public-health diseases and illnesses together, mentioned above, is greatly preferable to the piecemeal approach.
- *Wildlife and biodiversity.* Providing a logical arrangement of medium and large greenspaces, plus the connecting corridors and stepping stones linked to the surroundings, would provide urban habitat and biodiversity throughout the metro area, as well as enhance routes for continually incoming species.
- *Tree cover.* Doubling tree cover, for instance, in an urban area to have the greatest environmental benefit, would create a logical arrangement of woody vegetation that provides combined urban heat, wind, water, wildlife, and people benefits.
- *Water bodies.* Achieving 100% clean water bodies (e.g., streams, rivers, ponds), for example, would improve pipe and drainage systems, sewage treatment facilities, natural water-cleaning, and land-use patterns, a combined benefit for fish, aquatic ecosystems, flood reduction, wildlife, and recreationists.
- *Vehicle traffic.* Cutting ground-level traffic by, say, 50–75% in urban areas would provide alternative and new ways to move, as well as mixed-use areas with human needs in proximity, thus significantly reducing driving, traffic disturbance, and air and water pollution, plus improving mobility and more-livable neighborhoods.
- *Prime footprints.* Integrating urban areas with their near and distant footprint impacts, plus the transportation network linking residents to sources and sinks, would provide spatial, environmental, and economic clarity for reducing impacts and accelerating benefits.
- *Nature's services or ecosystem services.* Nature's services to society have been categorized as provisioning, regulating, cultural, and supporting services (Costanza *et al.*, 1997a; Daily, 1997; Norberg, 1999; Millennium Ecosystem Assessment, 2005; National Research Council, 2005b; Adams *et al.*, 2006; Brauman *et al.*, 2007; McDonald and Marcotullio, 2011; Sagoff, 2011; Levin, 2012; Termorshuizon and Opdam, 2009). All are familiar as "traditional" ecological knowledge (Huntington, 2000). Increasing these services, say, ten-fold, in urban areas (Bolund and Hunhammer, 1999) would provide a major increase in, and logical arrangement of, greenspaces and green cover. Benefits go directly to residents and shopkeepers, as well as government.

Other big-picture objectives with big solutions underpinned by urban ecology exist, such as environmental resistance, resilience, and adaptability, or having, e.g., 90% of incoming resources being local or from that geographic region. Of these big urban challenges and opportunities, which would have the greatest cumulative benefit for nature and us?

I will return to urban ecology solutions, both specific and broad, at the book's end. Meanwhile the chapters ahead reveal and develop the urban ecology principles themselves.

Part I **Framework**

Chapter 2

Spatial patterns and mosaics

> Towns don't want to be suburbs, suburbs don't want to be cities, and cities don't want to be wastelands.
>
> *Michael Dukakis, quoted in* Conservation and Values, *1978*

> If you know one of their cities, you know them all, for they're exactly alike ... Amaurot lies up against a gently sloping hill; the town is almost square in shape. From a little below the crest of the hill, it runs down about two miles to the river. ... the water runs clean and sweet all the way to the sea ... another stream, not particularly large, but very gentle and pleasant, which rises out of the hill, flows down through the center of town ... The town is surrounded by a thick, high wall ... The streets are conveniently laid out for use by vehicles and for protection from the wind. Their buildings are by no means paltry; the unbroken rows of houses facing each other across the streets through the different wards make a fine sight. The streets are twenty feet wide. Behind each row of houses at the center of every block and extending the full length of the street, there are large gardens. Every house has a door to the street and another to the garden ... They raise vines, fruits, herbs, and flowers, so thrifty and flourishing that I have never seen any gardens more productive or elegant than theirs. ... the founder of the city paid particular attention to the siting of these gardens.
>
> *Thomas Moore,* Utopia, *1516*

Between bedrock and stratosphere, a city lies uneasily on a sheet of wet soil and is covered by a sheet of air. Industrious lilliputians in the city keep creating pattern, upward, downward, inward, and especially outward.

The surface of the earth or wet soil is the benchmark. Upward pattern is largely formed by earthen fill, vegetation, low and tall buildings, and engineered structures (Figure 2.1). Combinations and arrangements create the vertical patterns familiar to urban residents – skyscrapers, street trees, mounded dumps, bridges, residential areas, and so forth. A cluster of high-rise buildings serving as a central business district (CBD) commonly characterizes a city center. Deep canyons cut between tall buildings. High rises poke up, mainly supported by bedrock, and may be scattered across the city. Many of the 5000 or so high rises (10+ levels) in Sao Paulo, for example, remain a block or more apart, so residents have adjacent space for park relaxation, children playing, and car parking. Night lights glitter from thousands of windows creating a magical sense of place ... while drawing migrating birds to their death. Rows of trees line streets and boulevards. A city's upward dimension is extremely diverse.

Downward from the ground surface is no mirror image, though the assortment of human constructions may double the city's vertical pattern. Deep and shallow wells, which lower the water table, often dry out the upper soil. An anastomosing three-dimensional network of rails used by noisy subway trains.

Figure 2.1. A planned and built city center with streets and tree lines radiating from circular intersections. Low-rise buildings and no skyscrapers; blocks typically have central spaces with plants. Arc de Triomphe (top); Palais Galliera (lower right). Paris. R. Forman photo.

Uncoordinated shopping walkways and arcades with disorienting branches. A dendritic (tree-like) pipe system with seldom-clogged acute angles for ever-flowing human sewage. Maybe a catacomb to store excess skeletons and skulls, a 3D labyrinth of interconnecting passageways, small reservoirs to hold clean water, and underground rooms with unseen and unseemly uses over time. Together these diverse networks and cavities appear disjointed, uncoordinated, disoriented: a spatial pattern awaiting a pioneer or explorer. Hardly anyone in history has seen or grasped all the underground parts of a city. A true frontier lurking at our feet. Exploring the urban underground promises a Christopher Columbus-type voyage in the abyss, in this case, eternally lost within meters of the multitude above.

Scale, human, and nature patterns

The preceding perspective highlights the vertical dimension. Horizontal patterns are central to this book and are introduced in the following sections.

But first, curiously the oblique pattern remains a frontier. Consider the diagonal up or down bird's-eye view, the city-center view from a street corner or skyscraper, and a diagonal ground-penetrating technology image. Although gravity and sun provide a vertical orientation, these oblique angles are quite normal in nature – tree branches, roots, caves, corals, and beetle legs. Many birds and insects fly obliquely through space. Diagonal patterns and views through nature appear familiar. In contrast, humans normally build horizontal and vertical structures. Oblique views through these constructions often highlight odd patterns absent from engineering and architectural drawings. What indeed are the main oblique patterns in cities? Are they important in how things work? How can they be improved?

Spatial scale

Many discussions of scale conclude that "Scale matters." Both patterns and processes tend to differ at different scales. Also scale involves both time and space. Temporal scales vary, say, from minutes to millennia. Here we focus on urban spatial scales, especially from meters to hundreds of kilometers.

Imagine having a giant zoom lens that you progressively open at a constant rate. Starting at the microscopic level, one might see, for instance, soil particles and algae cells. As the lens progressively opens, a sequence of recognizable images appears, such as a lettuce plant, vegetable garden, house plot, housing development, neighborhood, residential area, city, urban region, geographic region, continent, and globe. These are *scales* or levels of scale. The sequence observed proceeds from "fine scale" to "broad scale." The space or area or distance varies from small to large. Consider two maps, one of a local spot with the "scale" marked as 1:100 (map distance divided by actual distance on the ground), and the other of a nation with the scale given as 1:100 000. One over 100 is a much larger scale than one over 100 000. For simplicity and clarity, this book uses fine scale for the former map (local spot) and broad scale or coarse scale for the latter (a nation).

Each scale in the zoom sequence shows a heterogeneous pattern. Heterogeneity could be simply a gradient, or series of gradients, where objects or attributes gradually change in density across space such that no distinct boundaries are evident. However, almost always the pattern observed is a *mosaic* composed of relatively distinct objects with boundaries. This is because underlying substrates tend to be patchy, natural disturbances tend to create patches with relatively sharp boundaries, and, especially important to urban areas, most human activities produce sharp boundaries. These "land mosaics," from small to large, have easily understood characteristics at the heart of urban ecology.

The giant lens and the relative distinctiveness or indistinctiveness of boundaries highlights another key characteristic. What is the resolution relative to the grain size of the mosaic? "Resolution," as in art and photography, is mainly determined by two attributes, the degree of difference and the abruptness of a boundary. A black object adjacent to a white one is easily differentiated, whereas light and dark gray objects may appear as one. An abrupt or sharp boundary adds contrast and thus increases resolution.

Grain size, the average area or diameter of constituent parts or patches, is also a useful way to analyze an area or mosaic (Forman, 1995). A prevalence of small patches produces a fine grain, whereas large objects produce a coarse grain. Grain size has considerable importance for the flows and movements in an area, how the area works, and what kinds of species, people, and activities prevail. Relating grain size to the "extent" or area being examined provides insight into the spatial scale of ecological processes (Wiens, 1999; McGarigal and Cushman, 2005; Farina, 2006).

Opening the giant zoom lens, or zooming out by computer from a spot on a satellite image, highlights an important but poorly understood ecological dimension. As the lens opens at a constant rate, does one see a gradual change in heterogeneous pattern, or relatively distinct images that persist and then change rapidly in kind of a stair-stepped process? One line of reasoning leads to the second option. A *domain of scale* refers to the range of distance or scale in which an object remains in focus (as the lens opens) (Wiens, 1989). For example, the individual plantings in a vegetable garden seem to remain clear until suddenly they are hard to distinguish, as the whole garden appears to be an element in the house plot design. Analogously with the opening lens, a farmer's bean field remains clear until it quickly disappears into its agricultural valley surrounded by wooded ridges.

This stair-stepped interpretation of changing scale patterns seems to result from the basic mechanisms producing pattern. The garden pattern results from planting seeds or seedlings of different vegetables in rows. The bean field pattern results from plowing by a large tractor (which could not effectively turn around in the vegetable garden or house plot). The ridge and valley pattern results from ancient geological processes. In these fine-scale and broad-scale examples, no clear intervening pattern-forming mechanisms are evident. The pattern of Paris streets and buildings results from urban planning (Figure 2.1). Each pattern is a level of scale, or simply a scale (McGarigal and Marks, 1995; Wu, 1999, 2004).

A domain of scales contains two or more scales with a similar spatial pattern determined by essentially the same process(es). Natural processes and human activities serve as mechanisms producing a scale pattern, and then maintaining or changing it. Each spatial scale in an urban region can be planned and sustained using a combination of incentives, regulations, laws, care/stewardship, social action, market investments, or government action.

Such a hierarchy of spatial scales is common in urban areas. A *hierarchy* or hierarchical system has higher levels with larger, slowly changing units (entities), and lower levels with smaller, fast-changing units (O'Neill *et al.*, 1986; Farina, 2006; Alberti, 2008; Wu, 2008). Large objects have more inertia and are harder to change than are small ones. Thus, stability and predictive ability are greater at the higher level or broad scale. Small units normally fluctuate more and are less predictable.

In addition, the upper levels typically constrain the lower levels, while the lower level initiates changes affecting upper levels, though in both cases the opposite may occur. Conditions of a unit at one scale thus are affected by inputs or signals from both the scale above and the scale below. Furthermore, the unit is affected by competitive or collaborative conditions of other units at the same scale. Thus a particular unit is controlled by three scales, upper, same, and lower levels.

Different ecological and human factors operate and are important at different scales. For example, a moss species was found to have distinct distributions over nine spatial scales from the globe to a tree stump (Forman, 1964). Moreover, the environmental factors controlling the distributions differed at each scale. Three or more scales are typically needed to understand a park or a house plot. But the same is true for understanding a metro area or urban region (Hardoy *et al.*, 2004; Pacione, 2005; McGranahan, 2006). A common urban hierarchy is: megalopolis, urban region, metro area, city, residential area, neighborhood, housing development, house plot, and vegetable garden. But other hierarchies are equally useful, such as urban region, urban-region ring, peri-urban (or exurban) area, commercial area, and building space.

The Washington-to-Boston region illustrates a *megalopolis* as a specific object, i.e., a group of large cities with their surrounding zones of influence (Gottman, 1961; Borchert, 1993; Benton-Short and Short, 2008). From 1950 to 2000 this megalopolis increased in: total population from 32 to 49 million; human density from 1580 to 2411 people/km^2 (610–931 people/mi^2); autos from 6 to 24 million; and water usage from 1027 to 1500 gallons/person/day. In the year 2000, the Region had: 1.4% of the USA land area; 17% of the population; 20% of gross domestic product (GDP); and 4.3 pounds of municipal solid waste/person/day, as well as very long journeys for trash-removal trucks. Together these probably constitute the greatest environmental impact per person in the history of the world.

Other megalopolises are present in The Netherlands (Amsterdam, Rotterdam, Utrecht, The Hague area), Japan (Tokyo, Yokohama, and perhaps southward) (Hanes, 1993), and China's Pearl River Delta (Guangzhou, Foshan, Dongguan, and perhaps Shenzhen and Hong Kong). With the possible exception of the Ranstad area linking the major Dutch cities, the megalopolis seems to have been little studied by ecologists. Spatial patterns appear to be of particular interest. For example, what ecological and human functions do the low-population locations play, and how are they distributed? How much sharing of resources and facilities occurs among a cluster of cities with overlapping urban-region rings? Where are the ecologically best and worst places for future development, and for future parks? Is the city cluster more stable or less so than the equivalent cities dispersed, or indeed a single megacity? In what ways is a megalopolis good, or bad, ecologically?

Like roses or people, there are fractals and then there are fractals (Milne, 1988; Batty and Longley, 1994). A *fractal* is a geometrical shape repeated over different scales of measurement.

Three fractals may have use in understanding urban areas:

1. The self-similar fractal, illustrated by a typical fern leaf or dendritic stream network, refers to a basic form repeated in similar form at a series of spatial scales.
2. A boundary or edge fractal is illustrated by a boundary composed of several large lobes alternating with large coves. Focusing in on one lobe reveals that it also is composed of similarly shaped smaller lobes and coves, and then focusing in on one of them in turn reveals roughly the same pattern. The sequential forms in nature normally are not identical but similar.
3. A route or trajectory fractal effectively measures the convolutedness of a route in geometrical dimensions. A straight line between two points has a fractal dimension, $D = 1$, a squiggly route might be, for instance, $D = 1.3$. A back-and-forth route that covers an entire surface area has $D = 2$, and a route extending throughout an entire volume has $D = 3$. Nature tends toward curviness or convolution, commonly illustrated by two-dimensional boundaries and routes with a fractal $D = 1.1$ to 1.5. Consequently, it requires energy or a maintenance budget to maintain straight routes against natural processes.

Such fractal patterns are generally considered to be scale-independent. Nevertheless, until we know more about possible scale-independence, current research should specify the particular scale(s) observed.

Nature's patterns and human patterns

Whether any pattern is present in all landscapes and urban areas remains a mystery. My candidates for universality are patches and corridors (and perhaps networks and a background matrix) (Forman and Godron, 1981; Forman, 1995). Apparently the only two-dimensional situations without patches are homogeneous, or a space with only corridors (like a photo of spaghetti), or gradients (gradual changes without boundaries). Technically, homogeneity does not exist on land. I have never seen an only-corridor case. Spatial ecological gradients are scarce on the land and may be absent in urban areas intensely molded by humans. Corridors (strips) and networks of interconnected corridors are widespread in the urban environment, but always intermixed with patches. The matrix is discussed below. Many other patterns introduced in this section are widespread and might be universal.

First, consider the *horizontal natural processes* that effectively produce nature's pattern on land (Figure 2.2a). Wind and water (including ice) flows, plus animal movement by walking and flying, are the processes. All trace curvy routes. High-velocity linear flows produce straighter routes, while convoluted routes result from slow linear movements or from turbulence. Groundwater flowing through sand, and fire moving in high wind, tend to trace relatively straight routes. Animals foraging for food normally follow quite convoluted routes.

A rich array of distinct interesting spatial patterns results from the combination of these natural processes (Figure 2.2b). Nature's patterns are predominantly irregular, curvy, elongated, composed of coves and lobes, fractal or dendritic, aggregated, variable in size, and/or finely textured. These patterns appear at all scales.

What aspects of nature do ecologists normally consider to be most important, and of relevance to urban areas? Typical priorities for natural vegetation protection are (1) a large area; (2) a remnant representing former conditions; (3) a rare example; (4) a representative example; and (5) a species-rich patch or area. The same priorities apply to vegetation-protected water bodies (stream, lake, estuary, etc.), as well as their headwaters and fish habitats and movement routes.

So, how do we affect nature (Forman, 2012)? We simplify. We linearize and geometricize. We attempt to control. We reduce variability, and therefore adaptability. We multiply, and we sprawl. We pollute, and contaminate. We eliminate, and impoverish. We degrade patterns. We disrupt processes. We perforate, and we dissect. We fragment, and we shrink. We consume, and over-consume.

Consequently, human-created patterns imprinted on the land appear dramatically different than the

(a) Natural processes across the landscape

Ground- Water Water Wind Wind- Fire Seasonal Animal Animal Pollinator
water in in erosion dispersed move- migra- dispersal foraging pollina-
flow eroding meande- seeds, ment tion for food ting
stream ring dust flowers
river

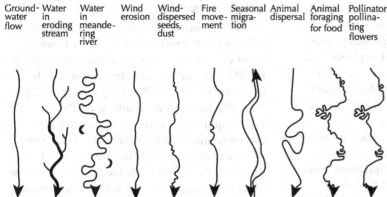

(b) Landscape patterns produced by:

Natural processes | Planning and design | Unplanned development | Long-term trial-and-error

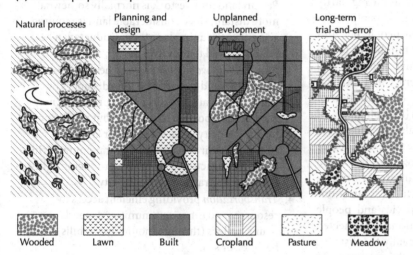

Wooded Lawn Built Cropland Pasture Meadow

Figure 2.2. Horizontal natural processes and four types of spatial patterns on the land. (a) Typical routes of flow or movement. (b) (left) Wooded areas surrounded by agriculture (barchan dune, left center); (center two diagrams) built areas with streams, woods, and lawn parks; (right) farmland with woods, crops, pastures, meadows, streams, and farmsteads (based on central Toscana, Italy). Adapted from Forman (2012); see also Forman (1995) and Forman and Hersperger (1997).

natural patterns. Motivations for creating our forms involve control, protection, access, efficiency, and community. Squares, grids, smooth curves, straight lines, dual lines, and circles with radiating lines cover built areas (Figure 2.2b). Rectangles with length-to-width ratios of 1.5:1 to 4:1 are especially widespread (Forman, 1995). These patterns or objects are mainly regular geometric forms, which together produce urban areas with a distinct Euclidian geometry.

Urban ecology focuses on the combination of natural and human patterns. The former predominate in outer portions of the urban region and the latter in city center. In outer areas the predominant natural forces are typically supplemented by planning and management of large areas for, e.g., wood harvest, water protection, and recreation. City center is often highly "planned" as a whole and "designed" in detail at spots.

Two types of land patterns, both unplanned as a whole, appear to be widespread in the complex intermediate suburban or peri-urban areas. One spatial arrangement results from long-term trial-and-error (Figure 2.2b), where the individual human processes have reached a dynamic equilibrium with natural processes based on "what works." Both natural processes and human activities continue apparently with few disruptions. The range of both natural and human patterns is truncated.

The second suburban pattern, in contrast, results from relatively new human activities imprinted on the natural land (Figure 2.2b). Early piecemeal activities with a short-term anthropogenic perspective inevitably degrade nature in spots. The subsequent sequence of such activities results in a landscape with severely degraded nature and a poor arrangement for people. In effect, the urban region contains planned built geometric landscapes, natural landscapes, and unplanned (as a whole) landscapes. The unplanned portion combines natural and built areas that are either stable products of trial-and-error or changing recent-development landscapes with markedly degraded nature.

Spatially measuring the varied ecological patterns in urban areas is a challenge indeed. One author suggests four particularly informative dimensions that also link ecology to human patterns (Alberti, 2008): form, density, heterogeneity, and connectivity. *Urban form* commonly indicates the centrality of objects (for instance, from compact aggregated to dispersed or polycentric; Anas *et al.*, 1998), as well as the regularity (or irregularity) of their distribution. Urban *population density* refers to the number of people per unit area, though density could refer to number of buildings, parks, industries, successional habitats, and so forth. *Ecological or habitat heterogeneity* (habitat diversity) refers to the degree and concentration of differences in vegetation or habitats (hydrological and socio-cultural heterogeneity may also be measured; Pickett *et al.*, 2001). *Connectivity* measures the degree to which species, people or resources are facilitated or impeded in movement across an area. *Corridors* or *stepping stones* are visible structural connections whereas, even in their absence, species and people may still perceive an area to be functionally connected (Tischendorf and Fahrig, 2000). Usually the distance between patches or habitats is a useful way to measure connectivity (Goodwin and Fahrig, 2002). Other ways to measure or model urban landscape pattern will be introduced below.

Geomorphic framework with cities added

People living for decades, plus buildings lasting decades-to-centuries, are aggregated in cities, which teeter on the mighty backbone of nature. The rocks of mountains and hills and ridges stand strong for millions of years. Even the unconsolidated earthen deposits of sand, silt, or clay cover the valleys and plains below for hundreds, thousands, or millions of years. Persistent microclimatic conditions also often last for centuries.

To reveal the *urban backbone*, i.e., the underlying geomorphic pattern for a metro area, consider the basic human needs as a village becomes a city. That should lead us to the special locations where cities develop, plus their underlying structure. Five human needs seem particularly important:

1. *Clean fresh water* is a daily human necessity. Water is heavy to transport so cities are close to water supplies. Commonly Roman aqueducts extended for kilometers and today's pipelines often extend for tens of kilometers. Restoring a water supply is normally the first priority after a disaster.
2. *Food*, required every day or few days, comes in diverse forms, is light weight, and many but not all forms can be cost-effectively transported long distances. Perishable foods and those with slow or expensive transport may be mainly grown locally in urban agriculture or market-garden farming. This requires ample good agricultural soil for cultivation within kilometers or tens of kilometers. Pastureland for livestock is normally somewhat more distant, since considerable land is required and either the animals or their products can be readily moved to the city.
3. *Minerals and wood fiber* are very heavy, non-perishable, and transported long distances by water, boat, train, or truck. Their proximity may have been important to a village, but usually not for the later city. Mineral and fiber are either transported directly to the city for manufacturing goods, or to factories elsewhere producing light-to-heavy goods transported to the city.
4. *Transportation* providing efficient access to resources and other communities depends on plains, valleys (through mountains or hills), rivers, harbors, or sea.
5. *Security or defense* against competitors focuses on distinctive arrangements of land and water, such as the base of mountains; wetlands; the sides, intersections or mouths of rivers; and coastal harbors, bays, peninsulas, and islands.

Villages exist in many locations across the land, sometimes appearing, other times disappearing. However, a place providing the five human needs in proximity illustrates where and why villages grow into cities.

In essence, most cities are located where three features come together: (1) clean freshwater, (2) good agricultural soil, and (3) a distinctive land–water interface location providing transportation and security (Platt, 2004; Pacione, 2005). The two prime city locations are (1) seacoast areas around bays, harbors, river mouths, and coastal mountains/hills; and (2) riversides, commonly where rivers join and near mountains/hills.

Designers have designed cities (Brasilia, Canberra) where the three central attributes did not come together. Another example are some recent Chinese cities (e.g., Ordos, Inner Mongolia) built mainly to attract investment and employment. Such cities depend heavily on importing resources. Future cities emerging away from clean freshwater, good agricultural soil, and a distinctive land–water interface location providing transportation and security may be unsustainable, at risk of major shrinkage.

Less common locations for cities are by large lakes [Toronto, Chicago, Salt Lake City (USA), Irkutsk (Russia), Baki (Azerbaijan)]. Also some cities have emerged at the intersections of major travel routes [Berlin, Indianapolis (USA), Denver (USA)].

The idea of a *convergency point*, where three or more habitats or land uses come together, is useful in understanding city location (Forman, 1995). Wildlife generally require food, water, a protected nest or den site, and escape cover. Some species prefer convergency points that provide ready access to diverse resources, and also stability during tough times.

Coastal cities have adjoining land and sea resources. A city by a river where mountain and plain come together benefits from the convergence of river, mountain, and plain resources. Such resources include crop and livestock production, water for transport and for hydropower, wood and mineral building materials, manufacturing resources, diverse recreational opportunity, and habitat diversity and rich biodiversity. Cities on boundaries between two land types, and especially at convergency points, probably also cause particularly high environmental degradation.

The land surface where cities appear is sculpted and deposited by flowing wind, water, and glaciers. The characteristic locations of cities on land mean that a distinctive set of nature's urban-backbone components is present (see Figure 4.3). Thus, a basic map can usefully differentiate:

1. Two *rock types*: limestone and silica rock (e.g., sandstone, granite).
2. Three types of *unconsolidated earthen deposits*: clay (commonly derived from limestone, and subject to shrinkage/expansion); silt (good for agriculture and septic systems when well-drained; also as wet alluvial river deposits); and sand (especially on a flat coastal plain or coastline beach/dune strip).
3. Four *topographic land surfaces*: mountain/ridge/hill top; steep slope; gentle slope; flat valleys/plain covered with unconsolidated earthen material.
4. Two *land–water interface locations*: riverside, river intersection, and river by mountains/hills; and seacoast bay/harbor/river-mouth often by mountains/hills.
5. At least two types of *wetlands*: marsh (grassy) and swamp (wooded).

These urban backbone components are long term. However, nature's short-term flows and movements – high wind, flooding, sinking, shaking, and slides – continually threaten the uneasy city.

Consider the key long-term microclimatic patterns (see Chapter 5), which are strongly linked to topographic land-surface forms and the urban backbone (Geiger, 1965; Moran and Morgan, 1994). Such microclimatic patterns are also readily mapped in urban regions:

1. Five *mountain/hill microclimates*: the top (with 10–20% higher wind velocity); north- and south-facing slopes (reflecting solar angle, energy received, and snow/ice conditions); and upwind (windward) and downwind (e.g., rain-shadow) slopes.
2. *Cool-air drainage on slopes and in valleys*: vegetated unbuilt slopes as sources of cool air moving downslope on still summer nights, with the moving air continuing along valleys (without high-rise buildings), result in significant cooling of the heated city and ventilating of the polluted air upward.
3. The *coastal-strip microclimate*: modified temperature, moisture, fog, windspeed, salt transport, and storm effects, usually within about 4–7 km of the sea.

Mapping these basic geomorphic and microclimatic patterns is fundamental for planning, building, pinpointing hazards, and providing solutions in metro areas. The map is an urban-backbone mosaic. Each piece of the mosaic provides different benefits and is subject to a different set of threats. Therefore, the map provides a clear view of suitable land uses, potential disasters, damage, maintenance/repair costs, public health, management, planning, and policy.

An array of geomorphic (and microclimatic) backbone patterns strongly affects the fine scale of buildings and neighborhoods within a city (Spirn, 1984; Waldheim, 2006; Marsh, 2010). For example, individual locations may have a high water table (Mander and Kimmel, 2008), be flood-prone from river or sea, be subject to high winds, be subject to mudslides or river-sediment deposits, or suffer subsidence such as sinkhole formation. Slope issues are especially well known, so optimum and maximum slope angles have been identified for different land uses (Berke *et al.*, 2006; Marsh, 2010): house sites (maximum 20–25%; optimum 2%); mowed lawns (max. 25%; opt. 2–3%); septic system drain-fields (max. 10–15%; opt. 0.05%); sidewalks (max. 10%; opt. 1%); 20-mi/h streets/roads (max. 12%); 40-mi/h roads (max. 8%); 60-mi/h roads (max. 5%); factory sites (max. 3–4%; opt. 2%); and parking lots (max. 3%; opt. 1%).

In effect, a sustainable city should fit within its broad geomorphic backbone and nature's processes. Furthermore the pieces of a city should fit within the geomorphic and microclimatic mosaic. An obvious example is market-gardening urban agriculture, which only works on good soil close to the city (Mougeot, 2005). Wise planning, or common sense, avoids placing a major sewage treatment facility in a tsunami-risk location, important buildings in a low site threatened by sealevel rise, or someone's home on or beneath a mudslide-prone slope.

Urban–rural gradient as spatial model

Models provide comfort in the face of a giant complex world composed of infinitely varying mosaics. Modeling focuses on a few key characteristics, simplifying and providing insight into this complexity. Thus, a *model* simplifies a complex system to gain understanding.

Consider an extremely simple model containing only two components, nature and people. Four flows are present in the *nature-and-people interaction model* (Figure 2.3): nature positively affects people; nature negatively affects people; people positively affect nature; and people negatively affect nature. The first two flows on the effects of nature are typically of moderate importance. The third interaction, people positively affecting nature, overall has been of minor importance. The giant interaction in the model is people negatively affecting nature. Conspicuous examples are everywhere.

The nature-and-people model is non-spatial. However, its spatial implications, e.g., for how to optimally arrange people on the land, are especially valuable. Thus, knowing that the negative effects of people on nature are the primary interaction highlights the importance of large natural protected areas, and compact rather than dispersed development. Economic models are largely non-spatial, one reason their predictions are so often inaccurate. The following models are spatial, and in most cases are simple for clarity and easy use.

City, suburb, exurb/peri-urb, farmland, natural land

Geographers have long highlighted the urban–rural divide, both the contrast and the linkage between an urban area and its surrounding rural area. Movements and flows of all sorts go both ways (Douglas, 2006; Kruger, 2006). People migrate to the city (Hardoy *et al.*, 2004; Pacione, 2005; Kruger, 2006; Tacoli, 2006). Agricultural products and other goods flow to the city. Natural resources flow to the city. Manufactured goods and recreationists flow to the rural area. Solid wastes, liquid wastes, and gaseous wastes flow to the rural (Hardoy *et al.*, 2004). With cheap oil and abundant vehicles, commuters move both directions. Change is especially prominent in the rural, where new development may be concentrated (Hardoy *et al.*, 2004; Browder and Godfrey, 1997). Also, natural resources are degraded or lost in the rural, such as depleted freshwater and woody vegetation lost to firewood use (Hardoy *et al.*, 2004).

Empirical observation of land-use patterns in different cities effectively highlights the value of spatial models. Thus, major land uses in many cities of the USA exhibit a similar distribution relative to four spatial areas: city center; inner half of city radius; outer half of city radius; and major transportation corridors (Hartshorn, 1992). While these patterns were described in 1992 for cities, they seem to apply relatively well for all-built metro areas.

1. General business: city center plus scattered locations mainly in the inner half of city radius.
2. Retail business: same as the preceding plus clusters near the city border.
3. Industry: along major transportation routes from city to border.
4. Public and semi-public land: small greenspaces in city center and inner half of radius, and larger greenspaces mainly near the border.
5. Apartments: concentrated in the inner half of radius, with fewer in the city center and outer half of radius.
6. Single-family homes: concentrated in both the inner and outer half of the radius.

Inner and outer suburbs (see Figure 1.4) within the metro area are typically distinguishable. Usually they

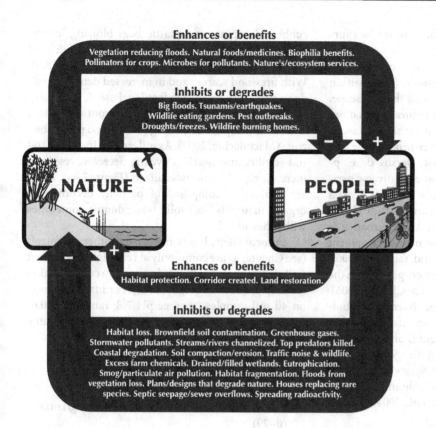

Figure 2.3. Nature-and-people interactions model. Examples listed of positive and negative effects. Adapted from Forman (2010a).

differ in the abundance of single-unit houses, multi-unit low rises (and high rises if present), greenspace, and number and size of shopping areas. In many ways, suburbs are a tension zone between city and rural. The urban–rural flows pass through suburbs going both ways. Meanwhile, suburbs are often densifying with buildings and population.

Some residents have a strong "sense of place" and community (Donahue, 1999), while many others are temporary residents for nearby employment. As suburbs rapidly change, some people stay, work to protect their place, and suffer, as seemingly endless densification changes the place. Meanwhile, others like a particular type of suburban milieu rather than a specific place, and simply move outward as urbanization rolls onward (Forman, 2008).

Suburban wildlife reflect the tension zone, being a combination of native biodiversity, non-natives and pests (e.g., too many deer, beaver, raccoons, wild boar, and so forth) (DeStefano and DeGraaf, 2003). Outer suburbs often have low-density housing sprawl with associated environmental degradation and public health problems (Frumkin et al., 2004). Single-use residential housing rather than multi-use land commonly creates a car-dependent transportation system. Water quality and air quality degrade. Physical inactivity, vehicle crashes, mental health problems, and more public health issues emerge.

Peri-urban refers to the zone containing some urban-related structures just beyond the continuous-built metro area (Allen, 2006; McGregor et al., 2006; Tacoli, 2006). Peri-urban areas, as widely analyzed in Africa, Australia, and parts of Europe, Asia, and Latin America, usually seem to have an agricultural matrix characteristic of most urban regions worldwide. In North America, where sprawl has sometimes spread beyond a city's surrounding farmland into forest or desert, the approximately equivalent term seems to be exurban. Thus, *exurban* seems to be a similar or slightly broader concept referring to a zone with urban-related structures just beyond the all-built metro area, and *peri-urban*, the predominant case, where the zone is mainly farmland. Generally in this book the concepts are considered to be synonymous. Commonly the structures are new groups of houses in a farmland landscape, though shopping centers, industries, and highways related to the city may characterize the exurban or peri-urban area.

Residential spread within a few kilometers of the Manchester (UK) metro area border (even within sight and sound of it) has fragmented the landscape, reduced cropland, reduced semi-natural vegetation, and increased recreational (leisure) sites (Ravetz, 2000). Some have called this peri-urban zone the "battleground" for conservationists versus developers. Examples of environmental issues in diverse peri-urban areas are instructive. A new highway bypass becomes the center of a peri-urban area outside Hubli-Dharwad (India) (Shindhe, 2006). A peri-urban area outside Accra (Ghana) focuses on new development near five existing villages (Gough and Yankson, 2006). Loss of irrigation water and certain crops occur in the peri-urban area of Mexico City (Diaz-Chavez, 2006). Plant colonization and succession have been studied in peri-urban woodlands in Germany and other European nations. Wildfire, logging, and conservation of exurban forests represent challenges in exurban areas of Florida (Vince et al., 2005). Sprawl and its environmental effects are especially visible in some peri-urban or exurban areas (Frumkin et al., 2004; Burchell et al., 2005).

An *urban-to-rural* concept has long been related to human population density. For example in the USA, the Census Bureau defines urban as having >3.86 persons/ha (1.56/acre). After reviewing literature on the subject, one author recommends the following (Theobold, 2004): urban, >3.86 persons/ha; suburban, 0.4 to 4; exurban, 0.02 to 0.4; and rural, <0.02 persons/ha (0.8 persons/100 acres). Such numbers would not work in South Asia, East Asia, and many other regions with extensive areas exceeding 4 persons/ha. Outside the metro area in North America, population densities vary in exceedingly complex patchy patterns, which are interesting but remain analytic challenges. Several authors have noted that population density is a poor measure of urban conditions. Nevertheless, the city-to-rural gradient remains a simple useful analytic framework.

A detailed analysis of soil and land use outside of Kano (Nigeria) is particularly interesting (Maconachie, 2007). Described as the overlap of urban and rural, or rural but under the influence of urban processes, the zone is effectively moving outward, as development occurs both adjacent to the metro area and dispersed in the peri-urban zone. Land degradation is conspicuous, with reduced water quality, degraded woodland, loss of especially valuable trees, sand removal, spreading rubbish dumping, and plastic bags blowing "everywhere." Twenty tree species provide economic and other functional values to the residents (see Figure 8.9). With firewood scarcer and in increased demand, residents progressively shift to fossil fuel use.

Soil texture, organic matter input from vegetation, and soil fertility are degrading in the Kano peri-urban zone (Maconachie, 2007). As soil erosion increases and soil fertility and nutrient cycling decrease, residents increasingly add various fertilizers (Figure 2.4a). These are commonly composed of partially decomposed organic materials from solid-waste dumps, sometimes with ashes added.

Unfortunately, heavy metals and other chemicals (see Chapter 4) are commonly at very high levels, often well above the recommended maximum concentration for crop production (Figure 2.4b). For instance, based on 40 soil samples (average pH 7.8, range 5.6–10.6) from three peri-urban agricultural sites outside Kano, scientists discovered the following:

- Pb (lead), 40% of the samples exceeded the 5.0 mg/l maximum recommended concentration (samples ranged from 0 to 46 mg/l).
- Cd (cadmium), 40% exceeded the 0.01 mg/l max. (0–29).
- Cr (chromium), 40% >0.10 mg/l max. (0–49).
- Ni (nickel), 85% >0.2 mg/l max. (0–272).
- Mn (manganese), 90% >0.2 mg/l max. (0–9).
- Co (cobalt), 93% >0.05 mg/l max. (0–23).
- Fe (iron), 98% >5.0 mg/l max. (1–45).
- Cu (copper), 100% >0.2 mg/l max. (range 0.5–3.5).

These consistently high levels suggest that toxic soils due to added fertilizer are widespread.

The use of ashes and decomposed dump material for fertilizer and crop production close to the city recycles wastes and may be cost-effective. However, perhaps unknowingly, consumers of the food produced are most likely getting a heavy dose of health-impairing chemicals. Soil animals and microbes, as well as herbivores consuming the plants, are doubtless severely impacted by such high levels of toxic chemicals.

A general city-to-natural-forest gradient around Seattle (USA) highlights relatively gradual changes in both human and ecosystem patterns (Alberti, 2008). As distance from the central business district increases: population density decreases, elevation increases, percent impervious surface decreases, vegetation cover decreases, land uses change, average parcel size increases, building density decreases, and the

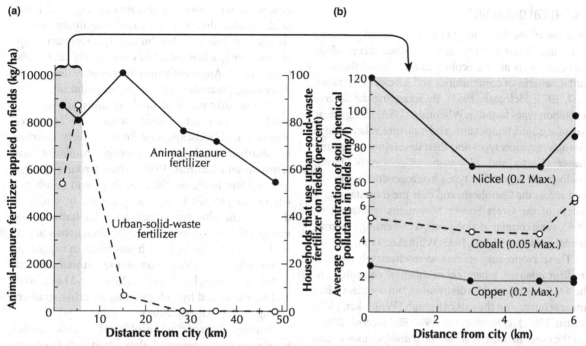

Figure 2.4. Fertilizers and chemical pollutants in soils of peri-urban/exurban farm fields. (a) Animal manure (taki) from local livestock. Urban solid-waste (shara) (composed of street sweepings, household refuse, and animal manure, with most stones, rubble, metal, glass and polyethylene bags removed) is sorted and transported. Animal-manure data at each site are based on three fields (plots) mostly 0.1–0.6 ha (0.25–1.5 acre). Urban-solid-waste data are based on a survey questionnaire. (b) Numbers in parentheses are the maximum soil pollutant concentration (phytotoxic threshold) recommended for crop production. Soil texture and 21 chemical components measured show considerable spatial and/or temporal variability but, with few exceptions, do not correlate with distance from city. Kano, Northern Nigeria. Based on Maconachie (2007).

aggregation of built areas decreases. Meanwhile, forest cover increases, aggregation of forest patches increases, biodiversity changes, microclimate changes, hydrologic flows change, and nutrient cycling increases. Such a sequential gradient provides a preliminary overview, which in turn leads to analyses of the essential mosaic or polycentric patterns on the ground (Alberti, 2008). Numerous ecological patterns and processes correlate with land use revealed in satellite images and GIS analyses (Luck and Wu, 2002), though many other processes operate at finer scales "invisible" in most GIS analyses (Hersperger et al., 2012).

Urban designers also have used the city–suburb–rural gradient in architectural and urban planning conceptions (Calthorpe and Fulton, 2001). A detailed and highly constrained example used by a group of "new urbanists" relates essentially everything planned to a "typical rural–urban transect" with six sequential "transect zones," plus a "special district" category (De Chiara and Koppelman, 1984; Watson et al., 2003; Duany et al., 2008). These zones are idealized and considered to be representative or average, though they may not exist in a linear sequence in a particular city. Although the transect idea was built on, or parallels, that in ecology, the numerous characteristics related to it are overwhelmingly human, with puzzlingly few ecological dimensions.

Finally, all the areas from city center to exurb/peri-urb fit the dictionary definition of *urban* (of or pertaining to a city), and therefore are considered as *urban areas* in this book. The urban-to-rural concept now should be more clearly specified, such as city-center-to-exurban, city-to-farmland, and so forth. The most typical gradient for a large city is: city center; outer city district(s); inner suburb; outer suburb; exurban or peri-urban; farmland; and natural land (forest, woodland, shrubland, grassland, or desert) (see Figure 1.4). One or more of these specific sequential areas may be missing on a particular radius or direction. For an urban region, the most general gradient concept would be metro area to inner urban-region ring to outer ring.

Ecological gradients

Since the 1950s, the concept of environmental gradients indicating a linear sequence of environmental conditions has been widely used in ecology to understand the horizontal patterns of communities and species (Whittaker, 1967, 1975; McIntosh, 1967). By arranging the diverse vegetation types found in Wisconsin (USA) along linear axes of, e.g., soil, temperature and moisture, relationships among vegetation types and their development became clearer (Curtis and McIntosh, 1951; Curtis, 1959). Analogously vegetation types heterogeneously distributed across the Caribbean, and over the mountains and valleys of the Great Smoky Mountains (southeastern USA), were compared along axes representing environmental gradients (Beard, 1955; Whittaker, 1956, 1962).

These pioneering studies were effectively *indirect gradient analyses*, where environmental conditions in the field may be patchily distributed, but are placed in a linear sequence on the axis of a graph (Whittaker, 1975; Austin, 1987; McDonnell et al., 1993; Pickett et al., 2009). On the other hand, a *direct gradient analysis* uses measurements taken at regular (or random) intervals along a straight line or transect crossing a series of environmental conditions, such as from dry upland to wetland, or proceeding up a mountain (McDonnell et al., 1993). Direct gradient analysis with regular linear sampling removes observer bias, both in selecting sample sites felt to be representative or to support a hypothesis, and in ordering samples along an axis. Selecting and arranging samples from the vegetation mosaic in urban regions should be done objectively and carefully described.

Sometimes an environmental gradient is considered to be "simple" if a single factor, such as temperature or soil moisture, is the major change (Whittaker, 1975; McDonnell and Pickett, 1990). More typically, a "complex" gradient involves a number of factors that change together, or co-vary, such as temperature, precipitation and soil when going up a mountain (McDonnell et al., 1993; Medley et al., 1995; McIntyre et al., 2000).

In recent decades, an *urban-to-rural gradient*, paralleling ecologists' gradient analysis techniques, has been widely used. Apparently the first major use was for lichen distribution around cities. From 1866 to 1972, a period including the industrial revolution, city-to-rural lichen distributions were analyzed around 84 cities, mostly in Eastern and Western Europe, but also for Caracas, Christchurch (New Zealand), and New York, as well as Montreal, Arvida, Sudbury and Wawa in Canada (Le Blanc and Rao, 1973; Schmid, 1975). In general, from center city to surrounding rural farmland, species diversity increased, dominant species changed, and the very few "urban" species often disappeared. The gradient patterns were usually curvilinear (non-linear). Apparently the patterns were due to both decreasing urban desiccation and air pollution.

In the early phase of modern landscape ecology, patch and corridor characteristics were compared along a broad landscape modification axis from urban to suburban to cultivated to managed to natural land (Forman and Godron, 1986). Thus, striking changes were evident in the origin, size, shape and number of vegetation patches along the gradient. In addition, different corridor types, networks, habitations, and matrix attributes change markedly from urban to natural land. A similar approach was taken in suggesting patterns for naturalness, human use, human and ecosystem functions, biodiversity, climate and land cover, and nutrients and hydrology along an urban gradient (Hough, 1990; Alberti, 2008).

Numerous specific ecological characteristics have now been measured along gradients from city to rural or natural land for many cities (Figure 2.5). Apparently no synthesis of the results yet exists. Thus, the following list highlights the range of characteristics, along with some illustrative studies as entrees into the literature.

1. *Ecosystem*. (McDonnell and Pickett, 1990; Pouyat and McDonnell, 1991; Pouyat et al., 1995; McDonnell et al., 1997; Zhu and Carreiro, 1999; Carreiro, 2008)
2. *Soil and chemicals*. (White and McDonnell, 1988; Pouyat and McDonnell, 1991; Pouyat et al., 1994, 1995; McDonnell et al., 1997; Groffman et al., 2004; Wollheim et al., 2005; Carreiro, 2008; Carreiro et al, 2009)
3. *Air*. (McDonnell et al., 1997; Gregg et al., 2003; Ziska et al., 2004)
4. *Water*. (Limburg and Schmidt, 1990; Arnold and Gibbons, 1996; Forman et al., 2003; Walsh et al., 2005; Kang and Marston, 2006)
5. *Plants*. (Le Blanc and Rao, 1973; Kowarik, 1990; Medley et al., 1995; Sukopp, 1998; Porter et al., 2001; Burton et al., 2005; Ochimaru and Fukuda, 2007; Zipperer and Guntenspergen, 2009)
6. *Animals*. (Jokimaki and Suhonen, 1993; Blair, 1996, 1999; Blair and Launer, 1997; Clergeau et al., 1998; Zhu and Carreiro, 1999; Germaine and Wakeling, 2001; Marzluff, 2001, 2005; Crooks et al., 2004;

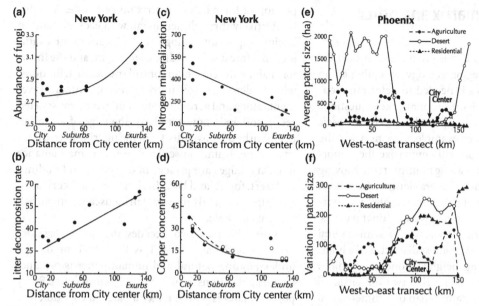

Figure 2.5. Landscape ecology and soil patterns along urban-to-rural gradients. (a) to (d) Forest soil variables from New York City (city center = Central Park) northward to exurban (rural) sites in Connecticut. Based on Pouyat and McDonnell (1991), McDonnell et al. (1993), Pouyat et al. (1995). (a) Measured as logarithm of meters length of hyphae per gram dry weight of litter ($R^2 = 0.82$). (b) Measured as percent loss of original litter mass placed in nylon mesh bags in the forest litter for 24 weeks ($R^2 = 0.71$). (c) Net mineralization in micrograms of N per gram dry weight of soil, based on laboratory incubation ($R^2 = 0.73$). (d) Heavy-metal Cu concentration in mg/kg. Dashed line = organic litter/humus layer; solid line = mineral soil. (e) and (f) Phoenix (Arizona, USA). West-to-east gradient from natural land (desert) to city center to natural land (desert). (f) Coefficient of variation of patch size plotted. Based on Luck and Wu (2002).

Desender et al., 2005; Huste et al., 2006; Hodgson et al., 2007; Evans et al., 2009; Garaffa et al., 2009; Magle et al., 2010)

The urban-to-rural framework has been particularly useful for understanding ecological patterns around New York (Figure 2.5a), Phoenix (Figure 2.5e and f), and Baltimore (McDonnell and Pickett, 1990; Pickett et al., 1997, 2009; Luck and Wu, 2002; Carreiro, 2008; Carreiro et al., 2009). A Seattle study suggests several spatial measures that may be especially useful for understanding ecological patterns along city-to-rural gradients: percent cover of a habitat or land use type, average patch size, perimeter-to-area ratio, edge density, fractal dimension, interspersion, juxtaposition, and aggregation (Alberti, 2008).

Analyzing urban gradients is a simple and useful way to gain broad insight. Yet they offer challenges. For example, measuring enough radii or samples in different directions is important to gain an average and an estimate of variability around a city. The area of a city affects the shape of curves generated along a gradient (Garaffa et al., 2009), as for example comparing the very vertical Shanghai with the extensively spread-out Sao Paulo. Oddly, urban gradient studies rarely include villages, towns, and satellite cities that are abundant and important in most urban-region rings (Forman, 2008; Prados, 2009).

The end points chosen have a strong effect on final results. For example, starting in city center or in one of the districts around it has an important effect on any overall curve along the gradient. Ending the gradient in a peri-urban area, cropland, satellite city, forest or desert also has a major effect on the shape of curves along the gradient. In addition, the specific land uses chosen, and their order, strongly affect indirect gradient studies. For example, one study highlighted specific land uses with decreasing daytime human population (business district; office park; residential; golf course; open-space recreational; and nature preserve) (Blair, 1999). The end points were polar contrasts, whereas each of the intermediate types may vary widely in vegetation and other bird-related features, so the overall results reflect the particular sites selected. Indeed, the ecological characteristic studied may primarily be determined by finer-scale patterns within the broader levels chosen for an urban gradient. These methodological issues are familiar to researchers in the field, but are introduced to improve and widen the base of urban gradient studies.

Patch–corridor–matrix and other spatial models

The urban-to-rural gradient is often a useful start for understanding urban region ecology. Sampling along many radii or directions is needed to determine representativeness of results. Still, many key questions require a two-dimensional approach, i.e., a *land mosaic model* for urban areas. For example, viewing a satellite image of an urban region will emphasize the importance of comparing things along multiple radii. Moving through space along a distance gradient, such as down a mountain or from center city to remote area, normally crosses a mosaic with relatively distinct patches, corridors, and boundaries. Thus, a large semi-natural park may adjoin a central business district, and a high-rise residential/shopping neighborhood may adjoin an outer forested landscape. The two-dimensional land mosaic perspective, rather than one-dimensional radius, adds the importance of patterns perpendicular to the radius. For example, suburbanites commonly commute, shop, and recreate in many surrounding directions. Water and wildlife move in all directions.

Land as patches, corridors, and matrix

Patches, as wide areas or "blobs" differing from their surroundings, appear to be universal in landscapes. Also *corridors* (strips), and the background *matrix*. Theoretically it is possible to have areas covered only by patches, but in practice corridors always seem to be present. An area could resemble spaghetti – only corridors – but this may not exist on land. A relatively homogeneous matrix without patches or corridors may also be non-existent.

The basic characteristics of individual patches are simple (Figure 2.6a). They vary in type (e.g., pond, dump, housing development), size (large, small), and shape (squarish, elongated, convoluted) (see equations, Appendix B). Corridors are wide-to-narrow, long-to-short, straight-to-convoluted, and so forth (Figure 2.6b). The matrix is connected-to-disconnected, extensive-to-limited, and perforated-to-continuous (Figure 2.6c). Definitions of these terms fit the dictionary so this spatial language of landscape ecology readily enhances communication among diverse interests.

All three elements described have boundaries or edges, also ubiquitous. The boundary may be considered to be a line. A narrow strip of *edge*, usually recognizable on each side of the line, differs from the *interior* portion of the patch or matrix on each side. An edge is effectively three-dimensional, with each dimension having important ecological significance. For example, for a forest, edge width, edge height, and the linear form along the edge help control species distributions, wildlife densities, soil nutrient levels, animal movements along and across an edge, and wind-transported seeds, soil and snow. Similarly, the edge of an urban neighborhood may differ from the interior in tree density, crime, traffic noise, land use, and maintenance intensity. Edges are present in objects from buildings to desert, forest, and lake. In short, patches, corridors, and edges are nearly universal in landscapes and urban areas at any scale.

Over the past two decades the *patch–corridor–matrix model* has emerged as the prime model for understanding natural and human patterns on land, and in applications for societal problem-solving. Every point on land is within either a patch, a corridor, or a matrix. Their attributes introduced above are handles for study and for application. The model permits direct comparison of diverse mosaics and their components.

The patch–corridor–matrix model has been widely and successfully used for scientific analysis, scholarly interpretation, and society's environmentally related solutions. An enhanced version of the model is needed to address two land patterns, though they are rare in urban areas: (1) gradual environmental gradients (no distinct patches, corridors, or boundaries across an area); and (2) distinct patches or corridors but indistinct boundaries (various "soft" boundary types) (McIntyre and Hobbs, 1999; McGarigal and Cushman, 2005; Lindenmayer and Fischer, 2006).

The land mosaic, like any system containing life, has structure, function, and change (or pattern, process, and dynamics) (Figure 2.6d, e and f). An area's *structure* is its spatial pattern, i.e., its components and their arrangement. *Function* or *functioning* refers to the flows, movements and interactions present of wind, water, animals, people and transport. *Change* refers to the alterations in structural pattern (and functioning) over time, as in a changing mosaic.

These three central characteristics of a land mosaic are intimately linked in feedback systems. Structure very much determines function. In fact, a key way to improve how an area works is to change patches or corridors in some way. But functioning in turn often alters or creates structure. Furthermore, structure and function both cause changed conditions, which consequently alter structure as well as function. Those are

Patch–corridor–matrix and other spatial models

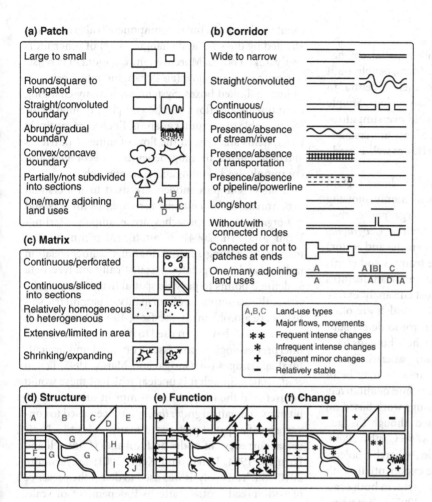

Figure 2.6. Patch–corridor–matrix and structure-function-change characteristics of a land mosaic. See Turner (1989), Forman (1995), Burel and Baudry (1999), Ingegnoli (2002), Farina (2006), Coulson and Tchakerian (2010).

feedback loops, where one thing affects a second, which affects the first. Feedbacks provide for both stability and dynamics. They also illustrate the old adage, "everything is connected." A change here and now quite understandably leads to a change there and then.

Corridors may exist singly but connected corridors of a particular type, forming a network, seem almost universal. Road networks, stream networks, hedgerow networks, trail networks, and pipe networks are nearly ubiquitous (Figure 2.6d and e). All function as systems, with flows from node (intersection) to node along their corridors. Furthermore, networks are barriers and filters that substantially interrupt and disrupt other flows across the land.

Three types of networks characterize urban regions. Nature creates tree-like *dendritic networks*, such as stream and river systems with flowing surface water. Ridge networks in rugged mountains, underground rivers in limestone, and the reverse-network of a mountain stream flowing out onto an arid valley are natural dendritic networks. Humans make dendritic forms such as irrigation and drainage canals, and pipe systems for water and sewage. We also make *rectilinear networks*, including street and hedgerow networks, with predominantly straight lines and many right angles. *Anastomosing networks*, typically with gently curving lines and sharp acute angles, are also common, such as most railway systems and animal-trail networks.

Many attributes, easily understood but little studied ecologically, describe networks (Forman, 1995). Consider, for instance, the effects of attached node size, corridor hierarchies, linkage density, connectivity, circuitry (abundance of loops), linkages per node, and mesh size. Numerous ecological phenomena should be highly sensitive to such network attributes.

Richness of spatial models

The models following tend to be highly spatial, emphasizing that the arrangement of organisms, people, built structures, and the physical environment is central to understanding urban areas and urban ecology. Simple spatial models (which, however, avoid oversimplification) generally have the advantage of easy understanding and communication among parties, as well as ready application or implementation.

Maps are models. The map-maker shows certain features and filters out others, such as only showing major highways and medium-to-large parks, thus eliminating small roads and small parks. *Geographic information system images* (GIS, both raster and vector) are also models that include some features and eliminate others. Hundreds of small but important features in a landscape are filtered out by most GIS analyses (see Figure 1.6). *Random* (or stochastic) models are often used by ecologists as a type of control to detect subtle spatial patterns. In urban areas where humans have used energy to create and maintain patterns, indeed a heavy human imprint, random models may be of limited use. "*Mental models*" where one qualitatively or quantitatively estimates how things work based on previous experience are widely used, though the lack of objectivity limits their usefulness (McCarthy, 2009). A *distance-decay* pattern or model, whereby objects dispersing from a source decrease exponentially with increasing distance, is particularly useful in both ecology and urban studies (Pacione, 2005). Dispersing seeds, animals and people commonly decrease in number with the square of the distance (d^{-2}) from a source location.

Input–output or "*black-box*" models are conceptually simple and useful. Though usually non-spatial, they can readily be made spatial thus increasing their value. In essence, a component or black box contains X amount of the prime item of interest. Input to the box is Y, and output from the box is Z. If the rates Y and Z are equal, the content of the box remains stable. If $Y < Z$, then X shrinks. However, if the internal production of X in the box increases and Z increases at the same rate, X in the box remains constant. But Z could change little, so X in the box grows. Even the growth (or shrinkage) of X could alter Y.

Input–output models are the normal foundation of *systems* models, which effectively are composed of a few or many components interconnected by flows. Thus, the size of a particular component is dependent on production or loss in the box, as well as the direct inputs and outputs. But the component is also normally affected by the size (and changing size) of other indirectly linked boxes. Moreover, the component is sensitive to the rates of flow (and changing rates) between indirectly linked boxes. Systems with many changing components and flows can be complex, though readily amenable to computer analysis. Engineers routinely use systems analysis. In the flow of mineral nutrients and other chemicals (biogeochemistry), ecologists also commonly use a systems analysis approach.

Lots of complex models are used to understand urban areas. *Cellular automata*, *complexity* models, and *self-organization* approaches are examples (Berling-Wolff and Wu, 2004). *Geostatistical* techniques or models (including semi-variograms, autocorrelation indices, and interpolation) statistically analyze data at defined points to provide spatial insight (especially where the assumptions of normality in parametric statistics are a problem) (Gustafson, 1998). Many "*landscape metrics*" have been used to measure patterns and interpret ecological processes in natural and agricultural landscapes (McGarigal and Marks, 1995; Leitao et al., 2006), though it is unclear which of these would be useful in the heavy-human-imprint urban area. A bit of mathematical *graph theory* has been used to convert specific spatial patterns at many scales to a "universal currency" of nodes and linkages (Cantwell and Forman, 1994; Urban and Keitt, 2001; Freyermuth, 2010). Such an analysis was used to detect the presence of widespread spatial patterns independent of scale, land use, population density, and geographic region. Some *economic land-pricing/value* models use a similar approach. Also see equations in Appendix B.

The following "models" are listed in four somewhat overlapping categories: (1) early models; (2) donut model; (3) ecological pattern models; and (4) urban form models.

1. Early models
   (a) *Zones of influence*. Also known as von Thunen bands, this early spatial model highlighted three major concentric bands around a city, intensive cropland, pastureland, and woodland (which provided wood products and game) (Losch, 1954; Cronon, 1991).
   (b) *Central place theory*. Largely an economic approach, this model highlighted population centers as markets, with goods being sold less at increasing distances and proportional to population size (Christaller, 1933; Hartshorn, 1992; Pacione, 2005). Population centers

across the land competed, which resulted in a hexagonal zone of market influence around each center. Villages, towns, and cities had different-sized zones of influence, hence conceptually producing a hierarchy of hexagonal patterns on the land, though actual hexagons were very hard to detect in aerial photographs. Many modifications of the model have been made.

2. Donut model

   The *donut model* highlights the urban region (as the donut) composed of the all-built metro area (the central "hole"), plus the surrounding urban-region ring upon which the metro area mainly depends (Figure 2.7) (Forman, 2008). At this broad scale, one readily recognizes the major types of city: (1) with no major water body (complete donut); (2) on a river (sliced donut); (3) at a river intersection (Y-sliced donut); (4) on a straight coast (half donut); (5) coastal city with bay or harbor (indented half donut); (6) coastal city with river and bay/harbor (indented sliced half donut), and (7) city with river set back from a coast (asymmetric sliced flattened donut). Looking more closely within the donut provides additional insights into key urban region patterns and processes (Figure 2.7). These include (Forman, 2008): (1) zone with coastal microclimate (onshore breeze, etc.); (2) zones subject to coastal storms, cyclones/hurricanes, and tsunamis; (3) overall degree of interaction between metro area and urban-region ring; (4) transportation corridors (radial, around a coastal city, and connecting satellite cities); (5) topography and water flows; (6) flood-prone areas; (7) industrial location and plume zone; (8) nearby competing city; (9) satellite cities and their zones of influence; (10) probability of cropland/former cropland; (11) probability of natural land, including near the metro area; and (12) recreational zones. In effect, the donut model provides a convenient visual tool in wise planning for the future of a city in its urban region.

3. Ecological pattern models

   (a) *Dispersion* model for the distribution of individuals of a species indicates that most species have an aggregated or clumped distribution, some are regularly or evenly distributed, and very few approach a random distribution (Smith, 1996; Cain *et al.*, 2011).

   (b) *Habitat selection* model in its spatial form highlights home range characteristics around

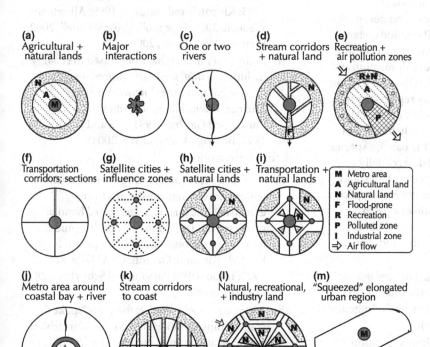

Figure 2.7. Donut models highlighting major land-use patterns in urban regions. Metro area = continuous built area containing city and (usually) suburban land. Agricultural land = active or recent farming, usually cropland. Natural, recreational, and industrial lands = hypothesized optimal locations. Narrow arrows = major water flows (river, streams, windstorms, tsunami). Wide arrows = air flows. (m) Elongated urban region squeezed by a coast, mountain range, or another adjoining urban region. See Forman (2008).

a site selected for nesting or denning by an animal, and probably also relates well to people choosing an apartment or home. For example, in farmland around Oxford, UK, the nesting site of most bird species (and the highest species diversity) correlated with landscape pattern at two spatial scales, i.e., the 5 ha (12.5 acres) right around the nest and a broad 250 ha (625 acres or 1 mi^2) area (Arnold, 1983; Forman, 1995).

(c) *Habitat arrangement* models that emphasize the interspersion of habitat, total boundary length between habitats, the relative lengths of each type of edge present, and convergency points where three or more habitats converge (Hunter, 1990; Forman, 1995).

(d) *Environmental gradient* or *species continuum* model, both in the direct and indirect gradient-analysis form, was described above (Curtis and McIntosh, 1951; Whittaker, 1975; Austin, 1999).

(e) *Ecosystem* models focus either on ecological energetics, i.e., the one-way flow of energy through feeding or trophic levels, or on mineral nutrients (biogeochemistry), which may cycle within or more often flow through an ecosystem (Odum, 1971; Odum and Barrett, 2005).

(f) *Island biogeography* theory or model graphically and mathematically relates species richness on islands to island size and distance from the mainland (species source), resulting from the processes of species colonization and extinction (MacArthur and Wilson, 1967). On land the model was gradually replaced by the patch–corridor–matrix model because (1) a patch is surrounded by diverse habitats, each being a source of effects on the patch, a species source in its own right, and differentially suitable for movement between patches, and (2) the roles of specialist and generalist species plus the interior-to-edge ratio of a patch seem to be quite important. Urban habitats normally are much closer to the land model than the island model.

(g) *Patch-corridor-matrix* model or *land mosaic* model was described above (Forman, 1979a; Forman and Godron, 1981, 1986; Collinge and Forman, 1998; Farina, 2006).

(h) *Boundary* or *edge* model relates vegetation density, herbivore density, predator foraging, species richness, and other ecological attributes to the three-dimensional form (width, height, and longitudinal heterogeneity) and the hardness/softness (straight abrupt, curvy/convoluted, gradient, or strip of micro-mosaic) of edges between habitats (Forman, 1995).

(i) *Metapopulation models* analyze the dynamics of local colonizations and extinctions on patches for a species with sub-populations on separate patches, but connected by occasional movements of individuals between the patches (Cain *et al.*, 2011; Morin, 2011).

(j) *Fractal* models were introduced above (Milne, 1988).

(k) *Ecological network* refers to the distribution and suitability of habitats as perceived and used by animals in movement through a land mosaic (Jongman and Pungetti, 2004; Opdam *et al.*, 2006; Dalang and Hersperger, 2012).

(l) *Emerald network*, as a group of large natural or semi-natural patches interconnected by corridors or effective stepping stones, may be the optimum ecological network (Forman, 2004b, 2008; Massa *et al.*, 2004).

(m) A suite of *landscape ecology* models is summarized in various reviews (Forman, 1995; Klopatek and Gardner, 1999; Alberti and Waddell, 2000; Ingegnoli, 2002; Waddell, 2002; Farina, 2006; Alberti, 2008).

(n) Many other models have been used in ecology including *spatial stochastic*, *process-based*, *fuzzy-logic*, and *neural-network* models (Turner, 1989; Sklar and Costanza, 1991; Turner and Gardner, 1991; Wu and Levin, 1997; Berling-Wolff and Wu, 2004).

4. Urban form models

(a) *Greenbelt* model and *urban growth boundary* pattern are circular constraints to city expansion, the former a designated band of protected land (e.g., around London) and the latter a line beyond which development is limited (Portland, Oregon, USA) (Howard, 1902; Elson, 1986; Parsons and Schuyler, 2000; Forman, 2008).

(b) *Metro area border* models focus on "spokes" (radial transportation routes), circular/orbital transportation routes, or edge cities/nodes (Garreau, 1991; Ravetz, 2000; Pacione, 2005).

(c) *Land market value* model relates land uses and urbanization to distance from city center and other key locations (Alonso, 1964).
(d) *Environmental constraints and urban form* emphasizes how metro areas do or do not fit well with natural patterns, and how mountains, rivers, water bodies, and so forth shape cities (Whitehand and Morton, 2004; Pacione, 2005).
(e) *Urban-to-rural gradient* model was described above (McDonnell and Pickett, 1990; Hahs and McDonnell, 2007).
(f) *Concentric, sector,* and *multiple nuclei* models, which have long been standards in urban geography, indicate major land-use (land cover) distributions around a city center as either concentric zones, pie-section-shaped sectors, or dispersed districts (Hartshorn, 1992; Pacione, 2005). Many modifications and elaborations have been made, recognizing differences in geography and mechanisms.
(g) *Orthogonal grid* models often reflect the central imprinting role of transportation within a city (Platt, 2004).
(h) *Pattern perception* models highlight generic features perceived and used by people in an urban area, such as lines, nodes, edges and so forth (Alexander *et al.*, 1977; Lynch, 1981; Lynch and Hack, 1996; van Bohemen, 2005).
(i) *City as ecological system* model goes well beyond specific ecosystem characteristics to provide a broad multi-dimensional approach for understanding an urban area (Alberti, 2008; Newman and Jennings, 2008). The model may include: principles of form and function; diversity; adaptiveness; interconnectedness; resilience; regenerative capacity; symbiosis; holism; systems interactions between parts; processes; complexity; hierarchical and context factors; flows of energy, materials and information (Grimm *et al.*, 2000); city-shaping relative to transportation patterns; self-organization (zero waste, self-regulating, resilient/self-renewing, flexible) (Bossel, 1998); and ecological succession. This array of characteristics of course represents a wish list or a recommendation for future study and modeling.

The preceding list of spatial, mainly ecological, models highlights the richness of models available, and offers ideas for selecting appropriate modeling approaches for urban ecology research. Some are major and others minor; some descriptive, others quantitative; some lie dormant, others are actively used. Some investigators recommend using at least two model types to generate valuable hypotheses and to approach understanding.

In a search for universal spatial patterns present in any landscape, a bit of graph theory from mathematics provides interesting insight (Cantwell and Forman, 1994). Twenty-four aerial photographs of areas varying in size by five orders of magnitude, and also varying widely in population density, land use, climate, and worldwide geography were examined. These aerials were directly compared (à la graph theory) by converting all photographed patterns into simple networks of nodes, and linkages between nodes. Perhaps surprisingly, five specific spatial patterns were found in 85% or more of these extremely diverse land mosaics. One pattern, for example was an area of matrix with scattered patches embedded within it, like a residential area with dispersed schools or a coniferous forest containing scattered bogs. Although only five simple patterns were widespread, other repeated patterns were detected in the diverse set of landscapes. Probably more widespread basic patterns await discovery. Perhaps even universal patterns exist. Tailoring societal solutions to such patterns offers promise.

Another way to evaluate universality is to convert everything to the same currency or units. Some economists use market values to give everything a monetary value permitting direct comparisons. Analogously, everything from toothbrushes and labor to soil and religion has been converted to the currency of calories or "emergy" (Odum, 1983).

Neighborhood mosaics and their linkages

To create a stunning patchwork quilt that lasts a family for generations, the quilter envisions clusters of colors and shapes, and sews the pieces of cloth together tightly. A unique "crazy-quilt," composed of diverse patch sizes, shapes, materials and textures, requires particular attention to linkages along the irregular boundaries. So it is with neighborhood mosaics. Land uses (land covers) and habitat patches are strongly interconnected.

We explore this idea in four steps: (1) land mosaic concept; (2) effect of the surroundings; (3) adjacencies and dependent pairs; and (4) neighborhood mosaics. To glimpse the big picture at the outset, we consider a land mosaic to be the spatial pattern of elements in a large area, an interwoven mosaic refers to elements tied together by strong interactions, and a neighborhood mosaic focuses on a cluster of nearby elements with mainly positive interactions.

Land mosaic concept

As two-dimensional patterns of diverse usually small adjoining elements, "mosaics" occur at any scale from sub-microscopic to the globe. *Land mosaics* on the Earth's surface generally represent the broader range of scales from perhaps meters-wide upwards (Forman, 1995). Thus, habitats and land uses predominate as the key elements or components present. A perfect grid has all elements the same, equal-sized squares. But nature's processes and human activities always produce mosaics with elements varying in size, shape, and type. Indeed, many ecological factors have been related to the properties of land mosaics (Turner and Gardner, 1991; Bennett *et al.*, 2006; Farina, 2006; Lindenmayer and Fischer, 2006; Haslem and Bennett, 2008). Although corridors or strips are normally present in land mosaics, patches predominate. The arrangement of the patches and corridors is a unique structure, a core characteristic of each land mosaic.

Furthermore, whole landscapes have properties that can be quantified and compared (Bennett *et al.*, 2006; Radford and Bennett, 2007). Structural properties of particular interest in these studies are: extent of habitat; composition of the mosaic; spatial configuration of elements; and geographical position.

The network of interactions tying the elements together is the other core attribute. Flows and movements of objects, materials and energy link the group or cluster of patches. In effect, these processes determine how the land mosaic works, how it functions. Flows among the elements also indicate how stable or long-lasting a mosaic is.

Thus, the strength of interactions, i.e., amount and rate of flows/movements, among mosaic components is a key (Figure 2.8). Strong rather than weak interactions indicate: (1) a tightly interwoven mosaic, with dependence of one element on another, or interdependence between or among patches; (2) an actively functioning mosaic; and probably (3) stability for the future.

Figure 2.8. A center city urban mosaic. Skyscraper area (lower right and right center); adjacent parks (Boston Common and Garden; lower center); low-rise residential areas (left and center); Charles River above dam (left); Cambridge business and residential areas (upper left); transportation and industrial area (upper center); mixed-use area (upper right). Road, rail, and river corridors interconnect the land-use patches. Boston. R. Forman photo.

An *interwoven mosaic*, as a group of landscape elements tied together by strong interactions (flows and movements), highlights an active functioning unit of tightly interacting land uses and habitats. Stronger interactions suggest flows or movements that cover much of the surface of patches, rather than simply in a narrow zone along patch borders. Indeed, although some flows go along borders, strongly interweaving a mosaic implies mainly perpendicular flows across boundaries.

The strength of an interwoven mosaic is analogous to the concept of a natural plant or animal community (Austin, 1999; Cain *et al.*, 2011; Morin, 2011). Most ecologists consider a *natural community* to be a group of species coexisting at a site. The species in the community may be tightly interlinked with abundant interactions, e.g., of nutrient flow, herbivory, predation, competition and symbiosis. However, if few or low-level interactions occur among the coexisting species, the group may be referred to as a "species assemblage."

Corridors within the mosaic typically play especially important functional roles. As conduits they channel flows, both positive and negative, along their length (Forman, 1995; Bennett, 2003). As barriers or filters, flows and movements across corridors are blocked or reduced. Consequently, the arrangement of corridors within and around a land mosaic strongly affects the direction and rate of flows. For instance, corridors parallel versus perpendicular to a boundary, and gently curving versus convoluted corridors, have quite different functional effects.

These patterns and principles permit us to estimate the size or extent of an actively functioning, tightly interwoven mosaic on land. One approach is to identify the major barriers or filters in the land. Corridors such as rivers, railways, ridges, and busy highways are familiar barriers separating both animal home ranges and human neighborhoods. Another common case determining the edge of an interwoven mosaic is a hard abrupt boundary between highly contrasting land uses. In both the corridor and hard-edge cases, estimates of flows crossing are valuable, since more interactions than expected may occur. For example, informal squatter settlements (favelas) in Sao Paulo and Rio de Janeiro are often quite close to, and actively tied by employment to, wealthy neighborhoods. Or, wildlife may regularly move in foraging and seasonal migration between north and south sides of a ridge.

A second approach to determining the extent or boundaries of an interwoven mosaic results from identifying the organizing force, if any, for the area. This may cover an area somewhat equally, such as government, religion, or ethnic identity, with a relatively distinct drop-off at the outer edge. The edge-to-interior ratio may be usefully estimated, since small mosaics have mostly edge conditions affected by the surroundings, while a large mosaic may have a significant interior portion with few interactions with the surroundings.

Alternatively, the organizing force may be at a somewhat central location, such as a major resource, employer, civic/cultural plaza, pond, or transportation hub. With a central location, the distance-decay relationship helps determine the extent of an interwoven mosaic. Thus, an exponential or d^{-2} decrease in flows or interactions typically occurs with distance from a source.

"Self-organizing principles" have been used to understand spatial patterns, and especially their similarity across the land. Farms and villages in a landscape often appear surprisingly similar. Housing developments around Orlando (Florida) or Calgary (Canada) do as well. An organizing force (such as farmers or developers), a mix of elements, and ample time often lead to similar results in a landscape.

Urban regions are large land mosaics where the city or metro area is the central organizing force (Forman, 2008). In effect, the city and its ring-around-the-city functionally interact, and seem to be interdependent. The mosaic pattern and conditions of the urban-region ring are strongly determined by flows from and to the central organizing force.

At a finer scale, a *locality-centered mosaic* or "region" was determined in a different way for a suburban town near Boston (Forman *et al.*, 2004; Forman, 2008). Surrounding towns were involved in a significant portion of the issues regularly faced by the town (e.g., education, airport, clean-water supply, wildlife movement, railway, rare species, vernal pools, large natural habitats, hospital, biking/walking trails, state forest, national park, and much more). Based on an extensive list of such issues, the towns involved were listed, and it was clear that 14 other towns regularly interacted with the central town. From the central town perspective, the group as a whole functioned in a town-centered or locality-centered region.

The third approach for estimating the size of an interwoven mosaic is to focus on the set of basic needs (or uses) of the species and humans present. The "home range," i.e., area used in daily movements by an animal, is helpful here (Hunter, 1990). The basic needs or uses of the animal are food, water, nesting/denning cover, and escape cover (from predators or danger). Analogously, urban people typically require or frequently use a residence, food store, goods shopping, school, employment, park or other leisure space, restaurants, and entertainment spots. Mapping the set of frequent-use locations helps identify not only flows and movements, but the extent of an interwoven mosaic.

Interactions also indicate much about the persistence or stability of a land mosaic. A stable organizing force provides a long life. Dependence of one component on another, such as homes depending on a well or other water source, adds stability. Interdependence, such as homes depending on a grocery store, and vice versa, adds still more stability. The other side of interdependence is that if one major component is degraded or eliminated, the mosaic system as a whole may degrade, even unravel. Industrial interdependence, or symbiosis, where one industry uses the output of a second, and vice versa, is an example. Providing redundancy, diverse resources, multi-modal transportation, alternative routes, and so forth provides stability (Forman, 2008). A sense of place by residents, resulting from familiarity and affinity, leads to care and stewardship. That also provides mosaic stability.

Adjacencies

Rarely can we escape being strongly affected by what is next to us. "Adjacency" refers to an object (e.g., land use or habitat) that adjoins (is in contact with) another object. An adjacent land use may surround the entire boundary of a patch, or at the other extreme, be in contact only at a point (point adjacency). Most adjacencies adjoin a portion of a patch's boundary. An *adjacency effect* is where one element significantly affects an adjoining element.

In a study of 40 urban/suburban woods in Springfield (Massachusetts, USA), bird species richness decreased as the density of buildings adjacent to woods increased (Tilghman, 1987). In a Finland study, adjacent building density surrounding a park affected avian nesting outside the park, as well as foraging by the birds inside the park (Jokimaki, 1999). Adjacency effects are doubtless widespread, such as factory air pollutants degrading an adjacent semi-natural park, fires in a fire-prone woodland destroying an adjoining housing development, and stormwater runoff and summer heat from a large tarmac/asphalt parking area affecting an adjacent downslope and downwind area. However, adjacent land uses may also have little interaction (Moran, 1984; Mesquita et al., 1999).

In the Barcelona Region, where wooded patches are adjacent to urban areas, the species richness of plants thriving in human-disturbed habitats (synanthropic species) is higher in large woods than in small woods (Guirado et al., 2006). Also the richness of common forest species, as well as all forest species, is higher close to the edge, and lower in the interior, of the woods. Edges of woods are visited by people more frequently than are interior portions. However, where woods are adjacent to crop fields, the synanthropic plant species richness is higher in small woods than in large woodland patches. Overall, forest size and adjacent land use type are the two most important variables affecting the plant species diversity.

In exurban/peri-urban areas, especially in sprawl areas of North America, several terms have been used to describe the mix of housing and natural or semi-natural (non-agricultural) land – interface, interdigitation, intermix, interspersion. The natural land has sometimes been called "wildland," which is wild relative to the center of a housing development but typically is much used, degraded, and not very wild relative to more-distant natural land. People use the adjacent semi-natural areas for recreation/leisure, dumping, resource extraction, and other activities that degrade the ecological value (Matlack, 1993; Radeloff et al., 2010). Meanwhile, nearby natural areas are sources of water flows, clean air, biodiversity and, in dry climates, wildfire entering the adjacent housing development.

The interface has been described as the area where natural vegetation and urban encroachment coexist, but neither predominates (Ewert et al., 1993; Theobold, 2004). "Interspersion" suggests land uses being intermixed or scattered at intervals, whereas "interdigitation" suggests an interfingering of lobes. One study, related to exurban wildfire, estimates that 1.9% of the contiguous USA has housing [>1 unit per 16 ha (40 acres)] near large continuous vegetation areas (Vince et al., 2005). However, housing intermingling with small areas of non-agricultural vegetation may cover about 7.4% of the land.

In the fire-prone Mediterranean Basin, often the most hazardous fire conditions are where patches of built area cover 20–40% of the land and are intermixed with large areas of natural land (Forman, 2004; Francisco Rego, personal communication, 2003). Large vegetation areas tend to have fewer but bigger hotter fires, whereas smaller wooded areas intermixed with housing normally have more but smaller fires, which are more easily extinguished. Fire management that protects buildings, protects forest growing wood products, and protects the set of rare or fire-adapted species is especially challenging in such intermixed areas full of adjacencies.

A closer look at adjacencies reveals the importance of *adjacency arrangement effects*, where the configuration of nearby elements significantly affects a particular element (Hersperger and Forman, 2003). South of Calgary (Alberta, Canada), woodland patches of quaking aspen (*Populus tremuloides*) are often surrounded by grassland and/or shrubland. Along a gradient from 0% to 100% adjacent shrubland (i.e., all to no grassland), plant species composition in the woods changed from many weedy and introduced species to many moist-environment species as adjacent shrubland increased (Figure 2.9b). The proportion of native species, perennial species, and shrubland species in the woods increased linearly, while the proportion of grassland species decreased linearly. Total plant species richness in the woods remained essentially constant as the adjacent habitat types changed.

A modeling study using graph theory attempted to see if adjacency arrangements were predictable for different types of land uses, irrespective of spatial scale and worldwide geography (Cantwell and Forman, 1994). House plots usually had four (or two) adjacent land uses irrespective of landscape context. Crop fields and hedgerows each typically had four adjoining land uses anywhere. In contrast, roads and rivers normally had

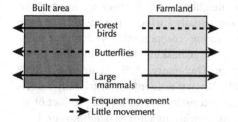

(a) Animal movement across different land uses

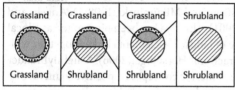

(b) Adjacency arrangement: plants in a wooded patch surrounded by different amounts of two land uses

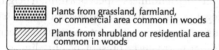

(c) Expected biodiversity of wooded park in exurban/peri-urban area

Park supports:
Many butterflies
Few forest birds
Some large mammals
Small area of house plot/garden plants

Few butterflies
Many forest birds
Some large mammals
Large area of house plot/garden plants

Few butterflies
Few forest birds
Some large mammals
Small area of house plot/garden plants

Figure 2.9. Biodiversity patterns in differing land-use adjacency arrangements. (a) Based on measurements and modeling of 3 butterfly species, 12 forest-bird species, and 1 large-mammal (deer) species in The Netherlands (Knaapen et al., 1992; Forman, 1995). (b) Based on measurements in aspen (*Populus tremuloides*) patches (averaging about 0.4 ha = 1 acre) in a rural area south of Calgary (Canada) (Hersperger and Forman, 2003). (c) Expected patterns based on (a) and (b).

many adjacent land uses (8+ in the study). Five adjacency arrangements of different land uses were found in ≥85% of the landscapes studied. In short, adjacency arrangements appear to be an important and promising basis for understanding land and neighborhood mosaics.

Dependent pairs and the surroundings

Animals that regularly use two or more habitats are effectively *multi-habitat species*. Common examples are a deer or hawk moving between woodland and meadow, geese between water body and grassy area, and common pigeons between built structure and open park. Active frequent usage of both sites suggests that the animals may depend on or require the sites to survive or thrive. However, without an experimental study, the usage may simply reflect preference. That is, if a site disappeared, the animal is genetically flexible enough to switch usage to a different site or type of site.

In many cases the presence or survival of a component in a site depends on another site, which may be adjacent to the first. A reservoir's water quality and quantity depend on the surrounding drainage basin. Wetland or stream level depends on the degree of pumping by an adjacent well. The well may lower the water-table, thus drying out any water body in its "cone of influence." A beaver pond and dam depends on the beaver having sufficient suitable woody plants to cut in the adjacent woodland. Bat activity (probably proportional to bat number) was found to be greater in industrial/commercial areas when they are next to greenspace (Gehrt and Chelsvig, 2003). On the other hand, bat activity in industrial/commercial areas was less next to farmland, but greater next to farmland containing a water body. More familiar is a city apartment or shop that depends on the adjacent street.

Some apparent dependencies are for a non-adjacent element. A market-gardening area depends on, but is often slightly separated from, the nearby city. A cranberry-growing bog depends on a water body, either adjacent or nearby. An area of warehouses for storage and trans-shipment of goods often depends on, but may be separated from a municipal airport. In some well-functioning "metapopulations," the sustained presence of a species on a small patch depends on a nearby large patch as a source of dispersing individuals (Forman, 1995; Magle et al., 2010; Morin, 2011).

Indeed, some patch pairs are *interdependent elements*, or at least mutually beneficial to each other. A sewage-treatment facility pipes its outflow to a water body, which in turn depends on the water and effectiveness of the facility. Visitor/tourist sites and hotels benefit each other. A small shopping area depends on the surrounding residential area, and vice versa. An elementary school and the residential area are symbiotic. Blackbirds foraging in residential areas depend on a wooded park for night roosting, while the park-roosting birds depend on residential areas for feeding. Rather than being clumped together, the thousands of high-rise residential buildings in Sao Paulo are often separated by small greenspaces (see book cover). These spaces, required by regulation, around high rises provide playground, parkland, walkways and meeting places for the residents above.

Other linkages between elements or land uses occur in mosaics. A matrix surrounding a patch is a major source of flows affecting the patch. A surrounding residential area depends on a park, but sometimes not vice versa. As described above, central place theory highlights the mutual interactions between a population center and its surrounding land or zone of influence (Pacione, 2005). A strategic point is mainly dependent on characteristics of the matrix, either the rarity of a site or its location relative to shape of the matrix (Forman, 1995).

Various empirical ecological studies highlight the important effect of "the surroundings" on a site (Luck et al., 2004). Invertebrates in a UK shrubland (*Calluna* heath) differ and change according to patterns in the surroundings (Webb and Hopkins, 1984; Webb et al., 1984). Intensive conifer plantations increase predation on bird nests in adjacent habitats (Villard, 2012). Surrounding land uses affect forest birds (Dunford and Freemark, 2004). Avian diversity and movement depend on the proportion of woodland in a built area within 500 m of the edge of a park (Ichinose, 2005). In each case, further study might reveal that a specific element in the surroundings, rather than the surroundings as a whole, is mainly responsible for the effect reported.

Degrading a non-adjacent area in the surroundings of a park in the Greater Yellowstone Region produces four major results (based on known mechanisms and ecological theory) (Hansen and DeFries, 2007). (1) The "effective size" of the protected park area is reduced. (2) Ecological processes and flows are disrupted. (3) Critical habitats, such as species sources and movement routes, in the surroundings are lost or degraded. (4) Edge effects from the degraded site ripple outward into the park.

Neighborhood mosaics

The idea of "neighborhood" focuses on nearness or proximity, and usually includes the idea of positive interactions among components (*Webster's College Dictionary*, 1991; Chaskin, 1995; Moudon, 1997; *Oxford English Dictionary*, 1998; Fainstein, 2000; Steiner, 2002). Nearby residents with friendly associations are characteristic. Often this is a product of, or leads to, similar conditions across a neighborhood.

A neighborhood mosaic links the human and the ecological dimensions by focusing on land uses or habitats as the components (Addicott et al., 1987; Forman, 1995; Engstrom and Mikuminski, 1998; DeAngelis and Petersen, 2001). The idea has been pinpointed as a local assemblage of landscape elements linked together with strong interactions (Hersperger, 2006). A slightly broader *neighborhood mosaic* concept is a positively interacting cluster of nearby land uses or habitats. Positive interactions highlight the friendliness aspect of a human neighborhood. Negative interactions could disrupt the cluster of sites. Thus, although negative interactions between components may normally be present, positive linkages predominate in a neighborhood mosaic.

Interactions tying a neighborhood together

The strength of interactions measures the tightness of a neighborhood mosaic or site cluster (Figure 2.8) (Addicott et al., 1987; Todd and Todd, 1994; Chaskin, 1995; Hersperger, 2006). Flows and movements of wind, water, heat, seeds, animals, noise, people and other objects among the land uses may vary from high to low. The contrast or difference between adjacent habitats and the porosity of boundaries affect flow rates tying the mosaic together. In effect, an "interwoven mosaic" refers to elements tied together by strong interactions, while a "neighborhood mosaic" focuses on a cluster of nearby elements with mostly positive interactions.

The preceding sections have highlighted the roles of adjacencies, adjacency arrangements, dependent habitat pairs, interdependent habitat pairs, and the surroundings (Figure 2.9). These spatial arrangements help integrate the neighborhood mosaic. Other types of spatial arrangements inter-linking the site cluster doubtless exist and warrant study. Also, as described above, several types of flows are spatially dependent. Distinct corridors, and often networks, are typically present in a neighborhood mosaic. Such linear features represent conduits and barriers for flows. Perceptual or functional connectivity emphasizes that animals or people may move through varied habitats or land uses without using visible structural corridors. In Australia, some Aborigines create a fine-scale mosaic on the land using "fire-stick" burning, and women gather food by ever-changing routes through the finite well-known mosaic (Forman, 1995). Other examples of flows and effects on flows in a neighborhood mosaic are "boundary-crossing frequency," "diverse-habitat crossing frequency," "follow-the-leader traplining," and "species dispersals in a metapopulation."

Guidelines for the typical distance that flows go have been determined in certain regions, as follows. More flow types could be added, but such data would seem to vary by region and locality.

- Most people will walk up to 1 km or 0.5 mi in commuting to work.
- Elementary schools in suburbs average about 1.5 km or 1 mi apart.
- Urban parks are also commonly planned for that distance, with residence to park 0.5mi apart.
- Zones around streams and rivers are typically mapped for the average 50-year and 100-year floods.
- Traffic noise disturbance from a busy highway often extends outward 1+ km.
- The cooling effect of a medium-large urban park typically extends from the edge 0.5–1.0 km (see Chapter 5).

Suppose, instead of mapping land uses and habitats, we mapped the flows between them (Figure 2.6d and e), and made the thickness of arrows proportional to flow rate. This rarely done flow-map would highlight areas of both major and little interaction among sites. A cluster of nearby sites is likely to be a concentrated area of flows. In cropland of Southern Ontario, rather little movement of small mammals and birds occurs between woods and fields (Wegner and Merriam, 1979). Also little movement occurs between hedgerows and fields. Most inter-site movement of the animals is between woods and attached hedgerows, highlighting a hot spot of movement at that point of attachment. Such studies of flows and movements in a neighborhood mosaic or site cluster would be valuable for spatial planning.

The home ranges of terrestrial vertebrates helps tie together neighborhood mosaics. In "habitat selection," animals choose a nesting/denning spot where the immediate surroundings are suitable, away from disturbance, and with ample cover to hide from predators and people. But in selecting the spot the animals also look for suitable broader surroundings, e.g., with ample food available, access to water, seasonal changes, and perhaps a refugium to survive severe disturbances such as fire or intense hunting (Lake *et al.*, 2007). Habitat selection by birds in Oxford (UK) farmland mainly focuses on both the immediate 5 ha (12.5 acres) and broader 250 ha (1 mi^2) (Arnold, 1983).

"Multi-habitat species," those that regularly use or require two or more land uses/habitats, move in a cluster of sites. If several or many such species use the same or similar habitats, the neighborhood is likely to be tightly integrated by these frequent movements.

Another widespread case (occasionally called "landscape complementation") is where many species, especially some insects, require different habitat types to complete their life cycles (Dunning *et al.*, 1992; Taylor *et al.*, 1993; Pope *et al.*, 2000). Where the habitats are in proximity, the movements contribute to the tightness of a cluster. Yet another example is short-distance seasonal migration, illustrated by many birds and ungulates moving between summer high-elevation forest/meadows and winter grassy valleys in the Alps or Rocky Mountains. In Southern California, deer migrate between winter south-facing slopes and summer north-facing slopes.

Many neighborhood mosaics seem to be centered on an organizing force, such as a central plaza, small park, shopping area, or government/cultural center. Probably this central place is the epicenter of flows, and radial flows and movements may predominate in the site cluster. Yet the organizing force may also be a corridor such as a riverside park or coastal walkway.

A *convergency point*, where three or more habitats or land uses converge, is a particularly interesting case of a central organizing force contributing to a neighborhood mosaic (Forman, 1995). In the US Midwest and Rocky Mountains, the optimum location for resident quail (bobwhite, a small game bird) is reported to be where four habitats converge: woodland, shrubland, meadow, and cropland (Leopold, 1933). A combination of the same four land uses is also considered to be optimum for cottontail rabbit (*Sylvilagus*), ringneck pheasant (*Phasianus*), wild turkey (*Meleagris*), and white-tailed deer (*Odocoileus virginianus*). Indeed, quail populations seem to be highest when these four habitat types each cover about a quarter of the area. Three close-by land uses may be optimum for moose (*Alces americana*), mule deer (*Odocoileus hemionus*), elk (*Cervus canadensis*) and grizzly bear (*Ursus arctos*).

The different land uses in proximity provide a diversity of food, and also provide stability in the face of disturbances. Those are the same main reasons so much of the human population lives near seacoasts. Yet two other functions occur at convergency points. The point is effectively the tip of a lobe for each adjoining land use or habitat (Forman, 1995). This means that much movement occurring in the habitat may be funneled to the lobe, with some movement continuing across the point and into the next habitat. In effect, the point is a narrows through which some animals are funneled. Furthermore, the convergency point is an ideal spot for a predator to hunt for food. Thus, a hawk perched on a tree at the point has different habitats visible in perhaps a 270-degree arc. In short, the convergency point is another key component of neighborhood mosaics.

A mixed use neighborhood may have residential, shopping, food-producing, small manufacturing, and recreational sites intermixed. Such a place is likely to be effectively an interwoven neighborhood (see Figure 11.1) (Yokohari et al., 2000).

Patch-centered and corridor-centered mosaics

A "patch-centered mosaic" has a key central patch interacting with the surrounding land uses. Several spatial characteristics familiar in landscape ecology and conservation biology affect, even control, the central element (call it patch X) in the mosaic: size; shape; distance to nearest large patch of type X; and the distribution of stepping stones (Primack, 1993; Forman, 1995). However, an *adjacency arrangement model* or pattern highlights the surroundings and their configuration as potentially of equal or more importance in affecting patch X (Hersperger and Forman, 2003). Adjacency patterns focus on the: (1) number and boundary lengths of adjoining patches; (2) number of types of adjoining patches; (3) sizes of the adjoining patches; and (4) number and type of corridors alongside, or attached at a point.

Thus, aspen woods in Alberta (Canada) differed in plant species richness and composition depending on whether one or two habitat types were adjacent, as well as the relative boundary length of the two types along the woods margin (Figure 2.9b) (Hersperger and Forman, 2003). For woods in Southern Ontario cropland, increasing the number of attached hedgerows (connected to other patches) from zero to four progressively improved habitat conditions for a small mammal species in the woods (Fahrig and Merriam, 1985; Henein and Merriam, 1990; Merriam *et al.*, 1991; Forman, 1995). Also the quality of the corridors affected movements of the woods animals.

Empirical literature on the other components of the adjacency arrangement model is scarce. Thus, we hypothesize a series of patterns. Adjacent patches may be a source of excess heat, floodwater, wildfire, or herds of overgrazing herbivores, or alternatively, a source of rare species, pollinators, much-needed water or scarce nutrients. Although some interactions in a neighborhood mosaic are negative, positive interactions predominate. Therefore, species richness in patch X may increase with the number of adjoining patch types. The rate of increase would progressively lessen as the species pool (total number in the landscape) is approached. Thereafter species richness can be expected to decrease, as only generalist species remain when adjoining patches are simply slivers touching patch X at points. Somewhat analogously, patch X species diversity increases and then drops as the number of types of adjoining patches increases. In this case of patch types, the peak is higher since each adjoining patch contains a different set of species. Increasing the size of adjoining patches also raises patch X species richness, because the larger patches typically contain more interior species. Points of contact, or short boundaries, with patch X, as mentioned above, indicate the presence of lobes of adjoining patches and thus funnels of species to patch X.

A "corridor-centered" neighborhood adds its own attributes to the adjacency model variables. Again familiar variables from landscape ecology and related fields include corridor width, connectivity, habitat quality, and straightness/convolution (Primack, 1993; Forman, 1995). The key adjacency-related characteristics seem to be (Figure 2.6b): (1) curvilinearity; (2) directionality of movement along a corridor (i.e., upwind/downwind, upslope/downslope, toward/away from high quality habitat or land use); (3) density and type of convergency points along the boundary; and (4) barrier or filter effect. Corridor-centered neighborhoods apparently have not been studied ecologically, but may be important in many urban areas.

Convoluted boundaries tend to have considerable movement between adjoining habitats, but almost no movement along the boundary (Figure 2.6a). This is illustrated in the USA by plants and small mammals around Great Smoky Mountains National Park (Tennessee/North Carolina) and by ungulates at grassland-woodland boundaries in New Mexico (Ambrose and Bratton, 1990; Forman, 1995). In contrast, straight

Figure 2.10. A distinct urban neighborhood with low-rise residential housing and shops on the ground floor. Extensive walking, plus tiny "squares" with outdoor tables and chairs for eating and sitting. Las Letras, Madrid. R. Forman photo.

boundaries have relatively limited movement between adjoining habitats, but tend to channel the movement of animals and people along boundaries.

Flow directionality is illustrated in the "road-effect zone," where three factors control flows outward from roads (Forman and Alexander, 1998; Forman et al., 2003). (1) Wind transports airborne dust, road salt, noise, and so forth. (2) Gravity underlies water-borne transport of water, sediment, pollutants, and floating plankton. (3) Habitat quality highlights locomotion and animal or human behavior in moving from low- to high-quality habitats. Convergency points occur in lines along a corridor, reflecting every change in type of adjoining habitat. Finally, a barrier or filter adjoining the boundary of a corridor eliminates or reduces interactions in that direction.

Clearly the land uses and habitats surrounding a patch or corridor are of major ecological importance in a neighborhood mosaic. The adjacency arrangement model or hypothesis is currently based on scraps of evidence. The relative importance of the familiar landscape ecology variables versus the adjacency model variables is unknown, and awaits research.

The perceptive eye sees mosaics everywhere, a near-universal pattern. They dazzle, and they draw in the inquisitive mind. A first glance reveals mosaics to be simply a combination of patches, corridors, and matrix ... little more. Yet no twins live in the world of mosaics, each view is unique. We have to, and can, model mosaics, but still the basic widespread types, presumably encountered repeatedly, remain a puzzle to solve.

Two neighborhood mosaics

Finally, some neighborhood mosaics seem to hang together by numerous tight interactions rather than by a central organizing force (Figure 2.6e). For an urban riverside area, the river is an organizing force. Yet other forces in four somewhat-parallel strips also tie the wide zone together: river, floodplain, infrastructure, and adjacent terrestrial land uses (see Chapter 7). (1) The river flows at normal flow stage, it floods, and it has low flows typically in drought periods. (2) The floodplain has rich habitat diversity, soil, vegetation, and often agriculture; it is periodically covered with water that deposits rich sediments; and it may have a low water-table during drought. (3) Infrastructure, such as oil, gas, sewer and water pipelines, electric powerlines, railways and highways, typically connect city center with rural areas by running along riversides. These structures are often in the floodplain, riverbank and adjoining upland, and frequently cross over or under the river. (4) The adjoining riverside upland frequently has a sequence of sections with residential types from single-unit homes to high-rises, somewhat linear parkland, and large and medium industry.

This sequence of land uses in the upland is usually affected by the river, floodplain and infrastructure along most of the corridor length. However, often access from upland to river and floodplain is limited to relatively distinct locations. Urban areas away from the riverside are quite different and seem relatively unconnected to the four river-related strips.

Many key characteristics of an urban human neighborhood are encapsulated as potentially instructive for understanding an ecological neighborhood mosaic. The neighborhood of Las Letras in city-center Madrid is about 2.5 km^2 (1 mi^2), roughly triangular, and bordered by busy streets (Figure 2.10). The surrounding five neighborhoods are each quite different. Las Letras has narrow streets, little vehicle traffic, much pedestrian usage, frequent shops on ground level of mostly 4–5-level residential buildings, a few religious structures and government buildings, about two schools, relatively few children outside playing, many restaurants and ample bars, and a relatively distinct history as home to writers and artists. Two open paved plazas and a few more tiny ones have outdoor/indoor restaurants and serve as magnets for tourists, city residents, and neighbors. The plazas, schools, religious structures and government buildings are well used, but do not appear to be major organizing forces for the neighborhood. Most daily shopping is walkable, especially along the busy border streets. Two adjoining neighborhoods, one of which is a major commercial shopping area, are also much used by Las Letras residents. Also two adjoining neighborhoods are major magnets for city tourists who spill into the neighborhood for restaurants and entertainment.

Ecologically, relatively small trees line a few of the streets. Three shops sell flowers and small nursery shrubs. Most buildings have small window balconies, with the first-level balconies averaging perhaps three or four plants of 2–3 species, and with fewer balcony plants higher up. Based on my observations during a typical spring, birds show relatively little interest in the balcony plants, which seem to have almost no insects. A few buildings contain some roof plants. Buildings usually have small interior courtyards, some containing a few planted and maintained plants. One of the two largest courtyards has extensive ivy cover, a handful of small trees, planted flower beds, and tiny spots of successional habitat succession. Six bird species and rather few individuals of each were observed in the courtyard. Two species, magpie and house sparrow, seemed to be residents in the vegetation.

City personnel sweep up trash and dirt from streets every day and frequently sweep-wash the streets with a truck. Constant sweeping seems to eliminate plants from the numerous cracks in sidewalks and cut-stone streets. Rainwater rapidly infiltrates through the cracks to the substrate below, though infrequent heavy rain causes water to flow into stormwater drains. Garbage from residents, restaurants and hotels is removed every day. Restaurant dumpster bins and their surroundings are kept noticeably clean. Large street-side bins for recycling materials are frequently emptied. Rodents must be present but were not observed by the author. Cockroaches were occasionally seen at ground level but not in an upper-floor kitchen area. Outdoor restaurant tables and food seemed dust-free, and leg-watching there revealed no sign of rat nibbling. The characteristic sprinkle of urban bird droppings was barely noticeable. The preceding array of human and ecological characteristics tied Las Letras together as a neighborhood mosaic, which contrasted with its surrounding urban neighborhoods.

Urban regions, metro areas, cities

Finally we focus on whole urban regions, metro areas, and cities as ecological mosaics. Drawing heavily on the preceding companion book, *Urban Regions: Ecology and Planning Beyond the City* (Forman, 2008), this key subject is briefly explored with three perspectives, from broad to specific: (1) regions: geographic, coastal, mountain, urban; (2) urban regions and metro areas as ecological mosaics; and (3) cities and ecological characteristics.

Regions: geographic, coastal, mountain, urban

Geographic regions cover 100% of a continent or the land surface. Two broad features characterize a *geographic region* (Forman, 1995, 2008): a common macroclimate and cultural-social pattern. Atmospheric "cells" form in the global air circulation due to solar energy differences and the arrangement of continents, mountain ranges and seas. Each cell exhibits a common *macroclimate*, that is, the history of weather patterns covering a relatively large area and differing from that in surrounding areas. A geographic region such as the USA Southwest or Mediterranean Region spatially corresponds with a macroclimatic cell, whereas the Southern England and Northeastern China regions are subsets of larger macroclimatic areas.

Culturally determined human activities on the land, mirrored in the idea of regionalism, also help determine a geographic region (Forman, 2008). A common culture or cultures covers the land, as evident in town and village forms, architecture, language and arts. Also a transportation network often connected with a single large city hub ties a region together socially and economically. The combination of the macroclimate and cultural-social dimension means that geographic regions (e.g., Canada's Maritime Provinces, Southwestern Australia) may be smaller, or larger (the Balkans, Central America), than a nation.

Coastal regions increasingly represent the global conflict between people and nature. Here human populations squeeze in, recreational and industrial demands skyrocket, distinctive biodiverse ecosystems concentrate, rich seafood resources are emptied, and climate change most threatens (Grove and Rackham, 2001; Forman, 2009). For instance, all 29 of the world's "ecoregions" more than one-third urbanized are coastal or on islands, and these areas contain 213 endemic terrestrial vertebrates, all threatened with extinction (McDonald *et al.*, 2008). A coastal region is an elongated strip of land and adjoining sea tied together by land transportation and marine activities, both usually linked to a single large coastal city and port.

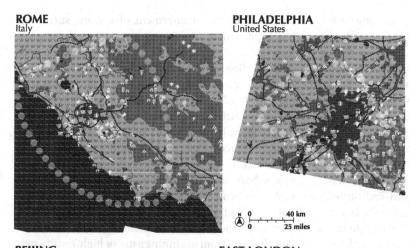

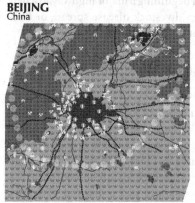 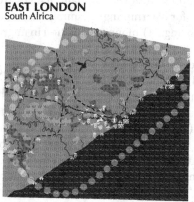

Figure 2.11. Urban regions and metro areas for four cities. Dotted line = boundary of urban region; dark blob near center = essentially all-built metro area. Black with lines = sea; medium-dark areas = wooded; lighter-gray area around East London = grassland; light-speckled area = agriculture. Metro-area boundary determined from satellite images; urban-region boundary enclosing a functional area based on many criteria. Tiny icons mainly represent diverse small features of importance for natural systems and their human uses. See Forman (2008).

Some three-quarters of the land-and-sea spatial patterns are generally parallel to, and at relatively predictable distances from, the coastline (Forman, 2009, 2010b). Most of the remaining patterns are terrestrial and perpendicular to the coastline. These patterns form a mesh or grid-like pattern with diverse corridors and highly diverse enclosed cells across the region. The main diagonals radiate from the city (Figure 2.7k and l). In contrast, the coastline (normally hundreds of meters to a few kilometers wide) has a fine-scale highly heterogeneous pattern of small sites packed together, such as towns, convex rocky headlands, concave beaches, stream and river mouths, and coastal wetlands. Often the sites of each coastline type are relatively equidistant.

Flows and movements in the coastal region also are overwhelmingly parallel or perpendicular to the coastline (Forman, 2010b). Most are parallel and go both ways. Most perpendicular flows on land are one way, and on sea either landward or both ways. On land, each flow or movement usually appears as a narrow band that in effect targets specific sites. Land-and-sea flows far from the coastline are mainly long distance, whereas those originating in the coastline are predominately short distance. With these coastal-region patterns, flows, and changes, conflicts between people and nature are targeted in the narrow coastline zone (Forman, 2009; Marsh, 2010). That is where biodiversity and natural flows are concentrated, and where human recreation and diverse economic activities are centered.

Mountain regions exhibit the common macroclimate and cultural-social dimensions of a geographic region, but are characterized by an abundance of high ridges and peaks. The region may include a number of large flat valleys and large cities, or none. Major transportation networks are limited in extent, and relatively isolated small communities are common. The mountain region is usually fine-grained, with small rather sharp-boundary landscapes, such as alpine, coniferous forest, and valley bottom (Swanson *et al.*, 1990; Forman, 1995). Gravity carries groundwater and piped water downward, as well as water and sediment over the surface and in streams. Wind carries seeds and spores downward, upward, and along ridges. Animals carrying seeds also move downward, upward and along ridges. Fires race up and creep down slopes. Mountain regions, like coastal and agricultural regions, are quite distinct.

An *urban region* is the area of active interactions between a city and its surroundings (Chapter 1) (Geddes, 1914, 1925; Mumford, 1968; Forman, 2008). This functional concept focuses on the flows and movements both toward the city and outward from the city. Indeed, the outer boundary of an urban region is determined by a drop in the rate of inward-and-outward flows, as one proceeds along a radius from the city.

Many criteria were used to map urban regions for 38 small-to-large cities (>250 000 population) worldwide (Forman, 2008). The primary criteria for determining urban-region boundaries ended up being: (1) mountain ranges and the sea; (2) major political boundaries; (3) another nearby major city (with distance scaled to the population sizes of both cities); (4) distance to one-day recreation and tourism sites; (5) major biodiversity areas; and (6) drainage areas around major water supplies. Radii were usually 70–100 km (about 40–60 mi) for the urban regions analyzed.

Urban regions with a dominant central city therefore differ markedly from geographic, mountain, coastal, and other region types including "bioregions" and "ecoregions" (Forman, 2008). Quite predictable radial patterns and flows predominate. Simple spatial models, such as urban-to-rural gradient, land mosaic, and donut models, are especially promising for ecological understanding and for urban planning (Figures 2.5 and 2.7).

Four urban regions, Rome, Beijing, Philadelphia and East London (South Africa), illustrate many of the patterns present (Figure 2.11). From the eye of a satellite, all have a conspicuous central all-built metro area, within which city and at least some suburbs exist. All four urban regions are dominated by agricultural land, especially close to the metro area. Natural or semi-natural land is mostly in the outer portion, and coastal regions include the near-shore marine zone. Rome has a ring highway mainly outside the metro area. Philadelphia's urban region is small, squeezed by those of surrounding cities. The Beijing Region contains two large reservoirs near the boundary and is sliced by a mountain range. East London is a small somewhat-isolated coastal city at a river mouth and surrounded by grassland. All the urban regions have major highways radiating outward to other regions.

Consider the arrangement of regions surrounding a particular urban region and the consequent interactions (Forman, 2008). Typically two to four regions adjoin an urban region, fewer for coastal cities and more where broad-scale topographic diversity is marked. If adjoining regions are mainly other urban regions, much competition and some collaboration may be prevalent. If nearby regions are mainly remote and rural, access to resources and room to expand the urban region boundary may be important. Issues near urban-region boundaries may cause rapid effects within, or in an adjoining, region. Inputs from, or outputs to, an adjoining region also may be beneficial or damaging. Indeed, an urban region is affected by distant changes, from beginning a major highway to a new immigration policy, livestock disease spread, or fluctuation in migratory bird populations.

Metro areas and urban regions as ecological mosaics

Human population density is especially important ecologically, in part because concentrated people have major impacts on adjoining and nearby areas. The highest densities of course are cities, such as: Shanghai, with 30 328 people/km^2 on average (7.4 million people in 244 km^2); Moscow, 18 085/km^2 (8.5 million in 470 km^2); Paris, 8847/km^2 (7.9 million in 893 km^2); London, 6215/km^2 (6.6 million in 1062 km^2); and New York, with 4039 people/km^2 (10.8 million in 2674 km^2) (Watson *et al.*, 2003). Unlike most cities, Shanghai, Hong Kong, and Singapore mainly grow upward. All-built metropolitan areas include a considerable area of lower-density suburbs. The population density of an urban region is much lower still, though quite heterogeneous or patchy, since city, suburbs, exurban/peri-urban areas, satellite cities, and numerous towns and villages are included along with extensive unbuilt land.

For 38 cities (>250 000 population) analyzed worldwide, both area and perimeter length of the metro area ranged from tiny Abeche (Chad) to huge Chicago (Forman, 2008). The *perimeter-to-area ratio* is also ecologically important, because more convoluted boundaries normally indicate more human effects on the surroundings. Some cities in the Amazon Basin (population about equal to that of a Shanghai apartment complex, though much more spread out) have quite convoluted margins with lobes and coves (Browder and Godfrey, 1997). Residents, especially in the built lobes, doubtless have extensive impacts on

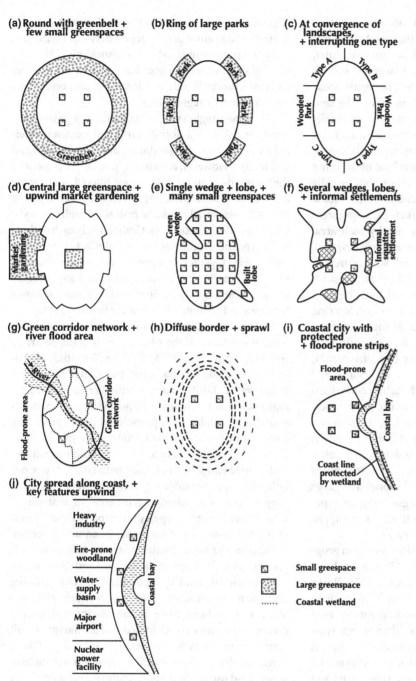

Figure 2.12. Ten alternative forms of metro areas. Six forms are conceptual based on many cities; four are illustrated in (a) London, (b) Seoul, (f) Brasilia, (g) St. Paul/Minneapolis (Minnesota, USA). Based on Potter and Salau (1990), Bosselman (2008), Forman (2008), and other sources.

surrounding rainforest. The perimeter-to-area ratio for a metro area tends to be inversely proportional to city population size. Large population cities tend to be more compact than small ones.

Ten contrasting forms or types of metro areas seem to represent the range of those worldwide (Figure 2.12). To ecologically compare these, I qualitatively evaluated each one for 23 ecologically related variables (i.e., analogous to Table 7.1). Negative roles or functions (relative to habitats, plants, animals, soil, air, water) strongly outweighed positives in two cases (Figure 2.12h and j). Positives predominated in two cases (e and g). Each metro area form contains both positive and negative variables, and their overall distribution between habitat/plant/animal variables and soil/air/water variables is about equal. Ecologically the two best metro area designs provide potential airflow into city center, plus cooling by greenspaces, and ready access of species from the surroundings, into and across the metro area.

Consistent with the first spatial models of the land (e.g., von Thunen bands), today's urban regions typically have agricultural land surrounding the metro area and natural lands in the outer portion (Figure 2.11) (Cronon, 1991; Forman, 2008). The agricultural land of the inner urban-region ring is mostly for crops, but usually includes some little-used former farmland. Often some of the area is irrigated, especially market-gardening areas with high food production for the nearby built area. Extensive floodplain areas may be extensively irrigated, such as for rice culture around Bangkok. Pastureland is typically limited in urban regions because livestock require large areas and land prices are relatively high.

Large natural lands in the inner urban-region ring are of prime ecological importance as sources for the "species rain" across a metro area. Thus, if biodiversity is to be sustained throughout a city or metro area, it depends on a continuous input of seeds, spores, and animals from the surroundings. Nearby rather than distant large natural lands are by far the most effective species sources (as indicated by d^{-2}, decrease in dispersal with the square of the distance). Natural lands in an urban region may be desert/grassland [e.g., Cairo, Tehran, Phoenix (USA)] or woodland/forest [Iquitos (Peru), Seattle (USA), Samarinda (Indonesia)]. Large natural lands in the outer urban-region ring are especially important for nature protection, water supply, wood production, wildlife, and recreation.

Urban form has long been of interest to urban geographers and planners (Lynch, 1981; Hartshorn, 1992; Whitehand and Morton, 2004; Pacione, 2005; Berke et al., 2006). Ecologically form is most important in affecting function, the flows and movements, or how a place works. For instance, as described above, typically little movement crosses hard boundaries, whereas considerable movement crosses various soft boundaries between two habitats or land uses. Thus, a compact metro area has less impact on or interaction with its surroundings [e.g., Bucharest (Romania), Edmonton and Winnipeg (Canada)]. On the other hand, a diffuse border indicates considerable movement from metro area to surroundings [Atlanta (USA), Milan (Italy), San Antonio (USA), San Jose (Costa Rica)].

The degree of convolution (abundance of coves and lobes) of a metro area border, represented by the "inner edge and hole" in the donut model (Figure 2.7b), also suggests the amount of interaction between metro area and its surroundings. Moscow, Stockholm, Brasilia, Santiago and Kuala Lumpur have several major lobes and coves that indicate considerable impact on the surrounding urban-region rings.

The urban-region ring contains many large land uses (land covers), such as forests, farmland areas, and satellite cities that form the predominant cover in maps and GIS images. However, certain tiny objects are present by the hundreds or thousands, yet are filtered out of maps and GIS images (Forman, 2004b; Hersperger et al., 2012). Tiny streams, small roads, farmsteads, hamlets, maybe villages are typical examples. Cumulatively each of these has a large and widespread ecological effect on the urban region. Also together they provide rich habitat diversity in the region. Furthermore, because they are unevenly spread, these tiny objects in abundance mold many of the flows and movements across the urban region.

A detailed analysis elucidated spatial patterns for natural systems and their human uses in the 38 urban regions (Forman, 2008). More than 75 spatial correlations were made for characteristics related to: nature; food; water; built systems; built areas; and whole regions. Many ecologically important spatial patterns and also principles were pinpointed in the preceding companion book (Forman, 2008; also see Forman, 2009). Interestingly, very few of the numerous nature-and-people variables correlated with either city population size or geographic region worldwide. This result suggests that the "inherent geometry" or spatial pattern created by a growing city is the major determinant of natural systems and their uses in an urban region. An expanding node creates a strong radial and secondarily circular imprint on the land and its ecology.

While local planning is common and city planning often done, surprisingly little urban-region planning seems to have been done. In the 38-city analysis, ten diverse attributes were identified that are likely to result from planning beyond the metro area (Forman, 2008). Attributes described the metro-area form and the urban-region land use configuration. Only two urban regions, both with planned cities (Canberra, Brasilia), exhibited four of the suggested urban-region planning attributes. Four more regions (London, Moscow, Beijing, Rome) exhibited three planning attributes. Seventeen cities exhibited one or two attributes. Surprisingly, 15 cities (40% of the total) (Tehran, Chicago, San Diego/Tijuana, Philadelphia, Nairobi, Kuala Lumpur, Bamako, Iquitos, Cuttack, Kagoshima, Erzurum, Ulaanbaatar, East

London, Rahimyar Khan, Abeche) showed no evidence of regional planning.

At a finer scale, lots of spatial features within an urban region have been pinpointed for having overall positive, or negative, effects on natural systems and their human uses (see Appendix A). These attributes are grouped into four categories: (1) city, metro, and region; (2) nature, forest, and food; (3) water; and (4) transportation, development, industry. Additional negative attributes are listed under (5) hazards. The positive spatial features can be fit together, and negative ones avoided, in creatively developing much-needed urban-region plans. Indeed, the ten alternative metro areas (Figure 2.12), and their ecological comparison described above, highlight the importance of planning for the big picture.

A city's region of course is also linked to other regions, both adjoining and distant. The arrangement of other regions strongly affects the flows and movements across the urban-region boundary (Forman, 2008). Some flows originate in the boundary zone, some mainly affect the outer edge portion of an urban region, and perhaps most affect much of the urban-region area. While distant changes are commonly ignored, periodically they strongly affect an urban region, such as transmission of a disease, major change in immigration policy, proposed new railway, new economic markets or incentives for nature-based tourism.

Certainly a metro area has a strong impact on nearby and distant landscapes, as highlighted by the *ecological footprint* concept (Rees and Wackernagel, 1996, 2008; Beatley, 2000a; Luck *et al.*, 2001). Indeed, this concept leads to a promising way to consider urban sustainability, or at least a better land and city. In essence, the *prime footprints* of a city or metro area are the primary specific landscapes providing inputs and receiving outputs, and are tied to the city by transportation corridors (Forman, 2008). This integrated giant network (represented by the prime-footprints model) is composed of city, corridors and landscapes, and is the object to plan for sustainability or a better future for nature and people.

A principle emerges from the preceding patterns and concepts. Strong reciprocal interactions between the metro area and its surrounding ring sustain both human and natural conditions in the urban region.

Towns, cities, ecology

A few key spatial and ecological insights are selected here from this huge topic. Nearly 50 nations have no city with a population >500 000 (Hardoy *et al.*, 2004).

Still, each small city has to protect its water supply, provide for sewage wastewater and solid waste, deal with stormwater runoff and pollutants, provide public health services, manage parks and public spaces, protect valuable natural land, and avoid settlement on hazardous land. Environmental dimensions permeate towns and small cities.

Towns and small cities are home for much of the world's low-income population, and are also expected to be 25 years hence (Satterthwaite and Tacoli, 2006; Forman, 2008). In China, towns/small cities are the main centers for the extensive farm population, government and service functions, and centers for trade (Kirkby *et al.*, 2006). Half of the world's urban population (and a quarter of the total population) lives in urban centers of less than a half million people. Megacities of >10 million contain <5% of the world's population, and are concentrated in nations with the largest economies.

The ecology of towns is surprisingly little known (Thayer, 2003; Forman *et al.*, 2004; Forman, 2008; McDonnell *et al.*, 2009). A study of human-dispersed plants in Finnish villages may be indicative (Hanski, 1982). Most species are either regionally common, or are rare and localized. Species richness increases with village size, and decreases with village isolation. The similarity in species composition decreases with distance between villages. Overall these patterns suggest that species composition is rather stable in small communities.

Much has been written about planned towns (Duany *et al.*, 2000, 2003; Bohl, 2002; Forman, 2008; Forsyth and Crewe, 2009). The Woodlands (Texas) remains my favorite, especially in its solutions for stormwater and flooding (Morgan and King, 1987). But overwhelmingly, planned-town plans focus on the human dimension, with minimal concern for ecological dimensions (Steiner, 2000; Forman, 2008). For example in descriptive materials for the plans, look at the key features largely ignored, overlooked, minimized or missing: (1) habitat, species diversity, and rare species; (2) environmental monitoring, management, and improvement; (3) hydrologic groundwater protection and habitat restoration; (4) habitat connectivity for regional wildlife movement; (5) ecological impact of traffic noise and pollutants; and (6) adaptive management for water conservation, stormwater runoff control, energy use, and water and air pollution. Existing communities have to address such issues by retrofitting. But when planning a town from scratch, how could the giant on the site remain unnoticed?

Cities provide protection, industry, employment, specialized features, and "the place to be." Cities expand outward and densify or transform inward (see Chapter 3). Outward expansion is largely unplanned as a whole, though may be intensively planned locally. An extreme case may be a new satellite city, Aguas Claras, close by Brasilia. Numerous dense high rises, apparently individually built by developers with little overall government planning, have created a chaotic place.

Rarely cities use central planning to transform an existing district. Two famous examples are a large residential and commercial district with circles and radial streets in Paris (Figure 2.1), and a large residential/commercial area of square-block buildings with courtyards and truncated corners for neighborhood shops in Barcelona (see Figure 11.5f). Bombing in Tokyo destroyed large areas that were rebuilt. A 1996 bomb devastated about a quarter of city-center Manchester (UK), which continues to be rebuilt (Ravetz, 2000). Recent abandonment of a military base in Seoul generated plans for on-site building, as well as linkages widely extending through the city. Gradual degradation of a large area in city-center Detroit (USA) awaits its next planned stage, hopefully not missing the "on-site giant" as in the planned towns above.

Several ecologically important patterns emerge from constructing new districts. The land use eliminated is normally either low- or high-density housing for the poor, and may include some small-plot farming. The new district generally does not provide for the people displaced who must move elsewhere (Hooper and Ortolano, 2012). The preceding small-farm urban agriculture provided some food locally, though was not intensive high-production characteristic of market-garden farming.

A new regular geometry of streets and buildings replaces the mainly unplanned mixture of nature, agriculture, streets, and buildings. New or much-improved infrastructure covers the area. New transportation may connect the city with its region (as in New York). Nature's backbone or framework is almost entirely covered with built structures. Wetlands are filled, and low-elevation areas subject to periodic flooding or sea-level rise sometimes dominate new districts (Boston's Back Bay; Pudong in Shanghai). Streams are eliminated by channeling water into pipes (Eixample in Barcelona). Ecologically important nearby natural land may be eliminated [desert by Albuquerque (USA); grassland by Calgary (Canada)]. Even large city-center areas change (Berlin's Tiergarten park completely replanted following 1940s wartime devastation; Detroit's downtown "urban removal" awaiting a new future).

Yet normally new districts have not created new large greenspaces. None of the cases has focused on connecting greenspaces. None restored a floodplain, though a city-center stream restoration daylighted a small river. None primarily reduced disaster risk, such as from mudslide, flood, or fire. Cities need new visions seriously recognizing nature.

Part I Framework

Chapter 3

Flows, movements, change

Thanks to the automobile, cities no longer have to dispose of tons of horse manure every day.
Virginia I. Postrel, Reason *magazine editor, speech for City Club of Cleveland, Ohio, 19 June 1990*

And then the tidal wave appeared as a high wall of foam rushing towards them, and soon it was on them, on everyone, crashing onto the land, crushing houses, sweeping huts away, drowning cattle and people ... Yet, a kind of miracle, most trees – the palms, the bunches of pandanus with great stalking roots, the sweeps of mangroves – were left undisturbed by the same wave that swept away fortress-like walls and paved roads.
Paul Theroux, Ghost Train to the Eastern Star, *2008*

The nature of flows and movements

Imagine a walk in the city and suddenly reaching a spot where nothing moved. People, leaves, birds, air, water, vehicles, and clouds are absolutely still. Weird, even scary. Better stop and look sharp. Flows and movements around us are universal, never stop.

Processes in urban areas are in effect flows, movements and transport through space (Forman, 1999; Ball, 2009). Some are mainly vertical, including rainfall, evapo-transpiration, tree falls, and ecological succession. Horizontal flows, the focus here, generally cross heterogeneous space thus linking different land uses and habitats (Wegner and Merriam, 1979).

Some flows and movements are basically human driven, such as cycling, motor vehicles, trains, electric transmission, and piping of water, sewage and oil. Most human-driven processes tend to be constrained to relatively straight lines. Straight routes have two theoretical advantages: efficiency for getting from here to there; and protecting the surrounding matrix from degradation by transport.

Natural horizontal flows and movements occur everywhere in urban areas – wind, dust and gaseous transport, surface water runoff, subsurface and groundwater flows, pollination, seed dispersal, sheet flow of water, erosion, sedimentation, fish movements, and animal foraging, dispersal and migration (Turner, 1987; Harris *et al.*, 1996b; Forman, 1999; Fischer *et al.*, 2006). Unlike human-driven flows, the routes of natural processes are overwhelmingly curvilinear (see Figure 2.2). Groundwater flow may be slightly curvy, whereas animal foraging and dispersal usually trace highly convoluted routes. Still, our rectilinear networks tend to straighten natural flows and movements around cities.

Flows and movements essentially describe how the urban area works. In effect, they are the interactions, the linkages, the connections, and the processes that link urban habitats and land uses. They tightly tie the urban mosaic together. Let's explore the subject from three perspectives: (1) flow and movement patterns; (2) animal and plant movements; and (3) system and ecosystem flows.

Flow and movement patterns

Four basic types of flows and movements are especially important in urban areas: (1) air flows; (2) water flows; (3) self-locomotion; and (4) motor-powered movement using outside energy. These flows carry or transport energy, materials and objects. Transported items vary in the direction, route, rate, amount, and distance carried.

Airflows are based on energy differences in the atmosphere, that is, from warm to cool areas (see Chapter 5) (Geiger, 1965; Ball, 2009). *Streamline airflow* moves in smooth parallel layers, such as wind moving across an extensive corn field or flat open golf course (Forman, 1995). *Turbulent airflow* has circular eddies normally with small up-and-down flows, characteristic of wind moving through a housing development with scattered trees and buildings. *Vortex airflow* is typically in a cylindrical form, such as on the downwind side of a flat-roof building or beyond the downwind edge of a high-rise building (see Figures 5.8 and 5.9).

65

Breezes due to temperature differences at a finer-scale local level are also important in urban areas (see Figure 5.1). Warm air normally rises to the colder upper atmosphere over a city on still nights. Cool air on hills and mountains flows downslope at night, forcing the warm valley air to rise vertically. An onshore breeze from the sea occurs in early fall when the water is warmer than the land. Similarly an offshore breeze in spring moves from the warmer land to the cooler sea.

Nearly horizontal water flows are based on gravity, going from upslope to downslope. Surface water flow over extensive impermeable surfaces, especially of roofs and roads, predominates in urban areas (see Chapter 6). Some of the water that infiltrates through surface cracks and soil with vegetation then runs nearly horizontally in subsurface water flow. However, the typical fine-scale heterogeneity of soil and fill in urban areas (see Chapter 4) often interrupts subsurface flow. Water that infiltrates deeply, e.g., in sandy fill areas, and encounters the water-table then moves horizontally, and often slowly, as groundwater flow. Especially in hilly and mountainous terrain, drinking water, stormwater, and sewage wastewater normally flow by gravity in pipelines and ditches.

Self-locomotion by animals and people is based on energy derived from the sun and then transformed by photosynthesis into organic compounds. These in turn are consumed as food and then metabolized producing energy for movement. People walk and bicycle. Terrestrial animals walk and run, flying animals fly, and locomoting aquatic animals swim. To sustain movement, the animal requires food, water and rest.

Movement by outside energy is characteristic of human-constructed and maintained infrastructure. Most trucks, buses and cars run on roads currently using petroleum from oil (Figure 3.1). Trains, subways, and trollies run on tracks using energy directly or indirectly from fossil fuel, i.e., oil or coal. Water, sewage, oil, and natural gas are pumped through pipelines, mainly using fossil fuel, especially oil. Electricity is transmitted in powerlines, often high above the ground. Electric generation depends on hydropower, or on oil-, natural gas- or coal-fired power facilities, which may be centralized or widely distributed over the urban area. Renewable energy, especially from wind and solar sources, provides a small but growing portion of the outside energy for movement.

These flows of wind, water, locomoting animals, people walking, motor vehicles, and infrastructure conduits fall into three groups based on energy. (1) Wind and water flows due to temperature differences

Figure 3.1. A major radial road connecting city center and rural area. Vehicle headlights heading inward at dusk; skyscraper lights a hazard to migrating songbirds. Street tree corridor lines the road. Calgary, Canada. R. Forman photo.

or gravity represent "mass flow" (mass transport). (2) Animals and people walking are self-locomotion based on energy from the food chain. And (3) transportation and infrastructure movements today are mainly based on fossil fuel, which represents long-ago solar and fossilized plant energy.

The energy-driven flows and movements are *vectors* that transport objects (including materials and energy). Trains carry people, freight, and seeds (Figure 3.2). Motor vehicles carry people, goods, seeds, insects, and even vertebrate animals. Wind transports heat and moisture. It also carries noise, dust, aerosols and gases, including heavy metals, sulfur dioxide, and toxic organic substances. Furthermore, airflows carry seeds, spores, algae, and "parachuting" spiders. High winds spread wildfire.

Flowing water transports floating algae and zooplankton. It frequently carries tiny clay particles to the sea, silt to river bottoms and deltas, and sand for short distances. Water flows carry a wide range of water pollutants from nitrogen and phosphorus to sediment, toxic chemicals, and rubbish. Floodwater transports nearly anything from rolling stones to rolling cars, washed-out bridges, and toxic-waste accumulations from industries. The mass flows of wind or water, plus the objects they transport, accelerate through narrow spots (the Venturi effect).

Animal and plant movements

Almost all of the preceding mechanisms also operate for the movement of plants and animals. Yet some

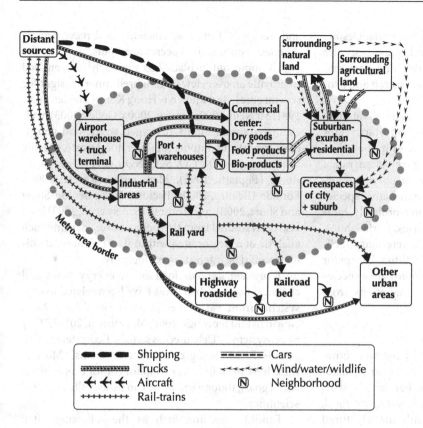

Figure 3.2. Probable major routes and vectors for non-native and native species movement. Dotted oval = a metro area.

additional mechanisms are important for understanding species movement in urban areas.

Most terrestrial vertebrate species have a small area close to the nest or den called the *territory*, which is defended against intruders, especially of the same species. The routes taken by an animal in its territory radiate out from the relatively central den, often tracing short somewhat-linear loops. Overall, territoriality is less prominent in urban than in natural land. The animal's *home range* is a larger area typically used in daily movements, especially foraging for food. These movements trace larger irregular loops radiating out from the nest. A third movement type, *animal dispersal*, refers to sub-adult males and females that leave the den or nest, heading off to find a mate and establish their own home ranges at some distance away. Finally, many species migrate seasonally to access suitable food conditions and avoid unsuitable environmental conditions. This cyclic *migration* may be latitudinal between tropical and temperate areas, altitudinal between high and low elevation in mountains, or locally between north and south sides of a ridge. Except for animal dispersal, the routes taken in the diverse movements are relatively predictable.

Urban areas both inhibit and facilitate these movements. Residential development disrupts patterns of many species, particularly those of butterflies (Knaapen et al., 1992; Forman, 1995). Busy roads and highways also block movements (Forman et al., 2003).

On the other hand, the network of stormwater drainage pipes is especially effective in facilitating the movement of some animals, such as rats around restaurant, hotel, and dump areas. Stormwater pipe networks are also great for cockroaches to enter apartment buildings and many other buildings with basements (Robinson, 1996). Pipes are well connected and grates or grilles over the pipes in basements have plenty of room for fat cockroaches to enter and leave. Such animals can move rapidly in this unobstructed system to find the best and oft-changing food sources. Termites readily move from building to building due to the abundance of wooden foundations and buildings in many urban areas, plus readily traversable sandy fill next to the buildings (Robinson, 1996).

Non-native species introduced into urban areas at ports, rail yards, and warehouse areas may then spread widely (Figure 3.2) (Trepl, 1990; McNeill, 2000; Song et al., 2005). Railways appear to be major, though spatially constrained, routes through the city. Despite severe water pollution, marine and aquatic species doubtless spread directly from a shipping port into marine ecosystems, though recreational craft and fishing boats may also play a role.

Trucks loading and unloading materials at storage warehouses may be an especially important vector for widely spreading non-native species in the urban region. Such trucks trace irregular yet often radial routes outward from warehouses near ports and airports. The other major spread of non-native species to "everywhere" in the urban region probably radiates from sites such as plant nurseries, florist shops, pet stores, botanical gardens, zoos, veterinary facilities, and some biological research facilities. Except for railways and certain highways, the non-native species spread probably mainly represents radiations from nodes at different spatial scales.

System and ecosystem flows

The metaphor, *urban metabolism*, has been used sometimes in referring to the overall flows of energy, materials and chemicals between an urban area and its surroundings. For instance, oil, gas, water, air, food, and raw materials enter a city, while manufactured goods, heat, air pollutants, sewage wastewater, water pollutants, and solid wastes leave. Much like an economic or financial budget, inputs and outputs are calculated providing insights into ways to improve the system (Wolman, 1965; Rees and Wackernagel, 1996; Hardoy *et al.*, 2004; Brunner, 2007; Wittig, 2008). Although mainly of socioeconomic interest, the analysis can be ecological in the case where organisms significantly affect, or are affected by, the input/output flows. For example, some inputs are affected by upwind or upslope vegetation conditions. Considerable urban vegetation or activity of nitrogen-cycle bacteria could affect certain outputs, and outputs often have strong effects on surrounding ecological conditions.

Ecological engineering or systems ecology provides key insights for inputs/outputs, internal system attributes, network structures, flow rates, and stability (Odum, 1983; Ma, 1985; Mitsch and Jorgensen, 2004; van Bohemen, 2005; Pincetl *et al.*, 2012; Adler and Tanner, 2013). As perhaps the simplest systems model, a metro area (or city) is considered to be a "black box," with inputs and outputs of energy, water, materials and people. Such simple systems highlight four ways that may increase the stability or sustainability of an urban area and its surroundings: (1) increase internal production and storage; (2) increase efficiency and recycling; (3) decrease inputs; and (4) decrease outputs.

The input/output black-box approach has been used to provide an overview or gain preliminary insight for several cities. Best known is Hong Kong (Boyden *et al.*, 1981; McNeill, 2000; Beatley, 2000a; Golley, 2003), with estimated annual inputs (of oxygen, water, food, petroleum, cargo) and outputs (of water, food waste, organic solids, cargo, CO_2, CO, SO_x, NO_x, C_xH_x, lead, dust). Rome (Pignatti, 1995), Manchester (Ravetz, 2000), London (Beatley, 2000a; Pacione, 2005; Benton-Short and Short, 2008), and Toronto (Sahely *et al.*, 2003) have also been studied from this perspective. Focusing such analysis of a city, or area within it, on plants and animals would be interesting ecologically.

Ecosystem analysis focuses on energy flows and chemical flows. Urban areas have been related to ecosystems from varied perspectives (Bradshaw, 2003; Newman and Jennings, 2008; McDonnell, 2011). "City as ecosystem." "Urban ecosystem." "Ecosystems of or in cities." "Modeling cities on ecosystems." Most of these have been used as metaphors or for economic or social goals, though each case can be usefully analyzed scientifically.

Familiar maxims such as the following often accompany the reference to ecosystems. Everything is connected. Materials don't disappear but go somewhere. Recycle like nature. The second law (of thermodynamics) is after us. Entropy will win. Like Mother Earth or motherhood, ecology is good. And so forth. Such phrases limit understanding if blindly used, but can be useful if critical analysis follows.

Overall, "urban energy flows" are mainly characterized by inputs of fossil fuel, followed by its use and dissipation to heat across the urban area (Pignatti, 1995). Meanwhile "urban chemical flows" are the incoming streams of materials, goods and water of varied chemical composition that are mainly dispersed into smaller units, often chemically changed, and then reaggregated into outgoing air, water, and transport waste streams. Normally little cycling or recycling occurs within the urban system.

Biogeochemical or mineral nutrient flows of chemicals within or through an ecosystem have been examined in urban areas, generally based on the traditional ecological model of nutrient cycling within a natural ecosystem (Smith, 1996; Odum and Barrett, 2005; Cain *et al.*, 2011). The urban carbon "cycle" has been described (Ravetz, 2000; Carreiro *et al.*, 2009), though the urban nitrogen "cycle" is most studied (Craul, 1992; Alberti,

2008; Carreiro, 2008; Carreiro et al., 2009; Pouyat et al., 2009). So many characteristics of urban areas differ from those in natural land that a new more useful chemical-flow model or paradigm is likely to evolve (Kaye et al., 2006). Chemical flow-throughs (rather than cycles) involving organisms may be a highlight.

A future "urban chemical-flow model" will probably have foundations in network patterns, spatial heterogeneity contrasts, spatial hierarchies, chemically subdividing-altering-reaggregating patterns, and linkages with the heterogeneous surroundings. Another research frontier lurks.

Linking input–output ecological flows to the spatial pattern within a metro area seems to be an especially valuable next step. The urban region is distinctive due to its strong radial source-sink flows of people, vehicles, goods, seeds, wildlife, microbes, dust, gases, water, chemicals, and so forth. In early evening, serpents of white lights and red lights moving in opposite directions along radii link city and surroundings (Figure 3.1). Still a hierarchy of secondary non-radial ecological flows and movements remains little studied. These varied flows and movements mainly occur in prominent anthropogenic networks, which are designed for both human and ecological flows in and around urban areas. Road, rail, pipe, powerline, and waterway systems channel, accelerate and interrupt urban chemical flows.

Flows around boundaries and mosaics

Flows and movements by boundaries

The flows and movements of objects in effect may be along, parallel to, or perpendicular to a boundary. Straightness facilitates movement along a boundary, including wind, water, and animal movements (Forman, 1995). Some animals move parallel to a boundary, as suggested by a deer path in a Swiss beech (*Fagus*) woods ca. 8 m in from the boundary, or raccoon (*Procyon*) tracks paralleling a New Jersey (USA) hedgerow. However, parallel movements are little studied and may or may not be common. Movements perpendicular to a boundary may reflect patch-interior animals moving toward or returning from the edge of a patch. More frequent though may be perpendicular flows crossing the boundary from one land use to another, and therefore represent interactions or connections between the elements.

An animal approaching a boundary basically has three options: continue on across; turn and go along or parallel to the boundary; or turn back. The first and third options seem to be most common, and probably depend largely on habitat/land use suitability on the other side of the boundary.

The edge portion of a natural habitat typically has denser vegetation with more species packed together, often referred to as the *edge effect* (Figure 3.3) (Hunter, 1990; Forman, 1995). Such edge conditions may inhibit movement across a boundary. If nothing crosses, the edge is a barrier but that is rare. Typically the edge is a filter, which slows and reduces flows across. This filtration effect changes over time, just as do flows through a semi-permeable membrane. In urban areas, edges tend to be heavily affected by people (Matlack, 1993; Guirado et al., 2006). Park edges may be quite open, encouraging people to enter. Alternatively, a dense shrub layer ("mantel," meaning overcoat in German) may be present that helps in protecting valuable resources within the park. A small nature reserve near Cambridge (UK) was ringed by a wide dense thicket of thorny plants (*Rubus, Crataegus*) and cut branches around its edge, thus nearly eliminating human access except at a designated entrance. Numerous options exist for minimizing human overuse of a protected area by managing the strip outside, the boundary itself, and especially the edge portion of the protected area (Forman, 1995).

Movement across a boundary also depends on its form (Figure 3.3) (Forman, 1995; Andreassen et al., 1996; van Bohemen, 2005; Hodgson et al., 2007). *Hard boundaries* are abrupt and relatively straight. *Soft boundaries*, which are gradual and/or curvy, seem to occur in four forms: (1) a gradual transition or "gradient" from one habitat to the other; (2) a "convoluted" or curvy boundary with "lobes and coves"; (3) a heterogeneous "patchy strip" between habitats; and (4) fractal or fine-textured. The form of many boundaries changes over time.

Observations of tracks and scats of elk (*Cervus canadensis*), mule deer (*Odocoileus hemionus*), and coyote (*Canis latrans*) around straight, curvy, and convoluted boundaries between woodland and grassland in New Mexico (USA) suggest important movement patterns (Forman, 1995). For the two herbivores, hard boundaries are major conduits for movement along, whereas convoluted boundaries have essentially no animal movement along them, either from tip-to-tip of lobes or following the

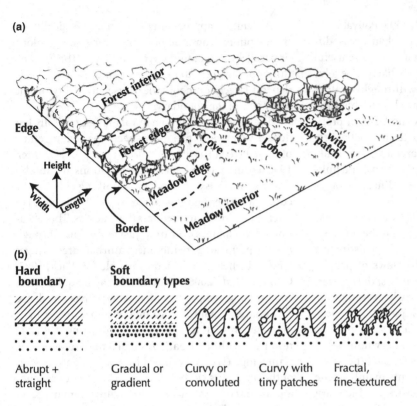

Figure 3.3. Habitat edge and interior, plus hard and soft boundaries. An edge, as the outer portion of a forest or meadow, has width, height, and length characteristics. Based on Forman (1995).

convolutions. In contrast, convoluted soft boundaries are the prime areas for animals moving between habitats. Rather few animals seem to cross straight boundaries. Boundary-crossing frequency might be proportional to the ratio of actual boundary length to straight-line length, or the boundary fractal dimension. The predator, coyote, also mainly moves along the straight boundaries. A few sections of the convoluted boundaries contained tiny wooded patches in the grassland lobes, reminiscent of a patchy-strip soft boundary. Based on track and scat densities, these boundary sections are among the highest in herbivore usage and in herbivore crossings between woodland and grassland habitats.

Individual lobes and coves are important in channeling movements around boundaries (Forman, 1995; Belisle and Desrochers, 2002). Flowing water and wind with transported sediments or snow seem to mainly cross boundaries at the ends of coves. Also, turbulence patterns around boundaries are relatively predictable from the lobe-and-cove pattern. Microenvironmental heterogeneity, such as cool, warm, windy and protected spots is high in lobe-and-cove boundaries, suggesting considerable localized wildlife movement. Animals in a suitable habitat or land use often seem to depart from it at the tip of a lobe. However, animals approaching a convoluted boundary may head toward a projecting lobe of the habitat ahead, but enter diagonally somewhere between the ends of the lobe and cove. These patterns are suggested by limited evidence and remain a research frontier.

Flows in mosaics of land-use pattern

Direction, route, and rate are three central dimensions of flows and movements in heterogeneous land mosaics. Understanding these flows portrays how any area, from neighborhood to urban region, works.

Certain spatial patterns particularly indicate flows (Figure 3.4; see also Figure 12.5). Specialist species from surrounding natural and agricultural lands arrive, but relatively few survive or thrive. Most species present are and will be generalists. For the goal of maintaining a high diversity of wildlife throughout the urban area, apparent priorities are: (1) to enhance continuous movement of species from surrounding habitats into and across the urban area; and (2) to enhance habitat diversity and amount within the urban area.

Several principles provide insight into directionality of flows. "Diffusion" of objects such as industrial and commercial products tends to go from higher concentration to lower concentration (Tacoli, 2006).

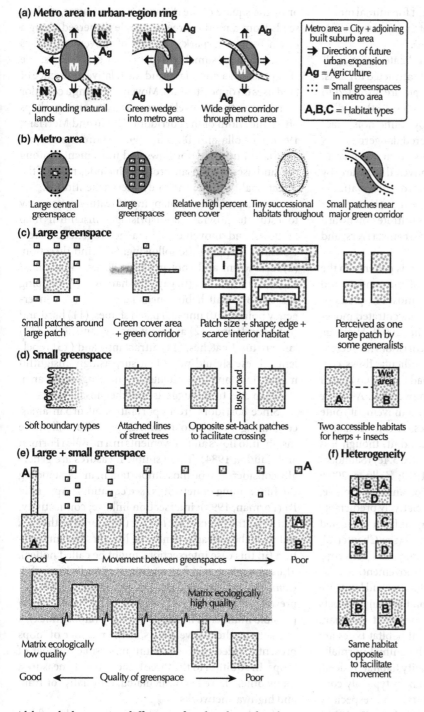

Figure 3.4. Suggested ecologically optimum greenspace patterns for a city. (a) to (f) Key spatial attributes affecting urban flows and movements. M = metro area; N = natural or seminatural land; I = interior habitat; E = edge habitat.

Although the passive diffusion of molecules is familiar in science, in urban areas diffusion is mainly a social-science concept (Hartshorn, 1992; Pacione, 2005). A net movement of products goes from urban to rural. Yet raw materials for those products go in the opposite direction, from rural to urban. Inputs and outputs indicate directionality, as do sources and sinks (Vink, 1983; Pulliam and Danielson, 1991). Regional "goods-and-services flow models" estimate flows into, within, and out of cities.

In metapopulation dynamics (see Chapter 9), a net movement of species from large natural patches to small natural patches occurs (Figure 3.4c) (see equations, Appendix B). A small habitat patch near a large

71

one seems to function as an attractant for animal movements in and out of the large patch (Opdam, 1991; Verboom et al., 1991; Forman, 1995).

Attractants and repellents also indicate directionality in the landscape. Thus, wildlife are attracted to parks, water, and food sources, while people are attracted to museums, restaurants, and tourist sites. Both wildlife and people are typically repelled by traffic noise, factory air pollution, and sites considered dangerous.

As described above, air moves from warmer to cooler areas, and water mostly moves downward by gravity. Regional winds blow toward cooler atmospheric areas, and local breezes toward cooler air over different land uses. Rainwater may infiltrate downward from soil surface to groundwater. Streams, rivers, and floodwaters flow downslope.

In addition to directionality, various patterns in the land indicate the primary routes of movements and flows (Forman, 1995). Corridors are most conspicuous, and act as conduits sometimes for concentrated movements in both directions (Tigas et al., 2002). Highway corridors for commuters and trail corridors for recreationists are characteristic. For perpendicular flows, corridors such as fencing and busy roads are often partial barriers, reducing and deflecting movement. A gap or narrows in a corridor is commonly a movement route for crossing between two land uses. A row of habitat "stepping stones" functions as a conduit, though less effectively than does a corridor (Figure 3.4e) (Saunders et al., 1987; Zollner and Lima, 1999; Zollner, 2000; Ichinose, 2005). Also the routes of animals, people, and wind-transported objects are relatively predictable around various patch shapes, ranging from elongated to convoluted. "Convergency points" (see Chapter 2), where three or more habitats or land uses converge, tend to be attractants or funnels for movement.

Since animals and people have habitat and land-use preferences, the distribution of habitats strongly affects the directionality and route of movement (Forman, 1995; Hough, 2004). Thus, if several habitat types are at equal distance from a source, the animal normally moves to the preferred or high-quality habitat. Indeed, the sometimes quite-curvy route taken typically concentrates on high-quality habitats, and especially avoids low-quality or dangerous sites (Prevett, 1991). Follow-the-leader or "traplining" is a particularly efficient way to move (Ball, 2009).

In addition to directionality and route, the rate or amount of movement is important in a land mosaic. The amount of movement typically decreases exponentially or by the square of the distance from a source, such as seeds from a mother tree or people dispersed along a beach from a carpark. The rate of movement across a boundary presumably depends on two factors: the abruptness of a boundary, and the relative difference of land uses on opposite sides. Movement along a corridor increases with the level of connectivity and, perhaps often, the width of the corridor (Henein and Merriam, 1990; La Polla and Barrett, 1993; Andreassen et al., 1996). Connectivity for flows and movements among the land uses of a mosaic probably includes most of the directionality, route, and rate patterns mentioned.

Even a preliminary mapping of features often easy to estimate provides a surprisingly insightful map of flows and movements in a heterogeneous area. Consider mapping the following: (1) wind direction, (2) local warm-and-cool areas, (3) slopes, (4) streams/rivers, (5) corridors, (6) gaps and narrows, (7) stepping stones, (8) small habitat near large one, (9) convergency points, (10) unusual patch shapes, (11) hard and soft boundaries, (12) apparent sources and sinks, (13) large natural patches, (14) attractants, and (15) repellents. These should provide a surprisingly insightful map of flows and movements for a heterogeneous area. The map effectively suggests how the mosaic works.

Since corridors are important in all urban areas, combining them into networks provides still more insight into flows and movements in a mosaic (Forman and Baudry, 1984). Three simple measures are generally considered to be most important in understanding the functioning of networks (see equations, Appendix B) (Forman, 1995): intersection linkage; connectivity; and circuitry. *Intersection linkage* (measured by the beta index) is the average number of linkages per intersection (in this case intersections might be called nodes). Thus, a perfect grid has an intersection linkage of 4. *Connectivity* (gamma index) is the number of linkages present between intersections divided by the maximum possible number of linkages. *Circuitry* (alpha index), a measure of alternative routes, is the number of loops present divided by the maximum possible number of loops. In addition to the three basic network measures, many others have been calculated, especially for train and highway networks.

Another major approach to understanding flows and movements in networks is the *gravity model*, which adds the size of nodes, such as cities, at intersections (see equations, Appendix B) (Hartshorn, 1992; Forman, 1995; Berling-Wolff and Wu, 2004; Pacione, 2005; Alberti, 2008). This is analogous to a bit of Newtonian physics and assumes a net diffusion of objects from large nodes to small nodes. Consider a typical distribution of large and small cities connected in a grid network. In its basic form, the gravity model simply uses the population sizes of nodes and their distance apart to estimate the amount of interaction, i.e., flows and movements, between nodes. Gravity models emphasize that distance is more important than node size in determining interaction.

A suite of "traveling salesman" models has evolved in part from the gravity model to estimate the optimum route through a network connecting population centers. Similarly, "Lowry journey-to-work" models have been used to estimate the proportions of a work force traveling various distances and routes to a workplace. "Graph-theoretic" models create conceptual networks to reflect connectivity, for example, based on common boundaries between land uses in a mosaic (Cantwell and Forman, 1994; Urban and Keitt, 2001).

Change-over-time principles

Is it better to live in a stable unchanging urban area, a gradually changing area, or one with a seemingly unending sequence of major disturbances? Which is better ecologically for vegetation, wildlife, and water conditions? A constant environment is only temporary. A strong government or a surrounding wall that tries to keep outside influences out at some point comes crashing down, typically with a convulsion. In contrast, unending major disturbances lead to continual behavioral adjustments and probably numerous adaptations. But such an urban area is only suitable for a limited number of widely tolerant people and generalist species. The intermediate case of a gradually changing urban area may change at the same rate (or faster or slower) as the surrounding land and society. In this case, occasional disturbances with ongoing adaptations and behavioral adjustments theoretically may continue forever.

The sections following first outline basic concepts of change, especially as used in ecology. Then changing ecological components – soil, water, air, vegetation, biodiversity – are highlighted.

Useful concepts from ecology

Change has been a core subject in ecology at least since the early 20th century studies of succession in the Chicago, Nebraska, and Minneapolis-St. Paul (USA) areas. Many of the concepts (Smith, 1996; Ricklefs and Miller, 2000; Cain *et al.*, 2011) briefly introduced here are useful in urban areas (Ramalho and Hobbs, 2012).

For a gradually changing ecosystem or site, we measure the rate, amount and direction of change. Growth is often modeled as linear, exponential (J-shaped), or logistic (S-shaped, sigmoid) (see equations, Appendix B). Stresses, such as limited water, high air temperature, or lawn-mowing that may change little or seldom over time, often continue for long periods. Terms such as stability, equilibrium, sustainability, and persistence describe aspects of a gradually changing system (van Bohemen, 2005).

Five alternative mechanisms seem to maintain stable trajectories: (1) inertia of a large system; (2) a strong control hierarchy; (3) negative feedbacks; (4) patch dynamics; and (5) mosaic stability. Thus, a large system or area is simply more stable than a small one, which may be readily disturbed or transformed. A strong hierarchy keeps components under control, limiting the opportunities for alteration (O'Neill *et al.*, 1986). Negative feedbacks, where one component stimulates a second, which in turn inhibits the first, maintain a cyclic trajectory, and tend to prevent major alteration (Wang *et al.*, 1992). Patch dynamics indicates that disturbed patches occur in different locations over time, yet the whole system remains stable (Pickett and White, 1985). Mosaic stability results from strong interactions among neighboring patches and corridors that in turn limit the intensity and spread of disturbance (Forman, 1995, 2008). Some inter-patch interactions are effectively symbiotic. Also, mosaic stability may increase with more heterogeneity present.

Disturbance is an event that causes a significant change in an ecosystem or urban area (Figure 3.5a). The events or driving forces causing such a response of course are diverse (Forman, 2008), and urbanization often changes the disturbance regimes over time (Rebele, 1994). If the cause of change is protracted or continuous, it is usually considered to be a *stress* rather than a disturbance or perturbation.

Some changes are cyclic, such as tides, seasons, and commuter traffic (Figure 3.5c and d). Many species easily adapt genetically to frequent cyclic changes, whether large or small. An alternative response is for individual animals or people to change their behavior to survive and thrive. Such animals tend to be generalists with wide plasticity or tolerance. Infrequent disturbances such as rare fires or cyclones are the hardest to adapt to, because the species may not have encountered and responded to them before.

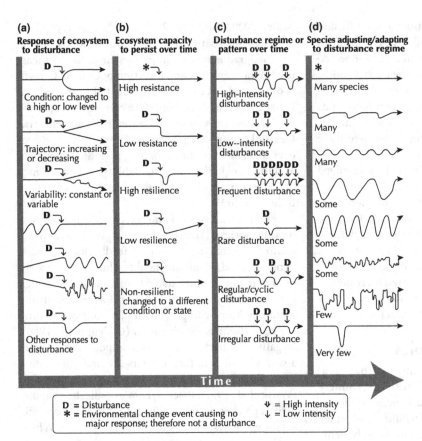

Figure 3.5. Responses of ecosystems and species to disturbance patterns.

Some infrequent disturbances are intense or acute, causing major changes. So-called "disasters" (especially for people) include: earthquakes, volcanic eruptions, tsunamis, cyclones/hurricanes, floods and wildfires. In urban areas these may be disastrous for people, but also typically cause major alterations in air quality, erosion, water quality, trees, wildlife movement and survival, semi-natural communities, and much more. Dysfunctional wastewater sewer systems, building rubble, spills of industrial chemicals, persistent stagnant water, and urban landslides produce an environmental nightmare. These dwarf the benefits of an increase in successional habitats following disasters.

Disturbances also may be intense or acute enough that the system or area does not return to approximately its preceding condition, but rather develops into a different, even new, condition or state (Figure 3.5b). Old park trees dying in an air-polluted city may have begun growing in clean air, but are then replaced by quite different vegetation tolerant of air pollution. The inner-suburb neighborhood destroyed by an earthquake originated as single-family homes on the outer city edge, but is replaced by apartment buildings to fill a housing shortage for the current city and inner suburb.

To persist in a world of potential disturbances, ecosystems and urban areas seem to have one of two mechanisms, either high resistance or high resilience (Figure 3.5b). In effect, *resistance* refers to the mechanisms that prevent an environmental or human factor from causing a significant alteration in the system. A high level of any of the five mechanisms listed above provides resistance against change. These mechanisms create a high "threshold," which the disturbance intensity does not overcome (Bennett and Radford, 2003). *Resilience*, in contrast, is the ability to bounce back or quickly recover from a major alteration (Vale and Campanella, 2005). Rapid plant recolonization, growth and reproduction, as well as behavioral change by animals, and adaptation through natural selection, all provide resilience to an ecosystem.

Semi-natural areas seem to have higher resistance than do urban areas. Also, natural areas are usually more resilient, bouncing back quickly from alteration, than urban areas. Urban recovery is often a slow process, depending more on the confluence of political, economic and social conditions.

A broader perspective on "resilience" has been proposed that combines the traditional resistance and resilience concepts and adds an anthropogenic dimension (Holling, 1996; Alberti and Marzluff, 2004; Cumming et al., 2005; Newman and Jennings, 2008). For this approach, resilience is effectively the ability of a system to maintain its identity in the face of and following disturbances. The system is described as going through phases of exploitation ("birth"), conservation ("growth"), release ("death"), and renewal ("reorganization"), and changing to a different state or condition. Complex social and ecological conditions are coupled (Liu et al., 2007; Alberti, 2008; Niemela et al., 2011). Putting aside the confusing and mushrooming terminology, useful insight for urban ecology might emerge from the discussion.

Urbanization occurs synchronously, though often differently, at different spatial scales (Harms, 1999). Thus, a neighborhood may either expand or contract while the city is spreading outward. Conversely, the neighborhood may spread or shrink while the city loses population. Densification and de-densification of a site also occur while the surrounding area changes. For instance, changing ecological conditions at a site may be significantly affected by increasing or decreasing vegetation cover and wildlife movement in the surroundings. Analogously the site may be a significant source of seeds and animals for the surroundings. The relationship of changes between scales seems to be little studied, at least ecologically.

Finally, the *legacies* of former ecological conditions, or the "time lag" between a disturbance and its major effect, are increasingly important in ecology (Foster et al., 2003). Building foundations and stone walls constructed 150–250 years ago in New England (USA) affect today's soil conditions and wildlife movement (Foster and Aber, 2004). The collapse of today's fish stocks in some cases correlates with overfishing patterns decades ago (Jackson et al., 2001). Plant and vertebrate biodiversity in wetlands correlates better with density of the surrounding road network of three decades ago, than with today's road-density network (Findlay and Bourdages, 2000).

Time scales

Twice a day commuter vehicle traffic between suburbs and city peaks, mainly flowing inward in morning and outward in afternoon/evening. During these periods an endless mass of moving vehicles creates barriers to wildlife movement. Associated traffic noise creates wide bands of land with little animal activity, since animals may not hear predators or communicate with young. Particulate matter raised from road surfaces and varied pollutants from the vehicles themselves bathe adjoining areas twice a day (Carpenter, 1983). Associated parking-lot activity peaks twice a day, so cats and rats adjust their behavior to be active at different times. Noisy polluting truck traffic tends to precede the "rush-three-hours" of car commuters.

However, urban changes vary markedly by time scale. Thus, urban traffic also varies on a weekly cycle. Commuter movement is scarce on weekends, yet the Friday and Sunday afternoon traffic of recreationists between city and natural land often clogs highways. Weekend traffic and people in nearby recreational areas may be dense, with wildlife doubtless adjusting behavior to the weekend cycle. Furthermore, weekend traffic and people also vary seasonally. For example, many people head to mountain ski areas in winter and coastal beaches in summer. Diverse seasonal changes in urban plants and animals are familiar to all urban residents (Houck and Cody, 2000; Saley et al., 2003; Murgui, 2007; Mitchell, 2008).

Time scales vary from very long to nearly instantaneous, with ecological responses equally varied. Evolutionary processes of forming species and gradually changing them vary over essentially the whole range, from millions of years to perhaps hours for many microbes. Some urban plant and animal species are genetically altered in centuries, decades, or less (Palumbi, 2001; Yeh and Price, 2004; Shochat et al., 2006; Evans, 2010).

Paleoecological studies of pollen deposits over about 11 000 years since the last glaciation highlight major changes in the vegetation of the Berlin area (Brande et al., 1990). Hence significant increases and decreases in the relative importance of wetlands, grassland, shrubland, forest, and ruderal (so-called human-disturbed wasteland) plants occurred over these millennia. The city itself only appeared some 700 years ago at the crossroads of two major trade routes. Vegetation changes since then have been equally striking and presumably much more rapid (Sukopp, 1990).

Many slowly changing (from a human perspective) characteristics of urban areas seem to mainly occur at the scale of decades or human generations (Hanes, 1993; Forman, 1995; Brenner et al., 2001; Hong et al.,

Figure 3.6. Volcano towering over city in a zone of earthquakes and volcanic eruptions. An Australian eucalpt on the right. Volcan del Agua, Guatemala City. R. Forman photo.

2008; Brantz and Duempelmann, 2011; Forman and Sperling, 2011). Biodiversity, major irrigation systems, farmland soil depth, desertification, transportation infrastructure, technological change, culture, sedimentation of reservoirs, and urbanization around reservoirs are examples. All of these human-generation changes also occur at longer and shorter scales.

Tides normally rise and fall twice a day, and very high tides occur on a lunar cycle. Saltwater moves in and out across mud flats, salt marshes and mangrove swamps. High-tide saltwater moves up streams and rivers, and freshwater tidal fluctuations occur farther up the rivers affecting coastal cities. These freshwater tidal areas, often impacted by urbanization, normally contain a concentration of rare species able to thrive in such an unusual environment.

Many so-called disasters mainly occur in days, hours or minutes. Hurricanes/cyclones are usually days-long events. Floods and lava flows from volcanic eruptions often persist for days or weeks (Figure 3.6). Earthquakes typically are minutes-to-days events. In desert cities, dust and sand storms are hours-to-days events and flash floods minutes-to-days (Allan and Warren, 1993). Urban bombing typically is a minutes-to-days event. For Manchester (UK), such bombing in 1940 destroyed 4 ha (10 acres) of city center, including 30 000 houses, and caused 1300 fires and 363 human deaths (Ravetz, 2000). Later, a 1996 bomb devastated a quarter of the city center. A nuclear-power facility explosion, such as at Chernobyl in 1986, seems to be an hours-to-weeks event, which causes effects lasting centuries or more (Yatsukhno and Kozlovskaya, 1998).

Urbanization

We begin with the spatial and temporal patterns of urbanization, both internal and external. Then a brief section follows on the effects of urbanization on habitat loss, habitat degradation, and habitat fragmentation.

Spatial patterns of urbanization

Urbanization is the process of internal and external urban land-use change. For the typical city growing in population, the characteristic internal change is *densification*, while usually the outward change is *expansion* over agricultural or natural land. Urbanization has been measured in many ways useful for ecological analysis (Zonneveld and Forman, 1990; Theobold, 2004; Frumkin *et al.*, 2004; Scheer, 2004; Ellis *et al.*, 2006; Xu *et al.*, 2007; Forman, 2008; Fialkowski and Bitner, 2008; Schneider and Woodcock, 2008; Mortberg, 2009).

We explore urbanization from four perspectives: (1) outward expansion; (2) outward urbanization models; (3) internal changes; and (4) planning and optimal expansion patterns.

Outward expansion

In the fourth century BC, Plato pointed out that a city should not exceed 50 000 people, and that the next person should found a new city. In the late 20th century, a leading urban-design scholar indicated that the optimum size of a city has a population between 25 000 and 250 000 (Lynch, 1981). Today numerous cities exceed a million and some exceed 10 million people, each covering a large land surface.

A four-decade (1936–76) study of the Southern Ontario Region observed that the number of cities tripled (from 11 to 33) (Dorney and McLellan, 1984). The number of largest-area cities changed from four with a radius >50 km, to seven (>50 km) plus one with a >80-km radius (Toronto). The number of separate non-overlapping-radii cities changed little (from 5 to 4), but the number of cities with overlapping radii sharply increased from 6 to 29. In short, regional urban growth produced more, larger, and more-overlapping cities. Effectively this created a *megalopolis* of coalescing urban areas.

Usually cities are not round, and of course would not be round in such a megalopolis of overlapping population centers. In the USA, city shapes have been described as radial-centric, rectilinear, branching, linear or ring-form. These shapes basically result from geomorphology plus urban growth patterns (Hartshorn, 1992;

Pacione, 2005; Forman, 2008). Cities tend to be abundant and elongated in coastal areas. Urbanization there seems to be especially damaging ecologically, because endangered species are often concentrated along coastal areas (McDonald et al., 2008; Forman, 2010b).

A landscape modification gradient, ranging from urban to suburban to cultivated to managed to natural, was used to understand many characteristics of patches, corridors, and other features on the land (Forman and Godron, 1986). For example, from urban to natural along the gradient, patch density decreased (curvilinearly), regularity of patch shape decreased (linearly), variability in patch size increased (curvilinearly), number of wide green corridors peaked in the middle, number of stream corridors increased (curvilinearly), and connectivity of the matrix increased. In landscape ecology such changing spatial patterns reveal much about ecology outside a city.

Along a similar gradient from urban to suburban to exurban to rural to "wildland," it was found that authors of different studies used the terms quite differently (Theobold, 2004). The sequence of land-use types was consistent and useful. But differentiating the types by human population density was extremely variable, and seemed not to be useful.

The types of outward urbanization (suburbanization) expansion have been described in different, though overlapping, ways. One study recognizes four types of spread (Schneider and Woodcock, 2008): compact with infill; fragmented development; widely dispersed development; and "frantic" outward land conversion. Another study recognizes six somewhat-overlapping types (Scheer, 2004): urban fringe growth, squatter settlement, planned communities, edge cities (Garreau, 1991), sprawl, and rings of growth. Some informal squatter settlements occur at the edge of cities; also "edge cities" form as nodes near the fringe of a large city. Planned communities are commonly well separate from a city.

A two-decade study (1979–2003) of urbanization in the Nanjing Region (China) identified infilling, edge expansion, and spontaneous growth beyond the edge (Xu et al., 2007). Edge expansion accounted for the greatest population growth, whereas spontaneous growth covered the largest area. Approximately 80% of the growth occurred within 1.4 km of the urban fringe. Spontaneous growth beyond the urban fringe was often followed by coalescence of the developed sites. Spontaneous growth sites showed a negative exponential decline with distance from the edge of a large urbanized patch.

Five outward urbanization models

A study of 38 small-to-large cities in geographic areas worldwide initially highlighted four urbanization models of outward expansion (Figure 3.7) (Forman, 2008). A fifth model, as a variant of one, also seemed to be important.

Bulges. Outward urbanization occurs in planned or unplanned patches adjoining the urban fringe, and in different directions over time.

Concentric rings. Urbanization from the urban fringe occurs at equal rates in all directions, forming an equal-width ring, and over time growth occurs in spurts, producing concentric rings.

Satellite cities. No growth occurs outside the main city, but instead concentric growth occurs synchronously around a few satellite cities in the urban region.

Transportation corridors. Urbanization progressively extends out radial transportation routes forming ribbon or strip development, which progressively widens.

Dispersed patches. As a type of sprawl, growth occurs in separate small sites that progressively become both denser and further out from the urban fringe.

The bulges model describes the expansion of London from medieval time to 1830 (Figure 3.7). The medieval city was oblong, with most of the subsequent expansion (excluding ribbon/strip development) by bulges, each with a length-to-width ratio of about 1.5:1 to 2:1. Compact cities (Beatley, 2000a) as present in Northern Europe reflect the bulges and/or concentric growth models. Recent satellite-city growth has been important in the Barcelona Region, and led to several somewhat-compact cities (Forman, 2004b; Catalan et al., 2008). Planned satellite-city growth is particularly interesting, because urbanization can be targeted to particular satellite areas where environmental impacts would be limited.

Transportation corridor growth is conspicuous outward from small cities in the Amazon Basin (Browder and Godfrey, 1997). Dispersed-sites urbanization is widespread in North America, but increasingly appearing in Europe and elsewhere, perhaps in part mimicking the North American pattern. Urban-related dispersed patches far from the city that cause environmental effects on valuable parkland and other protected areas ("naturbanization") are an extreme example of the dispersed site pattern (Prados, 2009). Since about 1980, these dispersed residential spots,

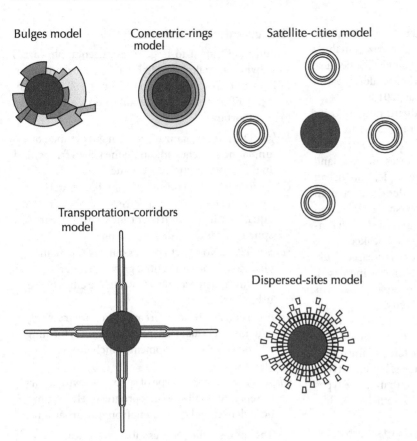

Figure 3.7. Five models of urbanization spread. Central circle = metro area. Three stages of outward expansion illustrated. Adapted from Forman (2008), largely based on analyzing the urban regions of 38 small-to-large cities worldwide.

often developing from recreational second-home nuclei, have become common in Europe and the Mediterranean Basin, as well as in North America.

Probably no city mimics only one of the models. Rather, outward growth in an urban region appears to be a combination of the patterns, often with one or two patterns predominant. Four decades (1960–98) of Cairo expansion were mainly by adjacent (concentric) growth, plus dispersed patches early and satellite growth later (El Araby, 2002). The urbanization of Phoenix (USA) over eight decades (1912–95) is particularly revealing. Early 1912–34 expansion was mainly by outward bulges, plus some adjacent (concentric) growth (Jenerette and Wu, 2001). Then in 1934–55, concentric growth, several new dispersed nodes, and one strip development corridor describes the expansion pattern. Next during 1955–75, numerous dispersed nodes and two strip-development corridors appeared, plus growth of a few satellite cities/towns. Finally from 1975 to 1995, urban bulges, satellite city expansion, and some dispersed nodes describe the urbanization of Phoenix. In this eight-decade period of urban expansion, all five model types are represented.

To determine which of the models above is best and which worst environmentally, the last four models (Figure 3.7) were applied to (superimposed on) the actual land-use patterns in the 38 urban regions studied worldwide (Forman, 2008). The effects of urbanization according to each model were estimated for 14 variables involving natural systems or human uses of natural systems (such as streams/rivers, tourist sites, large natural patches, and wildlife-and-walker corridors). Based on 2128 urban-region patterns (4 models times 38 urban regions times 14 environmentally related variables), the concentric-rings and satellite-cities urbanization patterns produce much less environmental impact than do the transportation-corridors and dispersed-site patterns. Little difference in overall environmental effect was evident between the first two models, and also between the latter two models. The bulges model was not evaluated, though its effect is probably almost the same as that for the concentric-ring pattern of outward expansion.

The dispersed-patch model is particularly useful because it reflects the predominant pattern of sprawl, characteristic of North America, increasing in Europe, and present in developing nations (Potter and Salau, 1990; Browder et al., 1995). A study comparing the ecological effects of compact versus dispersed growth around Brisbane (Australia) recommended high residential density with small yards and large connected greenspaces (Sushinsky et al., 2013). While most studies focus on patterns and problems within sprawl areas, the big-picture problem is poignantly highlighted by Janice Pearlman (1998):

> If the entire land mass of the planet were divided into individual household plots, there would be no space left for either agriculture or natural wilderness areas.

In essence, *sprawl* (unsuitable) is urban expansion with a low density of people and/or buildings (August et al., 2002; Squires, 2002; Frumkin et al., 2004; Soule, 2006; Forman, 2008). Since urban means city-related, the expansion of dispersed villages and towns here is not considered to be sprawl, nor is rural development far from a city. Typically either housing developments or houses are dispersed (sometimes called urban dispersion or leap-frog development) on farmland or natural land in a zone beyond the metro area or urban fringe. Many features are associated or correlate with sprawl, including strip (ribbon) commercial development along roads, large single-use rather than mixed-use residential areas, auto dependence by residents, loss of farmland, loss of natural habitat, fragmented disconnected land, diverse public-health problems, and commercial rather than civic and cultural centers of communities.

The distribution of different housing-unit plot (or parcel) sizes may be a good descriptor differentiating city, suburb, and exurb areas, irrespective of geography (Fialkowski and Bitner, 2008). Thus, determining changes in the total urban area strongly depends, particularly in North America, on what housing density is included as urban (Theobold, 2001). Over eight decades (1919–99), the approximate percent of area of all single-unit house plots in the USA changed over time as follows (Gordon et al., 2005):

Year	Plot size (acres)			
	≤0.5	0.5–1	1–5	5–22
1919	5	5	5	85
1959	6	4	10	80
1979	7	5	20	68
1999	9	6	16	69

(1 acre = 0.4 ha)

Throughout the eight decades very-large house plots predominated. Rather little change in plot sizes occurred in the first 40 years, 1919–59. But from 1959 to 1979, the proportion of largest house plots (5–22 acres) dropped sharply, while 1–5-acre plots doubled in percent. Then in the last 20 years, the proportion of large 1–5-acre plots decreased while the smaller house-plot sizes increased. A different study indicated that over the four decades (1960–2000) the relative proportion of housing in different plot-size categories (≤1, 2–10, and 11–40 acre plots) changed little (Theobold, 2001).

The bulk of these house plots represents outward urbanization described by the dispersed-site model, i.e., sprawl. Most of the sprawl has occurred on farmland, which is usually cheaper to develop and has soil particularly suitable for septic systems normally used in sprawl areas.

Where woodland or forest is close to a city, sprawl development often occurs in the forest (Figure 3.8). If the woodland is on hill- or mountain-slopes facing the city, development degrades the cool-air drainage process, useful for reducing urban heat and ventilating air pollution (see Chapter 5). In the USA about 65% of this sprawl-prone forest (wildland–urban interface) is considered to be subject to severe wildfires (Theobold and Romme, 2007). In 2000, this sprawl- and fire-prone forest covered 466 000 km^2 containing 12.5 million housing units. These threatened housing units are 52% more numerous than in 1970.

Internal changes

Internally an urban area may become denser or less dense in population and buildings, or may be transformed from one type to another without a change in density. Densification results from many processes such as conversion from low-rise to high-rise apartments, informal squatter settlement, infilling of vacant lots, and urban redevelopment. Classic examples of continued densification are Shanghai, Singapore, and Hong Kong, where dense populations are packed together in high-rises covering a limited total area. Most cities in the USA, known for outward sprawl, are also densifying (Katz and Land, 2003). Steep slopes, floodplains, wetlands and other greenspaces are characteristic sites for densification by informal squatter settlements in many parts of the world (Matthews and Kazimee, 1994). Cities in the USA have several hundred thousand homeless people who regularly use urban greenspaces.

Flows, movements, change

Figure 3.8. High-rise buildings, low rises, patch of single-family units, and an isolated house spreading up wooded and agricultural hill-slopes on the edge of a metro area. Barcelona. From R. Forman photo.

Infill, by building on vacant sites, removes greenspaces and tiny vegetated sites. But there is a limit to the amount of infill appropriate for an urban area. That *infill threshold* is the number and arrangement of parks and other greenspaces that optimize conditions for both people and nature. For people in low population-density areas, infill may lead to enhanced public transport, local services, even a sense of place (Frey, 1999). But many greenspaces are essential and green corridors and stepping stones important to sustain vibrant urban nature. Some infill eliminates valuable remnant wetlands (Dale, 1997; Mitsch and Gosselink, 2007) and may cover contaminated-soil brownfields.

The threshold for infill also highlights the issue of a compact versus spread-out city. Compactness greatly reduces the total amount of land consumed by development and its associated environmental degradation. But an all-built city without greenspaces or tiny vegetated sites is nearly devoid of nature, and arguably does not sustain a healthy population of people. Thus, the infill threshold is a key measure for urban compactness.

Urban areas also may become less dense (de-densification) in population and buildings (Haase and Schelke, 2010). Population shrinkage is typically accompanied by property abandonment and the crumbling and removal of buildings, leaving vacant plots or areas. In the late 20th century, population loss and land abandonment in the USA occurred in Detroit, St. Louis, Cleveland, Youngstown, Pittsburgh and Buffalo (Langner and Endlicher, 2007; Rink, 2009). War often leads to an exodus of people and a less dense city. Even in the more common case of growing cities, the inner suburbs may de-densify. Long ago these suburbs were outward expansion areas of the city, often with industries providing jobs. But now the areas may feature poor-quality housing areas and contaminated-soil brownfields.

In some cases, land is transformed or converted from one use to another without a major change in population density (Altshuler and Luberoff, 2003; Lukez, 2007; Bosselman, 2008; Brantz and Duempelmann, 2011). Or, while transformation is prominent, the population may increase. Portions of Paris in the late 19th century and Barcelona in the early 20th century were transformed from low-income to more-expensive planned neighborhoods.

Over history, Budapest has suffered floods from the big Danube River, often destroying major portions of the city then followed by rebuilding. In 1990, after decades of foreign occupation, the military forces left, leaving most of their equipment, almost all of their buildings, and many contaminated-soil areas (Enyedi and Szirmai, 1992; Sailer-Fliege, 1999; Dent, 2007). Since then the city has yet again been transformed, this time for the extensive traffic of a car culture, associated air pollution, shopping malls, and so forth.

Internal change may be affected by outward expansion, and vice versa. Thus, extensive internal land abandonment approximately correlated with the outward suburban expansion of Detroit. Analogously at a finer scale, a neighborhood may densify or de-densify partially in response to growth or shrinkage of the broader city or suburb. Ecologically this is probably particularly important, since context is so important for the flows of species and water in urban areas. Ecological studies of such hierarchically changing urban areas would be valuable.

Time-lapse photographs of a city over its history, as if taken from a cloud or flying carpet, would be magical, and quite informative. For instance, Amsterdam's center was transformed from a dense medieval structure dissected by narrow ways: to a rich traders' city with larger markets, buildings, streets and canals; then to countless four-level workers' apartments in the 19th century; next to an abundance of architect-designed block-size apartment buildings; then to numerous post-war high-rise buildings; and more recently Amsterdam includes many mega-block apartment buildings set within relatively continuous lawn (Habraken, 2000).

The underlying pattern of roads and blocks tends to structure transformation patterns. City-center New York is a rectangular grid and Boston an irregular network (folklore says laid out by the early settlers' cows), so sections of a city being transformed fit the regular or irregular network (Krieger and Cobb, 1999). Thus, sections of a city being transformed fit within the existing regular or irregular street network. Over centuries the internal form of some cities has markedly changed (Hall, 2002; Busquets, 2003). In the late 19th century, radical planning changed the underlying road network for wagons and carriages in a large section of Paris to a pattern highlighting circle intersections with radiating streets. In the early 20th century Barcelona also underwent a major spatial transformation of transportation.

A particularly informative approach is to map the rates of land-use change across the urban area, as done for Denver (USA) (Berke et al., 2006). Such an approach in rural areas of Denmark over a century found that the most transformed features were ditches, paths, hedgerows, and dry patches (Agger and Brandt, 1988). Thus, areas of rapid change contrast with stable areas. Again, the land-use change comparison is ecologically important because of adjacency and context effects on a site (see Chapter 2).

Because vegetation is normally so limited in a city, vegetation change during internal urbanization is especially conspicuous and important. The increase, decrease, and rearrangement of urban parks, community gardens, vacant lots, tree corridors, and other tiny vegetated sites are seen and felt by nearly every resident. The changes have a major effect on ecological conditions. During two centuries in Sydney, two of the nine original forest types remain only slightly changed in area, whereas another two types almost disappeared (Benson and Howell, 1990). Over a generation (1973–97) in the Southern USA, urban tree-canopy cover decreased 8–9% in Houston and Roanoke, but dropped 20–26% in Atlanta, Chattanooga, and Fairfax County (Virginia) (Hermansen and Macie, 2005).

For the five decades (1948–94) following atomic-bomb destruction of Hiroshima City, the farmland and woodland of the city changed markedly (Nakagoshi and Moriguchi, 1999). Of this total greenspace, farmland area decreased from 19% to 10%, and woodland from 77% to 67%. Residential cover spreading on the greenspace areas increased from 4% to 23%. In 1998, the city's farmland and woodland cover was composed of: rice paddy fields 5%; dry fields 2%; "natural" old-growth forest 0.3%; second-growth pine-oak woods 66%; and conifer plantations 20%. During the half-century (1948–98) of internal growth and densification, half the residential growth was on woodland and half on farmland. Yet since there was less agricultural area initially, half the farmland was lost.

Vegetation on greenspaces themselves is changed as part of internal urbanization. For example, in a 1940s urban-shrinking phase within Tokyo, existing greenspaces were transformed and new ones constructed for wartime defense (Ishikawa, 2001). Some greenspaces were converted to air defense anti-aircraft sites and other military uses. After earlier extensive hard-to-stop fires, open corridors were constructed to limit the spread of fires. Analogous "air strike defensive fire breaks" were constructed in Seoul about 1940 (Rowe, 2010). Today's Tokyo, again transformed, has removed many traces of wartime greenspace uses. Now the city is characterized by a large central Imperial Palace grounds, a fair number of medium-large parks, and numerous tiny vegetated spots and symbols of nature associated with individual buildings or residents.

A closer look at 11 residential areas in Merseyside (UK) over 25 years reveals different patterns. Only tiny barely noticeable changes occurred at individual spots. Yet overall, 5% of the total area was converted from plant cover to impermeable surfaces of buildings and pavement (Pauleit and Breuste, 2011). This change was greatest in wealthy neighborhoods. The maximum surface temperature for the total area increased an average 0.9°C. During a rainstorm of 1 cm/h, the surface runoff increased 4%. Biodiversity (at least as measured by the Shannon index) decreased over the 25 years. In short, the cumulative effect of tiny, barely noticed changes in residential areas appear to be environmentally significant.

Planning and optimal expansion patterns

Finally, a relatively small proportion of urbanization, especially outward spread, directly results from planning. Planning and design can create a much wider range of changing spatial patterns than is normally produced by lack of planning, which breaks up the land into relatively fine-scale human and natural processes. Ecological restoration in urban areas mainly focuses on small spaces but can address large areas (Harris et al., 1996a; Lee and Cho, 2008a; Hobbs and Suding, 2009).

Although planning is well beyond the scope of this book, a few examples that include ecological dimensions are mentioned for illustration. At a broad scale, the extensive Dutch planning traditions are particularly useful for, and in, a megalopolis (Kuhn, 2002; Opdam et al., 2002, 2006; van der Valk, 2002;). Urban region planning is especially useful because, unlike a huge megalopolis or the limited area of a city, much of the public regularly depends on an urban region (Mumford, 1968; Yaro and Hiss, 1996; Forman, 2008). Also, planning could make a semi-sustainable urban region. Metro area planning, for example in Asia, could build on the traditional mixture of urban and rural land uses, where farmland and woodland provide both key ecological functions and cultural services (see Figure 11.1) (Yokohari et al., 2000; Rowe, 2005).

Plans for whole cities typically focus on housing, jobs, transportation and economic development. Very few such city plans have environmental dimensions as a leading goal (Ravetz, 2000). New cities such as Canberra and Brasilia have been planned and constructed with varying environmental success (Le Gates and Stout, 1996; Reps, 1997; Forman, 2008). Greenbelts, urban growth boundaries, and greenways are familiar planned patterns around metro areas (Howard, 1902; Taylor et al., 1995; Pendall, 1999; Kuhn, 2002; Ahern, 2004; Forman, 2008). Transit-oriented development (TOD) highlights both urban and suburban development around public transport stations (Cervero, 1998; Berke et al., 2006).

Some broad-scale plans with valuable environmental dimensions have focused on the ring area of outward urbanization. Compact development contrasts with sprawl (Benfield et al., 2001; Selman, 2006; De Aranzabal et al., 2008). So-called smart growth plans have attempted to solve some challenging issues of exurban or peri-urban development (Garvin, 2002; Porter, 2002; Squires, 2002; Szold and Carbonell, 2002; Handy, 2005).

An analysis originally derived from logging patterns of old-growth forest in the USA Pacific Northwest (Franklin and Forman, 1987; Forman, 1995) provides especially valuable insight for outward urban expansion. Five spatial alternatives for cutting a forest over time were compared. The shrinking natural land is assumed to be more ecologically suitable than the expanding new land type. These are the five spatial-sequence models:

1. *Edge model.* Development parallel to and progressively expands from one boundary.
2. *Corridor model.* Development along a central corridor and spreading outward in both directions.
3. *Nucleus model.* Expanding concentrically from a central point.
4. *Few nuclei model.* Expanding concentrically and synchronously from a few points.
5. *Dispersed-patch model.* Dispersing developments in a regular distribution (minimizing the variance in distance between patches).

For forest cutting, the edge model (Figure 3.9a) has the least environmental impact, and is considered to be the ecologically best way actual forest cover is progressively decreased over time.

However, three changes significantly improve the edge model (Figure 3.9b). The late portion of the sequence is improved by a "jaws-like" urbanization process along two adjacent sides (Forman, 1995; Forman and Mellinger, 2000). The late portion is also enhanced by establishing at the beginning tiny protected patches and corridors that are maintained in place until finally being removed at the end. The middle potion of the sequence is improved by retaining a few large patches or "chunks," rather than a single very large one (i.e., spreading risk), that are then sequentially rather than synchronously eliminated. The resulting "jaws-and-chunks" model seems to be the ecologically optimum way to transform a more-suitable to a less-suitable land (Forman and Collinge, 1996). This model sequence may be optimum for outward urbanization.

A useful aspect of determining the optimum spatial sequence is that at any point in the time sequence, the best and worst location for the next change can be pinpointed. For example, at the midpoint (50%-transformed phase), the best location (least ecological degradation) for the next housing development would be along the edge of the large northwestern and southeastern green patches (Figure 3.9b). The worst location for a new development would be just southeast of the

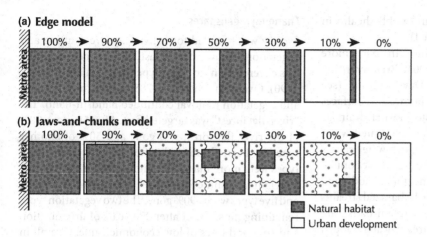

Figure 3.9. Proposed optimal models of land transformation. (a) Edge model describing real-world transformation of ecologically more-suitable to less-suitable conditions (Forman, 1995). (b) Ecologically improved "jaws-and-chunks" model. See text. (See Forman and Collinge, 1996). Adapted from original drawings courtesy of Kimberly Hill. Urbanizing from the outer rather than inner side of the large natural habitat better provides a "species rain" into the metro area (see Chapter 12).

center of the large southeastern green patch. Inversely, these locations would be the ecologically worst and best, respectively, for establishing the next protected natural area or park.

Rates and trajectories of ecological change

For convenience we present these in four groups: (1) soil, water, air; (2) vegetation and biodiversity; (3) urban land uses; and (4) habitat loss, degradation, and fragmentation.

Soil, water, air

Environmental and urban changes occur long-term but major alterations normally occur in discrete periods. The following illustrates the periods, their individual lengths, and the sequence of major human effects on the environment in the USA (Turner et al., 1990):

- Deforested area: 1690–1900.
- Terrestrial vertebrate diversity decrease: 1780–1900.
- Carbon releases to atmosphere: 1810–1960.
- Human population size: 1850–1970.
- Lead releases to air, soil, and water: 1910–1970.
- Water withdrawals from ground: 1930–1970.
- Sulfur releases to atmosphere: 1950–1970.
- Phosphorus releases to water and soil: 1960–1980.
- Nitrogen releases to air, soil, and water: 1975–1980.

Soil is degraded by urbanization in different ways (see Chapter 4) (Bennett, 1948; Brown et al., 2000; Sieghardt et al., 2005; Sauerwein, 2011). (1) Erosion increases, associated with the construction of roads, buildings and other structures. This also means that sedimentation of the eroded soil particles in water bodies increases. (2) Productive agricultural land is lost to urbanization. Since cities usually originated on or by good, even prime, agricultural soil, in many regions the best soils are buried by urban spread (Amundson et al., 2003). (3) Soil quality is reduced by chemical pollutants. Excess nitrogen blankets many urban areas, hydrocarbons from petroleum combustion reduce water infiltration into soils (White and McDonnell, 1988), and toxic substances from arsenic to organics limit the use of soil areas (Brown and Jameton, 2000; Hough et al., 2004). Soil-contaminated brownfields concentrated around cities are familiar examples.

Mineral nutrient flows (biogeochemical cycling) are much altered by urbanization (Groffman et al., 2004; Kaye et al., 2006; Carreiro et al., 2008; Alberti, 2008; Grimm et al., 2008). For example, nitrogen concentration may be higher than in many farmland areas. Phosphorus from stormwater pipes and septic systems eutrophicates urban streams and rivers. Heavy metals accumulate in soils sometimes to toxic levels.

Urban air is commonly altered by densification and outward spread (see Chapter 5). Microclimate, including urban heat, changes over local built sites as well as larger all-built metro areas. Air pollutant concentrations may increase or decrease. For instance, sulfur dioxide (SO_2) and particulate matter (PM) increase with coal burning, and decrease with a switch to cleaner coal and other fuels. Aerial nitrogen oxides increase with more vehicular traffic and other high-temperature-machine use. Urban air pollution also affects microclimate, such as more particles and aerosols leading to more rainfall (Chen et al., 2007). Combined microclimatic and pollution changes affect both environmental conditions and public health in the urban area (Frumkin et al., 2004).

Water is affected by urbanization in much more diverse ways (Paul and Meyer, 2001; Brenner et al., 2006; Baker, 2009; Butler and Davies, 2011) (see Chapters 6 and 7). Hard or impermeable surfaces accelerate stormwater flow into stormwater drainage pipes leading to water bodies. The accelerated water may cause erosion. The large increase in water causes flooding, which inhibits septic system functioning, fills basements, covers roads and neighborhoods, and washes away bridges and buildings. Urbanized streams are straightened and channelized, stream-banks are lined with rocks or concrete, and stream-water is piped underground. Streams in suburbia may disappear, reappear, and disappear again, the water eventually pouring into a nearby water body. Stormwater and sewage wastewater are combined in pipes and channels in many cities, while in others the flows are separated in different pipe systems, so the flows are only combined (CSOs) in heavy storm events. Water quality usually degrades, though planned projects such as wastewater sewage-treatment facilities improve water quality. Urbanization also leads to water shortages (Ma, 2004).

Six decades (1950–2010) of urbanization in China's Changjiang Region, which contains several large cities (Shanghai, Nanjing, etc.), is instructive. In 60 years the region lost a total 12.2% of four key environmental resources: forests, wetlands, freshwater, and cash-crop farmland (for vegetables and fruits; not grain production) (Kim and Rowe, 2012). The total resource loss around each of four large cities was: 10% for Shanghai; 20% Suzhou; 20% Wuxi-Changshou; and 20% Nanjing. Except for wetland near Shanghai, little variability from city to city and from resource to resource was evident. These four very large cities apparently contributed 35% of the total regional resource loss. However, 230 scattered small cities contributed a similar amount of total resource loss over the six decades. Per-person, the resource loss was much greater around the dispersed small cities than near the big cities. The loss of wetlands and freshwater to urbanization is especially significant, because they cannot be transported or replaced easily.

Vegetation and biodiversity

These key characteristics of urban areas are explored in two ways: (1) changing greenspaces; and (2) ecological succession and species change.

Changing greenspaces

Sydney was founded for convicts, so tree cutting and vegetation removal started fast. The area of today's city was covered by nine forest types (Benson and Howell, 1990). Over two centuries, 1788 to 1988, urbanization and vegetation removal continued hand-in-hand. The "river-flat forest" was targeted first. After 50 years, 90% of the river-flat forest was gone and 50% of another type had been eliminated. After a century, five forest types were more than half removed. After two centuries, eight of the nine forest types were >50% eliminated, and five types were >90% gone. The two vegetation types remaining most intact after 200 years of urbanization had soils and trees of low economic value. Overall in urban regions today, natural or semi-natural areas protected for biodiversity tend to be on low-quality soils (Hunter, 1990; Forman, 1995). Ironically, richer soils often support more species.

Considering specific features within an urban area, Boston's "Emerald Necklace" was very valuable for the city in the late 19th/early 20th century. A half-century later the green-corridor-connected large green patches were engulfed by urban land, cut in places, and modified by surrounding development activities. Today the Necklace is still valued regionally for its history, but its environmental and human features are mainly valued by the local surroundings. Unlike Boston, Canberra's wide green corridors connect to woodland surrounding the city (Forman, 2008). As the city grew for a century, urbanization has not greatly interrupted that connectivity.

Thus, some greenspaces are mainly of local value while others are of regional value. Protected vegetation around the somewhat-distant water supply for Canberra was planned and established from the beginning of urbanization. Maintaining market gardening areas, large parks, large water bodies, and greenway networks during an urbanization process provides ecological and societal value both locally and for the whole metro area.

In contrast, over five decades (1944–91) woodland in Rome doubled (from 4.3% to 8.3% of the city) (Attorre et al., 1998). Most was due to successional habitats, especially shrubland, growing into woodland. During the period, a few large vegetation patches changed to more and smaller ones (12 to 23; average 287 to 46 ha). Of this amount, woodland patches slightly increased in both number and size (83 to 99; average 3 to 5 ha). Meanwhile, separate urban built patches dropped sharply in number, and increased markedly in size (73 to 17; average 11 to 191 ha). During the half century, the predominant change patterns were: woodland remained woodland; farmland either remained farmland or became urban; shrubland became urban or woodland, or remained shrubland; meadow became urban or remained meadow; and urban land remained urban. The total boundary length between land uses increased 10%. Thus, considerable vegetation dynamics occurs during the urbanization process. Change represents a valuable opportunity to enhance ecological conditions across a metro area.

Shanghai over the past few decades also had a significant increase in urban parks, lawns, and street trees (Zhao *et al.*, 2006). Some increase occurred on former farmland. Negative effects of urbanization on greenspaces of course also occur. For instance, intensive firewood collecting for fuel around Lagos and other Central and East African cities has extensively eliminated trees. Loss of tree products, soil degradation, soil erosion, and increased atmospheric particles are among the results. Informal squatter settlement on urban greenspaces worldwide causes cascading environmental degradation.

Ecological succession and species change

For the short term of months or years, early *ecological succession* (successional habitats) on urban sites is especially important. Relatively few sites support succession for decades. Considerable green cover occurs in small sites on house plots. Suppose a house plot were abandoned and protected for 30 years. In Leicester (UK), it would probably become a mixed young birch-holly-ash-elder-hawthorn woodland ("scrub") (*Betula pendula, Ilex aquifolium, Fraxinus excelsior, Sambucus nigra, Crataegus monogyna*) (Owen, 1991). Beneath these tree and tall-shrub species would be brambles (*Rubus fruticosus*) and lots of herbaceous plants with diverse seeds continually arriving. Theoretically much later, a mixed deciduous woodland would be expected. This sequence almost never happens, because gardening by urban residents adds and removes plants, and manicures the area.

The later woodland stages continue to change, as seen in a 16-ha (40-acre) New York City woods over a half century (Rudnicky and McDonnell, 1989). In 1937 the wood was 70% covered by hemlock and oak (*Tsuga canadensis, Quercus rubra*), mainly in two large separated patches. In 1985, hemlock and oak canopies only remained in several small spots, while maple, birch and cherry trees (*Acer rubrum, Betula lenta, Prunus serotina*) dominated the canopy woods. The hemlocks and oaks were mostly large trees while the new dominants were small trees. In addition to tree growth and death, the prime reasons for the transformation were considered to be the stress of human trampling, plus hurricane, arson, and vandalism disturbances over a half century.

Now visualize the action on a "vacant lot" or abandoned parking lot over time. Rainwater infiltrates through cracks. Seeds germinate in cracks. Roots penetrate. Roots thicken, especially those of woody plant seedlings, forcing the cracks to widen. Dust including silt and clay particles accumulates. Leaves and stems die, producing litter. Microbes turn litter into humus. Roots also die, adding organic matter lower down. Taller plants increase the shade. Invertebrates colonize and multiply. Cracks form networks. Vertebrates feed on invertebrates, as well as plants. In this way a vacant lot ecosystem with distinctive spatial patterns has formed and ecological succession then accelerates.

Urban successional processes have been described more generally as follows (Berkowitz *et al.*, 2003; Bradshaw, 2003):

1. Plant immigration and establishment.
2. Surface accumulation of fine particles and mineral nutrients.
3. Increase in organic matter and microbial colonization.
4. Root expansion, soil changes, and some nutrient recycling.
5. Plant competition and facilitation.
6. Upward growth and more plant colonization.
7. Increase in herbivores and food web complexity.
8. Some species and growth limited by urban stresses.
9. Disturbances at spots within the site.

This sequence of processes produces a micro-mosaic over the site. Early succession seems similar across a city. However, since seed input, plant species success, urban stresses, and disturbances differ markedly, upon closer inspection successional habitats are quite different.

Furthermore, initial site conditions as well as the rates of succession differ from location to location (Figure 3.10). A high degree of unpredictability, though not necessarily randomness, is characteristic (Trepl, 1995; Niemela, 1999). Turbulence and vortices characterize urban winds (see Chapter 5). Sources of seeds and animals may be near or distant in the highly

Figure 3.10. Several years of ecological succession by an abandoned city building. Tall herbaceous vegetation with small tree of heaven (*Ailanthus altissima*) in front; trees and shrub in back remain from former yard plantings. Detroit, Michigan, USA. R. Forman photo courtesy of Jessica Newman.

fragmented urban area. The communities or assemblages of species are not tightly integrated as in many natural areas. Human disturbances are often unpredictable. Legacies of former stress or disturbance are often important. The trajectory of succession may be altered at any stage. Consequently at any time, successional sites across a metro area support an immense richness of species, patterns, and processes.

An alternative perspective on vegetation and species change in urban areas starts from the typical case of a site to be developed into a park or garden or lawn or other vegetated land use (Gilbert, 1991). Both the sequence and rate of ecological succession are more dependent on human activities than on plant colonization, competition, soil development, and so forth as described above. Five processes typically occur in stages:

1. Site clearing (removing unwanted objects and reducing heterogeneity).
2. Subsurface construction (digging, removing soil material, pouring foundations, adding sandy fill and topsoil).
3. Surface construction (laying walkways, constructing walls and any buildings, planting lawns, gardens, hedges and trees).
4. External effects (wind patterns, input of seeds, weeds and animals; diverse human impacts).
5. Maintenance (mowing, trimming, tree/shrub removal, chemical applications).

The first and fifth processes are considered to be most important in affecting vegetation.

The preceding sequence produces "human-molded vegetation" rather than vegetation from ecological succession. Costs are much higher. Predictability much higher. Species richness normally much lower. And stability, which depends mainly on a sustained maintenance regime, much lower.

Urbanization sprawl is widely known for reducing habitat cover and biodiversity (Theobold *et al.*, 1997; August *et al.*, 2002; McKinney, 2002; Squires, 2002; Johnson and Klemens, 2005; Ray, 2005; Jaeger *et al.*, 2010; Berland, 2012). Yet sprawl is a time process and species may adapt to it (Hitchings and Beebee, 1997; Vos *et al.*, 2001). Indeed, urbanization also creates habitats, widening the options for some species to survive (Chapman and Underwood, 2009). Some bridges are prime habitats for bats and birds. Walls are key habitats for some wall flowers. Lizards and flies, termites and cockroaches, cats and mice, house plants and spices, often thrive in buildings. Vegetable gardens with regular soil disturbance, plus fertilizers and pesticides, provide fine conditions for many rapidly reproducing herbaceous plants (often called weeds). Noise-tolerant animals can live near busy highways with fewer competitors to worry about. Even the urban concentration of non-native species provides novel habitats for other species, such as bats in holes of planted eucalypt trees or fleas carried by Norway rats.

The ranges or distribution limits of many species within cities have moved outward at the same time that urbanization has expanded outward from the urban fringe. Six medium-and-small mammal species that moved outward from the city-center area of Tokyo in a half century (1924–72) (Obara, 1995) provide a nice example. All species ranges (raccoon dog, fox, weasel, mole, hare, and bat) moved outward in spurts. This left a large central area dominated by mice and rats (*Mus musculus, Rattus rattus, R. norvegicus*). For the six native mammal species, three types of conditions were recognized in irregular concentric rings outside city center. Animals in the inner ring were dependent on the urban ecosystem. Animals in a narrow outermost ring were dependent on natural ecosystems. Animals in the variable-with middle zone were dependent about equally on urban and natural ecosystems. These patterns may be typical for urban animals in many metro areas.

Urban land uses

A glance at different locations in a metro area during urbanization is interesting. Here we will quickly hop from street trees and neighborhoods to transportation corridors and riversides, and to whole urban regions, plus the peri-urban or exurban zone at the epicenter of outward urban expansion.

Tree patterns along roads and streets in Rome are even more interesting than for greenspaces in the preceding section (Attorre et al., 2000). The tree species planted differed in, and sometimes symbolized, contrasting political and economic periods. In 1998 about 120 000 street trees representing 58 species were recorded in approximately 70 sections (urban units) covering the 350 km² area of Rome. Five planted-tree types along streets and roads were recognized:

- Elm (*Ulmus minor*): the traditional symbol of Papal ("the Pope's") Rome.
- Plane (or sycamore) tree (*Platanus* × *acerifolia*): symbol of the late 19th century Umbertine period.
- Pine (*Pinus pinea*): planted during the Fascist period, and symbolizing a cultural link with the Roman empire.
- Black locust (*Robinia pseudoacacia*): a non-native species from America planted during an intense urbanization period characteristic of the 1970s.
- Mix of small trees (*Ligustrum, Nerium, Hibiscus, Prunus*): a functional response to very small spaces, plus a wish for a rapid effect, and characteristic of the 1990s.

During the past century, 1898 to 1998, large trees (>20 m high) progressively decreased from 65% to 35% of the total, and small trees increased from 5% to 43%. Plane tree and elm dominance changed to a diverse assemblage of black locust, plane tree, pine, *Ligustrum* and *Nerium* dominance. Throughout the 100 years, tree species dominance in the individual city sections differed markedly, producing an ever-changing, broad distinctive mosaic across the metro area.

Over a similar 12-decade period (1867–1984) in Berlin, an analogous changing mosaic of herbaceous plant species occurred (Brande et al., 1990). At the outset the city had considerable greenspace in its center surrounded by rather densely built areas. During the period, buildings, lawns, flower gardens, vegetable gardens, rubble from war-destroyed buildings, and craters from explosions all appeared and disappeared. Diverse changing successional habitats characterized the urban mosaic.

Neighborhoods decline and are revitalized (Pacione, 2005). The neighborhood mosaic (see Chapter 2) often has a high degree of interactions, even interdependence, among its components. Changing a central component could trigger an unraveling and degradation of the neighborhood. More likely, however, the active interactions would provide compensation, limiting the spread of any disturbance effect. Altering the edge portion of a neighborhood, though, might change the whole area. Also, some boundaries including the urban fringe readily move, even going from concave to convex (Forman, 1995).

Transportation infrastructure typically is extensive with considerable inertia, and thus changes slowly. Individual linkages or nodes may be altered, but the network as a whole usually remains functional, especially if many loops (high circuitry) are present. On the other hand, urbanization changes along coastlines may be slow or rapid and, with a powerful sea adjacent, are relatively unpredictable (Marsh, 2005).

Riversides tend to change drastically with urbanization (see Figure 7.7). With more surrounding impermeable surfaces, river floods often get larger. More infrastructure lines the river connecting city center with areas outside the metro area. The infrastructure requires continual maintenance, causing continued habitat disturbance. Riverside land uses also change as old factories become less viable, commercial and hotel activities grow, and demand increases for replacing old housing with new apartments and condominiums. In Hiroshima City (Japan) over two decades (1976–97), riparian forest patches decreased in number (36 to 15) and especially in size (average 0.19 to 0.04 ha) (Tanimoto and Nakagoshi, 1999). The resulting woods were more elongated. Residential development on floodplain farmland caused the number of farmlands to double, while their average size plummeted from 1.26 to 0.14 ha. Parks tripled in number (from 5 to 17), though their average size and shape remained essentially constant.

Streams also change markedly with urbanization, since bulldozers, road building, and building construction are such powerful forces (Paul and Meyer, 2001; Korhnak and Vince, 2005). Yet more powerful flood times occasionally change that perspective.

Urbanization effects are just as striking at broader scales. The megalopolis of Dutch cities has a "green heart" of fragmented farmland and woods. Over decades the heart became even more fragmented (Harms, 1999). In a half century (1950–2000), the Seattle (Puget Sound) Region was fundamentally transformed, both in environmental and human conditions, by extensive urban expansion (Alberti, 2008). Such a pattern is replicated in many other North American metro areas and urban regions (Berger, 2006).

Finally, focusing on the epicenter of outward urbanization, a specific case of a new highway bypass around the city of Hubli-Dharwad (India) illustrates what happens at the urban fringe (Shindhe, 2006). Long-term residents interviewed repeatedly mentioned the following results of the highway bypass: (1) water-logged soil in the wet season; (2) deprived of water to carry away sewage; (3) fallow land due to water-logging; (4) blocked access to my land; (5) unable to grow vegetables and cash crops; (6) lost access to the water tank source; and (7) lost 35 mango trees, or 4 large shade trees, or 120 trees ....The peri-urban or exurban zone is normally the most altered, indeed transformed, area during urbanization.

Habitat loss, degradation, fragmentation

The key spatial processes causing habitat change are first outlined. Then habitat changes on agricultural and natural lands due to urbanization are introduced.

Spatial processes

Five *spatial processes*, i.e., mechanisms changing the pattern or arrangement of objects over time, predominate in outward urbanization (Forman, 1995; Lindenmayer and Fischer, 2006). *Perforation* makes holes in an object. *Dissection* slices an object, creating constant-width strips or corridors. *Fragmentation* breaks an object into pieces, at least some being widely separated. *Shrinkage* decreases the size of an object. *Disappearance* (attrition) is the loss or elimination of an object.

A general model highlights the relative importance of these spatial processes along a land transformation gradient from 100% more-suitable habitat or land use to 100% less-suitable habitat (Figure 3.11a) (Forman, 1995). Normally perforation and dissection predominate in the early portion of land-use change such as urbanization. Fragmentation and shrinkage are usually most prominent in the middle portion of the change gradient. Shrinkage and disappearance then predominate in the later portion of urbanization. Interesting exceptions to the model exist, but the general sequence of spatial processes paints the big picture.

Urbanization around a New Zealand city perforates, and then fragments, nearby forest (Norton, 2000). Habitat fragmentation is especially important ecologically in suburban and exurban/peri-urban areas, as discussed below (also see Chapters 8 and 9). Finally, as any city continues outward spread, nature's fragments continue shrinking and disappearing.

Gradual land-use changes seem to have predominated over eight decades (1912–95) in the dry Phoenix Region (USA) (Jenerette and Wu, 2001). Desert area decreased steadily. The total agricultural area changed little, but some moved outward with more irrigation. The urban area noticeably increased from about 1935 on, and steeply from 1975 on. The urban area increase was exponential and correlated with population growth ($R^2 = 0.99$). Fragmentation of the agricultural and natural desert lands by urbanization especially increased in the urbanization acceleration phase after 1975. Structural complexity of the land or mosaic heterogeneity also markedly increased as urban spread accelerated. Modeling the patterns for two decades (1975–95) suggests that the regional land-use trends can be understood by measuring four "landscape metrics" [patch density; edge density (boundary length); fractal dimension; and contagion (aggregation)].

Additional spatial processes are important for land recovery or restoration, and usually for environmental

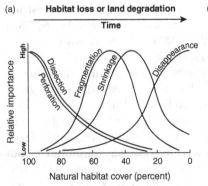
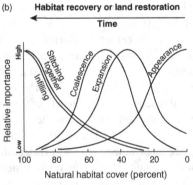

Figure 3.11. Spatial processes operating during habitat loss and habitat recovery. General models; local conditions may create variations. See Forman (1995).

planning, design, and management. These cases normally begin with less-suitable land, and the goal is to increase more-suitable land or habitat (Jordan et al., 1987; Harris et al., 1996a). Recovery spatial processes used in the reverse-of-urbanization scenario differ markedly from those just described (Figure 3.11b) (Forman, 1995): appearance; expansion; coalescence; stitching together; and infilling. Also proliferation, connection, aggregation or translocation might occur.

Thus, for example, the unplanned reforestation of mid-19th century agricultural land in New England (USA) probably involved most of these spatial processes of recovery (Foster and Aber, 2004). In an ecological planning project for the Greater Barcelona Region (Forman, 2004b), one large green patch appeared (was proposed in the plan), several large vegetated patches were expanded, some small-to-medium patches effectively coalesced, and connectivity among large green patches was greatly increased creating an "emerald network." Many urban plans, some implemented, propose increasing connectivity for both nature and people using major green corridors or greenways (Ahern, 2004; Jongman and Pungetti, 2004; Erickson, 2006). Finally, infilling occurs where an isolated built property (inholding) within a park is acquired and revegetated to establish continuity across the park.

Not surprisingly, the idea of spatial processes is useful well beyond the case of outward urbanization. Many features of urban areas, from an entire urban region to a neighborhood or cemetery or house plot, change with different spatial processes (Figure 3.11). Perforation, shrinkage, and expansion seem to be most frequent. The least common spatial processes are coalescence, stitching together, and translocation. In short, the analysis of spatial processes provides considerable insight into urbanization and ecological change over time.

Agricultural and natural land

Overall, habitat loss, degradation, and fragmentation by urban expansion mainly occur on agricultural land, and secondarily on natural land. *Habitat loss* refers to disappearance, *habitat degradation* to reduction in ecological quality, and *habitat fragmentation* to the breaking into pieces. Habitat loss doubtless has the greatest effect on ecological processes and biodiversity. Habitat degradation seems to have the second greatest effect, and fragmentation the third greatest effect (Forman, 2006).

The effects of habitat loss, degradation, and fragmentation specifically on vegetation and biodiversity are discussed in Chapter 8, and on wildlife in Chapter 9. This brief section is to highlight the agricultural and natural lands where outward urbanization most affects habitats.

Farmland, villages/towns, urban areas, and transportation corridors are the prime fragmenters of land. At a fine scale, almost any human activity on the land can alter habitats. At the broad scale, land or landscape fragmentation is an intense global phenomenon (Molnar, 2010). The only extensive, largely unfragmented areas remaining are the Amazon Basin, northern Canada, Gobi Desert, Arabian Desert, and Antarctica.

Near the big cities of Eastern China is some of the most intense habitat loss, degradation, and fragmentation in the world (Li et al., 2010). Urban land and good farm soil are highly correlated in China (Tan et al., 2005). In only a decade (1990–2000), urban land in the Beijing-Tianjin-Hebel Region increased 71%, and three-quarters of that was on arable cropland. Farmland loss (Brown, 1995) and vegetation loss (China Ministry of Transport, personal communication, 2010) are both important issues in China. Not surprisingly, the habitat disruption in Eastern China occurs close to remnant hot spots of plant and mammal diversity.

Land or habitat fragmentation has been analyzed with diverse spatial measures. "Landscape metrics" measure parameters such as patch sizes, shapes, density, aggregation/dispersion, boundary or edge length (density), corridor length and connectivity, and network connectivity and circuitry (Forman and Godron, 1986; Turner and Gardner, 1991; Forman, 1995; Klopatek and Gardner, 1999; McGarigal and Cushman, 2005; Leitao et al., 2006). "Effective mesh size" is a measure of fragmentation level (Jaeger et al., 2007, 2010). For example, over 75 years (1930–2005), land fragmentation mainly due to roads in Baden-Wurttemberg (Germany) increased by 40%, as indicated in effect by the smaller size of spaces enclosed by the road network.

Species richness is often high on rich soils, while rare species may be concentrated on both the richest and poorest soils. Urbanization on agricultural land commonly targets some of the most-productive agricultural soils (Imhoff et al., 1997). Urbanization not only causes farmland loss, but fragments agricultural land (Carsjens and van der Knaap, 2002), a pattern widely disliked by farmers. Market-gardening (truck-farming) areas just outside the urban fringe are an especially valuable source of fresh vegetables and fruits for a city, yet ironically are threatened by urbanization from that same city.

On Puerto Rico (a large Caribbean island) over four centuries, farmland expanded at the expense of forest land. Then in a short period (ca. 1977–94) the total urbanized area increased by 42%, almost all on agricultural land (Lopez *et al.*, 2001). That rapid urbanization not only significantly decreased food production for the people, but increased urban land from 11% to a very high 27% of the island.

Urbanization of natural land, i.e., forest/woodland and desert/grassland, is ecologically more serious than that on farmland, because habitat conditions have not been degraded by farming activities (Zonneveld and Forman, 1990; Browder and Godfrey, 1997; Stern and Marsh, 1997; Jongman, 2002; Prados, 2009). For 13 decades (1650–1780), the forest around Wilmington (Delaware, USA) was removed for grain production especially on good soils, and mainly for fuelwood on poorer soils (Matlack, 1997a). A vigorous agricultural economy then followed for 14 decades. During the subsequent two decades (1920–40), some fields were abandoned and woodland expanded. Finally from 1940 to 1995, the spread of suburban houses has competed with and outpaced woodland regrowth on the former crop fields.

A somewhat different scenario unfolded on much poorer soils in the Boston Region. The transformation of forest to cropland continued almost linearly for two centuries (about 1650 to 1850) (Whitney and Davis, 1986; Foster and Aber, 2004). Then 13 decades of forest expansion followed. Since about 1980, both forest and farmland have decreased, as residential sprawl spread.

In summary, key spatial processes elucidate the patterns of urbanization expansion, as well as land and habitat recovery. In agricultural land around cities, urbanization causes significant habitat degradation and loss, because semi-natural habitat is scarce. Where natural land is close to cities, urbanization normally eliminates semi-natural and natural habitat, degrades habitat, and fragments habitat, normally in that order of ecological severity.

Part II Ecological features

Chapter 4: Urban soil and chemicals

> The nation that destroys its soil destroys itself.
> *Franklin D. Roosevelt, letter to the governors,*
> *26 February 1937*

> ... microwildernesses exist in a handful of soil ... close to a pristine state and still unvisited. Bacteria, protistans, nematodes, mites, and other minute creatures ... A lifetime can be spent in a Magellanic voyage around the trunk of a single tree.
> *Edward O. Wilson*, Naturalist, *1994*

The essence of urban soil

Soil is the source of life. We build on soil, and nations rise and fall on their soil. Look closely and wondrous organisms appear, upon which we depend. In urban areas hard surfaces covering so much of the ground render much of the soil invisible. Consequently the remaining greenspaces and spots are that much more valuable for life and us. Most of the species central to urban ecology – trees and shrubs and flowers and wildlife and soil animals and soil microbes and us – depend on these remnant magical spaces.

Geologists, engineers, and land planners usually consider "soil" to be all the loose unconsolidated earth material (regolith) above the bedrock, thus focusing on the mineral component (Berke *et al.*, 2006; Marsh, 2010). In contrast, most agriculturalists and soil scientists consider "soil" to be the upper portion of this unconsolidated material that is modified by the biological activity of abundant microbes, soil animals, and plant roots. Both perspectives are important in urban ecology.

Thus, *urban soil* is the unconsolidated material above bedrock in cities and towns, and is composed of a lower mineral zone, plus an upper biological zone of mineral particles mixed with abundant organic matter and organisms. Minerals are also in the upper zone and organisms in the lower, but the bulk of biological activity is in the upper zone. *Topsoil* or surface soil commonly refers to the biological zone (combined A and B layers; or sometimes only the A) (Gilbert, 1991). *Subsoil* is the mineral-dominated zone below the topsoil. In urban areas, human-deposited sandy fill is widespread around construction projects, and normally contains no topsoil, only subsoil material. Earth material or subsoil material used as fill can be readily converted to biologically active soil by the addition of organic matter such as leaf litter (Bormann *et al.*, 1993).

Many human activities and natural processes have produced urban soil. Not surprisingly, these diverse processes have created a rich array of soil types in urban areas. The different types, in turn, serve a variety of extremely important functions for society, from engineering to ecology, farming, and recreation. So let us begin by introducing (1) some key types and functions of soils; this is followed by (2) core characteristics and (3) vertical and horizontal patterns.

Key types and functions

In the outer portion of an urban region, near-natural and agricultural soils are often widespread. "Natural soils," which developed on site without fill and with low levels of human chemicals, usually show an A layer over a B layer. The A horizon is mostly blackish due to the abundance of dead organic matter, and is a layer of leaching where rainwater infiltrating downward washes out nutrients. In the B horizon some of the leached nutrients accumulate. These soils may be better called semi-natural because of the aerial deposits of human chemicals associated with cities. Agricultural "crop soils," in contrast, usually show a relatively homogeneous layer down to a somewhat distinct line at about 30–40 cm depth, representing the depth of plowing or cultivation (Russell, 1961; Tivy, 1996).

In higher human-density areas "disturbed" soils of several types predominate (Wheater, 1999; Sieghardt *et al.*, 2005). Compacted soil results from compression by walking, vehicles or heavy machinery, and from vibrations by trains, vehicles or machines. Many activities produce low-organic-matter soils. Altered-nutrient soils are widespread. Rainwater leaching of calcium carbonate from concrete and mortar all around us raises the pH, making more alkaline soils. Acid rain acidifies soil. High levels of inorganic heavy metals create metal-contaminated soil, and high levels of organic materials such as hydrocarbons also contaminate soil.

"Improved" or "remediated" soils are scattered throughout urban areas. For some, organic matter derived from fallen leaves (yard waste), municipal solid waste, or (sewage) sludge is added. For others, nutrient enhancement is accomplished by the addition of fertilizer or the lowering/raising of pH. Perhaps most common is simply the addition of topsoil, or alternatively, *fill*, the usually sandy mineral material with very little organic matter used for building up the ground level (Knox *et al.*, 2000). Fill often includes various human-made materials, from brick-mortar-concrete rubble to metal, plastic, wood and glass objects.

Human designed-and-mixed soil creates a particular mix of mineral particles, often with organic matter. For example, using sieves and mixing, specific percentages of sand, silt, and clay are combined for use in containers or other special sites for horticultural plants. An extreme type of urban soil uses human-modified mineral particles, such as extruded (and fused) clay particles used in hydroponics. Plastic or other material may be added to hold the particles in place, as on green roofs.

The internal structure and the functioning of each soil type vary considerably. These urban soils from semi-natural to artificial are all blanketed with human chemicals. Heavy metals, petroleum hydrocarbons, nitrogen, and much more are deposited and accumulate. Such accumulations may reduce the rates of biological activity, water infiltration, organic matter decomposition, and so forth, and may be lethal to some roots, soil animals, and microbes. The intensity of effect varies by chemical concentration and soil type.

What are the functions or services provided to society by such an array of urban soils? Soils provide support for engineered objects on the surface. For high buildings, train tracks, and busy roads the mineral soil component has low compressibility. For low buildings, lightly used roads, paths, trees, and athletic fields, compressibility is moderately important. For green roofs, a lightweight soil is useful.

"Water drainage" is a key for several functions. A high level of drainage is important for all the engineering functions mentioned, as well as for paths, ball fields, and around foundations. An intermediate drainage level, neither too high nor too low, is optimal for septic systems, crops, and diverse plantings. In cases involving plants, organic matter or clay particles hold nutrients against excessive soil drainage. "Aeration" providing air in the soil generally correlates with drainage, and is valuable for trees, septic systems, and fill around foundations.

In short, a good topsoil provides air, water, and essential nutrients for roots, soil animals, and microbes in the soil (Sieghardt *et al.*, 2005). It helps moderate water quantity and quality problems in the surrounding environment. And it accumulates, breaks down, and recycles diverse chemicals, from mineral nutrients to toxic substances.

Core characteristics

Before considering the wide range of soil types, it is useful to visualize a typical or archetypal small area of urban soils (Figure 4.1). Eight patterns associated with human activities highlight the essence and distinctiveness of urban soils.

1. *Horizontal pattern*. Urban soils display a much finer spatial mosaic than do natural and agricultural soils (Craul, 1999; Blume, 2009). Both exposed soil and soil covered by the hard surfaces of buildings and roads commonly contain a number of different soil types, each a rather small patch or strip. Patch types appear fragmented and the soil mosaic highly heterogeneous. The horizontal distribution of fill sites and types of fill typically covers a significant portion of this mosaic. Boundaries differentiating soil types are typically abrupt and often straight, reflecting the predominant urban geometry.
2. *Vertical layers*. Soil profiles or vertical layering normally vary much more across an urban area than across a non-urban area. The topsoil with major biological activity is usually rather shallow. Few trees mean few deep roots. Initially little biological soil-development occurs on the widespread fill present. Chemical pollutants inhibit soil animals and the soil porosity they produce. The shallow soil contains a concentration of human chemicals, many toxic. Commonly beneath the topsoil are layers of fill, perhaps the

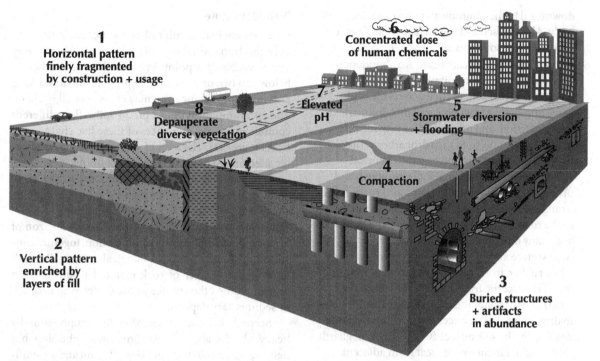

Figure 4.1. Eight major distinctive characteristics of urban soils. Natural and agricultural soils normally have none or few of these. See Craul (1992), Brown *et al.* (2000), Clement and Thomas (2001).

older the city the more layers on average. These layers are predominately sandy but may embed former topsoil. They usually include a range of human (anthropogenic) materials (Gilbert, 1991). Layers in the soil material tend to be quite distinct with sharp boundaries (Craul, 1992; Blume, 2009). Water infiltrates downward through these layers carrying some human chemicals from the topsoil.

3. *Human structures and artifacts.* The three-dimensional movement of water through soil material is accentuated by the complex accumulation of underground human structures over time (Clement and Thomas, 2001). Buried foundations, pillars, walls, floors, beams, and drainage structures are often on top of one another. Yet these may be dwarfed by the labyrinth of underground pipes. Formerly functional pipes may now be full of air or soil. Pipes carry drinking water, human sewage, stormwater, electricity, gas, oil, steam, or hot water. Many pipes leak, so water leaves or enters them at specific locations. Water commonly flows along outside pipes. Little subterranean dams and blockages occur. Subterranean rail lines, highways, walkways and shops may have shallow soil above them

(Clement and Thomas, 2001; Benedikz *et al.*, 2005). Demolition debris such as brick, mortar, concrete, and asphalt rubble is common. But glass, metal, plastic, wood, and other objects are also frequent. The net effect of buried structures is a fine-scale three-dimensional mosaic of water conditions, as well as of aeration, chemicals, and microhabitat conditions for soil organisms and roots.

4. *Compacted soil.* Repeated walking, games on athletic fields, and other human activities compress the soil particles together, reducing soil porosity and structure (Craul, 1992). Silty, clayey, and organic soils, and those with a mixture of particle sizes, are most sensitive to compaction. Compaction in turn degrades soil drainage, aeration, organic matter, and soil animal communities (Sieghardt *et al.*, 2005).

5. *Stormwater infiltration.* In built areas, much stormwater accelerates across hard surfaces into drains, along ditches, and through pipes that are widespread in a metro area. The result is a major reduction in water infiltrating into urban soil. Water from precipitation and snowmelt readily infiltrates into sandy soils, but only slowly in clayey and organic soils. Such infiltrating water may flow

downward to the groundwater or to a surface water body. Also surface and subsurface water flows often pass from one soil to an adjacent one.

6. *Concentrated human chemicals*. An abundance of pollutants from precipitation and dry fallout bathes the urban soil. Particles, aerosols, and gases originating from power plants, industries, transportation, and other sources provide this bath. Consider NOx, SO_2, hydrocarbons; the list goes on. Inorganic and organic chemical substances modify and mold all components of urban soils. Both vehicular traffic and roads release a range of chemical pollutants that accumulate in the road vicinity. Sodium chloride, phosphate, and many more pollutants originate from the road surface and roadside management activities, while rubber, iron, asbestos, chromium, zinc, and much more come from vehicles (Forman *et al.*, 2003). A hydrophobic surface crust of accumulated hydrocarbons that reduces water infiltration may develop on the soil surface (Craul, 1992; Sieghardt *et al.*, 2005). Furthermore, heat from adjacent structures raises the soil temperature and dries the soil (Craul, 1992).

7. *Elevated pH*. Water running over concrete and brick-and-mortar surfaces of buildings, roads, and sidewalks leaches out their constituent calcium and carbonate ions (Gilbert, 1991). The elevated levels of calcium reaching the soil raises the *soil pH* (lowers the concentration of hydrogen ions or acidity). Hydrogen ions are replaced by calcium ions on soil particles making the soil more alkaline. The elevated pH alters nutrient cycles and changes organism numbers and activity in the soil.

8. *Vegetation and its litter*. On top of urban soil typically only one or two vegetation layers is present, such as savanna-like grass cover with scattered trees. With limited shade, the heat causes rapid decomposition of soil organic matter, and a scarcity of tree roots limits the depth of the biological soil. In addition, fallen leaves and dead branches and trunks tend to be carted away (Benedikz *et al.*, 2005). This sharply reduces the amount of organic material incorporated into soil.

Vertical and horizontal pattern

A closer look at the vertical layering and horizontal pattern (Jenny, 1980) is a key to understanding urban soil ecology.

Vertical structure

Five characteristics with rather distinct attributes provide the framework for vertical soil structure. Only one is visible at a point on the soil surface, whereas below you two or three may be present, and in walking around some distance you may cross over all of them. The predominant types of urban soil reflect different combinations and sequences of these five characteristics (Figure 4.2). The level of soil-saturated groundwater affects most of the vertical soil patterns.

First, natural *soil stratification* refers to the A, B, and C soil layers that commonly develop over time by weathering on site under natural conditions of wind, water, temperature, and vegetation. The A horizon of leaching, and B horizon of accumulation, together constitute the zone of major biological activity. Beneath them is the C layer of rock material resulting from breaking down the surface of the underlying bedrock or sedimentary deposit.

Second, *compacted soil* due to compression by heavy objects above or to prolonged vibration has dense-packed mineral particles. This means a significant reduction in pore area, aeration, drainage, and soil structure. Organic and clay soils compact the most. Gravely and sandy soils, especially if the particles are of relatively similar size, are most resistant to compaction (Sieghardt *et al.*, 2005).

Third, *impervious or hard surface* is a covering (sealing) to support weight and human activities above the soil. Normally the building, road, parking lot, driveway and sidewalk is partially permeable to water infiltration, and without continued maintenance permeability may markedly increase. Typically a sandy-gravelly fill underlies the hard surface.

Fourth, *fill* associated with built structures is normally dominated by coarse sandy material (though any size particles may be present), often containing human-produced materials. Fill is primarily added for rapid water drainage and often for supporting heavy structures.

Fifth, *added topsoil* contains considerable organic matter, nutrients, and soil organisms. It may be topsoil with biological and mineral components that developed naturally elsewhere. Or it may be human designed and mixed soil, usually with varying proportions of sand, silt, and organic matter for plant growth in different locations (Gilbert, 1991).

Getting down on your knees with a small shovel, it is often easy to see interesting sublayers in the A

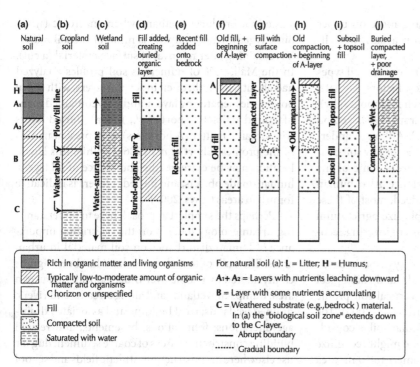

Figure 4.2. Ten soil profiles with layers of fill, water, and organic matter. Soil surface at top, and bedrock (or other parent material) beneath a profile. The first three profiles characterize non-urban soils, but are often present on small sites in urban areas. In contrast, the following seven soil profiles (containing fill) cover much of an urban area. See Craul (1992).

horizon of a natural soil (Figure 4.2a). At the top is the *litter*, where the origin of the dead organic matter is relatively clear: leaves, conifer needles, twigs, and so forth. Beneath the litter is *humus*, a more homogeneous black material where the origin of the organic matter is unidentifiable or unclear. Beneath these two all-organic layers appears the rather distinct *mineral soil* composed of sand, silt, and/or clay particles. The upper portion of mineral soil, or A_1 *layer*, contains considerable organic matter intermixed and is clearly blackish. The lower A_2 *layer* is much lower in organic matter and mineral nutrients. Under this is the *B layer*, where certain clay minerals and iron oxides leached from above often accumulate.

In urban areas the natural soil stratification is relatively clear where the other four attributes are absent (Figure 4.2). However, hard surfaces of buildings and roads are widespread and, where present, effectively eliminate the A and B biological zone. Fill is also widespread, especially in older areas where over generations, centuries, even millennia, construction activities have occurred repeatedly in the same locations and have spread outward with population growth. Fill reaching the soil surface has little organic matter and initially supports slow plant growth. But over time, organic matter accumulates, forming an A horizon that supports

good plant growth (Gilbert, 1991). If fill at the surface is compacted, colonizing plants grow very slowly, organic matter accumulates slowly, and topsoil develops slowly.

Compaction is often most acute at or near the soil surface (Figure 4.2), and thus plants on such soil surfaces usually grow poorly. However, a common urban pattern is the *buried compaction layer*, where construction machinery was operated on a layer of fill, which was then covered with topsoil. Precipitation water infiltrating through this topsoil reaches the relatively impermeable compacted layer, where it simply puddles or accumulates, thus drastically reducing aeration (Figure 4.2j). Woody plants may grow little or poorly.

Horizontal pattern

Soil type names, classification, and mapping look much like tongue twisters devoid of ready recognition, and thus largely remain in the domain of experts, especially soil scientists. An international classification system developed by FAO/UNESCO and used in some areas provides worldwide breadth, but highlights characteristics of the normally invisible soil material (beneath the A layer) (Short *et al*., 1986b; Fenton and Collins, 2000; Sieghardt *et al*., 2005; Sauerwein, 2011). Many nations, including Australia, Canada, China, France, Germany, Russia, and the USA, have their own classification systems, often focusing more on upper soil (A-layer) characteristics and on a narrower climatic range of soils.

Most *soil classification* systems recognize soil types as products of five interacting key factors: climate, parent material (e.g., bedrock), relief (topography), biological influences, and time (Craul, 1992; Fenton and Collins, 2000; Sauerwein, 2011). Although understanding these approaches is valuable for any soil classification and mapping, in urban areas some of the five key factors are of minor importance. Also, human influence, the giant, is missing. Indeed, most of these soil classification approaches emphasize agricultural soils, with much less emphasis on engineering, ecological, and urban dimensions.

Identification of a soil in these systems is based on measuring physical, chemical, and biological attributes at different depths in a vertical soil profile (Fenton and Collins, 2000; Sieghardt et al., 2005). The horizontal extent or size of a particular soil is considered hierarchically. For example, one might recognize a soil "individual" or "pedon" of some 1 to 10 m^2 area (which exhibits its natural variability), a "polypedon" as a group of similar pedons (e.g., at a location on a slope), and in similar manner a "series" and then an "order" (Craul, 1992, 1999; Fenton and Collins, 2000). Common soil maps, typically at the larger end of the hierarchy, are often less useful for areas smaller than about 10 ha (25 acres) (Berke et al., 2006).

As for most things mapped, lines represent boundaries between types. In natural areas a gradual gradient between types is more common than an abrupt boundary, whereas relatively sharp boundaries predominate in urban areas. Soil types on a natural-area map typically appear somewhat like giant amoebas packed together, even on top of one another, whereas urban soils often exhibit straight boundaries. In short, urban soils do not fit well with the existing classification systems, and some soil scientists are at work to modify or extend or replace the systems for urban areas.

For example, from a detailed soil analysis of the grassy "Mall" in central Washington, D.C., five urban soil types, called urbic, spolic, dredged, garbic, and scalped, predominate (Short et al., 1986a, 1986b; Fanning and Fanning, 1989; Evans et al., 2000). "Urbic" is mineral soil fill containing modern human-manufactured objects. "Spolic" is mineral fill without the human-made objects. "Dredged" is a less-dense (lower bulk density) mineral fill derived from wet areas. "Garbic" is an organic soil derived from organic waste that produces methane when anaerobic (without oxygen). And "scalped" is mineral subsoil material remaining after removal of surface material (a cut). In the Mall, 95% of numerous soil profiles analyzed had about 6 m of fill, often in several layers, with sharp boundaries separating layers, and 94% contained manufactured artifacts (Short et al., 1986a). Buried layers with organic matter (A horizons) were present in 42% of the profiles (Short et al., 1986b). While this iconic location may be extreme, the abundance of fill layers, human-made objects, and buried A layers is indicative for urban areas (Kays, 2000).

Perhaps the simplest approach here to understanding urban ecology focuses on the overriding importance of human disturbance on soil material in urban areas. The bulk of today's urban land has been significantly altered by people over time. Consider an urban area: dumps in wetlands and along streams and rivers; drained wetlands; road building and associated drainage alterations; foundations, basements, and related building construction; areas of coarse sediment dug for use elsewhere in construction; athletic fields and school yards compacted; demolition sites; sites with rich soil added for plant growth; flood scoured areas and major sediment deposits by floods; wells dug; septic systems built; straightened rivers and streams; ditches and canals dug; plowed fields for food production; industrial slag/spoil heaps and other waste sites; and areas covered with hard surface and later uncovered.

Urban soil types are closely linked to current and recent land uses. Thus, coarse fill, natural soil, compacted natural soil, buried compacted fill, added top soil, hard-surfaced soil, and so forth, as introduced in Figure 4.2, illustrate major soil types in urban areas. These are at a scale especially useful for engineering, planning, urban agriculture, and ecological understanding.

Physical, chemical, and biological characteristics of the types, both average and variability, are readily measured and compared. The bath of airborne human pollutants operates at a broader scale, e.g., differentiating urban and non-urban soils or industrial and residential areas. At much finer scales, the complex network of building foundations and overlapping pipe networks, as well as the components of rubble – bricks with mortar, tar-asphalt, roofing material, wood beams, iron pipes, plastic paraphernalia – may have significant physical, chemical, and biological effects.

In short, at present it is perhaps best to simply name and map soil types mainly based on the last major human disturbances or uses of the land. In remnant natural locations natural soils can be recognized. In this manner, soil types can become as widely familiar as plant and animal types, and thus highly useful to urban ecology and society.

In a later section we will add one key structural characteristic of soils that makes this approach especially useful. "Soil texture" refers to the proportions of sand, silt, and clay (large to small mineral particles). Several important soil characteristics including drainage and aeration typically correlate with texture. However, first we highlight the processes that make, mold, and change urban soils.

Key natural and human processes

These are conveniently presented in three groups: (1) geological, (2) ecological, and (3) human processes.

Geological processes and patterns

Over long long periods, "geological processes" mold the land on which cities form. Continents separate, move and collide. Mountains are pushed skyward by the Earth's tectonic pressures and heat (Figure 4.3). Mountains are eroded down to hills and plains (Forman, 1979b; Costa and Baker, 1981), which in turn may be uplifted and then eroded. Water and wind erosion leads to sediment deposits in valleys, as well as across plains. Seas rise and fall, covering coastal plains and inundating flat areas where corals, chemical processes, and compression produce limestone. Glaciers advance, grinding down hills and mountains, and then melt leaving a covering of sandy materials, some distinctive hills, and lots of low areas where wetlands form. These molding processes across the land endlessly recur. In seemingly

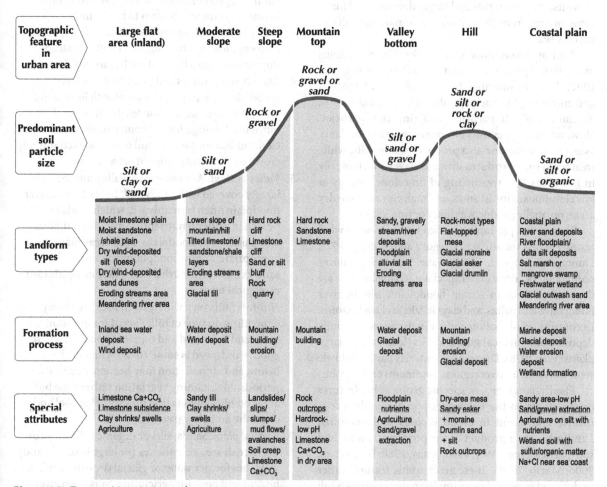

Figure 4.3. Topographic and substrate framework on which urban areas grow. Specific characteristics differ by region. Typically a topographic feature is dominated by a single landform type, formation process, and soil-particle size.

the last minute, humans in large numbers or with large machines have sculpted massive areas of the Earth's surface. We have become a geological process.

Rocks are fairly easy to recognize, as many "geology from the road" books emphasize. Three big types exist. (a) *Igneous rocks* come directly from the Earth's hot magma inside. They may cool slowly beneath the surface, forming large minerals as in granitic rocks, or, as in surface lava flows and volcanic eruptions, cool rapidly to form dark basaltic rocks. (b) *Metamorphic rocks* begin with any sort of rock that is later squeezed and bent by the Earth's internal pressures and heat. The different minerals in these highly diverse metamorphic rocks (e.g., slate, schist, and gneiss) tend to be aligned in distinct layers, which usually are non-horizontal and curved. (c) *Sedimentary rocks* are horizontal layers of deposited sediments that have been compressed by layers added on top. Sandstone develops from sandy deposits, shale from silty and clayey deposits, and limestone mainly from the calcium carbonate deposits in shallow seas.

Putting various processes to work on the rocks helps make the soil particles of urban areas (Way, 1978; Craul, 1992). The "weathering" (or breakdown) of igneous and metamorphic rocks produces large sand grains, medium-sized silt particles, and tiny clay particles. However, sandy deposits are often prevalent on land. Weathered sandstone also produces sandy soils, while weathered shale produces silty and clayey soil, typically in low areas. The weathering of limestone, usually at low elevations in moist areas, and high elevations in dry areas, normally produces a clayey or fine-silt soil.

These soil types can be eroded by water or by wind flowing across the surface (Gregory and Walling, 1973; Dunne and Leopold, 1978). Water "erosion" produces sediments or deposits. Especially characteristic are deposits of sands in stream floodplains, silts in river floodplains and deltas, and clays in lake and sea bottoms. In contrast, wind erosion, mostly in dry areas, generally deposits sands in local sites, and silts (loess) across large plains (Way, 1978). Tiny wind-blown clay particles slowly settle out of the air over regions, continents or the globe.

Finally, *landforms*, as specific geomorphic features or structures on the Earth's surface, are easily recognizable and useful in urban ecology and planning. Landforms are a product of the processes, rocks, and soil-particle types (Way, 1978; Jenny, 1980; Berke *et al.*, 2006; Marsh, 2010). These geomorphic features differ in different regions. For example, in dry climates a hill may be a flat-topped mesa or a high sand dune, whereas in glaciated areas a hill may be a moraine or drumlin. Yet seven "land-surface types" cut across such features and give considerable insight into the soils of urban areas (Figure 4.3).

1. *Large flat area*. Silty and clayey soils high in pH, Ca, CO_3, and other mineral nutrients predominate, so most of such areas are good for agriculture, plant growth, and septic systems. Sandy floodplains of eroding streams and silty floodplains of meandering rivers are common.

2. *Moderate slope*. Silts and sands, especially from underlying sandstone and shale or as deposits of eroded material from upslope, predominate. Eroding streams are common. Almost all urban land uses may work well here.

3. *Steep slope*. Rocks and gravel typically cover cliffs, though sandy or silty bluffs may be present. Semi-natural vegetation, rock quarries, and informal (squatter) settlements seem to be the most characteristic urban land uses. The unstable surface is an engineering challenge, and both the steep slope and the area below it are hazardous sites.

4. *Mountain top*. Rocks and gravelly sandy soils predominate. Except in dry areas with limestone, precipitation water tends to leach out mineral nutrients, leaving a low-pH nutrient-poor soil. Communications towers and tourist sites commonly crowd onto mountaintops in urban areas.

5. *Valley bottom*. A small stream valley may be largely covered with sandy soil. In contrast, a river floodplain often is mainly silt, which tends to be rich in mineral nutrients. Agriculture thrives and, outside the floodplain, built areas may be appropriate. Sandy floodplains sometimes include sand extraction for urban uses, despite significant ecological impacts.

6. *Hill-top*. Soils vary widely from rocky to clayey depending on type of hill. Except for limestone, soils tend to be acid and nutrient poor. Views are good, windspeed is some 10–20% higher than below, and air pollution may become bad in still periods. Maintaining vegetation rather than built areas on hill slopes facing the city helps cool and clean the air below (see Chapter 5).

7. *Coastal plain*. Sands, silts or organic soils may cover the flattish area near the sea (or large lake). A sandy soil from former water or glacial deposits tends to be acid and poor for agriculture. Silts from river floodplains and deltas are rich in nutrients, prime

for agriculture. Yet such sites are hazardous, due to large floods. Marshes and swamps, often over the silt base, have wet anaerobic organic soils, which are unsuitable for most urban uses. Aquifers and biodiversity are normally especially important in vegetated areas of the coastal plain.

Cities spread over all seven of these land-surface types. Before exploring ecological and human processes, *soil erosion* in urban areas warrants special mention (Figure 4.4) (Knox et al., 2000). Natural ecosystems have soil erosion, from moderate levels in dry climates to little in moist areas, for example, <1 ton/acre/year (2.2 metric tons/ha) on forest land (Craul, 1999). Cultivated agricultural land with extensive soil surface exposed to water and wind erosion has considerably higher annual erosion rates, e.g., 5–20 tons/acre (11–45 metric tons/ha). Erosion in urban areas varies much more from spot to spot. Stable urban areas almost all covered with roads, buildings, and vegetation seem to have only slightly more erosion than do vegetation-covered natural areas. Overall, urban streambank erosion is low. Indeed, many streams have disappeared, with their water flowing in buried pipes.

In contrast, construction sites usually have huge amounts of erosion. Individual sites studied in Washington, D.C. varied from 7 to 218 (average >33) tons of eroded soil/ha/year (Craul, 1992, 1999). About 94% of the material was carried away in only 6% of the time, emphasizing that most erosion of exposed soil occurs in occasional heavy rains. Moreover, for Washington as a whole, it was estimated that roads and roadsides contribute twice as much to erosion as construction sites (Craul, 1992). Several other characteristics may cause considerable erosion in urban areas, including steep slopes, a poor stormwater-drainage system, and active development with exposed soils.

Finally, eroded soil or sediment does not disappear; it is deposited somewhere. Sediments may accumulate in low land areas including wetlands. More frequently, sediments are deposited in water bodies such as pond, lake, estuary, or sea.

Figure 4.4. Steep eroded soil bank with parallel gullies and few plants. Eroded gullies in the silt and/or clay form and deepen mainly where plants are absent. Urban graffiti on wall. Madrid peri-urban area. R. Forman photo.

Ecological processes

Several key natural processes are external and affect the top of an urban soil (Craul, 1992). Precipitation falls. As solar energy is absorbed, the temperature of air in soil increases. Plants colonize, cover, and shade the soil. Under asphalt/tarmac, compared with a grass lawn, maximum soil temperature is higher down to 180cm (6ft) depth, and at 180cm minimum temperature is lower (Wessolek, 2008). Wind and sun cause evapo-transpiration, which pumps soil water to the atmosphere. Dead leaves and wood fall, forming the litter layer. In some climates, frost–thaw cycles or fires mold the soil. From below, the water-table rises and drops. Ecological succession changes the soil.

Some of these processes primarily affect the soil's physical and chemical conditions, while others affect biological conditions. Each natural process is significantly altered by urban conditions. Finally, the human effects on these natural processes represent handles for planning and improvement.

Another set of natural processes is primarily internal. Organic matter decomposes. Water flows downward, and in dry times or climates some is drawn upward by evapo-transpiration. Mineral nutrients are leached out of the A layer. Mineral nutrients accumulate in the B layer. Several specialized bacteria sequentially convert N_2 (in air) to NH_4 (e.g., ammonium in the protein of organisms), to NO_2, to NO_3 (nitrite and nitrate in the soil), to N_2. Bedrock is weathered, and the rock pieces are progressively broken down to smaller mineral particles (Russell, 1961; Craul, 1999; Fenton and Collins 2000). Earthworms and other large soil invertebrates move up and down, mixing organic and inorganic materials and increasing soil pores and permeability. In soil pores, oxygen diffuses downward and carbon dioxide diffuses upward. Bacteria and fungi decomposers are consumed by small detritus feeders, which are consumed by larger invertebrates, which are consumed by still larger ones, as well as vertebrates, in complex food webs. Roots absorb mineral nutrients.

Roots grow. Also roots die in the soil, providing organic matter as well as former root channels down deep.

These active external and internal processes represent the natural functioning of a soil, i.e., how it works. They mold physical, chemical and biological conditions in a soil. And together they highlight soil as a highly dynamic object.

Human processes

Whereas human activities may affect any of the soil's natural processes, a distinct set of anthropogenic activities directly altering the soil characterizes the soils of urban areas. Indeed, at a broad scale, urbanization itself highlights the process of forming urban soils (Craul, 1992). The placement and modification of soil material largely reflects the history and human activities of a city or urban area.

Some human effects originate from afar but are accentuated by local urban conditions. Floodwater that inundates, scours, and deposits sediment is often much worse because of residential development, engineered waterways, or the stormwater drainage system of a city. Particulate pollutants such as dust, soot, and heavy metals commonly originate in external agricultural and industrial areas, as well as from nearby industries, power facilities, construction sites, and busy roads (Spirn, 1984; Breuste et al., 1998; Pickett et al., 2001; Lee et al., 2008). Gaseous pollutants including NOx, SO_2, and hydrocarbons have distant and local sources. Such pollutants cover the urban soil. The hydrocarbons from vehicles can create a (hydrophobic) crust on the soil that reduces water infiltration (Craul, 1992).

Human activities adjacent to or near an urban soil may also have major effects. Tall buildings alter air temperature through prolonged shading, reflected solar energy, and radiation from the wall surfaces. Building foundations, sidewalks, and roads absorb heat and radiate some into the adjacent soil (Craul, 1992; Wessolek, 2008). Train and vehicle traffic causes repeated vibrations that compact soil. Surface and subsurface water flows from adjacent areas carry chemical contaminants into the soil.

On the soil surface numerous human activities alter soils. "Hard-surfacing" or sealing a soil for roads, parking lots, sidewalks and driveways alters almost all the properties of the underlying soil (Wessolek, 2008; Niemela et al., 2011). Soil compaction results from heavy objects, especially the repeated movements of construction equipment, truck and car traffic, well-used footpaths, and athletic fields (Craul, 1992, 1999; Sieghardt et al., 2005). Soil surfaces may be mechanically graded or plowed (Craul, 1999). Diverse soil materials are often added on the surface, including gravel, sand, topsoil, or a designed soil mix (Gilbert, 1991; Craul, 1992, 1999;). Fertilizers, organic or chemical, and pesticides are added. Leaf litter and fallen wood are often removed from the soil in urban areas, thus reducing important organic matter and habitat values (Benedikz et al., 2005; Gaston et al., 2005a; Jordan and Jones, 2007).

Finally and most brutally, soil material itself is removed, human objects are inserted, and fill, often from elsewhere, is added (Kays, 2000; Benedikz et al., 2005). Buried solid-waste sites are scattered throughout urban areas (Marsh, 2010), and their density may correlate with age of human occupation. Septic systems, cesspools, and pipes for human sewage are inserted in the soil (Craul, 1992; Anderson and Otis, 2000; Benedikz et al., 2005). Numerous human structures, from building foundations and walls to diverse functioning and non-functioning pipe systems perforate urban soils. Fill itself, which is largely sandy for good drainage, contains human materials ranging from demolition concrete, brick and mortar to wood, plastic and metal objects (Gilbert, 1991; Craul, 1992).

These diverse buried human structures and materials create three-dimensional heterogeneity widely affecting soil functions, from aeration and drainage to organic matter and soil fauna. The materials are also internal sources of contaminants affecting the soil as water and soil animals move through it.

Buried human structures in varying states of degradation may be differentiated in many ways, including (Hiller and Meusser, 1998; Sauerwein, 2011): (a) identifiable object; (b) odor; (c) color; (d) hardness; (e) surface structure; (f) internal anatomy of parts or fragments; (g) particle size; (h) carbonate content; and (i) diverse microscopic and chemical analyses. Archaeologists are experts at this.

In short, diverse human processes, most of which are external, alter natural processes, most of which are internal. Human inputs from the atmosphere, from adjoining areas, and from activities on the soil surface strongly modify soil conditions and how a soil works. However, the most disruptive human processes transform internal conditions by excavation, insertion of structures and materials, and fill. Clearly our urban soils are actively functioning and highly dynamic over time. Do not blink or something important may be missed.

Soil texture and associated properties

We begin with (1) soil texture and types, followed by (2) soil properties related to texture, and finally (3) the relative importance of soil properties.

Soil texture and types

When rock is weathered or broken down into small mineral particles, how big are they and what do we call them? The answers differ depending on the primary functions we ascribe to soil (Russell, 1961; Way, 1978; Craul, 1999; Hallmark, 2000). Thus, an engineer concerned with structural stability might focus on differences among gravels and coarse sand, plus the amount of fine-grained and organic soil material. The agriculturalist growing crops, in contrast, focuses on differences in sandy, silty, clayey, and loamy soils, and may lump all gravels together in one group. Ecologists usually use the agricultural approach. Although different classifications use different particle size ranges and names, they all agree on the general sequence of types from large to small mineral particles (Craul, 1992; Hallmark, 2000): (a) boulders or cobbles, (b) gravel, (c) coarse sand, (d) fine sand, (e) silt, and (f) clay.

Essentially all soils are mixtures of particle types, though typically the bulk of soil material is in the sand, silt and clay range. Thus, *soil texture*, the proportion of sand, silt, and clay (each a percent by weight), is an important and widely used soil characteristic (Way, 1978; Craul, 1999; Hallmark, 2000; Marsh, 2010). Sandy soils are mostly sand, such as a barrier beach or glacial outwash plain; silty soils mostly silt, e.g., a river floodplain or delta; clayey soils mostly clay, as for various wetland soils. Loamy soils have substantial proportions of all three particle types, illustrated by choice garden soils. In urban areas where the structural dimension of soil and fill is so important, instead of only the three-variable analysis (illustrated by a sand–silt–clay "soil texture triangle"), adding the sizes and amounts of gravel as a fourth variable seems particularly valuable.

Sand, silt, and clay proportions are readily determined in the laboratory with sieves of different mesh size (Marsh, 2010). Clay particles may be suspended in water, followed by drying and then weighing. However, a "hand test," which is useful in the field, provides a rough approximation of several key soils based primarily on soil texture (Figure 4.5) (Craul, 1992, 1999; Brady and Weil, 2002; Marsh, 2010). Consider a spectrum of six such soils, plus one soil added at each end

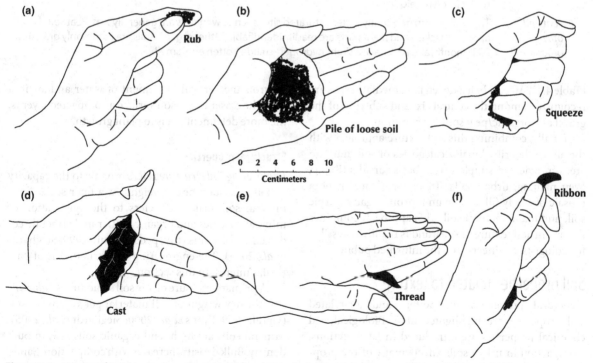

Figure 4.5. Hand test for soil texture type. See Table 4.1 for what can be learned. 1 centimeter = 0.39 inch. See Marsh (2010).

Table 4.1. Results of a hand test for rough determination of soil types. See Figure 4.5 illustrating the soil "cast," "thread," and "ribbon." Other attributes such as nuances in color (measured by a Munsel chart) provide further insight. Based on Craul (1992, 1999) and Marsh (2010).

1. *Gravel*	Viewing an intact handful of soil and using a millimeter ruler, most of the surface is covered by particles <8 cm (3 inches) and >2 mm (1/16 inch) in diameter (or >5 mm depending on the classification system used).
2. *Sand*	Individual grains of sand are readily felt and seen. When dry and squeezed in the hand, the soil falls apart as the hand slowly opens. When moist and squeezed, it forms a soil cast that crumbles when touched.
3. *Sandy loam*	Individual sand grains are readily felt and seen. When dry and squeezed, it forms a cast that readily falls apart. When moist and squeezed, it forms a cast that remains intact with careful handling.
4. *Loam*	Instead of feeling individual sand grains, the soil feels gritty yet fairly smooth. When dry and squeezed, it forms a cast that remains intact with careful handling. When moist and squeezed, the soil can be handled freely without breaking apart. When moist, the soil can be rolled to a pencil-like soil thread that readily breaks.
5. *Silt loam*	When dry, the soil may appear cloddy (composed of lumps), but the lumps are readily broken, and when pulverized the soil feels soft like flour. When wet, the soil readily forms a runny (muddy) puddle. When dry or moist and squeezed, the soil forms a cast that is freely handled without breaking. When moist, the soil can be rolled to a pencil-like soil thread that readily breaks. When moist and pinched out between thumb and forefinger, it does not form a flattened soil ribbon, but appears broken.
6. *Clay loam*	The soil usually breaks into clods or lumps that are hard when dry. When moist and squeezed, it forms a cast that remains intact with much handling. When moist, the soil can be rolled to a soil thread like a pointed pencil that does not readily break. When moist soil is pinched out between thumb and forefinger, it forms a thin soil ribbon that readily breaks under its own weight.
7. *Clay*	When dry, the soil forms hard clods or lumps. When wet, it is usually sticky. When moist, the soil can be rolled to a soil thread like a pointed pencil that does not readily break. When moist soil is pinched out between thumb and forefinger, it usually forms a long flexible soil ribbon that does not break under its own weight.
8. *Peat and other organic soil*	These soils from frequently water-saturated sites such as wetlands are generally wet. Peat soil appears blackish and plant parts are readily identifiable. Other organic soils are commonly grayish, feel smooth, and may have a slight hydrogen-sulfide "rotten egg" smell.

(Table 4.1). At least half a dozen intermediate types are recognized among the central six, and subtypes of the gravel and organic/peat soils are common.

Finally, combining this soil texture approach with the above-described major categories of soil in urban areas, provides a simple useful basis for identifying, even naming, urban soils. Thus, a park might be a mosaic of sandy fill, clay-loam topsoil, organic garbic soil, and loamy natural soil. This approach combines the origin and texture of soils, and is readily accessible to ecologists, engineers, and the informed public.

Soil properties related to texture

Structural properties and water- and air-related soil properties are highlighted here. Biological and chemical properties are introduced in later sections. Compaction in urban soils affects many of the properties, and the least permeable layer in a soil mainly controls the vertical movement of water and air in a soil. The closer to the soil surface a compacted layer is, the more detrimental it is to plant growth.

Structural properties

Engineering soil structure refers broadly to the capacity of soil to sustain heavy weight. At a finer scale, "biomineral soil structure" refers to the aggregation of mineral particles and organic matter in distinct three-dimensional forms (e.g., peds) (Craul, 1999; Sieghardt et al., 2005). The engineering soil structure is of particular interest in this section.

A primary measure of any soil structure is *bulk density*, the dry weight of soil material per volume of soil (Craul, 1999; Evans et al., 2000; Sieghardt et al., 2005). Mineral soils are high, and organic soils low, in bulk density. Bulk density increases with compaction. Sandy soils are lightweight and clay soils heavyweight.

Soils contain pore spaces, and *pore size distribution* refers to the total volume of pores of different sizes in the soil (Craul, 1992, 1999; Sieghardt *et al.*, 2005). Large structural pores, 10–50 μm in diameter, function differently in water flow and for plant growth than do the medium (0.2–10 μm) and small (<0.2 μm) capillary pores in soil, as noted below. Where aggregations of mineral particles and organic matter are present as distinct three-dimensional forms, pores occur between them.

Volume change, as a "deformation" of soil, occurs in several ways: (1) compression by an external load, (2) shrinking or swelling as proportions of water and air change in large soil pores, (3) absorption of water in small capillary pores or within many clay particles themselves, and (4) frost expansion (e.g., upward frost heaving). The *compressibility* of soil refers to its capacity to resist volume change (Way, 1978). In urban areas with heavy trucks, trains, and buildings, even skyscrapers, the properties affecting soil compression (or bearing capacity) are especially important.

Shear strength is the resistance to a diagonal stress or force (e.g., as measured in kg/cm^2) that disrupts the continuity of soil pores (Way, 1978; Costa and Baker, 1981; Hallmark, 2000; Sieghardt *et al.*, 2005). Shear strength is higher in a soil with similar-sized and similar-shaped mineral particles, such as certain gravels and coarse sands. Cohesive bonding among particles as in some clay soils also somewhat increases shear strength. In urban areas where buildings and busy roads are commonly adjacent to greenspaces, shearing of the soil is especially important.

Plasticity is the capacity of a soil to become and remain deformed without volume change or rupture (Way, 1978; Hallmark, 2000). Mainly important in clay soils, plasticity reflects (colloidal) slippages and collapses at grain-to-grain contact points. Finally, *elasticity* is the ability to rebound to an original volume after a change in an external applied load (Way, 1978). This is important in soil under transportation corridors when a heavy truck or train passes.

These engineering soil-structure characteristics are a key to the placement and arrangement of built structures, as well as the effects on adjoining or embedded greenspaces of urban areas. Such structure characteristics also play key roles in the fine-scale anatomy of soil and its functioning for water, air, and organisms.

Water- and air-related properties

Drainage usually refers to the movement of water over the soil surface (surface runoff) and within the soil (Russell, 1961; Craul, 1992, 1999; Marsh, 2010). Permeability and hydraulic conductivity (described below) affect the internal flows. A sequence of drainage classes is recognized, such as excessively drained, somewhat excessively drained, well drained, moderately well drained, somewhat poorly drained, poorly drained, and very poorly drained (Craul, 1992, 1999). Although these terms are mainly targeted to one soil function, plant growth, the measurements used for identifying each class could be applied for any of the several functions of urban soils.

Infiltration refers more specifically to the rate of water penetration by gravity into the soil from its surface (Craul, 1999; Sieghardt *et al.*, 2005; Marsh, 2010). In most mineral soils infiltration water flows through large structural pores. But in organic and some clay soils (Figure 4.6) considerable infiltration may take place in smaller capillary pores. Infiltration water that reaches the saturated-zone water-table recharges the groundwater.

Permeability is the capacity of a soil to permit either water or air movement through it, and is measured as a rate per volume of soil (Way, 1978; Craul, 1999; Marsh, 2010). Hydraulic conductivity is the capacity of a soil to conduct water, especially through the large pores (Sieghardt *et al.*, 2005). *Percolation*, as a more specialized term, is the downward movement of water within soil, often in the vicinity of the saturated zone. A "percolation test" is commonly used to evaluate the suitability of soils for septic systems (Craul, 1999; Marsh, 2010).

Aeration is the movement of air through the soil, primarily oxygen from the atmosphere downward and carbon dioxide upward and out of the soil (Craul, 1992; Sieghardt *et al.*, 2005). The bulk of the soil air flow and diffusion of gases occurs in the large pores. While carbon dioxide constitutes only about 0.04% of the atmosphere, at low depths in soils it often reaches some 150 to 300 times that level, i.e., about 5–10% of the air within soil. On the other hand, oxygen constitutes about 20% of the atmosphere. Air at lower levels of coarse-particle soils commonly drops to only 5–10% O_2. In wet, organic, and/or clay soils (Figure 4.6) oxygen commonly approaches or reaches zero percent, which indicates *anaerobic conditions*. Clogging up, for example, >90% of the soil pores may lead to oxygen shortage and root damage (Sieghardt *et al.* 2005).

An *oxidation–reduction* balance or threshold occurs at low oxygen levels (Craul, 1992). Positively and negatively charged ions "shift" (subsurface flow),

Figure 4.6. Black organic crop soil and gray roadside clay soil used for irrigation ditch. Note deep footprints in clay (lower left) and the lowered field level due to years of light-weight organic soil being oxidized and blown away. Valencia peri-urban area, Spain. R. Forman photo.

so that many chemicals in the oxidized form (presence of oxygen) are converted to their reduced form (insufficient oxygen), or vice versa. Thus, when oxygen drops below the oxidation–reduction threshold, soil may change color and the availability of several key mineral nutrients drops. The nitrogen, sulfur, iron, manganese, and heavy metals present become unavailable to plant roots and other soil organisms. Furthermore, below the oxidation–reduction threshold some elements, such as chromium and arsenic that are often present in urban-contaminated soils, become toxic to organisms.

While there is no need for an urban ecologist to know every concept outlined here, the big picture is important. Engineering soil structure is strongly determined by bulk density, compressibility, shear strength, and plasticity of soil. Water infiltration is mainly controlled by pore size distribution, permeability, and compaction of soil. And aeration for O_2 and CO_2 strongly depends on pore size distribution. With insufficient O_2, many chemicals in reduced form are unavailable or toxic to organisms. All of the soil properties can be negatively, or positively, affected by human activities in urban areas.

Relative importance of soil properties

To get a handle on which soil properties among the bewildering array present are most informative or important, lists from diverse specialists are informative. These lists are generated for different stated purposes and thus further emphasize the diverse societal roles played by soils (Kays, 2000; Amundson *et al.*, 2003).

1. *For engineering activities* (Hallmark, 2000). Urban engineering projects are concerned with many and different soil properties: (a) corrosion of metal (especially steel) is affected by soil drainage, texture, acidity, and conductivity (resistance); (b) corrosion of concrete by texture, Mg-Na sulfates, NaCl, pH, and high humus content; (c) slope stabilization by slope angle, soil strength, and distinct rock-layer discontinuities; and (d) shallow excavation by texture, slope angle, depth to bedrock or pan, high humus content, cobbles/stones, ponding/flooding, water table depth, shrinking/swelling, high bulk density, and frost/ice. Four project types, i.e., (e) dwellings with basements, (f) dwellings without basements, (g) fill for roads, and (h) streets/roads, have similar soil concerns, i.e., slope angle, depth to bedrock/pan, cobbles/stones, frost/ice (excluding dwellings with basements), and subsidence (excluding road fill).
2. *Usefulness rating for standard soil-chemical analyses* (Craul, 1992) (properties ranked in order of importance): (a) pH; (b) phosphorus; (c) potassium; (d) sodium adsorption ratio; (e) exchangeable sodium percentage; (f) organic matter; (g) cation-exchange capacity; (h) total nitrogen; (i) nitrate nitrogen; and others.
3. *Chemical analyses for suitability of urban forestry* (which emphasizes wood production) (Sieghardt *et al.*, 2005) (properties ranked in order of importance): (a) pH; (b) organic matter; (c) cation exchange capacity; (d) contents of phosphorus, potassium, and nitrogen; (e) magnesium; (f) aluminum; and others.
4. *For land development issues* (Marsh, 2010): (a) soil composition (relative amounts of mineral particles, organic matter, water, and air); and (b) soil texture.
5. *Suitability for urban "greening"* (adding vegetation, especially trees) (Sieghardt *et al.*, 2005): (a) texture; (b) structure; (c) pore size distribution; (d) bulk density (weight per volume); and (e) chemical parameters.
6. *Optimal characteristics for a highly plant-productive soil* (Craul, 1992): (a) clay content; (b) humus in the A layer; (c) effective rooting zone; (d) root zone capability for available water; (e) air capacity (pores >50 μm); (f) hydraulic conductivity; (g) pH ($CaCl_2$); (h) S-value of the root zone; (i)

earthworms; and (j) granular structure with high water stability.

Lists for other societal objectives such as agricultural production, septic systems, and ball fields would also differ. Perhaps most striking from these lists is their dissimilarity, both in properties included and in their relative ranking of important factors. Soil requirements for engineering are overwhelmingly structural plus a few chemical and no biological requirements. Lists for land use and plant-related activities emphasize physical (structural) and chemical attributes. The concept of bioassays in evaluating the suitability of soils remains a promising frontier.

Life in the soil

Soil "hums" with the activity of soil organisms. Food webs in urban soils vary from simple to quite complex, and remain a research frontier. We explore these key subjects from three perspectives: (1) plants and organic matter; (2) microbes; and (3) soil animals.

Plants and organic matter

The essence of this important topic is captured in (1) soil ecosystem development, (2) organic matter, and (3) roots (Jenny, 1980).

Soil ecosystem development

In ecosystem development on a bare soil, such as new fill or a topsoil-removed area, the soil changes just as much as does the vegetation overhead (Fenton and Collins, 2000; Berkowitz et al., 2003; Bradshaw, 2003). Moss, herbaceous plants, and some woody plants spring up from existing seeds and spores, or colonize from arriving seeds/spores. Species resistant to urban drought, scarce nitrogen, and chemical pollutants grow, compete, and form a vegetation cover. This provides some shade and cooling for the soil. Vegetation cover provides some friction against wind and water flows, and helps stabilize the surface against erosion. Dead leaves and stems provide a film of surface litter, while dead roots provide a bit of organic matter in the upper mineral soil. Nutrients including nitrogen begin accumulating.

Some microbes and soil invertebrates colonize the surface organic matter, accelerating its decomposition to humus. New plant species favored by these altered conditions colonize, outcompete the early colonists, form a higher canopy cover, and produce much more litter. From then on, the improved soil conditions favor more vegetation including shrubs and trees. In a positive feedback process, these shrubs and trees improve the soil. During this soil development, organic matter increases, water retention improves, water infiltration increases, aeration improves, shade and soil cooling increase, root growth is greater, decomposition microbes increase, soil animals become more numerous, and soil nutrients, especially nitrogen, accumulate.

The soil-development process typically continues until plant roots are inhibited by a layer of compacted or contaminated soil, or by a high water table. Or the process is arrested by above-ground human activities such as cutting woody plants, mowing, trampling, or being covered by fill. If the activity is a short one-time disturbance, such as cutting trees or a fire, soil development restarts. If the activity is repeated or chronic as in repeated mowing or continued trampling, the soil conditions degrade, at least until the next significant human activity occurs. Soil degradation may mean a significant increase in erosion, decrease in organic matter, or worsening of water and aeration conditions. In an urban greenspace, these diverse processes produce a mosaic of conditions for roots, organic matter, microbes, and soil animals. The soil mosaic mirrors both recent and current human activities.

Litter differs in quality, and soil development reflects the differences. Most conifer needles tend to acidify the soil. Some leaves such as maples (*Acer*) decompose rapidly, while others including oaks (*Quercus*) break down more slowly. Fallen branch and twig litter decomposes more slowly, reduces erosion, and provides more microhabitats for soil animals than does leaf litter. Trees that tolerate soil compaction, such as box elder (*Acer negundo*), honey locust (*Gleditzia triacanthos*), and willow (*Salix*), may contribute more to urban soil development than the more environmentally sensitive species, including Austrian pine (*Pinus nigra*), Japanese maple (*Acer palmatum*), and sugar maple (*Acer saccharum*) that grow poorly. Retaining rather than removing lawn cuttings noticeably increases soil organic matter, nutrients, and soil animals, which worm-eating lawn birds appreciate.

The widespread removal of fallen leaves, branches and trees in urban areas means less litter, humus, and organic matter in the A_1 horizon, as well as fewer nutrients and few soil animals (Benedikz et al. 2005; Gaston et al., 2005a; Sieghardt et al., 2005; Jordan and Jones, 2007). Standing dead trees and branches as important habitats for various wildlife species are often removed for public safety.

Organic matter

As the non-living carbon-containing material in soil, *organic matter* comes in many forms (Gilbert, 1991). In natural soils it is predominantly the litter and humus plus a prominent component of the A_1 mineral horizon, though organic matter is present throughout the A and B layers, and at very low levels beneath. In urban areas organic material may occur in layers buried by fill. On the soil surface, additions, such as peat moss, sawdust, woodchips, sewage sludge, and topsoil from elsewhere, are common in urban areas. Buried material from wooden buildings, animal waste, and garbage dumps is also organic matter in urban areas.

Organic matter at or near the surface plays lots of roles in soil development (Craul, 1992). It lightens the soil. It binds mineral particles into aggregates, improving soil structure. Organic matter increases water infiltration, and also water retention. It enhances aeration of soil. It facilitates root penetration. It stimulates microbial populations and decomposition. And it increases soil animal populations and soil functioning. Thus, not surprisingly, in adding topsoil or artificially designed and mixed soil, organic matter typically is 1–5% or more by weight. If peat moss is used dead organic matter may be 10% of the volume, or in compost 15% by volume, depending on location.

Basically all life in the soil – roots, microbes, and animals – depends on the energy and nutrients in organic matter (see equations, Appendix B). The carbohydrates and diverse organic compounds in leaves and wood are composed of carbon, hydrogen, and oxygen, plus nitrogen and other elements. Some soil animals physically break down litter to small pieces by chewing, and in the process get energy and nutrients for their growth (Figure 4.7). Microorganisms, especially bacteria and fungi, decompose these small pieces with their large surface area, and in the process get energy and nutrients for their growth and metabolism. The microbes decompose complex organic compounds to simple organic ones, and then to inorganic elements in the soil. Roots in turn absorb inorganic elements such as nitrate, phosphate, and magnesium for plant growth. These processes actively take place in the presence of oxygen, highlighting the importance of soil aeration.

In the absence of adequate oxygen the soil is anaerobic, with two major effects on decomposition (Craul, 1992). First, decomposition of organic material is much slower, meaning that little energy is released for microbial metabolism. Most of the energy and nutrients remain tied up in the organic matter that may accumulate. Second, the products of this slow organic-matter decomposition are different. Methane (CH_4) and hydrogen sulfide (H_2S), for example, are given off into the soil by microbes rather than CO_2 and H_2O as in aerobic decomposition. A high concentration of H_2S, a gas that diffuses through the soil pores and smells like rotten eggs, is toxic to plants.

A major factor producing anaerobic conditions is standing or ponded water in the soil at some distance below the surface. In urban areas this commonly results from a high water table, a buried compacted layer, or buried human structures such as foundations, drainage objects, and walls.

Roots

Root growth of course reflects above-ground conditions, but is especially dependent on the three-dimensional distribution of environmental conditions in the soil. At the surface, soil temperature contributes to, and closely tracks, urban air temperature. However, temperature about 20 cm (8 in) down changes little from day to night, though in sandy or dry soil it may vary a bit more (Jackson and Raw, 1966). Roots seem particularly sensitive to low-temperature damage, which occurs at night when root growth is more rapid (Craul, 1992). Oxygen is especially important for roots. For instance at 25°C, roots may daily absorb O_2 gas equal to about 9 times their volume. As suggested above, oxygen is primarily important for root metabolism and respiration, but also for nutrient absorption and to prevent root rot. Even short periods of anaerobic conditions due to standing water-saturated soil can damage root systems.

Roots absorb much of their water from the medium-sized soil pores (0.2–10 μm diameter), which store and make available most of the *capillary water* (held by cohesion next to soil particles) in the soil (Figure 4.7) (Sieghardt *et al.*, 2005). In effect, an unbroken stream of water extends from the soil through root, stem, and leaf to the atmosphere. Thus, when a molecule of water is evaporated (transpired) from a leaf to the air due to sun or wind, it puts the entire continuous water stream under tension. That tension moves the strongly cohesive water molecules upward, effectively pumping water from soil to atmosphere through the plant.

The requirement of both water and oxygen, the latter mainly inhibited by too much water, determines much about where and how plants grow in cities. Street tree plantings, for instance, on average only last a few to

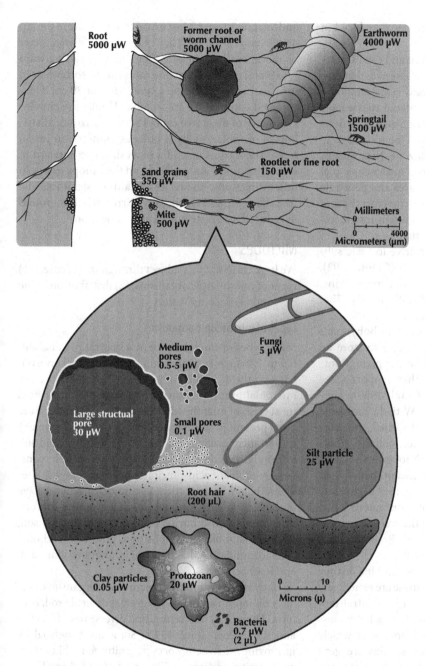

Figure 4.7. Sizes of major biological and physical components in soil. Larger objects above and smaller ones below. W = width; L = length; 1 micrometer = 1 micron = 1 μm. See Hallmark (2000), Schulze et al. (2005), Townsend et al. (2008).

several years. Excavation of dying trees typically reveals that the volume of uncompacted soil was inadequate for good root growth, and especially that the roots were inundated or had insufficient water (Craul, 1991, 1999; Benedikz, 2005; Sieghardt et al., 2005). A planted street tree optimally has a dependable source of water near the top of the root system or root ball, and good drainage near its bottom.

Growing roots tend to follow paths of least resistance, especially channels of former roots and routes of soil invertebrates (Sieghardt et al., 2005). In urban soils, additional routes of roots are along old pipes, sewer lines, cracks in roads and sidewalks, foundation walls, stormwater drains, and the like. Some of these routes reflect the fact that, except in wet soil, roots tend to grow toward water. Root penetration, i.e., the rate of downward growth, of course is reduced in compacted soil.

Large coarse roots anchor the plant and can lift sidewalks. Attached to them is a network of *fine roots* that generally densely perforate the organically rich humus and A_1 layers (Figure 4.7). In mineral fill material, coarse roots may protrude outward in asymmetric patterns, reflecting disrupted water flows in the soil. Fine roots tend to be concentrated in an upper shallow zone. Near the tips of elongating fine roots are numerous microscopic *root hairs*, which provide a large surface area and penetrate between soil particles. Water and nutrients enter the root through the root hairs. For rapid root growth, intermediate soil moisture conditions are optimal.

Root growth is also highly sensitive to toxic substances so prevalent in urban areas (Craul, 1992). Elevated levels of ammonia in solution, copper, zinc, or lead stop both root growth and water uptake. That leaves non-functioning roots, which die.

Roots play many roles. They store carbohydrates produced in leaf photosynthesis. They also store an array of other organic substances for the plant. Roots anchor the plant against erosion. They support a tree against being toppled by wind (Craul, 1999). They absorb water, and absorb nutrients. With the inorganic nitrogen, phosphorus, and sulfur absorbed, the roots synthesize a diversity of organic compounds. Roots even produce plant hormones for shoot growth.

Moreover, roots provide significant benefits to the soil. By binding soil particles together they create a bio-mineral soil structure. Roots continuously die, thus adding organic matter within the soil where roots grow, rather than on top of the soil. Roots increase water infiltration, and aeration as well.

So how big are urban root systems and what do they look like (Benedikz, 2005)? In part these are genetically determined. Typically, palms have thick central taproots that go straight down, grasses have a thick mass of fine fibrous roots, and willows (*Salix*) have widely branching horizontal roots. Some species are generalists, growing over a wide range of environmental conditions (e.g., black locust and red maple, *Robinia pseudoacacia*, *Acer rubrum*), some are specialists in very specific soil conditions, and some thrive in the relatively new, seemingly difficult urban environment (such as tree of heaven, *Ailanthus altissima*). Almost all specialists do poorly in cities.

Yet root systems for a particular species are highly variable, depending on the 3-dimensional distribution of environmental conditions in soil, especially too little and too much water. Most urban tree roots are in the top 45–60 cm (18–24 in). They may extend down a meter or more, but horizontally they often spread outward for several meters. Most fine roots that absorb water in urban areas are in the top 15 cm (6 in), or 30 cm if texture and compaction are suitable. Unlike the circular symmetry of most tree canopies, the horizontal distribution of roots usually is quite asymmetric in urban trees. These roots extend a short distance outward in some directions, and well beyond the canopy extent in other directions. The relative health or sickliness of a tree above ground is generally mirrored below ground in reduced root distribution and elongation.

Microbes

We look closely at this subject through three lenses: (1) types of microscopic organisms; (2) distribution in the city; and (3) decomposition.

Types of microscopic organisms

Many types of microbes live at a neutral pH of about 6.5 to 7, which is common in urban areas. However, bacteria thrive at this rather high pH and are the key microbial group present. As soil becomes more acid most types of microbes decrease. Although bacteria are often present in some abundance, fungi are of primary importance in acid soils, e.g., of pH 4 to 5.5. Acid soil locations include some woodlands (especially conifers), significant acid-precipitation locations, and sites where heavy metals, such as zinc or cadmium, have eliminated key soil animals leaving a poorly drained litter-humus. Also buried poor-drainage organic soil, where anaerobic decomposition produces carbonic acid accumulation, has a low pH, fungi predominant, and few bacteria.

Bacteria. As the most numerous soil organisms, perhaps billions in an ounce (28 grams) of fertile soil, bacteria (Figure 4.7) play central and diverse roles (Jackson and Raw, 1966; Craul, 1992). Some are beneficial by improving soil structure or cycling nitrogen, while others are parasites or diseases. Elongated or rod-shaped bacteria (about 1 μm diameter and a few micrometers long) predominate and swim around in soil. Many can form spores that resist heat and desiccation, a benefit near the urban soil surface. Some can reproduce in 20 minutes though most multiply more slowly. Certain bacteria are aggregated around root tips and presumably facilitate nutrient absorption by the plant.

Aerobic bacteria (those in the presence of oxygen) typically metabolize and reproduce much faster (e.g.,

20–30 times) than do *anaerobic bacteria* (no oxygen). Rapid bacterial growth can use up the oxygen present in the soil. That causes local anaerobic conditions even in a well-drained soil. Water-saturated soil is dominated by anaerobes. Most bacteria derive their carbon and energy from the decomposition of organic matter (heterotrophs).

In contrast, some *chemo-synthetic bacteria* gain their carbon from CO_2 in the soil and their energy by oxidizing inorganic compounds. Thus, with oxygen present, nitrifying bacteria gain energy from nitrates and nitrites, and sulfur-oxidizing bacteria from sulfates. Iron bacteria derive their energy by oxidizing ferrous to ferric salts, though these bacteria do not corrode iron objects such as pipes and debris buried in the soil.

Fungi. Mostly composed of one-cell-thick filaments, fungi in 28 grams (an ounce) of moderately acid soil may have hundreds of meters of filaments (Figure 4.7), together weighing more than the mass of bacteria present (Russell, 1961; Jackson and Raw, 1966; Craul, 1992). Overall three types of fungi predominate. Mainly "single-celled fungi" (Phycomycetes) include most of the molds. Root rots and plant parasites (e.g., potato blight) also fit here. Second, the "cup- or bowl-shaped fungi" (Ascomycetes) are particularly important in decomposing cellulose-rich plant material. Their spores are highly resistant to heat. Third, the "stalked fungi" (Basidiomycetes) are most familiar as mushrooms or toadstools, some edible, some tasteless, and some poisonous. Stalked species include various plant parasites.

Both the single-celled and stalked fungi include *mycorrhizae*, the fungi penetrating or attached to root tips forming a mutually beneficial symbiosis. The fungus receives carbohydrates from the plant and the plant receives nutrients from the fungus. *Penicillium*-type fungi that produce antibiotics against bacteria, and various yeasts, may be abundant in soils.

Although parasites and symbionts are present, most fungi ("saprophytes") decompose organic matter and are relatively effective at breaking down cellulose-rich material. Unlike bacteria, most fungi are particularly effective in decomposing lignin, a major component of wood, and hence these species are widespread in woodland soils. A few fungi can attack and parasitize soil animals, such as roundworms (nematodes). Gardeners and commercial flower-growers know that some seedlings and roots are quite susceptible to fungal disease, and may add fungicide to the soil. Finally, stored topsoil loses its fungi, including mycorrhizae, rapidly.

Other microscopic soil organisms (Russell, 1961; Jackson and Raw, 1966; Craul, 1992). (a) *Viruses*, as tiny entities of DNA and protein, are not organisms but rather multiply within cells. Some (bacteriophages) parasitize and inhibit the activity of bacteria. Plant viruses, such as tobacco mosaic, remain infective for long periods in the soil, and can be transmitted plant to plant by soil animals and fungi. (b) *Actinomycetes* are similar to bacteria though with fungus-like characteristics, and are mainly in higher pH soils characteristic of cities. Some produce antibiotics (e.g., streptomycin from *Streptomyces*) that inhibit bacteria and fungi, providing a competitive advantage in the soil. (c) *Algae*, as unicellular or filamentous photosynthetic organisms requiring light, primarily grow on the surface of moist higher-pH soils. Considering the many urban sites with little leaf litter, green algae and blue-green algae, the most abundant types, sometimes form temporary mats during wet periods. (d) *Protozoa*, as mobile single-celled organisms, are the primary consumers of bacteria. Amoeboid forms and cells that swim with flagellum or with cilia are typically common in higher-pH soils characteristic of cities. Most protozoa can form thick-walled heat-resistant cysts that readily blow in urban dust and are viable for long periods.

Distribution in the city

Urban soils are bathed in microbes from the air. For example, pollen collectors on hospitals collect astronomical numbers of fungus spores. Dust deposits from urban roads, traffic, and other sources are full of microorganisms of diverse types. Areas particularly subject to acid rain are likely to be rich in soil fungi.

Several types of sites are hot spots of microorganisms. Garden compost piles are usually designed for decomposition by numerous aerobic bacteria. Garbage dumps (tips) near the soil surface support high levels of diverse microbes, both aerobic and anaerobic, decomposing the variety of materials present. When a dump is capped with soil or other covering, anaerobic bacteria predominate in the decomposition, as the emission of methane indicates. Topsoil organic matter buried beneath fill may have anaerobic conditions, or may be acidic with many decomposition fungi.

The soil microbes around certain urban sites with specialized functions may be exceptional, but are little studied (Clement and Thomas, 2001). Consider: hospitals; cemeteries and burial crypts; sewage treatment facilities; commercial mushroom-growing operations; catacombs; military facilities; and smuggling contraband sites. Activities and materials associated with various infrastructural networks, e.g., subway system, underground highway, electric system, telephone system, and underground shopping arcade, facilitate the transmission of microbes throughout the urban area. Flowing water, including water supply, hot water for heating, stormwater drainage, and human sewage system, is particularly important in creating microbial concentrations, because leaks are present. Leaked water may stimulate or inhibit microbial activity. Especially in the case of human sewage, leaks (and CSOs) are significant sources of microorganisms. Finally, people moving underground in maintenance activities and for other purposes are both a source and a transmitter of microorganisms.

This complexity of subterranean human structures and functions not only affects microbes, but also creates microhabitats and routes for lots of animals below the surface (Clement and Thomas, 2001). Rats, crickets, moths, cockroaches, spiders, bats, beetles, flies, millipedes, mollusks, crustaceans, and much more may thrive. Even algae grow around underground lights. Such animals move about, die, and are eaten and decomposed. The hidden soil microbe story is rich.

Decomposition

The breakdown of organic matter to CO_2, H_2O, and heat by microbes sounds simple, but is exceedingly complex (see equations, Appendix B). The essence is interesting. Plants contain starches, fats, and proteins that are broken down to simple sugars, amino acids, etc., which in turn are recombined to make other proteins, fats, starches, and related compounds. All of these organic chemicals are readily decomposed by bacteria and fungi. Yet, two other sets of chemicals, mainly associated with the structure of plant cell walls, decompose more slowly: (1) cellulose and its relatives; and (2) lignin (more carbon and less oxygen than in cellulose), which is particularly prominent in wood.

To get a sense of the overall process and its constraints, consider seven key characteristics that accelerate decomposition of organic matter (Russell, 1961):

1. A low level of lignin.
2. Tiny organic particles, such as resulting from the activity of large soil invertebrates (e.g., earthworms).
3. An adequate amount of available nitrogen.
4. Neutral to slightly acid soil (microbial populations decrease and most soil animals die in a strongly acid soil).
5. Good aeration, and with adequate moisture (not anaerobic and not water-logged conditions, which reduce bacteria and their activity).
6. Temperature fairly high.
7. A mixed leaf litter of different species rather than of a single species.

Compared with litter and humus decomposition, the process in a compost pile has similar characteristics except that the temperature is higher and aeration is less. Urban areas generally provide good conditions for items (1), (4), and (6). However, several of the other items are often in short supply on urban sites. Furthermore, urban areas have patchy amounts of soil organic matter because fallen leaves and branches are frequently removed (Benedikz et al., 2005).

The "carbon-to-nitrogen ratio" in organic material has sometimes been used as an indicator of decomposition rate (Russell, 1961; Craul, 1999; McGregor et al., 2006). An optimal C:N ratio for decomposition is about 25–30:1. If one wishes to decompose sawdust (from wood), which has a C:N ratio of up to 400:1, the sawdust needs to be mixed with a low-ratio material such as chicken manure at 7:1. Leaves and straw usually have a moderately high C:N ratio. Too little nitrogen slows the decomposition rate, whereas too much nitrogen may lead to compaction, acid conditions, and an ammonia smell.

In urban areas, many organic materials other than leaves and wood are decomposed. In fact, the decomposition breakdown of varied human-made organic materials is considered to be one of the "services" provided by nature or ecosystems to society. Domestic waste including food scraps, some plastic bags, paper products, and so forth may be composted relatively rapidly on site (McGregor et al., 2006), or much more slowly in a large garbage dump.

As described above, fungi, especially cup-forming fungi in high-pH aerobic conditions, are particularly effective in decomposing cellulose-rich materials, including paper and cardboard products. Lumber and other wooden building materials are full of lignin, which decomposes best by fungi in acid conditions. Herbicides, insect pesticides, organic acids, alcohols, and hydrocarbons (including polycyclic aromatic hydrocarbons – PAHs) are widely used in urban areas. Some of these and other human-made organic chemicals decompose readily, and some slowly. Also some are highly resistant to microbial decomposition and hence accumulate to high levels. The idea of contaminated soils, or brownfields, often indicates the

Life in the soil

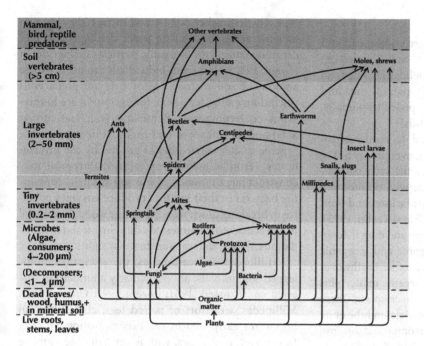

Figure 4.8. Basic food web in soil. Each group of organisms includes numerous species whose feeding patterns vary somewhat. Soil food webs in urban areas are less studied, and may normally be simpler with fewer groups present. 1 μm = 0.001 millimeter; 1 mm = 0.1 centimeter = 0.039 inch. Based mainly on Russell (1961), Jackson and Raw (1966), Odum (1971), Krebs (1972), Palmer and Fowler (1975), Smith (1996), Odum and Barrett (2005), Townsend et al. (2008).

accumulation of such industrially produced chemicals in the soil.

Soil animals

Three types of urban soil animals are easily recognized: (1) vertebrates (e.g., mice and snakes), (2) large invertebrates (earthworms, beetles), and (3) tiny invertebrates (mites, springtails) that the eye can just barely recognize (Figure 4.8) (Owen, 1991; Bolen, 2000). Of the huge number of species in soils, a relatively few from each group are pests (see Chapter 1), which therefore grab our attention. But the rest are performing important functions in our urban soils. The chewing of litter by many invertebrates results in their feces becoming a major component of clean rich black humus. By burrowing, the large invertebrates increase the porosity of soil, whereas the tiny invertebrates mainly move between soil particles without changing soil porosity.

Vertebrates

Many mammals, reptiles and amphibians, plus a few birds, live in burrows in the soil (fossorial species) (Bolen, 2000). In North America, kingfishers (*Ceryle*) nest in burrows along certain streams in urban regions. Snakes often live in dry, somewhat sandy soils, various toad and salamander species thrive in moist litter, and many urban turtles and frogs live in wetter soils.

Small mammals in burrows include mice, voles, shrews, and moles, which may feed on roots, litter, or large invertebrates such as earthworms (Figure 4.8). Soil texture is important to various burrowing animals including gopher tortoises (*Gopherus*) in sand, woodchucks (*Marmota*) and red foxes (*Vulpes*) in sandy loams, ground squirrel species in loams, and muskrats (*Ondatra zibethica*) in clayey soils. Some species, such as burrowing owls (*Athene cunicularia*) and cottontails (*Sylvilagus*), generally colonize the former burrows of other species. Burrows are particularly important in compacted soils.

Some burrows dug by a large animal serve as a habitat for several other vertebrate species, a common occurrence under shrubs in desert areas around cities. Just as "mineral licks" rich in sodium, calcium, or magnesium salts attract diverse wildlife in remote areas, the nutrient content of soil affects vertebrates in burrows. For example, in North America, the cottontail (*Sylvilagus*) is larger and reproduces better in soils with a pH >6 and rich in calcium and phosphorus, characteristic of many urban soils (Crawford, 1950). Cottontails may be common in cemeteries without dogs. Piles of rock or logs on the soil are prime habitats for chipmunks (*Eutamias*). Finally, various highly urban vertebrates, such as the raccoon (*Procyon*), thrive in the stormwater drainage system. Here they escape dogs and cars yet readily emerge to feed on widespread garbage and pet food in a city.

Large invertebrates

Six groups of large invertebrates are widespread and functionally quite important in urban soils (Figure 4.8) (Owen, 1991): (1) earthworms (annelids); (2) slugs and snails (mollusks); (3) beetles (coleops); (4) ants and termites; (5) millipedes and centipedes (diplopods, chilopods); and (6) spiders (arachnids).

Earthworms, essentially limited to moderate-to-high pH soils such as in urban areas, overall may be the most important animals in soils, as the early naturalists, Gilbert White and Charles Darwin, pointed out (Figures 4.7 and 4.8) (Russell, 1961; Jackson and Raw, 1966; Owen, 1991; Craul, 1992; Broll and Kaplin, 1995). Earthworms burrow. They chew litter and drag it down into the burrows, thus mixing organic and inorganic matter often far down in the soil profile. Earthworm manure, composed of undigested organic matter plus some earthworm secretion, creates *wormcasts*. These tiny aggregations or lumps of soil particles may be quite abundant on the surface of an urban or natural soil, and are much higher in available P, Ca, Mg, K, and nitrate than the soil beneath (Russell, 1961). Darwin took a special research interest in earthworms, including estimating the rate at which ancient buildings, pavements, and cinders were buried by wormcasts (Jackson and Raw, 1966). Earthworm burrows form large pores that aerate the soil and enhance soil-water drainage (Figure 4.7). Earthworms do poorly in acid soil, water-saturated soil, drought, or cold. A related group of worms, the enchytrids, is often abundant in wet organic soils such as present in many sewage treatment facilities.

Slugs and snails commonly thrive in moist soil and a cool temperature (Figure 4.8) (Jackson and Raw, 1966; Owen, 1991; Craul, 1992). They are especially important in the decomposition of plant material, because the gut of these mollusks contains microorganisms producing an enzyme (cellulase) that breaks down cellulose and related substances of plant cells. Like the earthworms, the slugs and snails do poorly in acid soil. Also these mollusks increase soil porosity and improve soil structure by creating aggregations of soil particles. Many of the mollusks are omnivorous, and some are predaceous, feeding on other large invertebrates in the soil.

Beetles, including their juvenile worm-like grubs, increase soil aeration and water drainage by moving through the soil (Russell, 1961; Owen, 1991). Many are predaceous (carabids) on other soil invertebrates.

Ants and termites, while evolutionarily unrelated, in some ways operate similarly (Figure 4.8). By typically nesting beneath the soil surface, they burrow, increase aeration, facilitate water drainage, and mix organic matter with inorganic particles further down (Craul, 1992). Numerous individuals can move a huge amount of soil material. Most ants are predaceous (including eating termites), though some are herbivorous or decomposers. Some ants (e.g., leaf-cutting ants) and termites feed on fungi that decompose the organic matter carried below ground by the invertebrates. Termites, mainly tropical and subtropical, may construct large mounds. These soil animals, containing bacteria in their gut that produce enzymes effective in breaking down wood, live near wood. In tropical urban areas, termite-infested crumbling wooden foundations, floors, and walls are familiar sights.

Millipedes and centipedes, the elongated multilegged soil animals, are particularly abundant in forested soil (Russell, 1961; Owen, 1991; Craul, 1992). Millipedes, with lots of paired legs, chew litter, an important early step in the decomposition process. In contrast, centipedes, with fewer paired legs, move faster and are predaceous mainly on other invertebrates in the soil.

Finally, spiders in the soil may form webs but mostly they forage through the organic layers of the soil as predators of invertebrates (Owen, 1991). Various other large invertebrate groups of importance burrow in soil, including pillbugs or woodlice (isopods) and bees and wasps (hymenops) (Russell, 1961; Owen, 1991; Wheater, 1999).

Tiny invertebrates

While numerous tiny invertebrates populate soil, two groups seem to be particularly abundant and important (Owen, 1991): (1) mites and springtails (acarines, collembolans); and (2) nematodes (roundworms).

Mites and springtails often predominate in acid soils, where few or no large invertebrates are present, and are present in huge numbers (Figures 4.7 and 4.8) (Russell, 1961; Craul, 1992; Broll and Keplin, 1995). Springtails feed mainly on organic matter and fungi. They in turn are eaten by mites, beetles, centipedes, and spiders. Typically mites are oval and barely visible to the eye, whereas springtails are cylindrical and a few millimeters long. Mites primarily consume organic matter, though some are predaceous on nematodes, springtails, and insect eggs.

Nematodes (Jackson and Raw, 1966) are microscopic, and in some soils are the most abundant of the animals considered. Roundworms are best known as pests and parasites, as well as vectors for various plant and animal diseases (see Chapter 1). However, the sheer abundance of nematodes suggests the important role they play in the decomposition of organic matter. They are also major consumers of soil bacteria, and some are herbivores or predators. Rotifers, an unrelated group of microscopic animals, are also frequently abundant in soil.

Finally, from knee height, gently push aside some soil litter and look sharply. Then with fingers or a scoop pick up some humus with black mineral soil beneath, and gently spread it out. As Pulitzer-Prize winner E. O. Wilson pointed out: Strange and wonderful and unknown organisms lie within meters of where we sit.

Urban soil chemicals

The soils beneath us in urban areas contain a cornucopia of chemicals, not just a long list, but spatially arranged and interrelated in intriguing ways. We absolutely depend on them for the trees, the gardens, the parks. Buildings and roads, and their longevity, depend on the chemicals present. Waste disposal and treatment and public health are affected by soil chemicals. Some of the chemicals are natural substances such as zinc and chloride, while others are from the >100 000 artificial substances made by humans (Ellis and Mellor, 1995; Stengel et al., 2006; Sauerwein, 2011). Some are good and others bad for us. Analogously, in ecology, some improve or increase plant growth, microbial communities, aquatic ecosystems, and so forth, while others degrade or decrease them (Rowell, 1994).

In essence the diverse urban chemicals originate from four major sources (Sauerwein, 2011). (1) The geological substrate: Ca and carbonate (CO_3) from limestone; silica (Si) from almost all other rocks, as well as sand and gravel deposits. (2) Buried human-made materials: Si from sand and gravel fill; Ca and Mg from rubble fill; sulfur (S) in dredged fill from bottom of water bodies (van Bohemen, 2005); natural carbon compounds and N, P, Ca, and Mg from added topsoil; existing built structures and foundations; former structures and foundations; CH_4, H_2S and CO_2 from a dump (tip); and numerous inorganic and organic compounds from human made artifacts. (3) Atmospheric inputs: SO_2 and NOx from acid precipitation; heavy metals from dust particulates and aerosols; heavy metals and hydrocarbons near roads; Ca and carbonate from cement factories; diverse chemicals from industrial areas; and hydrocarbons from some urban trees. (4) Human surface applications: Na and Cl in road salt; organic and inorganic pesticides in lawn and vegetation management; Ca and carbonate in runoff from buildings; heavy metals and hydrocarbons in stormwater runoff; carbon compounds, N, P and K in leaf litter and humus applications; carbon compounds in wood compost; carbon compounds, N, P and heavy metals in sludge from human wastewater treatment.

Natural soils beyond the urban area seem decidedly impoverished, though nature works smoothly with its chemical subset. Agricultural soils for crop production have a different subset. Perhaps seven characteristics best highlight urban, in contrast to natural, soil chemistry:

1. Widespread soil compaction and its effects on drainage, aeration, chemistry, microbes, animals, and plants.
2. High pH (high alkalinity or low acidity) due to the prevalence of calcium-carbonate-rich concrete and mortar (natural limestone soils also have a high pH).
3. Considerable leaf litter and wood transported off-site by people, wind and water.
4. Accumulated aerial deposits of diverse chemicals from the concentration of industry, transportation, power generation, and buildings.
5. Underground human structures, plus different types of fill often containing human objects/artifacts.
6. Human applications of diverse chemicals and materials on the surface.
7. Restricted aeration due largely to impervious surfaces.

To reveal our treasure chest of soil chemicals, we explore four key aspects: (1) first is an overview of the main types of chemicals present. This is followed by (2) underground human structures, rubble, and artifacts, (3) contaminated soils and organic wastes, and (4) chemical flows.

Types of chemicals

Inorganic elements

Nutrients. Eighteen chemical elements are considered to be "essential" for living organisms (Craul, 1992; Lambers et al., 1998; Smith, 1996; Schulze et al., 2005). Carbon (C), hydrogen (H), and oxygen (O) are the big three. These are the components of carbohydrates produced by plant photosynthesis, which are then converted into many other organic molecule forms. Animals and humans thus mainly eat the big three, C, H, O.

Six other chemicals called "mineral nutrients," or simply nutrients, are present in plants and animals in relatively large amounts: nitrogen (N), a component of all amino acids and proteins; phosphorus (P), important for energy conversion within cells and for plant growth; potassium (K), between cells and important in carbohydrate metabolism; calcium (Ca), in bone, teeth, and plant cell walls; magnesium (Mg), in the chlorophyll molecule and important in enzyme reactions; and sulfur (S), in proteins. These six elements (N, P, K, Ca, Mg, S) are called *macro-nutrients*, since they are needed in abundance and, together with C, H, O, compose 99+% of living tissue (Figure 4.9) (Ricklefs and Miller, 2000).

Nine other mineral nutrients are required at low levels (grams or kilograms per hectare, or ounces or pounds per acre of vegetation) for life, compared with 100–1000 times that amount for macro-nutrients. These *micro-nutrients* (trace elements) are: sodium (Na); chlorine (Cl); iron (Fe); boron (B); manganese (Mn); zinc (Zn); copper (Cu); cobalt (Co); and molybdenum (Mo) (Figure 4.9). Almost all urban soils contain sufficient amounts of these nutrients, though sometimes a nutrient is unavailable, that is, cannot be absorbed by organisms. For instance, at a high pH most of the Fe, Mn, Cu, and Zn present cannot be absorbed by roots (Lambers *et al.*, 1998). Warm dry soils often provide ample Ca, Na, K, and H for plant growth, while cool moist soils may additionally provide Mg and Al (Craul, 1992). Sandstone-derived soil has more H, while limestone-derived soil is higher in Ca and Mg. Silica (Si) is especially abundant in plants.

In contrast, the six macro-nutrients, especially N and P, are often in limited supply in soil. For example, N, P, K, and Mg are frequently inadequate for good street-tree growth. Common chemical fertilizers are various amounts of N, P, and K. In acid soils, available P, Ca, and Mg are in limited supply.

These nutrients are absorbed by organisms as "available" ions in solution in the soil. The macro-nutrient

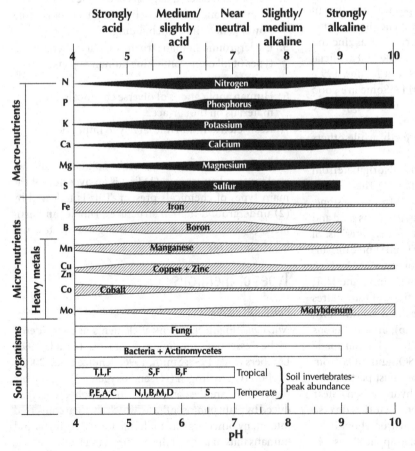

Figure 4.9. Soil chemicals and organisms relative to soil pH. Width of bars refers to the relative amount of nutrients available for absorption by plant roots (upper and middle portions). Letters at bottom indicate typical maximum abundance level for each group of soil organisms. T = termites; L = litter earthworms; S = soil earthworms; F = ants (Formicidae); B = beetles; P = protozoa; E = enchytrid worms; A = mites (Acarina); C = springtails (Collembola); N = nematode worms; I = isopods; M = millipedes/centipedes; D = fly larvae (Diptera). pH (measuring acidity or alkalinity) represents concentration of hydrogen ions; each pH unit = 10 times the next lower unit. Based mainly on Craul (1992), Lavelle *et al.* (1995), Lambers *et al.* (1998), Edwards (2004), Schulze *et al.* (2005).

ions (Craul, 1992) are nitrogen (NH_4^+, NO_2^-, and NO_3^-); phosphorus ($H_2PO_4^-$); potassium (K^+); calcium (Ca^{2+}); magnesium (Mg^{2+}); sodium (Na^+); and sulfur (SO_4^{2-} and S^{2-}).

Heavy metals. The term heavy metals usually refers to a set of 12 mainly "heavyweight" chemical elements: manganese, zinc, copper, cobalt, molybdenum, aluminum (Al), arsenic (As), cadmium (Cd), chromium (Cr), mercury (Hg), nickel (Ni), and lead (Pb) (Figure 4.9). The first five heavy metals listed are essential micro-nutrients (Co, Cu, Mn, Mo, Zn), but the other seven are apparently not required for the growth of organisms.

Heavy metals originate from rock material, but increase enormously in concentration in the soil due to industrial, transportation, and other human processes (Craul, 1992; Wessolek, 2008). Lead is often at high levels in soil near buildings with lead paint, and by roads where lead-containing gasoline is used (Evans, 2000). All the heavy metals are persistent in the soil, that is, highly resistant to being washed out by rainwater (Sieghardt et al., 2005).

Just as for the micro-nutrients, if a heavy metal element is in a salt or particulate form, it is unavailable for organisms. But when available in solution, high levels of heavy metals become toxic to organisms, inhibiting or killing them. For Co, Cu, and Zn, availability (and toxicity) increases with low pH (Figure 4.9) [acid conditions usually associated with low organic matter, low clay content, and low fertility (e.g., cation exchange capacity)] (Craul, 1992; Lambers et al., 1998). Molybdenum is more available at high pH. In contrast, low levels of heavy metals are tolerated by organisms, which may absorb the elements (e.g., Cr, Ni). Some heavy metals, including Cd, Cu, and Zn, *bioaccumulate*, that is, in passing through the food chain they become more concentrated and typically toxic (Sieghardt et al., 2005). Other contaminants are present in soils, such as asbestos near many demolition sites (Evans et al., 2000).

De-icing salt. Sodium chloride (NaCl) is the main salt used in cold climates for reducing ice and snow along roads, parking lots, and walkways (Craul, 1992; Forman et al., 2003). Some 20–60% of the salt applied to certain German roads was carried by airflow and deposited 2 to 40 m from the road, sometimes further (Sieghardt et al., 2005). In water, Na and Cl readily separate as ions and are highly mobile. They then permeate the soil, are absorbed by plant roots, or reach the groundwater where Na and Cl persist and accumulate.

In the presence of NaCl the pH rises. Several nutrients, including Fe, Mn, Cu, Zn, and Co, decrease (also N, P, Ca, Mg depending on how high the pH is) (Craul, 1992; Lambers et al., 1998). The pH rises, which decreases the availability of most heavy metals and thus their potentially toxic effects. Street trees are especially subject to and inhibited by high NaCl levels. Most tree species grow poorly or die, though some are tolerant of the salt. Marine aerosols carrying salt into coastal cities accentuate the problem.

Presently the primary alternative is calcium chloride ($CaCl_2$), a less persistent salt. Better still, Ca causes less degradation of soil chemical and physical properties than does Na (Sieghardt et al., 2005). KCl provides potassium, which is sometimes deficient in urban soils, but which is highly alkaline, causing high-pH problems. Calcium-magnesium-acetate (CMA), an organic salt that readily breaks down, is being studied as a de-icer for roads. Sodium and magnesium sulfates react with minerals in concrete, so that adding water causes expansion and crumbling of concrete structures (Hallmark, 2000).

Organic substances

Pesticides are widely used in urban areas to target certain groups of organisms: herbicides for unwanted plants; fungicides for fungi, especially the pests on plants; insecticides for many types of insects; and other less-used chemicals including rodenticides (for rodents), nematocides (for nematodes), and acaricides (for mites). While obviously a diverse group of chemicals, pesticides last a few days to years and can accumulate to high levels in soil (Sieghardt et al., 2005). In addition to being lethal for their target organisms, pesticides inhibit or kill many microbes and other soil animals. In some cases the by-products of pesticide breakdown are more toxic than the pesticide.

Although pesticides are widely used in agriculture, about a quarter of their use in the USA is in urban areas (Mizell and Hagen, 2000). Much of this is indoors for cockroaches and termites, though considerable amounts are used outdoors for pests on lawns and ornamental plants. In warm moist climates, insecticides are widely used for mosquitoes, midges, and flies, as both pests and public health threats.

Many pesticides used in urban areas accumulate in soils (Craul, 1992). Once applied, the chemical may: vaporize into the air; be adsorbed onto (especially small) soil particles; be washed to lower soil levels or groundwater; be decomposed by soil microbes or chemical processes; or be adsorbed on plant surfaces. Most pesticide applications kill target organisms and various related species. However, the repeated fumigation of soils can be lethal to most soil organisms.

Hydrocarbons, as long chains of H, C, and a few O atoms, particularly result from the use and incomplete combustion of petroleum products. Polycyclic aromatic hydrocarbons (PAHs) are widely produced by vehicles and in industrial processes (Forman *et al.*, 2003; Sieghardt *et al.*, 2005; Wessolek, 2008). Certain trees produce airborne hydrocarbons in abundance (see Chapter 8). PAHs often accumulate along roadsides, near tar and asphalt sites, and in contaminated rubble. Mono-aromatic hydrocarbons, such as benzene and toluene, emanate from crude oil and various petroleum products.

Other organic chemicals of enormous and little known variety, especially from industry and transportation, end up in urban soils (Evans *et al.*, 2000; Forman *et al.*, 2003). Polychlorinated biphenols (PCBs), methyl tertiary-butyl ether (MTBE, an additive to gasoline that readily pollutes groundwater), fatty acids, and alcohols are among the small percentage of organic products studied for, and known to have, environmental effects. Some inputs to urban soils are gradual and often dispersed in air, land and water. Others are concentrated accidental spills or purposeful dumping, typically with acute ecological effects.

Gases in soil

Leaks of methane (CH_4) from major gas pipelines, plus the thousands of kilometers of small distribution pipes to buildings, strongly inhibit many roots, microbes and animals in the soil. Considerable methane and CO_2 also result from the anaerobic decomposition of organic materials in major, as well as numerous tiny, buried dumps around a city. In some urban areas dump methane is captured for energy use. Not surprisingly, anaerobic decomposition of the diverse components buried in dumps produces several gases, including carbon monoxide (CO), hydrogen sulfide (H_2S), hydrogen cyanide (HCN), ammonia (NH_3), ethane, ethylene, and propylene (Craul, 1992). Toxic to many organisms, these gases surely have diverse and significant ecological ramifications.

Radon (a product of radioactive decay of uranium) is widely dispersed in certain rocks and sediments, and is liberated into the soil as a radioactive gas (Craul, 1992; Nielson and Rogers, 2000). Although it is inert with a half-life of a few days, it causes human lung cancer, and doubtless kills or inhibits soil animals and other organisms.

At high levels almost any chemical is toxic to organisms. Different species – microbes, soil invertebrates, wildlife, humans, plants – are inhibited at different concentrations. Widespread substances hazardous to humans in cities include (Nielson and Rogers, 2000; Sieghardt *et al.*, 2005; Sauerwein, 2011): (1) arsenic, cadmium, lead, zinc, nickel and copper; (2) chlorine chemicals (e.g., PCB, DDT, and volatile chlorinated hydrocarbons); and (3) radioactivity (e.g., radon, radionuclides).

More local but potentially hazardous substances include: (a) chromium, cobalt, uranium; (b) cyanide; (c) aromatic compounds (e.g., benzene, toluene, naphthalene); and (d) phenols. Many other substances could be added to these lists.

Underground human structures, rubble, and artifacts

Throughout human prehistory and history whole communities have been buried. Middle Eastern cities now lie beneath mounds called tells, and lost villages have been discovered across Britain. The buried materials in such sites are magnets for archaeologists.

Underground structures are composed of materials resistant to compression by weight above, to water penetration, and to chemical degradation related to pH, water, and anaerobic conditions. Such materials of course are diverse, but masonry and metals predominate. Corrosion of their surfaces liberates chemicals into the soil, facilitating or inhibiting the growth of soil microbes, animals, and roots.

Most important, however, to understanding urban soils are the materials and chemicals in fill (Craul, 1992; Evans, 2000). Major solid-waste dumps are considered in Chapter 12. Here the focus is on persistent materials, not organic materials such as paper products, cardboard, clothing, food garbage, leather, and brush piles that decompose rather rapidly. A vast array of chemicals is added to urban soil from vehicles, road dust, and other sources (Spirn, 1984; Breuste *et al.*, 1998; Forman *et al.*, 2003; van Bohemen, 2005). Also, although mining wastes are prominent in some cities such as Johannesburg and various Pennsylvania (USA) cities (Kovar, 2004; Berger, 2006), mining is not a characteristic urban phenomenon.

Several categories of human materials and *artifacts* (human-made objects) in fill are readily recognized:
1. *Rubble with cement*: masonry, mortar, bricks, building stones, chunks of concrete, iron reinforcing rods.
2. *Other building materials*: lumber, nails, roofing (tiles, slate, asphalt shingles), pipes (of Fe, Pb, Cu, Al), iron and steel structural supports.
3. *Road and railroad debris*: chunks of tarmac/asphalt, concrete chunks, diverse iron structures, railroad-bed cinders.
4. *Industrial waste products*: exceedingly diverse materials depending on the industries; furnace ash, slag, cinders (also from coal-fired power facilities).
5. *Domestic/home refuse/trash*: long-lasting human-made objects/artifacts, including metal cans, ceramics/pottery, diverse plastic objects, and glass (the most persistent artifact).

Dispersed artifacts, such as items lost or tossed away, are concentrated in fill, but have a similar composition to domestic refuse. For example, the soils around multi-unit housing for military families in New Hampshire (USA) contained the following dispersed human-made objects (Evans *et al.*, 2000):
1. *Housing used for ca. 40 years* (1950s–90s). Topsoil (down to an average 18 cm): pieces of plastic, wire, paper, brick, pantyhose, clothes pin. Subsurface soil (18–60 cm depth): toy block pieces, wood, nails, pottery, brick, woven material, cigarette filter.
2. *Housing used for ca. 20 years* (1970s–90s): Topsoil: toy car, piece of pottery. Subsoil: Styrofoam, plastic tape, paper, ribbon.

The chemical inputs to soil may be more diverse than the structures, materials, and artifacts (Craul, 1992). Cement-based materials give off Ca, CO_3 (carbonate) ions, Mg, and heavy metals. Corrosion of iron provides the nutrient Fe in a form available to organisms. Readily decomposable plastics often liberate chemical by-products, including gases, that are toxic to many roots and other organisms.

For example, in an analysis of some rubble in Berlin (Blume, 1982; Craul, 1992), particles <2 mm in diameter (sand-size and smaller) composed 60% by weight of the material. Ten percent of these small particles were calcium carbonate ($CaCO_3$), but significant amounts of boron, copper, manganese, and zinc were present. Large particles of rubble were 22% bricks, 12% mortar, 1% slag, a trace of coal, and 3% artificial products (artifacts). Mortar, artifacts, and the tiny particles contained most of the $CaCO_3$. Artificial products contained the greatest amounts of boron (B), copper (Cu), and zinc (Zn), and second most manganese (Mn) (Figure 4.9). Slag was high in Mn, and coal in Zn. Bricks and mortar both contained relatively low levels of the four heavy metals (glass contains considerable B and sometimes Pb). The sand-sized particles, predominantly from mortar, can aggregate with clay to form a relatively impermeable concrete-like material (Gilbert, 1991), which supports few plants and soil organisms.

An analysis of rubble soils from several building sites in Britain found an average pH of 7.0 (Figure 4.9), with very low levels of total nitrogen averaging 0.6% (Gilbert, 1991). Calcium levels were high, averaging 4597 ppm. Phosphorus (32 ppm), K (198), and Mg (233 ppm) levels were also usually high, largely reflecting their abundance in the clay used for making bricks. Although cinders, furnace ash, sulfur from gypsum plaster, and other materials may locally lower soil pH, a widespread high pH in urban areas is characteristic, due to water running over masonry and concrete structures. The de-icing salt NaCl in cold climates, as well as irrigation with Ca-rich water in hot climates, tends to raise the pH. Distinctive urban plants survive or thrive at a high pH.

Contaminated soils and organic wastes

Contaminated soils

Chemically contaminated soils, here called *brownfields*, are usually difficult to deal with for several reasons: (a) a continuing input of chemicals; (b) the diversity of chemicals present; (c) the relative immobility of many chemicals; (d) types and amounts of chemicals at different soil depths; and (e) the paucity of microorganisms, soil animals, and plant roots able to survive, indeed thrive, in the toxic chemicals (Craul, 1992; Hollander *et al.*, 2011). Contaminated soil may be removed and deposited at another site (Sieghardt *et al.*, 2005). Or uncontaminated soil can be added over the site at a thickness greater than roots penetrate.

Alternatively, one might try "bioremediation," a soil treatment or cleaning process to stabilize, mobilize, extract, and/or volatilize the chemicals. For example, if poplars (*Populus*) could thrive in the contaminated soil, they could extract certain chemicals from the soil. Or certain microbes, if tolerant, could stimulate the uptake or breakdown of certain chemicals. The poplars would have to be harvested and deposited on another site. A key limitation to the effectiveness of bioremediation is to find a mechanism where the rate of removal is sufficient to rapidly remove the chemical accumulation (plus ongoing inputs) (see Chapter 12). For many soil contaminants, regulatory standards have been established to help in the evaluation of contaminated land, as well as for human health (Craul, 1992; Maconachie, 2007).

In Rosario, Argentina, livestock manure, crop residues, and municipal organic waste were mixed to fertilize peri-urban soils. This fertilizer resulted in few earthworms and a high concentration of heavy metals. In an extensive study at Kano, Nigeria, water from an industrial area dominated by tanneries and textile mills was piped untreated into a stream used for irrigation of a peri-urban agricultural area (Binns and Maconachie, 2006; Maconachie, 2007). Forty soil measurements were made in the farmland (2 samples × 20 sites; average pH 8.1, Mg 12.1 mg/l, and Ca 13.9 mg/l). For the heavy metals, Hg, Cr, Pb, and Cd, 40% of the measurements exceeded recommended permissible levels for industrial use and for irrigation. Almost all measurements for Fe, Mn, Co, Cu, and Ni exceed permissible levels. The farmers knew of the toxic contaminants but apparently considered water and crop production to be more important.

Soils around heavy-industry areas may be illustrated by two sites on the southern coast of South Korea dominated by chemical, including petrochemical, industries, which give off high levels of sulfur dioxide pollution (Lee and Cho, 2008b). Soil pH averaged a very low 4.25. Organic matter was 11% and total nitrogen a low 0.7%. Phosphorus in the soils averaged only 19 ppm, K 100, Ca 148, and Mg 39, while Al averaged 462 ppm.

The varied levels of soil chemicals in the preceding examples are indicative but may not be representative. Certainly variability from city to city and site to site must be high. In short, chemically contaminated soils are hazardous for food production and other uses, and difficult to clean up.

Organic wastes and recycling

Massive amounts of organic waste are produced in cities, and interest in its recycling grows. The big three are: *yard waste* (leaf litter, brush, and wood); *solid waste* (often collected by municipalities, organic components include paper products, food, and plastic); and *human waste* (typically in wastewater of sewage and septic systems) (Kidder, 2000). These organic matters can be composted for a period, thus stimulating partial decomposition by aerobic bacteria, and then applied on the soil surface. Humus from yard waste is commonly spread onto gardens. Recycled solid waste is used in some urban agriculture, though heavy metals are often a problem (Maconachie, 2007).

Untreated or partially treated human waste is commonly inserted directly into the ground in holes, compost toilets, cesspools, and septic systems (Figure 4.10). The soil provides some organic-matter decomposition, though often considerable organic matter is carried into a local water body. "Sludge" (biosolids) from partial wastewater treatment may serve as a fertilizer on certain lawns, golf courses, and athletic fields, if essentially all spores of pathogenic organisms have been killed.

Chemical analyses of each organic-waste type are important to determine whether levels are safe for water supplies, food, and various ecological processes. Decomposed leaf litter is least likely to have toxic-level chemicals (Kidder, 2000). Sewage sludge often has high levels of heavy metals (Evans, 2000; Sieghardt *et al.*, 2005). Composted solid waste doubtless varies the most in types and levels of toxic chemicals because of the diverse materials included.

Chemical flows

Four major sources of chemicals in urban soils predominate: (1) minerals in the geological substrate (e.g., silica [SiO_2], calcium carbonate [$CaCO_3$]); (2) water flows, especially in floods (e.g., phosphate, nitrate);

Figure 4.10. Erosion and line of wet soil in road of informal squatter settlement. Soil erosion due to both water and wind. The wet soil line during a dry period usually indicates community drainage of sewage wastewater. Favela on west edge of Rio de Janeiro. R. Forman photo.

(3) atmosphere (e.g., dust in dry deposition, sulfur in acid rain); and (4) humans (e.g., hydrocarbons from transportation, organo-chlorides from manufactured products, heavy metals from solid waste, ammonium and phosphate from human wastewater). Some chemical inputs are spatially limited, concentrated, short-term, and may affect any soil layer (Pouyat et al., 2007; Sauerwein, 2011). Other inputs are relatively diffuse, extensive, low-concentration, and continuous, mainly affecting the upper portion of soil.

The concentration of diverse chemicals is striking because most of these urban chemicals disappear so slowly (Blume, 2009; Sauerwein, 2011). Cities are worsening mounds of toxic chemicals. Some organic chemicals are decomposed by microbes in a few days, but most persist, even for years. Some substances such as nitrates and chlorides are readily leached downward by rainwater to groundwater and surface water bodies. But heavy metals usually leach very slowly. Removal of contaminated soil and attempts at bioremediation of sites are basically expensive and local. Thus, day by day chemicals accumulate in cities and towns, poisoning organisms and us.

The law of conservation of matter indicates that chemical elements can neither be created nor destroyed, just combined in different forms. *Biogeochemical cycles*, such as the familiar water cycle, have built on this to highlight the movement of chemical elements and substances from component to component within a system (see Chapter 3) (Kaye et al., 2006). *Mineral nutrient cycles*, such as for carbon, phosphorus, nitrogen, and sulfur, focus on nutrients flowing, especially cycling from soil to organism to soil to organism and so on (Craul, 1992; Carreiro, 2008). However, concentrated heavy metals as in many urban sites may interrupt nutrient flows (Gilbert, 1991). Cycles may also involve the soil or atmosphere.

For a particular urban ecosystem such as a park or woods, chemicals cycle within the system, but also flow one way through it as a *chemical flow-through* (Forman, 1995). The flows are commonly driven by wind and water, such as N, Ca, or Zn arriving by air and being washed out by water to another ecosystem. People and wind and water combine to accelerate the inputs, flows, and outputs. Thus, urban areas normally have high rates of aerial deposition of chemicals, wind turbulence among buildings and other structures, extensive impermeable surfaces, extensive pipe systems, moving vehicles depositing and resuspending chemicals, and large numbers of people seemingly walking everywhere spreading chemicals. Chemical flow-throughs predominate in urban soils, sites, and ecosystems.

These vectors and massive horizontal flows operate across the urban fine-scale mosaic. As discussed earlier in the chapter, urban soils differ at the scale of meters, tens of meters, occasionally hundreds of meters across, rather than often being an order of magnitude (10 times) or more greater as in natural and agricultural soils. Similarly, sites and distinct ecosystems tend to be much smaller in built areas.

In effect, the large rapid flows of chemicals through small soils and sites means that typically an atom or chemical is "here today, gone tomorrow." In addition, the total amount of a chemical on site may remain relatively constant, but, relative to agricultural and natural soils outside the city, the amount is likely to change, even rapidly and drastically. Urban soils, ecosystems and sites with small size and rapid chemical flow-throughs are likely to be fluctuating rather than equilibrium places.

The urban underground

To sense the richness of structure underground we sequentially explore: (1) structures and their distribution; (2) organisms and habitats; and (3) networks and their forms.

Underground structures and their distribution

Imagine living in a big city for a period, absorbing the sights and sounds and aromas on sidewalks and streets, as well as in all kinds of buildings throughout the city. Familiarity and understanding, even an embryonic sense of place, develops. But suddenly someone says, "You barely scratched the surface ... half the action lies underfoot." Indeed, just climbing down a few feet through a manhole opens up an entirely new world, like the first time peering through a microscope or the first snorkeling on a coral reef.

Initially our impression underground is of darkness, moist air, a slightly unappealing aroma, silence except perhaps for distant trickling water, and maybe helplessness or fear with few familiar cues. Yet soon curiosity creeps in. A frontier found by few lies at your fingertips.

Some objects or structures are small and may be numerous, relatively easy to find on a treasure hunt. Manholes, entrances/exits, ladders, gates, stairs, elevators, light wells and air vents are part of getting around.

Urban soil and chemicals

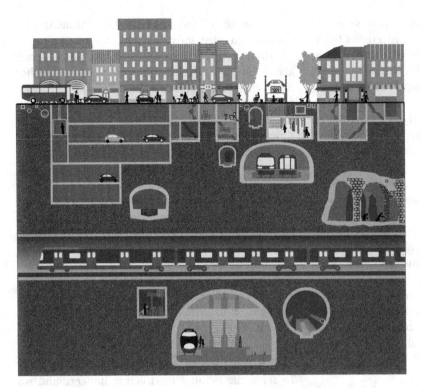

Figure 4.11. Underground structures at different levels in a city. Based mainly on the extensive, diverse, and longstanding underground in Paris (Clement and Thomas, 2001). Lower: inter-city train, stormwater, electric power system, subway. Middle: wastewater, stormwater, subway, garage, quarried spaces. Upper: garage, basement, elevator, wastewater, stormwater, telephone cable system, clean water supply, heating/cooling pipe system, natural gas, walkway shopping arcade.

Basements of buildings, wine cellars, nightclubs/entertainment spots, and rest rooms often contain people (Figure 4.11). Less familiar are building foundations, former pipes and foundations, pumps, flow-control equipment, sewer cleaning/flushing apparatus, leaking pipes, pockets of dangerous gas, and scarcely known black holes. These are just the small things in abundance (Curiel-Esparza et al., 2004).

Some large structures or objects loom underground but are few in number and may require more searching. Subway stations (as in Beijing, Berlin, Baltimore), train station(s) for suburban commuter rail (San Diego, London, Tokyo) and inter-city trains (Barcelona, Washington, Osaka), bus terminals (New York), and carpark garages (Boston, Madrid, Paris) are the key underground transportation spots. People are attracted to underground shopping nodes or centers (Toronto, Chicago), office buildings (Kansas City, USA), storage warehouses (Kansas City), light industry (Kansas City), military facilities (Bagdad, London; Stafford, 1999), and even a vault with France's gold reserves (Paris; Shea, 2011). Here too are less-familiar large structures, such as a water reservoir (Paris; Clement and Thomas, 2001; Shea, 2011), flooded quarries (Paris), collapsed rock piles with rubble from former above-ground buildings (Paris), flower-growing spaces (Paris), and fungus-growing spaces (Paris). Even catacombs (ossuaries) store the bones of former residents dug from overcrowded cemeteries, as in Rome, or in Paris (where bones of 6 million former residents are stored) (Clement and Thomas, 2001).

Much easier to find underground are the diverse intertwined networks, each another frontier inviting exploration. Transportation creates most of the large-tunnel networks (Figure 4.11), including the subway (e.g., dense systems in Tokyo and New York), suburban commuter rail (San Francisco, Moscow), inter-city railways (many large cities), multi-lane highways (Boston, Madrid), and streets (Kansas City, USA, and Leipzig, Germany; Fritsch, 1896).

But water-related networks are also the big story. Many types include: piped former streams (all moist-climate cities); underground aqueduct (Paris); clean water-supply (main) pipe system (all cities); hot-water heating pipes (London, many Chinese cities), stormwater drainage system (all cities); and wastewater sewage system (all cities). Heavy rainstorms above ground can send walls of onrushing water through stormwater pipes and tunnels. Many cities entirely or partially combine two dissimilar flows, stormwater and human sewage, into a single pipe system (e.g., of 1340 km length in Barcelona).

Important underground networks for people include pedestrian walkways (Chicago, Montreal, Toronto), tourist historical routes (Seattle and San Antonio, USA), and quarry tunnels (Kansas City, USA, and 290 km in Paris). Other key networks are gas lines (many cities), electrical conduits (most cities), telecommunication cables (many cities), and pneumatic tubes (Paris).

Fortunately, these explorations of the city are made slightly easier because of the vertical distribution or layering of structures (Clement and Thomas, 2001; Shea, 2011). Objects and networks may be roughly grouped into three layers that are logical though also complicated by historical development and urban growth (Figure 4.11). The upper layer just beneath the city's streets and greenspaces may contain basements of buildings, parking garages, electricity conduits, gas lines, telephone or telecommunications cables, small water-supply pipes to buildings, small hot-water heating pipes to buildings, stormwater drainage pipes going diagonally downward from street to a small connector pipe, and human sewage pipes going diagonally downward from basements to collector pipes (Cano-Hurtado and Canto-Perello, 1999).

A middle layer often contains more parking-garage space, a subway system, pedestrian walkways, shops, some little-used quarry space, clean water-supply (main) pipes, large hot-water heating pipes, medium-size stormwater-drainage collector pipes, and large human-sewage collector pipes (Figure 4.11). At a still deeper level may be more tunnels of the subway system, tunnels containing high-voltage electrical conduits, a large stormwater-drainage collector tunnel, and perhaps a large human-sewage collector tunnel. If an inter-city railway system is underground, it is often at the bottom level.

Plenty of features facilitate flows and movements up or down among the layers. Air and light penetrate in vertical shafts. Ladders and stairs are used by people. Stormwater flows diagonally downward, and the heavier more-viscous sewage wastewater does too. All the pipe systems leak from time to time and location to location. Leaking gas, oil, gasoline, clean water, stormwater, and wastewater-sewage move downward by gravity. The ever-aging complex of objects and network systems enmeshed in soil and water below ground promise lots of interactions, mostly negative, between systems. For instance, riding most subway systems normally highlights the abundance of leaks in the "waterproofed" tunnels. Occasional blockages and explosions disrupt the endless seemingly silent flows in underground systems.

Organisms and habitats

With a dense and diverse human population only meters away, the underground activities of people are exceedingly diverse. Perhaps most interesting are the "free-spirit" people (catophiles in Paris) who simply love the freedom present underground, an escape from the regimentation needed for a high population density above ground (Shea, 2011). The free-spirits roam widely through the networks, love discovery, and often are the most knowledgeable about the little-explored portions. Some tend to be loners who relish the freedom of nightclub parties, drugs, and artistic expression in murals, while others may be serious speleologists (cavers), scuba divers, mappers, and so forth (adventurers in New York). Also usually in low density are the "homeless," who may appreciate freedom but additionally welcome convenient shelter, especially during cold or wet times above.

More familiar are the regular travelers on subways and trains, shoppers in pedestrian tunnels, business people going to and from work, and some of the many key maintenance personnel who keep the underground city working (Fulford, 1995; Belanger, 2007; Brick, 2009). Specialized underground activities such as military uses, wine aging, fungus growing, and storage of valuables of course widen the range of people valuing underground resources.

People bring down food, carry dirt, microbes, spores, insects and human diseases, and transmit pathogens to rest rooms. Human food and diverse types of solid waste attract mice and rats (Feng and Hinsworth, 2013) and more. In effect, the urban underground has a permanent, abundant and rich input of organisms (Figures 4.7 and 4.8) from above. Conversely, these organisms endlessly move upward into our streets and buildings.

Microbes. Greenish algae are often present in moist tunnels around continuous lights. Many types of bacteria, from decomposers of organic material to pathogens in sewage, are present on all substrates. Fungi may also thrive, especially in more-acid locations. Sometimes edible mushrooms are commercially grown underground. Protozoa, the single-cell animals, feed on essentially all types of microbes.

Invertebrates. Lights attract moths (Lepidoptera), flies (Diptera), and other winged organisms. Thus, sometimes spider webs and spiders are present around lights. Cockroaches/cucarachas (Blattidae) and other scavenging invertebrates mainly feed on garbage and dead organic matter. Representatives of the rather distinctive fauna characteristic of caves may be present in the anthropogenic underground. In the groundwater, usually meters or tens of meters below the urban ground surface, is a still stranger or less-familiar hyporheic fauna (see Figure 6.4). These distinctive micro- and macro-invertebrate forms are densest just below the water-table (top portion of groundwater) and where chemical pollutants are not excessive.

Vertebrates. Manifold mice and rats (e.g., *Rattus norvegicus* or *R. rattus*; Boada and Capdevila, 2000) thrive on garbage and dead organic matter both above and below ground. Feral cats often forage and live on the abundant mice and other foods below ground. Raccoons (*Procyon*) frequently live underground and come up at night through stormwater drains to scavenge for food. Bats may have roosts underground and emerge at dusk to feed on flying insects in the air above ground at night. These and other rodents and predators are common carriers of fleas, rabies, and other diseases (Rossin *et al.*, 2004). In addition, many types of vertebrates have been reported from under cities, though populations often do not persist long term underground. These include fat fish in a pond beneath the old opera house in Paris (Shea, 2011); anguila (*Anguilla anguilla*) in Barcelona (Boada and Capdevila, 2000); and alligators in various North American cities.

Network flows and forms

Material and objects moving. Water in the underground soil, but outside of pipes and tunnels, is the most pervasive flow. Some water infiltrates through the hard surfaces and greenspaces above to become subsurface flow. Leaky pipes add to the amount. *Subsurface water* flowing in the upper underground layer (Figure 4.11) mainly goes horizontally but slightly downward to a nearby water body, or it may get deep enough to join the saturated-soil groundwater beneath. Flowing subsurface water normally carries a large number of microbes and invertebrates through the soil and into network conduits. Groundwater in porous rock or sandy material flows very slowly, whereas groundwater in limestone (or karst) rock usually flows rapidly. The deepest level of structures is commonly within the groundwater, which in some cities encloses the middle layer or even the upper layer (Figure 4.11). Pipes, tunnels, small objects and large objects are typically "waterproofed" so that subsurface water and groundwater do not penetrate them.

Moving people is the primary rationale and flow in the underground pedestrian walkways, streets, highways, subway, suburban commuter rail, and inter-city trains. Transported goods and waste materials, especially related to shops and storage, overall are minor. The transmission of electricity and telecommunications occurs in pipes containing air. Natural gas, oil, and gasoline fill and flow in their own separate pipes. Flows in pipes and tunnels for clean water-supply, hot-water heating, human sewage/wastewater, and stormwater runoff are all water-based, though both water and air fill the conduits.

Network forms. Connected corridors in a network function in several ways, especially as conduit for movement along and as barrier or filter against movement across. Watertight underground corridors generally have negligible barrier effect, since subsurface water and groundwater in the soil and rock can simply flow around and past such pipes and tunnels. However, if subsurface water in the soil encounters a leaky pipe or tunnel containing air, the subsurface water flows into the corridor, which then carries it elsewhere. Much more important though is the conduit effect, where water, microbes, invertebrates, vertebrates, and people are transported within pipes and tunnels.

Network form is thus central to movement patterns (see Figure 6.6). Three indices of network form provide the big picture (see equations, Appendix B) (Forman, 1995): (1) *connectivity* (measured by the gamma index) indicates how connected by corridors the network nodes or intersections are; (2) *circuitry* (alpha index) measures the relative abundance of loops in the network; and (3) *corridors per node* (beta index) is the average number of corridors radiating from an intersection. The direction, rate, route, and alternative routes available for movement are strongly determined by these three attributes. For example, attempts to eradicate a pest animal in a location are relatively ineffective if the animal's preferred network has many alternative routes available (high circuitry) to circumvent the eradication location.

But networks have many other attributes also affecting flows and movements, as illustrated by the following networks:

1. *Pedestrian walkways.* Toronto has the world's largest underground shopping complex, including

1200 stores (Fulford, 1995; Belanger, 2007). More than 30 km of underground shopping tunnels and scattered retail shopping nodes/centers are also connected to above-ground resources, including 50 office buildings, 6 hotels, 2 department stores, several tourist destinations, and many public transit stops. Overall the pedestrian tunnel network is rectilinear, roughly paralleling the above-ground streets, and covers an area with a length-to-width ratio of about 1.5 to 1. The underground network has a few main corridors or axes connected to underground shopping nodes, and contains numerous short spurs (no outlets, or dead ends), "T" intersections, "+" intersections, jogs (zig-zags), and diagonals. A few curves, loops, major shopping nodes, and disconnected small network sections are present.

Less extensive but similar-form networks are present in Chicago, Montreal, and Calgary. The abundance of underground shops and people, plus an extensive corridor length, means that the city center will long be plagued by diverse pests that thrive with these conditions. Underground shopping nodes or centers easily walkable by a dense population, with no parking spaces available or needed, is effectively an antidote to the suburban shopping mall.

2. *Quarried limestone tunnels.* Limestone is soft easily harvested rock. The quarried spaces are rainless and snowless, and, for cities with cold winters or hot summers, have pleasant constant temperatures (e.g., ca. 18°C or 65°F). A small-to-medium sized city, Kansas City (Missouri, USA), has an underground quarry area with about 9.6 km (6 mi) of roads and 3.2 km of railroad. These corridors apparently connect 55 businesses, storage warehouses, office space, and light industry.

The quarried-tunnel network of Paris is composed of mostly straight corridors, some quite long without intersections (Clement and Thomas, 2001; Shea, 2011). The long corridors connect areas of dense short interconnected corridors. Many diagonals are present so intersections have acute, right, and obtuse angles. Large and medium-size enclosures are rectilinear in form, whereas small and medium enclosures are typically in portions with convoluted (squiggly) interconnected corridors. The overall area has a length: width ratio of 1:1.

3. *Catacombs network.* A 1-km-long portion of the Paris quarry network with highly convoluted tunnels is lined with bones, sometimes neatly stacked, of some six million former residents (Clement and Thomas, 2011; Shea, 2011). Consideration of the number of people who lived in a city over time, versus the space or number of markers in aboveground cemeteries, highlights a little-recognized but long-standing disconnect in older cities. In Paris the catacombs network has numerous short spur tunnels plus many loops. A few large rooms are attached to the network. The catacombs network in Rome is similar in form.

4. *Sewage wastewater network.* A good way to understand a city's human sewage system is to descend into the Musée des Égoutes de Paris (Paris Sewer Museum). In addition to hearing the occasional rumble of a subway, the double-clanking of a vehicle passing over a loose "manhole," the soft splashing of water, or low voices, one quickly becomes accustomed to, and largely forgets, the moist somewhat pungent air. Intriguing history oozes from every direction: famous names on the tunnel "streets"; wartime resistance activities right under the aboveground offices of an occupying military force; Jean Valjean's famous escape into the infamous sewer system (in Victor Hugo's *Les Misérables*), when soldiers crushed a band of citizens behind street barricades in the French Revolution.

Sewage wastewater is viscous, flows slowly by gravity, and frequently experiences blockages in pipes. Flushing wastewater down kitchen and bathroom drains, and of course toilets, sends the liquid material rapidly down a vertical or steeply diagonal pipe to a more horizontal collector pipe for many drains and toilets (Clement and Thomas, 2001; Shea, 2011). This primary collector goes to a larger and somewhat-deeper secondary-collector pipe, and so on. Pumps are used periodically to keep the fluid moving. In Paris these larger pipes lead to channels down the center of tunnels with walkways along sides. The channels in turn lead to a "river" of sewage, essentially filling the width of a tunnel several meters wide. The sewage river leads northwestward to a huge sewage treatment facility on the outskirts of the city. Before reaching that facility, sewage is full of pathogenic bacteria and other microbes that are readily and widely dispersed by workers, scurrying vertebrates and invertebrates, leaking and flowing water, and air movement.

The network from first-collector-pipe to sewage river is *dendritic* (tree-like), with virtually only acute angles at intersections. Infrequent spur corridors are mainly for the varied equipment used to clean or flush the pipes and tunnels (large wooden balls have been used for pipes, and a wide flat boat for the river sewage). Several characteristics are scarce in the sewage network: loops (which are accessed by pumps to avoid blockages); large open rooms or nodes; sharp curves; jogs (zig-zags); and narrows.

5. Distinctive characteristics of many other underground networks stand out. (1) Subway networks commonly have loops and are attached to many subway-station nodes (at least two in Paris include delightful tiny tropical rainforests). (2) An aqueduct is basically an unbranched corridor leading to a reservoir in the city. (3) Piped former streams are a single corridor, or with a few tributaries connecting at an acute angle. (4) An inter-city train network typically has a single station with railway tunnels leading outward in two or a few directions. (5) Electricity, telephone, gas, and clean-water-supply networks often have right angles. (6) Stormwater drainage networks start with numerous small pipes (and drains) mostly on the edge of streets, and lead to a nearby water body. Conversely, (7) hot-water heating networks begin at a power station and end in numerous small pipes typically in the basements of buildings.

Paris is a large city and its underground has a long and rich history. Smaller and newer cities are likely to have only portions of such a complex of networks. In coastal cities the underground networks, and a city's dependence on them, are especially at risk from salt-water intrusion and sea-level rise associated with climate warming.

Part II Ecological features

Chapter 5 Urban air

... this most excellent canopy, the air, look you, this brave o'erhanging firmament, this majestical roof fretted with golden fire, why, it appears no other thing to me than a foul and pestilent congregation of vapours.

William Shakespeare, The Tragical History of Hamlet Prince of Denmark*, 1603*

We all live downwind.

Bumper sticker in USA, 1980s

We all live downstream.

Environmentalist's motto, quoted in The New Ecology of Nature*, 2002*

We daily bathe in urban air. So does everything else in built areas. The nature of this air overwhelmingly depends on microclimate and pollutants, which in turn are strongly determined by the built environment.

Consider the urban trees and other plants that bathe continuously outdoors. Individual trees, tree lines and woods, plus mowed grass, low spontaneous plant cover and ornamental plantings, are the prime vegetation types found repeatedly across urban areas. Individual shrubs are common, though shrubby areas tend to be uncommon despite their ecological importance (Forman, 2008).

The *roles of trees* are exceptionally diverse and important for urban air (Gartland, 2008). Trees cool surfaces by shading, and cool the air by shading and evapo-transpiring water. Trees may heat or cool the air by creating or disrupting streamline, turbulent, and vortex wind patterns. Trees give off water molecules and may increase relative humidity. Trees, especially those with extensive leaf surfaces, catch airborne dust particles, often containing heavy metals (Spirn, 1984). Trees absorb the greenhouse gas, CO_2 and other gases including SO_2, NO_2, and O_3 (Forsyth and Musacchio, 2005). Some trees emit quantities of hydrocarbons (a major type of VOCs) (Gartland, 2008). Many trees liberate lots of pollen, and some give off long-airborne seeds. Trees produce leaves, flowers, pollen, fruits, and seeds that attract insects, bats, and birds.

Yet trees also suffer from many of these same factors. Trees grow poorly, and may die, with excess heat, wind, and shade from buildings. Trees wither from excess dust, SO_2, aerial road salt, heavy metals, and other polluting aerosols and gases. Some trees die from frost, and excess light at night may stimulate growth that is sensitive to frost. Trees suffer from too much herbivory by insects, even excess bird droppings. Considering all these roles, trees will be key players in the urban air sections ahead (also see Chapter 8).

Human-built structures of seemingly infinite variety also predominate in urban areas. To gain understanding in such a complex situation, we simplify by developing models (and try to avoid false insights from oversimplification). In urban areas we recognize some 15 distinct types of structures to model. Each is present by the hundreds or thousands across the urban area, and together the structures represent the bulk of the area. Five groupings emphasize some common features present (Erell *et al.*, 2011):

1. *House lot, suburban block, city block*: typically a rectangular form, and building dominated.
2. *Urban street "canyon," suburban road, highway segment, rail corridor section*: linear transportation form separating two adjacent land uses.
3. *Shoreline section (by lake, estuary, sea), riverside section*: linear form by water body and differentiated by the adjacent land use
4. *Courtyard/patio, plaza/small greenspace, large open greenspace*: open vegetated patch surrounded by taller structures.
5. *Parking lot, roof, wall*: a patch of hard ("impervious") surface.

In the sections following in this chapter, urban air relative to these apparently basic structures will be discussed in varying detail.

Microclimate

Two perspectives are particularly valuable here: (1) air in and out of cities; and (2) layers of air and the air-dome.

Air in and out of cities

Urban and non-urban air

Some representative statistics provide insights into air differences between urban and non-urban areas (for cities of about 1 million population in mid-latitudes). As a general comparison with nearby non-urban areas, the microclimate of urban areas is characterized by the following (von Stulpnagel et al., 1990; Gilbert, 1991; Oke, 1997a; Hough, 2004; Sieghardt et al., 2005; Alberti, 2008):

1. *Radiation*: (a) much less UV radiation (25–90% less); (b) less solar radiation (1–25%); (c) greater infrared radiation input (5–40%).
2. *Heat and temperature*: (a) increased temperature (1–3°C annual average, and up to 12°C at spots on occasions); (b) more (upward) heat flux (50%); (c) greater heat storage (±200%).
3. *Moisture and water*: (a) less evapo-transpiration (±50%); (b) decreased humidity (summer daytime); (c) more thunderstorms; (d) less snow (some turns to rain); (e) more total precipitation, especially on downwind side of city.
4. *Airflows*: (a) greater turbulence intensity (10–50%); (b) decreased windspeed (horizontal streamline) (5–30% at 10 m above ground surface); (c) altered wind direction (1–10 degrees).
5. *Sky conditions*: (a) reduced visibility; (b) more haze in and on downwind side of city; (c) more clouds on downwind side of city; (d) more, less, or no difference in fog.

In contrast, for air pollutants, differences between urban and non-urban areas are usually much greater, even an order of magnitude or more (Spirn, 1984; Gilbert, 1991; Marsh, 2010): (a) gases 5–25 times greater in urban areas; and (b) particulates and aerosols 10 times more. However, such differences can rise or drop rapidly.

These broad comparative differences are useful as a first step in understanding, yet variability is the second big step. Note that the ranges given are rather large. Also, "non-urban areas" beyond a city may be, e.g., dusty cropland, extensive wetland, or mountainous, each with different climatic properties. Similarly, mid-latitude cities vary from desert to wet forest to island locations. Furthermore, human activities have created regional-scale smog over Los Angeles, nitrogen dioxide (NO_2) over Beijing, and dust over Mexico City. So, consider the broad trends above as generalizations, but expect each city, and each urban to non-urban radius, to be quite distinctive (Brazel and Heisler, 2009).

Energy, heat, airflow, and pollution are the big stories of urban air. Before diving into these key subjects we briefly consider atmospheric moisture.

Moisture in the air

Invisible moisture (or water vapor) from land and sea is carried upward by wind into the colder atmosphere. The moisture condenses (releasing heat), often forming visible clouds of droplets. Liquid falls as rain. If the temperature is below freezing, snow or ice form, and drop. Because cities are significant sources of pollutant particles that act as nuclei for moisture condensation, rain especially falls on the downwind side of metro areas (Kuttler, 2008). Overall, the total amount of water in the atmosphere is small, equivalent to about 2.5 cm (1 in) of water spread over the globe (Moran and Morgan, 1994).

Urban areas are typically covered by extensive hard impervious surface and limited vegetation cover, the opposite of surrounding natural and agricultural lands. Thus, much of the urban rainwater rapidly drains away in ditches and pipes to water bodies. Hard surfaces such as walls and roads are normally wet and evaporate water molecules for only a short time (Rosenberg, 1974; Kuttler, 2008). Combined with limited plant cover, this means that little moisture is released to the atmosphere in the city (little evapo-transpiration). Excluding the precipitation and associated evaporation, the main sources of atmospheric moisture are localized: (1) vegetation patches; (2) water bodies (e.g., ponds, rivers); (3) release from machinery combustion (e.g., industry, domestic heating, vehicles); and (4) horizontal movement (advection) from an adjoining source.

"Humidity" refers to the concentration of moisture or water vapor in the air (Rosenberg, 1974; Moran and Morgan, 1994; Kuttler, 2008; Erell et al., 2011). Higher humidity reduces the evaporation and transpiration of water molecules from surfaces. *Relative humidity*, in percent, is the moisture present compared with the amount present if the air were saturated (on the verge of rain). Relative humidity is strongly affected by rainfall patterns, though in general urban air is relatively dry. Overall, urban daytime relative humidity may be lower than that of the surroundings due to the extensive hard surface and stormwater runoff, plus limited vegetation. Nighttime humidity may be higher (despite less moisture from dew) because of higher temperature in the urban area. Dew condensed on surfaces tends to be limited in urban areas due to dry air (Rosenberg, 1974; Kuttler, 2008).

In streets lined with high buildings (forming "street canyons") and somewhat limited air movement, relative humidity can be increased with well-watered plant cover such as grass, flowers, and shrubs, as well as by trees (Erell *et al.*, 2011). Also pools, water channels, and fountains increase the relative humidity. Most tree evapo-transpiration liberates moisture in the upper canopy, where airflow tends to be greater. Tree shade cools and somewhat reduces evapo-transpiration from plant cover or pools beneath.

Several interesting implications follow from these moisture patterns. More urban air pollution, particularly of particles, generally leads to more foggy days (Kuttler, 2008). Low indoor humidity, as in winter-heated buildings, is increased by many indoor plants (Moran and Morgan, 1994; Kellert, 2005). Dry urban air means a high evapo-transpiration rate for a plant, and consequently the need for more water for plant roots. Dry air dries out soil, resulting in fewer microbes, fewer soil animals, less decomposition, and less mineral-nutrient cycling. Molds are scarcer in dry urban air. On the other hand, in arid climes the surroundings are typically dryer than the urban area which receives some irrigation.

Ventilating the city

Look at the air high overhead. Regional airflows or winds, resulting from macroclimatic often regional temperature differences on sea and land, carry heat, pollutants, and airborne organisms into and out of urban areas (Figure 5.1a). Clustered high buildings force the horizontal layers of *streamline airflow* across the land upward a bit. The structures also cause *turbulence*, somewhat random flows with up-and-down eddies. Turbulence (and vortex airflows) especially separates pollutants, including heat and particles, from urban surfaces. Thus, regional winds tend to ventilate and clean both urban surfaces and the air, always carrying the pollutants to somewhere downwind.

In most cities, the regional wind typically drops at night, leaving still air. Since heat rises from urban surfaces toward cold outer space, a modest amount of ventilation occurs naturally. Warm and polluted air moves upward, thus pulling in some cooler air from

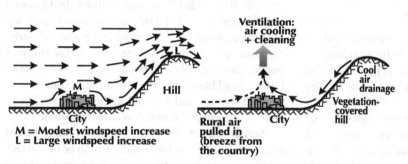

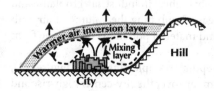

Figure 5.1. Airflows by city and hill. (a) Arrows indicate streamline airflow. (b) and (c) Warm air from city rises. (c) Inversion layer of warm air, with cooler air beneath, interrupts the normal upward airflow from warm substrate to cold upper atmosphere (troposphere).

surrounding cropland, natural land, or water bodies – effectively a "breeze from the country" (which may contain pollutants) (Figure 5.1b).

However, if nearby hillslopes or mountainsides are present, a stronger airflow, *cool air drainage*, takes place on still nights. Heavier cool air pours downward into the city pushing out the lighter warm air. This can produce a major cooling and cleaning of the urban air. Stuttgart (Germany) is a particularly well known case, where urban planning kept high-rise buildings out of valley bottoms so that cool air drained efficiently throughout the city (Thurlow, 1983; Spirn, 1984; Hough, 2004). At no cost, this mechanism both cleans the air by removing pollutants and reduces the urban heat effect by cooling the city. Cool air drainage is especially effective on slopes covered with vegetation (Forman, 2008).

Under certain microclimatic conditions with no regional wind, a layer of warm air forms a temperature *inversion*, and remains for a period above the urban area (Figure 5.1c). With this stationary warm layer, the natural upward ventilation of warm city air is interrupted. Thus, urban heat builds up beneath the inversion layer, gradually spreading outward, often heating suburbs and beyond. Since there is no upward ventilation through the inversion layer, pollutants also progressively build up under the layer, resulting in poor urban air quality.

Finally, a regional wind or storm blows the heat and pollutants away. Residents then enjoy crystal clear air.

Layers of air and the airdome

Analyzing the air above us reveals a number of different layers. Outer space is characterized by the near absence (extremely low density) of molecules (Figure 5.2). The "stratosphere" contains an atmosphere of molecules held by the Earth's magnetic pull (Ahrens, 1991; Moran and Morgan, 1994). Small meteorites encounter the friction of the molecules and burn up here. Commercial airliners commonly fly in the jet stream at the bottom of the stratosphere. An ozone layer (ozone shield) filters out most UV, X-ray, and gamma/beta/alpha radiation (the very short wavelengths) emitted by our Sun, thus protecting all organisms and us in the city.

Especially important for urban ecology is the *urban boundary layer* (UBL) in the lowest 1000–2000 m (3000–6000 ft) (Figure 5.2) (Oke, 1987; McPherson, 1994b; Gartland, 2008; Kuttler, 2008; Erell et al., 2011). This is the lower portion of the "troposphere" (between stratosphere and surface) with reduced windspeed due to friction from the Earth's surface. Temperature inversions (Figures 5.1c and 5.2) occur in the UBL. It also usually contains significant amounts of human-caused pollutants including heat. Small propeller planes often fly just above the urban boundary layer.

The bottom portion of the UBL is the *urban canopy layer*, a mix of streamline, turbulent, and vortex airflows determined by the heights of trees and buildings. Immediately above this is a "roughness UBL," often with considerable turbulence and eddies containing pollutants that have been removed from the urban canopy layer.

Above the roughness layer, a "surface UBL" is dominated by regional streamline airflow (Figure 5.2). Here the windspeed increases logarithmically with height until reaching 100%, above which no drag effect due to the urban area is present (Hough, 2004; Marsh, 2010; Erell et al., 2011). The thicker urban canopy and roughness layer of cities, compared with suburbs, creates a greater drag on regional airflow, and thus a higher column of reduced-speed streamline airflow above the city. Horizontal streamline airflow dominates near ground level in large open areas, whereas turbulence predominates in most urban areas.

The layer of significant human-caused pollutants – heat, particles, aerosols and gases – effectively forms a dome or *airdome* of varying thickness over an urban area (Balling et al., 2001; Hough, 2004; US Environmental Protection Agency, 2008b; Marsh, 2010; Erell et al., 2011). Somewhat like a giant amoeba in form (Figure 5.3), the thickest portions usually are over the central business district and major industrial or power-generation pollution sources. The thinnest portions are typically over large greenspaces and water bodies. In addition, the mosaic of urban and suburban/peri-urban areas, and the building density within them, affects the form of the amoeboid urban airdome.

Several factors affect the buildup of dome height over a land use. Most important seem to be (1) surface roughness, (2) aspect ratio, (3) percent vegetated area (or impermeable area), and (4) source of particles, aerosols, and gases (Alberti, 2008). Heat tends to be highest in a center-city high-rise area and in high-density urban-residential areas. Particles and chemical pollutants are usually highest in industrial, cropland, and commercial areas. Combining the volume of air with elevated heat and material pollutants creates the form of the airdome.

In the conceptual example illustrated (Figure 5.3), peaks of the dome are over the city-center high-rise and

Energy and radiation

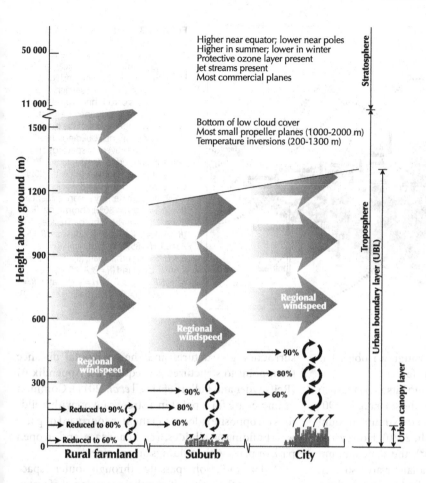

Figure 5.2. Layers of air above the land. Within a few hundred meters of the ground surface, the windspeed of regional streamlined airflow is reduced, while turbulence (size of the circular eddies) increases. 1 meter = 3.28 feet. Based mainly on Ahrens (1991), Moran and Morgan (1994), Gartland (2008), Kuttler (2008), Marsh (2010), Erell et al. (2011).

industrial areas. Dips in the dome are over the wooded parkland, institutional, and suburban-residential areas. Cropland has its own flattish dome of heat and materials adjacent to the amoeboid urban airdome. In the illustration, forest and water body have no overhead heat and pollutant buildup.

The different thicknesses of the airdome are most pronounced in short-term still air such as at night. A persistent warm-air inversion creates a relatively smooth-topped dome of polluted air (Figure 5.1c) (Geiger, 1965; Ahrens, 1991; Marsh, 2010). Light regional winds tend to lower the dome and extend it downwind into a plume. Strong winds may eliminate the heat-and-pollutant dome altogether. Afterward, with pollution emitted and relatively calm air, the dome gradually reforms.

Beneath the dome is a mosaic of warm and cool air over different land uses (Figure 5.3). Since heat moves toward cool areas, local horizontal airflows occur over land uses. Thus, air with pollutants moves, for instance, from cropland to suburban residential, and from city center to urban park. These local horizontal airflows tend to smooth the airdome thickness and surface.

Temporally the dome changes daily, weekly, and seasonally. A prolonged warm-air inversion with ample mixing of polluted air beneath produces a relatively smooth-topped dome. The more common case of heterogeneously distributed pollution sources and light regional airflows produces a variably thick dome. Strong regional winds change the airdome from large and thick to nearly nothing.

Energy and radiation

Solar radiation

Consider sitting by a campfire or in direct sun on a cool day and feeling warmth on your face. Radiant energy or radiation from the fire or sun has warmed you, not heat from the air. *Radiation* moves through air as electromagnetic waves that, upon reaching an object, simply release heat and warm it. Any radiant energy

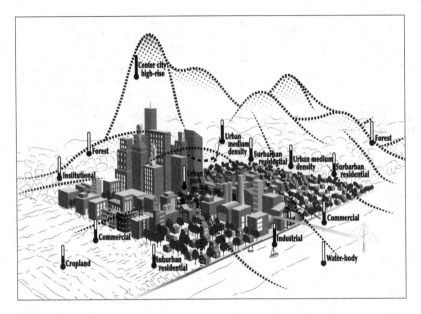

Figure 5.3. Airdome of heat and pollutants over different land uses of city and surroundings. General heat and pollution levels indicated by "thermometers": very high; high; medium; low; very low. For each land use, estimates of roughness (*R*) (effective terrain or surface roughness), aspect ratio (*A*) [average height of the main roughness elements (buildings, trees) divided by their average spacing], and percent hard surface (*H*) (buildings, roads, etc.) are as follows (Alberti, 2008): center city high-rise ($R = 8; A > 2; H > 90\%$); medium density urban residential ($R = 7; A = 1.0; H = 80\%$); commercial ($R = 5; A = 0.1; H = 85\%$); suburban residential ($R = 6; A = 0.4; H = 50\%$); industrial ($R = 5; A = 0.1; H = 85\%$); institutional ($R = 5; A = 0.3; H < 50\%$); urban park ($R = 5; A > 0.5; H < 50\%$); cropland ($R = 3; A > 0.05; H = 1\%$); forest ($R < 4; A > 0.05; H = 1\%$); water ($R = 2; A > 0.05; H = 0\%$). See Balling *et al.*, 2001; Hough (2004), Marsh (2010).

not absorbed (or transmitted through an object) is reflected directly away from the surface.

When solar radiation from the Sun hits a surface such as a wall or the city perpendicularly, the solar angle is 90° and each square meter receives the maximum amount of energy (Rosenberg, 1974; Marsh, 2010). However, if the wall is tilted (solar angle <90°), the same amount of Sun's radiation is spread over a larger area so each square meter receives less energy. Shading and shadows result from objects such as buildings and trees intercepting radiation as the solar angle changes throughout the day (see equations, Appendix B). Except in the overhead sun of a tropical city, these objects always produce shadows to the side, with the smallest shadows at midday. A row of street trees or row of buildings produces a "corridor of shade," which sharply reduces direct solar radiation to the ground (Figure 5.4) (Emmanuel *et al.*, 2007; Marsh, 2010). Such a shadow corridor may be unwelcome to a pedestrian on a cold day, but quite welcome on a hot summer afternoon.

The size of open areas exposed to the sky is a useful indicator of several microclimatic dimensions. Within an open area, the *sky view factor* (or view factor) is the exposed area visible from a point. An easy method to estimate the sky view factor is with a fish-eye lens pointed upward. This is particularly valuable in areas with many trees that tend to have irregular boundaries and holes in their canopy. Alternatively, using geometric modeling, the sky view factor is readily calculated with simple trigonometry from the height of surrounding structures and the horizontal distance from point to structures (see equations, Appendix B) (Reifsnyder and Lull, 1965; Erell *et al.*, 2011). Common examples are for a point in a street canyon with buildings on opposite sides, a rectangular courtyard or patio with building on all sides, a circular space, and an open park or plaza with buildings on one side.

Solar radiation passing through outer space encounters the Earth's atmosphere composed of water, nitrogen, oxygen, ozone, carbon dioxide, and other molecules. Above urban areas we have added lots of pollutants. Solar radiation encountering the molecule-laden atmosphere is divided into three. Some energy is reflected outward. Some is absorbed by the atmosphere. And some passes through the atmosphere to the urban surface beneath.

Solar radiation is composed of energy traveling in a spectrum of wavelengths, from short to long (see equations, Appendix B). Each type of molecule in the atmosphere effectively filters out a particular wavelength of incoming energy. For example, ozone (O_3) filters out most of the very short ultraviolet wavelengths (<0.4 μm), whereas CO_2 and H_2O filter out much of the very long infrared (IR) wavelengths (>0.7 μm) (Ahrens, 1991; Forman, 1995). We can only see in the visible wavelengths (0.4–0.7 μm), which also drive plant photosynthesis.

An important change or threshold appears at a wavelength of about 3 μm. Below that level *short-wave radiation* is mainly the UV, visible, and near-infrared

Energy and radiation

Figure 5.4. A corridor of shade where transpiring trees cool street, sidewalk, and wall. Tree roots receive oxygen from the air and water runoff from the sidewalk and street. Guatemala City. R. Forman photo.

(0.7–3.0 μm). From 3 to 25 (or 1000) μm is infrared *long-wave radiation*. At 8–10 μm is an "atmospheric window," where an abundance of IR is given off from the Earth's surface and passes through the atmosphere to outer space.

Surface energy balance of an area

In a broad sense, the incoming energy simply equals the outgoing energy plus any change (gain or loss) in the amount of energy stored in an urban area, a greenspace, or other object. Except over relatively long periods, energy input and energy output are not equal. In bright sun, energy input is higher than the output, so energy in the form of heat builds up in the city. At night with little energy input, the built-up stored energy is almost all emitted as output to outer space. Arithmetic and ecology work together.

The *surface energy balance* or budget of an urban area is basically determined by six energy flows, easily understood (Rosenberg, 1974; Kuttler, 2008; Grimmond *et al.*, 2010; Erell *et al.*, 2011; Parlow, 2011). This balance readily applies to an entire urban region and any piece within, such as commercial area, urban park, city block, or tree (Njorge *et al.*, 1999). For each energy flow (or flux) we will indicate below key characteristics affecting it in urban areas. This provides potentially useful ways to improve our urban air.

$$Q^* + Q_F = Q_H + Q_E + \text{change in } Q_S + \text{change in } Q_A$$

Energy inputs | Energy outputs | Can be inputs or outputs

Q^* is the net all-wave radiation (short- and long-wave energy), the combination of incoming direct solar and diffuse sky radiation. Q_F is the anthropogenic heat added by human activities (such as heating, cooling, and transportation). Those are the two inputs.

Q_H is the sensible heat (calories determining air temperature that we feel). Q_E is the latent heat (energy given off to the air by evapo-transpiration of water molecules, mainly by plants). Those are outputs. Q_S is net heat storage (gain or loss of accumulated heat from an object). And Q_A is the net horizontal heat advection (transfer of heat to or from an adjacent, e.g., non-urban, area).

The incoming *all-wave radiation*, Q^*, is composed of direct and diffuse solar energy. Direct solar radiation from the Sun predominates on clear dry days. However, in urban areas a significant portion of the solar radiation may be absorbed or "scattered" by water vapor, including clouds, and pollutants in the sky's atmosphere. Energy radiating downward from this atmospheric scattering is called "sky radiation" or diffuse radiation, an important component of the all-wave radiation reaching urban surfaces (see equations, Appendix B) (Kuttler, 2008).

Solar and sky radiation energy encountering the urban area is either reflected upward or absorbed within the system. The amount of incoming short-wave radiation (or solar irradiance) absorbed by the city is largely determined by the surfaces and three-dimensional patterns of buildings, streets, and vegetated areas beneath the urban canopy layer. The amount of incoming long-wave energy (or atmospheric counter radiation) absorbed in an urban area depends mainly on air temperature and water vapor in the lower atmosphere.

The surface reflection of energy is particularly important since in many cases people can have a major effect on it. Reflection is largely determined by surface characteristics and the combination of objects present. The percent of incoming radiation reflected, or *albedo*, varies widely (Ahrens, 1991; Dandou *et al.*, 2008; Gartland, 2008; Marsh, 2010; Erell *et al.*, 2011; Parlow, 2011):

1. *Built areas*: (a) concrete, 25%; (b) asphalt (tarmac) 13%; (c) overall urban area, 15%; (d) commercial city center, 14%; (e) area of buildings 12%.
2. *Vegetated areas*: (a) mowed grass, 23%; (b) crops, 20%; (c) meadow, 15%; (d) deciduous forest, 16%; (e) coniferous forest, 12%; (f) tropical rain forest, 8%.
3. *Other surfaces*: (a) fresh snow cover, 85%; (b) fresh white-painted surface, 80%; (c) rock quarry, 17%; (d) airport, 10%; (e) surface water (e.g., lake), 7%.

131

The percent figures given are general averages for comparative purposes. Considerable variation in albedo exists for most such features, especially for bare soil, concrete, and areas of buildings. Cities in Europe and the USA are generally in the 15–20% range, whereas some North African towns with mostly white buildings exceed 40% reflection (Erell et al., 2011). A high density of buildings and a uniformity in their height produce high albedos. Overall, light-colored smooth dry surfaces reflect the most radiation.

A *wavelength shift* is another key reason we separate short- from long-wavelength radiation. When short- and long-wave radiation encounters a structure, the non-reflected energy is absorbed. The absorbed radiant energy of both wavelengths is converted into heat in the structure. This heat, however, can only be emitted from the structure's surface as long-wavelength infrared energy, essentially heat. In effect, within the structure the short-wavelength radiation absorbed has been shifted or converted to long-wavelength heat. Its emission to the air raises air temperature.

The story is still more interesting because it leads to the *greenhouse effect*. As mentioned above, long wavelengths tend to be filtered out by common atmospheric gases such as H_2O and CO_2, even methane (CH_4), N_2O, and ozone (O_3). Airborne aerosols in relatively high concentration reflect some incoming solar radiation, typically <10%, skyward (Erell et al., 2011).

In clear air over a city, the outgoing long wavelengths of heat readily pass upward at night to outer space, thus cooling the urban air temperature. But in dirty air with an accumulation of chemical pollutants, especially airborne particles, the emitted long waves do not readily pass through. Rather, much of this long-wave energy is reflected, or absorbed and reradiated, downward by the atmospheric accumulation. Consequently, urban heat accumulates and temperatures rise.

The pattern is quite comparable to that of a greenhouse or glasshouse. Short and long-wavelength energy enters, is absorbed by soil and plants, and then the energy is emitted upward as long-wavelength infrared radiation, which cannot pass through the glass covering. Therefore, heat builds up inside the greenhouse, broiling or frying the plants.

Now consider the other energy flows in our urban area. *Anthropogenic heat* (Q_F) directly added by human activities in most cities is relatively low compared with the other heat sources resulting from solar radiation (Gartland, 2008; Kuttler, 2008; Erell et al., 2011; Parlow, 2011). Indeed, the metabolic heat given off by living organisms, including the human population, is negligible compared with solar-driven heat sources.

Transportation and heating or cooling buildings are the primary human sources of heat, and these vary enormously both in space and time (McElroy, 2010). For example, office blocks in central business districts of cold-climate cities may give off more than 1000–1500 watts/m^2 of anthropogenic heat in winter. But in mild climates, such heat output is in the range of a few watts in suburbs to 25 watts/m^2 in city center. Eight Australian and US cities had mid-day values from 7 to 62 watts/m^2, with summer averaging 18 watts/m^2 less than in winter (Erell et al., 2011). Even the relatively high-latitude cities of Vancouver (Canada), Fairbanks (USA), and Berlin (Germany) had average annual anthropogenic heat flows of 6–21 watts/m^2, although the value for Moscow was much higher. Fossil-fuel energy use per capita has generally risen in cities in recent decades, thus adding anthropogenic heat. Yet outward urbanization sprawl has lowered average population density in many metro areas, so the overall role of Q_F as an input in the energy balance probably remains minimal.

The outputs in the surface energy model for urban areas are equally interesting. *Sensible heat* (what organisms feel and respond to), Q_H, flows from materials or structures to the air, and raises air temperature. Within the heterogeneous urban structure, heat flows by convection from structures to the adjoining air. However, for a large urban area, air turbulence (in the surface UBL above the city) carries most sensible heat skyward.

In contrast, *latent heat*, Q_E, (the energy required to evaporate water from liquid to gaseous form), is effectively given off to the air in the evapo-transpiration process (Rosenberg, 1974; Ahrens, 1991; Grimmond et al., 2010). *Evaporation* liberates water vapor from non-living surfaces, such as soil, streets and roofs. *Transpiration* gives off water molecules from the surfaces of living organisms such as plants, especially trees with so much leaf-surface area. Both heat and wind accelerate this upward pumping of water to the air (Kuttler, 2008).

The latent heat given off in the evapo-transpiration process cools the surface of a roof, a leaf, or a suburb. Latent heat energy essentially ends up in the molecules of water vapor. Therefore, the evapo-transpiration process does not heat the air or raise air temperature. Also, unlike sensible heat, animals and people do not feel latent heat. Most importantly though, the cooling of a surface by evaporation or transpiration means that the cool surface radiates or emits less sensible heat.

With less heat given off, the adjoining air temperature is cooler.

Two heat flows, storage, Q_S, and advection, Q_A, can either gain or lose energy. This contrasts with solar radiation and anthropogenic heat, which are only gains, and sensible and latent heat, which only remove heat. Radiant energy absorbed by a structure such as a building or city is stored therein. *Energy storage*, Q_S, depends on the "thermal conductivity" and "heat capacity" of a structure (Gartland, 2008; Erell *et al.*, 2011; Parlow, 2011). Steel is highly conductive and can rapidly rise in temperature, and stone is also relatively high in both characteristics. In contrast, wood is relatively low in both thermal conductivity and heat capacity. Expanded polystyrene (e.g., Styrofoam) is extremely low in both, and thus very poor for storing heat. Typically in spring, or during the day, heat storage in the city is increasing, and in autumn or at night, storage decreases.

Advection, Q_A, the horizontal flow of airborne energy or material, results from temperature differences, such as between city and adjacent suburb, or buildings and a park. Incoming solar radiation varies rather little horizontally across an urban area, while anthropogenic heat sources do vary spatially but the amount of heat is relatively small. Thus, the major horizontal contrasts are normally in sensible and latent heat flows. These flows primarily mirror nearby differences in the amount of vegetation present.

On summer afternoons, advected heat flows from a warm built treeless area to a cooler wooded park. Similarly, hot air moves from a large carpark to an adjoining houselot with lawn. But late at night with cooled buildings and carpark, advected heat can flow in the opposite direction. Thus, like storage, advection can either add energy to or take energy away from an urban area.

Finally, surface emission (or upward long-wave radiation) is the primary microclimatic variable determining why non-urban areas are so different from urban spaces. Radiation emitted from a structure depends strongly on surface temperature (Stefan–Boltzmann law or black-body effect). A warmer surface radiates energy toward a cooler surface. Standing by a cold wall (or an ice sculpture) means that body heat radiates to the wall, and a person feels colder. In urban areas, the abundance of concrete, bricks, steel, and asphalt materials that readily hold and conduct heat means that daytime surface temperatures are higher than in non-urban areas. Consequently, built areas emit considerable long-wave-infrared energy, or heat, skyward.

In an urban area west of Tokyo, summer daytime maximum air temperatures (at 1.2 m above ground) averaged 32°C (90°F) (Asaeda and Ca, 1998). Meanwhile the surface temperatures were 55°C for asphalt and concrete, 40° for a mowed-grass area, and 38°C (127°F) for canopy foliage atop woods in a park. Therefore, much more heat would be emitted skyward at night from the impervious built surfaces than from the vegetated surfaces. The lawn and tree-canopy surfaces differed little in temperature.

The relative size of the different heat flows in the energy balance varies greatly from city to city and spot to spot, as well as from day to night and summer to winter. Overall though, Q^*, the radiant heat flow from sun and sky (solar radiation and diffuse sky radiation combined) often reaches ±500 watts/m², and greatly exceeds any other peak mid-day flow (Erell *et al.*, 2011; Parlow, 2011). Sensible heat flow, Q_H, which determines air temperature (often reaching a peak of 200–300 watts/m²) (see equations, Appendix B) tends to be larger than heat storage flow, Q_S, and latent heat flow, Q_E, combined (some 150–200 watts/m²), in urban areas. Anthropogenic heat flows, on the other hand typically only reach about 3–30 watts/m², and advection overall is probably often in the same range (Forman, 1979b). These relative diurnal peak flows, however, only provide a partial picture. Based on the total amounts of energy flow over 24 hours, radiant heat flow, Q^*, is by far the greatest factor determining the temperature and energy flows in a city, or a site within.

Urban heat

Except in hot dry regions, almost always non-urban areas are cooler than urban areas. The *urban heat island* refers to a built area with a higher air temperature than its surroundings (Landsberg, 1981; Klysik and Fortuniak, 1999; Mills, 2004; Gartland, 2008; US Environmental Protection Agency, 2008b; Erell *et al.*, 2011). This slightly odd but well-used term can be useful for ecological understanding, but often has a negative implication because the heat may be uncomfortable for residents. The concept has ecological implications such as less-cold nights and a longer growing season in cool climates, as well as human implications such as pollution-caused asthma and other airborne diseases (Endlicher *et al.*, 2008).

In today's world fewer and fewer cities could be considered island-like. A megalopolis, such as the Tokyo-Yokohama area, the connected big Dutch cities, and the Boston-to-Washington area, shows a complex pattern of connected elevated temperature. Indeed, most cities are connected with suburbs and nearby towns or other cities. The amoeboid dome concept (Figure 5.3) may be more useful and informative than the island metaphor.

Nevertheless, we start by highlighting the main characteristics of urban heat islands or airdomes, and then briefly explore their causes. Finally, the roles of trees and greenspaces in affecting air temperature, both in the city and its surroundings, are highlighted.

Characteristics of a "heat island"

The highest temperature or greatest heat intensity tends to be near the center, or just downwind, of the area of elevated temperature. *Heat intensity* of the "island" usually refers to the air temperature difference between the hottest location in the urban area and that in a "comparable" location in the non-urban surroundings. The second location poses measurement problems because the surroundings usually are quite heterogeneous, and comparable or extreme locations require interpretation. Still, North American cities typically have higher heat intensities than European cities of the same population, for several reasons related to the causes of heat islands (von Stulpnagel et al., 1990; Erell et al., 2011).

Urban heat islands with isopleths or lines of equal temperature have been mapped for numerous cities, including Mexico City (Erell et al., 2011), Washington, D.C. (Moran and Morgan, 1994; Marsh, 2010), Regina, Canada (Marsh, 2010), Seoul, Korea (Lee et al., 2008), Tokyo, Japan (Gartland, 2008), Berlin, Germany (von Stulpnagel et al., 1990), and Grenada, Spain (Gartland, 2008). In North America, cities in temperate broadleaf and mixed forest have the most pronounced summer urban heat islands (Imhoff et al., 2010). Nevertheless, a recent review of 190 heat-island studies reported methodological problems in most of them, and counseled caution in interpreting results (Stewart, 2011). Nineteen high-quality studies were cited, however.

London is a particularly informative example (Figure 5.5) (Spirn, 1984; Marsh, 2005). Apparently the first description of a heat island was made by a local climatologist, Luke Howard, in 1818 (Gartland, 2008; Parlow, 2011; Erell et al., 2011). Similar descriptions followed for European cities from 1855 onward and for US cities from 1953 on. In two centuries, London has densified and also expanded to become a megacity. Built lobes that have recently grown outward (Forman, 2008) presumably are also lobe-like fingers of heat projecting outward.

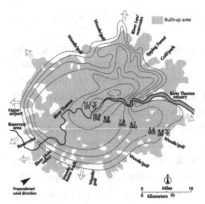

Figure 5.5. Urban heat zones relative to the built area of London. Land uses marked around the perimeter refer to generally short green wedges (white space) projecting into the built area (shaded). Arrows indicate apparent major areas of outward urbanization, 1959–2000. Adapted from Spirn (1984), Marsh (2005), Forman (2008).

The hottest spot is near the center of London with a heat intensity of 6.7°C (12°F) (Figure 5.5). Green wedges of this city project inward and built lobes project outward. The heat contours follow these coves and lobes rather well, though the temperature lines are less convoluted. Green wedges are typically dominated by woods, golf courses, and/or rivers. The steepest temperature gradient occurs where the Thames greenspace wedge reaches closest to the core nearly all-built area. Major green wedges and green corridors slicing through a city probably noticeably ventilate it, reducing buildup of hot air and pollution.

Also, in the last century London established a wide delineated greenbelt around itself, so in fact today's city is an "island" of elevated heat. The outer boundary of mapped development corresponds rather closely with the inner greenbelt boundary.

As cities expand outward, both the heat island area and the maximum air temperature within it increase (Gartland, 2008). The heat-island intensity seems to increase with diameter of metro area (Wong, 2009; Imhoff et al., 2010). But the heat intensity may decrease or increase, perhaps largely based on changes in the surrounding non-urban areas.

The heat island is most pronounced or intense in large areas dominated by impervious surfaces (roofs, streets, sidewalks), and least intense in large areas dominated by vegetated greenspace. But note that the spatial correlation is rough or general as suggested by the patterns in Figure 5.5. The temperature at a given location largely relates to urban conditions close-by, rather than to distance from a non-urban boundary.

Surface temperatures of human structures show a clear heat-island effect, with higher daytime temperature differences between urban and non-urban, and small differences at night (Erell *et al.*, 2011). In contrast, the air temperatures at about the average height of buildings in the urban canopy layer show the heat island to be marked on still nights, but with little urban versus non-urban effect by day. For large cities at least, a heat island effect is present upward to about 1000 (to 2000) meters in the urban boundary layer (UBL) by day, but only hundreds of meters or less at night (Figure 5.2). Thus, although urban surfaces are warmer by day than at night, people feel air temperature (sensible heat). Overall at night, air is warmer in the city than in surrounding non-urban areas. But by day, generally the temperature difference is small.

The urban heat island of elevated air temperature is normally present year round, winter and summer, and is most evident in calm clear air. In North American cities the average heat intensity or heat island amplitude is 1.3°C in winter and 4.3°C in summer (Imhoff *et al.*, 2010). Still, in some cases the heat island is more pronounced in winter than in summer (Gartland, 2008). Light regional wind blows the urban airdome into a temporary plume downwind. Stronger wind eliminates the high urban-boundary-layer pattern, and minimizes the lower urban-canopy-layer pattern (Figure 5.2). Nevertheless, island-like elevated daytime surface temperatures may remain.

The considerable upward moving heat over urban areas at night also contributes to the formation of temperature inversions (Gartland, 2008). Unlike warm-layer inversions at fairly low altitudes in rural areas, urban heat inversions are commonly up near the top of the urban boundary layer. The bottom of an urban temperature inversion is often about 100–300 m high (Ahrens, 1991), and below it the air tends to be well mixed (Figure 5.1c).

In hot arid areas, although city air temperature is extremely high, the urban heat island is less pronounced (Brazel *et al.*, 2000, 2009; Imhoff *et al.*, 2010; Erell *et al.*, 2011). This is because the temperature of the non-urban may be similar to or even higher than that of the urban, which may have considerable irrigation. In subtropical cities, the temperature-difference intensity is less than in temperate zones (Roth, 2007). Also the tropical/subtropical heat island effect may be more evident in the wet season than in the dry season. Larger cities on average have greater heat island intensities than do smaller cities (Oke, 1973). Small towns form urban heat islands, though hamlets may not (Landsberg, 1981;

Mills, 2004). Heat islands also form in non-urban areas such as a large sandy shrubland surrounded by moist forest (Moran and Morgan, 1994).

Before examining the causes of urban heat, consider briefly some of its ecological effects (von Stulpnagel *et al.*, 1990; Gilbert, 1991; Kaye *et al.*, 2006; Parmesan, 2006; Alberti, 2008). The growing season (and frost-free season) for plants is longer, and especially extends further into autumn. The phenology of plants, including earlier blooming and later leaf drop, is changed. Bird nesting and migration differ in timing. Many other natural rhythms are altered by the added urban heat.

Causes of urban heat

To understand urban heat differences we examine (1) energy flows, and (2) surfaces and structures.

Energy flows

Numerous factors affect the intensity of a city's heat island, and their cumulative effects are reflected in basic physical processes of energy flow. The difference in the surface energy balance, introduced above, of the urban and adjacent non-urban areas is central (Gartland, 2008; Erell *et al.*, 2011; Parlow, 2011).

In late afternoon and early evening, air temperature begins to decrease in both city and rural areas. However, with considerable stored heat being emitted from surfaces in the city, rural cooling takes place faster. That creates a larger city-rural difference in air temperature. A more intensive heat island is formed at night, a nocturnal heat island, usually most pronounced a few hours after sunset (Oke, 1981; Erell *et al.*, 2011). Note that if the rural area has humid overcast conditions, rural cooling is much less, and therefore the urban heat-island intensity is less. Shortly after sunrise the process is reversed, as the non-urban area heats up faster than the urban. Consequently, the daytime urban heat island is less intensive or non-existent.

These broad patterns are affected by geography, weather, season, and so forth. But most important are how the characteristics of the local urban surface area affect the variables in the surface energy balance: the inputs of net radiation and anthropogenic energy; the outputs of sensible and latent energy; and the changes in energy storage and advection.

Incoming net radiation, Q^*, is typically lower in the city than outside due to polluted air, and of course at night no net radiation arrives at either place (see equations, Appendix B). Anthropogenic (human contributed) heat inputs (Q_F) which overall are relatively small, may be large in certain locations, such as high-density building areas. Such local heat sources could affect air temperature contours within a heat island (Figure 5.5). Especially at night in cool climates, heating systems give off sensible heat, Q_H, raising nocturnal heat-island conditions somewhat. Sensible heat, especially from heated surfaces, is given off to the outer atmosphere day and night, in and out of the city.

Evaporative cooling in summer by air conditioning systems cools built structures, and gives off latent heat, Q_E. In both city and surroundings, latent heat is mainly given off in daytime when most plant evapo-transpiration occurs. However, latent heat does not raise air temperature.

Horizontally moving advected heat, Q_A, can be important locally, but normally has little effect on the heat island intensity. If anything, advection would reduce the urban–rural temperature difference by warming areas downwind of the city.

Thus, except for the obvious day–night difference in incoming net radiation from Sun and sky, the other four energy flows just discussed show relatively small urban-to-rural differences. However, energy storage, Q_S, is a key to the heat island intensity (Parlow, 2011). The mainly nocturnal heat island of higher air temperature in city than in surroundings depends primarily on heat emitted from built urban surfaces. That stored energy, absorbed during the sunny day, heats the nighttime air in cities more than in rural areas, and dominates the urban heat-island effect.

Surfaces and structures

We focus on the differences in structures and surfaces between non-urban and urban areas (Landsberg, 1981). Non-urban areas usually have considerable vegetation cover and moisture in the soil. Therefore, evapo-transpiration and latent heat flows are much higher than in the less vegetated city with little exposed moist soil. In the city, upward latent heat flow is typically small, and in non-urban areas considerable latent heat moves upward without warming the air. After rain, evaporation from wet built structures briefly gives off latent heat, which cools the surfaces of the structures but does not raise air temperature.

Extensive impervious surfaces in urban areas (e.g., some 75–95% cover in cities and 20–75% in suburbs), compared with very few in many non-urban areas, have major effects on the surface energy balance. Steel, concrete, asphalt (tarmac), and bricks-and-mortar readily absorb, conduct, store, and emit heat. Incoming net radiation energy by day is readily stored in these materials, and the warmer their surface, relative to the night air above, the more they radiate heat to the air.

Built structures using these impervious materials create urban surface forms that are also important in affecting the intensity of a heat island. A higher building density (for buildings of similar form and height) increases the albedo or reflected energy, and therefore has less energy absorption and storage. Lower building densities tend to have more vegetation, especially tree cover, which absorbs more energy and gives off more latent heat. A second major factor is the convolutedness or roughness of the urban surface (within the urban canopy layer). This relates to the variability in building form, height, and density; the relationship of streets and buildings; the distribution of greenspaces; and so forth.

One measure, the average "street canyon" height-to-width ratio (or sky view factor) separating blocks of buildings is particularly useful. The heat island intensity rises as the average height-to-width ratio (H:W) increases (sky view decreases) in the urban area, though the relationship is curvilinear (Oke, 1981; Erell et al., 2011). A steep increase in urban versus non-urban temperature difference occurs as H:W increases up to 1.0–1.5, above which the heat island intensity increases gradually. Thus, while the heat intensity is greatest in center-city high-rise areas, it is especially sensitive to road widths and building heights in the outer city and inner suburb areas.

Heat island intensity is decreased by many other factors, including: (1) wet rural areas; (2) cloud cover and atmospheric moisture; (3) regional wind; and (4) onshore and offshore breezes of coastal cities. The interaction between global warming and the urban heat island is explored later in this chapter.

A daytime "urban cool island" effect, whereby city air is cooler than that of surrounding non-urban areas, has occasionally been highlighted (Erell et al., 2011). This results from high city albedo plus shading of surfaces by built structures. Although the urban cool island is normally a small and/or temporary temperature difference, it may be important for cities in hot dry climates. This cool-island effect differs from the familiar "oasis effect" in hot arid land, where cooling largely results from shade and evapo-transpiration by vegetation on moist soil in the oasis.

Finally, the urban heat island and its heat intensity are familiar as problems for the comfort and health of urban residents. An alternative perspective is that high air temperature is simply a normal characteristic of cities, which are composed of concentrated buildings and roads that absorb, store, and emit heat. Incorporation of the considerable knowledge available on urban climate and heat into urban planning and design remains surprisingly scarce, given the expected extensive populations of cities ahead (Grimmond et al., 2010; Mills et al., 2010).

The island metaphor or model, with its core heat-intensity concept measured by subtracting two variables may be oversimplified and in need of a richer paradigm or broader perspective. The basic processes in the surface energy balance should serve well in such a paradigm. Key factors may include (1) the connected amoeboid airdome; (2) non-urban areas being more heterogeneous and diverse than urban areas; (3) combining the cover and diversity of vegetation types; and (4) integrating other horizontal flows, especially water, wildlife, and transportation. Building a model on such major observable patterns and assays could provide useful understanding of heat around urban areas.

Greenspaces and impervious spaces

Greenspaces

In the urban canopy, vegetation with trees and built structures with impervious surfaces are the two primary features of built areas. They also are the two prime determinants of urban heat buildup through energy flows (Klysik and Fortuniak, 1999; Grimmond et al., 2010). Here we separately consider the cumulative role of vegetated greenspaces and of built surfaces in affecting air temperature across an urban area.

The effect of the size of greenspaces is of obvious interest. Does it matter temperature-wise whether parks are large or small? A particularly interesting study compared the air temperature in 42 Berlin greenspaces of different size versus the temperature in their built-up surroundings on a calm summer evening (Figure 5.6) (von Stulpnagel et al., 1990). Small greenspaces <30 ha were on average approximately 1°C cooler than their surroundings. Medium-sized ones of about 30–500 ha averaged about 3°C cooler, and the four large urban greenspaces >500 ha were about 5°C cooler. Variability in temperature was rather low among the small unbuilt areas, as well as among the large ones. However, variability was high for the medium-size greenspaces, with a range from 0.5 to 5.4°C. This high variability may be due to different land uses included (e.g., lawn, woods,

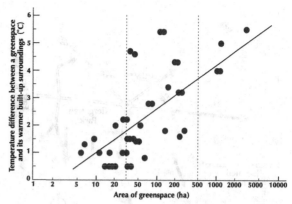

Figure 5.6. Air cooling relative to size of urban greenspace. In West Berlin. The coolest temperature measured in a park or other greenspace is compared with the temperature in its surrounding built area. Measurements at 2 m aboveground in midsummer (July 9, 1982) at 11:00 PM on a calm clear night. 1 hectare (ha) = 100 m × 100 m = 2.5 acres. Adapted from von Stulpnagel et al. (1990).

road, parking lot, ballfield, wetland, pond, and so forth), and/or the shape, topography, and location of a greenspace. Nevertheless, clearly larger parks cool the air more than do smaller ones.

An equally interesting question is how far outward, and in which direction, a cooling effect extends from a greenspace boundary into the surrounding built-up area. The cooling effect occurs in a "breeze from the park" mainly in the hours after sunset, and is related to warm air rising from built areas surrounding a greenspace.

Measurements in the surroundings of five Berlin greenspaces found cooling effects (significantly lower air temperature at 2 m) reaching 1500 m downwind, and 500 m upwind, of medium-size greenspaces (Figure 5.7) (von Stulpnagel et al., 1990). Overall cooling extended further from medium greenspaces (125–212 ha) than from small ones (18–36 ha), though differences were not significant within size categories. Cooling extended downwind on average 2 to 2.5 times further than upwind for both greenspace-size categories. The data also suggest that the cooling effect extends furthest upwind in calm conditions, whereas downwind cooling distances seem about the same in calm conditions and moderate wind. With stronger winds, the cooling effect beyond a greenspace commonly disappears, though in two cases it extended outward a considerable distance.

These Berlin studies suggest a modeling approach to estimate whether a single large greenspace or the same area subdivided into smaller greenspaces

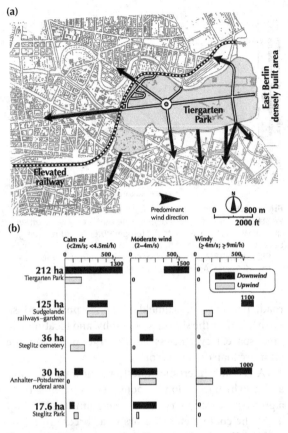

Figure 5.7. Distance a cooling effect extends outward from a greenspace. (a) Length of arrow = distance a lower temperature was measured. (b) Horizontal bar indicates the range of cooling distances (in meters) recorded outward from a greenspace. 0 = no cooling detected outside greenspace. West Berlin; temperature measurements at 2 m aboveground under varying weather conditions. 1 hectare (ha) = 100 m × 100 m = 2.5 acres; 1m = 3.3 ft. Also see Figure 5.6 caption. Adapted from von Stulpnagel et al. (1990).

provides more overall cooling for an urban area. One might compare the total greenspace area cooled by one 1000-ha park versus ten 100-ha parks versus a hundred 10-ha parks (Figure 5.6). We assume that parks are square, minimum temperature is in the park center, and temperature isopleths are equidistant from center to border. With this model, one large park has 1000 ha with an average cooling of 2°C, ten medium parks have 1000 ha averaging 1.25°C cooling, and a hundred small parks have 1000 ha averaging 0.5°C cooling.

Alternatively, we may assume that, say, the central 90% of the large park is 4°C cooler and its outer edge portion averages 2°C cooler, with the analogous 90%/10% areas of different temperature in the medium and small parks. In this alternative, one large park has 1000 ha of 3.8°C cooling, compared with 2.4° and 1.0°C cooling, respectively, in the same total area of medium and small parks. In both modeled alternatives, the large greenspace provides noticeably more overall cooling (both by area and temperature) than do the smaller parks of the same total area.

A more complete answer emerges by adding the outward-extending cooling effect (Figure 5.7) to these results. We assume a very rough average cooling distance of 300 m for a medium-size 100-ha greenspace, and 200 m for a small 10-ha park. Although the average temperature decrease surrounding the medium park is greater than that around small parks, the total area around small ones is considerably greater. Together these patterns would probably narrow the difference in cooling benefit based on within-park data, but the overall conclusion of greater total urban cooling by one large rather than several small parks seems to remain valid.

The representativeness of these results awaits studies in other cities. Studies in Northern Europe on hot summer days have found a 2–3°C cooling in "large" parks, 1°C in 10+ ha parks, no significant cooling in parks <1 ha (Kuttler, 1993), and cooling extending to 200–400 m in built-up areas downwind of large parks (Tyrvainen et al., 2005). In San Francisco, cooling extends some 2000 m downwind of the large 700-ha Golden Gate Park, where winds are typically strong (Schmid, 1975; Marsh, 2010). In Montreal, the more-typical 20-ha LaFontaine Park on a calm winter early-evening showed a downwind cooling of about 1250 m, and almost no upwind cooling (Schmid, 1975). In spring early-evening there, with calm-to-light wind conditions, the approximate downwind cooling extended to 900 m, and upwind to 300 m. Interestingly, the coolest spot in both cases was not in the park but about 450 and 100 m, respectively, downwind of the park boundary. This downwind "coolest-spot" pattern could be usefully factored into the simple modeling just above to provide a yet more useful picture of cooling by urban parks.

Impervious spaces

In contrast to greenspaces, patches of impervious surface such as roofs and parking lots tend to be small and distributed in huge numbers across an urban area. Even street and sidewalk segments (e.g., between intersections), while in linear form, mostly cover a limited area and are widespread. Airports, truck/lorry transport-and-warehouse terminals, and wide highway/boulevard segments are large and few in number in an urban area.

As indicated above, albedo, the percent of incoming net radiation reflected by a surface, is important because all non-reflected energy is absorbed by the structures. There it is converted to heat, which is stored and then re-emitted, especially at night, accentuating the heat island. Light-colored, dry, and smooth concrete surfaces reflect the most (Gartland, 2008; Erell *et al.*, 2011). White roofs and new concrete have high albedo and therefore contribute little to raising air temperature. Dark surfaces such as asphalt/tarmac roads and tar-based roofs have low albedos, thus absorbing most incoming energy, which mainly ends up raising the city's air temperature. As noted above, moist surfaces such as water bodies, vegetation, and impervious surfaces after a rain, also have low albedos. However, the latent heat given off in evaporation and transpiration from these moist surfaces does not heat the air.

Two strategies are the key to ameliorating hot temperatures in cities (Gartland, 2008): (1) orient and color surfaces to maximize upward reflection of incoming radiation, and (2) increase vegetation, especially trees. Paint roofs and walls white as in many Mediterranean towns and cities. Also use light-colored materials for roofing. On the sunny sides of buildings, use narrow downward slanting strips or shelves to reflect net radiation upward, and shade the vertical surfaces between strips. These urban surfaces are cooled by increasing evaporation of rainwater on them (thus also reducing stormwater flooding below). New concrete on roads, parking lots, and sidewalks has a high albedo, but hydrocarbons and other pollutants from transportation usually quickly darken it. General urban particulate and aerosol pollution tends to darken roofs, walls, roads and sidewalks, and thus increase urban heat.

The companion strategy of increasing urban vegetation can also be used to minimize heat buildup in impervious urban materials, and the consequent air-temperature increase. A key is the arrangement or juxtaposition of trees and other vegetation relative to the surfaces of built structures. A tree or tree canopy over an impervious surface, such as street, parking lot and sidewalk, provides triple benefit. (a) The tree provides shade, thus minimizing energy absorption by the surface and consequent heat buildup. (b) The underside of the tree canopy absorbs energy emitted upward from the impervious surface (Ali-Toudert and Mayer, 2007). And (c) the foliage transpires water upward, thus "dissipating" the incoming radiant energy into latent heat without raising air temperature. Plants covering green roofs and green walls provide the first and third temperature benefits.

So, shade the streets and sidewalks and parking lots with trees (Gartland, 2008; Gaston, 2010; Grimmond *et al.*, 2010). Indeed, shade the roofs and walls of homes, and of low-rise residential and commercial structures. A well-placed street tree can shade a building wall in addition to sidewalk and street. In choosing locations for trees, diurnal and seasonal solar angles and determining when surface energy absorption contributes most to raising air temperature are critical. Choose tree species according to whether they will grow or wither, and whether they have a leafless period. Consider whether trees are located to increase or inhibit cooling by winds, to be explored in the next section. An invisible factor is perhaps most important. Make sure that roots continue to have sufficient water to support the tree's considerable evapo-transpiration (ET) in the face of incoming radiant and wind energy. To provide a major cooling effect for urban areas, during the day ET pumps a lot of water out of the ground.

Also plant cover on roofs and walls (see Chapter 10) greatly reduces their surface temperature. That means much less heat emitted to the air. For instance, ivy covering a building may reduce the building surface temperature by 15°C (27°F), and air temperature 1 meter from the surface by up to 4°C (von Stulpnagel *et al.*, 1990).

Yet another approach to reduce urban heat buildup remains largely untapped. As highlighted in Chapter 6, cities use impervious surfaces and pipes to get rid of rainwater or stormwater as fast as possible, thus degrading local water bodies and causing periodic floods. Evaporation of water from those same surfaces reduces heat radiation to the air. Couldn't creative thinking and technology capitalize on the considerable rainwater and built surfaces to produce combined solutions for the urban heat, stormwater, and water body problems?

Diverse airflows

Urban airflows are explored from four perspectives: (1) local breezes; (2) winds and windbreaks; (3) street canyons; and (4) isolated buildings and trees.

Local breezes

The upward flow of warm air from an urban area helps to ventilate and clean urban air. The process also draws in cooler air to replace the rising air, and several types of this cool air movement are important. We typically call them breezes, as the windspeed is relatively low. A *breeze from the country* is cool air from a surrounding rural or natural area moving inward through the urban area (Figure 5.1a). Similarly a *breeze from the park* is airflow from a relatively large greenspace into its surrounding built areas (Figure 5.7).

Cool air drainage down a mountain or hillslope at night results in part from the cool air upslope being heavier than warm air downslope, and hence air moves downward by gravity (Figure 5.1b). Upward moving air from a nearby heated city also draws this cooler mountain air to the city. Vegetated slopes without development are particularly effective in generating cool air drainage. For this, the relative advantage of meadow or woodland on hillslopes and mountain slopes remains a subject of research (Forman, 2008). The cool air primarily flows downward in gullies, valley bottoms, and open strips between buildings or trees. The absence of high-rise buildings in these channels enhances ventilation of the urban area.

Onshore and offshore breezes occur in coastal areas by the sea or a large lake. These breezes are seasonally dependent and may occur by day or night. In spring following the cold season, solar radiation warms the land faster than the seawater. Warm air rising from the land then draws cool air from over the sea inland, creating an *onshore breeze* (Yoshikado, 1990). Heat rising from urban surfaces by the coast accentuates the onshore breeze. This cooling breeze typically extends a few kilometers (up to about 7 km) inland from the coastline. Conversely, in autumn following the hot season, the land surface cools more rapidly than the seawater at night, so heat rising over the sea draws cool air seaward, creating an *offshore breeze*.

In essence, these breezes and air drainage are local processes. All result from stored energy being radiated skyward, and several are strongly affected by the heat emitted at night from impermeable urban surfaces. These airflows contrast with winds generated by atmospheric temperature differences at the regional scale.

Winds and windbreaks

Regional wind across the land may be visualized as flowing in horizontal layers in the urban boundary layer, with higher windspeeds aloft (Figure 5.2). In a large smooth open area, such as cropland or pastureland in flat terrain, the air flows as *streamlines*. If streamlines encounter a hill with gentle slopes on both windy and downwind sides, the lowest air layers are squeezed upward against the upper layers of flowing air (Brandle *et al.*, 1988; Forman, 1995; Rampanelli *et al.*, 2004). At the hilltop, windspeed is commonly some 10–15% higher (due to the Venturi effect). Like wind passing an airplane wing, streamline airflow is maintained on the upslope, top, and downslope of the gradual hill, or a similarly contoured windbreak or town.

If the hill has a steep slope on either upwind or downwind side, the streamline airflow separates from the ground surface, creating a zone of *turbulence* composed of seemingly chaotic "eddies," typically with strong localized circular up-and-down motion. The steep-slope hill is a *bluff object* that disrupts streamline airflow, creating turbulence or a vortex. Turbulence extends downwind from the hill until streamline airflow is reestablished along the ground. A *vortex*, as the third basic type of airflow, is strong airflow rotating in the form of a cylinder, such as a twister or tornado. Vortices form by streamline wind flowing over or around a long object such as a long wall or the edge of a tall building.

Winds separate objects and energy from surfaces, such as dust from a construction site, butterflies from a butterfly garden, heat from a wall, or hats from pedestrians. The items are deposited downwind. The ability to separate objects from a surface increases in the following order: (a) breeze (as described above), (b) streamlines, (c) turbulence, and (d) vortex. Generally people try to minimize turbulence and vortices.

The minimum regional windspeed that eliminates the urban heat island was calculated for several cities (Schmid, 1975). For cities of 2 to 8 million people, the heat island disappeared when wind reached 11–12 m/s (ca. 25 mi/h). A 6–8 m/s wind did this in cities of 120 000 to 400 000, and a 3–5 m/s (7–11 mi/h) regional wind eliminated the urban heat island of small cities (33 000 to 50 000 population).

A treeline, wooded strip, or stone wall acts as a windbreak to decrease streamline windspeed (Brandle *et al.*, 1988; Forman, 1995). Three simple variables mainly determine the effectiveness and interesting wind patterns around a windbreak (see equations, Appendix B). Assume that the windbreak is on flat extensive mowed grass or cropfield, and that wind direction is perpendicular to the windbreak. Streamline airflow is reduced and turbulence present both upwind and downwind of the windbreak. The distance of reduced windspeed both upwind and downwind is proportional to *windbreak height*, H. Typically windspeed is reduced upwind some $3-6H$ (e.g., 6–12 m upwind of a 2-meter-high hedge, or 60–120 m upwind of a 20-meter-high treeline). Just beyond the windbreak is a "quiet zone," with little streamline airflow, that extends a distance of about 8H downwind. Beyond that is a "wake zone" of turbulence extending to 15–25H downwind of the windbreak before streamline airflow is reestablished.

Windbreak porosity has a major effect on these airflows. Downwind of an impermeable wall or building or nearly impermeable woods, strong turbulence is present, and relatively strong turbulence upwind as well. In contrast, a porous treeline or hedgerow permits some streamline air to flow through the windbreak (bleed flow). This is sufficient to prevent or reduce turbulence both upwind and downwind. Relating the length of quiet and wake zones to windbreak porosity produces a bell-shaped response curve.

In addition to windbreak height and porosity, location is the third key variable. Locating the windbreak in an extensive open area produces the above wind patterns, but urban areas are mainly characterized by numerous buildings and trees that create turbulence and vortices. If the wind encountering a windbreak is already slightly or strongly turbulent, the distance, H, in which wind patterns are affected both upwind and downwind of the windbreak is reduced. In addition, if wind encounters a windbreak at an angle rather than perpendicularly, H is reduced, and the windbreak is less effective in reducing windspeed (Erell *et al.*, 2011).

Street canyons

The idea of a valley or canyon with flat bottom and vertical sides has been used to help understand the microclimate of a street lined with buildings on both sides (Oke, 1981; Erell *et al.*, 2011). Consider a *street canyon* with a height-to-width ratio (aspect ratio) of

1. If streamline wind above the roofs is in the same direction as the street, the air also flows along within the canyon creating minimal turbulence or vortices. On the other hand, above-roof airflow crossing over the street perpendicularly creates a secondary flow in the form of a horizontal rotating vortex within the canyon (Figure 5.8). Streamline wind flowing diagonally over a street also creates a horizontal canyon vortex. However, instead of airflow "rotating in place," the air both rotates and flows along the street. Air with pollutants picked up along the canyon spills out at the end of the street.

Wind crossing perpendicularly over a deep street canyon, e.g., with $H/W \geq 2$, creates complex secondary airflow patterns (Erell *et al.*, 2011). The vortex formed beneath the roof level rotates in a cylinder as in a shallow canyon, but this rotation may cause another vortex to form below it that rotates in the opposite direction. Counter-rotating vortices can also be produced in wider canyons ($H/W < 1$). The relative temperature of the three canyon surfaces, as well as the over-roof windspeed, strongly affects such airflow patterns.

At city street intersections, wind moving along one street reaches the corner of a building and forms a vertical vortex in the end portion of the intersecting street canyon (Figure 5.8). However, windspeed is lowest at street level and higher above, so the complex vortex and turbulent airflows present may be simply viewed as a "street-corner eddy."

These generalized airflow patterns provide the framework for understanding wind in urban areas with street canyons. Variability in the heights and continuity of buildings creates turbulence at the above-roof level, and this turbulence disrupts the geometric streamlines

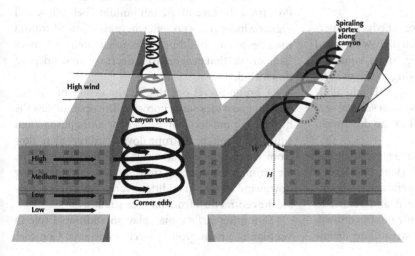

Figure 5.8. Air flows in street canyons. Decreasing W (street width), or increasing H (building height), tends to decrease or eliminate the canyon vortex at street level. High, medium, low refer to windspeed. Adapted from Erell *et al.* (2011).

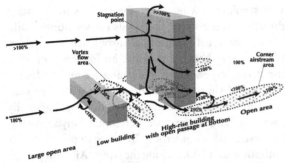

Figure 5.9. Airflows by low and tall buildings. * = Windspeed of streamline airflow in a long upwind, relatively smooth-surface open area. Increases and decreases in windspeed given are idealized, and vary according to the presence of other structures such as buildings, bridges, trees, and parked trucks. Nevertheless, the locations and relative amounts of windspeed change illustrate the principles and patterns to be expected in urban areas. Based on Spirn (1984), Oke, (1987), Brandle *et al.* (1988), Forman (1995), Hough (2004), Erell *et al.* (2011).

and vortices. Also, frequent large objects such as trucks and street trees at the bottoms of canyons produce turbulence where most organisms move or live. Building walls with balconies, ornament, and vegetation create fine-scale temperature differences, all of which create some turbulence. East-west oriented street canyons (H/W ratio = 1–2) feel hotter than north-south ones, due to prolonged relatively direct solar radiation on walls facing the sun. Therefore, east-west streets can be expected to have more upward heat movement at night (Ali-Toudert and Mayer, 2007; Erell *et al.*, 2011). Indeed, diverse street-canyon designs varying airflow types, built radiation surfaces, tree distributions, and more can produce a rich array of microclimatic effects (Yamaoka *et al.*, 2008).

Isolated buildings and trees

An elongated isolated building, like a hedgerow perpendicular to the wind, has elongated high-windspeed zones at each end, where moving air is squeezed between the structure and other flowing air (Figure 5.9) (Spirn, 1984; Hough, 2004; Erell *et al.*, 2011). In addition, eddy effects, as at street corners, form vertical vortices. These may affect much or all of the downwind side of a building (Erell *et al.*, 2011).

The higher a building is the longer the sheltered reduced-windspeed zone downwind (Gut, 1993). For a very thin building or stonewall, perpendicular streamline flows are forced upward over it and return to ground level at some distance downwind. In contrast, for a large flat-topped building the streamline airflow is forced upward at the edge of the building, and then is quickly reestablished along the flat roof. Immediately downwind of the building, strong downward turbulence occurs, which extends only a short distance downwind before streamlines are reestablished at ground level. Also, airflow over woods has some turbulence due to canopy roughness, so immediately downwind the zone of turbulence and reduced-windspeed is usually very short.

To reduce streamline airflow and increase farmland production, the distance between parallel windbreaks is partially determined by the height of windbreak trees, and partly by the "porosity" of the windbreak. A medium-porous windbreak is commonly used in order to limit turbulence yet provide long downwind windspeed reduction (Brandle *et al.*, 1988).

With buildings, height and distance apart are the main variables determining airflow patterns, since almost always buildings have zero porosity. For a series of buildings, a H/W of 1 or less produces "skimming flow," where streamlines continue over streets with minimal turbulence present (Gut, 1993; Erell *et al.*, 2011). When buildings are relatively far apart or isolated (small H/W ratio), streamline airflow may touch ground level at spots between buildings, thus creating an overall rough pattern of flow (isolated roughness flow).

The airflow patterns around buildings of different size and shape of course are diverse. Consider wind flowing perpendicular to one low long building separated from a second flat high-rise building behind (Figure 5.9) (Spirn, 1984; Hough, 2004; Erell *et al.*, 2011). At ground level, wind is reduced upwind of the low building. Downwind of the first building, windspeed is increased (rather than decreased as for an isolated building), due to additional downward airflow from the face of the tall building behind. A still higher windspeed is present where air is forced around the sides of both buildings. This accelerated air forms a "side streak" that progressively decreases in windspeed downwind.

A ground-level opening in the center of the tall building provides some porosity. The opening has the highest ground-level windspeed, as air is pushed and squeezed through. A strong downward turbulent flow from air passing over the tall building prevents the air through the opening from forming a central long downwind streak. In addition, vertical vortices separate the central flow from the side streaks.

Skimming airflow may play an additional role in ventilating urban ground-level conditions. Consider the so-called "prairie dog effect" named for a burrowing mammal (*Cynomys*) in North American grasslands (Butler, 1981). Prairie dogs have two or more entrances to their underground homes. The upwind hole is at the top of a mound built by the animal, while the downwind entrance is closer to ground level. Streamline airflow across the grassland is forced upward and slightly accelerated by the mound, a process that also draws air upward from the burrow. Air enters the burrow system in the second hole where streamline airflow is slightly downward toward the ground. The prairie dogs are well adapted to ventilating their burrow system with fresh air in this way. Analogously, funnels are used on ships to ventilate lower levels, and wind-towers funnel air downward to ventilate many Middle Eastern buildings.

Urban trees appear repeatedly as key players in this chapter. It is useful to summarize their roles relative to air flow (also see Chapter 8) (McPherson *et al.*, 1994b). By evapo-transpiration "cooling," trees affect the breeze from the country and breeze from the park. Tree abundance on mountain slopes and paucity in the nearby city increases cool air drainage. Their cover on the land affects both onshore and offshore breezes. In the street canyon, trees reduce the along-street airflows, both streamline and vortex. Trees may shade the street and sidewalk, as well as the sunny-side, windward, and/or leeward wall, in all cases affecting radiation and upward airflow. Trees "cool" the street canyon with evapo-transpiration. They create turbulence from streamlines. Trees disrupt street-corner eddies. They may accelerate airflow beneath their canopies. Trees may accelerate or inhibit airflow and cooling along wall surfaces, with different implications for wall temperature in summer and winter (Gartland, 2008; Erell *et al.*, 2011). In essence, trees are a key to urban microclimate.

Air pollutants and effects

The air over natural land contains plenty of gases, aerosols, and particles, some of which are essential to life and its diversity. Consider the general composition of air for organisms (Ahrens, 1991): 78% nitrogen gas (N_2), 21% O_2, 0.039% CO_2, 1% argon, plus traces of many other gases such as neon, helium, and methane (CH_4). But water vapor may range from zero to 4%.

Lots of natural microbial activities add and remove chemicals, including CO_2 and NOx (nitrogen oxides), in significant amounts from the air. Volcanic activity adds SO_2 (sulfur dioxide) and PM (particulate matter, or simply particles). Sea spray adds NaCl (salt). Natural lightning-caused fires add CO_2, CO (carbon monoxide), and PM. Various trees and other plants produce hydrocarbons (HCs, a major type of volatile organic compounds or VOCs). Wind erosion in dry land contributes dust particles to the air, especially silt. Wetlands contribute CH_4. And plant evapo-transpiration contributes water vapor. These natural materials may be scarce or dense, but are not (unwanted) "pollutants."

Urban air contains these chemicals in addition to pollutants from human activities unrelated to urban areas. Livestock and rice paddies produce CH_4; pastureland and cropfields give off soil particles (PM); smelters, SO_2 and heavy metals; unpaved roads, particles (PM); paper mills, SO_2; refineries, HCs; and anthropogenic fires, CO_2, CO, and PM. Although mainly originating elsewhere, most of these airborne materials are transported by wind into urban areas, where they may be an important component of urban pollution.

Ten urban air pollutants

Ten air-pollutant types are most important in urban areas (Figure 5.10). The gases are CO_2, CO, SO_2, NOx, and HC. The particulates and aerosols are PM, HM (heavy metals), O_3 (ozone) smog, and CFL (chlorofluorocarbons). Toxic substances, the tenth type, are a heterogeneous group of gases and aerosols/particulates. Gases are lightweight and invisible, while particles are heavier and visible. Airborne particles and aerosols are often grouped together, because particles are "dry" and *aerosols* are particles in droplets of water.

Pollutant variability from city to city and place to place within an urban region is great, and pollutant levels sometimes rise or drop sharply and quickly. Furthermore, countless other pollutants mainly produced by industry and motor vehicles enrich urban air, usually at very low levels but sometimes at high levels (Benton-Short and Short, 2008). Radioactive isotopes from distant or local sources are in urban air normally at "low" levels (basically, no radioactivity is safe for living organisms).

Fossil fuels – gasoline, diesel, coal, oil, and natural gas – produce somewhat different polluting chemicals and are the primary indirect or underlying sources for urban air pollution (Figure 5.10). Biomass fuel, solid waste, surface materials, and manufactured products are also indirect underlying pollutant sources.

The direct urban-region sources produce extremely different pollutants. Motor vehicles and industries

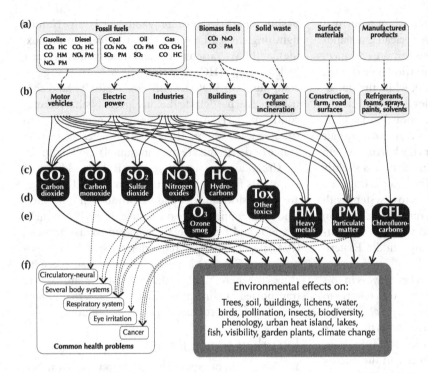

Figure 5.10. Ten key urban air pollutants and their primary sources. (c) Gases; (d) aerosols and gases; (e) particulates and aerosols. Ozone smog (O_3) is produced in the atmosphere from the combination of nitrogen oxides, hydrocarbons, and solar radiation. Based mainly on Hartshorn (1992), Moran and Morgan (1994), Puliafito et al. (1998), McNeill (2000), Ravetz (2000), Frumkin et al. (2004), Alberti (2008), Benton-Short and Short (2008).

produce by far the most number (seven each) of air pollutants (Figure 5.10). Some sources produce only two major pollutant types.

Particulate matter, PM, is produced by six of the seven urban-region sources. CO_2 emanates from five, and NOx from four. CO and CFL are each produced in quantity by a single source type.

The effects of urban air pollutants range from human health to ecology and the environment (Figure 5.10). The predominant health effects are to the respiratory system, and secondarily in causing cancer (Curtis et al., 2006). However, some of the cancers are in lungs, e.g., from very fine particles, so in effect damage to the respiratory system is the overwhelming health problem. For instance, Manchester, the so-called pollution capital of the UK, has particularly high rates of respiratory disease (Ravetz, 2000).

In brief, the major ecological and environmental effects of elevated pollutant levels are:

1. *CO_2*. Associated with anaerobic (no oxygen) conditions in soil, and therefore little root growth, few soil animals, and presence of anaerobic bacteria and decomposition. Also, a major greenhouse gas leading to global warming.
2. *CO*. Reduces the transport of O_2 in blood of vertebrates, leading to death.
3. *SO_2*. Damages leaf tissue, leading to death of plants. Lowers pH. Acid precipitation chemically erodes limestone, concrete, mortar between bricks, and some statues (Moran and Morgan, 1994).
4. *NOx*. Nitrogen dioxide (NO_2), the primary problem; nitrogen oxide (NO) not so; nitrous oxide (N_2O) mainly from biomass burning, including wood-fire cooking/heating and forest/savanna burning. NOx may lead to smog.
5. *HCs*. Hydrocarbons, or volatile organic compounds (VOCs), include many derivatives of petroleum, including polycyclic aromatic hydrocarbons (PAHs). Hydrocarbons may lead to smog.
6. *O_3 smog*. NOx and HCs in the presence of sunlight, heat, and O_2 produce smog, with the rate apparently increasing with temperature. Dominated by ozone (O_3), smog may contain PAN (peroxyacetyl nitrate), formaldehyde, ketones, and other organic constituents (Moran and Morgan, 1994). Corrosive to metals and rubber.
7. *Tox*. Toxic or hazardous substances include organic compounds, especially benzene, formaldehyde, chloroform, methyl chloride, polychlorinated biphenols (PCBs), dioxins, pesticides (e.g., DDT), and compounds containing cadmium (Cd)

(Benton-Short and Short, 2008). Other toxics are inorganic, including mercury, lead, and arsenic. The worst industrial air pollution case for humans was in 1984 when a cloud of methyl isocyanate gas from a major industry in the city of Bhopal, India spread across the urban area killing 4000 people and injuring some 200 000.

8. *HM*. Heavy metals inhibit various microbes and decomposition processes, as well as root growth. Also they reduce aquatic biodiversity. Lead (Pb) remains in gasoline for vehicles in many developing nations, where urban air, e.g., in Cairo, Lagos, and Cape Town, is polluted by Pb, which damages the nerve system (Benton-Short and Short, 2008). After the removal of lead from gasoline in Athens, Pb dropped sharply in the air, and in urban parks gradually moved down into the mineral soil A-horizon (Michopoulos et al., 2005).

9. *PM*. Particulate matter (or simply particles) damages leaves and plant growth, and may smother lichens and mosses. Airborne PM reflects incoming solar and sky radiation outward. It increases the turbidity of water bodies, inhibiting clear-water fish. Large particles >10 μm in diameter (PM_{10}) include pollen, fly-ash from combustion furnaces, and dust from soil erosion, cement, and coal (Marsh, 2010). $PM_{2.5}$ (>2.5 μm) includes smaller coal dust and fly-ash particles, and is particularly important in damaging the respiratory system. Very fine particles <1 μm include photochemical smog, tobacco smoke, auto exhaust, oil smoke, and sea-spray salt. Plants may act as partial filters of PM, as particles accumulate on leaf surfaces and are later washed to the ground by rain. Vertical vegetation layers, a diversity of species, and certain leaf attributes increase the filtering effect (Spirn, 1984).

10. *CFL*. Chlorofluorocarbons breakdown O_3 molecules in the stratosphere (Figure 5.2) (a cause of the "ozone hole"), increasing the penetration of UV radiation, which damages living organisms (Moran and Morgan, 1994).

Spatial distribution of pollutants

Broad plumes of materials reaching the stratosphere, such as radioisotopes from nuclear explosions, are carried around the globe and slowly settle downward. In contrast, plumes of materials moving in the troposphere below it are often carried to the ground in precipitation events. Such plumes, however, are commonly more extensive than weather fronts and may be rapidly replenished by pollution sources after rain. Thus, a huge layer of elevated NO_2, mainly related to vehicular traffic, covers much of East Asia and surroundings, bathing Beijing, Shanghai, Hong Kong, Seoul, and even Osaka and Tokyo (Parlow, 2011). A similar cloud of particulate matter originating from soil erosion in dry pastureland and cropland spreads far eastward into the air of the same cities.

Many of the severely and long-polluted cities of history owe much of their murkiness to coal combustion (McNeill, 2000). London in the 17th–20th centuries, Pittsburgh (USA) in the 19th–20th centuries, the Ruhr Region cities (Germany) in the 19th–20th centuries, Calcutta (Kolkata) in the 20th–21st centuries, and today Baotou (China's blast-furnace center) are notable. Mexico City, Los Angeles, and Athens suffer from severe ozone smog, especially in summer (Benton-Short and Short, 2008). Yet air quality may dramatically improve, as in Ankara, Pittsburgh, and Cubatao (Brazil).

Some cities have air dominated by a single pollutant type, such as NOx in New York and Tokyo, or PM in Delhi (Laakso et al., 2006; Benton-Short and Short, 2008; Kuttler, 2008). On the other hand, several pollutant types dominate some cities, such as CO, NOx, and O_3 in Buenos Aires, or SO_2, NOx, PM, and O_3 in Beijing (Figure 5.11). Mexico City's air is high in SO_2, CO, O_3, NOx, and PM. Some major cities have low or relatively low levels of all major pollutant types, including Auckland, Montreal, Vancouver, Vienna, Berlin, and Edinburgh. San Francisco, constantly swept by ocean winds, has the cleanest major city air in the USA (Mayer and Provo, 2004), quite a contrast with the air of its neighbor, Los Angeles.

Differences in urban versus suburban areas are usually visible for particulates and present for gases. For instance, overall in the USA, airborne particulate matter in suburban areas may be 60% lower than in urban air, NOx 99% lower, and SO_2 17% lower, though of course this varies by city and by radius measured (Marsh, 2010). The upwind radius typically shows a sharper difference than in the downwind direction. Pollution sources are more likely to be concentrated on the downwind side of a city, and in addition, polluted urban air is blown downwind (Madronich, 2006; Moran and Morgan, 1994). In dry air, heavier particles tend to settle out nearby, while lighter pollutants are carried further downwind.

Figure 5.11. Visibility commonly limited to about three blocks due to diverse air pollutants. Street trees including *Sophora japonica* apparently pollution-resistant; school children (foreground) doing exercises. Beijing. R. Forman photo.

Urban industrial areas typically are sources of condensation nuclei that spur cloud development and potential precipitation, a case of air pollution affecting weather (Moran and Morgan, 1994). Precipitation may be greater within a city (e.g., Detroit, USA) or greater downwind of it (Chicago, St. Louis) (Marsh, 2010).

Within an urban area, however, pollutant levels correspond to land uses at finer scales. For example, in Hyderabad (South-Central India) particulate matter is a primary air pollutant (Sekhar, 1998). Over a 4-month measurement period, the rate of dust fall was consistently highest in commercial areas, intermediate in industrial zones, and lowest in residential zones. The author hypothesizes that dust in commercial areas largely comes from heavy traffic using poorly maintained road surfaces. Industrial operations produce dust in the industrial areas, and construction activity mainly contributes dust in residential zones.

Particulate matter, being heavier than gas, tends to be densest in small airdomes or plumes around local sources. In Rio de Janeiro, a concentration of industry is the major PM source, and a plume across the city occurs during certain wind conditions (Habbel *et al.*, 1998). Unpaved roads and uncontrolled open burning of waste are dispersed secondary PM sources. In Leipzig (Germany), industries, rail yards, and train stations with associated PM concentrations were at somewhat dispersed locations (Geiger, 1965).

In the USA as a whole, cars and trucks annually contribute 77% of CO, 56% of NOx, 47% VOCs (HCs), 31% air toxins, 30% CO_2, 25% directly emitted PM_{10} (28% $PM_{2.5}$), and 7% SO_2 (Frumkin *et al.*, 2004). Of course traffic varies enormously throughout the day, with commuters coming and going, as well as through the week, as weekday and weekend travel differ. The role of vehicle congestion is emphasized by the relationship of pollutants produced at different driving speeds. For CO and VOCs, by far the greatest emission rate occurs with vehicle movement below 24 km/h (15 mi/h), and the curve rises again above 60 mi/h. In contrast, NOx emission is relatively low at lower speeds, and rises above about 80 km/h (50 mi/h). Within residential areas, vehicle emissions per housing unit increase with lower housing density. Lower connectivity of streets also results in more emissions. In effect people drive and pollute more in sprawl areas.

Intersections of busy streets also have high PM concentrations, due to the concentration of vehicular exhaust and the continual lifting of particles from street surfaces by moving vehicles and people. In Kuala Lumpur, PM in a street intersection peaked during morning and afternoon commuter-traffic times (Carpenter, 1983). Street dust contains a long list of chemical elements. For example in Urbana (Illinois, USA) the primary elements measured were manganese, zinc, barium, nickel, strontium, and chromium. Yet arsenic, uranium, calcium, cadmium, cesium, potassium, sodium, silver, lead and mercury were also common. Other elements in the street dust were at lower levels (Spirn, 1984). Noses and lungs on street corners are bathed in such a mix.

Air quality inside vehicles and buildings not surprisingly can be much worse than in outside urban air, due to sources of concentrated pollutants, limited volume, and limited ventilation (Spengler and Chen, 2000; Frumkin *et al.*, 2004; Benton-Short and Short, 2008). Cooking and heating with wood, animal dung, or kerosene give off considerable PM and CO, the latter being especially dangerous in an enclosed space. Similarly, in cars and other vehicles the range of vehicle-related pollutants (Figure 5.10), including CO, may accumulate to high levels. Respiratory disease is particularly prevalent with polluted inside air.

Some air pollutants, such as NOx and O_3, increase with temperature and are especially problematic in intense urban heat islands (Sarrata et al., 2005; Kuttler, 2008). In Taiching (Taiwan) with a 0 to 4°C urban-heat-island intensity, CO_2 concentration did not change with temperature, CO increased linearly, SO_2 showed a steep increase between 2 and 3°C, PM (both PM_{10} and $PM_{2.5}$) steeply increased between 3 and 4°C, and NO_2 sharply increased between 4 and 5°C (Lai and Chengb, 2008). Thus, as air temperature rises, different pollutants are sequentially added to the problematic mix.

Finally, cities in arid land are special cases for several reasons. Usually regional air temperature is high, humidity is low, and considerable PM arrives from upwind soil erosion. Silt is the predominant incoming particulate matter, since heavier sand grains drop out quickly and lighter clay particles are carried higher and long distances (Knox et al., 2000). Most such urban areas, such as Phoenix and Albuquerque (USA), developed in valleys where water was available (Parlow, 2011). Ventilation is somewhat limited, so pollution accumulates in the city. Also, many people live and drive further out. With abundant solar radiation, plus NOx and HCs from vehicles, this is a formula for ozone smog development in the city's valley.

Local construction sites provide lots of PM to the arid air. Continuous dust input, with a paucity of rain, covers and damages leaves of plants. Trees from moister climates are often planted and irrigated across arid urban areas. Pollen levels rise, causing respiratory problems. In the past half century, air conditioning has become widespread, and chlorofluorocarbons in refrigerants have markedly increased the damage to our protecting stratospheric ozone layer (fortunately a recent drop in use of these chemicals has rapidly reduced the problem). Wind also carries urban-generated pollution to areas out of the city, such as pollutants from Los Angeles covering the downwind Mojave Desert.

Heavily polluted Cairo air highlights further dimensions of arid cities (Benton-Short and Short, 2008; Parlow, 2011). With vehicles using leaded fuel, lead levels are high in Cairo's air. The burning of wastes producing black-particle soot, diesel traffic exhaust, and unpaved or poorly maintained road surfaces, all in the urban region, seem to dominate the PM present. The smokiness of the air also results from biomass burning for cooking and heating.

Global warming and urban air

Is global climate warming a major cause of the urban heat island, or is urban heat a major cause of global warming? Neither of the above seems to be correct (Peterson, 2003; Parker, 2004; Parmesan, 2006; IPCC, 2007; Alberti, 2008; Alcoforado and Andrade, 2008; Rosenzweig et al., 2011).

Certainly urban areas, as contrasted with the heat directly given off from them, affect global climate. For example, perhaps about 85% of the human-produced CO_2, CFC, and ozone (smog in troposphere) is generated in or near urban areas (Oke, 1997b). Also considerable particulate matter and aerosols are produced by urban areas. CO_2 is a prime greenhouse gas leading to higher global temperature. Meanwhile PM and aerosols block some incoming net radiation, effectively cooling the urban air.

Over a century or so, global atmospheric CO_2 has increased several percent and temperature has risen a few degrees centigrade. Urban air temperature typically correlates well with area of urbanization, or distance inward from edge of urbanized area. Thus, as many cities have expanded outward over a few decades, urban temperature has risen at a considerably faster rate than global warming (Alcoforado and Andrade, 2008). Urban areas are expected to greatly expand in the upcoming few decades, promising really hot cities.

Urban areas cover < 1% of the global land surface, and directional heat plumes only extend kilometers or tens of kilometers downwind of cities (Alcoforado and Andrade, 2008). Thus, direct urban heat is considered to be essentially a negligible factor in producing global warming. [Note that considerable uncertainty exists in diverse global-climate models for the regional scale (especially in incorporating city effects) (IPCC, 2007; Alcoforado and Andrade, 2008; Oleson et al., 2008; Rosenzweig et al., 2011). Many other conditions may have greater effects on model predictions, such as anticyclones, urban heat dissipation with vertical instability, changes in heating/air conditioning, and altered snowmelt/cover.]

Many indirect effects may be important in evaluating the urban and global heat relationship. Global warming seems to heat the non-urban surroundings more than the urban area (Alcoforado and Andrade, 2008; Erell et al., 2011). That decreases, rather than increases, the heat intensity difference and the urban heat island.

Decreasing the temperature difference between city and surroundings also means less breeze from the country, and less cool air drainage from adjoining hills/mountains. Consequently, at night less incoming cool air means higher urban temperature. Similarly, increasing ocean temperature would reduce, and alter the timing of, sea breezes and offshore breezes. This may lead to slightly higher temperature in coastal urban areas. More extreme weather events, such as severe droughts, floods, freezes, and cyclones, predicted by some global models could produce both temporary cooling and warming for cities.

In short, cities are funnels of concentrated greenhouse gases, heat, and other pollutants rising into the global atmosphere. The total amount of urban CO_2 is highly significant compared with other sources. Yet the urban heat directly given off is limited or minuscule compared with heat from the Sun, land, and oceans (IPCC, 2007; Erell *et al.*, 2011). Thus, urban CO_2, not urban heat, increases global warming (Alcoforado and Andrade, 2008). Cities worldwide are expanding, so the picture could change.

Meanwhile, global warming seems negligible in raising the urban heat intensity. Overwhelmingly, energy from the Sun absorbed, stored, and emitted as heat by extensive urban impervious surfaces creates the urban heat-island intensity. In essence, heat moves both from city to "globe" and globe to city, but the relative amounts are tiny.

Part II: Ecological features

Chapter 6
Urban water systems

We never miss the water till the well runs dry.

James Kelly, A Complete Collection of Scottish Proverbs, *1721*

Tortuous, fissured, unpaved, crackling, interrupted by quagmires, broken by fantastic bends, rising and falling illogically, fetid, savage, wild, submerged in obscurity, with scars on its pavements and gashes on its walls, appalling, seen retrospectively, such was the ancient sewer of Paris. Ramifications in every direction, crossings of trenches, branchings, multiple forkings, stars as if in mines, cul-de-sacs, arches covered with saltpeter, foul cesspools, a cabby ooze on the walls, drops falling from the ceiling, darkness; nothing equaled the horror of this old voiding crypt, Babylonian digestive apparatus, cavern, grave, gulf pierced with streets, titanic molehill in which the mind seems to see prowling through shadows, in the excrement that has been splendor, that enormous blind mole, the past. … Today the sewer is neat, cold, straight, correct. … Do not trust in it too much, however. Miasmas still inhabit it.

Victor Hugo, Les Misérables, *1862*

No place even resembles the water flows of an urban area. We have straight, fast predictable flows over hard surfaces and through pipes. Streams are channelized, even disappearing into pipes and "magically" reappearing. Rain brings down the rich chemical mix arcing over us. Stormwater quickly washes the city clean, carrying away dust from streets and sidewalks and an even richer cornucopia of chemicals from buildings and other surfaces. A water-supply pipe squirts a small addition of hyper-clean water (except for chlorine, fluorine, and other chemicals) into the system. Another pipe or ditch system carries off and partially cleans the eternal concentration of human wastes so they don't build up in the city. River water bulging with sediment and countless pollutants often squeezes through a city, sometimes with roaring floods, which clean out the accumulated toxics from industrial sites and carry off inappropriately located urban structures.

In contrast, out in natural land water generally flows more slowly, less predictably, and in crooked routes full of objects creating friction. The chemical mix washed down by rain and carried over surfaces is typically rather depauperate, often dominated by inorganic macro-nutrients. No pipe squirts in hyper-clean water, and no pipes carry off tubfuls of human wastewater. Clear water fills most rivers, while streams appear curvy, heterogeneous, often nutrient-poor, and seem to flow unendingly.

Urban areas also display an array of special water structures absent or scarce in both natural and agricultural lands. Water fountains and fire hydrants are familiar. Yet a closer look may reveal a green roof, rain garden, stormwater basin, vegetated swale, grey-water, aquaculture, serpentine cleaning-channel, riparian pond, daylighted stream, and biofilter. Engineering predominates, but ecological engineering appears (Ma, 1985; Mitsch and Jorgensen, 2004; van Bohemen, 2005).

On the subject of water, two quantity-and-quality questions always arise. How much? How clean or polluted?

Most wetlands in the land around today's cities were long ago drained or filled by farmers, so today's remaining small wetlands tend to be scarce habitats with uncommon species. In dry climates during much of the year, stream and river water may be invisible, as it slowly flows in the oft-sandy sediment. In wet climes, streams and rivers may be considered as hydraulic conduits to drain the land, and thus stream-corridor trees, which fall and disrupt flows, are discouraged.

Flooding greatly depends on the land upstream/ upriver of the city, where land-use change may cause so-called 100-year floods to occur twice a decade. Alternatively, upstream water diversions may leave

Urban water systems

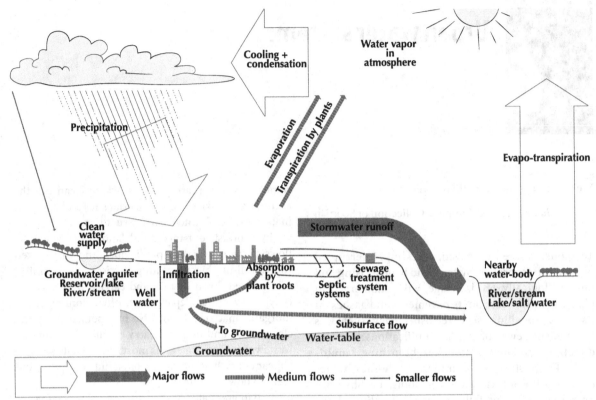

Figure 6.1 Urban water flows in the global hydrologic cycle. Precipitation on an urban area results overall in both major and medium water flows, while clean water piped into an urban area produces smaller flows. In addition to climate, the amounts of water flow vary according to percent impervious-surface cover, subsurface human modifications, and effectiveness of urban-infrastructure functioning. Distinctive urban water flows contrast with those in natural and agricultural lands.

barely any visible flow in an urban river. Recent dramatic flooding of coastal cities worldwide emphasizes that walls of water also come from the sea in hurricanes/cyclones, monsoonal storms, and tsunamis. Rising sea level promises worse things ahead for coastal metro areas.

Meanwhile clean freshwater supply is running short worldwide, creating scarcity and raising costs for cities, which then ponder environmentally damaging alternatives (Ma, 2004; Ghassemi, 2006). Ironically, these cities seem oblivious to the abundant water flowing through them, an obvious water source when kept clean, or cleaned up. Even migratory river-and-sea fish would welcome renewed passage through a city from cleaning up river pollution.

Clearly water runs through the core of urban ecology. In this chapter we explore the subject in five big steps: (1) urban water flows and cycling; (2) groundwater; (3) clean water supply; (4) sewage and septic wastewater; and (5) stormwater and pollutants.

Urban flows in the water cycle

Global water is distributed in five major places: atmosphere (mainly as water vapor); land surface (lakes, rivers, vegetation, etc.); subsurface (groundwater and shallow subsurface water); ice; and sea. Of the liquid water on land, <2% is in lakes and rivers while 97% is subsurface water, overwhelmingly groundwater (Gibert *et al.*, 1994a).

As part of the "global water cycle," urban areas contain a number of relatively distinctive flows (Hall, 1984; Korhnak and Vince, 2005; Todd and Mays, 2005; Shanahan and Jacobs, 2007; Welty, 2009). Hydrologic (or water) budgets of inflows and outflows are usefully determined for a whole urban area, or any of its components (see equations, Appendix B). A very small amount of urban water is cycled or recycled in the urban area, normally using pumps.

In the water or hydrologic cycle, atmospheric water vapor cools and falls as precipitation (mainly rain in the warm air of cities) (Carpenter, 1983; Alberti, 2008). Some falling water is "intercepted" by buildings, roads, and soil, where it directly evaporates back to the atmosphere (Figure 6.1). Some water infiltrates through cracks in the hard-surface cover and into the soil of greenspaces and smaller vegetated spots. The infiltrated water may be absorbed by roots and pumped upward by plants in transpiration to the atmosphere. *Evapo-transpiration*, the combination of *evaporation* from non-living surfaces and *transpiration* from plants, in urban areas typically correlates with the percent of vegetation cover and its extensive leaf surface (Berthier *et al.*, 2006).

Alternatively, water infiltrating into the soil may move by subsurface flow nearly horizontally to a stream, river, or other water body. Any remaining infiltrated water flows further downward into the groundwater. However, most of the water falling on urban hard surfaces, especially in heavy rains, quickly flows as "surface runoff" across the surfaces and into stormwater drainage systems composed of pipes and/or ditches. This stormwater runoff mainly empties into water bodies, sometimes causing floods. Evapo-transpiration from land and water body then transfers water back to the atmosphere as water vapor.

In addition, usually a constant flow of clean drinkable freshwater is piped into urban areas (Figure 6.1). Except in arid cities, this addition is typically tiny compared with the amount of incoming precipitation. An extensive, usually old and leaky pipe system carries the clean water to all users (Galloway, 2011). Urban residents then rapidly convert much of the piped clean-water to wastewater flushed down drains and toilets. Some of the clean water is used for irrigating parks and lawns and for other uses. In many outer urban areas, the wastewater commonly passes through septic systems or cesspools (e.g., holes filled with gravel) directly into the ground, where the liquid becomes subsurface flow. With denser populations using sewer systems, instead the wastewater is rapidly piped to a sewage treatment facility. From it cleaner water is channeled to a nearby water body.

The extensive pumping out of groundwater from the sandy soil of much-urbanized Long Island, New York illustrates additional dimensions of urban water (Todd and Mays, 2005). Over time hard surfaces spread, and stormwater drainage systems carried water coastward. Sewage systems expanded, carrying wastewater to sewage treatment facilities, and the water coastward. Wells reaching the groundwater had to be deeper. Stream flows decreased. Wetlands dried out. Clean groundwater supplies became scarcer. A familiar story.

Overall, the most distinctive features of the urban water flows are: (1) the extensive impervious/hard surface; (2) the stormwater drainage system; and (3) alternative targets for the supplemental piped-in water (Figure 6.1). All three features accelerate water flow to local water bodies. The hard surface cover also evaporates considerable water and limits plant evapo-transpiration. Furthermore, hard surfaces add a range of pollutants, including heat, to both the surface-runoff water and the soil-infiltrating water. The piped-in water carries off wastewater pollutants and pathogens. Natural water cycles, i.e., those without these core urban attributes, have noticeably less evaporation and surface runoff, but more infiltration, plant evapo-transpiration, subsurface flow, and groundwater recharge. The distinctive urban characteristics permit a concentration of people, buildings and roads to thrive with minimal flooding and without drowning in pollutants.

Groundwater

Groundwater normally refers to water in the water-saturated zone (phreatic zone) in the ground (Dunne and Leopold, 1978; Todd and Mays, 2005; Shanahan, 2009). The upper surface of this zone is the *water-table* (Figure 6.1). Most of the time, a relatively small amount of water is present in the aerated soil zone between ground surface and water-table. Nonetheless, this aerated zone holds the bulk of the roots, microbes and soil animals, and consequently is extremely important for water absorption and growth by plants.

In natural land, groundwater and surface water, such as streams and ponds, are normally connected as a single resource (Winter *et al.*, 1998). We see the top of the resource that emerges, iceberg-like, above ground level. However, in urban areas the surface waters and groundwater are often "disconnected." This results from extensive pumping, which lowers the water-table, leaving a non-saturated soil zone between groundwater and stream, river, or lake.

Now let's delve into the invisible. First, we explore groundwater flows, heat, and pollutants. Then groundwater habitats and animals are highlighted.

Flows, pollutants, heat

Three-quarters of Europeans drink groundwater, while half of the residents in the USA do (Gibert *et al.*, 1994a; Margat, 1994). For urban areas of concentrated people this clean groundwater comes mainly from *aquifers*.

Normally these are large sandy/gravelly underground volumes or porous rocks such as sandstone that store water. Envision an underground lake full of sand.

We use aquifer groundwater supplies for three major objectives: (1) urban and rural drinking water; (2) industrial uses; and (3) agricultural irrigation. Of the three, drinking water is the major use in many European nations including Italy, Austria, Denmark, Hungary, and Russia. Industrial uses predominate in Japan, South Korea, The Netherlands, and Norway. Agricultural use of aquifer water is predominant in India, China, Australia, Saudi Arabia, Greece, Spain, South Africa, Argentina, Mexico, and parts of the USA.

Groundwater flows

Today's cities commonly have higher land in the surroundings, so both surface water and groundwater generally flow into, and then out of, the urban area. With hills and valleys, the land surface varies considerably. In contrast, the water-table surface beneath normally only undulates gently (Winter *et al.*, 1998; Marsh, 2005). Thus, the water-table often reaches the ground surface in valleys and depressions, where flooding is most likely.

To visualize water flows more clearly, consider a stream flowing between two low hills with a typical soil mainly composed of small particles, i.e., silt and/or clay. The water-table in the hills is at a higher elevation than the surface of the valley stream (Winter *et al.*, 1998; Marsh, 2005). Rain falls on the hills, where some water infiltrates down to the groundwater and then flows horizontally downward to the stream. In this way the stream is "recharged" with water and continues flowing. But suppose the soil in the hills is predominantly sandy, with large particles and considerable pore space through which water readily drains. In this case the water-table in the hills typically is lower than the stream level. As the stream flows over the sandy area, the stream loses water, which flows downward into the sandy soil, "recharging" the groundwater. Most cities on clay soil began with groundwater near the surface and frequent floods, whereas those on sand or other porous material started with a relatively low water-table.

Irrespective, the pumping of groundwater by wells in and around cities, especially for industrial uses, commonly lowers the water-table (Figure 6.1). The amount of groundwater pumped from a porous rock or unconsolidated soil material depends on the pore space present (Chapter 4), plus the ability to pull water through it (i.e., hydraulic permeability or porosity, plus conductivity). For example, groundwater can be rapidly pumped from gravel, clean sand, (karst) limestone, and permeable (igneous) basalt (Todd and Mays, 2005; Marsh, 2005). But water is only slowly extracted from low-porosity shale, marine clay, and unfractured igneous and metamorphic rock.

In the London Region, the distribution of porous rock and soil causes groundwater from the northwest, southwest, and south to flow (according to Darcy's law) (see equations, Appendix B) toward the center of London (Trafalgar Square) (Shanahan, 2009). Yet, unlike pre-London (pre-Roman) times with wetlands in the area, today the city's water-table is mainly >10 m (30 ft) down, and in some locations >40 m deep. Groundwater pumping caused that.

Lowering the water-table by groundwater pumping may lead to *land subsidence*, such as the ground-level drop of perhaps a meter for Houston (Texas), and several meters (15–20 ft) in central Mexico City (IPCC, 2007; Shanahan, 2009). In the latter case, during a 7-yr period the ground surface dropped 1 m (3 ft) while the water-table dropped an average of 1 m each year (World Resources Institute, 1996). Naturally, urban land subsidence causes problems for buildings (especially on clay soils), and may lead to deeper and larger wells for further water extraction (Shanahan, 2009). Even in Las Vegas (Nevada, USA), where surrounding mountains provide considerable groundwater, pumping has lowered the water-table and dried out springs (Todd and Mays, 2005). Moreover a near-surface underground reservoir of polluted stormwater has formed. Cities in moist limestone areas typically have "sinkholes," where a rounded area of land, often with some buildings, drops meters rather quickly, as in Orlando and Tampa (Florida) (Maire and Pomel, 1994; Mangin, 1994; Marsh, 2005).

The withdrawal of groundwater by pumping (Figure 6.2) for urban uses worldwide is minor compared with that for agriculture. Europe and North America though use a significant proportion of groundwater pumping for clean-water supply and industry (Todd and Mays, 2005; UN-Habitat, 2005). Much of the groundwater pumped for urban use is returned to local water bodies, but commonly in polluted form due to industrial processes, surface runoff, or human wastewater. Thus, urban groundwater pumping (withdrawal) especially leads to degradation of aquatic ecosystems and fish populations surrounding and downriver of the urban area.

Figure 6.2. Community water supply pumped from groundwater. Steel pillars reduce problems with elephants that also wish to use the water. Peri-urban area of Livingston, Zambia. R. Forman photo.

A lowered water table may dry out wooden-foundation pilings or pillars under buildings that have long been protected from decay by being immersed in essentially anaerobic (oxygen-free) groundwater. Pumping lowered the water-table under important buildings with wooden foundations along St. James Avenue in Boston (Shanahan, 2009). The wood of the ancient trees was rapidly oxidized, rotted by microbes, and chewed by invertebrates, requiring expensive replacement of building foundations. The pattern is highlighted in the following case of a 14-centuries-old city built on essentially oxygen-free mud.

> Wherever you walk in Venice, not far beneath your overheated feet is one of over 22 million wooden stakes, the majority of which are as sound as the day they were driven into the soft silts of the lagoon.
>
> (David Bellamy, 1976, cited by Hough, 2004)

Yet urban groundwater can also rise. Sharply reducing industry and its pumping, or abandoning old city wells, may cause the city water-table to rise (Foster *et al.*, 1998; Shanahan, 2009). Thus, since 1980 the water-table in the center of London has been rising. This results in ponding, sewer malfunction, concrete and masonry degradation, and flooded underground transportation. An abundance of leaky pipes adds to groundwater rise. If a reduction in groundwater pumping in a city is accompanied by increased pumping in exurban/peri-urban areas, the polluted city groundwater may then flow outward to the exurban areas.

Furthermore, relatively water-tight structures such as underground garages and subway stations projecting down into the groundwater have buoyancy, like the lower portion of a ship. Rising groundwater puts upward pressure on the structures, which may crack. Such problems are the harbingers of sea-level rise effects on coastal cities.

Urban groundwater may also rise from the extensive removal of exurban woodland, either upslope or downslope of a city (Thaitakoo *et al.*, 2013). This removes the evapo-transpiration pumping-power of trees, leaving more soil water, which raises the water-table. More groundwater from upslope flows toward the city, while a rising water-table downslope on a plain may decrease groundwater flow from the city.

Groundwater around coastal cities has two additional distinctive flows. First, the groundwater flowing from the higher land to the lower sea normally reaches the ground surface in a narrow *groundwater emergence* zone, commonly within 100–200 m of the coastline or shoreline (Todd and Mays, 2005; Forman, 2010b). On the landward side of the coastline the freshwater rises to the surface in low areas, particularly floodplains, deltas, swamps and marshes. Urban areas on these locations may have ample freshwater for pumping, but are especially subject to flooding, an acute problem with sea-level rise. On the seaward side of the coastline, the fresh groundwater flows directly through the soil or rock to enter the sea underwater, just offshore. If not severely polluted, this near-shore (littoral) zone with arriving fresh groundwater is especially rich biologically, in some areas characterized by mangroves, sea-grass, or coral reef.

The second distinctive groundwater flow of coastal cities is *saltwater intrusion* (Acebillo and Folch, 2000; Todd and Mays, 2005). Excessive urban pumping, even in this zone where groundwater normally rises toward the ground surface, can lower the water-table. With less pressure downward due to less groundwater, the deeper seawater seeps or intrudes landward into the ground. Since freshwater is lighter than saltwater, the saltwater intrusion extends inland sometimes hundreds of meters or more, with freshwater on top. In this case wells are commonly contaminated with salt, and further or deeper pumping simply causes greater saltwater intrusion.

Although groundwater levels are normally quite low under cities due to pumping, even in desert cities flooding may occur. The commonest cause in arid land is the "flash flood" from heavy rains, whereby surface runoff over the land far exceeds water infiltration, and sheets of water wash down slopes into the city. A more specialized case is where a "hardpan" (somewhat impermeable layer of soil particles partially cemented together) has formed usually from previous land use (Shanahan and Jacobs, 2007; Shanahan, 2009). Thus, as in Kuwait City, rainwater plus water from urban greenspace irrigation, leakage from pipes, and other sources can only infiltrate downward very slowly (Al-Rashid and Sherif, 2001). Therefore, the rainwater simply puddles or floods.

Groundwater pollution

Cities are epicenters of diverse groundwater contaminants. Consider the major sources and the pollutants (Fetter, 1999; Sharma and Reddy, 2004; Marsh, 2005; Todd and Mays, 2005):

1. *Industrial wastes.* A wide array includes heavy metals, organic compounds, petroleum products, and radioactive materials. Groundwater in contaminated-soil brownfields reflects this array.
2. *Urban stormwater.* Transportation is a major source and, though most stormwater goes to streams and other water bodies, some contaminants infiltrate into the soil. Especially in sandy soils, groundwater receives heavy metals, organic compounds, petroleum residues, nitrates, and road salt.
3. *Human wastewater.* Septic systems, cesspools, and outhouses directly incorporate nitrogen, phosphorus, sodium, and chlorinated organic compounds into groundwater. Functioning wastewater sewer systems reduce the input of these chemicals to groundwater, but overflows (CSOs) and leaks pollute the groundwater.
4. *Solid-waste dumps.* Numerous chemicals such as solid and hazardous wastes, as well as methane and benzene from residential garbage, reach groundwater even in dumps (tips, landfills) meeting government regulations. Small buried dumps from former use are often abundant next to urban water bodies.
5. *Construction fill.* The pervasive use of fill from varied sources, including building rubble and mining debris, creates a vast, highly heterogeneous complex of chemical sources enriching urban groundwater. The constituents of roads and roadbeds are illustrative (Forman *et al.*, 2003).
6. *Farmlands.* Perhaps least important in urban areas, nitrogen, phosphorus, and pesticides may reach groundwater, though legacies of former agriculture persist.
7. *Spills and leakage.* Numerous chemicals from roads, railways, industrial sites, and other sources leak from underground storage tanks, pipelines, and chemical storage sites. Also, residential areas provide paint, cleaning compounds, oil, and gasoline to groundwater. At night certain trucks illegally pour chemicals or wastewater into stormwater drains.

A waste site such as a dump or leaking tank is a "point source," which creates an elongated chemical "plume" in the groundwater pointing in the direction of groundwater flow. Since, except in limestone or karst areas, groundwater moves very slowly, the groundwater remains polluted for a long time. Consequently, a small amount of pollution may contaminate a large area.

Normally groundwater can only clean itself slowly. Cleaning basically depends on four processes (Todd and Mays, 2005): filtration of the chemicals, adsorption onto soil particles, chemical breakdown, and dilution. Since microbes and oxygen are usually rather scarce in groundwater, microbial decomposition, an important process in cleaning surface water, is overall of limited importance in cleaning.

However, most of the pollutant sources are dispersed "non-point sources," so the urban groundwater as a whole is polluted. For example, approximately 324 leaking underground storage tank sites are known in the 47 km^2 (18 mi^2) area around a small Michigan city (Genesee County, USA) (Marsh, 2005), yet many more are unrecorded. The groundwater is doubtless a persistent "witches' brew" with gradually changing mixtures from location to location. That doubtless characterizes the groundwater of most cities (Todd and Mays, 2005), and illustrates why city drinking-water supplies seldom come from beneath the city. Even industries are motivated to move outward where cleaner groundwater can be pumped (and then normally polluted).

Relatively few studies are available on the effects of pollutants on groundwater organisms (see section below) (Notenboom *et al.*, 1994). Nitrogen and phosphorus compounds are used by groundwater microbes. Inorganic and organic toxic compounds, heavy metals, and pesticides seem to alter groundwater faunas. Acute toxicity levels are known for very few groundwater species. Slight evidence exists for the "bio-accumulation" of heavy metals, that is, the increasing concentration of a pollutant in organisms higher in the food web. In some locations groundwater contains radioactivity, which may bioaccumulate to high levels in the ecosystem (Brenner *et al.*, 2006).

Organic compounds from human sewage, wastewater sludge, and other sources may be of particular concern, though the ecological effects of leaking wastewater sewage pipes on groundwater seem to be little studied (Barrett et al., 1999; Shanahan, 2009). For example, one sewage-polluted site had an abundance of surface-water organisms mixed with the groundwater ones (Gibert et al., 1994a). Typically, increased organic matter leads to more microbial decomposition, which lowers dissolved oxygen levels, but in this case groundwater organisms thrive in the near absence of oxygen. Overall, evidence suggests that the groundwater animals are less sensitive to pollutants, including toxic substances, than are comparable surface-water species.

As a postscript, a common approach recommended for stormwater pollution management is to channel stormwater to small ponds or basins, in part so that the pollutants can move into the soil rather than be carried onward directly to water bodies such as streams and lakes. However, a pond itself has considerable weight, often increased or maintained by rain, and thus the bottom water and pollutants tend to be pushed down into the ground, even into the groundwater. As noted above, groundwater moves very slowly and pollutants there, in the absence of oxygen, persist for long periods. Urban stormwater management remains a challenge for creative solutions.

Heat in groundwater

In many northern European cities with an aquifer within a few meters of the surface, considerable heat associated with the heat-island effect enters groundwater (see Chapter 5) (Allen et al., 2003; Taniguchi et al., 2005). Basically, solar (or net) radiation heats constructed objects such as roads and buildings, which then radiate heat to the air but also transfer heat by conduction into the ground. In summer, Winnipeg (Canada) groundwater temperature may be elevated as much as 5°C, with heat penetrating to 130 m depth (Ferguson and Woodbury, 2004). The groundwater temperature increase seems to be most related to heat loss from buildings, with the effect extending outward in the ground a few hundred meters (several hundred feet) from a heated structure.

Electric-power generating facilities scattered across an urban area are another source of heat into the soil and groundwater (Figure 6.3). Scattered smokestacks often mark their locations. Thus, the combustion of coal, oil, gas, organic solid waste, and other energy

Figure 6.3. Smokestacks indicating power facilities heating water that is piped underground to buildings. White roof in center reflects considerable solar radiation (high albedo). City in Jilin Province, China. R. Forman photo.

sources heats water, which in turn is piped underground to government buildings, skyscrapers, high rises, and other locations. Heat loss from the combustion process at the power-generator site and from the diverging pipeline network is then conducted into the urban soil and groundwater.

Moreover, the core of the Earth itself is a heat source for groundwater, soil, and urban areas. This *geothermal heat* maintains groundwater at a rather constant temperature, often about 16°C (62°F), and prevents winter freezing of underground water pipes. In areas with many hot springs and spas such as Iceland and Yellowstone Park, so-called "high-temperature" geothermal energy is readily available for human use. However, normally the high-temperature energy is at a considerable depth and is expensive (if not hazardous) to drill and use. Nonetheless, "low-temperature" geothermal energy, such as that keeping the soil somewhat warm, is widely available and relatively inexpensive to drill and use. Unlike high-temperature geothermal heat, the low-temperature heat is a modest heating source, though increasingly used for heating buildings and homes in some urban areas.

All three of the heat sources – solar-heated constructions on the surface, power-generation facilities and pipe systems, and geothermal energy – could be used for urban winter heating, and perhaps summer cooling. Ecologically, the heat combined with water provides great growing conditions for microbes, soil animals, and plant roots. The relatively constant warm groundwater conditions buffer the organisms from

Urban water systems

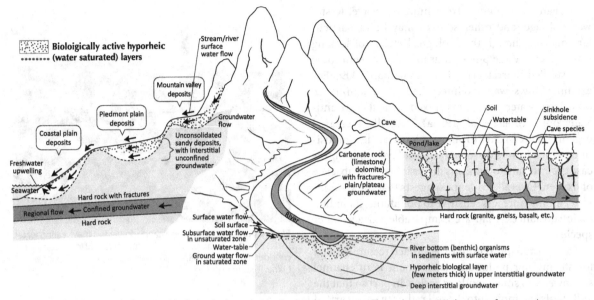

Figure 6.4. Water-saturated zones with distinctive hyporheic animals in groundwater. Three characteristic locations for groundwater animals: (left) surface-water sediments from mountains-to-sea; (center) river sediments in floodplains; (right) carbonate (karst) rock with fractures and groundwater. These rarely seen, and hence strange-looking, animals live with microbes in very-low-oxygen groundwater beneath, e.g., the oxygenated water in flowing rivers and river-bottom sediments in urban areas. Based on Creuze des Chatelliers et al. (1994), Gibert et al. (1994b), Maire and Pomel (1994), Mangin (1994).

environmental fluctuations and some human effects. Urban vegetation and ecosystems are enriched by organisms benefitting from the heat. Shut-downs of a power-generation facility, however, may cause considerable mortality of mostly underground organisms, followed by recycling of nutrients and growth of the next generation of organisms. Heated groundwater may also flow into local water bodies, with analogous enhanced and disruptive effects on aquatic ecosystems and fish.

Groundwater habitats and animals

Groundwater is essentially an invisible zone, unseen by ecologists or anyone else. Strange and amazing animals lurk, awaiting discovery. Some of us share the excitement of exploring caves and mines, and sense that the animals there are quite different from what we see on land. Unfortunately, as yet few ecologists study groundwater.

Groundwater animals and microbes seem to function, move, and vary at a wide range of time and space scales. These are estimated at 1 to 100 years, and 1 to 100 m³ (Gibert et al., 1994b). Aquifers ranging from 100 to 10 000 years in duration, and 100 to 100 000 m³ in volume, seem to be particularly important locations or habitats for groundwater organisms.

Habitat conditions

Groundwater microbes include bacteria and fungi, plus protozoans as primary consumers (Gounot, 1994). All the microorganisms are fed on by multicellular animals in groundwater, i.e., micro- and macro-invertebrates, which have a generalized diet.

Groundwater animals are concentrated in the upper zone of groundwater, especially in the sediment under and near flowing water (Gibert et al., 1994a). Here microbes and dead organic matter (detritus) filter downward from above and are food for the groundwater animals. This interaction zone between groundwater and surface water, the *hyporheic zone*, is prominent in two types of substrates, limestone-karst areas and porous rocks and sediments. Limestone-karst rock readily dissolves in water, and therefore, in moist climates, contains caves, sinkholes, numerous crevices, and rapidly flowing water as underground streams and rivers (Figure 6.4) (Gibert et al., 1994a). Groundwater organisms beneath karst cities have as yet received little study. So here we focus on habitats and organisms in the groundwater of porous rocks and sediments (Figure 6.4).

This porous substrate or medium (interstitial space) usually has elongated, rounded, varied-size, and intermixed soil particles or rock minerals (Zilliox, 1994).

Groundwater flow is slow with curvy routes. "Porosity" defines the percent volume of water in an underground reservoir or aquifer. "Permeability" describes the flow or flux through this underground reservoir, and "dispersivity" defines the dispersion of, e.g., organisms or pollutants, in the reservoir. Oxygen levels may vary significantly from spot to spot, though some evidence suggests that at ≥2 m (6 ft) depth below surface water, relatively constant "hypoxic" (very low oxygen) conditions are present, with <0.2 mg/l dissolved oxygen (Pospisil, 1994; Strayer, 1994).

Habitat heterogeneity is low, habitats are small, and habitat fluctuation is limited with high predictability (Gibert et al., 1994b). Eternal darkness prevails.

Animals and microbes

Overall the total population density of groundwater animals at a location is low, especially at greater depth (Gibert et al., 1994b). Species diversity (richness) is low. Seldom does one species dominate an area. Organisms are heterotrophic, predominantly feeding on dead organic matter. Productivity is low. Food webs are short and simple.

Species exhibit many adaptations for this distinctive environment (Gibert et al., 1994b). Organisms generally lack pigmentation and have limited development of ocular and other sensory organs. Appendages are typically long, numerous, and with highly developed mechanical and chemical receptors. Organisms such as Annelids (roundworms), Planaria (flatworms), and Crustacea are often longer and thinner than their surface-water relatives. Slow metabolic rates, infrequent reproduction, and long life spans characterize these groundwater organisms. The species must be adapted to surface-water flooding that temporarily disrupts subsurface species assemblages, as well as to particular water temperatures and the amount of incoming organic matter, but these are apparently little studied. Basically a tub of intriguing animals lies beneath our feet.

Excluding the surface-water species that feed on, or are fed on by, groundwater species, two types of groundwater organisms may be recognized (Gibert et al., 1994b): (1) those that occasionally use the surface-water environment ("stygophiles," including Plecoptera, nematodes, oligochaetes, mites, copepods, ostracods, cladocerans, tardigrades, and larvae of stream-bottom aquatic insects); and (2) those restricted to groundwater ("stygobites," such as amphipods and others only in deep groundwater).

Around rivers, groundwater animals vary in all three river dimensions, longitudinal, lateral, and vertical (Figure 6.1) (Gibert et al., 1994b). Longitudinal studies from 1 to 500 km along rivers suggest that the groundwater fauna does not correlate with the patterns of surface-water stream/river species, but rather seems to be distributed in smaller patches. Laterally across a floodplain, the density of groundwater organisms and their species assemblages apparently vary at scales from meters to tens of meters, but do not correlate well with observed floodplain boundaries. Vertically, striking differences in total density and species composition are evident at the scale of tens of centimeters, even at >1 m depth, below the surface-water level in river-bottom sediments. The vertical pattern may largely reflect progressively lower dissolved-oxygen levels in groundwater beneath a river.

Apparently the only relatively detailed urban study is of groundwater organisms just downriver of Vienna with nearby seepage from a solid-waste dump (landfill), plus perhaps pollutants from a construction site (Pospisil, 1994). Three sites were studied here by the Danube River and associated floodplain wetland. The groundwater species present were widespread, including species characteristic of both Eastern and Western Europe. The upper groundwater reached 8°C higher in mid-summer than that at deeper levels. The "specific conductance" (a measure of the concentration of certain pollutant ions) overall was slightly higher in the groundwater than in the surface water. Dissolved-oxygen levels were high in the upper groundwater zone, quite variable at 0.5 m depth, and very low and stable at ≥2 m below surface water.

One site, relatively close to oil seepage from a harbor, had 86 groundwater animal species, 7 of which were stygobites restricted to groundwater. Measurements at a second site with deeper groundwater recorded 21 species, almost all stygobites. The third site had 35 species. In the first two sites the most abundant organisms were ostracods and cylopoids. In contrast, the third site had an occasional amphipod and an abundant snail species. Finally, it seems likely that an altered but persistent groundwater fauna exists in slightly polluted sites on the upslope side of urban areas and away from industries. In contrast, the diverse array and heavy dose of pollutants under the central portion of a city must severely degrade or eliminate the groundwater fauna.

Clean water supply

Worldwide most cities get their water supply from reservoirs, though some extract drinking (potable) water from a lake, river, stream, deep groundwater, or even shallow groundwater (Forman, 2008). Relatively small amounts of water are derived from coastal desalinization facilities or from rooftop and other cisterns catching rainwater. To some people, water in Northern Europe has a different taste than that in the USA. The former mainly uses clean deep groundwater while American cities overwhelmingly use surface water, which is therefore treated with chlorine and often other chemicals. Most towns and major industries get their water supply from a convenient local inexpensive source.

Surface-water supply almost always comes from upslope or upriver of a city, thus avoiding the pollution produced by the city. Damming a river to create an adequate-sized reservoir can produce a dependable clean water supply only if the upslope river and stream areas are sufficiently protected with natural or semi-natural vegetation. With insufficient land protection, usually day-by-day and year-by-year eroded soil is carried and deposited as sediment in the reservoir. However, the bulk of the sediment normally arrives in one or a few mega-storm events. Sedimentation of the reservoir bottom progressively decreases the capacity or water volume of the reservoir, thus reducing the available water supply for the city. New York and Boston both have good land protection around their rather distant reservoirs, which provide dependable water with little water-treatment cost (Platt, 2004).

In the erosion process, fine clay and organic particles tend to be eroded first, even in light rains. These are the most valuable soil particles for holding nutrients and maintaining fertility for plant growth. The fine particles tend to stay suspended in solution in the water, like tiny clay particles keeping a glass of water murky for hours. Therefore, most of the eroded clay particles continue on downstream/downriver to the sea, where they slowly filter to sea-bottom. Silt particles are eroded by somewhat heavier rain and readily transported to the reservoir, where water velocity is minimal and the silt filters to the bottom. Reservoirs tend to fill up mostly with silt. Sand is still heavier, requiring heavy rainfall for erosion. But also a significant portion of the sand may be deposited in the stream/river floodplain before reaching the reservoir.

In wet periods reservoir water backs up a bit into tributaries, where some water moves laterally into the nearby soils as "bank storage" (Winter et al., 1998). Droughts or upstream water diversions that lower the reservoir level also slowly draw out bank-storage water as a kind of stabilizing mechanism. Development or other cutting of the stream bank reduces the effectiveness of bank storage as a clean-water supply. Land protection around a reservoir also limits stream/river flooding and its degrading effect on water quality (UN-Habitat, 2005; Sekercioglu, 2011).

Suburbs and small cities commonly have wells pumping shallow groundwater, which comes from rainwater infiltrating through the suburban land. Each well has a "cone of influence," i.e., a cone-shaped soil volume beginning at the bottom point of the well pipe and widening to the ground surface. The surface of this cone is roughly the area where pumping lowers the water-table (Winter et al., 1998). A lowered water-table tends to dry out wetlands, ponds, vernal pools, and streams. Development with impervious surfaces accentuates this drying out of surface water resources. Therefore, natural vegetation and land protection are especially important atop a well's cone of influence, which is thus called the "wellhead protection area."

Most users of the water supply are urban residents who require clean pathogen-free water for drinking and cooking. They also typically use the clean water for other uses such as flushing toilets, washing, cleaning, watering plants, and so forth that do not require such clean water. A separate water supply is typically provided for other major users, i.e., electric-power-generation facilities, some large industries, and urban-agriculture irrigation, which do not require rigorous water-cleaning treatment (Westerhoff and Crittenden, 2009). Power production and most industries use water for cooling, and liberate warm or hot water into the environment. Unusual aquatic ecosystems and fish populations capitalize on this added heat, but large fish kills and other ecological disruptions typically occur during the inevitable facility break-downs and shut-downs.

Over four decades (1955–95) in the USA, water withdrawal from the groundwater by industry has decreased in part due to increased industrial efficiencies (Todd and Mays, 2005). However, tighter pollution regulations mean that less acceptably clean water from industry is recycled back into natural systems. Urban water uses during the period have changed considerably. The total urban public water-supply increased 2.3 times, electric-power-generation water use increased 2.6 times, and industrial water use decreased by 25%.

Figure 6.5. Aqueduct to carry clean water a few kilometers from a source to a Roman city. For the intermittent channel at bottom, water rushes over sand in wet periods, and slowly flows beneath the surface in dry periods. Built by Augustus Caesar ca. 25 BC; cut stones fit together without mortar. Tarragona, Spain. R. Forman photo.

As cities expand outward, even becoming megacities, the problem of obtaining sufficient clean-water supply becomes serious (IPCC, 2007). Roman aqueducts brought clean water kilometers or tens of kilometers usually from sources upslope and within the drainage basin of the city (Figure 6.5). Today such water sources are usually scarce and/or polluted.

"Inter-basin or inter-regional water transfers" are one solution, but with big problems (Wang, 1999). The Romans and Incas both sometimes transported water in canals, tunnels, and aqueducts from one drainage basin to another. The heavily urbanized San Diego-Los Angeles-San Francisco region mainly transports water supplies from far-northern California, the Sacramento Valley, and the Colorado River, all hundreds of kilometers distant (Hartshorn, 1992; Thayer, 2003; Ghassemi, 2006). Volga River water is pumped a similar distance to Moscow, the megacity (Golubev and Vasiliev, 1978).

Such water transfers sharply reduce water in the source areas. Streams, rivers, wetlands, lakes, ponds, and groundwater either have lowered water levels or disappear altogether. Ships lie on their sides in today's almost waterless former Aral Sea. Habitat loss and species disappearance occur on a massive scale in the dried-out source areas. Ironically, the cities receiving the distant water have lots of water, but it is polluted and needs cleansing, either chemically or by being channeled through natural ecosystems.

Extensive and diverse infrastructure is required for water supply in a metro area (Westerhoff and Crittenden, 2009). Each component of the system has considerable ecological effect, either positive or negative, on water cleanliness. Thus on balance, a dammed reservoir can be extremely beneficial. Its protected area of natural vegetation is one of the largest, if not the largest, natural area in the urban region. The large green area typically sustains high habitat diversity, high species diversity, many rare species, viable populations of interior species, large-home-range vertebrates, clean groundwater, connected streams, a semi-natural disturbance regime favoring certain species, a buffer against flooding, and a buffer against urbanization and climate change effects (Forman, 1995). The reservoir itself supports a population of warm-water fish and often contains some deep water that may support cool-water fish. Diverse aquatic and semi-aquatic habitats are often prominent around the reservoir margins.

Water is commonly piped from reservoir to water-treatment facility where chemicals such as chlorine and fluorine are often added to purify the water. The facility is frequently located on a protected greenspace of moderate size, typically supporting species important in its surrounding urbanized area. Often wet spots are visible even during droughts, though also relatively toxic chemicals (when concentrated) are transported to and stored on site. Small cities often have distinctive water tanks on towers or tanks on nearby hills. These tanks provide water pressure throughout the urban area. Relatively small pump buildings, sometimes on small greenspaces, are often scattered over the built area to keep water moving.

Water mains (the large water pipes) periodically burst, causing water distribution problems, urgent repair activity, and news stories. The unruly water bubbling up mainly flows into the storm drainage system leading to a local water body, which is promptly altered by both more and clean water. The aquatic ecosystem and its species are temporarily altered. In 2010 a major water-supply pipe (3 m = 10 ft diameter) for Boston ruptured, causing 2 million people in the city and 29 surrounding suburbs to have to boil their drinking and cooking water for 4–5 days. Water pressure was apparently maintained for fire hydrants and toilets during the period.

Another huge but almost invisible ecological effect results from the hundreds or thousands of kilometers of secondary to tiny pipes leading from the water mains to almost every building and other water user (Westerhoff and Crittenden, 2009). Water is maintained under pressure in these pipes. Imagine how many of these pipes and joints of different types, some old and some new, leak. No-one knows how many. A relatively big leak often results in repair, but trickles, gradually increasing, go unnoticed. The water-supply system endlessly contributes small amounts of water to countless spots across the metro area. Benefits of the added water may flow to deep tree roots and upward tree growth, to soil animals and microbes, to polluted ground water, to urban stream-flow, and to wetlands and ponds. On the other hand, if the soil is already rather wet, the added water seeping or squirting out of water supply pipes is bad news for many of the preceding soil and water components.

Worldwide a much higher proportion of urban residents than rural residents has "safe" drinking water (UN-Habitat, 2006). But in numerous cities extreme differences in water quantity and quality exist from section to section within a city. Thus, water-related illnesses are extensive, as illustrated by 65% of hospital patients in India reportedly being treated for illnesses from water. Water piped to taps (faucets) in housing units predominates, though public taps, e.g., at wells or fountains, predominate in some areas (Figure 6.2) (UN-Habitat, 2005). In Johor Bahru (Malaysia), Hong Kong, and Beijing, 100% of the population has house taps, while in Colombo (Sri Lanka), Almaty (Kazakhstan), and Chennai (India), 26–29% of the people are served by public taps. However, in Jakarta (Indonesia) and Cebu (Philippines), three-quarters (73–77%) of the population is without piped water to either housing unit or public tap. Residents in such urban areas get water from tube wells, dug wells, rainwater collectors, ponds, rivers, and other sources.

In North America, total indoor household water use averages about 265 l/year (70 gallons/year), and varies rather little from region to region (Westerhoff and Crittenden, 2009). On the other hand, outdoor water use varies widely, for instance from ca. 5 gallons/year in Waterloo (Ontario) and Cambridge (Massachusetts) to 225 gallons/year in Phoenix and Scottsdale (Arizona). In the desert city of Phoenix, warmer night temperatures result in more outdoor water use by residents (Brazel et al., 2000; Guhathakurta and Gober, 2007). In these dry Arizona cases, the added water supports trees, shrubs, flowers, lawns, birds, mammals, insects, and other animals that would quickly perish without the water. The species dependent on supplemental water would be largely replaced by native dry-country species from the surroundings.

Water-supply pipes lead to fire hydrants, fountains and wells for public use, fountains and ponds in parks, irrigation systems for greenspace lawns, and of course people's kitchens, bathrooms, and water closets. Spills and leaks occur in all of these locations, and many kinds of species take advantage of these key bits of added water. Our water-supply system permeates the city and provides a cornucopia of mostly positive though tenuous ecological benefits.

In short, the water-supply addition seeps into an urban area in numerous spots associated with the water-supply infrastructure permeating the area. In most spots the added water supports microbes, invertebrates, vertebrates and/or plants that would otherwise die and be replaced by drier-adapted species, or not be replaced. Overall the added resource and associated microhabitat diversity sustains an enrichment of urban biodiversity.

Finally, *potable-water reuse* is a special form of treatment that, in several steps, transforms human wastewater into potable drinking water (Law, 2003; du Pisani, 2006).

> Water should be judged not by its history but by its quality.
> *Lucas Van Vuuren, 1970s, quoted by Law (2003)*

Using a sequence of technologies, including reverse osmosis, ozone with activated carbon, microfiltration, and a membrane bioreactor, Windhoek, the arid-land capital of Namibia, has had potable-water reuse since 1968. Singapore, Brisbane (Australia), and Orange County (California) also apparently reuse some water. In many cities of the world where clean freshwater-supply is increasingly scarce and expensive, converting clean water to wastewater and back to clean water becomes an increasingly compelling idea.

Sewage and septic wastewater

> Is there a toilet in the home and a tap for hand-washing? If not, is there a well maintained toilet in easy reach? If this is a public toilet and there is a charge for using it, is it kept clean, can low-income households afford to use it and is it safe for women and children, especially after dark?
>
> Is there provision to remove human wastes and household wastewater?
>
> *(UN-Habitat, 2005)*

No-No-No-No are the emphatic answers for the majority of urban residents in Asia, Sub-Saharan

Africa, and Latin America. Variation from city to city in the proportion of the population connected to the city's sanitary sewage system ranges, for instance, from <30% in Buenos Aires, Belem (Brazil,) and San Juan (Puerto Rico) to >90% in Medellin (Colombia), Monterrey (Mexico), Caracas (Venezuela), and Greater Santiago (Chile) (UN-Habitat, 2005).

Yes-Yes-Yes-Yes for the above questions describes conditions in high-income sections of most cities. But even in these sections the piped water supply may be intermittent and of poor quality. The percentage of households with flush toilets (in cities of >100 000 population) ranges from about 95% in North Africa to 83% in Southeast Asia, 70% in South and West Asia, 68% in Latin America to 24% in Sub-Saharan Africa.

This section begins with human sewage systems, wastewater aquaculture, and the ecological effects of sewage systems. Then septic systems and other approaches for dealing with household wastewater mainly on-site are introduced.

Wastewater sewage systems and associated effects

We explore this intriguing subject with three lenses: (1) wastewater and pipe systems; (2) wastewater sewage treatment; and (3) wastewater aquaculture for food.

Wastewater and pipe systems

Sanitary sewage or wastewater is generally a combination of human wastes (feces and urine), household wastewater (from sinks, basins, tubs), commercial wastewater, industrial wastewater, and stormwater infiltrating into a sanitary-sewage pipe system. Human wastes are full of microorganisms, many being pathogenic or disease-causing, and thus are of exceptional ecological and public health importance.

Cities normally began by having wastewater carried away to "somewhere" by streams or rivers, or being diluted by the sea or estuary (Metosi, 2000; Anderson and Otis, 2000). With urban population growth, sewage pipe systems (Figure 6.6d) with "manholes" and clean-outs continued to expand, as elegantly portrayed by the Musée des Égoutes de Paris (Paris Sewer Museum) (Clement and Thomas, 2001). The pipe systems mainly have acute-angled connections to reduce blockages, though blockages have been cleared with pipe-sized wooden balls and other technologies, even interesting boats in tunnels. Gravity for moving the viscous sewage was seldom sufficient, so pumping facilities were added. Interconnections among pipes were added to circumvent blockages and keep the liquid moving. The amount of sewage became excessive for simple disposal into the environment. Therefore, sewage-treatment facilities for cleaning wastewater were added to break down the diverse sewage components for recycling back into natural systems.

Usually all wastewater comes from the clean-water supply. Typical water uses, and hence components of the outgoing household wastewater are (e.g., in the UK): 31% toilet (WC) flushing; 26% washing/bathing; 15% food preparation/drinking; 12% laundry; and 10% washing-up (the remainder is mostly used outdoors) (Butler and Davies, 2011).

Wastewater from human toilet use is primarily composed of water, organic matter, bacteria, nitrogen/phosphorus/other nutrients, and the disposal of various solid materials (Westerhoff and Crittendon, 2009; Butler and Davies, 2011). Bacteria are decomposing the organic matter, while coliforms (e.g., *E. coli*) and streptococci may be pathogens. Wastewater from household sinks, basins, and tubs contains various organic and inorganic chemicals from food waste, soap/detergent, dirt, and other sources. Salt and pharmaceutical compounds from households are also environmental hazards warranting treatment.

The inputs from commercial areas may be rather similar to those from residential areas, whereas inputs from industrial facilities are quite different (Butler and Davies, 2011). In addition, chemicals leach into wastewater from sewage pipes and connections that are composed of lead, copper, and other materials. The heterogeneous mixture of materials in wastewater requires a set of contrasting approaches for treatment and cleaning. Even with normal sewage treatment, challenges persist for some chemicals that are toxic or persistent and/or bio-accumulate in the food chain.

Industrial chemicals, supposedly treated separately at the industrial site or elsewhere, find their way into sewage systems. A truck-farming area irrigated with municipal sewage water (presumably partially treated) from Paris contains high levels of copper, mainly associated with organic matter (Kirpichtchikova *et al.*, 2003). Also present are high levels of zinc and lead, mainly bound to mineral nutrients in the productive farming soil. Combined sewer overflows (CSOs) (discussed below) contain industrial pollutants (Alberti, 2008). Asbestos is present in significant amounts in the municipal sludge of two-thirds of cities in the USA (Manos *et al.*, 1991).

Urban water systems

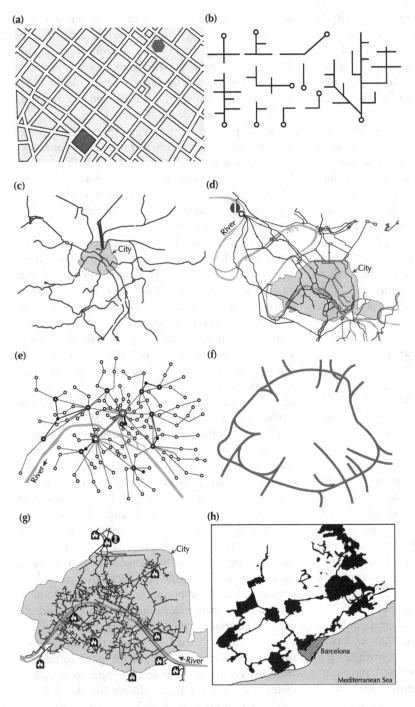

Figure 6.6. Eight network and flow patterns in urban areas. (a) Street network grid in city center (Lima, Peru). (b) Stormwater drainage system in suburban Boston. Traditional form on right; reduced flow amounts and velocities on left. (c) Commuter rail system, Paris Region. (d) Sewage wastewater system, Paris Region. (e) Main telephone-cable system, Paris. (f) Ring-rail network (Petite Ceinture), early 20th century Paris. (g) Heating and cooling pipe-system for buildings, Paris. (h) Emerald network of large connected green patches, proposed plan for Barcelona Region. Adapted from Clement and Thomas (2001), Forman (2004b).

In streets we see the familiar manholes, essentially the top tiny circles of a gigantic connected network (Figure 6.7). But sewage pipe systems beneath our feet have problems. Growth and expansion of tree roots cause a major portion of the sewer system backups and overflows in Los Angeles (Sklar, 2008). Rainwater infiltrating through the soil seeps into holes and joints of the pipes. Below the water-table, considerable groundwater often squirts from saturated soil into the sewage pipe system. Indeed, if stormwater pipes are directly connected to the sanitary sewage system, stormwater pours in.

Figure 6.7. Electric-trolley public transit and manhole access to pipe system under street. Grass- and-block- covered surface to increase water infiltration and reduce heat buildup. Zurich, Switzerland. R. Forman photo.

Above the water-table, cracks and holes permit some "exfiltration" of wastewater from the pipe system into the soil. So far apparently little evidence exists that this pollutes the groundwater, which in urban areas tends to be quite polluted itself. Nonetheless, toxic chemicals and pathogenic bacteria from wastewater would be bad in a sensitive water-supply area or a rare natural ecosystem.

Many cities have a *combined sewage/stormwater system* composed of pipes that carry both sanitary wastewater and stormwater runoff directly to ditches, streams, river, and/or estuary (Benton-Short and Short, 2008; Baker, 2009). The combined system thus pours chemicals and solids from households, commercial buildings, and industries, as well as from urban building and road surfaces, to local water bodies. The wastewater volume heavily pollutes areas near pipe ends. Combined systems may also flow to sewage treatment facilities.

Separate stormwater and human sewage systems have separate pipes for the two flows (US Environmental Protection Agency, 2007; Butler and Davies, 2011). The sanitary wastewater sewage may flow directly into a local water body. Alternatively and preferably, the flow leads to a sewage treatment facility, which partially or largely cleans the wastewater. In contrast, the urban stormwater system is usually composed of many shorter separate pipe-systems, often with right-angle connections, and with many pipe ends (Figure 6.6b). Therefore, as described below, stormwater with its pollutants and debris pours into many depressions, ponds, gullies, streams, rivers, estuaries/sea, and even into underground temporary storage spaces.

Combined systems predominate in cities of developing nations, though they are also widespread in developed nations, including several hundred cities in the USA (Benton-Short and Short, 2008). In dry weather, essentially only sewage wastewater flows in the pipes. In light rains, stormwater typically somewhat exceeds the flow of wastewater in the pipes. In heavy rains, however, stormwater flows may be 50–100 times greater in the system. If the pipe system leads to a sewage treatment facility, the facility is unable to treat the heavy flow.

Combined sewer overflows (CSOs) refer to the temporary direct outflow of combined wastewater and stormwater into a natural system or environment (CSO also refers to the pipe structure that permits this overflow) (Paul and Meyer, 2001; Benton-Short and Short, 2008; Butler and Davies, 2011). In the USA some 75 000 CSO events occur annually, most discharging into rivers (US Environmental Protection Agency, 2007). Even though CSOs are predominately composed of stormwater, it contains pollutants from urban surfaces. Combined with some sewage wastewater, the CSO outflow pours considerable pollution into the environment.

A flushing of chemicals from urban surfaces occurs early in a rainstorm ("first flush") as water flowing over surfaces picks up and cleans the accumulation of particles and chemicals. Water from light rains, and especially from a series of light rains, carries a relatively concentrated flush of particulates and dissolved chemicals. The chemical flush is also evident early in heavy rainstorms and CSO overflows.

The separate wastewater/stormwater system offers significant ecological advantages over the combined system (Butler and Davies, 2011), including: (1) less serious pollution of local water bodies; (2) smaller less-intrusive infrastructure (pipes, pump facilities, sewage treatment facility), meaning less environmental degradation from construction and maintenance; (3) no road grit (particulate material) carried to a treatment facility and to its outflow; and (4) in floods, only stormwater, without sewage wastewater, bubbles up from manholes. However, some ecological advantages exist for a combined sewage/stormwater system, such as: (1) fewer pipes, meaning less construction-and-maintenance effect; and (2) except in heavy rains, the sewage treatment facility provides some cleaning treatment of

Figure 6.8. Sewage treatment facility by a river carrying considerable sewage wastewater. Large dark circles are secondary treatment tanks mainly for oxygenating wastewater to reduce organic matter with aerobic bacteria. Upriver the population far outstrips sewage treatment capacity. Small vegetable-garden plots in floodplain; sometimes midge populations (chironomids) from the rivers are pests. During wet periods stormwater may fill or overflow the channelized-river concrete banks, sometimes causing huge floods. Note small access roads used by infrastructure maintenance vehicles and local residents. Besos River, Barcelona. R. Forman photo courtesy of Mark Montlleo.

stormwater. Overall, separate systems, though more expensive, are ecologically optimal for urban areas.

Some decomposition processes occur in the sanitary sewage pipes even before wastewater reaches a treatment facility (Anderson and Otis, 2000; Butler and Davies, 2011). "Hydrolysis" in the presence of water and enzymes decomposes large organic molecules to smaller ones. "Aerobic bacteria" in the presence of oxygen decompose large and small organic molecules to simple stable end-products, including CO_2, H_2O, and various inorganic ions. Some "nitrification" by nitrifying bacteria produces nitrites and nitrates, and "denitrifying bacteria" convert nitrate to nitrogen gas (N_2). "Anaerobic bacteria," essentially in the absence of oxygen, slowly decompose organic molecules to simple stable end-products. This last process, however, produces smelly hydrogen sulfide (H_2S) and/or combustible methane (CH_4). Sewage pipes themselves are active places, though the volume of urban wastewater requires a sewage treatment facility or a natural system for cleaning.

Sewage treatment facilities

Wastewater sewage treatment facilities are typically built around the urban fringe, which then expands outward, often leaving a moderately large greenspace in the metro area (Figure 6.8). Both water and nutrients are abundant. Consequently, sewage-treatment-facility greenspaces are often valuable for the abundance and species richness of both plants and animals. Bird-watching may be good around sewage treatment facilities.

The effectiveness of a sewage-cleaning process can generally be measured in four ways (Anderson and Otis, 2000). (1) *BOD* (biological or biochemical oxygen demand) estimates the quantity of biodegradable organic material. (2) *TSS* (total suspended solids) measures the concentration of undissolved solid materials. (3) *TN* and *TP* measure the concentration of the nutrients, total nitrogen and total phosphorus. (4) The *fecal-coliform-bacteria level* indicates the potential occurrence of pathogenic bacteria of enteric (intestinal) origin. Three levels of sewage treatment are usually recognized in reducing and removing these components of wastewater (Anderson and Otis, 2000; Kalff, 2002; Baker, 2009): primary, secondary, and tertiary treatment.

Primary sewage treatment removes large solids and most other solids that settle to the bottom of a tank, as well as greases, oils, and varied solids that float at the top. The remaining effluent is composed of suspended solids, dissolved organic materials, and other soluble pollutants.

Secondary treatment removes much, commonly >85%, of the dissolved organic materials (BOD) and suspended solids (TSS), and significantly reduces bacterial numbers. Typically disinfection, e.g., by adding chlorine to reduce microbes, especially fecal-coliform bacteria, occurs at the end of the process.

For the majority of centralized sewage treatment facilities in the USA, secondary treatment effluent is discharged into water bodies. Some facilities with advanced secondary treatment achieve a >95% reduction in BOD and TSS. The effluent from secondary treatment contains relatively low levels of organic material, suspended solids, and live bacteria, but still contains most of the original wastewater nitrogen and phosphorus.

Increasingly an advanced or *tertiary wastewater treatment* is included to reduce total phosphorus and total nitrogen. This is particularly important because these two mineral nutrients are the primary causes of eutrophication of water bodies, especially phosphorus in freshwater and nitrogen in saltwater. Typically TP and TN are reduced by >90%, and BOD and TSS by >95%, in tertiary treatment. In wetland soils, treatment of phosphorus may be more effective in drier conditions and nitrogen in wetter conditions (Fischer and Acreman, 2004). Disinfection to further eliminate microbes commonly occurs where effluent from the tertiary process is discharged aboveground or into surface waters.

Of course many variations and technologies are used in the primary-secondary-tertiary treatment process. For instance, screens, aerators, sludge recycling, drying, incineration, reverse osmosis, chemical additions of iron, aluminum or calcium, or ultra-violet radiation may be used (Kalff, 2002). Because people are pouring into cities, wastewater solutions are a global priority.

Since municipal sewage-treatment facilities are expensive to construct and maintain, smaller *neighborhood* (or package) *treatment facilities* sometimes are built for dense apartment, condominium, or neighborhood locations. Too often these small sewage-treatment facilities have soon malfunctioned, causing significant environmental pollution and public health problems. To sustain neighborhood wastewater treatment facilities seems to require permanent municipal resources.

Sludge that builds up at the bottom of secondary treatment facilities is sometimes used as fertilizer for agriculture and even lawns, because of its richness in nitrogen and phosphorus. However, sludge also normally contains heavy metals, other pollutants, and living spores of bacteria that are then both diluted and widely distributed in the environment. As discussed below, natural systems instead of built treatment facilities are sometimes used for secondary and tertiary treatment. Almost always, sewage systems deposit their effluent water downslope or downriver of a city, and hence upslope or upriver of the neighboring municipality.

Recycling of the water in sewage through groundwater recharge is of increasing interest and promise (Todd and Mays, 2005). This is only considered in exurban/peri-urban areas that still have, and are likely to continue to have, clean groundwater, normally a scarce commodity in urban areas. Basically the organic carbon, nitrogen, pathogens, and in some cases heavy metals in wastewater have to be intensively treated (cleaned) before the effluent water enters the groundwater. Eliminating parasites, bacteria, and viruses, including resistant spores of the first two types, is particularly important for public health. After cleaning, the water enters (recharges) the groundwater, which then is pumped upward as clean-water supply for drinking and other household uses.

Natural-systems wastewater treatment, especially using marshes and ponds, may clean wastewater from whole communities (Vymazal *et al.*, 2006; Mitsch and Gosselink, 2007). This approach depends on natural physical, chemical, and biological processes, such as just described. Primary, secondary, and tertiary treatment goals are met by having wastewater flow through marshes and/or ponds. The total area required, usually many hectares or acres, is related to the human population size served.

The amount of treatment cleaning is mainly determined by the (1) length of route and (2) rate of flow through wastewater systems, so routes tend to be convoluted. Generally pumps are required to keep the flows moving, since the wetland terrain is relatively flat. Many other technologies may be added to complement the natural processes.

For decades, a pond/wetland complex has successfully treated the wastewater for >15 000 people in Arcata, California (Figure 6.9) (Gearheart, 1992). Primary treatment occurs in a building, secondary treatment in 18 ha (45 acres) of pond and 2.4 ha (6 acres) of marsh, and tertiary treatment in a similar-sized marshes-and-pond area. The site provides two major additional benefits. The site supports dense wildlife populations. It is also a significant feeding and rest stop for migrating birds, especially waterfowl, in a state that has drained or filled 90% of its wetlands. Secondly, the natural system sewage-treatment facility is a major recreation area, with loop trails, viewing towers, ponds with waterbirds, and valuable educational facilities.

Wastewater aquaculture for food

Adding fish food to ponds and other water bodies to gain impressive production of one or a few fish species for human consumption has a long history (Costa-Pierce *et al.*, 2005; van Bohemen, 2005). In the present case of aquaculture, urban wastewater is used as both nutrient and food input. The addition of nitrogen is especially important in catalyzing growth in estuaries and other habitats of coastal cities. Huge growing urban

Urban water systems

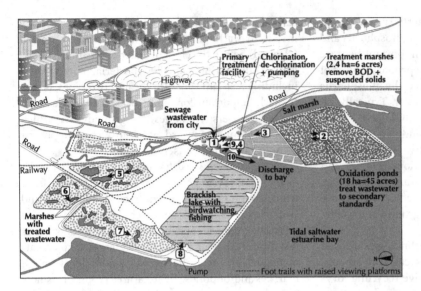

Figure 6.9. Pond-marsh wastewater treatment, wildlife, and recreation area. Sewage wastewater treatment serves the >15 000-population of Arcata (California). Numbers trace the route of flow during treatment. Water is twice chlorinated and de-chlorinated. The estuarine Arcata Bay is a center of oyster production. Adapted from City of Arcata brochure (ca. 1991), Stewart and Streshinsky (1990), France (2003).

populations have a major need for food, and produce an enormous amount of wastewater rich in nitrogen. The wastewater is thus used for local food production, a central goal in achieving food security for an urban area (Koc et al., 1999). In addition, the production process helps to treat or clean the wastewater, thus reducing seawater pollution effects, especially due to nitrogen. Since the rate of population growth exceeds that of constructing tertiary treatment facilities, seawater nitrogen pollution is expected to increase. For instance from 2000 to 2050, nitrogen output is expected to increase by 21–27% in Los Angeles, New York, and Sao Paulo; 100% in Jakarta; and 117–120% in Mumbai and Kolkata (Calcutta) (Costa-Pierce et al., 2005).

Various experimental and technological systems for aquaculture are used in North America, including tanks, pumps, inputs/outputs, and recycling processes. Europe grows several species aquaculturally, including Atlantic salmon (*Salmo salar*) and blue mussel (*Mytilus edulis*) in saltwater, and rainbow trout (*Oncorhynchus mykiss*), European eel (*Anguilla anguilla*), and African catfish (*Clarias gariepinus*) in freshwater.

However, Asian cities have been at the forefront of aquaculture using urban wastewater (Costa-Pierce et al., 2005; McGregor et al., 2006). Kolkata (India) has wastewater aquaculture pond-systems growing Indian vegetables and fish, as well as fish polycultures, especially of three carp species. Hanoi (Vietnam) pond systems grow fish polycultures of tilapia and carp species, as well as rice and fish. Particularly interesting are the diverse wastewater aquaculture ponds around Ho Chi Minh City (Vietnam) (Van and De Pauw, 2005). Crops harvested in different ponds include: water morning-glory (*Ipomea aquatica*); lotus (*Nuphar lotus*); lotus plus duckweed (Lemnaceae); water mimosa (*Neptunia oleracea*); duckweed and tilapia; and mixed fish species.

Lots of problems exist for such wastewater aquaculture systems, including pollution types and levels, pests, public health, economics, sustainability of the systems, and effects on the surroundings. If wastewater from industry enters the system, toxic substances such as heavy metals and organic chemicals are probably present. Perhaps public acceptance of food grown with urban sewage is the greatest hurdle. Future proposals, designs, and products for urban wastewater aquaculture of course exist (Edwards, 2005). Nevertheless, the greatest benefits of the approach is in addressing several major worldwide urban issues concurrently, i.e.: the increasing need for food, especially locally produced food; growing amounts of wastewater produced; huge excess of nitrogen put into natural systems; and reduction of pollution in local urban water bodies.

Effects on local water bodies

In general three types of wastewater are discharged from sewage systems (Butler and Davies, 2011): (1) normally functioning sewage treatment facilities produce continuous low-level pollutants; (2) separate wastewater and stormwater systems produce intermittent large loads of suspended solid material and/or ammonia;

and (3) combined wastewater and stormwater systems produce intermittent CSOs, which consist of mixed stormwater, residential-commercial-industrial wastewater, and deposits from the sewage pipes. Leaks and illegal discharges along the sewage pipe system generally add pollutants to the soil rather than directly to water bodies. The major wastewater discharges enter into and pollute local water bodies, especially streams, rivers, lakes, and estuaries in and around metro areas. Urban rivers are more likely to receive CSOs, because the volume of flowing water can dilute the pollution (Paul and Meyer, 2001).

Consequently, four primary ecological problems are particularly prominent in these local receiving waters (Kalff, 2002; Butler and Davies, 2011):

1. Insufficient dissolved-oxygen.
2. Excessive nutrients, especially phosphorus and nitrogen.
3. Excessive sediment in both the water body and on its bottom.
4. An excess of toxic substances, both inorganic and organic.

In general, low dissolved-oxygen is likely in slow streams, large rivers, lakes, and small estuaries, and highly likely in small rivers. Nutrient excess is likely in deep lakes and small estuaries, and highly likely in shallow lakes. Excessive sediment is likely in all local water bodies except fast-flowing streams. Excessive toxic-chemical concentrations are also likely in all local water bodies except large estuaries. In addition, pathogens, including parasites, bacteria, and viruses, are highly likely in all local water bodies. Disease-causing pathogens not only affect humans, but perhaps all animal species in and associated with the water bodies.

The processes in water bodies that "treat" or clean up wastewater effluent are nearly the same as those operating in sewage treatment facilities (and in the soils in and under septic drainfields). Physical processes include dilution, mixing, flocculation, sedimentation, thermal breakdown, and aeration (Kalff, 2002; Butler and Davies, 2011). Chemical or biochemical processes include aerobic oxidation, anaerobic oxidation, nitrification, and adsorption of metals and other toxic substances. Biological or microbiological processes include decomposition by aerobic and anaerobic bacteria, blooms or massive growth of green algae and blue-green algae, and massive die-offs of algae and other microbes. An aquatic biologist can usually estimate the degree of wastewater pollution in a water body simply by looking at the species present, a surprisingly effective bioassay.

A low level of dissolved-oxygen creates *anaerobic* or "anoxic" conditions, commonly resulting in fish kills. Almost all organisms of the natural aerobic aquatic ecosystem also die. Consequently, anaerobic bacteria and other microbes predominate in low- or no-oxygen conditions, with CH_4, H_2S and bad smells commonly produced.

An excess of phosphorus and nitrogen nutrients commonly leads to *algal blooms*, i.e., "eutrophication" (Kalff, 2002). The dense floating algae produce shade, so algae photosynthesis (which produces oxygen) is essentially limited to a thin upper layer of the water body. In freshwater, excess phosphorus commonly catalyzes blooms of green algae, though nitrogen may also. A very high level of nutrients often causes blooms of blue-green algae, which in turn commonly produce toxic substances in the water. In salt water, added nitrogen is more likely to cause the bloom. An extensive growth of algae has additional big ecological consequences.

A concurrent massive death of algae cells occurs, and the cells gradually filter downward in the water. With oxygen present, aerobic bacteria then massively multiply and decompose the organic matter of these dead cells. However, this bacterial decomposition quickly uses up the dissolved-oxygen present, thus creating anaerobic conditions. Anaerobic bacteria take over, and decompose both the dead algal cells and the dead aerobic bacterial cells. Most fish and other aquatic animals also die in the anaerobic conditions. In this way, the input of wastewater phosphorus and nitrogen triggers a sequence of ecological changes leading to anoxic conditions in most of the water body. Anoxic "dead zones," some covering hundreds of square kilometers, are characteristic in the sea off of many coastal cities.

Sediment particles from wastewater also produce varied ecological effects in the local water bodies. Heavy sediment such as sand and most silt quickly accumulates on the bottom (Paul and Meyer, 2001; Kalff, 2002). In large streams and small rivers the sediment smoothes the bottom surface, greatly reducing habitat diversity. Sediment smoothing also renders the bottom unsuitable for successful reproduction by some fish, including salmon and trout. Sediment accumulations in flowing water commonly alter the water-flow patterns. Lighter-weight sediment such as clay, fine silt and some organic matter remains in the water a long time before settling to the bottom, and thus clogs the gills of fish and feeding structures of filter-feeding animals. The suspended sediment also reduces light penetration and algal photosynthesis.

Toxic substances come in an array of forms (Kalff, 2002). Inorganic compounds including ammonia, arsenic, and heavy metals may be directly toxic and kill aquatic organisms. Some inorganics also "bio-accumulate," that is, increase in concentration at each step in the food chain. Thus, predators such as many fish, and herons and raptors feeding on the fish, get a heavy dose of heavy metals, which may be toxic to the predators. Toxic organic compounds, such as hydrocarbons from transportation and chemical by-products from industry, are exceedingly diverse. Some quickly break down while others are quite persistent in the water body. Both acute and chronic effects on the aquatic ecosystem occur from wastewater inputs. Almost all the inputs reduce biodiversity of the natural water body.

Microbial densities, especially of coliform bacteria, tend to be high in almost all urban water bodies, especially those receiving water from sewage treatment facilities and CSOs (Paul and Meyer, 2001). Protozoan pathogens including the protozoan *Giardia* may also be widespread. Wastewater pollution by organic matter seems to sharply decrease invertebrate biodiversity. This often results in water bodies dominated by midges (Chironomidae, Diptera) and oligochaetes (worms on the bottom).

In general, urban streams have low species richness in response to toxic substances, sedimentation, and organic compounds. In contrast, the abundance of invertebrates typically decreases with more toxins and siltation, but increases with added inorganic nutrients and organic compounds. Wastewater effluent discharged to local water bodies can be expected to degrade aquatic habitats, degrade food sources, have direct and indirect toxic effects, and eliminate most organisms by organic-matter enrichment. As in the case of algal blooms described above, considerable added organic matter stimulates a massive growth of aerobic decomposing bacteria, which quickly deplete the dissolved-oxygen, resulting in fish kills.

Some of the wastewater microbes in water bodies are of considerable public health importance in urban areas (also see Chapter 1) (UN-Habitat, 2005, 2006; Butler and Davies, 2011). "Water-borne diseases" include cholera, typhoid, roundworms, and hookworms. Often most prevalent are diarrheal diseases resulting from viruses, protozoa, and coliform bacteria such as *E. coli*. "Water-related diseases" are even more diverse. These include mosquito-transmitted malaria, dengue, and yellow fever. Worm diseases involve tapeworms, roundworms, schistosomiasis, and varied eye and skin infections. Slow-moving water is often optimum for public-health disease microorganisms.

Typically very few surface water bodies in and around metro areas retain natural aquatic ecosystems. Some urban water bodies are slightly or somewhat polluted by wastewater, stormwater, and other inputs, and might be considered as having semi-natural aquatic ecosystems. These are often categorized as "not swimmable or fishable." Swimmers would be at high risk of pathogenic infection, and fish are likely to contain high levels of toxic substances. Many other urban water bodies are highly polluted, and have aquatic ecosystems with a considerable anaerobic component, short food webs, and very low species richness.

Septic and other systems

High-density residential areas and adjoining low-density areas are normally best served by a sewage wastewater system of pipes leading to a treatment facility. A lower-density housing area might be well served by a small government-maintained neighborhood sewage-treatment system. Most common, however, is for each isolated house to handle its wastewater on-site.

Various types of *latrines*, basically holes in the ground for human waste, are the longest serving method of human-waste disposal (Anderson and Otis, 2000; Hardoy et al., 2004; UN-Habitat, 2005). To minimize water-supply contamination, latrines are placed downslope of wells (McGregor et al., 2006). A "cesspool" as a hole, e.g., filled with gravel and covered, takes both human waste and household wastewater. Or a hole may contain a bucket for human waste that, when full, is transported for disposal, composting, or treatment off-site. For centuries, "night soil" (human feces) has been recycled as valuable fertilizer for urban agriculture. The latrine may contain a tank, often partially filled with water, that is periodically emptied by pumping. The "outhouse" or "privy" is usually a latrine with roof and walls, an object made vestigial in many areas by water closets or toilets within residences. Latrines normally provide very little treatment of waste.

In contrast, the *septic system* is designed to provide on-site waste treatment (Anderson and Otis, 2000; Marsh, 2010; Butler and Davies, 2011). The system depends on belowground processes to convert most of the wastewater from toilets, basins and sinks to simple stable end-products including CO_2, H_2O, and inorganic compounds. Three major components constitute the system: (1) a septic tank (generally >1 m^3); (2) a drainfield (leachfield) containing subsurface trenches with perforated pipes or jointed tiles; and (3) aerated soil beneath the drainfield.

Wastewater from a building is piped to the septic tank, where most solids settle to the bottom as "sludge." Anaerobic bacteria slowly decompose some of the sludge. Grease, oils, and other materials float on top of the wastewater as "scum." The septic tank is regularly pumped out to remove both sludge and scum.

Liquid from the septic tank then is spread in the soil by flowing through the perforated pipes in the drainfield. Finally, the fluid flows downward through the aerated (non-saturated) soil beneath. A richness of physical, chemical, and biological processes in soil treats the wastewater (Anderson and Otis, 2000). Physically the soil disperses and filters solid particles. The soil chemically removes dissolved pollutants by adsorption, cation exchange, chemical precipitation, and forming chemical complexes. Aerobic soil bacteria decompose organic molecules. Nitrifying and denitrifying bacteria transform nitrogen compounds. Plant roots absorb nitrogen, phosphorus, sulfur, calcium, and other nutrients including heavy metals.

Thus, in a well-functioning septic system, the tank and soil treat (clean) most of the wastewater on-site, leaving harmless natural substances in the soil or nearby water body. Pathogenic bacteria from the wastewater apparently are virtually eliminated. Almost all organic matter is broken down to simple inorganic substances. Nitrogen and phosphorus are largely absorbed by the soil, and hence do not eutrophicate the local water body. In short, the septic system accomplishes primary, secondary, and tertiary treatment on site. In this way water is also released to the soil on a site.

Significant constraints limit where a septic system is effective or legal (Anderson and Otis, 2000; Marsh, 2010). A sufficient soil area is needed for the drainfield. A sufficient depth of aerated soil [e.g., 1.3 m (4 ft)] beneath the drainfield is needed. A "percolation test" measures how readily or fast water moves downward through soil (Marsh, 2010). Percolation of water through a sandy soil is too fast and too much pollution reaches groundwater or a local water body. Percolation through a clayey soil is too slow, so the wastewater and pollutants tend to accumulate, puddle, and smell. Just as for most agricultural crops, the best soils for septic treatment are loamy or silty soils (Chapter 4).

Indeed, apparently a significant proportion of septic systems do not accomplish the wastewater treatment goals. Many reasons exist: built on inappropriate soil; water-table has risen due to nearby development, dam, or beaver activity; soil compaction due to trucks, construction or a temporary swimming pool; tree- or shrub-root disruption of the drainfield; toxic chemicals killing the important aerobic bacteria of the drainfield and soil beneath; regular pumping of the septic tank interrupted; and other factors. Any of these factors results in an excess of pathogenic bacteria, wastewater organic matter, and/or nitrogen and phosphorus reaching groundwater or local water bodies. A rotten-egg smell of hydrogen sulfide indicates that the septic system is seriously malfunctioning.

Grey-water recycling is another approach that significantly reduces the amount of domestic wastewater needing treatment, thus enhancing the efficiency of human-waste treatment (van Bohemen, 2005; Butler and Davies, 2011). Grey-water is the drainage from sinks, basins, and tubs. One major use is for flushing toilets (WCs), thus recycling or using the same water for two different uses in sequence. The other grey-water use is for watering plants, as in gardens and lawns. Grey-water contains considerable phosphorus (and organic matter). The phosphorus typically stimulates plant growth and reduces the need for other fertilizers.

A further variant on recycling that enhances wastewater treatment is "urine recycling" (van Bohemen, 2005; Butler and Davies, 2011). Urine is rich in nitrogen and low in heavy metals, thus being good as a fertilizer for plant growth. Using special no-mix toilets, the liquid fertilizer is readily collected and eliminated from the wastewater stream and its treatment. Finally, numerous other interesting technologies and approaches and their combinations provide variations on the three basic approaches for disposing of wastewater, i.e., latrines, septic systems, and sewage treatment systems (Anderson and Otis, 2000; Marsh, 2010).

"Composting toilets" are still another option for some urban conditions (Van der Ryn, 1978). For these, suitable conditions of temperature, moisture, chemistry, and air are maintained to facilitate the natural decomposition of human feces. Composting toilets have the great advantage of recycling wastes on site and not using water to transport wastes. The combination of avoiding water use and minimizing the movement of pollutants into water bodies is highly significant ecologically.

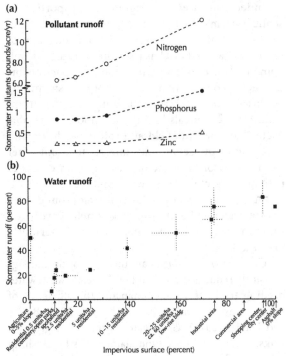

Figure 6.10. Impervious surface cover, stormwater runoff, and pollutants. Patterns from urban sites in USA. (a) Presumably nitrogen and phosphorus = TN and TP. 1 pound/acre/year = 1.2 kg/ha/year. Based on Marsh (2010). (b) Dotted lines roughly indicate common range of variation (where data were available). At 10% impervious surface, low to high points = residential, cemetery, sports-field; at 60%, the point = 20–25 units/ha; at 75%, low and high points = light and heavy industry. 1 ha = 2.5 acres. Based on US Environmental Protection Agency (2001), Forman et al. (2003), Dunnett and Kingsbury (2004), Vince et al. (2005), Berke et al. (2006), Wessolek (2008), Marsh (2010).

Finally, which is environmentally better for an exurban/peri-urban area or outer suburb: septic systems or a sewage treatment system? Both have advantages and disadvantages (Anderson and Otis, 2000). A sewage treatment system: (1) minimizes water-quality pollution problems in water bodies, groundwater, and soil across the landscape; and (2) reduces the number and dispersion of untreated or poorly treated wastewater releases into the environment. Numerous septic systems: (1) put water into the soil across the land rather than hydrologically funneling it to a single downriver spot; (2) reduce the drying out of wetlands, ponds, and streams; (3) have no CSOs (combined sewer overflows); and (4) provide tertiary treatment that minimizes eutrophication-causing phosphorus and nitrogen inputs to water bodies. Therefore, when a residential population becomes sufficiently dense, it is normally best to connect houses to a sewage treatment system. Then old septic tanks can be pumped, crushed, and covered.

Stormwater and pollutants

Cities get rid of rainwater as fast as possible, yet pump in scarce freshwater. Is rainwater a waste or a resource?

Hard or impervious surfaces, plus a storm-drainage pipe system, concentrate and accelerate stormwater flows. Flooding in and downslope of the city is a frequent result. The rapidly flowing stormwater picks up and carries off pollutants. That effectively cleans the city and dirties the local water bodies.

To provide suitable conditions for both people and nature, we would keep water levels below flood stage, minimize pollutants entering water bodies, and keep adequate flows in streams during dry periods. In essence, four treatments or approaches are used to accomplish these goals: reduce, slow, infiltrate, and filter stormwater. Both water quantity and quality are addressed at different spatial scales (Walesh, 1989; Loizeaux-Bennett, 1999; Alberti, 2008), from sprawl areas (Frumkin et al., 2004; Korhnak and Vince, 2005) to dry-climate urban areas (Westerhoff and Crittenden, 2009).

Impervious surfaces and water flows

Rainwater falling on an urban area in effect either evaporates upward from impervious surfaces, is transpired upward by plants, infiltrates downward into the soil, or flows nearly horizontally over or under the urban surface as *stormwater runoff*. The percent runoff (amount of runoff divided by rainfall) normally correlates with the percent of *impervious surface* (or hard or impermeable or sealed) cover (Figure 6.10) (Butler and Davies, 2011). Roofs, driveways, sidewalks, streets, highways, and other such surfaces are considered to be impervious (Lee and Heaney, 2003; Wessolek, 2008; Breuste, 2009). Imperviousness of an individual surface type ranges from <10% for grassland or woodland to 80% for roof surfaces and 100% for very high-quality roads.

Many approaches and equations are used for determining percent stormwater runoff (see equations, Appendix B) (Butler and Davies, 2011). Urban areas in the UK composed of many surface types have the following stormwater-runoff percentages on average: 5–30% parks and gardens; 30–50% low- and medium-density housing; 50–70% high-density housing, apartments, and suburban business areas; 50–90% industrial areas; 65–100% city commercial areas; and 70–95% city center.

Residential areas in North America on average have approximately the following percent impervious surfaces (Figure 6.10) (US Environmental Protection Agency, 1993; Schueler, 1995; Arnold and Gibbons, 1996; Paul and Meyer, 2001; Forman et al., 2003) (1 acre = 0.4 ha): 20% impervious surface for 1-acre house-plots; 25% for ½-acre plots (about 12% for cluster development); 30% for 1/3-acre; 38% for ¼-acre; and 65% for 1/8-acre house-plots. Other urban land uses approximate 75% impervious surface for an industrial area; 85% commercial area; and 95% shopping center.

Woods may produce 5–10% stormwater runoff, while lawns and other planted areas normally produce 10–20% runoff (Schueler, 1995; Arnold and Gibbons, 1996; Lynch and Hack, 1996; Paul and Meyer, 2001; Alberti, 2008). A 20% impervious surface produces about 20% stormwater runoff; 35–40% imperviousness produces about 30% runoff; and 85–90% imperviousness produces about 55% stormwater runoff. Thus, adding impervious surface in an exurban area generates more runoff than does the same surface area added to a more developed inner suburb.

This increase in runoff with more impervious surface also means a decrease in evapo-transpiration and infiltration. Evapo-transpiration is reported to drop only slightly (from 40% to 38%) in going from 10% to 90% impervious surface. However, infiltration into the soil drops sharply from some 50% to 15% along that imperviousness gradient. Along the gradient, the shallow infiltration of water, which supports vegetation growth and some stream flow, drops from about 25% to 10%. But deep infiltration potentially reaching the groundwater drops more, from 25% to 5% along this increase in impervious cover.

These broad percentages of impervious surfaces or stormwater runoff provide a preliminary overview, but understanding stormwater flow patterns depends on the types and permeability of individual surfaces and their spatial arrangement (Wessolek, 2008). Making a smooth surface rougher and adding objects to reduce water flow increase "surface depression storage" (see equations, Appendix B). This reduces stormwater runoff and increases evapo-transpiration. This solution is especially effective for light rains. Adding soil and plants increases both infiltration and transpiration, thus reducing percent runoff.

A more informative indicator of stormwater conditions is referred to as *drainage connection* or *connected*

Figure 6.11. Small stormwater detention basin where runoff from roofs and parking lot accumulates. The basin volume may hold drainage from a 1 inch (2.5cm) rainstorm. City of Lake Placid, Florida. R. Forman photo.

impervious cover. This is the percent of an area with impervious surface directly connected to a water body (Taylor et al., 2004; Ladson et al., 2006). A roof or carpark for instance, may be directly connected by a pipe system to a pond or river. In this case essentially no reduction, slowing, infiltration, or filtering of the stormwater occurs, so the water body receives maximum water quantity and quality impacts. On the other hand, a low drainage connection normally means that water from the roof or carpark drains to an adjacent (Figure 6.11) or distant vegetated soil. Such soil is often in a basin or depression where some or all of the four treatments (reduce, slow, infiltrate, and filter) occur. Reducing drainage connection provides many ecological and societal benefits.

Since greenspaces and spots of green cover typically have very low runoff percentages, the spatial arrangement of impervious and vegetated surfaces is a key to stormwater flows. For example, in an urban area gradually sloping down to a river, impervious areas alternating with green corridors parallel to the river could have almost all stormwater treated in the corridors. Virtually no surface runoff into the river occurs; only cleaner subsurface flows enter the river. Alternatively, stormwater runoff from impervious surfaces could be funneled to a reasonable density of scattered greenspaces across the urban area. Spatial pattern remains an opportunity for stormwater research.

Wooded greenspaces are particularly effective at reducing stormwater runoff (Lynch and Hack, 1996; Gartland, 2008). Most urban impervious-surface cover, however, replaces former farmland with its furrows, ditches, and roads. This surface transformation only modestly increases stormwater runoff, though it considerably changes the pollutants carried in stormwater. The highly pervious sand and gravel used in and around impervious surfaces, and covering railways, tends to absorb water and reduce runoff. Transportation structures including roads and carparks often compose more than half of the urban impervious surface (Southworth and Ben-Joseph, 1996). A 0.4-ha (1-acre) carpark may produce 16 times as much stormwater runoff as does a meadow (Schueler, 1995), an irony, since commercial carparks are often only 30–40% used (Benfield et al., 1999). Narrow streets have significantly less impervious surface than do the usual wide streets of suburbia (Frazer, 2005; Girling, 2005).

Drainage ditches or swales with grass or other plant cover provide several functions – friction, infiltration, transpiration, and retention – that reduce stormwater runoff (Spirn, 1984; Hill, 2009). Tiny "check dams" in ditches further reduce runoff (Marsh, 2005). Indeed, attempting to mimic nature by using a decentralized stormwater-infrastructure system to disperse human effects is a core of "low-impact development" (Richman et al., 1997; Forman et al., 2003; France, 2003). The use of vegetated rather than impervious infrastructure can greatly reduce stormwater runoff (France, 2003).

For instance in a small residential watershed or catchment, instead of funneling most stormwater in a pipe network to a large water-detention basin, in low-impact development a decentralized collection of small water collectors distributed across the watershed sharply reduces runoff. These collectors, variously called rain gardens, swales, depressions, drainage ditches, wetlands, biofilters, detention ponds, and retention ponds, effectively catch, hold, and treat rainwater. In low to average rainfall periods, such an approach results in little to no stormwater runoff at the bottom of the watershed. Runoff from heavy rains is sharply reduced. The decentralized vegetation-dominated system typically increases in efficiency over time as vegetation matures, and is less costly, less subject to variations in government budgets, and less likely to fail in heavy rains.

Another approach for reducing stormwater runoff is to make pavement surfaces more porous or permeable. *Porous pavements* are composed of a material such as tarmac/asphalt or concrete that is full of 3-dimensional spaces or voids (Ferguson, 1998; Scholz and Grabowiecki, 2007). The surface pores permit some runoff water to infiltrate into the voids and downward into the sandy soil beneath. In a light rain most water may infiltrate through porous pavement, whereas the proportion infiltrating is much less in heavy rain. Porous pavements are prone to clogging and reduced effectiveness within a few years, though they may be cleaned by expensive vacuuming or pressure washing, plus pollution collection and disposal. Technology research attempts to reduce or eliminate the clogging problem. At present, porous pavement may be most useful in sidewalks and light-traffic areas such as driveways and some carparks.

Permeable pavements, in contrast, are concrete blocks or plastic structures with large voids (holes) containing gravel or soil and grass (Figure 6.7) (Scholz and Grabowiecki, 2007). Thus, rainwater readily infiltrates downward and evapo-transpires upward. Occasionally the soil is inoculated with decomposition microbes to aid in treatment (cleaning) of stormwater pollutants. Permeable pavements reduce, treat, and increase infiltration of stormwater. In so doing, the pavements increase water recharge to the soil (and potentially groundwater), decrease water input to the stormwater pipe network, and reduce pollutants in the stormwater.

A study of flows through asphalt/tarmac and four types of permeable pavements in a Seattle (USA) carpark found virtually no runoff from the permeable pavements (Booth and Leavitt, 1999; Brattlebo and Booth, 2003). Compared with runoff from the asphalt, all the permeable pavements' runoff had lower levels of copper and zinc, plus no motor oil, a characteristic of carpark pollution. Furthermore, less atmospheric heating and evaporation of hydrocarbon pollutants (VOCs) occurs over carparks with permeable pavements than over tarmac/asphalt carparks (Asaeda and Ca, 2000).

A *hydrograph* typically portrays the amount or discharge (cubic meters or feet per second) of stormwater runoff in a stream or ditch or pipe, before and following a storm event (Hartshorn, 1992; Marsh, 2005; Vince et al., 2005; Butler and Davies, 2011). Compared with a non-urban area, several hydrograph patterns are characteristic of stormwater flows from an urban area.

Peak flow, the maximum height of water, occurs earlier in an urban area, due to more rapid runoff. Peak flow is higher, meaning that the potential flood level is higher and flood damage greater. Peak flows, and hence potential floods, are more frequent. The total amount of stormwater flow is greater, reflecting less infiltration and evapo-transpiration, further increasing flood potential (see equations, Appendix B). Low flows following the storm event are lower, and therefore streams and other water bodies are prone to drying out, with loss of fish. The daily variation in streamflows may be greater (Konrad and Booth, 2005). These patterns of urbanization effects on stormwater runoff highlight the hydrologic value of reducing peak flow and maintaining adequate flow (base flow) in streams during low-flow periods.

The lag time between rainstorm and peak-flow urban runoff is commonly a few to several hours. Frequent spikes in a hydrograph record indicate short-duration high-peak discharges of stormwater following storms, highlighting the *flashiness* of urban runoff. Flashy discharges over time tend to degrade stream channels and widen rivers.

In Singapore, a highly efficient storm drainage-pipe system accelerates stormwater flows to rivers, resulting in severe river flooding. Cities in developing nations typically have less-efficient stormwater-drainage systems, and therefore more water infiltration into the soil that tends to somewhat limit river flooding. Indeed, channeling stormwater runoff into ditches or "swales" containing grass, herbaceous vegetation, or woody vegetation produces an array of generally useful results: more friction, more infiltration, more subsurface water flow, more groundwater recharge, a higher water-table, less subsidence of the surface, more evapotranspiration, less surface-water flow, less erosion, less sedimentation, longer lag time to peak flow, lower peak flows, and less flooding. That's a bundle of benefits.

Dutch studies illustrate how stormwater discharge rate relates to percent impervious (sealed) cover (Tyrvainen *et al.*, 2005). Increasing impervious surface cover from 0% to 5% to 10% to 20% to 30% to 40% had increasing stormwater discharges of 14 to 16 to 24 to 42 to 60 to 80 m^3/s. Runoff rate changed little at the outset (0% to 5% impermeable surface). But from 10% upward, discharge increased linearly with increasing impervious surface cover (Schueler, 1995; Arnold and Gibbons, 1996). Naturally most "impervious" surfaces become more permeable as cracks form over time (Wessolek, 2008).

Stormwater pollution and local water bodies

One estimate suggests that 70% of the urban-related water pollution in the USA is due to stormwater runoff, far exceeding that directly from industry or human wastewater (Loizeaux-Bennett, 1999). Not surprisingly, stormwater pollution, i.e., the excess materials, chemicals and heat carried by runoff, originates from many sources and is quite diverse. The pollutants come from rainwater, dry aerial deposition (including wind-borne material), vehicles (leaks, wear, exhaust), commercial waste, industrial waste, construction sites, rubbish from people, animal wastes, road de-icing, urban agriculture, and lawns and parks. Pollutants are dissolved from, and picked up by, stormwater running over urban surfaces, including concrete, bricks-and-mortar, tarmac/asphalt, metals, roofs, and vehicles. For instance, water running over concrete or mortar between bricks dissolves and picks up calcium carbonate. Water running over tarmac/asphalt roads and carparks picks up hydrocarbons (including PAHs, e.g. pyrene, naphthalene, anthacene, fluoranthacene, benzine compounds) (Frazer, 2005).

The *first flush* of stormwater runoff after a rainstorm normally is richest in pollution, because the water quickly washes off much of the material accumulated on surfaces since the previous rainstorm. The longer the time between storms, the richer the first-flush pollution. Stormwater, especially the first flush, effectively cleans the city's surfaces.

The array of stormwater pollutants in turn includes oil, fecal coliform bacteria, nitrogen and phosphorus from lawn fertilizers and other sources, plus sediment, road salt, numerous chemicals from leaks and spills, and heat. Pesticides mainly come from domestic lawns, gardens, parks, and golf courses. Carparks are normally key sources of heavy metals and hydrocarbons (Mielke *et al.*, 2000). Stormwater pollution is often separated into two components: (1) suspended solids that can be filtered or, in still water, settle to the bottom; and (2) dissolved substances that are chemically in solution in the water. To give a sense of the characteristic composition of stormwater, the following are averages for a single stormwater event in the urban UK (ranges from place to place and event to event are wide) (Butler and Davies, 2011):

- 90 mg/l total suspended solids (TSS).
- 85 mg/l chemical oxygen demand (COD, a measure of organic matter that is decomposable by a strong oxidizing chemical).
- 9 mg/l biological oxygen demand (BOD, a measure of organic matter that is decomposable by microbes in the presence of dissolved-oxygen).
- 0.56 mg/l ammonia nitrogen.
- 3.2 mg/l total nitrogen (TN).
- 0.34 mg/l total phosphorus (TP).

- 0.30 mg/l total zinc (Zn).
- 0.14 mg/l total lead (Pb).
- 1.9 mg/l total hydrocarbons.
- 0.01 mg/l polycyclic aromatic hydrocarbons (PAHs).
- 400 to 50 000 ("most probable number" MPN/100 ml) fecal coliform bacteria (*E. coli*).

Analogous stormwater concentrations occur in the urban USA (Schueler and Holland, 2000; Marsh, 2005). Total suspended solids include organic matter as well as soil particles and other inorganic particles. The COD and BOD organic matter comes from dead leaves, garbage, combined sewer overflows, and other sources. Nitrogen and phosphorus come from almost all types of urban area including lawns, transportation, industry and commercial areas (Bernhardt *et al.*, 2008). Heavy metals originate from many sources, especially industry, transportation, and flows through metal pipes (see Figure 5.10). Hydrocarbons primarily originate from petroleum products, and fecal coliforms from human wastewater.

Local water bodies such as streams, rivers, lakes, and estuaries are the main recipients of stormwater pipe or ditch flows. The impervious surface cover of the urban area drained is sometimes considered to be the primary factor determining conditions for aquatic ecosystems and fish in these water bodies (Paul and Meyer, 2001; Lee and Heaney, 2003). For example, increased accelerated water flows and flooding strongly alter erosion/sedimentation patterns and fish populations (Frazer, 2005; Konrad and Booth, 2005). The area of a water body close to the ends (outfalls) of stormwater pipe systems is usually strongly polluted and altered by periodic heavy stormwater flows.

Water bodies, including wetlands, contain five important processes that tend to reduce the ecological effects of stormwater pollutants:

1. *Settling* to the bottom by suspended solids or particulates in calm water.
2. *Filtration* of debris and particulates by underwater plants, stems, dead leaves and branches.
3. *Assimilation*, the uptake of nutrients (including metals) by growing rooted plants and algae.
4. *Absorption* (or adsorption) by humus and mineral soil on the bottom.
5. *Decomposition* of organic matter by microbes.

These pollutant-cleaning processes, however, are highly sensitive to, and inhibited by, pollution itself. Furthermore, some of the stormwater pollutants are extremely toxic, or quite persistent, or accumulate through the food chain to become toxic to fish predators.

The combination or interaction of stormwater quantity and quality produces most ecological effects in local water bodies (Paul and Meyer, 2001). For example, rapid flows and flooding pick up and carry more pollutants to water bodies than do slow flows. Typically in polluted streams, the quantity and diversity of algae present are low. Likewise, compared with more-distant natural streams, the quantity and diversity of invertebrates are low. Fish communities normally are degraded. A study of midges (chironomids), the tiny biting pest insects common in urban waters, found similar densities in streams along an urban-to-rural gradient, even though species composition changed along the gradient (Gresens *et al.*, 2007). Particularly detrimental to fish populations in urban streams are low flows, which have high water temperature, concentrated pollution levels, semi-isolated pools or deep holes, and frequent fish kills.

Finally, stormwater pollutants often affect clean-water supplies. In exurban/peri-urban and suburban areas, typically stormwater rushes through many short pipe-networks to local wetlands, ponds and/or streams. Water in these water bodies commonly recharges the groundwater, which in turn serves as the source for water-supply town and private wells. Infiltrating polluted water through the soil partially cleans the water, as illustrated by 30 cm (1 ft) of soil significantly reducing heavy metal (zinc) levels in stormwater (Remmler *et al.*, 1998). Analogously for a city, major stormwater pipe-systems often funnel stormwater pollutants into a river, which serves as water-supply for a downriver urban area. Both arresting the water flows and cleaning the water pollutants enhance potential water supplies for urban areas.

Part II Ecological features

Chapter 7 Urban water bodies

There are no fixtures in nature.
The universe is fluid and volatile.
Ralph Waldo Emerson, Circles, 1841

Ye nymphs that reign o'er sewers and sinks,
The river Rhine, it is well known,
Doth wash your city of Cologne;
But tell me, nymphs, what power divine
Shall henceforth wash the river Rhine?
Samuel Taylor Coleridge,
The City of Cologne, 1800

Virtually all cities began by a water body. After centuries or decades of expansion, today a city can claim lots of urban water bodies of many types. We now dive into six key topics: (1) urban wetlands and ponds; (2) constructed basins, ponds, wetlands, biofilters; (3) urban streams; (4) urban rivers; (5) flooding by river and stream; and (6) urban coastal zones.

Urban wetlands and ponds

Wetlands

The types and definitions of wetlands vary widely. We refer to *wetlands* as vegetation-covered areas where water is at or above the ground surface for an extended period most years (Keddy, 2000; Marsh, 2005; Mitsch and Gosselink, 2007). Such a water regime produces three major characteristics of wetlands:

1. *Hydrology* – water in sufficient quantity flows into and maintains the wetland.
2. *Soil* – wetland soil is often saturated, contains considerable organic matter, and mainly exhibits anaerobic decomposition.
3. *Vegetation* – plants adapted to wet soil predominate.

In exurban/peri-urban and suburban areas where regulations limit development within a fixed distance of a wetland, these three key characteristics are important in determining the *wetland boundary* or limit (Marsh, 2005). Generally wetland vegetation covers the smallest area, wetland soil a somewhat larger area, and wetland hydrology the largest area. Thus, a developer may prefer using wetland vegetation, and a conservationist using wetland hydrology, to delimit a wetland boundary. In locating septic systems, the presence of wetland soil is a clear sign of a wetland. However, a percolation test for water flow through soil typically provides a more conservative measure for determining suitable non-wetland conditions.

Thus, with freshwater, swamps often have a visible water surface for some 1–3 months ("hydroperiod"), and marshes perhaps 2–5 months, during a year. Coastal saltwater wetlands are tidal, with saltmarshes mainly in temperate zones and mangrove swamps in the tropics (Figure 7.1) (Mitsch and Gosselink, 2007). Ponds may be present within wetlands.

Human perceptions of wetlands have traditionally been negative – places full of mosquitoes and flies, diseases, odors, and evil spirits, places to get lost in, useless waste space. Consequently, wetlands have been extensively eliminated by filling or draining, especially by early farmers in the land today urbanized.

In recent decades, perceptions have changed in many regions so that wetland functions and values are more important than the negatives. Accordingly, wetland protection, even restoration, has become a priority (Middleton, 1999; Mitsch and Gosselink, 2007). Wetlands as habitats in natural land are well studied ecologically (Keddy, 2000; Mitsch and Gosselink, 2007). Therefore, we simply introduce a handful of wetland characteristics that are particularly relevant to urban areas.

"Surficial wetlands" normally are shallow and form over a buried hard layer such as clay, concrete, or tarmac/asphalt, or form where flowing water is blocked, e.g., by a road or building construction (Marsh, 2005).

Figure 7.1. Planting small mangrove trees by tall marsh grass along edge of sewage-polluted brackish water in river mouth. Red mangrove (*Rhizophora mangle*) and reed grass (*Phragmites*) at high tide. Rio de Janeiro. R. Forman photo.

Such sites often dry out during dry seasons and are sometimes called seasonal wetlands. Surficial wetlands themselves tend to be temporary, appearing and disappearing. In contrast, "groundwater wetlands" receive water from the groundwater and the wetland surface water is essentially at water-table level (see Chapter 6). Pollutants in urban groundwater reach the wetland. "Riparian wetlands" form alongside streams and rivers, and thus are especially sensitive to annual and periodic fluctuations in flowing water. Basically wetland species are adapted to, and some "require," fluctuating water levels (Middleton, 2002).

Finally, coastal wetlands (Figure 7.2) are especially complicated because they often receive stream or river flows, plus freshwater from groundwater (upwelling), but also receive saltwater in varying amounts from daily tides and periodic storms. Coastal wetlands frequently appear sequentially in saltwater, brackish, and freshwater zones, though the types often intermix in complex patterns.

The *freshwater tidal wetland* is a particularly relevant one for coastal cities by rivers (Figure 7.2) (Mitsch and Gosselink, 2007; Guntenspergen *et al.*, 2009). These distinctive wetlands appear along rivers with daily saltwater tides rising and falling at the river mouth. The tides cause flowing river water to also rise and fall for some distance upriver. On a flat plain, the vegetation and fauna in freshwater tidal wetlands extend a considerable distance upriver. The species are adapted to these highly unusual fluctuating conditions, and thus many rare species are normally present. Upriver pollution and riverside development extensively degrade these distinctive wetlands.

As suggested, urban wetlands are generally characterized by being relatively small and isolated, having highly variable water levels, receiving considerable pollution input, and being heavily disturbed by the surrounding concentration of people. Urban wetlands are especially dynamic sites where biodiversity reflects diverse ongoing alterations (Middleton, 1999; Keddy, 2000). The seasonal fluctuation in water level means that wetlands normally have both an upper aerobic zone and a lower anaerobic zone. Wetlands are usually covered with dense vegetation, which may or may not be species rich. Large patches each dominated by a single species are common.

Wetland functions are especially rich, creating the classic multi-functional ecological habitat. Even tiny urban swamps and marshes perform most of the functions. Wetlands reduce flooding. They absorb many pollutants. They decompose pollutants. They improve water quality. They recharge groundwater. They pump water upward in evapo-transpiration. They protect shorelines and harbors. They provide habitat for diverse wildlife. They serve as roosts for local birds, and rest stops in the urban matrix for migrating birds. They support rare species in the metro area. So, in addition to providing various values for people, urban wetlands are of greater ecological importance than their abundance would indicate.

A closer look at a few of these urban functions is useful. Wetlands are sometimes compared to a sponge, because of an ability to absorb and hold water. But when wetland soil is saturated, the wetland can absorb very little water. So the sponge analogy is only valid in the months when the water level is well below the soil surface. Wetland microorganisms decompose many common organic and inorganic pollutants into harmless by-products or mineral nutrients (see Chapter 4) (Campbell and Ogden, 1999; Mitsch and Gosselink, 2007). Microbes in the aerobic soil zone have a high rate of decomposition compared with the low decomposition rate of anaerobic microbes beneath.

Much of the phosphorus entering a wetland is basically absorbed and held in the wetland soil, though some is used in vegetation growth. In contrast, many transformations of nitrogen occur in a wetland (Keddy, 2000; van Bohemen, 2005). In the water, nitrogen fixation, ammonification to NH_4^+, volatilization, nitrification to NO_2 and NO_3, de-nitrification to N_2, and decomposition of organic compounds containing nitrogen occur. Assimilation of ammonia (as NH_4^+) occurs in sediments, while the air contains

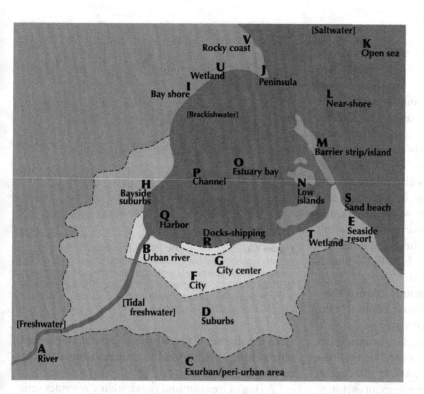

Figure 7.2. Land and water patterns around a coastal-estuary bay or lagoon of a river-mouth city.

N_2, N_2O, and NH_3 gases. Wetlands may reduce phosphorus, nitrogen, and other pollutants in water when functioning well. But high levels of pollution tend to decrease the effectiveness of, or eliminate, the water-cleaning function. Therefore, to provide sufficient wetland water-cleaning requires having a sufficient wetland vegetation cover and distribution relative to the incoming level of pollution.

A wetland often has considerable habitat heterogeneity, for instance with tree areas, shrubby areas, varied zones or patches of herbaceous vegetation, and even open ponds (Keddy, 2000; Mitsch and Gosselink, 2007). Frequent diverse disturbances mean that several successional stages are often present in patches. Plant diversity may be rich. Waterbirds such as herons, egrets, rails, and ducks are often conspicuous (Parsons, 2002). Amphibians and reptiles thrive in wetlands without too much pollution. Invertebrates are abundant and diverse. Since most of these species are scarce in a metro area, urban wetlands may not only be biodiversity "hot spots," but also valuable educational and recreational sites.

A glimpse of a few urban wetlands highlights additional key characteristics. San Francisco Bay was originally ringed with salt marshes, many of which over time have been filled and built on, a common pattern for many coastal cities, including Tokyo, Manila, New York, New Orleans, and Mumbai. Other San Francisco marshes have been diked, harvested, degraded, eliminated, and/or restored.

The East Kolkata (Calcutta) wetlands contain the world's largest (125 km²) wastewater ecosystem (see Chapter 6) (Costa-Pierce *et al.*, 2005). Located on river deltas, the area was long ago canalized and partially drained, with ponds created for harvesting salt. Today, sewage wastewater feeds fields, paddies, and ponds that annually produce thousands of tons each of vegetables, rice, and fish for food. Furthermore, the area is a rich reservoir of biodiversity.

Finally, Indonesia's capital, Jakarta, is one of the world's largest metro areas, covering some 650 km² of delta and other land surrounding the mouths of approximately 19 rivers emptying into the sea (Marshall, 2005a). Much of the area was former wetland, about half of which has been eliminated by filling for buildings and roads. Much of the remaining wetland, however, is home to informal squatter settlements. Residents create their own homes, lanes, schools, boat transport, electrical lines, and so forth in the mangrove and other swamp vegetation. Although many of the swamp species remain, much of the wetland habitat is intensively degraded.

Ponds

Ponds appear to be scarce in urban areas, yet the diversity of pond types is typically quite high. "Ponds" are small water bodies encircled by land. The combination of smallness, apparent scarcity, and heterogeneous types indicates that, in total, ponds harbor a large number of species. However, the small habitats are relatively isolated and have rather different species assemblages, and species movement among ponds is probably limited. The relative scarcity of ponds also means that by themselves they are of limited use in absorbing and treating urban pollutants. However, pollutant-cleaning by ponds can be important where stormwater or wastewater is channeled to a pond for filtering and decomposing pollutants (Costa-Pierce et al., 2005; Bernhardt et al., 2008; Butler and Davies, 2011).

Four major external factors mainly control characteristics in an urban pond (Colburn, 2004): (1) the surrounding land use; (2) surface-water and stormwater runoff to the pond; (3) subsurface groundwater flow to the pond; and (4) shoreline conditions. If the surrounding land is the lawn of a park or golf course, the pond may receive considerable phosphorus, nitrogen, and pesticide, whereas surrounding roads provide considerable hydrocarbons and heavy metals. Surface-water runoff typically comes from the immediate surroundings, whereas piped stormwater runoff often drains from several square kilometers of urban area. Normally groundwater slowly brings water and urban pollutants from a large area in one direction, and flows underground into the pond. Shoreline conditions of an urban pond also range widely, from natural woodland (see Figure 1.5) to a ring of trees, lawn, eroded soil, rock-pile (rip-rap) border, or concrete wall.

Pond shape varies from round to having a convoluted margin (Kalff, 2002). Although ponds are usually only up to a few meters (several feet) deep, a rounded pond may be deep enough in the center to support some relatively large fish. On the other hand, a convoluted pond has an extensive edge or shallow "littoral zone," which provides important aquatic-habitat heterogeneity, and consequently pond biodiversity. The littoral zone may drop down steeply or may deepen gradually from the pond margin. A gradual littoral zone, in addition to being safer for people, provides a series of habitats (Kalff, 2002). Typically one sees a sequence of emergent, floating, and submerged vegetation, each type being a different microhabitat for aquatic insects, amphibians, small fish, and larger fish. If far enough

Figure 7.3. Park pond with floating green algae and bordered by cattail marsh. Algal eutrophication probably due to excess phosphorus from park management, stormwater, and waterfowl. Nesting boxes added to enhance insect-eating birds; cattail (*Typha*) and weeping willow (*Salix babylonica*) behind. Toronto. R. Forman photo courtesy of Jessica Newman.

from the shoreline, tiny islands in a pond may be suitable nesting habitats for some animals threatened by terrestrial predators.

A ring of trees around the shoreline provides some water cooling, while fallen branches and leaves provide aquatic habitat (Figure 7.3). Dense shoreline vegetation offers cover for water-dependent frogs, salamanders, turtles, snakes, and alligators or crocodiles. Some ponds eutrophicated with excess phosphorus contain abundant rooted aquatic-vegetation as well as floating algae in the water. Other ponds may have a dense cover of floating plants such as duckweed (Lemnaceae) or water hyacinth (*Eichhornia*). One or two species of plant often predominates in a particular microhabitat. Small fish and some larger fish are usually present. With relatively high microhabitat heterogeneity present, the overall biodiversity in a pond is usually relatively high.

Ponds being small and surrounded with people and intensive land uses means that the aquatic ecosystems are frequently disturbed. On average that further enhances biodiversity. However, pond-repair mechanisms vary from very good (e.g., beavers repairing a leaking dam) to minimal (a worn-out pump for water input is not replaced).

As suggested, urban ponds are highly subject to incoming pollutants from surrounding lawns, roads, buildings and people, as well as distant pollutants entering in stormwater pipes and groundwater. Urban ponds frequently look green due to phosphorus and nitrogen eutrophication (Figure 7.3). Pond water nearly still, so sediment and pollutants such as phosphorus readily settle to the bottom. Eutrophication and extensive algae production in a still pond not infrequently leads to rampant microbial decomposition, loss of dissolved-oxygen, anaerobic conditions, and death of fish and other aquatic organisms. Pathogenic coliform bacteria from wastewater-pipe leakage, or CSOs (see Chapter 6), or from feces of ducks, geese, and pets, live in most urban ponds. To counteract some pond-pollution problems, chlorine is sometimes added to kill most pathogens, and water may be oxygenated with an air bubbler.

At least 16 types of ponds are present in urban areas:

- *Natural pond.* Surrounded by woods; scarce in urban area; biodiverse.
- *Vernal (ephemeral) pool.* Water body dries out in dry season; often supports rare species.
- *Oxbow on floodplain.* Scarce (little river migration in urban area); biodiverse.
- *Beaver pond.* On dammable stream; usually temporary in urban area; biodiverse.
- *Dammed pond.* For recreation, power or flood control; repairs keep it going.
- *Pond in former gravel pit.* Often quite acid, deep, steep banks, low biodiversity.
- *Pond in golf course.* Receives fertilizers/pesticides; eutrophicated; often used for irrigation.
- *Pond in city/suburban park.* Fertilizer/pesticide/animal inputs; ringed with people.
- *Public swimming pool.* Chlorine added; intensively used; nearly void of species.
- *Swimming pool in backyard house plot.* Chlorine added; nearly void of species.
- *Backyard pond in house plot.* Often changing/temporary; many types of species.
- *Institution or business-center pond.* Fertilizers/pesticides; some biodiversity.
- *Blocked-drainage pond.* Surrounding inputs; no maintenance; may be biodiverse.
- *Constructed stormwater-runoff pond.* Peak flows and pollutants enter; see below.
- *Pond at wastewater-treatment facility.* High in microbial pathogens and/or nutrients.
- *Industrial-waste holding and/or treatment pond.* Rich in toxic chemicals.

This array of pond types is equally diverse in their locations within the metro area. The broad functions of removing pollutants and of supporting biodiversity also vary enormously. Some species movements between ponds are known, such as amphibians moving through suitable habitat between vernal pools (Colburn, 2004). However, linking the diverse scatter of ponds together to provide integrated functions for the metro area remains a research frontier and planning opportunity.

Constructed basins, ponds, wetlands, biofilters

A distinctive feature of most urban areas is the array of tiny ponds, basins, wetlands, swales, and biofilters constructed to deal with stormwater and its pollutants (Rowney et al., 1999; Pitt and Voorhees, 2003; France, 2003; van Bohemen, 2005; Davis et al., 2009; Hurley and Forman, 2011). Some of these *constructed basins* or depressions (or "best management practices") (Jing et al., 2006; Hogan and Welbridge, 2007) end up as ugly rubbish collectors, while others with attractive plantings are welcome contrasts with the surrounding built area. Usually all of the basins are ringed by a shallow shelf for safety, and often with a narrow strip of aquatic or wetland vegetation. Basins have relatively steep sides with a weir or weirs or an outlet pipe near the top.

All constructed basin types are designed to receive and at least temporarily hold stormwater so that it does not quickly pour into a local water body. Effectively these basins reduce peak flows (cut the tops off hydrographs) and flooding. Some also treat and clean stormwater pollutants. Furthermore, designs differ depending on whether the underlying urban soil is contaminated or not. If the soil is chemically contaminated (e.g., a brownfield), it is best to avoid having stormwater flow through it, due to the probability of leaching out chemicals and carrying them to a local water body. Bioretention basins are typically used on uncontaminated soil.

Bioretention or *retention basins* are small constructed ponds or wetlands that retain (hold) and treat (clean) stormwater runoff (Hammer, 1997; Thompson and Solvig, 2000; Hsieh and Davis, 2005; Davis et al., 2009; Hurley and Forman, 2011). If the basin is above the water-table, usually a clay or other rather impermeable layer is placed at the bottom, on which a more-porous soil mix is added (Campbell and Ogden, 1999). A key to constructing wetlands is to get the incoming water flow right (Kusler, 1990; Larm, 2000; Braskerud et al., 2005).

Wet-tolerant grasses and other herbaceous vegetation are common in bioretention wetlands, though some shrubs and trees provide shade, habitat diversity, and more evapo-transpiration. Some bioretention wetlands above the water-table dry out during dry seasons or dry periods. Bioretention ponds, though often eutrophicated by incoming phosphorus and nitrogen, typically also retain good levels of dissolved-oxygen and aerobic decomposition partly because of mixing by incoming stormwater (Mallin *et al.*, 2002; Tanji *et al.*, 2002; Lavielle and Petterson, 2007; Vollertsen *et al.*, 2007).

Stormwater entering the bioretention depression is stored, gradually evapo-transpired upward, and infiltrated downward into the soil, potentially recharging the groundwater (Larm, 2000). Furthermore, stormwater pollutants are treated (cleaned) by several processes (see Chapter 6) (Mallin *et al.*, 2002; Barrett, 2005; Braskerud *et al.*, 2005; Dunnett and Clayden, 2007; Hogan and Welbridge, 2007): settling, filtration, assimilation, adsorption, and degradation/decomposition.

Suspended solids (particulate matter), including attached pollutants such as phosphorus, settle in the soil or pond bottom. Aerobic microbes decompose organic matter. Anaerobic microbes beneath also decompose organic matter at a slower rate. Other microbes oxidize inorganic compounds, including nitrogen compounds. The soil filters and absorbs/adsorbs many pollutants out of stormwater infiltrating downward. Consequently, bioretention basins, which typically require little maintenance, are the flagship of constructed stormwater structures because they accomplish so many key functions for society.

Swales, as wide shallow ditches covered with grass or other low herbaceous plants, are designed for stormwater flows, with the grass providing some friction compared with smooth flow in a pipe (Thompson and Solvig, 2000). In light rains much of the stormwater ends up in the swale, where evapo-transpiration and infiltration occur as well as the varied processes of pollutant treatment. In heavy rains, most of the stormwater flows along the swale to its end, though in the days following, some water and pollutants infiltrate into the soil under the grass-covered swale. Because of this water infiltration, although limited, swales are most appropriate on non-contaminated soil.

Two structures are particularly used over contaminated soils: detention basins or ponds, and biofilters (biofiltration cells). *Detention basins or ponds* (sometimes called catch basins), basically detain stormwater runoff, letting it slowly continue onward. This water detention reduces flooding (cuts off the peak flows of hydrographs). Between storm events, detention basins remain ponds if fed by groundwater, but often dry out over time if the water-table is deeper. A small amount of evaporation, infiltration, and settling of suspended solids occurs (Mallin *et al.*, 2002), but unlike bioretention ponds/wetlands, almost all the water from detention basins and ponds keeps flowing onward. Hardly any of the stormwater reaches the contaminated soil beneath. Detention basins are often seen near the lower end of large carparks, where they accumulate trash and require continued maintenance. In effect, detention basins and ponds slow stormwater flow but do not really treat (clean) it.

Biofilters, usually tiny enclosed wetlands designed to both slow water flow and treat pollutants, are used over both uncontaminated and contaminated soils (Welch and Jacoby, 2004; Dietz, 2007; Weiss *et al.*, 2007; Hurley and Forman, 2011). Normally these are tanks containing engineered soils and wetland plants that mainly depend on incoming stormwater flows. Evapo-transpiration occurs and water infiltrates through the internal engineered soils, where the various pollution-cleaning processes take place. If the bottom of a biofilter is above the water-table and an underdrain or subdrain pipe is present (Dietz and Clausen, 2008; Davis *et al*, 2009), water flows out the bottom and onward in the pipe. If the water-table is higher, water typically flows out in a pipe near the top, and onward. Thus, tiny biofilters both slow stormwater flow and treat stormwater pollutants.

"Street swales," instead of stormwater drain pipes, alongside suburban streets in Seattle and Portland (Oregon, USA) are effective in both slowing and cleaning water (Dunnett and Clayden, 2007; Hill, 2009). Also, stormwater running off the street flows directly to the roots of street trees. With street swales, the stormwater functions are combined with, and enhanced by, attractive plantings and other features. Indeed, biodiversity can be noticeably enhanced with street swales.

The appealing term, "rain-garden," usually refers to a tiny vegetated depression that absorbs stormwater (Dunnett and Clayden, 2007; Dietz and Clausen, 2008). This idea can apply to a tiny bioretention wetland as well as a biofilter. But it also applies to simple depressions created in parks and house plots, for example, to absorb stormwater from the downspout of a roof drain. Other tiny stormwater structures include a "rain barrel" that captures roof runoff, which in turn is used to water a garden. A "roof cistern" or roof surface catches rainwater that is stored in a tank and available for household uses. An underground empty structure or simply a hole filled with gravel will temporarily hold stormwater. Green roofs, porous pavements, permeable pavements, and green roofs do too (see Chapters 6 and 10).

As small structures in the urban environment, these basins and other solutions are highly subject to disturbances (Middleton, 1999). Droughts, heavy rainfall events, chemical spills, and human vandalism only hint at the range of disturbances that must be withstood for continued effectiveness in dealing with stormwater. Designing for change, for example using diverse rather than monoculture wetlands, is valuable. Minimizing maintenance and repair costs and effort sustains the array of stormwater treatment structures.

Usually the various small constructed stormwater-structures outlined are not considered appropriate for wastewater treatment, because of the high levels of organic matter and pathogenic microbes (Vymazal et al., 2006). However, large ponds and wetlands fed by wastewater are used in various metro areas for growing vegetables, rice, and fish for food (see Chapter 6) (Costa-Pierce et al., 2005).

Soil *remediation* commonly refers to the use of one or a few species for treating or reducing the pollutant levels in a contaminated soil, though many other definitions exist (Field et al., 1993; Terry and Banuelos, 1999; Kirkwood, 2001; Hollander et al., 2011). "Phytoremediation" uses a rapid-growing plant such as a grass or tree species, while "bioremediation" usually uses bacteria or fungi. Two or more plant species may be combined (Kadlec and Knight, 1996), and a range of technologies and materials are commonly used, especially in bioremediation (Margolis and Robinson, 2007). Even polluted stormwater could be remediated (Carleton et al., 2000). However, remediation is especially considered for the contaminated soils of brownfields. Typically such soils contain the products and by-products of former industrial or manufacturing processes, and consequently contain a range of chemicals that degrade ecosystem processes and are toxic to many organisms.

Perhaps most relevant and useful here in the context of urban water is to simply pinpoint the key issues in evaluating when remediation, in this case phytoremediation, is promising for effectively cleaning a contaminated soil.

1. What chemical pollutants are of prime concern, at what concentrations, and at what levels in the soil? What is the continuing input rate (e.g., from the atmosphere), if any, of the pollutants? What is the output rate (e.g., in surface water or groundwater flow) of pollutants?
2. With the chemical concentrations present in the soil, will the plant species selected grow rapidly for a sufficiently long period? Will it survive droughts, heavy-rain periods, and other disturbances?
3. Are the soil organic matter, available water, main root mass, and any necessary microorganisms sufficient and at the right level in the soil, so that the plants effectively absorb the target pollutant(s), or otherwise lead to its breakdown or detoxification?
4. If the plants absorb the pollutant, how often will they be harvested and where will the harvested plant material go?
5. Is the expected rate of pollutant reduction in the soil, or time until achieving a clean soil, cost effective, ecologically appropriate, and reasonable for society?

Such questions emphasize the challenges of brownfields and remediation, so continued research and pilot projects are important.

Many studies show the effectiveness of individual small constructed basins, ponds, wetlands, and so forth (Rowney et al., 1999; Tilley and Brown, 1998; Jing et al., 2006). However, much less is known about their cumulative effectiveness for an urban area, say, of tens or hundreds of hectares or acres.

A modeling study evaluated the amount of phosphorus reduction in stormwater using bioretention ponds and biofilters distributed across 80-ha (200-acre) industrial and institutional areas close to Boston's Charles River (Hurley and Forman, 2011). For each area, 1 to 40 detention ponds covered 5–15% of the total surface, and ca. 900 to 4300 biofilters covered 5–10% of the surface. The greatest phosphorus reduction was achieved when all stormwater flowed through a constructed treatment basin, irrespective of basin types, numbers, sizes, and locations. For basins covering 5% of the total land surface, a single pond achieved a 65% P-reduction, while biofilters with no underdrain achieved a 75% P-reduction. Doubling the total surface covered by basins only slightly improved the results. The arrangement of treatment basins was more important than their total area.

The combination of swales and bioretention ponds/wetlands in urban areas is designed to mimic stormwater processes in nature, accomplishing some of the key functions of streams, wetlands, and ponds (Thompson and Solvig, 2000). As much as possible, the constructed water features are fit to the form and contours of the watershed or catchment. If swales are curvy or convoluted, water flows are further slowed and reduced, and pollutant treatment increased. In a convoluted swale, typically considerable phosphorus settles out in the early portion and much nitrogen is absorbed and metabolized in the latter portion. The local downstream water body receives relatively little stormwater and pollution from a swale-and-bioretention system.

Low-impact development (see Chapter 6) uses an integrated array of small stormwater solutions, such as swales, bioretention ponds/wetlands, detention ponds, biofilters, green roofs, permeable pavements, and so forth to minimize the amount of stormwater and its pollutants flowing off a site or out of a watershed (US Environmental Protection Agency, 2000a; Dietz, 2007; Hood *et al.*, 2007; Booth and Bledsoe, 2009). Sometimes called a "begin-at-the-source" solution, this contrasts with an "end-of-pipe" solution in which most stormwater is funneled off-site to a large detention basin or local water body.

One study found no difference in the total stormwater runoff, total nitrogen, and total phosphorus flowing out of a low-impact development compared with that from a semi-natural area (Dietz and Clausen, 2008). However, total runoff, nitrogen, and phosphorus were a hundred times (two orders of magnitude) less from a low-impact development than from a standard housing development (subdivision). The array of small stormwater solutions was exceedingly effective.

For the pollutant levels typical in urban stormwater, the best solution is probably the conservation, recovery, and protection of natural systems including wetlands and ponds. However, at present the main alternative seems to be treatment in constructed basins, ponds, wetlands, and biofilters. Otherwise local water bodies will continue to be polluted or, more likely, become worse.

Urban streams

Unlike most of the preceding water features, which are unique to or characteristic of urban areas, the next three water-body types, i.e., streams, rivers, and coastal zones, are primarily non-urban. An extensive ecological literature exists for each of the three. Rather than attempting to encapsulate that knowledge, instead here

Figure 7.4. Stream directly receiving stormwater and pollutants from roofs and streets. Large rocks and logs, which survive the high-velocity spring flows, cover much of the stream bottom, thus causing splashes that oxygenate the water and providing potentially good fish habitat. Rascafria, peri-urban town north of Madrid. R. Forman photo.

we briefly highlight characteristics of each that make the urban versions unusual or distinctive.

Traditionally, and well before the rise of modern ecology, urban streams were basically considered to be a component of the stormwater drainage system (Figure 7.4) (Malcom *et al.*, 1986; Carpenter *et al.*, 2003). The "pipe-like" urban stream carried stormwater away as fast as possible to a local "receiving water" (Butler and Davies, 2011). Furthermore, stormwater was the essential cleaning mechanism for the city or town. The water carried off commercial wastes, manufacturing wastes, solid wastes, surface accumulations, and human wastewater. How else could they have been efficiently removed from a city?

> Engineered streams as lifeless sewers, designed to reduce flooding in one location [but which] compounds the problem downstream
>
> *Michael Hough*, Cities and Natural Process, *1995*

Several characteristics of urban streams, including stormwater, flooding, processes treating pollutants, and low-impact development, have been introduced in Chapter 6 and the preceding section. Here we present urban stream-related concepts in two groups: (1) watershed, floodplain, stream; and (2) water quantity, quality, biology.

Watershed, floodplain, stream

Water quantity, water quality, and stream-channel habitat conditions are three major direct controls on the ecology of urban streams (Rabeni and Sowa, 2002). In general, water quantity is strongly affected by conditions of the broad watershed or urban land. Water quality is primarily affected by watershed conditions but also by some local site conditions. Channel habitats are mainly affected by local site conditions but also by some broader watershed conditions. Thus, a stream site receives water and pollutants from the large upslope urban area but also from a single stormwater pipe just upstream. Both inputs strongly determine the stream vegetation, algae, invertebrates, and fish, indeed how the aquatic ecosystem functions at that site. In essence, stream ecology reflects the effects of urban land at both broad and site scales (Walsh et al., 2005b; Marsh, 2010; Wenger et al., 2009).

At the broad scale, the amount and arrangement of impervious surfaces, stormwater pipe systems, greenspaces, and the varied basins, ponds and wetlands greatly determines how much water flows in the stream (Jennings and Jarmagin, 2002). As noted above, much less water flows out of a low-impact development than from a standard housing development. A "small-scattered-patches" perspective of a landscape applies well in urban areas, and is useful in understanding urban streams (Forman, 1995; Corry and Nassauer, 2002). Even stream restoration sometimes focuses on the concept that "it takes a watershed to make a stream" (Williams et al., 1997a; Walsh et al., 2005a). A viable urban stream requires greenspaces of both urban area and stream site.

Since the urban groundwater water-table is often low, perennially flowing streams usually begin lower on the slope or valley, and hence almost all urban streams are on a previously formed floodplain. The streams theoretically could migrate back-and-forth across the floodplain, but typically human structures such as bridges and rock-lined banks keep the stream channel in a fixed position. Throughout history, urban floodplains, like wetlands and pond sides, have been magnets for dumping rubbish and debris. Hundreds, even thousands, of small mostly buried dumps line the water bodies of a metro area. All kinds of chemicals, some toxic, enter the stream water from such adjacent buried sources.

More familiar and conspicuous is the infrastructure on, over, and alongside urban floodplains (Carpenter et al., 2003; Forman, 2008): wastewater pipes, stormwater pipes, water-supply pipes, oil/gas pipelines, electric powerlines, roads, and railroads often pass in and out of cities along floodplains. Bridges, culverts, and even dams cross these periodically flooded surfaces. All the structures require maintenance and repair work. In short, the streamside strip is a dynamic place (Resh et al., 1988).

Riparian vegetation, particularly woody plants along urban floodplains, is another central feature that affects stream ecology (Groffman et al., 2003; Binford and Karty, 2006). The vegetation reduces streambank erosion, shades and cools the water, serves as habitat and wildlife corridor, is well used by birds (Knopf et al., 1988), drops branches that provide aquatic habitat, and plays many other ecological roles. In effect, riparian vegetation is important functionally and for biodiversity (Naiman et al., 1993). Protecting existing riparian vegetation (Ozawa, 2004) seems best, restoring semi-natural vegetation is almost as good, and covering the riparian zone with lawn and rows of trees is better than impervious surface. Still, much remains to be learned about the ecology of urban-stream riparian zones.

More than 130 000 km (80 000 mi) of streams and rivers in the USA bear the marks of urbanization (Paul and Meyer, 2001). Given the branching or converging form of stream networks commonly leading into a city, it is likely that most of the total could be called "urban streams" (e.g., with >10% impervious cover in the surrounding area; Arnold and Gibbons, 1996). Probably rivers constitute a small portion of the total length, and semi-natural streams predominate in exurban/peri-urban areas (with, say, <10% impervious surface cover).

A *stream* is a narrow water body with perennially flowing water on the surface, though periodically it may dry out in severe droughts (Giller and Malmqvist, 1998; Cushing and Allan, 2001; Kalff, 2002). An "intermittent channel" or gully only has visible surface flow during wet seasons, and "ephemeral channels" only have surface flow during some storms. In dry climates water usually flows beneath the ground surface in ephemeral channels. Much is known about urban stream ecology (Booth and Jackson, 1997; Paul and Meyer, 2001; Wang et al., 2001; Booth and Bledsoe, 2009), though variability is so great that discoveries doubtless lie ahead.

Headwaters of a stream or river refer to the beginning network of tiny streams (e.g., 1st to 3rd-order) (Cushing and Allan, 2001; Kalff, 2002; Allan and Castillo, 2007). Roads cover about 30% of American cities, and it is said that "streets are the headwaters of urban streams" (Stephanie Hurley, personal communication, 2010). This may increase "stream density" (see equations, Appendix B), and is especially evident

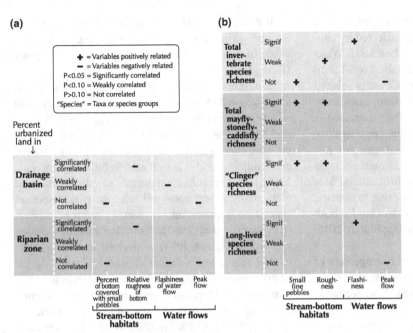

Figure 7.5. Invertebrate species richness, urbanization, and stream characteristics. (a) Stream characteristics relative to urbanization (percent urbanized land) at two spatial scales. Local riparian zone = 200 m on both sides of the stream, and extending 1 km upstream of a sample site. (b) Species richness relative to stream characteristics. Mayflies-stoneflies-caddisflies are especially characteristic of natural streams. Clingers typically cling to stones and rock, often on the underside. A 10-variable "benthic index of biological integrity" was significant for roughness and flashiness; not significant for bottom fine-pebbles and peak flow. Based on Spearman rank correlation coefficient (r); 7–18 streams in the Seattle Region (USA). Adapted from Morley and Karr (2002).

where rainwater runs from road surface to roadside ditch to stream. The roads and roadside ditches thus serve as ephemeral channels. But in a more built-up area, the road-surface water typically flows into a stormwater drain, and then continues through a pipe network to a local water body, which might or might not be a stream.

Indeed, in suburban areas of moist climates, it is not uncommon for a small semi-natural stream to disappear into a pipe. Then downslope the water reappears in a straightened ("channelized") urban stream, which further downslope also disappears. The channelized stream is frequently *armored*, that is, lined with rock (rip-rap) or concrete (Vince et al., 2005). Hydrologic equations are used for water flow in the heterogeneous semi-natural stream; engineering hydraulic equations for water in the pipe; and something in between for the channelized-stream water (Marsh, 2010; Butler and Davies, 2011).

The bottom of a stream is especially important ecologically (Figure 7.5) (Cushing and Allan, 2001; Kalff, 2002). Water flowing past large rocks and logs usually scours out deep holes where big fish (and little) often stay, as fishermen know. Tiny headwater streams (first order) may have a "step dam" appearance, a series of tiny dams and pools formed by rocks, logs, branches, accumulated leaves, and diverse dumped debris. Larger semi-natural streams (second to ca. fourth order) may have a "pool-and-riffle" bottom, a sequence of alternating deep pools with slow water and fast-flowing splashing water being oxygenated. Various combinations of rocks, gravel, sand, and silt often make a mosaic-like stream bottom. Each of the bottom surfaces, and many more, is a habitat used by different aquatic species. Stream bottom heterogeneity means habitat diversity, which supports biodiversity. A familiar way to increase stream habitat diversity, biodiversity, and fish is to simply add logs and rocks.

Urban streams are relatively straight because of human straightening or channelizing. Water velocity is greater in straight (sinuosity ratio <1.5; see equations, Appendix B) than in curvy streams. Fast-flowing water washes away small particles such as silt and sand, leaving a mainly rocky, gravelly stream bottom. Slower-moving water permits silt and sand to settle and cover the bottom, greatly reducing its habitat heterogeneity.

In the Seattle Region (USA), the relative roughness of stream bottoms decreases with more urbanized land in both the broad drainage and the adjacent riparian zone (Figure 7.5a). The species richness of mayflies-stoneflies-caddisflies and clinger-invertebrates in the stream, both groups characteristic of natural conditions, increases with stream-bottom heterogeneity (Figure 7.5b). However, total invertebrate diversity and long-lived-species diversity increase with the flashiness of water flows (Chapter 6).

Normally, straight streams are incising or cutting the land. Yet also sediments, from stream banks, construction sites, development-covered eroding hillsides, and stormwater that cleaned urban surfaces, accumulate on the stream bottom. Periodically a heavy storm causes a high-velocity high-peak-flow to wash away most of the sediment, leaving a wider, deeper, steep-sided stream channel (Neller, 1988; Paul and Meyer, 2001; Vince et al., 2005; Marsh, 2010). This urban stream channel is left with low habitat diversity.

Water quantity, quality, biology

The typical low water-table of urban areas means that streams are discharging water that infiltrates into the soil, rather than being sustained by water entering from the groundwater (to support base flow). Also, an extensive impervious surface-cover greatly reduces the shallow infiltration of rainwater, which would (via subsurface flow) provide water to the urban stream. Thus, urban streams are typically water-starved, except for periodic flushes by stormwater from pipes and surface runoff (Carpenter et al., 2003; Walsh et al., 2004).

Considerable variation in day-to-day streamflow is common (Brown, 2005). More distinctive is the alternating regime of brief high-velocity high-peak-flows and prolonged sluggish low-flows. The flashiness of peak flows, and the alternating peak- and low-flows, characterize almost all urban streams. The pattern tends to be most accentuated in small headwater urban streams (Cushing and Allan, 2001; Roy et al., 2005; Allan and Castillo, 2007). Few species are adapted to such a fluctuating environment, though as noted below, some urban species are (Figure 7.5b).

Consider the water flows around a stream in a small suburban watershed (catchment, basin) (Marsh, 2010). In a small storm event, visible surface water may slightly widen the stream, which mainly remains within its streambanks (if present). In a medium storm event, the stream may widen some distance across the floodplain, but also, surface water is often visible for a considerable distance up the intermittent and ephemeral channels/gullies. In the occasional extreme storm event, rushing water may cover the floodplain, and will normally be visible further up the ephemeral channels.

Furthermore, in valleys and gullies where surface water is virtually never seen, considerable (invisible) subsurface-water flows to the ephemeral channels. The subsurface flows do not clean surfaces or cause erosion, but they do flood basements and cause septic systems to malfunction. From the perspective of water quantity, peak flow, and flooding, the protection of ephemeral and intermittent channels with greenspaces and green cover is especially effective.

Water temperature in urban streams tends to be high and the dissolved-oxygen level low, especially in low-flow periods. Several factors contribute to the warm water, including urban "heat island" air temperature, scarcity of streamside riparian trees providing shade, typical lack of inflowing cool groundwater, and stormwater absorbing and carrying heat from urban surfaces (Paul and Meyer, 2001). In Roanoke City (Virginia, USA), reduced shade and increased channel width were found to be more important than the urban-surface heat in stormwater for increasing stream-water temperature (Krause et al., 2004). A slight temperature increase may occur from the input of sewage (CSO) wastewater (Kinouchi et al., 2007).

Water quality, as discussed above, is normally poor in urban streams (Novotny and Olem, 1994; Paul and Meyer, 2001; Miltner et al., 2004; Walsh et al., 2005b). A rich mix of heavy metals and hydrocarbons from transportation, organic matter from wastewater, nitrogen and phosphorus from many sources, and diverse toxic and other inorganic and organic substances from industry is characteristic. Typically, the changing mix mainly reflects conditions of a local site rather than the whole watershed (Rabeni and Sowa, 2002). For instance, local woodland cover increases ammonium-nitrogen in a stream, while local residential land and stormwater inputs strongly affect levels of dissolved-oxygen and chemical oxygen demand (COD). Nitrate-nitrogen concentrations in small suburban streams may be higher than in larger suburban/urban streams, due to runoff from lawns and other sources (Pouyat et al., 2009). Total-nitrogen levels in urban streams are important in affecting the abundance of midges (chironomids) and other invertebrates, as well as the complexity of aquatic food webs (Lawrence and Gresens, 2004).

In a city-to-exurban gradient study, herbaceous vegetation lining streams varied enormously in density, species diversity, and species composition (Urban et al., 2006). The vegetation patterns seemed to reflect characteristics of the stream, as well as vegetation and built-area patterns in the surrounding urban area.

Herbaceous vegetation along "wet ditches" tends to be dense and species rich (Geertsema and Sprangers, 2002; Blomqvist et al., 2003; Williams et al., 2003). Water flows along the ditch, especially in wet periods. Wildlife such as amphibians, reptiles, and raccoons

(*Procyon*) often use it as a movement corridor. The vegetation appears to be highly dynamic, changing in density and species, and particularly sensitive to ditch bank conditions such as aspect, angle, height, and erosion.

Invertebrate diversity in urban streams, compared with streams out in natural land, is usually quite low at a site, though may add up considerably along the length of a stream (Paul and Meyer, 2001; Fletcher *et al.*, 2004). In most cases population levels are low, probably resulting from heat, toxins, siltation, organic matter, low flows, and other factors. However, a few species thrive, the "urban-stream species," including types of chironomids, isopods, amphipods, and oligochaetes. Indeed, some may be adapted to periodic dry conditions.

A common riparian-zone salamander species (*Desmognathus fuscus*) in Atlanta (Georgia, USA) was scarce along urban streams where scouring from urban runoff had eroded streambanks (Orser and Shure, 1972). Also the reduction or removal of herbaceous groundcover in the riparian zone along urban streams reduced salamander populations.

As expected, the fish community of urban streams appears to be characterized by few species at a site, and modestly more along a stream (Karr and Chu, 1999; Paul and Meyer, 2001; Wang *et al.*, 2001; Fletcher *et al.*, 2004). Very few species are characteristic of riffles (stony stretches with splashing/turbulent flow) (Roy *et al.*, 2005). The fish community appears to change seasonally in response to the prevalence of low flows and flashy peak flows.

Canals, generally on a floodplain paralleling a river and with a steady slow water flow, are a distinctive variant of a stream/river (Braithwaite, 1976; Gilbert, 1991; Compagnie National du Rhone, 1996). Lots of characteristics make canals a distinctive ecological feature in urban areas: (1) eutrophication and plankton communities; (2) steep banks and a relatively homogeneous silt-covered bottom; (3) flat towpath and a sequence of small structures alongside; (4) generally less pollution than in urban rivers; (5) locks facilitating fish movement and the transfer of aquatic species between watersheds; and (6) usually a sequence of elongated wetland-vegetation patches. In cross-section, the canal corridor may support several fairly distinct plant communities (Gilbert, 1991). Lengthwise the corridor serves for movement of commercial or recreational boats, recreational walkers, and doubtless lots of birds, mammals, and other species.

Finally, *stream restoration* usually focuses on some combination of stormwater management, bank stabilization, channel reconfiguration, and riparian planting (Brown, 2000; Carpenter *et al.*, 2003; Walsh *et al.*, 2005a; Bernhardt and Palmer, 2007). As noted above, stream restoration may be a component of urban watershed improvement (or restoration) (Williams *et al.*, 1997a; Walsh *et al.*, 2005a). A review of 24 urban stream restoration projects concluded that problems and failures were greatest where an attempt was made to re-create natural channel geometry (Brown, 2000). Basically, surrounding land-use patterns have greatly changed over time, and today's streamflows respond to today's urban patterns, not to former natural or other land-use patterns. Recovering major characteristics of stream, floodplain, and vegetation cover, rather than attempting to restore detailed forms, vegetation types, and species, is a more promising solution.

Stream daylighting, the exposing of a previously buried (piped) waterway, has emerged as a component of recovery (Riley, 1998; Pinkham, 2000; Gleick, 2002; Bernhardt and Palmer, 2007). Hydrologically, daylighting relieves inadequate-capacity culverts and choke points, and reduces downstream flooding by using the stream's floodplain to hold and slow flowing water. Ecologically, the goal is to recover semi-natural stream-related functions and biodiversity. The enhanced sun, air, and soil conditions support a significant increase in infiltration, evapo-transpiration, aquatic species, riparian vegetation, and associated wildlife populations. The daylighted stream corridor provides a wildlife movement route, and may reconnect fragmented greenspaces of the urban area. Local residents seem to relish daylighted streams.

Urban rivers

Perhaps half of the world's cities are located by a significant river, and a fifth of those at the intersection of two rivers (Forman, 2008). Rivers run down a crease in the land, so tributaries and pipe systems draining land and built surfaces on both sides carry water and pollutants to the crease. Urban rivers commonly provide boat transportation of both people and goods, as well as waste disposal of stormwater, human wastewater, and industrial pollutants. Air movement along the river limits urban heat buildup and also helps clean out the city's air pollutants. Rivers make memorable cities.

In an urban region, the river not only connects the land and people in these ways, but forms a barrier to movement across. Bridges, especially near city center, provide connectivity for vehicles and trains but not for wildlife. The water and pollutant characteristics of river water mainly reflect the urban region or broad watershed (catchment) (Barber et al., 2006; Binford and Karty, 2006; Bryant, 2006; Galster et al., 2006).

Only a fifth of 38 urban regions analyzed worldwide have >80% natural vegetation covering the land within 2.5 km of streams and rivers (Forman, 2008). Most urban regions have less than a third of the land near streams and rivers in natural vegetation. Two-fifths of the urban regions have considerable built land (10–40% cover) present near waterways. Cropland dominates the land near streams and rivers in the great majority of urban regions, and is a major source of sediment and agricultural chemicals in urban rivers.

River channel and water

A natural river normally migrates on its floodplain, that is, has its channel moved from side to side during major floods (Gregory and Walling, 1973; Dunne and Leopold, 1978). This *river migration* leaves small mounds, depressions, wetlands and ponds (oxbows), which provide rich habitat heterogeneity and biodiversity on the floodplain. Near cities, people effectively almost eliminate river migration by construction in the floodplain, including bridges, piers, abutments, causeways, levees, dams, buildings, and various engineered banks (Brooks, 1998). Some of these structures squeeze or narrow the river through the city, so that water flow is accelerated (by the Venturi effect) (Forman, 2008). Also the river may be straightened or channelized through the city, further speeding up water flow.

The Los Angeles River is an extreme illustration of some of these attributes (Gumprecht, 1999; Orsi, 2004). Before the city, the river and its mouth migrated back-and-forth across a broad plain and delta with lots of wetlands. Such an area is highly heterogeneous, dynamic, and species rich (Acebillo and Folch, 2000; Keddy, 2000). Over time, almost all the water was diverted for other uses and groundwater feeding the river was pumped elsewhere. Therefore today, except in flood times, the river trickles with barely any wetland. Also urban construction initially constrained any migration of the seemingly unruly river, and then progressively much of its length was engineered into a wide concrete trough. The water trickle is mostly fed by stormwater pipe systems from across the city. Since the natural mechanisms of flood reduction were essentially eliminated and more peak flows were funneled to the still-called river, flooding can be severe. Residents and government have restored bits of the water flow and wetland in short sections of the channel. River restoration often focuses on the river channel and floodplain (Laenen and Dunnette, 1997; Nolan and Guthrie, 1998; Palmer et al., 2005), but as described below for Curitiba (Brazil), also includes solutions for headwater areas, riverside parkland, and riverbanks.

A worldwide survey suggests that in humid and temperate areas, urban river channels now have been enlarged on average 2 to 3 times, and some rivers up to 15 times the pre-city size (in cross-section) (Chin, 2006). Tropical urban rivers, however, may be somewhat smaller. A channelized river has a steeper slope, flows faster, has more high-peak flows, has lower low-flows, and has a more homogeneous bottom (Brooks, 1998). The lower low-flows mean that solar radiation heats the shallow water, reducing its dissolved-oxygen and most fish populations.

In some urban regions, a huge ongoing amount of silt is eroded from cropland and deposited on river bottoms. This sediment accumulation may require regular dredging to maintain boat traffic, though dredging also tends to undercut the urban riverbanks, as in the Brisbane River in Brisbane, Australia. Urban hillside development, construction sites, and riverbanks may be supplementary erosion sites and sediment sources to the river.

Riverbanks represent a sequence of habitats themselves, containing upland, riverside, and aquatic hyporheic species (see Figure 6.4) (Gilbert, 1991; Postel and Richter, 2003). A study of vegetation along riverbanks with fluctuating water levels by the River Thames (London region) highlighted the importance of riverside microhabitat heterogeneity (Francis and Hoggart, 2009). An abundance of plants and animals lives on river walls with various materials (stone, brick, concrete, wood, steel), angles, surfaces, cracks, depressions, drainage, sediment accumulation and exposure, and subject to varied maintenance and disturbance regimes. Two widespread plant species grew on most walls sampled, whereas almost all other species were only found on one or two walls. No invasive species were recorded. Vegetation cover and species richness increased with more wall cracks and heterogeneity. Downriver species richness was low, whereas upriver richness and vegetation cover were higher (many environmental conditions vary from upriver to downriver).

The vertical dimension from river-bottom to "water column" to surface layer is also ecologically

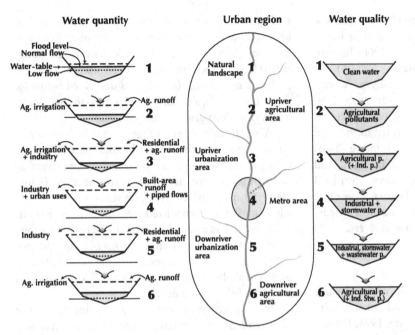

Figure 7.6. River water quantity and quality in urban region. Water runoff into the river decreases with loss of agricultural land, and increases with urbanization expansion. Tributary stream-water is commonly piped through the all-built metro area. Ag = agriculture; Ind = industrial; Stw = stormwater; p = pollutants.

distinctive in the three-dimensional urban river (Giller and Malmqvist, 1998; Allan and Castillo, 2007). Rivers mainly with low-flow conditions may have a diversity of rooted vegetation scattered across much of the river bottom (Gilbert, 1991). Otherwise, rooted vegetation is mainly limited to the riverside, and attached algae cover underwater structures such as bridge piers and logs.

Floating algae (phytoplankton) are abundant in the upper water column with abundant light, and are much less dense deeper down. Numerous tiny herbivores feed on the algae, and abundant aquatic insects feed on both tiny herbivores and fine organic-matter from upriver vegetation. Deeper into the river-bottom sediment are an array of highly distinctive low-oxygen-requiring groundwater (hyporheic) animals and microbes, as described in the Danube River sediment downriver of Vienna (see Chapter 6) (Pospisil, 1994).

A river slicing through an urban region exhibits a relatively predictable changing pattern of water quality (Figure 7.6), based on the sequence of pollution sources along the water-flow route. Also, water quantity patterns follow somewhat predictably (Figure 7.6). In the hydrologic case, four water levels are ecologically important: (1) normal flow; (2) water-table; (3) flooding; and (4) low flow.

In flood time, water rushes to the city. Then the many stormwater drainage-pipe systems lining the river pour in some more water, so that the downriver urban area may suffer large and many floods. The effects of big urban floods on society are well known. But the important ecological effects, as described below, are much less known (Thaitakoo et al., 2013). Such big urban floods are expected to increase with global climate change (IPCC, 2007).

On the other hand, most cities have "adapted" to the frequent regular seasonal increases in river flows, e.g., by building riverside embankments, providing depressions in adjoining parkland to hold floodwater, constructing a secondary greenspace route for floodwater flows, and protecting headwater areas. In urban areas that have not been narrowed or channelized, seasonal flooding provides access to rich floodplain habitats for fish and aquatic insects (flood-pulse model) (Junk et al., 1989; Kalff, 2002). Nutrients are available for rapid growth, young fish have an assortment of nurseries to avoid predation by big fish, and nutrients are carried back into the main channel. Even individual heavy-storm events provide some of these benefits. I recall the Sarapiqui River in Costa Rican rainforest rising 7.3 m (24 ft) in about 18 hours due to heavy rain in the distant mountains. The muddy floodwater briefly spread out across a forested and pastured floodplain, dispersing seeds and covering everything with nutrient-rich tree-growth-stimulating sediment.

Urban rivers also have low-flow times. In non-urban areas groundwater entering the river from below keeps an adequate water level (base flow) for supporting fish populations (Orsi, 2004). However, most cities have a lowered water-table due to pumping. In addition some, such as the Los Angeles River mentioned above, have diverted water or its supporting groundwater away for irrigation, industry, and/or urbanization. Low urban-river flows, especially when combined with water pollution, are highly degrading to the aquatic ecosystem. Cities, particularly in dry climates, may pump greywater or sewage-treatment effluent to support riverside floodplain wetlands.

River pollution and fish

In general, river high-flow or flooding time is especially characterized by water high in turbidity (in this case, muddiness) and sedimentation (Hartshorn, 1992; Laenen and Dunnette, 1997). Low flows particularly show the detrimental ecological effects of thermal pollution, low dissolved-oxygen, pollutant toxicity, salinity (where relevant), and excessive algal growth. Indeed, where low flow is mainly fed by groundwater, urban groundwater is usually quite polluted. Any flow level may have elevated microbial pollution, including fecal coliform bacteria such as *E. coli*.

It seems that if a river species is primarily limited by hydrologic water flow, the percent of impervious cover is a useful predictor (Allan, 2004). On the other hand, if the species is mainly limited by pollution, the percent of urbanized land in the watershed may be a better predictor.

Urban rivers, as the traditional major cleaning structures, carry away sewage wastewater and industrial wastes, as well as stormwater-drainage pollutants from residential, commercial and industrial areas. Industries and sewage-treatment facilities are often concentrated on the city's downriver side (Figures 6.8 and 7.6). In flood times, combined sewer overflows (CSOs) and cleanouts of industrial-waste sites are common. In effect, downriver urban areas (and further downriver) are both flood-prone and subject to a large dose of water pollution.

Conspicuous cases of river pollution are illustrated by the Cuyahoga River in Cleveland (Ohio, USA) catching fire in 1936, and again in 1952 (Wohl, 2004). In 1990, mainly untreated sewage from 12 million people, chemical pollution from >1000 industries, diverse pollutants from upstream tributaries, and tons of rubbish was dumped into the Tiete River of Sao Paulo. After considerable cleanup, the river today is less polluted even though the city has grown much larger. Still, many cities of the world are squarely on the trajectory of the Tiete before clean-up.

Two pesticides, DDT and chlordane, heavily used in agriculture have been recorded at high concentrations in New York's Hudson River (Wohl, 2004). Hundreds to thousands of tons of heavy metals (Cd, Cu, Pb, Hg) during the 1960s were annually dumped into New York's Hudson/Raritan river areas. Long after four decades of dumping polychlorinated biphenyls (PCBs) in Connecticut's (USA) Housatonic River, fishing is still banned. Pesticides – at least 8 herbicides, 8 fungicides, 17 insecticides, 2 nematocides – and industrial organic compounds (including dioxin, PCBs, and five others) in rivers are known to cause reproductive and endocrine-disrupting effects.

The ecological effects of various pollutants have been outlined above under streams as well as in Chapters 4 to 6. Thus, simply a few river-focused dimensions are mentioned here (Novotny and Olem, 1994; Laenen and Dunnette, 1997; Korhnak and Vince, 2005). Mainly sediments and agricultural chemicals wash right through the city, and have effects in downriver lower-flow areas. A legacy of industrial-waste dumping in rivers is mirrored in today's decades-later degradation of the aquatic ecosystem (Grimm *et al.*, 2008). Hazardous waste spills, with toxic chemicals commonly washed into the river, are concentrated in urban areas, along with industries and transportation (Wheeler *et al.*, 2005). Heavy metals and polycyclic aromatic hydrocarbons (PAHs), especially from transportation, degrade aquatic invertebrate populations (Paul and Meyer, 2001; Wheeler *et al.*, 2005). Many chemicals, including some heavy metals, bio-accumulate up the food chain to toxic levels (Wheeler *et al.*, 2005). Mercury channeled into the Sudbury River in suburban Boston more than 70 years ago accumulated in river-bottom sediments at levels such that the fish are still unsafe to eat for 50 km downriver. Partially treated wastewater entering a river produces a well-known downriver sequence of species types, ranging from highly tolerant to highly intolerant of sewage (Carpenter, 1983; Gilbert, 1991). Nitrogen at elevated levels was recorded down the Seine River for 100 km from Paris' major sewage-treatment facility (Chesterikoff *et al.*, 1992). Stormwater-runoff pollution seems almost innocuous compared with these river inputs, but, as described in Chapter 6, it has significant ecological effects on the aquatic ecosystem.

Even moderately clean urban rivers are generally not "swimmable or fishable." For example, the Charles

River slicing through the Boston Region has had excessive phosphorus, mainly originating in stormwater runoff from almost all urban land uses (Hurley and Forman, 2011). Nutrients do not recycle much in urban rivers; mainly they simply flow through the city. Reducing the phosphorus source is the main key to making the river more suitable for recreation.

Few fish thrive in urban rivers (Gilbert, 1991; Laenen and Dunnette, 1997; Karr and Chu, 1999; Allan, 2004). Muddy sediment from upriver agriculture clogs up gills, tends to homogenize the river-bottom, and degrades food sources. Then add all the urban pollutants, and urban-river fish survival is difficult. Most resident fish tend to be slow-water fish in little patches of slow water, while species of fast-water riffles are especially scarce. The relative abundance of impervious surfaces and scarcity of riparian vegetation in riversides may have a significant effect on fish populations (Karr and Chu, 1999; Binford and Karty, 2006).

The importance of vertical and horizontal structure within the river is illustrated by heat pollution from an electric-power facility located by Boston's Lower Charles River. The government permit for the facility apparently allows water heating of deeper river-water to extend only a short distance from shore, while in the upper water-level, heated water can extend outward about two-thirds of the river width. In this way, river fish have a considerable area of bottom for feeding, but more importantly, can swim past the polluting power facility in normal-temperature river water near the far shore.

Indeed, structural habitat heterogeneity seems to increase fish populations and diversity even in somewhat polluted urban rivers and other waters (Wolter, 2008). Suitable heterogeneous and semi-natural shorelines are particularly valuable for the urban fish community.

River water pollution close to cities blocks *fish migration* up and down river, such as (anadromous) fish mostly living in the sea and breeding in freshwater (Lucas and Baras, 2001; Wolter, 2008). Different types of river fish often migrate following large water discharges or floods. This is illustrated during especially high flows in a Manchester (UK) ship canal, when pollutants are diluted and dissolved-oxygen levels high (McCleave *et al.*, 1984; Lucas and Baras, 2001). Fish movement in a local area of a river may track food availability, while longer-range movement may be mainly a search for suitable food sources.

For a city in Brazil, migratory fish were not recorded within 300 km downriver (Alex Bager, personal communication, 2010). But after the city built a sewage treatment facility, migratory fish moved up close to the city on the downriver side. In the river network of tributaries, apparently forested parks with natural floodplains serve as "fish refugia," where the fish can stay until suitable water conditions occur for long-distance movement (Carolsfeld, 2003).

Riverside structures and riparian vegetation

City riversides are typically lined with low rises, high rises, old industries and infrastructure, along with scattered parks. Promenade walkways along the river are particularly attractive to tourists and residents. Upriver, more greenspaces and low-density residential areas typically appear. Downriver, more industries, a sewage treatment facility, and more flooding are characteristic. About half the riverside cities get a major portion of their water supply from the river, mostly within 15 km upriver of city center (Forman, 2008).

Close human interactions with an urban river are considered to be important for residents, tourists, and river commerce, as illustrated in Paris, Firenze (Italy), Moscow, Melbourne, St. Louis, San Antonio (USA), and other cities. Generally this means not separating the people and river by a railway or highway. Sometimes a walkway or tree-lined promenade is a solution (Figure 7.7d). A wide linear greenspace or greenway is a favored solution for residents, as well as ecologically for the river corridor (Cole, 1993; Baschak and Brown, 1995; Asakawa *et al.*, 2004; Bryant, 2006; Erickson, 2006; Hellmund and Smith, 2006).

A diverse, heavily used *riverside infrastructure* parallels rivers in urban areas. Gas pipeline, oil pipeline, electric powerline, water supply conduit, sewage wastewater pipeline, railway, and highway are common. These all require local maintenance roads, some of which are often on the floodplain. Also, sand and gravel mining for construction creates pits in floodplains.

Small objects lined up along the urban floodplain are common. Stormwater pipe outlets, ditches, and tributaries pour water into the river. Fill on the downslope sides of roads is particularly prone to erosion. In exurban and suburban areas, wildlife moving between land and river cross the infrastructure barriers through occasional suitable bridges and culverts, such as along Washington's Potomac River. Specifically designed wildlife underpasses would facilitate movement.

Typically, urban riversides have a dock area for recreational boats, ferryboats, and shipping (Figure 7.7g). Though heavily polluted, the aquatic habitat

Urban rivers

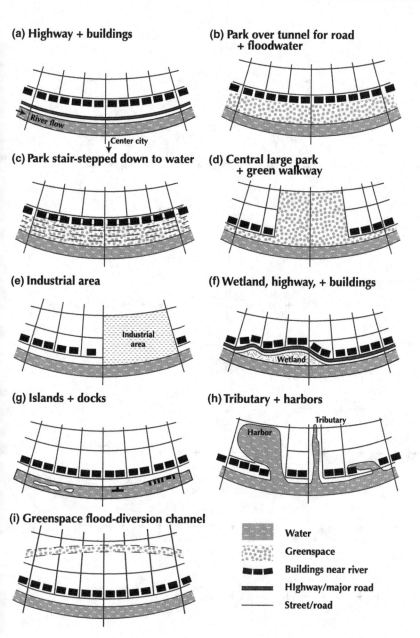

Figure 7.7. Nine alternative forms of urban riversides. Based on Wrenn et al. (1983), Holzer et al. (2008), Brown (2009) and other sources.

heterogeneity provided by piers and docks often supports some relatively uncommon urban species. In some river cities, harbors or lagoons have been dug out for boats, e.g., in a former wetland, and doubtless contain high pollutant concentrations plus some uncommon aquatic species along edges (Figure 7.7h).

The more natural land uses on riversides contrast with the hard engineered borders. Small or tiny wetlands typically appear scattered along the shore (Gilbert, 1991; Giller and Malmqvist, 1998). These collect a tiny amount of sediment and pollutants. The riverbank sometimes has stretches with a line of shrubs, trees, and other vegetation. Such narrow strips do little for arresting runoff and pollutants from the land, but do provide beneficial conditions for riverside aquatic species, as well as terrestrial vertebrates moving along the river. Floodplains, especially in suburban and exurban areas, may contain productive community gardens (see Figure 6.8), and, if extensive, may provide market-garden production for the city.

Nine contrasting forms or types of riversides seem to characterize urban areas worldwide (Figure 7.7). A

Table 7.1. Evaluation of environmental conditions on diverse urban riversides. See Figure 7.7. Overall evaluation for each environmental variable: + = positive; − = negative; . = not relevant, or positive and negative dimensions about equal

	(a) Highway, buildings	(b) Park over road/flood tunnel	(c) Park stair-stepped to water	(d) Central large park and green walkway	(e) Industrial area	(f) Wetland, highway, buildings	(g) Islands, docks	(h) Tributary, harbors	(i) Flood diversion channel
Air, soil, water									
Summer heat	−	+	+	+	−	.	−	+	.
Wind	−	+	+	+	−	.	−	.	−
Air pollution	−	+	+	+	−	.	−	+	.
Soil organisms	−	+	+	+	−	+	−	.	+
Erosion, deposits	+	.	−	.	−	+	−	−	.
Pollution to river	−	+	.	+	−	.	−	.	.
Flood-prone	−	+	+	−	.	+	−	+	+
Wetland, aquatics	−	.	.	.	−	+	.	+	.
Fish & migration	−	.	.	.	−	+	+	+	−
Habitats, species									
Vegetation cover	−	+	+	+	.	.	−	.	.
Natural vegetation	−	.	+	+	.	+	−	+	.
Habitat diversity	−	+	+	+	−	.	−	.	+
Habitat connects	−	+	+	+	−	+	.	+	.
Biodiversity	−	.	+	+	.	.	−	+	.
Riverside trees	−	−	+	+	−	+	.	.	−
Water-birds	−	.	.	.	−	+	+	+	.
Positives	1	8	12	11	0	8	2	8	3
Negatives	15	2	1	1	10	0	11	1	3

comparison of the riverside types based on 16 ecologically related variables suggests several patterns of interest (Table 7.1). Essentially all of the riverside forms have both ecologically positive and negative roles relative to air, soil, water, habitats, species. Negatives strongly outweigh positives in three cases (Figure 7.7a, e, and g). Positives predominate in five cases (Figure 7.7b, c, d, f, and h). The relative riverside patterns for soil, air, and water overall are similar to those for habitats and species. Finally, individual variables or features play key positive, or negative, roles on riversides.

Riparian vegetation, i.e., natural or semi-natural plant cover alongside a river or stream, is the most ecologically important component of the riverside. The vegetation provides dead organic matter in both particulate and dissolved form, as well as terrestrial insects, to the river. These are important food sources for aquatic insects and the river food web (Baxter *et al.*, 2005). In addition, the riparian vegetation provides friction, filtration, riverbank protection against erosion, some water storage in the soil, shade, and fallen logs and branches. The wood creates valuable habitat for fish and other organisms in an urban river (Naiman *et al.*, 1993; Binford and Karty, 2006; Naiman, 2009).

The riparian vegetation may be essentially continuous on one or both sides of the river, as in Washington, Edmonton (Canada), and Pamplona (Spain). Even passing under bridges, as in Washington, the woodland connectivity, with few breaks for lawns, roads or buildings, is useful for wildlife movement. Some migrating birds over a large metro area move and disperse seeds along riparian corridors (Tabacchi *et al.*, 1998). However, a three-decade (1966–97) comparison of urban riparian-woodland patches along the Furukawa River of Hiroshima City found that the riparian patches decreased in size, and became longer and narrower (Tanimoto and Nakagoshi, 1999).

A study of river riparian vegetation in an extensive, mainly built area roughly centered on Manchester (UK) found more non-native plant species in degraded river sites and in woodland, shrub, and tall-herb sites, compared with those in grassy sites (Maskell *et al.*, 2006). Sites with higher fertility and pH seemed to favor native species. Many later-successional riparian sites (see Chapter 8) were dominated by a non-native species. Not surprisingly, given the scarcity of urban riparian vegetation, rare species may be present. Thus, a vireo species (small insectivorous bird) in the San Diego (USA) metro area may survive at least temporarily alongside an urban river (Beatley, 1994).

Flooding by river and stream

Floods strike fear in people due to the high risk of property damage and drowning. Flooded basements, malfunctioning septic systems, and blocked roads and driveways also cause inconvenience. Yet long-term floodplain farmers may like the deluge, the overflow or expanse of water covering their land. Following the powerful rushing flows, receding waters normally leave wet depressions and cover the land with an uneven layer of silt rich in nutrients.

So it is with urban stream and river flooding. Puddles and ponds and wet depressions seem to proliferate. Wood and debris piles appear. Silt and mud paint the place, filling cracks, lightly covering smooth surfaces, and accumulating to impressive depth in spots. Plants and animals quickly respond and cover these new rich water and soil surfaces. The flooded urban area is ecologically rejuvenated, while people are cleaning up and rebuilding.

Flooded coastal cities instead get painted with salt, and sometimes sand. These surfaces quickly dry out and rather few plants and animals colonize and thrive. Overall, most of this new post-flood life in the city is gradually snuffed out by a long cleanup process. Indeed, pest species such as mosquitoes and midges (chironomids) may stimulate people to accelerate the cleanup. Nevertheless, the flood process provides a successional-habitat toehold for some plants, butterflies, diverse invertebrates, birds and other animals that spread in the urban area, or are ready to spread with the next flood.

Floods repeat. People in flood-prone areas can compare the present flood with those of the past. Society often estimates or maps the area covered by a 50-year flood and a 100-year flood, and may have regulations against building in those zones. These are zones where there is a 2% chance, or 1% chance, of a flood each year (see equations, Appendix B). But such flood zones change over time, depending largely on changing land use. For instance, Tulsa (Oklahoma, USA) apparently had four "100-year floods" in 14 years (1970–84). Since the flood zones had not been re-mapped for some time, each flood was considered an extreme case. Like many flooded cities, each time the urban area seemed to rebuild toward pre-flood conditions, as if suddenly floods were expected to stop.

Generally, *flood risk* can be measured as peak flow, that is, the maximum water depth from a storm or flood event (Butler and Davies, 2011). In estimating risk, some propose to link peak flow with its undesirable consequences. For instance, flood damage to a building is usually minimal if peak flow remains below ground-floor level, whereas considerable damage occurs if peak flow reaches 0.5 m above floor level.

But such a measure of risk would need to incorporate damage to natural systems in urban areas. Floods may flatten park trees; pour salt into freshwater ponds and lakes, killing the aquatic flora and fauna; clog up wastewater sewage systems, causing serious pollution; disrupt storm drainage; wash out small dams; spread toxic wastes from industrial treatment ponds; and spread mosquito-borne diseases. Natural systems underlie and permeate all cities and are at risk of significant alteration by floods.

Four major sources or causes of urban flooding are usefully recognized: (1) land upriver or upslope; (2) floodplain conditions upstream and upriver; (3) conditions in the city, especially stormwater drainage; and (4) coastal storms (even tsunamis). Rainwater, sometimes combined with snowmelt, is central to the first three, while seawater mainly represents the fourth source. The last two, urban stormwater flooding and coastal flooding, are explored in a section below. Here we introduce flooding from land and floodplain.

Floods from the upriver or upslope land normally result from rain or storm events. In a light rain over a long period, water infiltrates into and saturates the soil (fills the soil pores), which increases *base-flow* levels in streams. Slight prolonged flooding may follow. In a heavy, short rainstorm, relatively little water infiltrates and most water quickly runs off the surface (see equations, Appendix B). This commonly produces a large powerful short-duration flood that may cause damage. The problem is accentuated if the ground is frozen, or if it is a "rain-on-snow event" when a significant amount of snowmelt water is combined with rainwater runoff. A long-duration heavy rain saturates the soil, and runoff produces prolonged severe flooding. The last case is illustrated by a 1995 mega-flood along most of the Mississippi River (Marsh, 2005). Cities and towns on riversides or bisected by flood-prone rivers and major tributaries are frequently flooded, as illustrated by New Martinsville (West Virginia, USA) on the Ohio River.

The downtown area flooded so regularly that there was a local law requiring that motorboats cruising on Main Street not exceed 15 miles per hour – a law intended to minimize breakage of shop windows!

(Baker, 2009)

Land use (land cover) is a major determinant of how much water runs off the land. Natural vegetation greatly reduces and slows water runoff, even on rather steep slopes, thus minimizing flood risk. Converting forest or woodland to pastureland greatly increases downslope flood probability. Converting the surface to cropland, especially with straight drainage or irrigation channels, further increases downslope/downriver flooding (and additionally accelerates soil erosion and downriver sediment deposits).

Converting the woodland surface to built land often results in the most severe flooding. Although it is possible to design development to limit flooding, the typical built area is full of structures that channel or funnel water downslope. Roads, ditches, and pipes accelerate water runoff down gullies and streams. Periodically in the much-developed Maresma hills by Barcelona, water and mud crash down gullies, rolling cars, wiping out structures, and leaving mounds of debris at valley bottoms.

Thus, development on hillsides leads to a significant flood risk due to rapid short-duration, high peak flows, often called *flashy* or *flash floods*. Such floods, also characteristic of some desert areas, may be especially damaging, both because of their quick and oft-unexpected appearance and the high water level briefly reached. In general, urban flash floods tend to be especially severe where impervious hard surface covers 30% or more of the land (Walsh *et al.*, 2005a). Typically, housing development on the slopes of hills and mountains surrounding a city directly results in flooding parts of the urban area.

Several human activities in the floodplain valley itself cause or accentuate flooding. Removal of upstream/upriver woody vegetation in the floodplain reduces the significant *floodplain friction* benefit, by which riparian vegetation slows water flow. Eliminating most floodplain friction leads to higher peak flow in the downriver urban area. Agriculture, including cropland and pastures, benefits from the high water-table and nutrient-rich soils of a floodplain. Indeed, wet rice production essentially depends on the frequent floods. Irrigation channels in floodplains are a special challenge because during dry periods they must carry sufficient water to all fields, and in wet periods drainage channels must prevent too much water in any field. This combined irrigation-and-drainage

Figure 7.8. Narrowed channelized portion of a river through a city center where history has recorded huge floods. Continuous tree lines along both sides of the River Seine and on the island. Floodwaters draining a huge agricultural land, formerly forested, have inundated urban roads alongside the river. Low-rise buildings with large central courtyard (lower right) and a series of tiny courtyards (lower left); recreational fields and courts adjoining. Paris. R. Forman photo.

system is more-or-less adapted to annual high flows, whereas big floods require considerable subsequent repair.

Traditionally, industrial development close to a river used the water for power, cooling, and waste disposal, functions that now can usually be provided away from a river or stream. Commercial and residential development in the floodplain is basically misplaced, and regulations often prohibit it. Narrowing of the river channel or floodplain by rock and concrete barriers raises and accelerates water flows, too often leading to more and worse flooding. Narrowing is especially characteristic of the city-center stretch (Figure 7.8).

The Cheonggyecheon Stream/River in city-center Seoul had severe flooding due in part to urbanization of the surrounding land plus floodplain narrowing (Rowe, 2010). An urban restoration project in the 2000s ripped out a congested elevated highway over the river, added recreational parkland and walkways on both sides, and created buried compartments and tunnels under and alongside the river to handle heavy water flows (Figure 7.7b). Although an input of pumped water is required, the objective of reducing or eliminating city-center floods there seems to have been met.

An even bigger picture is useful or essential in water-flow issues. Thus, a flood solution may improve conditions at one location but worsen them downstream. Alternatively, it may improve conditions in both places.

Sometimes *levees* (dikes built on the floodplain somewhat paralleling the river) are added to help floodwater rush past a city without covering the floodplain. Development then sometimes creeps onto the floodplain. Levees work, until they are occasionally overtopped or otherwise rupture, causing major flood damage to the inappropriate floodplain development.

Many of the preceding human structures and developments tend to be just upriver of or in the urban area, and consequently tend to increase flooding in the area downriver of city center. However, a narrows or dam or other blockage of water flow also raises the water level for a short distance upriver, which may send floodwater into adjoining urban areas. Of course the infrequent rupture of a dam instantaneously fills the downriver valley with floodwater and mud that may spread over both land and built area.

Stream/river *channelization* straightens the stream (reducing its natural curviness) and reduces heterogeneity in the channel (removing large rocks and logs that produce turbulence and decrease the downstream flow rate) (Figure 7.8). Europe's big Rhine River was straightened and shortened in the early 1800s (McNeill, 2000). In the late 1600s beaver managed the river floodplains of Illinois (USA), but two centuries later a quarter of the total river length in the state was channelized.

Numerous solutions worldwide for stream/river-related urban flooding have been implemented. In The Woodlands, an outer suburb of Houston (Texas,

USA), perceptive designers successfully used the air spaces within soil, especially sandy soil, as "temporary basins" to hold heavy rainwaters, and essentially prevent floods in a flood-prone area (Spirn, 1984; Galatas and Barlow, 2004). In Denver (Colorado, USA), several parks with low, mostly grassy areas along the flood-prone Platte River and tributaries serve as temporary holding basins for floodwaters originating in the nearby mountains.

In the late 19th century, F. L. Olmsted designed a large Fens-and-Riverway project on the edge of Boston with one-third of the area for flood control and sanitary improvement (Spirn, 1984). The design later became a major recreation area. Large storms dumped water on the created wetlands, which absorbed much of the water, gradually over weeks releasing it downstream and to the air. This water-focused solution contrasted with the adjoining lower basin, where extensive hard surfaces and straightened waterways channeled the water downstream within hours and days causing floods.

In the 1970s–90s, Curitiba (Brazil) created a remarkable multi-dimensional solution for flooding that also addressed a range of key urban problems (Schwartz, 2004; Irazabal, 2005; Moore, 2007). Five rivers essentially converge in the city and produce damaging floods. Headwater areas far upriver were protected against development. Informal squatter settlements and other development were removed from the highest flood-risk areas in the river valleys. Major riverside parks with green elongated depressions to temporarily hold floodwater were established. Habitat diversity in parks greatly increased, both from the varied surface topography created, and from diverse plantings including fruit trees, ornamentals, and native species. Several flood-control reservoirs were constructed with protected land around them, and were used for recreational fishing. Channelization was minimized so that the natural curves of streams and rivers would slow water flows and reduce downriver flooding. Throughout the process, the major goals were flood mitigation and recreation for all segments of society. Yet by-product benefits were considerable, including valuable wildlife habitat, improved air quality, decreased illegal waste dumping, and informal squatter settlement on less-threatened sites.

Finally, extreme cases of flood-prone cities such as Bangkok and Venice offer insights into human adaptation to existing conditions and sea-level rise. A rising water-table tends to disrupt underground drainage systems and transportation, cause surface ponding,

Figure 7.9. Saltwater fish that live near freshwater of a river mouth. St.-Malo, Bretagne, France. R. Forman photo.

and flood buildings built in earlier lower-water-table times (Shanahan, 2009; Thaitakoo et al., 2013). In severe-flood areas residents mainly live on the second and higher levels of buildings, leaving the ground floor empty or as warehouse storage space. A zonation of algae and other species thrives on diverse moist building surfaces. A range of fish and other aquatic species (Figure 7.9) that ordinarily could not survive in urban-polluted waters thrive in the cleaner floodwaters and diverse aquatic habitats available in and around buildings, bridges, and other structures.

Streets where sidewalks and ground-level shops are frequently flooded are good candidates for establishing a second level of sidewalks and shops, elevated over the existing ground-level ones. A third upper-level sidewalk/balcony could serve businesses and residences above.

Low-flow conditions in urban streams and rivers are essentially the opposite of flooding. Low flows may occur during drought or low-rainfall times, but urbanization causes or exacerbates low flows. In urban streams, low flow is mostly a result of humans removing groundwater rather than meteorological drought. Groundwater wells are pumped for industry and other uses, including clean-water supply, thus lowering the water-table and drying out wetlands and streams. Lowering the water-table often leaves an urban stream or river with a thin saturated soil beneath it, but these are "disconnected" from the lower groundwater by a zone of non-saturated soil. Being disconnected from the groundwater means that the riparian zone with its stream/river is "perched." The perched riparian zone continually loses water to the drier layer beneath and the stream typically has long-term low-flow conditions.

Low flows have wide ecological ramifications (Lake, 2011). Scattered deep spots along the stream or river are pools serving as confined refuges for some fish. Water temperatures tend to be extreme, e.g., from all ice in shallow portions in winter to high temperature in summer. High water temperature means low oxygen content in the water, which eliminates most fish species and many other aquatic organisms. Reservoir, lake, and pond levels drop in droughts, often with similar ecological effects for their aquatic species. Wetlands and small channels dry out, i.e., no longer have a visible aboveground water surface. Unlike sand, clay noticeably shrinks when dried out, so cities built on abundant clay may have continual problems with cracked building foundations. During drought many animals move further in search of water. Thus, throughout the urban area, rats, mice, cockroaches, ants and much more appear in unexpected places. As in the desert, spots with water in the metro area are hot spots for animals.

Urban coastal zones

City facing sea ... urban advances, ocean in charge. The ecology of the interaction of these two giant forces is enormously complex, important, and intriguing, a subject for whole books. So, following a glimpse of the big picture, we will highlight a few threads of particular urban-ecology importance and provide references for further exploration.

To sense the place where city encounters sea (Figure 7.2), the richness of structures, flows, and changes are listed in staccato fashion.

1. *Natural structures.* Sea. Estuary. Lagoon. Bay. Barrier island. Beach. Rocky shore. Coral reef. Mud flat. Salt marsh. Mangrove swamp. Lagoon island. River mouth. Tidal river. Tidal stream/creek. Upper water layer. Lower water layer. Bottom habitat. Sea-grass bed. (Boada and Capdevila, 2000; Keddy, 2000; Mitsch and Gosselink, 2007; Marsh, 2010).
2. *Human structures.* Harbor. Port. Dock (pier/wharf/quay). Pole (pile/piling). Filled land. Seawall (and jetty, groin, bulkhead). "Armored" shoreline. Dam. Building. Road. Impervious surface. Stormwater pipe system. Sewage system and treatment facility. Combined sewer overflow (CSO). Industry. Recreation/tourist area. Marina. Recreational boat. Commercial area. Highway. Fishing dock area. Fishing boat. Passenger/ferry terminal area. Passenger boat. Cargo-ship dock area. Cargo ship. Warehouse. Railway along harbor. Urban park. Promenade/walkway along harbor. Dredged channel. Shoreline residential area. Beach resort/recreation area. Shore greenspace. Wetland greenspace. Dead (hypoxic/anoxic) zone. (Wrenn et al., 1983; Benton-Short and Short, 2008; Brown, 2009).
3. *Natural flows and movements.* Onshore wind, breeze. Offshore breeze. Alongshore current. Sand and rock erosion. Sand transport and deposition. River/stream sediment flows and deposition. Nitrogen and phosphorus flow. Organic matter and animals moving from wetland to lagoon/estuary, and to sea. Tidal and storm-driven saltwater up streams and river. Wind-blown salt inland. Waterbirds moving among diverse habitats. Birds stopping in migration. Terrestrial animals moving. Aquatic invertebrates moving. Fish foraging (Figure 7.9). Fish migrating between sea and river. Marine mammal and turtle migration.
4. *Human-related flows and movements.* Beach recreation. Recreational boat cruising, fishing, sailing. Commercial fishing. Passenger ferry transport. Cargo shipping. Park recreation. Train with freight. Truck with cargo. Vehicle traffic. Walker, local commuter. Industrial air and water pollutant dispersing. Stormwater pipe-system flow. Stormwater sediment, pollutant dispersing. Sewage wastewater-system flow. Elevated nitrogen flow from sewage system, residential development, recreational boating. Ship ballast pumping of water and non-native species. Solid waste dumping.
5. *Natural changes.* Storm, cyclone/hurricane effects. Tsunami effects. River/stream flooding. Sea-level and land-level rise or drop. Species colonization and extinction. Species population increase, decrease. Seasonal ice, monsoon. Species migration.
6. *Human-related changes.* Seasonal recreation, blue-green algae blooms. Oil spills. Invasive species spread. Pathogenic bacteria increases. Wetland degradation, elimination. Beach replenishment. Seawall construction. Shoreline armoring. Beginning/ending of industry or shipping. Offshore dead zone expanding. Bridge, dam, little-lagoon construction. Dune destruction, restoration. Coastal development. Anthropogenic climate change effects.

Although incomplete and containing overlapping concepts, the lists highlight the striking natural and human

Urban water bodies

Figure 7.10. Dock and warehouse area of a coastal city port. The city expanded outward from the skyscraper area by covering extensive salt marshes and sea-bottom habitat with added fill and created the port. Sea-level rise threatens such an area. Boston. R. Forman photo.

differences in the urban coastal zone. Natural and human structures are intimately tied together. Indeed, natural flows, movements and changes strongly affect human structures, while human flows and changes affect natural structures. Envisioning the interactions among these patterns emphasizes the complexity of the system. It also suggests the enormous biodiversity and ecosystem flows present.

The urban coastal zone topic is introduced in three sections: (1) city, waterfront, harbor; (2) estuary lagoon area; and (3) outer coastal area. Lakes are briefly mentioned near the end.

City, waterfront, harbor

Coastal cities normally develop at uncommon sites, such as a river mouth, delta, or estuary (Figure 7.2). Cardiff (Wales, UK), Miami, and Sydney, as examples, inherently have an abundance of uncommon species and high biodiversity (Wheater, 1999). The familiar impervious surfaces and stormwater pipe systems predominate in the city, yet the systems near coasts frequently become overloaded or plugged, especially in flood times. Stormwater carries pollutants to the harbor but also provides a bit of turbulence, locally raising the dissolved-oxygen level. Furthermore, just as adding a tiny pond or bird feeder to increase house-plot wildlife, human structures in the water are numerous and diverse, potentially supporting a considerable biodiversity (Kowarik, 1990).

A harbor, with the waterfront by city center, usually contains small parks, multi-unit residential buildings, and a commercial area including hotels, restaurants and shopping, all for people's use on land (Figure 7.10) (Wrenn *et al.*, 1983; Clark, 2000; Brown, 2009). In contrast, four urban land uses along the harbor explicitly draw people into the water area (Wheater, 1999). A recreation and tourist area usually has a marina with recreational boats (Eriksson *et al.*, 2004; Thai *et al.*, 2007). A ferry or passenger-ship terminal and dock area supports a regular movement of people to locations across the water. A fish (fishing, fisherman) dock area with fishing/shellfishing boats normally has a strong aroma, lots of waterbirds, and perhaps many terrestrial rodents (Wrenn *et al.*, 1983; Brown, 2009). A cargo shipping area typically has large piers, warehouses for goods, a railway and highway often along the harbor, and large seagoing cargo ships. In addition, some urban harbors contain active and former industries, which may use the water for waste disposal and may have docks and ships. Boat docks/piers in urban harbors are mostly on filled land (Figure 7.10) (Sien, 1992).

Most harbors and ports are periodically dredged to keep ship channels open. The dredged and highly contaminated sediments may be deposited in a "confined disposal facility" in the urban region. Soil berms, a clay liner beneath, and sometimes an added soil layer on top of the disposal site are designed to minimize the spread of contaminants.

Estuary lagoon area

Saltwater *lagoons* (or *open-water estuaries*) are relatively protected from the ocean waves, typically by a sandy or

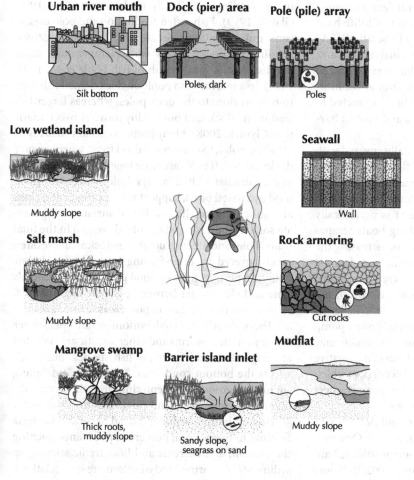

Figure 7.11. Edges and habitats of the open-water urban estuary. Each edge type supports a rather different set of aquatic species (including algae, fish, shellfish, and other invertebrates) and terrestrial species (including flowering plants, birds, and mammals). Salt marsh is in the temperate zone, mangrove swamp in the tropics.

stony barrier of islands, and usually only reach a depth of a few meters (several feet) (Figure 7.11) (Kjerfve, 1994; Lasserre and Marzolo, 2000; Valiela, 2006; Chapman *et al.*, 2009). Lagoons are common locations for seaport cities, including Tokyo-Yokohama, New York, Karachi, Kolkata, Dhaka, Shanghai, Cairo, and Lagos. Wetlands originally ringed an open estuary, salt marshes in temperate climates and mangrove swamp in the tropics (Keddy, 2000; Mitsch and Gosselink, 2007). Tidal streams near wetlands bring in freshwater and associated species from the land. The wetlands are enormously productive and support high wetland species diversity. Equally important, organic matter and invertebrates from the wetlands are carried or move into the lagoon, contributing mightily to the high density of fish, shellfish, phytoplankton, and other aquatic organisms in the lagoon.

A few usually low islands are often present near the edge of a lagoon. Such harbor islands have been intensively used over time for military activities, prisons (Howell and Pollak, 1991), waste disposal/dumping, excavation fill, parkland recreation, and nature conservation. On some 30 islands in the Boston Harbor, hundreds of species are present (Elliman, 2005; Paton *et al.*, 2005): 521 plants (54% native); 10 mammals; 3 reptiles; 1 amphibian; and 136 birds (50% probably breeding). Species richness seems to reflect the duration and intensity of land use and the island size, more than distance from the mainland. Harbor islands traditionally were valuable sites for breeding and roosting herons/egrets, because their terrestrial predators were largely absent (Parsons, 1995; Parsons *et al.*, 2001).

Although the habitat quality of a lagoon tends to reflect conditions of the broad urban area (Van Dolah *et al.*, 2008), several features of the open-water estuary are also important (Figure 7.11). The relative amount of bordering wetland, eroded-soil edge, or hard edge armored with rocks, concrete, and the like may be especially important for lagoon conditions. Without the wetland vegetation, wildlife habitat is sharply reduced. A hard edge in Seattle (USA) was "softened" by constructing a gradual underwater slope with different-sized rocks that was colonized by a sequence of aquatic plant and animal communities (Hill, 2009). Low islands can be constructed to absorb some energy of large storms and protect harbors (Aldhous, 2012).

The bottoms of lagoons are normally covered with silt and sand from river and stream flows. In relatively clear water, sea-grass beds rich in aquatic animals may cover a sandy bottom. A deep channel for shipping usually cuts through the lagoon and is periodically dredged. Equipment of certain fishing boats scrapes or sucks up sediment from the bottom, increasing the water turbidity. Recreational boat traffic may degrade the surrounding aquatic vegetation, create turbidity, and pollute the estuary with wastewater, nitrogen, and hydrocarbons (Murphy and Eaton, 1983; Eriksson et al., 2004; Thai et al., 2007). Cargo ships often pump out ballast containing species from afar, which may spread in the enclosed lagoon. Hundreds of non-native species are established in this way in US ports, of which some 85 species are considered to significantly affect ecological conditions (Wasson et al., 2002).

Coastal wetland vegetation filters and cleans water, dissipates wave energy (e.g., outside New Orleans), reduces shoreline erosion, and is a major wildlife habitat, especially for waterbirds (Parsons, 2002; Callaway and Zedler, 2004; Adams et al., 2006). The coastal marshes and open estuaries are among the most productive ecosystems worldwide, and the edge portions of coastal wetlands may be especially productive (Lee et al., 2006; Bernhardt and Palmer, 2007). Convoluted wetland edges accentuate these ecological benefits. A single greenspace wetland in a harbor, such as the 62-ha (152-acre) Belle Isle Marsh in Boston's Inner Harbor, attracts a large number of species, including a diverse resident fauna, many foraging species from the lagoon area, and migratory birds (Kelly, 1997). Mangrove swamps play an equally important role in tropical harbors, and also provide cool shade in the tropical midday sun (Murthy et al., 2001; Vannucci, 2004).

Boat landing areas along the harbor feature *docks* (piers, wharves, quays) supported by *poles* (piles, pilings) projecting down through the water into the sediment (Figure 7.11). Beneath docks, the algae, crustacea, various filter-feeders, and other organisms attached to the poles are relatively different from those in surrounding areas (Duffy-Anderson and Able, 1999; Glasby, 1999). Fish, often feeding on the pole organisms and on bottom species, are also fairly distinct. Shading by docks lowers water temperature, algae photosynthesis, and growth of fish (Able et al., 1998, 1999). In a marina with pontoon docks, small fish tend to remain close to the dock poles, whereas larger fish feed in the dock area but readily move between marinas (Clynick, 2008). Many harbors contain an array of standing poles, like an area of dead trees, where former docks existed. If poles are close together sediment usually accumulates. Pole arrays (pile fields) are much used by coastal birds, support numerous aquatic algae and animals along the roughened surface, and generate some turbulence and dissolved-oxygen in the tidal water. Poles and other structures in dock areas are often much-covered by filter-feeding invertebrates. If low pollution permits near-optimal growth, these animals filter and clean water between periodic disturbances. Fish tend to be abundant in pole arrays.

The sediment-covered bottom of an open-water estuary is rich in worms and other aquatic animals that are important food for many fish species. But much affects the bottom conditions: "upwelling," including submarine groundwater entering from the land; sedimentation from rivers and tidal streams; stormwater stirring up bottom sediment; dead algae and bacteria filtering to the bottom; numerous pollutants reaching the bottom; and currents and boat traffic stirring up sediments. Estuarine sandy-bottom areas of relatively clear water may have *sea-grass beds* (e.g., *Zostera*), rich in fish and other species (Figure 7.11) (Bowen and Valiela, 2001; Bradley and Stolt, 2006). However, sea-grass beds are usually scarce in urban areas, in part because peak flows of stormwater runoff carry sediment and also scour the bottom. These processes raise the turbidity or murkiness of the water (Gilbert, 1991; Lee et al., 2006).

A river and streams carrying sediment to the estuary may create a delta or mud flats that are much used by diverse waterbirds. Depending on water pollutant levels, migratory fish may move up the river past the city. The seaward portion of the river for some distance through the urban area is tidal freshwater, which inherently supports unusual plant and animal communities. Various small dams, artificial waterways, lakes, and small lagoons are often constructed around an open-water estuary. Together these support an impressive array of fish species, which are especially distributed according to salinity levels and distance from open-estuary water (Figure 7.9) (Waltham and Connolly, 2007).

Chemical pollutants from numerous dispersed sources (non-point pollution) flow through and accumulate in the water and sediments of estuaries (US Environmental Protection Agency, 2000a; Breen and Rigby, 1996; Valiela, 2006; Bernhardt et al., 2008). Indeed, heavy metal contamination levels (especially lead and copper) in estuarine sediments are apparently correlated with the amount of nearby land development (Hollister et al., 2008). Stormwater runoff systems seem to be the major pollutant contributor in urban estuaries of the USA, while wastewater is probably the prime source for cities with limited sewage treatment. Boats of various sorts contribute oil spills, chemical discharges, wastewater discharges, and ballast-pumped water. The types of pollutants, and some of their treatment or breakdown mechanisms, are familiar (see Chapters 4 and 6, and section above):

1. Particulate matter from industries, power facilities, river/stream sediment, stormwater runoff.
2. Heavy metals from industries and stormwater runoff; nitrogen and phosphorus from stormwater runoff, river/streams, wastewater.
3. Hydrocarbons from stormwater runoff, boats.
4. Toxic chemicals from essentially all sources.

Estuarine pollution not only degrades the natural communities and kills organisms but may cause genetic differentiation in populations. Polychlorinated biphenols (PCBs) in urban Southern New England (USA) apparently created genetically different populations of an estuarine fish (*Fundulus heteroclitus*) at distances of 2 km (1.2 mi) or less (McMillan et al., 2006). Some harbors apparently have been polluted so long that with water-quality improvement, marine borers (e.g., shipworms) begin to destroy the wooden poles and structures (Levinton and Drew, 2006).

A horizontal line of stormwater pipe openings pours the surface accumulations of a city into an estuary. This stormwater and pollution has major effects on estuarine water salinity, turbidity, temperature, and dissolved-oxygen, as well as the distribution and abundance of aquatic native and non-native species. Human wastewater from boats, septic systems of low-density coastal residential development, combined sewer overflows (CSOs), and incomplete sewage treatment fills estuaries with bacteria, including pathogenic coliforms (e.g., *E. coli*) (McKinney, 2004). Bacteria, water turbidity (or clarity), and nitrogen-nitrate levels in the estuary water seem to correlate with the percent of impervious cover in the urban area (Mallin et al., 2000). Even low levels of estuarine nitrogen from wastewater near Boston apparently increased eutrophication, altered the phytoplankton species composition, increased macroalgae abundance, and decreased or eliminated seagrass beds (Bowen and Valiela, 2001). But estuarine nitrogen can originate from tens of kilometers (miles) inland. Groundwater from sugarcane and other agricultural areas carried nitrogen to the Miami (Florida) coast in sufficient concentration to degrade coral reefs (Finkl and Charlier, 2003). Finally, the wide assortment of environmental impacts on urban estuaries markedly alters the fish populations present (Valiela, 2006).

Outer coastal areas

So-called *dead zones* (anoxic or hypoxic or oxygen-starved areas), kilometers to tens of kilometers wide and kilometers to tens of kilometers offshore of many coastal cities, such as New Orleans and New York, seem to be mainly caused by excess nitrogen. In essence, as described in sections above, excess nitrogen stimulates algae blooms (eutrophication), algae rapidly grow and die, dead cells filter downward, aerobic bacteria decomposing the dead cells explode in numbers, the bacteria use up the dissolved oxygen, fish swim away from the dead zone, and organisms of most other species in the hypoxic zone die. In Narragansett Bay offshore of Providence (Rhode Island, USA), one anoxic area is apparently mainly due to the major sewage-treatment facility. A second smaller dead zone is largely due to nitrogen from coastal residential septic systems, perhaps supplemented by nitrogen from small sewage treatment systems and numerous recreational boats (Andrew Altieri, personal communication, 2011). The population densities and mortality of several mollusks (bivalves), shellfish, and fish species seem to be correlated with nitrogen concentrations and the dead zones.

Narrow *barrier islands* or strips, usually at least partially covered with sand and containing one or more gaps (inlets), typically separate the estuary from the sea (Figures 7.2 and 7.11). Walking across a natural barrier island from sea to estuary traverses beach, primary dune, inter-dune trough (sometimes with small narrow freshwater wetland or pond), larger secondary dune, back-dune area, and bayshore saltmarsh or mangrove swamp (a small freshwater wetland may precede the saltmarsh/mangrove zone) (McHarg, 1969; Beatley et al., 1994; Platt et al., 1994; Pilkey and Dixon, 1996). Wind and salt spray from the sea, plus saltwater on the edges and freshwater in the center, are primary factors structuring the barrier island vegetation. Periodically major storms cut through a barrier strip.

Urban beaches are sandy shoreline magnets for crowds of recreationists and tourists (Benton-Short and Short, 2008). Dunes quickly become flattened and the vegetation degraded. As a powerful natural process, major storms often shrink or widen, or wash away, beaches (Pilkey and Dixon, 1996; Marsh, 2010). In urban areas these are sometimes "replenished" by pumping in sand from an offshore location, and later, sand is again eroded away.

Beaches extending beyond the estuary area are strongly affected by residential coastal development. Built areas may worsen near-shore water quality, beach width, dune height, beach bird populations, bird and sea-turtle nesting habitat, and habitat fragmentation. A model evaluating options for how to best combine beach recreation and coastal natural-habitat protection evaluated three variables (Forman, 2010b). Most important was the number of human-access points to the beach; second the recreational quality (including facilities) of an access point; and third the number of recreationists.

Seawalls (and jetties, groins, bulkheads, and other armoring structures) are mainly constructed of large rocks, sometimes with concrete or brick, to protect a beach or other coastal area from waves of major storms (Figure 7.11) (Pilkey and Dixon, 1996; Bulleri et al., 2004; Martin et al., 2005; Bertasi, 2007). The rocks provide new habitat colonized by algae and diverse invertebrates, and surrounded by fish and birds. However, the structures disrupt the normal movement of sand along the coast, as well as the habitats of many coastal animals. In The Netherlands, huge seawalls have been built to produce lagoons, and then water pumped out to create farmland in below-sea-level "polders" covering a third of the nation (McNeill, 2000; Steenbergen and Aten, 2007; Deinet et al., 2010).

Rocky coasts, ranging from cliffs to gradual slopes, support quite different arrays of species. The *intertidal zone*, containing horizontal sub-zones related to low, medium, high, and extra-high tides, is especially rich, because quite different species adapt to being repeatedly inundated and exposed to the air for different lengths of time. Gradual-slope rocky coastlines may cover a considerable area and be pockmarked with small intertidal pools. The pools contain their normal intertidal fauna, plus sea species temporarily trapped by a receding tide. Rock cliffs are pounded by ocean waves and thus full of cracks and crevices and ledges, many supporting various algae, invertebrates, and sea birds (Nybakken, 1997; Larson et al., 2000). For example, crevice-favoring mussels, chitons and limpets are especially sensitive to roughness of the rock face, whereas barnacles and oysters are more sensitive to the slope of a rock surface. Natural stone and brick seem to be better habitats than concrete.

Typically tropical *coral reefs* are slightly submerged limestone-like rock ridges covered by an enormous richness of algae, coral animals, filter-feeders, small tropical fish, large fish, and more. Such reefs are offshore from the beach, and optimally require clear water. Even a relatively low level of nitrogen can cause algae to overgrow and degrade a coral reef (Ward-Page, 2005). Not surprisingly, live coral reefs are scarce near cities and the mouths of rivers, due to water turbidity and pollutants.

In a coastal region, the terrestrial flows and movements of air, water, wildlife, and people are overwhelmingly parallel or perpendicular to the coastline (Figure 7.12) (Forman, 2010b). The same pattern appears on the seaward side of a coastal region. Although important exceptions exist, most flows go both ways, most are narrow bands, and most extend a long distance. Flows originating in the coastline mainly extend only a short distance. Flows and movements around a coastal city port are dense, often curved, and extend in many directions.

Flooding of urban areas by rivers and streams was discussed above, but *floodwater from the sea* is also ecologically important. Most dramatic of course is a "tsunami," though its impact on the human-built structures is usually worse than on natural systems, especially in zones where tsunami frequency is sufficient for species adaptation. For example, many palm trees, large multi-stemmed pandanus trees, and mangrove trees/shrubs survive tsunamis.

"Cyclones/hurricanes," particularly if large or moving slowly, may dump lots of rain over the land, which then pours into estuaries, just as monsoons and other heavy rains do (without such heavy winds). Yet cyclones and other major storms approaching a coast also push a lot of seawater toward the urban estuary. Flooding is severe when this added water coincides with high tide or extra-high tide. Southern Bangkok, for example, is periodically inundated by floodwater from the sea, and then northern Bangkok by river

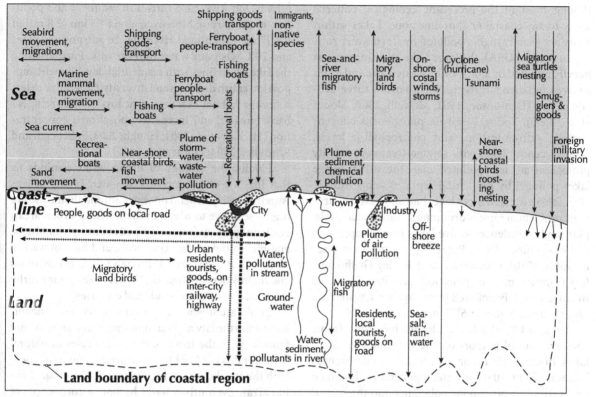

Figure 7.12. Characteristic flows and movements in a coastal region. Central coastline area is hundreds of meters to a few kilometers wide. Horizontal arrows in the sea portion indicate flows at different distances from coastline. From Forman (2010b).

floodwater from the land. Major floods such as these commonly inundate significant portions of a metro area (UN-Habitat, 2006). Much of Mumbai is on filled former mangrove swamp, especially the city's extensive informal settlements, so repeated extensive flooding in the city itself produces numerous tiny spots of sediment, plants, and animals (Murthy *et al.*, 2001).

Flooding also reconfigures the estuary (Figure 7.11) – destroying docks, buildings and stormwater systems; cutting through the barrier strip; washing away beach; rearranging or reshaping low islands; destroying some wetlands; creating new habitat for wetlands; filling the estuary with turbidity; killing countless organisms; and much more. Yet lots of new habitats appear; recolonization and rebirth follow. The urban estuary is dynamic indeed.

Sea level rise associated with anthropogenic climate change seems to promise more of the above, severe storms and flooding. Flood frequencies and risks are expected to rise ("10–30 times") in the upcoming future around UK cities and Sydney (Ashley *et al.*, 2005). Clogging stormwater drainage pipe systems and inundating infrastructure are familiar in large floods (McGranahan *et al.*, 2007). Additional ecologically important patterns are expected from sea-level rise (Douglas *et al.*, 2001; Valiela, 2006; Nicholls *et al.*, 2007; IPCC, 2007):

1. Converting low urban land to wetland or water body.
2. Erosion of coastal beaches.
3. Increased severity of flooding, and storm alteration of human and natural structures.
4. Increased salinity of surface waters (particularly up rivers and streams) and groundwater (saltwater intrusion).
5. Altered water-tables.

With the hard surfaces and armored shorelines of coastal development, the salt marshes and mangrove swamps may be narrowed or eliminated (Hartig *et al.*, 2002; Anthony *et al.*, 2009). Naturally these sea-level-rise-related patterns are likely to produce major changes in urban plant and animal communities (Wilby and Perry, 2006; IPCC, 2007; Aldhous, 2012).

Lakes and reservoirs (see Chapter 6) (Gilbert, 1991) have no saltwater, salt marsh, or mangrove swamp, though most of the preceding ecological patterns apply to any coastal or shoreline zone. Lakes within a city are usually highly polluted by stormwater runoff. In Seattle (USA), the large Lake Washington was heavily polluted by stormwater, wastewater and other sources, but then largely cleaned up over a three-decade period (Edmonson, 1991; McNeill, 2000; Moore et al., 2003). Today, based on phosphorus concentration, eutrophication, and chlorophyll-*a* levels, Seattle's central lakes with sewage-treatment-facility protection are less polluted than the urban-fringe lakes polluted by residential septic systems (Moore et al., 2003). Mexico City's lakes had and have a hydrologic problem of rapidly dropping groundwater levels, plus land subsidence, so five lakes have disappeared in five centuries (John Beardsley, personal communication, 2010). Canberra's Lake Burley Griffin suffers from stormwater pollution, and Brasilia's Lake Brasilia is heavily polluted by sewage wastewater and other sources (Chapter 6) (Forman, 2008).

Toronto, Cleveland, and Chicago have waterfronts and constructed harbors on the shorelines of major lakes. In an 11-km linear stretch of Central Toronto's lake waterfront, 10 stormwater pipes and 13 combined sewer overflow pipes pour pollutants into the harbor water (Royal Commission, 1991). However, the city-center Don River also carries in sediment and pollution to the harbor. Moreover, along 4.5 km (2.8 mi) of the Don's urban-land riverside, 26 stormwater pipes and 20 CSOs pour water into the mix. Further along the lake, shorelines with residential development support lower fish densities and diversity, largely due to less shoreline habitat heterogeneity, loss of "fish refuges," bank erosion, and increased phosphorus concentration (Hickley et al., 2004; Hough, 2004; Scheuerell and Schindler, 2004; Elliott et al., 2007).

Finally, the ecology of urban water reminds us that life is >98% composed of water. Living organisms intimately depend on water around them, and are highly sensitive to what's in the water. In urban areas, H_2O runs over surfaces or is piped in through buildings, and picks up particles, chemicals, and human wastes. These in turn are carried through pipes and channels and mainly deposited in local water bodies, our nearby streams, rivers, lakes, ponds and estuaries.

Look carefully, and an amazing array of microhabitats covers the city and surrounding urban area. We are familiar with the treasure chest of species in natural land. Yet it is hard not to be surprised, and impressed, with the rich bounty of biodiversity in the urban-habitat array. Even urban water harbors a cornucopia of species.

Part II Ecological features

Chapter 8

Urban habitat, vegetation, plants

There's a tree that grows in Brooklyn. Some people call it the Tree of Heaven. No matter where its seed falls, it makes a tree that struggles to reach the sky. ... the only tree that grows out of cement. It grows lushly ... there are too many of it.
  Betty Smith, A Tree Grows in Brooklyn, 1943

The requirement to maintain species richness sets the most stringent limits on many forms of land use.
  Gordon H. Orians, Environment, November 1990

Suppose a chemical was accidentally released into the air and killed all the plants in a city. Would it matter? Would the city look slightly different, or fundamentally different? Would almost everyone leave?

Or perhaps only certain types of plants died – trees, flowers, or native species. Or, in only certain habitats – lawns, street trees, or woodlands. Or, it prevented flower formation, seed germination, or shrub growth. Or it cut biodiversity by 90%, or tripled the number of species. Or vegetation change slowed, or accelerated, tenfold. Or areas were hit at different intensity, reconfiguring the vegetation patterns of the urban landscape. While these are extreme examples, they point to many of the key patterns, processes, and changes to be explored in this chapter.

We begin with (1) urban vegetation and habitat, followed by (2) the types of plants present. Next we introduce (3) urban plant biology. Then (4) trees and shrubs, and (5) plant community structure and dynamics are explored. Finally, we broaden to (6) plants and urban habitat fragments.

Urban vegetation and habitat

After briefly examining the different types of habitat and vegetation in urban areas, we focus in on habitat diversity or heterogeneity in this section.

Types of habitat or vegetation

Simple classifications

An afternoon drive around a city reveals scores of vegetation or habitat types. The plant species are not randomly distributed among habitats, nor are they in non-overlapping groupings. Although the species mixtures or assemblages overlap from site to site, it is easy to see the relative distinctness, e.g., of pond margin, cemetery, and street-tree vegetation.

To enhance easy recognition and communication (and avoid a morass of vegetation-type terminology), we refer to generally familiar urban sites such as lawn, marsh, railway, and flower garden. These are habitat types that differ in vegetation, especially species composition.

How can they be grouped into nice simple informative categories? Groupings or classification of urban vegetation types range from complex to simple. "Phytosociology" (or Zurich-Montpellier) classification, using Latin names to suggest hierarchical relationships among groups of coexisting species, has a long history in parts of Europe (Braun-Blanquet, 1964; Mueller-Dombois and Ellenberg, 1974). "Biotope mapping," sometimes using phytosociology as a conceptual basis, remains active in Northern Europe (Murcina, 1990; Sukopp, 1990; Pysek, 1995a) and scattered other cities, e.g., Durban (South Africa) (Roberts, 1993), Antalya City (Turkey) (Mansuroglu et al., 2006), and Chonju (South Korea) (Zerbe et al., 2004).

Seventy-five habitat types are recognized in the small city of Dusseldorf (Germany) (Godde et al., 1995). Thirty-eight of them are considered primary or widespread: sealed (paved) parking place; mixed forest; intensive grassland (small lawn); native woodland;

deciduous forest; inner city (city center); field (for crops); high-density block (buildings) site; ancient city; nursery garden; market garden; parking place; farm; scrub (shrub area); residential area; avenue; farmland verge (edge); wood-edge (edge portion of woods); allotment (community) garden; river bank; sports ground; shore of pond; harbor; moist meadow; hedge; dry meadow; extensive grassland; roadside verge; low-density block site; trade (business/commercial) site; waterworks (water supply facility); railway site; industrial site; gravel pit; cemetery; parkland; swamp woodland; and wasteland (unmaintained early-to-mid succession site). The bulk of these habitats are conspicuous in most metro areas.

A more general grouping of habitats in the UK recognizes 11 characteristic land uses (Gilbert,1991): urban common; industrial area; railway; road; city center; city park; allotment and leisure garden; cemetery; garden; river-canal-pond-lake-reservoir and water-main (pipeline); and woodland. Conditions in these land uses have been related to disturbance, stress and competition characteristics (Gilbert, 1991; Grime, 2001). Similar categories have been recognized and mapped in Chicago, Akron (Ohio, USA), and Tianjin (China) (Schmid, 1975; Whitney, 1985; Hu et al., 1995). High-definition remote sensing has been used with some success for urban vegetation mapping in Dunedin City (New Zealand) (Mathieu et al., 2007a, 2007b).

Six other vegetation classifications of urban vegetation have used still broader groupings: (1) cultivated sites with introduced plants; versus uncultivated sites (Spirn, 1984); (2) cultivated plant group; native plant community; and naturalized urban plant community (Hough, 2004; Forman, 2008); (3) native (remnant); managed (constructed); and ruderal (adaptive) (Del Tredici, 2010); (4) trampled; along roads, walls and fences; industrial; nutrient-poor; and nutrient-rich (Pysek and Pysek, 1990); (5) natural/semi-natural; managed; abandoned; and bare ground (Rieley and Page, 1995); and (6) natural vegetation remnants; agricultural-land vegetation; ornamental, horticultural and designed urban vegetation; and spontaneous urban vegetation (Pauleit and Breuste, 2011).

Other conceptual approaches to classifying or comparing urban vegetation exist (Cadenasso et al., 2007). For instance, vegetation has been grouped according to degree of human modification (hemeroby) of natural conditions (Ziarnek, 2007). Viewing the origins and subsequent changes of urban vegetation provides insight. For instance, the vegetation of Halifax (Canada) is dominated by plants from rocky habitats, secondarily from grassland-floodplain habitats, and thirdly from other continents (Lundholm and Martin, 2006). Sydney's current vegetation mainly results from two centuries of progressive removal of vegetation on economically valuable sites, which has left vegetation characteristic of low-value sites embellished with non-native species (Benson and Howell, 1990). Habitats can also be grouped by the percent of native species (Zerbe et al., 2004).

Mechanism grouping of vegetation

A somewhat different vegetation *grouping-by-mechanism* approach portrays or classifies urban vegetation or habitat types as a product of three hierarchical factors in a time sequence. First, drawing from ecology in natural areas, vegetation is distributed along a gradient of site moisture conditions, i.e., from dry to wet (Figure 8.1). Second, for each moisture level, vegetation is either largely planted by people or largely *spontaneous*, that is, from natural plant colonization. Third, for each plant origin, the vegetation reflects degree of human maintenance or natural disturbance, i.e., intensive, medium, or scarce/none.

Thus, a city's vegetation or habitat type at the dry and wet ends of the framework would likely be most different (Figure 8.1). Overall, within a moisture category, planted and intensively maintained/disturbed vegetation would differ most from spontaneous little-disturbed vegetation, and so forth. The most similar vegetation would likely be within one of the bottom categories of Figure 8.1. In this way sites requiring considerable human maintenance are grouped together, and contrast with sites with the most natural ecological succession grouped together. This grouping-by-mechanism system for classifying habitats provides ample flexibility, for instance, to add an intermediate level of planting/spontaneous vegetation, separate human maintenance from natural disturbance, and add vegetation from cities in different regions.

Nonetheless, two central messages emerge from the mechanism-grouping diagram (Figure 8.1). Three key mechanisms produce an urban vegetation or habitat type. Second, vegetation types from different sites are readily compared and potentially most similar within a group.

Terminology quickly becomes inhibitory in many classifications, so the approach here only uses terms familiar to the public. General terms for urban

Urban vegetation and habitat

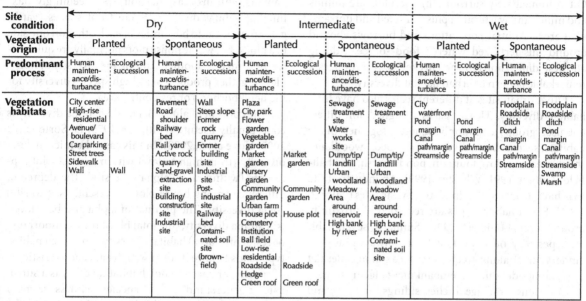

Figure 8.1. Urban habitat-vegetation types based on site condition, vegetation origin, and predominant process. Site conditions along a gradient from dry to wet are determined primarily by water in substrate, secondarily by microclimate. Vegetation origin refers to a gradient ranging from all human planting (trees/shrubs/flowers/grasses) to all natural spontaneous. Predominant process ranges from intensive human maintenance/disturbance (e.g., mowing, cutting, herbiciding, traffic) to all natural processes. Illustrative worldwide vegetation-habitats mainly based on Spirn (1984), Gilbert (1991), Sukopp et al. (1995), Wheater (1999), Zerbe et al. (2004), Pauleit and Breuste (2011).

vegetation or habitat used throughout the book, such as natural, semi-natural, planted, spontaneous, early succession, meadow, shrub, woody, plantation, woodland, marsh, and swamp, are familiar or defined in Chapter 2. The term, "ruderal," sometimes applied to vegetation but not used here, usually refers to disturbed wasteland. However, so-called wasteland (a subjective term of society) includes many site types, often with interesting and important ecological characteristics. "Disturbed" also seems awkward since basically all urban sites have been disturbed, often many times.

Habitat diversity

Fine-scale habitat diversity within land uses

Habitat diversity or heterogeneity is a key to understanding urban vegetation. One approach is to compare differences among habitats of a certain type dispersed across the city. For example, in Taipei (Taiwan), 164 tree species are found in 30 tree communities, representing 10 vegetation types times 3 landscape types (urban park, riverside park, and street verges) (Jim and Chen, 2008). Few tree species are in all three landscape types, and none in all ten vegetation types. In this case habitat diversity, as indicated by tree species, appears to be high for these widely separated habitats. This pattern may also characterize non-urban natural land, though perhaps less so, agricultural land.

More interesting and informative is the fine-scale habitat diversity in urban areas. Consider perhaps the extreme case, a community garden (allotment), where small plots are dug each year and annual plants, including vegetables, thrive (Gilbert, 1991). The area is gridded into tiny squares or rectangles and each person or family grows a different set of plants, e.g., from all beans to diverse vegetables to luxuriant weeds. The community garden resembles a checkerboard with 100+ intermixed colors instead of two colors. The size of the plots, number of types, and contrast of types present, all squeezed together, creates a habitat diversity perhaps exceeding any in nature.

Yet this community garden pattern is still richer, essentially an "enriched 100-color checkerboard." Paths between plots, mainly covered by perennial plants, are trampled at different intensities by people. Plot borders are weeded with different diligence. Humus and mulch are patchily added and spilled. Fertilizers and pesticides are spread around in very different amounts. Water is added at greatly different times and amounts. Fences, poles, and other creative structures are dispersed about.

Often immediately surrounding the plots are humus and mulch and waste piles, plus trampled and untrampled areas. Animal populations and burrows abound around this fertilized watered food-source. Adding all these dimensions to the checkerboard produces a remarkable pattern of urban habitat diversity.

Other urban sites at different scales also have high habitat diversity. Urban parks designed for different user groups commonly squeeze together many habitats, such as lawn, diverse flower beds, woodland, shrub area, meadow, fountain, pond, marsh, and ballfield (Gilbert, 1991; Wheater, 1999). An industrial site may have many types of industrial waste piles (Gilbert, 1991). Seven habitat types are recorded in proximity around a canal (Gilbert, 1991). Similarly, seven habitat types may be important by a railway: gravel or cinders (cess) along tracks; cut banks alongside; flat areas alongside; rail bed embankments facing opposite directions; drainage ditches; sidings (or yard); and masonry structures (see Chapter 10) (Muhlenbach, 1979; Gilbert, 1991; Wheater, 1999). Little overlap in characteristic plant species is present among the habitats. Analogous habitat diversity is present along roads (Forman et al., 2003). Plant species even differ markedly from the top to the base of a wall (Wheater, 1999; Dunnett and Kingsbury, 2004). Finally, a botanical garden or zoo may display the greatest habitat diversity and biodiversity.

House-plot urban diversity at two scales

Perhaps the most striking examples of high habitat diversity in urban areas are the widespread house plots (lots, gardens) of residential areas (see Chapter 10). A house plot typically includes many microhabitats, such as building, foundation planting, lawn, different flower gardens, vegetable garden, different tree species, wall, street trees, side shrub-or-fence lines, vegetated backline, driveway, and walkways. Many other objects including pond, brush pile, shed, and so forth may be present. Each habitat tends to have a somewhat different set of species, or "species pool" (Gilbert, 1991; Owen, 1991; Zobel et al., 1998; Smith et al., 2005, 2006). Adding them together creates quite a diversity of species (see equations, Appendix B).

But equally prominent is the diversity of house plots in a block or neighborhood or residential area. Adjacent or nearby plots may be contrasts, by being predominantly lawn, or woodland, or flower garden, or early-succession meadow, or stone xeroscaping. Or each house plot may be a different combination of habitats, with an occasional house plot seemingly maximizing habitat diversity. Some front spaces (yards, gardens) appear to be mimics of each other on one side of a block, but whole house plots normally are quite different from one another. Adding the diversity of house plots together produces very high species diversity for the neighborhood or residential area.

The concept of alpha, beta, and gamma diversity is conceptually useful (Whittaker, 1975; La Sorte et al., 2007; Hope et al., 2008). In this case, the microhabitat diversity (from high to low) within a house plot represents (within site) "alpha diversity." The degree of difference among house plots represents (site to site) "beta diversity." The product of alpha and beta levels is "gamma diversity," the total block (or neighborhood or residential area) habitat diversity. Since each habitat type tends to have its own somewhat characteristic set of species, or species pool, habitat diversity is a surrogate or general indicator of species richness (number or diversity). Many small habitats within a house plot contain many species. Many different or contrasting house plot designs produce a species-diverse neighborhood, as a study in Sheffield (UK) found (Thompson et al., 2003). Multiplying the within-site and site-to-site levels produces habitat diversity and species diversity higher than that found in most natural land.

Finally, urban species diversity seems to be a product of the fine-scale (small size) spaces or habitats, plus the contrast among them. This diversity is enriched by the enormous species pool available for planting (e.g., tens of thousands of plant species in UK nurseries), plus the influx of species from nearby areas outside the city as well as from far-off regions. Most of the planted species are in low abundance and often present as "singletons" in house plots, but they noticeably add to species totals (Smith et al., 2006).

Urban plants

This near-infinite topic is simplified into three dimensions: (1) types of plants; (2) taxonomic groups of plants; and (3) native, non-native, invasive, and naturalized species.

Types of plants

Many familiar ways can effectively divide up the millions of plants, and the hundreds or thousands of plant species, in a metropolitan area. Stand at the edge of a park, and doubtless trees, shrubs, grasses, and other herbaceous plants are in visible abundance. Looking

Urban plants

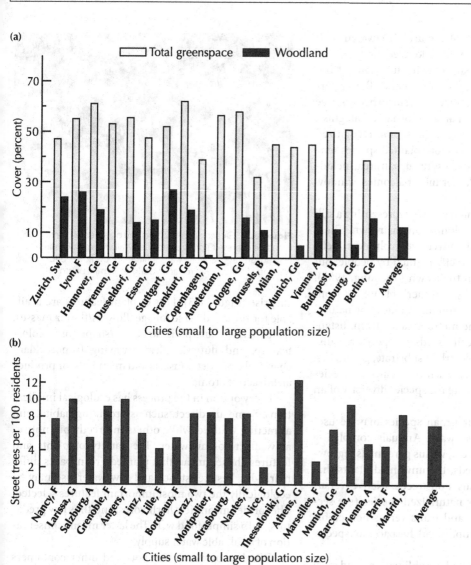

Figure 8.2. Greenspace, woodland, and street trees relative to city population size. A = Austria; B = Belgium; D = Denmark; F = France; G = Greece; Ge = Germany; H = Hungary; I = Italy; N = Netherlands; S = Spain; Sw = Switzerland. Adapted from Pauleit et al. (2005); population data from Turner (2010).

more closely within woods often highlights canopy trees, subcanopy trees, understory trees, a shrub layer, and a herbaceous layer. Also street trees, park trees, and yard trees are differentiated by location.

Larger cities have a higher percent of woodland (including taller closed-canopy forest, if present) than do small cities, at least in Europe (Figure 8.2a) (Pauleit et al., 2005). Zurich and Stuttgart are about one-quarter woodland, whereas four of the 18 cities compared (Milan, Amsterdam, Copenhagen, and Bremmen) have barely any woodland at all. On the other hand, the total percent of greenspace is somewhat lower in large cities. On average, about half the area of the 18 cities is greenspace. Brussels is less than a third, and Frankfort and Hannover (Germany) more than 60%, greenspace.

Street trees are proportionately more abundant in small cities than in large cities (ca. 8 versus 6.5 trees per 100 persons) (Figure 8.2b). However, street tree abundance in cities varies widely, from about 2 to 12 trees per 100 persons. Athens has the most street trees per capita, and Nice and Marseilles in France the fewest.

One estimate suggests that about 60–70% of urban greenspace is planted vegetation (Gilbert, 1991). That also means that 30–40% of the area is covered with spontaneous vegetation that colonized long ago or recently. The planted urban cover includes lawn (from golf course and institution to mini-park), flower gardens, vegetable gardens, street trees, shrub hedges, tree plantations, and more. An array of ornamental, grass, woody, and vegetable species is present in recognized greenspace areas. Yet even more planted species may live in countless tiny sites across the urban area, including backyards, patios, planter tubs, balconies, window boxes, and so forth.

Nevertheless, unplanted species representing a still more extensive palette colonize on their own. These *spontaneous* (unplanted, adventive, volunteer, self-sown) species are not directly planted by people, but rather colonize a site on their own (Thellung, 1905). Plants come from nearby "mother" plants, or from other sites in the city, or from agricultural or natural habitats surrounding the metro area, or from distant cities or habitats. Typically seeds or spores are dispersed, land on a suitable urban substrate, germinate, and produce new plants. Such spontaneous species perhaps normally dominate the species diversity of an urban area.

Not surprisingly our urban species may be usefully grouped in diverse ways. Annuals completing their life cycle within a year, versus perennials regrowing year after year. Weeds, the unwanted plants that plague gardeners, versus non-weeds. Native versus non-native, invasive, or naturalized species (see section below). Vegetables and fruits versus non-edible species. Ornamental woody and herbaceous species versus the rest.

Almost all plants planted in public spaces, and most planted in other urban spaces, were grown in nurseries. *Nursery plants* in turn are mainly species, varieties and forms that have been horticulturally selected because they are effective in accomplishing certain goals (Brickell, 2003; Cullina, 2009). Genetic varieties are often selected for rapid growth, for an abundance of flowers or fruits, for cold hardiness, for minimal toxic chemicals, for pest resistance, for pollution resistance, or for low water use.

A second selection is made by the planter in choosing species from the array available. The prime goal may be an "ornamental" to embellish, decorate, or beautify (Nassauer, 1988, 1997; Reed and Hilderbrand, 2012). Or plants to provide privacy, produce food, cut wind, add shade, enhance a structure, facilitate access, evoke a mood, thrive in a container or flower bed, or create

Figure 8.3. Garden of horticultural and ornamental plants including cypress (*Cupressus*) topiaries. Walkways of soil to limit heat buildup and facilitate water infiltration; low flower beds before planting of annuals. Retiro Park, Madrid. R. Forman photo.

surprise (Figure 8.3). Lots of plant features are available for these – diverse foliage, flowers/flower masses, fruits/berries/nuts/pods, structure/shape/form, color/texture, and fluttering leaves/waving fronds. Plant chemicals may deter insects and mammals, or produce an attractive aroma.

Every option in the process has ecological implications. Some are direct, such as providing habitat and attracting wildlife, while others indirectly but noticeably affect soil/air/water. The variety of plants and features chosen mean that planted greenspaces differ markedly. Pest resistance and low water use may be ecologically the two most important features selected. The former means few herbivores, and thus few predators and a simple food web. The latter means conservation of a valuable water supply.

Window boxes, flower pots and other containers support ornamental flowers, as well as cooking and medicinal herbs (Rapoport, 1993). Submerged plants, floating plants and emergent plants line most aquatic ecosystems. Vines wrap around or climb over plants. Epiphytes grow on the bark or branches of trees, and in moist tropical areas also on roofs, posts, wires, and even the leaves of shrubs. Genetically bred horticultural varieties thrive in gardens. Other plants live on gravestones, around lights in underground infrastructure, on ocean seawalls, or the backs of turtles. In effect, the number of plant species in cities is huge, and numerous ways are used to usefully subdivide them into groups.

Taxonomic groups of plants

Normally in this book "plants" simply refers to "vascular plants," those with internal tubes that conduct water and carbohydrates. Flowering plants predominate, but ferns and cone-producing woody plants are included. Ferns are usually uncommon, except in the wet tropics or where some limestone-growing species thrive in high-pH urban habitats. Cone-bearing trees including pine, fir, spruce, hemlock, and larch are widely present in cities, though many of the species are damaged by air pollution.

Among the *flowering plants*, the most abundant plant families, e.g., in the industrialized city of Sheffield (UK), are composites (Compositae), grasses (Graminae), and mints (Labiatae) (Hodgson, 1986; Gilbert, 1991). Many of the urban species in these families have intermediate-weight seeds, rapid germination, rapid seedling growth, and extensive lateral spread. Together these attributes provide flexibility for the stresses and disturbances of urban life. Species of three other families – pinks (Caryophyllaceae), mustards (Cruciferae), and scrophs (Scrophulariaceae) – are intermediate in abundance. Normally few species of legumes (Leguminosae), buttercups (Ranunculaceae), roses (Rosaceae) and parsleys (Umbelliferae) are present.

A detailed 15-year study of a single 0.1-ha (1/4-acre) house plot in Leicester (UK) documented 384 native flowering-plant species (Owen, 1991). The leading plant families by number of species are: Compositae (56 species); Labiatae (24); Rosaceae (21); Cruciferae (20); Liliaceae (16); Leguminosae (15); Solanaceae (14); Caryophyllaceae (11); Ranunculaceae (11); and Scrophulariaceae (11). In both UK studies (Sheffield, Leicester) only native species are included. Planted cultivated species and non-native spontaneous plants are predominantly in different sets of plant families. The Leicester study also reported 38 non-vascular plant species [25 macro-fungi or mushrooms; 11 bryophytes (9 mosses and 2 hepatics or liverworts), 1 lichen, and 1 alga].

Bryophytes are tiny leafy plants reproducing by spores. Two species, *Ceratodon purpureus* and *Bryum argenteum*, sometimes called sidewalk mosses because of their tolerance to drought, pollution and trampling, often predominate in cities worldwide (Gilbert, 1970, 1991; Le Blanc and Rao, 1973). The silvery *Bryum* thrives in more nutrient-rich sites compared with *Ceratodon* on more nutrient-poor sites. These and other widespread moss species, *Funaria hygrometrica*, *Bryum* spp., *Pohlia annotina*, and *Tortula muralis*, reproduce abundantly, grow rapidly, and are relatively pollution resistant. These species characterize somewhat dry and polluted sites. But some urban greenspaces in wet climates, such as in Portland (Oregon, USA) and Seattle (Washington, USA) and tropical rainforest are draped with a variety of moss species (Houck and Cody, 2000).

In West Berlin, 265 bryophyte species are reported, including 219 mosses and 46 hepatics (Schaepe, 1990). The predominant urban habitats (biotopes) and number of bryophyte species are: forests (115); cemeteries (110); mesotrophic-oligotrophic moors (shrublands) (102); walls (92); parks (87); flowing waters (73); lowland moors (wetter shrublands) (70); trees (68); ruderal sites ("disturbed wasteland") (67); construction and gravel pits (63); ponds (43); backyards (32); lawns and meadows (29); arable land (22); and dry grassland (11). Most of these bryophyte species are generalists, growing in a number of urban habitats.

Lichens, a symbiosis of fungus and alga together, are noted for being assays for air pollution level. A combination of desiccation, SO_2, and perhaps particulate matter may be the prime reason that lichen abundance and species diversity drop sharply from rural area to city center. Hundreds of such rural-to-urban lichen studies have been done since the early 1800s (Le Blanc and Rao, 1973; Schmid, 1975). Metro areas in the UK, for example, exhibit four concentric zones based on (epiphytic) lichens growing on tree bark (cited by Wheater, 1999): (1) central "desert," with zero species; (2) inner transition zone, with two crustose-lichen species present, *Lecanora conizaeoides* (a non-native species) and *Lepraria incana*; (3) outer transition zone, with some crustose lichen species and several foliose species, especially *Parmelia*, present and higher up on tree bark; and (4) "clean" zone, with a wide range of foliose- and crustose-lichen species growing from base to canopy of trees. A highly pollution-tolerant crustose lichen, *Lecanora muralis*, is reported to have spread progressively, and increased its growth rate, from asbestos roofs to cement-tile roofs to sandstone to tarmac to wood substrate (cited by Gilbert, 1991). Apparently this pattern was related to decreasing SO_2 levels in the air.

The 476 *macro-fungi* (mushrooms, etc.) species recorded in a 10-year study in Lodz (Poland) exceeds the number of native vascular-plant species in the surrounding city (cited by Gilbert, 1991). A central zone of the city has few "mycorrhizal" symbiotic fungus species, and no species with perennial woody reproductive structures (hard "shelf" species). A middle zone contains many more mycorrhizal species and some species with woody fruiting bodies. Outward, the suburban zone has a fungal flora slightly more diverse, perhaps due to habitat heterogeneity, than in a more-natural nearby forested landscape.

In Warsaw, the macro-fungi seem to grow best in urban woods, though the fungi are also present in partially open grassy habitats with humus or around temporary bare soil (Skirgiello, 1990). Some common macro-fungus species seem to have disappeared from the city, perhaps due in part to road salt. The fungus flora includes non-native species that may have arrived in the root-ball soil of horticultural plants.

Finally, green algae, and blue-green algae especially associated with certain types of water pollution, are abundant in almost all urban water bodies. Green algae also grow on moist urban surfaces of many types, including walls, sidewalks, and tree trunks. Marine brown, red, green, and other algae may survive or thrive in the seacoast and estuarine environments of cities.

Native, non-native, invasive, and naturalized species

Origins of urban plants

After land appeared from the sea, plants arrived, adapted and spread over the surface. Long-distance, in addition to nearby, dispersal of seeds and spores has been the rule ever since (Elton, 1958; Davis, 2009). Ocean currents, cyclones/hurricanes, migrating birds, and river flows continually carry propagules hundreds or thousands of kilometers. Species ranges expand and shrink with climate changes. People also purposely and inadvertently carry plants long distances. Two thousand years ago Cleopatra in Egypt received silk from China, and doubtless plenty of valuable seeds. The Inca and Maya peoples transported valuable gems, gold, dyes, and most probably valuable seeds, to and from far-off locations. In short, floras are constantly being intermixed and enriched.

When covered with massive ice thousands of years ago, the future sites of Boston, Toronto, Denver, Vancouver, Sapporo, Moscow, Warsaw, and Edinburgh were plantless. Afterward tundra plants colonized and spread. Soon many more species from warmer climes took over. Then somewhere a farmer settled on well-drained land by a river or coast, or by intersecting transportation routes, and quickly replaced the natural vegetation and wetlands with farmland. Other settlers arrived and did the same. A village formed, followed by a town ringed with extensive farmland. Town became small city, which densified and expanded to be a large city. Most cities are still mainly ringed by cropland (Forman, 2008), though low-density residential sprawl sometimes spreads beyond it into natural land. In essence, agriculture wiped out the natural vegetation and species, whereas city basically eliminated agriculture and its species.

Attempting to restore an urban site to pre-city conditions or to conditions outside the city normally means to farmland and its species. But today's urban environment renders success unlikely. Restoring an urban site to conditions close to long-ago nature seems impossible.

The origins of the cornucopia of urban plants are exceedingly diverse. In Canberra, seeds filtered from a car-wash facility for 26 months, planted, and grown for 9 months (Wace, 1977; Forman et al., 2003) produced 18 500 seedlings, representing 259 plant species. Seven percent of the species apparently came from outside the region, mostly more than 170 km (105 mi) distant. Thirty-one percent came from native vegetation surrounding the city, and 40% came from the city and surrounding agricultural land. In Rome, nine native plant species widespread on walls, wasteland, fields, trampled sites, and gardens are very rare in natural habitats (limestone and other rocks, tuff, shady mud, gravels, and snow beds) outside the city (Pignatti and Federici, 1989). In short, introduced plants are from both near and far.

Purposeful human planting covers urban gardens, plantations, window boxes, and so forth, whereas spontaneous plants colonize and grow on their own. Various planted cultivars, cultigens, varieties, and forms of plants are horticulturally developed by genetic breeding, grafting, and other methods. Even some of these human-modified plants turn feral and grow spontaneously (Sukopp and Sukopp, 1993; Thompson et al., 2003).

Weeds are unwanted plants, and, like art, determined by human preference (Baker, 1974; Stein, 1988). For a farmer, a species that competes with the crop plants is a weed. For one gardener, plants that compete with the planted roses or lettuce are weeds, yet the same plant species may be treasured by the gardener growing a richness of flowers. Over 15 years, a 0.6 ha English house plot had 80+ weed species, and over 25 years, a 0.5 ha plot in the same city had 95 weeds (Gilbert, 1991; Owen, 1991). Even though golfing began on highly diverse grassland/shrubland, a golfer trying for a low score may not want dandelion (*Taraxacum officinale*) on the green to create heterogeneity. In contrast, dandelion is welcome in lawns where homeowners appreciate the bright flowers on monotonous grass, and may even eat the young leaves in salad.

Some species transported from a far-off site are simply described as *non-native*. Non-native species are widespread, even predominant, where human activities have been extensive, as in many agricultural landscapes and tropical islands. For a city, the major sources or origins of non-native species seem to be other metropolitan areas, plus nearby or distant dry or rocky areas (Sachse *et al.*, 1990; Lundholm and Martin, 2006). Within a metro area, major sources of non-natives are probably shipping and railroad sites, commercial and industrial warehouses, and plantings in suburban/exurban residential areas.

Distribution within urban areas

Military terminology – invade, attack, wipe out, aggressive, and even racial purification, alien and exotic – has permeated the non-native-species literature (Kendle and Rose, 2000). Urban ecology studies help get past that phase (Davis *et al.*, 2011). Few native species thrive on recent novel human-created surfaces, such as walkways, walls, road shoulders, industrial waste piles, vacant lots, and the base of buildings. In such locations where few natives survive, vegetation dominated by non-natives seems much better ecologically than bare urban surfaces.

However, in addition to native and non-native, two terms seem useful, invasive and naturalized. *Invasive* species (non-native or native) successfully colonize, compete, reproduce and spread in an area. Some are invasive in human-dominated sites, such as a construction sites (reducing soil erosion), cracks in old pavement (providing early-succession vegetation), along a mowed roadside (normally adding to species diversity), or in farmland. Others are invasive in natural or semi-natural land (sometimes significantly altering vegetation structure or species composition).

After some short or long period of spread the species is no longer invasive, but rather is *naturalized*, a successful reproducing component fitting into ecosystem processes (Sukopp and Trepl, 1987; Sukopp and Sukopp, 1993; Peterken, 2001). Mainly species become naturalized in sites of frequent human disturbance (Muhlenbach, 1979; Dunn and Heneghan, 2011). But some species become naturalized in natural or semi-natural areas.

In the two centuries of humans introducing species to Australia, big economic disruptions have resulted from certain non-natives. Nevertheless the naturalization process, involving adjusting or genetically adapting to new conditions, is illustrated as follows (Fox, 1990): (1) some non-native herbivores feed on native plants, while native herbivores feed on non-native plants; (2) some non-native predators feed on native herbivores, while native predators feed on non-native herbivores; and (3) non-natives and natives of each feeding level coexist in habitats. Also in Australia and New Zealand, sometimes species "native" to a nation have been differentiated from species "indigenous" to a smaller region. Still, in urban areas most native species cannot keep up with the rate of environmental changes, so a continual "rain" of new non-native species helps keep the place green.

An extensive non-native-species literature focuses on the characteristics of the plant (including where it came from), and of sites susceptible to colonization or invasion (Sukopp and Trepl, 1987; Godefroid and Koedam, 2003; Alston and Richardson, 2006). Cities have more non-native species, typically some 70% of the flora in a European city, than do natural areas. The proportion of non-native plant species apparently does not differ significantly between European and US cities, though the latter have more dominance by a few of the species (Muller *et al.*, 2010).

Cities also have more disturbed, bare, nutrient-rich, and invadable sites (and perhaps empty niches). Non-natives seem to be mainly generalists with broad tolerance, and more resistant to urban conditions than are most native species. Occasional local extinctions of native species in urban areas due in part to non-natives seem likely (Sukopp and Trepl, 1987).

The packing together or density of plant species along a city-center to outer-suburb gradient of Berlin is lowest in the city center (inner city), and highest in the intermediate zones (immediate ring around city center and the inner suburb) (Figure 8.4a). Non-native species slightly decrease in the innermost three zones (from 50% to 43% of the flora), and then drop sharply to the outer zone (Figure 8.4b). Native species richness in the outer suburb is only a bit higher than in the surrounding urban-region ring (29% versus 22%).

Dissecting those non-natives into two groups is useful in Europe. "Archaeophytes" arrived before 1500 (when ships began bringing numerous seeds from distant lands), and "neophytes" colonized since 1500 (Figure 8.4). Archaeophytes gradually decrease outward along the gradient. Neophytes are considerably more frequent in center city and outward decrease

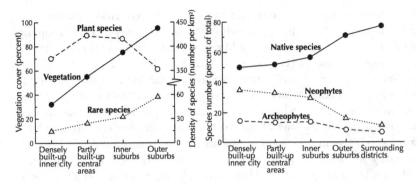

Figure 8.4. Plant species types and vegetation along city-to-rural gradient. Berlin (West); surrounding districts = average of five rural Brandenburg districts. Archaeophytes colonized before, and neophytes after, the year 1500, when ship explorers began bringing numerous plants from far-off lands to Europe. 1 km^2 = 0.39 mi^2. Based on Sukopp and Werner (1983) and Kowarik (1990).

sharply, so the two groups are in similar abundance outside the suburbs.

Few rare native plants (17) live in city-center Berlin, gradually increasing to 58 in the outer suburbs (Kowarik, 1990). Although no information is given, rare non-native species are probably most abundant in the city center. Some 75% of the urban flora in European cities has been introduced in the past 200 years (Sukopp, 2008). Per unit area, probably most new non-natives first colonize in the city center.

In residential areas, institutional areas, and vacant lots representing 95% of the tree locations in Chicago, the most abundant tree is buckthorn (*Rhamnus*), a non-native species (Whitney and Adams, 1980). Still, 70% of the Chicago trees are ash (*Fraxinus*), box elder (*Acer negundo*), willow (*Salix*), and cottonwood (*Populus*), four resilient native species that have a short life span and readily colonize open sites.

At a finer scale, the proportion of non-native species varies by habitat, as in Chonju (South Korea) (Zerbe et al., 2004): city center, 29% of the flora; urban residential areas, 31%; rural residential areas, 34%; industrial areas, 32%; commercial areas, 28%; railway and roadway sites, 27%; institutional sites, 24%; pre-1990 greenspaces, 14%; post-1990 greenspaces, 28%; rice paddy fields, 17%; dry fields and orchards, 28%; rivers, lakes and ponds, 23%; fallow land, 23%; and forests, 3–13%. The highest proportion of non-native species (37%) is for rural residential areas on a plain, and the lowest (3%) for coniferous tree plantations. Except for forests, old greenspaces, and rice paddies, the percent of non-native species is relatively similar (23% to 34%) in vastly different types of areas across the metro area.

Three familiar processes or filters tend to eliminate native species from agricultural and natural sites in cities (Forman, 2006; Williams et al., 2009). Habitat loss, as the land is transformed, has the greatest effect. Habitat degradation from environmental and human effects is a close second, and habitat fragmentation is important for some species with limited seed production and dispersal. Two other processes are especially important in urban areas. Planting and maintaining a site such as garden or lawn eliminates native species. Secondly, the concentration of successfully colonized non-native species seems to provide a formidable set of competitors that inhibit native species.

Urban species mixtures ahead

Where does all this lead? Mixtures of native and non-native species have characterized urban habitats since the origin of cities, some 6000 years ago in the Middle East. Hurricanes, ocean currents, migrating birds, river flows, and humans carrying seeds long precede the first city, and continue today. Human transport has accelerated the mixing process. Over ecology's 140-year history, ecologists have focused on forest, desert, grassland, farmland, aquatic and marine systems, but until recently few have focused on the equally, or more interesting, urban ecosystems. Terms such as new vegetation types, new habitats, novel ecosystems, and synthetic ecosystems have increased with the new interest (Whitney and Adams, 1980; Bridgewater, 1990; Fox, 1990; Rapoport, 1993; Peterken, 2001; Davis et al., 2011).

The diverse-origin plant communities or vegetation or ecosystems or habitats of urban areas add meaning to the idea of species diversity. What species? How many species? How much dominance and rarity? What types of origin are present? What rates of change?

The urban species mixtures are inherently intriguing, little studied, and frequently changing. A suburban New Jersey (USA) woods has non-native Norway maple (*Acer platanoides*) dominating canopy, subcanopy, understory, shrub, and herbaceous layers. This pattern reminded me of the same situation for a native southern beech (*Nothofagus* sp.) stand well west of Christchurch (New Zealand), that had almost no other plant species present. A non-native *Phragmites* reed area in northern New Jersey seemed to be an extensive

Figure 8.5. Tiny urban spot with intermixed planted, crop, and spontaneous plant species. Maize plant in center; two white dots to its right are butterflies pollinating flowers, an uncommon sight in such a small relatively isolated urban spot. Zacatecas, Mexico. R. Forman photo.

plant monoculture, though all ecosystems contain some habitat heterogeneity suitable for other species.

Local native species generally do not keep up with the rate of urban change. The sorting of continuously arriving non-native species from diverse environmental conditions and evolutionary histories probably can keep our cities green and diverse. But the area of green and the richness of species is probably greatest with mixed-origin communities (Figure 8.5). Cities have "tough" plants that colonize, survive, spread, and may adapt to granite-faced skyscrapers, salty roadsides, and belching smokestacks. How many ecosystem services do these non-native plants provide? How many people relish the benefits? As yet we have few answers.

Some non-native species are so disliked for their invasiveness that they are banned. One cannot sell Japanese knotweed and giant hogweed in the UK (Wheater, 1999), or Norway maple, Japanese barberry (*Berberis thunbergii*), and European honeysuckle (*Lonicera* spp.) in Massachusetts (USA). Yet other non-natives seem to have little or no effect on native species or vegetation.

The mixed-origin communities, constant new species input, and habitat heterogeneity (fine-scale and broad-scale) of urban areas suggest that there is little threat of biotic homogenization within a city, or among cities (McKinney, 2006; La Sorte *et al.*, 2007). House sparrows, rock doves/pigeons, and European starlings are common in perhaps most cities, yet each city has many more bird species. In dispersed regions, cities boast an abundance of crows, parakeets, magpies, Indian mynahs, collared doves, swifts, nighthawks, storks, gulls, and the list goes on. Big population changes can be expected, such as the dramatic drop in house sparrows when motor-power replaced horsepower in Paris, or today's sparrow drop in urban England.

Pulling out non-native plants near a massive "mother" source of dispersing seeds (Davis *et al.*, 2011), such as a botanical garden or residential area covered with non-natives, seems analogous to using a flyswatter among billions of mosquitoes. Or tilting with windmills à la Don Quijote.

Perhaps the 11-hectare Sudgelande Railway Park in southern Berlin best portrays urban vegetation of the future. For decades seeds from distant parts of the Soviet Union leaked from railway cars/wagons in this former rail yard (Kowarik and Langer, 2005). Many seeds germinated, adding an exceptionally wide assortment of non-native species to the flora of rail-yard-resistant native plants. Then the area was fenced for a decade to keep people out. Doubtless more natives and non-natives colonized and ecological succession continued. After that, landscape architects made modest modifications, and a limited-access park has been open to the public for more than a decade. During the post-rail-yard 20 years, the vegetation grew into a woodland. But this "mixed-origin" woodland is altogether distinctive, displaying an extremely high plant diversity composed of a rich and unique mixture of native urban species and non-native species mainly from many regions of Asia and Europe.

Urban plant biology

Several important disparate subjects are introduced here in three groups: (1) physiological plant ecology; (2) herbivory, pollination, and dispersal; and (3) plant genetics and adaptation. All are extensively analyzed in other books on horticulture, gardening, and botanical gardens, as well as texts on physiology, plant genetics, and ecology. However, urban plant-biology patterns, as distinct from those for natural and agricultural land, have yet to be highlighted. Thus, rather than summarizing general patterns, diverse studies and concepts are introduced to stimulate thought and research.

Physiological plant ecology

Environmental conditions and plant responses

Good health is the norm, or is it a temporary phase between illnesses? For the urban plant, "illness" typically results from not enough or too much of something. A gradient from insufficient to excess amount produces a sequence of plant responses: death; survival;

poor growth; good growth; good growth with successful reproduction; good growth; poor growth; survival; and death. Effectively this describes a bell-shaped or normal response curve.

The city is much richer than natural or agricultural land in things that will cause such a pattern. A plant must deal with competitors, herbivores, and parasites that limit growth, resources and environmental stresses that also limit growth, and disturbances that damage or kill (Grime, 2001). In urban areas, consider the concentrated arrays of heavy metals, toxic organic substances, micro-nutrients, microclimatic conditions, types of waste, building types and forms, road types, microhabitat heterogeneity, and human activities. With so many resources varying from insufficient to excess, relatively few plants exhibit good growth. Fewer still have good growth and successful reproduction.

Substrate and microclimate are illustrative. Soils in urban areas vary drastically, from none to ample, compacted to porous, dry to waterlogged, heavily polluted to less so, soil animals abundant to scarce, and so on (see Chapter 4). Industrial wastes themselves are extremely different. Thus, for different wastes, the following characteristics severely inhibit plant growth (Wheater, 1999):

1. Colliery (coalmine) waste-material: compaction, temperature, nutrients, acidity.
2. Pulverized fuel ash: nutrients, alkalinity.
3. Calcareous rock-quarry waste-material: nutrients.
4. Metal mining waste: nutrients, acidity, toxicity, salinity.

For microclimatic conditions, comparing urban versus non-urban areas highlights the following general patterns in a temperate climate (see Chapter 5) (Gilbert, 1991; Sieghardt et al., 2005)(Note that such summary information should be interpreted with caution, since differences within non-urban areas may be greater than those between urban and non-urban):

- UV radiation: much less (25–90%) in urban than in non-urban.
- Solar radiation: less (1–25%).
- Duration of bright sunshine: less (5–15%).
- Infrared radiation input: more (5–40%).
- Gaseous air pollution: more (500–2500%).
- Visibility: less.
- Heat flux: more (50%).
- Air temperature: higher (1–3°C annual average; up to 12°C on occasion).
- Evapo-transpiration: less (50%).
- Summer relative humidity: less (8–10%).
- Rainfall: more (5–10%).
- Snow: less.
- Streamline windspeed (at 10 m above ground): lower (5–30%).
- Wind turbulence: more (10–50%).
- Wind direction: altered (1–10°).

Within urban areas, plant growth is affected by differences in each soil and microclimate factor. Varied combinations of factors may be more severe, but are certainly less predictable, for plant growth.

So to thrive, urban plants must be *generalists*. These species have considerable genetic variation, providing a wide range of tolerance to stresses and disturbances. In a cool winter or spring the plants may benefit from urban heat. But in summer they are stressed by heat, lack of water, and in some cases, by heavy metals in the substrate and hydrocarbons in the air. Note also that some older long-lived plants, especially trees, may have grown well under former, different environmental conditions, but now simply survive in the present altered conditions.

In the face of disturbance, urban plants typically have either high *resistance* or high *resilience* (see Chapter 3). Resistant plants are tolerant and survive disturbances, whereas resilient species recover rapidly after a disturbance. Many large slowly reproducing trees ("K-selected" species) have an extensive root system, thick bark, relatively few large fruits, and so forth, and can withstand or survive in the face of many, diverse, and/or severe disturbances. On the other hand, many grasses and annuals ("r-selected") with rapid growth, rapid flowering and fruiting, and numerous tiny wind-dispersed seeds are killed by most disturbances, but readily bounce back by rapidly recolonizing sites. Urban floras are presumably a combination of both resistant and resilient species, though the relative proportions, e.g., for trees, grasses, non-natives, or planted perennials, is apparently unknown.

Summer temperature is highly dependent on shade, and trees are typically prime shade-makers. Usually residential areas have more tree cover than non-residential areas (Gartland, 2008), and high-income neighborhoods more than low-income ones (Jenerette et al., 2007). Shade mainly falls on three surfaces, grass, pavement, and roof. In residential areas of four US cities, a third of the grass present is shaded on average, 16% of the pavement, and 6% of the roof

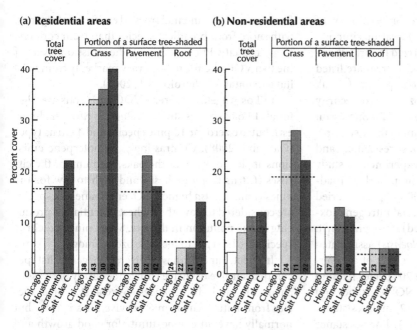

Figure 8.6. Tree-shaded surfaces in residential and non-residential areas. Horizontal dashed lines are averages. Numbers at bottom are total percent cover (shaded plus unshaded) for each surface type. Cities in USA; C = City. Based on Gartland (2008).

area (Figure 8.6a) (Gartland, 2008). But variation is striking. Salt Lake City (Utah) has a noticeably higher, and Chicago a noticeably lower, portion of the grass and roofs shaded. For shaded pavement, Sacramento (California) is highest, though percentages differ little from city to city. Non-residential areas have lower percentages of all three surfaces shaded (Figure 8.6b). Differences among cities are less in non-residential areas, though Chicago has hardly any shaded grass and shaded roofs. The low shaded values in cities suggest opportunities for increasing tree cover and shade, thus decreasing summer temperatures.

Trees of course produce different intensities of shade depending on their foliage. "Leaf area index," the total amount of leaf surface per unit of land area, is a widely used measure. For instance, in Toronto the leaf area index (m^2/m^2) varies from about 2.0 in "open" areas to 0.1 in industrial areas and 0.0 in commercial areas (Kenney, 2008). Institutional areas have a leaf area index of about 1.2 and residential areas about 1.0, the latter differing little according to housing density. The study suggests that average leaf area index could be most increased in industrial areas, but also increased in institutional and open areas.

Species functions

Some tolerant plant species with high resistance also have the characteristic of *assimilation* or uptake of material. For these species, absorption of chemicals does not kill the plant. If the plants die, the species does not have high resistance to those chemicals. Plants with the characteristic of high uptake/assimilation thus effectively contribute to cleaning the air or soil (phytoremediation).

Carrying this characteristic one step further, the added chemicals may not only be assimilated and accumulated in the plant, but may benefit or stimulate plant growth. Fertilizing the garden increases production but also adds nitrogen to plant leaves. In the UK, at the base of certain urban trees regularly used by dogs as signal posts or markers, a green alga, *Prasiola crispa*, thrives apparently with the added nitrogen in urine (Gilbert, 1991).

The patterns of response to urban stresses may be examined in more detail and suggest causative mechanisms. Consider an urban site dominated by horse chestnut trees (*Aesculus hippocastanum*) but degraded by road salt (Oleksyn *et al.*, 2007). The elevated salt level is a key factor in low nitrogen concentration in foliage, low photosynthetic rate, and greater leaf senescence. Three years of adding mulch and fertilizer reversed all three of these responses. In addition it lowered the soil pH. Heavy metal concentrations in leaves increased, thus providing some uptake or assimilation effect. Also phenolic compounds in the leaves decreased, which makes foliage potentially more palatable to insect herbivory. Increased herbivory could result in tree defoliation, even death. Or it could mean higher insect abundance and diversity, more food for birds, and higher avian populations and diversity.

Not surprisingly, the tolerance or sensitivity of some urban plants, especially trees, to various important urban stresses has been summarized for practitioners and the public. For example, urban trees are listed according to their sensitivity to soil compaction (Craul, 1992), road salt, sulfur dioxide, ozone, and security lighting at night (Grey and Deneke, 1992). Successful tree planting chooses species resistant to these stresses.

The difference between species resistance and assimilation is illustrated by an experimental study of 70 woody plant species (taxa) along urban roadsides in Japan (Takahashi et al., 2005). Species varied widely in both their resistance to aerial nitrogen dioxide (NO_2) and the uptake of (reduced) nitrogen from it. For instance, a cherry (*Prunus yedoensis*) assimilated 122 times the amount of nitrogen as that of an evergreen (*Cryptomeria japonica*). The 70 species fell into four recognizable groups: (1) high NO_2 resistance and high NO_2 assimilation (13 species); (2) high resistance and low assimilation (11 species); (3) low resistance and high assimilation (11 species); and (4) low resistance and low assimilation (35 species). Broadleaf trees in the first category that also have fast growth and high biomass (*Robinia pseudo-acacia*, *Sophora japonica*, *Populus nigra*, *Prunus lannesiana*) are thus considered to be the most promising species for limiting or reducing nitrogen oxide in the urban air.

Many tree leaves also assimilate polycyclic aromatic hydrocarbons (PAHs), which are organic compounds mainly emitted by motorized traffic (Kuhn et al., 1998). Eleven PAHs were measured in the air and in leaves of poplars (*Populus nigra*) along a busy main road in Frankfurt/Main (Germany). In the air, one (phenanthrene) was in much higher concentration than the others. However, in the leaves, four (phenanthrene, pyrene, fluoranthene, chrysene) were in relatively high concentration, indicating uptake by the trees (uptake varies with water vapor pressure). Overall, tree vegetation is considered to be an important sink or absorber of aerial PAHs.

In addition to resistance/tolerance, resilience, and uptake/assimilation, one more attribute bears mention for urban plants. Plants are *emitters* or sources of chemicals. Biogenic volatile organic compounds (VOCs) are emitted by many plants into the air. In cities these hydrocarbons can combine with nitrogen oxide in the presence of sunlight to form ozone smog. Two types of VOC are important here, isoprenes and monoterpenes (Li et al., 2008a; Noe et al., 2008). VOC emission is reported from urban trees in Shenyang, China. In Barcelona, 11 ornamental trees all emit monoterpenes, with more from broadleaf species than from conifers. In an arid city, Phoenix (USA), the emission levels of the two VOC chemical types vary markedly from habitat to habitat (Godefroid et al., 2007).

In Los Angeles, isoprene VOC emissions are high for all 13 oak species studied (*Quercus garryana* highest), but are zero for 10 pine species and 4 citrus types (Gartland, 2008). Contrastingly, monoterpene emissions are low or zero for the oaks, medium for the citruses (*Citrus limon* highest), and high to low for the pines (*Pinus clausa* highest). Overall, nine oak species are considered to have the highest potential for producing ozone pollution in the city. Seven pine and citrus species are least apt to increase ozone smog.

In effect, our cities have a diversity of generalist species that survive or thrive with a huge array of stresses, chemicals, and disturbances. Specialist native species from natural areas may survive for a period, but normally have little opportunity for good growth and successful reproduction. Non-native generalists commonly colonize and some thrive. Many urban species are resistant to disturbance; others are resilient, rapidly rebounding from disturbance. Some species assimilate and "passively" store certain urban chemicals, while others benefit from and grow better with added chemicals. Finally, some species produce and emit chemicals into the urban environment.

Herbivory, pollination, dispersal

Herbivory

Although herbivores consume some plant tissue in urban areas, overwhelmingly leaf cells simply die in place. Out in natural and agricultural areas herbivory is normally much higher, as suggested by simply observing the frequency and size of insect-chewed holes in leaves. Several hypotheses for the usual relatively low urban herbivory seem logical. Severe environmental conditions limit leaf insect diversity and abundance. Non-native species typically arrive without their herbivores, and local herbivores acclimate or adapt slowly to the new plants. Insecticide use is widespread in some urban areas. And of course much of the area is covered by hard surface.

Other hypotheses, such as predation limiting herbivores, seem to play a limited role. In some locations mammal herbivores such as deer, kangaroos, or rabbits have dense populations, but the urban structure and scarcity of a shrub layer limit the suitable area for these species. Some plant leaves contain high levels of organic chemicals that limit herbivory, e.g., by being unpalatable or toxic. But such plant species do not seem to dominate most urban areas.

The heterogeneous distribution of herbivory in greenspaces is illustrated in a study of the extent of herbivory, that is, leaf chewing and surface "damage," in three urban locations: tree canopy of small woods; edge of large woods; and interior of large woods (Christie and Hochuli, 2005). No difference was found in the frequency of leaves with herbivory evidence among the three sites. However, the amount of herbivory was higher in the small woods than in the interior of large woods (forest edge herbivory was intermediate).

Herbivores of course have feeding preferences, so the combination and distribution of plant species has a strong effect on herbivory. In the 1940s and 1950s in London, a huge increase in a flowering herbaceous plant, rosebay willowherb, occurred taking advantage of recently bombed sites (Gilbert, 1991). This was accompanied by a sharp increase in conspicuous elephant hawk-moths (*Deilephila elpenor*), which in the caterpillar stage feed on the plant. Later the population of these large moths decreased, presumably in part because of fewer "wasteland" sites (especially suitable for the plant) remaining in the city. But also predatory flies and wasps, perhaps from surrounding rural areas, became more abundant in the city.

Plants in the city are not like those in a florist shop, where evidence of herbivory is hard to detect because of horticultural breeding, insecticide use, and removal of holey leaves. Even the non-native species both inside and outside the city have herbivores, holes in leaves, and other evidence of herbivory (Tello *et al.*, 2005; Benedikz *et al.*, 2005).

Five prominent non-native species introduced over time into the UK are instructive (Wheater, 1999). (1) "Sycamore" (*Acer pseudo-platanus*) arrived in 1578 or earlier from Central/Southern Europe, and today has 43 insects associated with the tree, including nectar-feeding bees. That is about 10% of the insect diversity on native willows (*Salix*) which have 450 insect species [oaks (*Quercus*) with 423, and birches (*Betula*) with 334 insect species], but similar to the diversity on native hornbeams (*Carpinus*) and field maples (*Acer*), each with 51 insect species. (2) *Rhododendron punticum* was introduced in the late 1700s from Turkey, and today has few associated insects; its nectar is toxic. (3) Japanese knotweed (*Polyganum* sp.) introduced in 1825 from Japan and East Asia now harbors several bugs, beetles (weevils and leaf beetles), and caterpillars of moths and butterflies. (4) Indian balsam arrived in 1839 from the Himalayas and today supports two aphids, elephant hawk-moth caterpillars, and five nectar/pollen-feeding bumblebees. Finally, (5) giant hogweed introduced in the late 1800s from the Caucasus Mountains in Asia now has slugs, snails, and 30 herbivorous insects, including true bugs, two-winged flies, moths, and beetles.

These few urban examples for non-native plants do not show a clear increase in insect diversity over time. Nevertheless, considerable non-urban evidence indicates that insect diversity correlates with the time since a tree species arrived in the UK (Kennedy and Southwood, 1984).

Plant pathologists have extensive data on the effects of pest and disease species on urban trees (Benedikz *et al.*, 2005). For example, the following urban trees are most damaged by a type of insect or other organism [listed by number (high to low) of European nations reporting damage] (Tello *et al.*, 2005).

- Insect leaf feeders: *Quercus, Pinus, Aesculus, Robinia.*
- Insect wood borers: *Populus, Acer, Salix.*
- Insect bark borers: *Pinus, Ulmus, Picea, Fraxinus.*
- Insect sapsuckers: *Tilia, Picea, Acer, Platanus.*
- Mites: *Tilia, Fraxinus, Picea.*
- Fungi and bacteria (leaf and shoot diseases): *Platanus, Pinus, Populus, Aesculus.*
- Fungi (stem and branch diseases): broadleaf trees, *Ulmus.*
- Fungi (root and collar diseases): broadleaf trees, conifers.
- Fungi (wood decay): broadleaf trees, conifers.

Thus, different types of herbivores and diseases tend to target different trees. The explosion in abundance of a particular herbivore may defoliate a tree species over a wide region. A diversity of plant species reduces the chance of severe insect defoliation in a greenspace or a city.

In addition, herbivore populations may rise and fall rapidly, and seem to track the phenology or seasonal time sequence of changes in plants. An urban study of 34 common woody ornamental plant species and 33 insect pests found that the abundance of insects closely followed the sequence of flowering by the plant species (Mussey and Potter, 1997). Over three years, the timing for herbivores tracked plant phenology better than calendar date, and the sequence of herbivores was consistent from year to year.

Pollination and seed dispersal

Pollen and seeds are both transported by wind and animals (Murray, 1986; Wilson, 1992), though their dispersal in urban areas remains an important research frontier. Wind-dispersed pollen is typically produced in massive amounts, and must arrive at certain flowers at certain times for pollination to be effective. In animal pollination, bees, butterflies, hummingbirds, and certain bats fly preferentially to specific flowers where pollination occurs (Figure 8.5). Animal pollinators tend to be scarce in most urban areas, though common in some (Matteson, 2008) (see Chapter 9).

English cities are known for their abundance of gardens and flowers, especially ornamentals. Flower pollination activity by bees continues for 8–9 months of the year, in part due to the urban heat island. Beehives are moved among various parks, thus enhancing fruit and seed production with the floral luxuriance. Also, London makes 10% of the honey sold in its shops (see Figure 12.2). Suburban/urban honey largely avoids the chemicals of today's agriculture.

On the other hand, to be effective, seeds dispersed must reach a suitable substrate for germination and plant growth (Cheptou et al., 2008; Dornier and Cheptou, 2012). Such substrates are scarce in some urban areas but often are in moderate abundance. The dispersal of seeds is important for most spontaneously colonizing species, though relatively unimportant where almost all plants are human planted. Seed dispersal by animals may be facilitated by vegetation connectivity (Sarlov Herlin and Fry, 2000), and thus is commonly limited in the fragmented urban areas. Compared with the persistent fruits on many horticultural plants visually favored by the public, seed-enclosed fruits favored and rapidly consumed by bird and mammal dispersers tend to be uncommon. Indeed, the dispersers themselves are usually in limited abundance. Widespread buildings and roads and a low density of most plants further dictate against seed dispersal by animals.

Nevertheless, urban seed-dispersal mechanisms are considerably more diverse than in natural and agricultural land. Wind dispersal is especially important in urban areas. Turbulent and especially vortex airflows in urban areas readily lift seeds and pollen off surfaces. The seeds spin upward and encounter more horizontal streamline airflows, which often accelerate down roads and street canyons (Chapter 5). Moving vehicles and trains also generate airflows that transport seeds (Kowarik and von der Lippe, 2011).

Vehicles carry seeds stuck in tires, enmeshed in mud, and on inside floors and motors (Gilbert, 1991; Forman et al., 2003). The spread of non-native plants within a city may be greater than from city to suburb (Botham et al., 2009). Railway freight cars transport and distribute seeds along railways and in rail yards. Transporting horticultural plants with soil carries numerous seeds throughout an urban area. Indeed, people in huge numbers walking "everywhere" transport seeds. In effect, the distinctive characteristics of the built environment and its human activities greatly facilitate the movement of some pollen and most seeds.

A large urban woodland is likely to be particularly important for seed dispersal by animals. Fruit-eating bird populations may be relatively high, with the birds feeding in woodland and moving outward across the built areas dispersing seeds. Also migratory birds flying over an urban region may feed and rest in large greenspaces. Upon resuming flight, birds characteristically lighten their load, often defecating viable seeds.

The heterogeneous suburban landscape is also probably a key source of seeds, including of non-native species, spreading outward into surrounding agricultural and natural lands. For example, suburban gardens of Cape Town (South Africa) are a major source of non-native plant seeds (Alston and Richardson, 2006). Large numbers of people move back and forth between suburbia and the surrounding natural and semi-natural areas. The richness of non-native species in these surrounding areas correlates with the distance from the suburban gardens (the assumed seed source), as well as with habitat disturbance by people. Fewer non-natives are present under the cover of a tree canopy.

Plant genetics and adaptation

If genetic change and adaptation are fast enough, they provide a good way to survive and thrive in the face of climate change, urbanization, and other urban-related effects. *Genetic change* refers to the change in frequencies of genes in a population over generations. *Adaptations* (resulting from the process of adaptation) then are genetically determined attributes than provide an advantage or increased fitness to an individual or population. Adaptation contrasts with *acclimation* (acclimatization), the adjustment of an individual (one generation) in response to changing conditions. Large populations with ample reproduction tend to have little genetic change and few associated adaptations.

Although little is known directly relating plant genetics to urban attributes, several characteristics of the urban area suggest that many species may have adapted and continue to adapt to urban conditions (Cheptou et al., 2008; Muller et al., 2010). Most habitats are small and isolated. Most species are small or very small in population size. Demographic variability over time in the size of small populations means a greater chance of disappearing. Inbreeding tends to be high in small populations. That produces inbreeding depression, the loss of genetic variation. Inbreeding depression means a greater chance of weak or sterile offspring. Fewer seeds produced and lower seed germination rates, plus less competitive ability, may be expected (Kery et al., 2000). Demographic fluctuation plus genetic inbreeding greatly increases the probability of local population extinction. Also pollinators of pollen and animal dispersers of seeds tend to be limited in cities. The intervening spaces between plant populations are often covered with hard surfaces unsuitable for vegetation. Overall these characteristics favor genetic change and adaptation.

Conversely, other urban area attributes tend to limit genetic change and adaptation. The abundance of linear features including greenways, railways, street-tree rows, and hedgerows favor the movement of animals dispersing seeds. Large parks in the urban area, and especially the surrounding urban-region ring, provide a continual "rain" of pollinators and dispersers over the area. For instance, pollen from populations surrounding a city increased seed production and germination of an urban scarlet-gilia population (*Ipomopsis aggregata*) (Heschel and Paige, 1995). While the regularity of the urban grid fragments the area, it also tends to provide widespread similar habitat and hence the same species widely distributed. Many urban plants are wind-dispersed and thus enhanced by widespread streamline and turbulent airflows.

A literature review reports change in genetic diversity for one moss (*Leptodon smithii*) and six herbaceous plants (*Taraxacum officinale, Primula elatior, Saxifraga tridactlites, Epipactus helleborine, Viola pubescens, Grevillea macleayana*) (Evans, 2010). Four species showed change in urban populations, three divergence within urban populations, and three divergence of urban and rural populations. Short-term evolution may occur, such as in about five to 12 generations for seed dispersal of an urban weed (*Crepis sancta*; Asteraceae) (Cheptou et al., 2008).

A study of violets (*Viola pubescens*) in the Cincinnati (Ohio, USA) Region provides interesting insight (Culley et al., 2007). Genetic variation (and genetic distance) was measured (3 markers and 51 loci) for violet populations across the region, plus four populations in surrounding agricultural land and one distant population in Michigan. Urban populations had high and similar levels of genetic variation (plus similar genetic distances and no unique alleles), which differed markedly from populations in agricultural and distant areas. The results suggest that spatial fragmentation does not impede gene flow in the urban region for the violet species. It is hypothesized that insect pollinators for the plant are abundant enough to overcome fragmentation and isolation effects.

In southeastern Australia, a rare native shrub (*Grevillea macleayana*) is cultivated in residential house plots (gardens). The shrub grows nearby in remnant vegetation patches within the urban area, as well as in the surrounding urban fringe (Roberts et al., 2007). Numbers of inflorescences, visits by pollinators, and fruits per plant were similar in all three habitat locations. Urban plants had higher genetic variation than those in the urban-fringe bushland. All three locations had moderate genetic variation (differentiation), and the remnant and fringe populations were most similar. Thus, the authors note that cultivated native plants may exchange genes with plants in nearby more-natural locations and help to sustain the species there. However, the potential exists for outward gene flow altering the genetics of surrounding native populations (Whelan et al., 2006). In the *Grevillea* shrub study, 19 non-native plant species in urban gardens were also genetically analyzed, along with seeds from the same species in remnant and bushland fringe sites. No evidence was found for gene flow outward to bushland plants.

A broader approach focuses on adaptations or traits. Human-dominated cultural areas ("habitats") in Central Europe are broadly separated into cropland with weedy vegetation and built areas with ruderal ("disturbed wasteland") vegetation (Lososova et al., 2006). Croplands tend to have predictable, frequent, regular, and broad-scale disturbances. Meanwhile built areas have unpredictable irregular disturbances creating varied-scale mosaics with successional vegetation. The authors hypothesize that these drastically different disturbance regimes should lead to different "plant traits or adaptations." Using two large data sets, they found that: (1) cropland plants are more often annuals, insect- or self-pollinated, seed-reproducing,

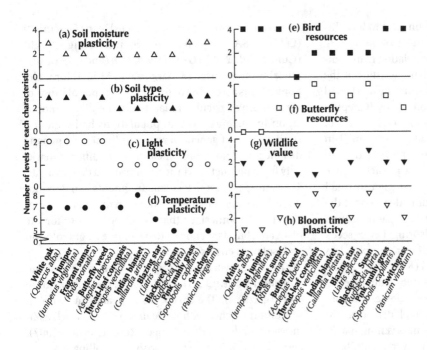

Figure 8.7. Characteristics of plants with high plasticity. (a) Number of soil moisture conditions for acceptable growth; (b) number of soil types; (c) number of light conditions; (d) number of overwintering hardiness zones; (e) number of seasons plant provides food or habitat; (f) number of months plant provides food or habitat; (g) number of major wildlife types using plant for food or habitat; (h) number of months in flower. Threadleaf coreopsis = moonbeam form; red juniper = grey owl form; switchgrass = Shenandoah form; fragrant sumac = gro low form; blackeyed susan = Indian summer form; pink muhly grass = *Sporobolus capillaris/Muhlenbergia capillaris*. Based on Hunter (2011).

and archaeophytes (Figure 8.4), and have overwintering green leaves and a persistent seed bank; whereas (2) built-area plants are more often biennials or perennials, wind-pollinated, flowering in mid-summer, reproducing both by seeds and vegetatively, dispersed by wind or humans, and neophytes. The contrast in plant adaptations between the old cropland area and newer built area is striking.

Preadaptation, the development of genetically based beneficial traits in one environment that are advantageous in a different location, provides further insight into urban plants. Most non-native species arriving in a city presumably die quickly. However, some from similar natural or urban environments have preadaptations that permit survival, even good growth and successful reproduction. Urban environmental conditions are not severe for such preadapted species. Some of these species spread into or invade disturbed sites in the city, while a few invade semi-natural areas in urban greenspaces and outside the metro area. Preadaptations enhance the chance of a recent invader becoming reclassified as a naturalized species, which "fits into" a new ecosystem.

A review of literature on climate-related genetic diversity within populations and the potential evolutionary responses of plant populations to future climate change points to sobering patterns (Jump and Penuelas, 2005). The authors suggest that in fragmented landscapes rapid climate change could overwhelm the ability of plant species to adapt. Genetic composition could be drastically altered and species extinction risk high. These patterns may apply to urban areas.

A useful way to develop vegetation resistant or resilient to disturbances and stresses, including climate change, is to plant generalist species. Such a plant has high *plasticity* (phenotypic), i.e., a genetically determined wide amplitude of tolerance to environmental conditions (Saebo *et al.*, 2005).

For example, ten species, including tree, shrubs, herbaceous perennials and grasses, with high plasticity are compared in Figure 8.7 (Hunter, 2011). Many of the species also provide ecological values. The tree, white oak (*Quercus alba*), has high plasticity levels for five of the eight variables. The two shrubs, juniper and sumac, each have high levels for four variables. Urban vegetation planting could emphasize highly plastic generalist species in the face of disturbances, urbanization, and climate change.

Other clues to plant genetics and adaptation in urban areas exist. In wetland ecology, native, non-native and hybrid cordgrass (*Spartina*) species compete in saltmarshes around San Francisco Bay (Ayres *et al.*, 2004). Apparently native and non-native dandelion (*Taraxacum*) species have successfully hybridized. Horticulture and crop science routinely select or genetically modify varieties and forms of plants planted in urban areas (Saebo *et al.*, 2005). Many of these species

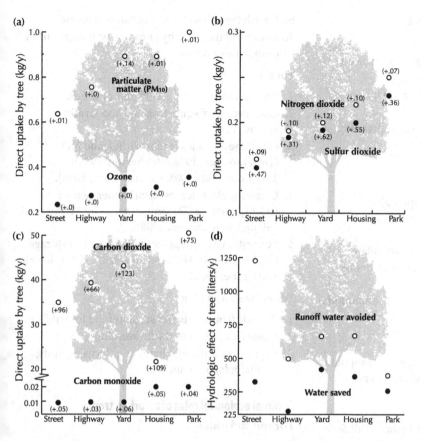

Figure 8.8. Projected annual benefits of a tree in different urban locations. Benefits are 30 years after planting green ash (*Fraxinus pennsylvanica*) saplings in typical locations in Chicago. Tree heights and diameters (dbh) are: street, 9.8 m and 31 cm; highway, 10.4 m and 33 cm; yard, 11.0 m and 36 cm; housing, 11.3 m and 37 cm; park, 11.9 m and 41 cm. Numbers in parentheses are additional benefits from "avoided emissions" (Chen and Jim, 2008).1 m = 3.28 ft; 1 cm = 0.4 inch; 1 kg = 2.2 pounds; 1 liter = 0.26 gallons. Based on McPherson (1994b).

are somewhat specialized, with little ability to adapt naturally. Nevertheless, it should be clear that plant genetics and adaptation of urban plants, currently represented by scattered studies, remains a frontier to explore and understand.

Trees and shrubs

Trees are ecological players on the stage of virtually every chapter. So here we focus on trees themselves, and to a lesser extent, shrubs in the urban environment. First the ecological or functional roles of trees are highlighted. Next we encapsulate the distribution patterns of urban trees, and finally consider the special importance of shrubs.

Ecological roles of trees

Individual trees dispersed over an urban area are the focus, rather than trees in the continuous cover of woods, forests, or plantations. This tree pattern is somewhat analogous to that in a grass-dominated savanna or an area of small farms. For convenience, the ecological or functional roles of trees are roughly grouped into three categories of overall relative importance: (1) major, (2) minor, and (3) minimal. Most of the ecological roles could be considered important out in natural land, but urban areas have quite different patterns. Readers may puzzle over or disagree with the classification of some functions. That's fine. Each reader has different experience in different cities, and to reach a conclusion requires serious thought about urban patterns.

Clearly the ecological roles of a tree vary from city to city and location to location within a city (Figure 8.8). Thus, planting a tree in a park provides on average much more uptake of air pollutants (particulate matter, ozone, NO_2, SO_2, and CO) than planting the tree by a road. The same is true for absorption of carbon dioxide. However, a street tree reduces stormwater runoff more than does a park tree. Overall, trees planted by highways and in yards and multi-unit housing sites provide intermediate benefits.

The ecological importance of an individual tree varies according to surrounding urban conditions, as well as the inputs from and outputs to those surroundings. Moreover, ecological roles change over time.

Major ecological roles of an urban tree

For soil and water

1. Shade and cool the soil surface and herbaceous plant cover.
2. Produce leaf litter that accumulates on the soil surface.
3. Reduce and aerate compacted soil by spreading roots (see Chapter 4).
4. Reduce stormwater runoff.

For air

1. Cool the air by shading (see Chapter 5).
2. "Cool" the air by evapo-transpiration (see Chapter 5).
3. Clean the air, as foliage filters out particulate matter (DeSanto et al., 1976; Grey and Deneke, 1992; Tyrvainen et al., 2005).
4. Emit biogenic VOCs (volatile organic compounds) (see section above).

For animals and plants

1. Make carbohydrates and other organic compounds consumed by herbivores and decomposers.
2. Produce pollen that disperses.
3. Produce seeds that disperse.
4. Provide cover used by birds and other wildlife.
5. Provide nest/den sites used by wildlife.
6. Provide food (leaves, nectar, fruits, seeds, etc.) used by wildlife.
7. Produce layers of foliage and microhabitats.
8. Support the biodiversity and abundance of insects.
9. Enhance avian biodiversity.
10. Provide stepping stones or corridors used in wildlife movement.
11. Inhibit competing plants by shading and root absorption.
12. Survive and grow on contaminated soil.
13. Support non-vascular-plant epiphytes (lichens, bryophytes, algae) on trunks.

Minor ecological roles of an urban tree

For soil and water

1. Transport carbohydrates to roots in the soil.
2. Add deep soil organic matter, as roots die.
3. Harbor mycorrhizae associated with fine roots.
4. Harbor nitrogen-fixing bacteria in root nodules.
5. Change the soil pH.
6. Enrich the soil fertility with mineral nutrients.
7. Lower soil moisture by pumping water upward in evapo-transpiration.

For air

1. Protect against strong, cold, or hot winds.
2. Cool the air by squeezing and accelerating airflows.
3. Clean the air, as leaves absorb SO_2 and NO_2 (DeSanto et al., 1976; McPherson, 1994b; Tyrvainen et al., 2005; Chen and Jim, 2008).
4. Clean the air, as leaves absorb other gases, e.g., CO, O_3, fluoride (DeSanto et al., 1976; McPherson, 1994b; Chen and Jim, 2008).
5. Sequester CO_2 by uptake/absorption and storage (McPherson, 1994a; Bradshaw et al., 1995; Nowak and Crane, 2002).

For animals and plants

1. Provide fruits, nuts and seeds as food for wildlife.
2. Colonize bare ground with seedlings.
3. Support vascular epiphytes (e.g., ferns, orchids, bromeliads) on branches.

Minimal ecological roles of an urban tree

For soil and water

1. Reduce soil erosion by wind.
2. Reduce soil erosion by water.
3. Decrease soil nutrients (macro- and micro-) by root uptake.
4. Reduce (phytoremediate) organic substances in soil.
5. Intercept rain and snow, preventing them from reaching the ground.
6. Lower the water table by pumping water upward in evapo-transpiration.
7. Reduce levels of floodwater.

For air

1. Cool the air by reflecting incoming solar and sky radiation.
2. Add oxygen to the air in photosynthesis.
3. Cool the area just downwind.
4. Increase relative humidity in the area just downwind.
5. Reduce noise levels (Bradshaw et al., 1995; Chen and Jim, 2008).
6. Produce sounds.
7. Produce odors/aromas.

Trees and shrubs

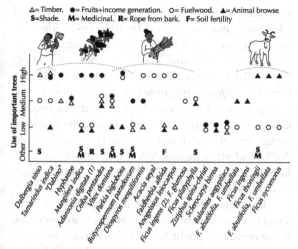

Figure 8.9. Use of economically important trees in peri-urban area of Kano, Nigeria. "Close settled zone" within about 80 km of Kano. Common names of trees (left to right): darbejiya; tsamiya; dabino; goriba; mangwaro; kuka; rimi; dinya; dorawa; kandanya; kanya; dushe; gawo; marke; kawuri; gangi; kurna; danya; aduwa; yandi; shirinya; cadiya; durumi (differs from yandi); baure. Notes: (1) Leaves are the income generation. (2) Var. *tomentosa*. Adapted from Maconachie (2007).

For animals and plants

1. Attract pollinators.
2. Attract fruit-eating birds that disperse seeds.
3. Provide visual cues with waving branches/fronds or shimmering leaves.
4. Drop branches and logs that provide wildlife habitat.
5. Drop branches and logs in water bodies, providing fish habitat.
6. Support vines.
7. Support epiphytes (lichens, bryophytes, algae) on leaves.
8. Produce organic chemicals (allelochemicals) that inhibit other organisms.

More ecological roles of individual trees, mostly minimal or minor, can surely be added to the list. Note that the list does not consider the possible indirect value of avoidance or mitigation (Nowak and Crane, 2002; Chen and Jim, 2008). Perhaps surprisingly, the bulk of the major ecological roles of trees in urban areas are also of major importance in non-urban areas. A few, such as cooling the air, cleaning the air, emitting VOCs, and aerating compacted soil, may be primarily important in urban areas.

Of course urban trees also provide a wide range of benefits to people (Grey and Deneke, 1992; Bolund and Hunhammar, 1999; Tyrvainen *et al.*, 2005; Chen and Jim, 2008; Jones, 2008; Del Tredici, 2010). For example, individually distributed trees may: reduce energy costs; enhance property values; provide privacy; enhance health; relieve stress; produce fruits; produce wood; and enhance aesthetics. Trees also create hazards.

Benefits provided strongly depend on the species of tree planted. In the peri-urban area of Kano (Nigeria), at least 24 types of tree provide significant value to residents (Figure 8.9) (Maconachie, 2007). Six tree species provide (high and medium use) timber products, while a largely different seven species provide browse for livestock. Ten species provide fruits and other income, while a mainly different 11 species provide fuelwood. Several other values (shade, medicinal, rope, and soil fertility) are also provided to residents by the peri-urban trees. Elsewhere, in Curritiba (Brazil) urban trees have been widely planted to provide a richness of values for residents.

Various economic analyses estimating the values of trees have been done (McPherson, 1994b; McPherson *et al.*, 1999, Tyrvainen and Miettenen, 2000; Chen and Jim, 2008). As expected, estimates vary depending on assumptions and variables included. These analyses emphasize market values, though some attempt to also estimate the non-market values of urban trees.

Most of the ecological roles listed above could be called *nature's services* or *ecosystem services*, that is, the natural patterns and processes that benefit society (Daily, 1997; National Research Council, 2005b). The above list of ecological functions for trees seems extensive. Yet it is but a portion of the total range of natural patterns and processes, generated by soil, water, air, animals, plants and microbes, that provide benefits to people.

Tree distribution and arrangement

The first question is to see where trees are in urban areas. Hence we briefly examine different habitat types in a series of cities to grasp patterns of distribution for numbers of trees and types of species. The second question is how the trees are arranged or aggregated within a land use. For this we highlight six basic types of tree arrangement.

Where are the trees?

In Guangzhou (South China), most vegetation occurs in patches of 100 to 10 000 m², that is, up to 1 hectare (2.5 acres) or about the size of a football field (Guan *et al.*, 1999). A study of 6527 trees in different land uses across Nanjing (China) found the greatest number of trees in roadsides, and the fewest in parks (roadsides 32% of the total, neighborhoods 21%, factories 20%, institutions 17%, and garden parks 10%) (Chen and Jim, 2008). The leading dominant species in the different land uses differed considerably. A tree health index indicated good health for institution, garden park, and factory trees, and poor for roadside and neighborhood trees.

For Hefei (Anhui Province, China) on the other hand, 62% of the trees were in residential areas and only 6% in roadsides (16% on institutional land, 9% parks, 8% in a ring park along the former city wall immediately surrounding the city center) (Wu et al., 2008). In this case, tree density was high in the ring park (651 trees/ha) and the parks (459 trees/ha), and much lower in the other three land uses (139–169 trees/ha). Tree species richness showed yet another pattern, with 63 species in institutional land, 52 in parks, 34 in ring park, 28 in residential areas, and 16 species in roadsides. Thus, Hefei roadsides are low in all three tree categories (tree number, density, species richness). However, the highest numbers differ by category (total number of trees in residential areas; tree density in ring park; and species richness in institutional areas). Also in Guangzhou, tree species richness is lower in roadsides than in other land uses (Jim and Liu, 2001). The Hefei results emphasize that observing one tree pattern would be misleading; the three patterns (tree number, density, and species richness) paint the picture of trees in the city

A study of 441 front yards (gardens, spaces) in Auckland (New Zealand) recorded 4700 trees (including tall shrubs and "tussocks"), 71% of which are non-native species (Meurk et al., 2009). Overall front-yard tree density is 464 trees/hectare, overwhelmingly planted trees. The most abundant three native species grow spontaneously at a density of only 1–2 trees/ha. In Auckland, 5 native and 7 non-native species predominate, each being present in more than a fifth of the front yards.

Woody plants cover 42% of a suburban residential area of Milwaukee (USA) (Dorney et al., 1984; Dunn and Heneghan, 2011). Backyards, which include both planted and spontaneous trees, had 36% of the trees. Front yards (28%) and street trees (25%) had lower but similar numbers. Tree species richness of front yards (30 taxa) was nearly triple that for street trees (11).

A closer look at street trees highlights change over time. Based on sampling 6496 trees in a small city (Manhattan, Kansas, USA), distinctive age cohorts were present: 55–60-year-old trees; 40-year trees; and 10–20-year trees (Grey and Deneke, 1992). The oldest street trees were elm (*Ulmus americana*), hackberry (*Celtis occidentalis*), and black walnut (*Juglans nigra*). The medium cohort was mainly Siberian elm (*Ulmus parvifolia*) and pin oak (*Quercus palustris*). The youngest cohort was composed of eight tree species. Changing street trees may reflect changing fashions, sapling availability/cost, disease/pest effects (Zipperer, 2002), and other factors. Street tree cohorts today in Rome in major part reflect changing political times over many decades (Bruno et al., 2006).

Observers often note the high mortality rate of street trees, commonly due to inadequate soil for roots, soil compaction, too little or too much water, road salt, high temperature, disease, and damage by vehicles and people (Bradshaw et al., 1995; Konijnendijk et al., 2005). For example, planted trees often last on average only about 7–15 years. This means that trees along a street, unless all recently planted, are likely to be of different ages and have different heights, tree-crown diameters, and root development. A Chicago study found that trees of streets, highways, housing areas, and parks had similar and high mortality rates (McPherson et al., 1994b). However, trees in house yards lasted twice as long as those in the other land uses.

Less common urban land uses may also have characteristic species. Cemeteries often have pendulous (weeping) varieties, columnar cypress trees (*Cupressus*) pointing to the sky, and other symbolic ornamental trees (Gilbert, 1991). Dumps/rubbish sites, railways, vacant lots, industrial waste piles/heaps, and inactive quarries are normally dominated by spontaneous trees, which are usually quite different from the planted species (Kunick, 1990).

Urban-to-rural gradient studies indicate that urban woods may have lower stem densities, fewer understory species, more non-native species, and higher tree species richness (McDonnell et al., 1997; Muller et al., 2000; Porter et al., 2001; McDonnell and Hahs, 2008; Berland, 2012). Based on an Ohio (USA) city study, woods intensively used for recreation may have especially high tree diversity (Porter et al., 2001).

Urban tree species also change in ecological succession. Over a few decades in a New York City woods, oaks (*Quercus*) and hemlock (*Tsuga canadensis*) significantly decreased, while birch (*Betula*), cherry (*Prunus*), and other species increased (Rudnicky and McDonnell, 1989). In 32 woods of a small city (Worcester, Massachusetts, USA), ash (*Fraxinus*) and oak (*Quercus rubra*) are apparently sharply decreasing, while a naturalized maple *(Acer platanoides)* is increasing (Bertin et al., 2005).

Major environmental disturbances also cause change in urban tree density and diversity. Over many decades Kyoto has experienced typhoons (hurricanes), floods, and fires (Sakamoto, 1988). Also humans have removed trees for new developments and have planted trees in different areas. Despite these major alterations, the total number of elm-related Ulmaceae trees (the predominant tree group composed of three species) has fluctuated around a rather constant level over the time period. All three species are planted, while two of the species grow spontaneously. The author concludes that the city residents' appreciation for trees and their functions helps stabilize the urban tree cover.

How are urban trees arranged?

In natural land, the *dispersion* or spatial arrangement of individuals of a plant species is generally aggregated into clumps. Much less common is a somewhat regular distribution of individuals, and in rare cases individuals seem randomly distributed. Also patterns may be polycentric, monocentric, or dispersed (see equations, Appendix B).

Rather than learning from diverse examples, here we highlight six basic tree arrangements found in almost all cities, and briefly explore the characteristics of each. The basic arrangements are: (1) single trees; (2) tree rows; (3) narrow wooded corridors; (4) wide wooded corridors; (5) small wooded patches; and (6) large wooded patches.

1. *Single trees*. These are especially common in yards, parks, cemeteries, industrial areas, vacant lots, brownfields, and so forth. Single trees are open grown, often with mid- to low-level branches present on all sides. Typically no shrub or tree understory is present. Both planted and spontaneous (adventive) trees are common. Although the tree's canopy is separate from neighboring trees, the root systems may extend more widely and be in competition, especially in dry areas. The form of the planted tree, and hence the species may be particularly important, as for street trees, a Japanese garden, or a cemetery. Lists of species and the availability of trees and their varieties are commonly used in selecting plantings (Benedikz et al., 2005; Saebo et al., 2005). Planting urban trees includes selecting location, species, and variety, as well as the processes of planting and maintenance (Millard, 2000). Overall, the success rate for plantings is considered to be moderate to low. In contrast, the natural colonization process itself selects the species that are often (pre-)adapted to survive in the urban environment. Spontaneous trees may have a higher success rate, and are established at a lower cost. Arranging separate single trees provides opportunity for creativity (Bell et al., 2005; Gary Hilderbrand, personal communication, 2012). For example, a regular grid of trees, as in a large parking lot or a fruit orchard, is often present in urban areas. The parking lot may have several species, whereas a fruit orchard typically has but one species. In the absence of maintenance, the regular tree pattern can change quickly into an irregular pattern or into regularly distributed clumps. In the latter case, trees may produce root sprouts around the "mother." Or for trees producing fruits, birds often feed on fruits and defecate the seeds upon flight, so tree seedlings and saplings are clustered around the mother plant.

2. *Tree rows*. In urban areas these are often single rows, such as on one side of a street or along a boundary. But double rows are also common, as on both sides of an avenue or an allee in a park. Except for some boundaries with spontaneous trees, tree rows are overwhelmingly planted. Shrub and tree understory layers are normally absent. Walkways may have tree rows on one or both sides. Rows of trees are familiar in some commercial and institutional areas. Tree rows tend to be of a single species, though multi-species tree rows provide somewhat different functions. Species selection often focuses on the form of trees.

Trees on one side of a street are often planted to reduce summer afternoon heat, and perhaps secondarily to facilitate snow and ice melt in winter (Bell et al., 2005). Some wildlife readily moves along tree rows, but the usual single tree species and the absence of a shrub layer greatly limits the effectiveness for wildlife movement. Windbreaks normally are of a single species, whereas visual-screen tree rows often have at least a shrub layer also. If canopies of two tree rows form an arch over a street, the road is all shaded, so, for instance, lizards may readily cross the road on hot days. The double row of plane trees in European cities, and of palms in tropical cities, is iconic. In London more than half the street trees are a single highly resistant hybrid species, the London plane tree (*Platanus*) (Wheater, 1999). As noted in the preceding section, newly planted street trees are subject to many stresses and disturbances, and thus usually have high mortality rates.

3. *Narrow wooded corridors.* These strips of tree-dominated vegetation are mostly of spontaneous plants, though each corridor may have originated as a planted tree row. Boundaries between fields, and along the back property line of medium-size house plots, often have vegetation strips. Such narrow strips are common along railways, on steep banks, and by a stream, canal or ditch. In contrast to tree rows, most of these wooded corridors receive little maintenance. Many wildlife species move along, either inside or alongside, narrow wooded corridors in urban areas (Forman, 1995). Narrow spots and breaks in the corridor are common, and may reduce effective wildlife movement. Such narrow corridors are highly subject to conditions on both adjacent sides, so the trees present are mostly generalists. People tend to dump many types of materials in the corridors, thus also affecting their use for wildlife and human movement.

4. *Wide wooded corridors.* Normally rather few of these major corridors are present in an urban region. They are basically strips of spontaneous semi-natural vegetation or woodland, typically remaining along a river, stream, or pipeline. If used in a major way for recreation, the corridor is sometimes called a "greenway." A narrow greenbelt, such as proposed for Toronto, could be a (semi-)circular wide wooded corridor. Also, many of these wide corridors tend to be major infrastructure routes for pipelines, electric powerlines, railways, and roads into and out of a city. This means the presence of unending maintenance and disturbance. Wildlife movement and human trails are prominent functions (Forman, 1995). Species also move and disperse between these wide wooded strips and the surrounding residential and commercial lands.

5. *Small wooded patches.* Such woods or woodlands of varied size and shape tend to be common, and of major ecological importance, throughout the urban area. Some are planted; most are of spontaneous vegetation. House plots, city parks, town parks, institutional areas, industrial areas, dumps, infrastructure areas, farm areas (Grey and Deneke, 1992; Attorre *et al.*, 1997), steep hillsides, and somewhat-high water-table areas are typical locations containing small wooded patches. Some small patches are surrounded by a particular land use such as residential land, while others are between two types, or even where three land uses converge (analogous to a convergency point) (Forman, 1995; Kowarik and Korner, 2005). Maintenance levels in such woods vary widely from intense, as in some Kyoto gardens, to none, where ecological succession is the predominant process. The shape of small wooded patches is especially important (see equations, Appendix B). Tiny patches of two or three trees together often have a noticeably different microclimate than their surroundings, and may contain semi-natural understory, shrub, and herbaceous-layer vegetation. Larger aggregations of trees progressively modify the interior environmental conditions and hence the species present. In a New Jersey (USA) study, woody vegetation patches up to about 1.5 to 2 ha had a much higher density or packing of both tree and bird species than did patches >2 ha (Forman *et al.*, 1976; Forman, 1995). Such small patches bulging with species thus are of particular interest in parks, where people wish to see wildlife. Most of the tree species are relatively tolerant generalists, and fast-growing successional species are common (Kowarik and Korner, 2005).

These small patches often have trails, are used for dog walking, and hence contain dog scents and waste that inhibit terrestrial wildlife. Dumping of varied materials is common in small wooded patches. In urban areas the woods are often disturbed and seem open, with the shrub layer suppressed to enhance human visibility and security. Tokyo and Berlin seem exceptional in maintaining abundant shrubs in wooded parks. A series of small or tiny patches serving as stepping stones is considered to be the primary route for most wildlife movement across urban areas, though this is little studied. Such a pattern of course means that each wooded patch in the sequence, as well as each intervening space, is very important for effective movement. A continuous wooded corridor is far better, and an elongated cluster of small patches is better, for wildlife movement (Forman, 1995, 2008). However, both patterns are in limited supply in an urban area.

6. *Large wooded patches.* These woodlands (or forests) are the least common of the six forms

and, when present, are of exceptional ecological importance in urban areas. Table Mountain Park in Cape Town and Tiergarten in Berlin are examples. Seoul has a ring of large wooded parks on its edge, representing the remnants of a former greenbelt (Forman, 2008). Semi-natural vegetation with mainly native species dominates, and is enriched by lots of non-native species present. Large woodland or forest patches, as around some German cities, provide many ecological functions and services to society, including wildlife protection, clean-water protection, climate modification, recreation, wood products, and mining (Forman, 2008). Large forested areas with similar functions also protrude as green wedges into some cities, including Portland (Oregon, USA) and Stockholm.

Commonly, even in these large patches, the shrub layer is suppressed in many areas for visibility and security. An ample shrub layer of *Lonicera* and other species in Berlin's Tiergarten is exceptional. Large wooded patches also have considerable horizontal heterogeneity, including scattered openings or glades. Such openings in a woodland are rare in an urban area and thus of considerable ecological interest. Ponds and streams may be present (see Figure 1.5). Dumping primarily occurs along the edge of the wooded patch or along wide trails accessible by vehicles (Matlack, 1993). A considerable trail network criss-crossing the urban woodland is characteristic of large wooded spaces. The trees and other plants are overwhelmingly spontaneous in origin. The species are mainly generalists, though some rare specialist species are often present in low numbers. A large forested patch in relatively new cities may be mainly a remnant of previous widespread forest habitat. However, the changed environment due to surrounding urbanization typically means that the remnant species reproduce less successfully, and are being replaced by generalists and non-natives (Rudnicky and McDonnell, 1989; Zipperer, 2002). Only large wooded patches contain many interior species (see equations, Appendix B).

Plantations, usually of a single tree species, are occasionally found in urban areas. Thousands of birch (*Betula pendula*) planted by the Amsterdam airport to reduce bird–airplane collisions is an example. Such monocultures are generally of low ecological value and low recreational interest. A

Figure 8.10. Small tree, shrub-height plants, and pond in urban traffic circle. Showy grass clumps; bottlebrush (*Callistemon*) tree on left. Few animals thrive in such a stressful hazardous site. Barcelona. R. Forman photo.

large mangrove swamp by a tropical coastal city may also be dominated by a single tree such as red mangrove (*Rhizophora mangle*). However, in this case, the tree and ecosystem are natural and of major ecological value.

Shrubs in urban areas

Shrubs and shrubby areas appear to play exceptionally important roles in urban areas because of their dense foliage relatively close to the ground (Figure 8.10). The foliage commonly provides dark shade and cool moist conditions at soil level, cover for both wildlife habitat and movement, and visual screening for people. Despite these roles, urban shrub ecology remains a research frontier.

Spontaneous shrub growth is common along boundaries, hedgerows and railways, at the edge of woods and quarries, on rubble and abandoned sites, along riverbanks, and in swamps and large wooded patches. In contrast, planted shrubs predominate in hedges, foundation plantings, flower gardens, side boundaries of house plots, cemeteries, entranceways, and narrow highway-median strips. The highest diversity of shrubs may be in certain cemeteries (see Chapter 12) (Gilbert, 1991). Here a diversity of planted shrubs around many markers is supplemented by spontaneous shrubs in less manicured areas.

However, shrubs tend to be scarce under single trees and planted tree rows. Many paths and walkways are devoid of adjacent shrubs. Small wooded patches heavily used by people generally contain spontaneous shrubs, but in very low density.

A study of spontaneous vegetation in ten German cities found four shrub species to be predominant (Kunick, 1990): black elder (*Sambucus nigra*), goat willow (*Salix caprea*), bramble or blackberry (*Rubus fruticosus*), and butterfly bush (*Buddleja davidii*). The first two often grow together, but all four produce dense thickets, which provide good habitat and cover for urban wildlife. Butterfly bush is mainly in city centers, apparently favored by urban heat. Sandy/gravelly railways favor bramble, often along with shrubby birch (*Betula pendula*) or black locust (*Robinia pseudo-acacia*). Butterfly bush, birch, and black locust thrive on building rubble. Black elder and goat willow seem to favor partially shaded soil with ample organic matter. Bramble grows on abandoned gardens, and black locust thickets on quarry piles. Spontaneous vegetation in certain cemeteries, including yew (*Taxus*), lilac (*Syringa*), and ivy (*Hedera*), form distinctive areas of spontaneous urban shrubland in the cities (Figure 12.7).

As just illustrated, tree seedlings and saplings may be a significant component of shrubby areas, and help determine the future succession of vegetation. Disturbance in woods may favor more generalist species (Guntenspergen and Levenson, 1997). Most spontaneous shrubs readily spread vegetatively and thus may cover a considerable space, unless cut back. Some shrubs are thorny or spiny, such as bramble, cactus, and *Yucca*. The spines deter people, but normally have little effect on urban wildlife movement. Maintenance personal tend to avoid areas covered with spiny shrubs, so such sites usually accumulate trash and debris.

Shrubs produce fruits in greater proportion than do urban trees, which more likely produce wind-dispersed seeds. Thus, fruit-eating birds and mammals (frugivores) are often attracted to shrubby areas. People also pick fruits from these convenient-sized plants such as blackberry and blueberry (Houck and Cody, 2000).

For almost all urban wildlife, a shrub can provide cover. A line of shrubs creates a promising wildlife movement route. An adequate-sized patch of shrubs provides habitat for nesting or denning. Of course very low shrubs, effectively ground cover, may provide none of these ecological benefits.

Buckthorn (*Rhamnus cathartica*), a non-native from Europe, was found to be the most common tree (essentially a small tree or tall shrub) in Chicago, 13% of the total trees present (Whitney and Adams, 1980). Thus, the leading woody plant is not planted, but colonizes, sprouts, and spreads spontaneously. One study suggests that the survival rate for birds nesting in non-native shrubs is lower than that in native shrubs, because the plant structure is more easily penetrated by predators (Moorman and DePerno, 2006).

An interesting study of understory plants in the urban woods of Barcelona provides insight into the distribution of shrubs (many Mediterranean understory species are shrubs) (Guirado *et al.*, 2006). The richness of human-disturbance-dependent (synanthropic) species is higher in small than large woods, if the small woods are adjacent to crop fields. Also, within larger woods these plants are most frequent near the edge. However, rare forest species are less frequent in small woods if the woods are adjacent to an urban area rather than to cropland. Furthermore, common forest species, total plant species richness, and the presence of people are more frequent close to a forest edge if the edge adjoins an urban area rather than cropland. Ecologically this emphasizes the importance of adjacent land use on species patterns. More people enter (presumably carrying seeds of human-disturbance-dependent species) and disturb woods from an urban border, than from a cropland border.

Finally, *vines* are also present and play ecological roles in urban areas, though the ecological literature on urban vines is scarce. Ivy (*Hedera helix*) and Boston ivy (*Parthenocissus*) readily colonize and grow on walls. Boston ivy and Virginia creeper (*Parthenocissus quinquefolia*) grow on tree trunks. Clematis (*Clematis* spp.) vines planted in gardens may grow into large masses that attract an abundance of pollinators. A naturalized non-native vine, Asian bittersweet (*Celastrus orbicularis*), is bird-dispersed and abundant in northeastern USA suburban areas. In tropical cities, thick woody vines or lianas are usually present at low frequency. Some birds nest in these urban vines, though overall low insect abundance limits breeding bird numbers.

Plant community structure and dynamics

Even a glimpse of vegetation highlights the two key aspects of plant communities, spatial structure and species pattern. Spatial structure includes vertical layers and horizontal patchiness. In contrast, species pattern emphasizes the diversity and composition of species making up the spatial structure.

Yet a glimpse misses the all-important change and dynamics over time, our second topic here. For this, we explore vegetation-related patterns over different time scales from hours to millennia, and then focus in on ecological succession.

Spatial structure

We first examine the vertical layers in vegetation and then the plant species present in the vegetation.

Vertical stratification

Ecologists commonly recognize up to five layers or strata in a forest: canopy, subcanopy, understory, shrub layer, and herb (or herbaceous) layer. The *canopy* is the more or less continuous uppermost layer of tree foliage, although some tropical forests have scattered emergent trees above the canopy. Just beneath the canopy may be a *subcanopy*, mostly of trees that grow upward when canopy trees die. An *understory layer* beneath is often dominated by smaller trees that only reach that height, though species that grow into upper layers are commonly present. A *shrub layer* of about 1–3 m height may be dominated by shrub species, but often contains many young trees that grow taller. The *herbaceous* or *herb layer* below about 1 m height is mainly of herbaceous plants, though usually many tree seedlings and/or low shrubs are present.

Urban woods or forests typically have two, three, or four layers of foliage, largely determined by the history of human activities on a site (Nowak, 1994). Normally a four-layer woods has little subcanopy, and three layers indicates the scarcity of either an understory or shrub layer. Usually two strata are only the canopy and herb layers. While recognizing that the presence of stratification and layers is convenient, some ecologists emphasize that these are but a model, and that really there are no layers. Instead foliage is distributed in gradually but unevenly changing amounts from ground to canopy top.

Consider the important Mongolian oak (*Quercus mongolica*) woods in Seoul (Lee *et al.*, 2008). One wood (stand) in city center is compared with three in the suburbs (inner urban boundary), and three in the exurban area (outer urban boundary). All woods are dominated by the oak, though small mountain ash (*Sorbus alnifolia*) trees are common in the city-center woods. Plant species richness is highest in the city-center woods and lowest in the exurban woods. The tree canopy in city center is thin and 15–18 m (49–59 ft) high, whereas the suburban and exurban woods are similar with a thick 8–18-m high canopy. In all three cases no distinct subcanopy is present. The city-center wood has a thin dense understory at 6–8 m height, compared with a thick medium-dense understory at 3–8 m in the other woods. The shrub layer is thin at 1–2 m in the city center and thick at 1–3 m in the suburban and exurban woods. All three woods have a similar-thickness herb layer at 0–1 m height. Thus, for these four-layer woods, the city-center wood differs markedly in vertical structure from the suburban and exurban woods, which are similar.

Two-layer woods are common in urban areas where the shrub layer has disappeared due to heavy human use or is removed for visibility and security. Also, unlike woods in natural and agricultural land, the edge portions of urban woods commonly have much less shrub and understory foliage than in the woods' interior (Johnson and Klemens, 2005). However, where people do not remove the shrubs, urban forest edges have somewhat denser foliage than in the interior.

Many ecological functions and benefits are lost in a two-layer rather than a multi-layer wood (Tyrvainen *et al.*, 2005). These stratification patterns occur in both remnant woodland (a site where woods were never cleared) and in regenerated woods, i.e., regrown from a cleared site (Hamabata, 1980; Zipperer, 2002; Bell *et al.*, 2005). Remnant woods persist, for example, in Amsterdam, St. Petersburg, Syracuse (New York, USA), and Portland (Oregon, USA). Regenerated woods dominated by spontaneous species commonly appear along rivers and on railway land (Gilbert, 1991), and are usually younger with fewer layers.

The horizontal pattern of urban plant communities is equally important though less studied directly. Parks are multifunctional and hence normally quite heterogeneous (Gilbert, 1991). Indeed, horizontal microhabitat heterogeneity characterizes urban areas, as emphasized in a section above. Even within a recognizable plant community or habitat, the density of vegetation varies from high to low. Openings in the canopy are normal and play important ecological roles. In addition to the presence of somewhat-distinct tiny patches, linear features such as trails are common in urban vegetation. Thus, the degree of horizontal heterogeneity or patchiness is an important characteristic of urban plant communities, and mainly reflects the history of human activities.

As discussed below, plant community boundaries or edges are overwhelmingly linear and abrupt. This pattern contrasts with edges out in natural land that are mainly curvilinear or convoluted, and sometimes in the form of a strip of tiny patches. Some urban vegetation

Urban habitat, vegetation, plants

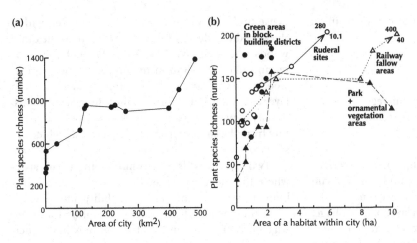

Figure 8.11. Species–area relationship for city and for habitats within city. (a) Cities in Germany and Eastern Europe; two largest cities = Warsaw and Berlin (West). From Klotz (1990). (b) Sites in Berlin (West). 1 km² = 0.39 mi²; 1 ha = 100 m × 100 m = 2.5 acres. (Ruderal is often described as "disturbed wasteland"). From Sukopp and Werner (1983).

has slightly curvy boundaries created by design, or due to set-backs from buildings, signs, towers, and other human objects (Kenney, 2008).

Species composition and biodiversity

A *natural community* (or species assemblage) is considered to be all the species present in an area, and is normally restricted to a relatively homogeneous area. Thus, the species in a meadow and in an adjacent woods would be measured separately as two communities, not combined as one. The degree or "tightness" of interactions among the species is of considerable ecological interest (Austin, 1999). Thus, a high degree of predation, parasitism, symbiosis, mutualism, and intra/interspecific competition indicates a group of species that have coexisted for a considerable period, and may have evolved together. A high degree of interaction suggests a more stable community.

Urban natural communities overall seem to have a low degree of species interaction, with species arriving, spreading and disappearing at a relatively rapid rate. However, species in urban communities are not random mixtures. Many species combinations are readily recognizable as repeated plant communities across urban areas. Planted areas such as flower and vegetable gardens may have the lowest degree of species interactions and stability. Plantings normally change drastically when human maintenance stops.

Biodiversity is sometimes described as the variety of life, and composed of three levels, communities, species, and genetic types. Since those are such different characteristics, ecologists usually simplify the concept of biodiversity to be species richness (number). Some ecological studies exclude non-native species. But, as noted above, this perspective seems to be of limited use in urban areas, where non-native and native species coexist in abundance and both play important ecological roles. Genetic types of almost all species are unknown and difficult to determine, so operationally these are normally excluded in ecological studies of biodiversity.

Thus, while recognizing that biodiversity exists at different levels, we consider the primary "measure of biodiversity" to be the richness (or number or diversity) of species present, no matter how long they have been in the area. A second informative ecological measure is sometimes added, i.e., habitat or vegetation or community diversity, the number of plant communities present in an area.

Of course, species number masks the diverse types and roles of the species, including dominant, rare, keystone, native, etc. species (Whittaker, 1975; Smith, 1996; Cain *et al.*, 2011). The degree of *dominance* (or inversely, evenness) in a community is especially informative (see equations, Appendix B). A community has high dominance if, for instance, one or two species compose more than half of the individuals present, and low dominance if, for instance, the most abundant 5–6 species compose half the community. The first case typically has low species richness and few rare species, whereas the second case has more diversity with many species in low abundance.

Plant species richness commonly correlates with the area or size of a habitat. One study of city size suggests that species richness rises sharply in cities up to about 125 km² (50 mi²), does not differ from 125 to 400 km², and again increases sharply in cities above 400 km² (Figure 8.11a) (Klotz, 1990). An alternative, more likely interpretation of the Figure 8.11a data is that species richness increases steeply in cities of up to

Plant community structure and dynamics

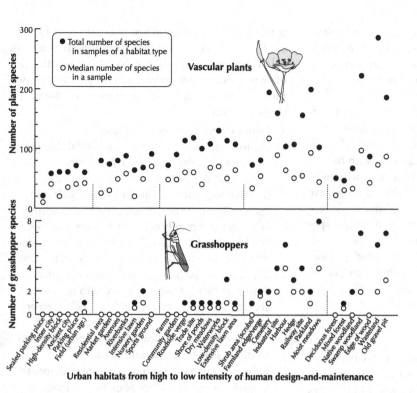

Figure 8.12. Species richness of plants and grasshoppers in urban habitats varying by human impact. Vascular plants; grasshoppers = Orthoptera. Five groupings of habitat types, plus finer-scale ordering of habitats within a group, are done by the author, and can be expected to vary in different cities. Habitat and plant data for Dusseldorf, Germany (Godde et al., 1995).

about 10 km² (4 mi²), and then constantly but gradually increases in cities of 10 to 500+ km².

Considering this *species–area relationship* for different habitats within a city is perhaps more ecologically informative. Plant species richness in general increases with area in all four habitat types studied (Figure 8.11b) (Sukopp and Werner, 1983). The rate of species increase is lowest for park and ornamental vegetation areas, and second lowest for railway sites. For tiny vegetation habitats ≤2 ha (5 acres), the steepest increase may be for "green areas in block-building districts" though variability is high and no large areas were sampled.

Other urban studies also show plant species richness increasing with area within a city (Pysek, 1998; Dunn and Heneghan, 2011). These include urban woods in Minneapolis-St. Paul (USA) (Hobbs, 1988), vacant lots in Chicago (Crowe, 1979), and spontaneous vegetation in Birmingham (UK) (Angold et al., 2006). Although species richness typically correlates with habitat area, other variables such as site conditions and human activities normally are more important in determining biodiversity (see equations, Appendix B).

Species richness of course varies by habitat type. Thus, the median plant species richness for plots within the 38 primary habitat types of a small city varies by a factor of 10 times (Figure 8.12) (Godde et al., 1995). About the same degree of variation exists for total species richness by habitat type.

Plant diversity seems to correlate inversely with degree of human design-and-maintenance effort (Figure 8.12). More intense levels, as for parking places, inner city, and high-density blocks, have fewer species. Also low plant richness occurs in forest habitats, perhaps because of dense shade. In contrast, many neglected areas, including old gravel pits and "wastelands," have high species richness.

The same human-impact pattern was found for the diversity of grasshoppers (Figure 8.12), land snails, wood lice, and butterflies (Godde et al., 1995). These results suggest that people are designing and managing against biodiversity in urban areas. Yet this is also a ripe opportunity for improvement.

Plant species richness or biodiversity is an important variable in many sections of this chapter. Diverse insights are added for perspective. Harvard University's Arnold Arboretum in Boston is one of the most plant-diverse areas anywhere outside the tropics. Covering 80 ha (200 acres), the collection of 15 467 planted plants represents 4030 species (taxa). If the spontaneous plants present were included, several hundred more species would be added to the total plant biodiversity (Peter Del Tredici, personal communication,

2012). Indeed, spontaneous species in diverse habitats of Rome constitute 41% of the flora in urban parks, in contrast to 73% in suburban areas, 83% in peri-urban habitats, and 98% in a protected urban nature reserve (Faggi *et al.*, 2008).

Biodiversity also correlates with various human patterns. In residential areas of an arid urban area (Phoenix, USA), plant species richness correlates positively with the family income of residents, negatively with the age of housing, and negatively with previous farming of a site (Hope *et al.*, 2008). Older residential areas have more irrigated low-diversity grass and trees, whereas newer areas have more xeroscaping with diverse species. Farming decades ago, which eliminated most native species, left a biodiversity legacy on today's land. Overall, within-site alpha diversity is high (equivalent to that for nearby desert), site-to-site beta diversity is low, and total regional gamma diversity is very high.

Change and dynamics

Change here refers to differences in spatial pattern over time, and dynamics adds insight into the mechanism or causes of change, typically with correlations. Rates of change vary widely so we start with time scales, and then focus on ecological succession.

Time scales from hours to millennia

For various ecological patterns and processes a rough correlation exists between space and time (Delcourt and Delcourt, 1988). That is, things that extend over a long time typically cover or affect a large area (e.g., glaciation), whereas quick changes occur on small spaces (e.g., tree-fall gap in a forest).

Numerous natural processes affecting the land seem to be almost instantaneous (earthquake, tornado, tsunami), while others seem eternal (corrosion, fungus rot, evapo-transpiration). Relatively few natural processes mainly operate at the human lifetime scale of years to decades (Forman, 2012). Thus, the natural world gradually changes, punctuated with powerful perturbations. Meanwhile, our powerful changes in the land are mainly at the years-to-decades (sometimes centuries) scale. Consequently we continually fight the inexorable slow changes with maintenance budgets and energy, while also having to respond to big disturbances ("surprises") with big repairs.

To envision what changes, as well as the rate, it is useful to group changes into different time scales. For urban plants and vegetation, four scales are particularly important: (1) short time scales including hours, days, weeks, months, and seasons; (2) years; (3) decades; (4) centuries and millennia. For each time scale below, the processes or mechanisms operating are often suggested. Some urban patterns change over two, three, or four scales, though most are predominant at a single time scale.

1. *Hours, days, weeks, months, seasons.*
Photosynthesis, transpiration, and tidal changes occur on a regular daily basis. Bombing, explosions, demolition, tsunamis, cyclones (hurricanes), and earthquakes typically alter sites and vegetation within hours or days. Pollination of species commonly occurs over days to weeks. Seed dispersal normally occurs in hours to weeks. Plant phenology follows a seasonal cycle. However, differences in phenology, such as for *Forsythia*, a flowering shrub, from site to site within an urban area may extend over days and are easily seen or mapped (Gilbert, 1991). Disturbances such as combined sewer overflows (CSOs) and woodland wildfire cause effects in hours to days.
Informal squatters commonly create settlements in weeks or months. Plant responses to land abandonment are usually striking within weeks to months. Such responses may include species expansion, disappearance, and colonization. Habitat loss is mainly a short-term event, though it may continue over years or more. Rapid environmental effects can cause species extinctions (Williams *et al.*, 2009).
Various vegetation types such as lawns and meadows are designed and intensively maintained for stability, but change rapidly because of fluctuations in the intensity of management (Ahern and Boughton, 1994). In contrast, some urban gardens and xeroscaping are designed for stability and low maintenance (Quigley, 2011).
2. *Years.* A 14-year study in Argentina documented plant colonization of a residential area (Rapoport, 1993). Seed banks perhaps typically change significantly over years. Vegetation responses to large disturbances such as demolition, cyclone, and tsunami typically extend over years. Species migration, range contraction, and range expansion often occur over years for rapidly reproducing species or those readily transported by people. A newly arrived non-native species may spread over years (Whitney and Adams, 1980; Sukopp, 2008). Building and road construction mainly extends over years. Habitat degradation and much

vegetation change commonly occur over years (Kunick, 1990).

Over three years a small nature reserve in a Prague suburb started with 99 plant species, lost 7, and gained 20 (Kubikova, 1990). Garden management and maintenance, such as planting, trimming, pruning, mulching, watering and harvesting, produce effects on plants over months and years (Owen, 1991). Many such activities or disturbances have *ripple effects* on other species (a species cascade).

3. *Decade changes*. Increasing air temperature associated with urbanization over decades is considered a prime reason for changed plant phenology. In the eastern USA from 1990 to 1999, the growing season increased an average of 7.6 days, mostly due to an earlier start in spring (White *et al*., 2002). An earlier start rather than a later finish was also reported for altered phenology in Beijing (Luo *et al*., 2007).

For seven cities on four continents, urban versus rural phenologies differed in all cases (Gazal *et al*., 2008). In spring the flush of new leaves was 1 to 23 days earlier in urban than in surrounding rural areas. Urban "bud burst" was earlier in three of four temperate cities, but in only one of three tropical cities.

The invasion of non-native species is prominent over decades, though invasives causing extinctions of urban native plants may be rare. Species extinction may be mostly a years-to-decades phenomenon. Extinctions of urban non-native species apparently remain unstudied. The naturalization of a non-native species seems to be most prominent over decades (Muhlenbach, 1979; Sukopp and Trepl, 1987; Peterken, 2001). Atmospheric warming related to global climate alters phenology, decreases native species, and increases non-natives over decades. Windstorms associated with climate change may be more frequent and severe, thus changing vegetation patterns.

The elimination or severe degradation of shoreline vegetation is typically an over-decades period, involving different processes section by section (Sukopp and Markstein, 1989). Invasion and hybridization in *Spartina* salt marsh is evident over decades (Ayres *et al*., 2004; Silliman *et al*., 2009). Change in urban land-use cover is often striking over a few decades (Pauleit *et al*., 2005). The legacy of decades-ago farming is evident in today's vegetation in residential areas (Hope *et al*., 2008). Major change in street trees seems to be a decades characteristic though it may extend over centuries (Grey and Deneke, 1992; Bruno *et al*., 2006).

Four prominent outward urbanization patterns – concentric rings, satellite cities, transportation corridors, and dispersed patches – seem to form over decades from cities (Forman, 2008). A "bulges model" whereby a city expands over time by bulges in different directions may be most characteristic of decades change. Suburbanization and associated species loss commonly occurs at this time scale (Moorman and DePerno, 2006). The fragmentation of urban habitats is typically an over-decades process (Luck and Wu, 2002). As a specialized case, greenspaces (canyons) in San Diego seem to lose most of their native birds and mammals in the 50-year (or 80) period after urbanization surrounds and isolates a greenspace (see Figure 9.2) (Soule, 1991).

Over about 8–10 decades, both native and non-native species spread and decline, and the relative abundances of species change noticeably (Kowarik, 1990). The spontaneous flora in parks changes considerably (Gilbert, 1991). Tree plantations change markedly. Over decades in Kyoto, the vegetation cover of the predominant Ulmaceae trees fluctuated around a rather constant level, despite major natural disturbances and human activities (Sakamoto, 1988).

Over 12 decades in Pizen (Czech Republic), 805 plant species remained present, 368 disappeared, and 238 new species arrived (Pysek *et al*., 2004). Species richness in the city increased, but decreased in the surroundings. The similarity of the flora from beginning to end was greater in the surroundings than in the city. From the 1960s to 1990s in Pizen the richness and proportion of archaeophyte species dropped, while the neophyte proportion did not change.

The effects of human activities such as management, maintenance, conservation, reclamation, restoration, development, historic preservation, and flower garden/vegetable garden planting are especially evident over decades. The same is true for naturalistic plantations (as in The Netherlands and UK) characterized by clumps of trees, diverse shrubs, intense management, and planting of the herbaceous layer after some 5 years (Gilbert, 1991).

4. *Centuries and millennia*. Plant adaptation through genetic change is perhaps most frequent over centuries, though it occurs over decades as well as much longer time frames. Species migration, range contraction, and range expansion for species with low reproduction rates occur over varied time frames, though perhaps typically over centuries. This rate is probably too slow for many species to survive climate change in a city (Jump and Penuelas, 2005).

Cultivated plants becoming feral (changing enough to thrive in the wild) seems to be mainly a decades-to-centuries phenomenon. Feralization may occur when a domesticated plant or cultigen hybridizes with a closely related wild plant, or when the cultivated plant genetically reverts to the wild type (Sukopp and Sukopp, 1993). Plant families with many natural hybrids include the sedges (Cyperaceae), composites (Asteraceae), grasses (Poaceae), and willows (Salicaceae). Cultivated plants becoming feral in rural areas seem to have been frequent in Europe since the 1200s, though apparently little is known specifically about plants in urban areas.

Deforestation and the spread of livestock and cropland occurred over centuries where many of today's European cities grow (Kowarik, 1990). Very few cities have existed for more than a millennium, and few over-millennia studies (e.g., pollen analyses) are done in cities. Nevertheless, both centuries-long (AD 750–1150) and millennia-long (13 000 years) studies show significant changes in vegetation, presumably largely due to climate change supplemented by human activities (Brande *et al.*, 1990).

Plant extinction rates over centuries have been calculated for 22 cities on five continents (Hahs *et al.*, 2009). Cities >400 years old that spread over cropland long after deforestation lost an average of 22% of the flora per century. Young cities (<200 years old) that expanded over native vegetation or a mixture of cropland and vegetation lost an average of 2% per century, and intermediate-age cities lost 12% of their plant species per century. Much of the difference in extinction rate correlated with the age of a city, as well as the current amount or proportion of native vegetation remaining.

Land use changes within a city, even within a park, are striking over a few centuries (Aey,

Figure 8.13. Ecological succession on a vacant lot. Tall herbaceous plants including grasses, sprouting catalpa trees (*Catalpa bignonioides*), and tree of heaven (*Ailanthus altissima*) between them. Philadelphia. R. Forman photo.

1990; Gilbert, 1991). Over centuries, European urban woodland and urban parks added rides (open strips), coppicing, extensive gardens, and more (Forrest and Konijnendijk, 2005). Public greenspaces were designed for public health. Plazas (piazzas, squares, etc.), cemeteries, botanical gardens, and street trees were added.

Ecological succession

A sequence of plant communities or vegetation stages is the simplest and most familiar concept of succession or vegetation dynamics. Five stages are common: (1) low herbaceous vegetation is replaced by (2) grass-dominated vegetation, which is replaced by (3) shrubby vegetation, which is replaced by (4) a small-tree cover, which is replaced by (5) taller-tree woodland or forest (Figure 8.13). In urban areas one or more stages are often skipped and variability in colonization pattern is conspicuous (Muller *et al.*, 2010). Small trees sometimes directly colonize quarries and mine-spoil sites, or shrub cover immediately follows the low colonizing plants (Wheater, 1999). These five successional stages are usually easy to see, yet several interesting patterns following provide some understanding of succession.

First, a particular pattern of changing vegetation very much depends on initial site conditions (Pickett *et al.*, 1987). The type, size, isolation, soil depth/moisture/fertility, and heterogeneity of a site especially affect the early phases of succession. Second, species availability strongly controls early phases, but also affects later stages. Thus, the seed bank (seeds in the soil), distance from plants producing seeds, wind patterns, abundance of animal seed-dispersers, and proximity of various human seed-dispersal mechanisms play prominent roles. Third, the roles and responses of species themselves strongly shape later phases, but also affect earlier ones. Plant life-history characteristics, changing environmental stresses, eco-physiological responses, competitors, allelochemicals (chemicals produced that inhibit other species), and herbivory are keys to succession.

Other patterns help interpret vegetation dynamics. In some cases, almost all the species appear essentially at the beginning of succession. Thus, the sequence of stages mainly represents the life span of different species, from annuals in the first year to long-lived trees in the last stage.

Also, the species in succession may be usefully grouped into three categories: tolerants, inhibitors, and facilitators. "Tolerants" grow in spite of difficult conditions. "Inhibitors" exert influences that reduce competitors and hence prolong the persistence of a species. "Facilitators" change the environment in ways that reduce persistence and enhance replacement by other species.

Although little-studied, it is likely that all of these mechanisms operate in urban areas, though their relative importance probably differs from that out in natural land. For example, seed banks, animal seed-dispersers, and herbivory seem low in many urban habitats, whereas wind dispersal, seed dispersal by people, site disturbance, and microhabitat heterogeneity are often high. Also different urban habitats are colonized by quite different species (Wheater, 1999).

Additional insights emerge from examples of urban succession. For spontaneous species in a 0.6-ha garden site in Kyoto (Japan), species turnover (percent of the species that disappeared) was 31% in the first year, and dropped to 19% by the ninth year (Imanishi *et al.*, 2007). By the ninth year few new species appeared.

Succession on a former nutrient-rich community garden (allotment) had highly heterogeneous mosaics in the early herbaceous stages (Gilbert, 1991). In years 4–6, tall herbs, especially garden weeds, and some bramble/blackberry (*Rubus*) dominated. At 10+ years, tall grasses, bramble, garden weeds, and more woody species formed a patchy mix.

Successional vegetation in cemeteries appears highly variable and sometimes quite distinctive (Gilbert, 1991). In London's Highgate Cemetery, a classic with overgrown spontaneous vegetation, the stage of dense small trees (pole stage) eliminated grassy areas. Soon afterward the small-tree cover began to "open up," with more light penetration and more herbaceous and shrub-layer vegetation.

On urban coal-mining waste spoils, plant species richness increased from the initial herbaceous stage to the following shrub stage 2 years later (Haeupler, 2008). On four mine sites the increase was: 50 to 70 species; 20 to 40; 10 to 15; and 10 to 15 species. Also, species disappeared each year.

In a former railway yard, plant species richness continued to increase as woody vegetation doubled and herbaceous vegetation cover dropped (Kowarik and Langer, 2005). A railway site abandoned for 35 years was easy to walk through with a 7-m-high closed canopy of goat willow (*Salix caprea*) (Gilbert, 1991). Some herbaceous plants and shrubs persisted from earlier phases. But saplings of sycamore (*Acer pseudoplatanus*) and ash (*Fraxinus*), indicating the future, were abundant, and a number of woodland herbs were slowly colonizing under the canopy.

Woodlands on industrial sites tend to be a patchwork or conspicuously mosaic-like (Weiss *et al.*, 2005). A 40-year successional stage dominated by birch (*Betula pendula*) had a small seed bank, patches of early and mid-succession stages, and abundant seedlings of the sycamore tree. An 85-year stage originally planted with black locust (*Robinia pseudo-acacia*) had tall locusts, dead locusts, canopy gaps favoring young locust growth, and lower levels dominated by bramble/blackberry and ferns. A mature woods in New York City decreased markedly in oaks (*Quercus*) and hemlock (*Tsuga canadensis*), while increasing in cherries (*Prunus*), birches (*Betula*), and other tree species during a 30-year period (Rudnicky and McDonnell, 1989).

These examples emphasize the different trajectories on different habitats, and indeed on a single habitat type. The familiar 5-stage successional series mentioned at the beginning is actually a "sequence of patchy mosaics." One stage typically contains small patches of the previous stages, as well as patches of plants representing the next stage. The patches in the mosaic change at different rates. Ecological succession is a changing vegetation mosaic with dynamics within dynamics. Of course, all the successional sequences are strongly affected by the concentration of human effects, from trampling and dumping to soil removal/addition and air pollution.

Some additional perspectives are useful in considering some successions. *Arrested succession* refers to a stage that persists for a long period, due to the prevalence of inhibitor species, or due to human maintenance or disturbance activities. For instance, semi-stable shrublands often grow under electric powerlines in exurban Connecticut (USA) (Niering and Goodwin, 1974). *Retrogressive succession* (retrogression) refers to a simplification and loss of vertical structure of the plant community over time, such as by a persistent stress or by repeated disturbance (Whittaker, 1975). *Cyclic succession* describes the trajectory from herbaceous to woody vegetation and back to a herbaceous stage. This may occur with periodic repeated disturbances such as fire or brush clearing by people in exurban areas. At a finer scale, old trees in a woods die, thus permitting tiny patches of herbaceous plants and subsequent shrubs and/or small trees to thrive alongside the surrounding large trees (a "gap dynamics" process).

Finally, it should be emphasized that vegetation succession focuses on the changing plants, but really *ecosystem development* is occurring. Soil, microclimate, microbes, vertebrates, and indeed all aspects of the ecosystem change. Succession is an especially rich subject in urban areas.

Plants and urban habitat fragments

Although small dispersed habitats are perhaps the most conspicuous ecological pattern of urban areas, surprisingly little research has explored the responses and roles of plants in the pattern. To paint the picture we begin with some broad patterns of the city. Next, patches and edges are the focus, and then corridors and species movement are briefly considered.

The urban area of the contiguous USA doubled during 1969–94, and in 2001 covered 3.5% of the total surface area and contained 3.8 billion trees (Nowak et al., 2001). The average tree cover per city was 27%, higher in forest regions and lower in desert regions, but with considerable variation within a region. Tree cover also varies widely according to habitat type, e.g., in European cities from 95% in woodland (forest), 35% in cemeteries, and 27% in freshwater (swamp) to <5% on railway corridors, meadows and cropland (Pauleit et al., 2005).

The "floristic similarity" (similarity in plant species present) differs markedly from city to city. For instance, ten large parks in or bordering major cities from Boston to Washington, D.C. have a total of 1391 plant species, of which 490 are non-native species (Loeb, 2006). Fewer than 1% of the species, both for natives and non-natives, are in all 10 parks. Native species decrease, while non-native species increase, in cities with larger human populations. Also for five Italian cities from Palermo to Milan, floristic similarity was highest among wooded habitats within a city. The similarity in plant species present was intermediate when comparing urban versus surrounding semi-natural vegetation, and lowest when comparing wooded habitats of different cities (Celesti-Grapow and Blasi, 1998). The similarity of floras in different types of habitat within a city would doubtless be even lower.

Understanding the *species pool*, that is, the total set of species that could reach a site, is useful (Zobel et al., 1998). The pool of species varies by regional vegetation type (or biome), and includes, for example, species arriving from agricultural land surrounding a city and natural land beyond it (Wace, 1977; Forman et al., 2003; Breuste, 2004). Cities often formed in especially species-rich areas (Kuhn et al., 2004; Hahs et al., 2009). The species pool includes the species now existing in habitats of the urban area. Note that the species pool is affected by the distance from species source to site, as well as dispersal ability or seed-dispersal mechanism. Increasingly species from other continents are part of the species pool (La Sorte et al., 2007). Thus, many North American species grow in European cities, and many European species in North American cities.

A study of 22 tree species in Birmingham (UK) suggests some ways that a city's overall vegetation affects individual habitat fragments (Bastian and Thomas, 1999). The proportion of habitat patches occupied by a species correlated with the density of available suitable patches. For many species, the occupancy of patches increases with three variables: age, area, and habitat diversity of a site. Occupancy by species decreases with distance from a patch containing the species. These results for urban habitat patches are consistent with the role of colonization and extinction in "island biogeography theory," which related species richness on islands in the sea to island size and distance from mainland.

The *patch–corridor–matrix model* gradually replaced the earlier island concept, partly because a patch is surrounded by a mosaic of diverse habitats. Each of these habitats: (1) may significantly affect the patch, (2) is a source of species with its own distinctive species pool, and (3) is differentially suitable for species movement between patches. In addition, the patch interior-to-edge ratio, and the prevalence of generalist species in an edge and specialist species in the interior, are important determinants of species richness on patches of different size. The interior-to-edge ratio

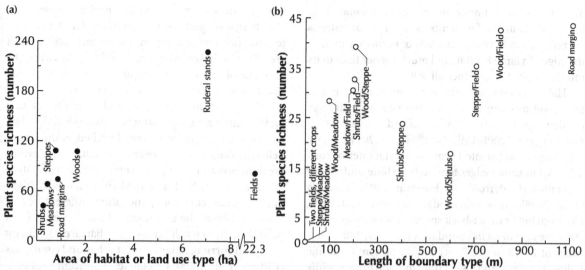

Figure 8.14. Plant diversity relative to area of habitat type and length of boundary type. Based on sampling a 580 m × 580 m plot (0.37 km²) in the peri-urban (exurban) area of Prague (Trneny Ujezd, "a mosaic of semicultural and still agriculturally used suburban landscape ... in the Prague external zone"). Boundary types named by the adjacent patches. No plant data for the following five boundary types (boundary lengths in parentheses): (1) Stone-pit/Wood (100 m); (2) Manure deposit/Field (100 m); (3) Manure deposit/Meadow (100 m); (4) Stone-pit/Field (200 m); (5) Stone-pit/Steppe (350 m). 1 ha = 100 m × 100 m = 2.5 acres; 1 m = 3.28 ft. Based on Kovar (1995).

and generalists/specialists distribution are particularly important for the ecology of patch shape (see equations, Appendix B).

Perhaps the heterogeneous and highly impervious city center functions as an intermediate matrix condition between sea and land, and a modified island biogeography model would be useful. However, most of the urban area seems more analogous to natural and agricultural land, with patterns fitting the landscape ecology model.

The *adjacency effect* (an adjoining area affects conditions in a site) (Hersperger and Forman, 2003) may be especially important in urban areas. Thus, in Springfield woods (Massachusetts, USA), avian species richness decreased where adjacent building density increased (Tilghman, 1987). In the Birmingham study above, the presence of a tree species in a habitat also correlated positively with the degree of similarity of adjacent habitats (Bastian and Thomas, 1999). The abundance of shrubs in suburban Tokyo decreased when surrounded by more urbanized area (Hamabata, 1980). In the Barcelona Region, many patterns of plant species beneath the canopy of urban woods depend on whether built area or cropland adjoins the woods (Guirado et al., 2006).

Plant species richness typically correlates with area of a city, as well as habitat size within an urban area (Figure 8.14a). This has been found for cities (Klotz, 1990; Pysek, 1998; Dunn and Heneghan, 2011), parks, railway sites, green areas in block-building districts, urban woods, and vacant lots (Crowe, 1979; Sukopp and Werner, 1983; Hobbs, 1988; Angold et al., 2006). Even comparing different types of habitat suggests an increase in plant diversity with increased area (Figure 8.14a) (Kovar, 1995). In these species-area studies, variability in response is relatively high, indicating that other variables are important.

Large woodland patches commonly contain planted ornamental trees in parts, and not only contain more species, but the interior tends to be criss-crossed by trails and people (Hamberg et al., 2008). Trampling eliminates some species while permitting others to colonize in the forest interior. Edge environmental conditions sometimes extend 50 m or more into a woodland. Thus, patch size, shape, and degree of trampling affects plant species within the woods.

Small patches are entirely composed of *edge effect*, i.e., the ecological conditions near the perimeter that are significantly affected by the surroundings (and differ from interior conditions, if present). Demographic variations in population size are expected to be high in small patches, and probably account for most disappearances of species, especially specialists (Byers and Mitchell, 2005). Inbreeding in small populations of small patches leads to weak or sterile offspring (Roberts et al., 2007), though this occurs over generations, and may not be very important in urban habitat fragments, particularly for plants. Small habitat patches are reported to have higher levels of herbivory than in the edge of a large habitat, and much higher than in its interior (Christie and Hochuli, 2005).

Habitat edges are relatively abundant in urban areas, and are often abrupt next to roads and buildings (Cadenasso et al., 2003; Godefroid and Koedam, 2003; Hamberg et al., 2008; Cilliers and Siebert, 2011). Dense edge vegetation (mantel) protects habitat interior conditions, but urban edges tend to have little understory vegetation (Godefroid and Koedam, 2003; Hamberg et al., 2008). Also, urban edges are mainly straight, although buffer or set-back spaces around objects create some curviness in boundaries (Kenney, 2008). Setbacks of 12–23 m (40–75 ft) may be present for traffic lights, stop signs, and power transmission towers, while 1.5–4.5 m (5–10 ft) buffers are typical for buildings, sidewalks, railway beds, fire hydrants, and so forth.

Habitat edges themselves differ markedly and thus contain different species. In the peri-urban (exurban) area of Prague, nine *edge types* (named for the two adjoining habitats) are present (Figure 8.14b) (Kovar, 1995). Road margins and wood/field edges have the greatest length, while the wood/meadow edge is shortest. The number of plant species seems to vary mainly by edge type, rather than by length of edge.

Corridors and stepping stones are discussed in Chapter 9 on urban wildlife. Yet seeds of some plant species are dispersed by animals, so the spatial arrangement of urban habitats is especially important for such plants. For instance in Seoul, the highest proportion of non-native species (28–37%) is in industrial and commercial areas, as well as in built-up areas with mixed land use (Zerbe et al., 2004). The lowest proportion (7–18%) is in wooded sites and parks. It seems likely that the industrial-commercial-built-up areas act as sources of non-native species for the wooded sites and parks. Thus, the arrangement of habitats between them is important for dispersal. Forest songbirds tend to travel in or close to woody vegetation, so large gaps between habitats may be rarely crossed (Belisle, 2002).

One study hypothesizes that to sustain a bird species, more than 20% of the landscape should be suitable breeding habitat, irrespective of the spatial arrangement of the habitat (Fahrig, 2002). This suggests that few bird species would be sustained in urban areas. However, with a lower proportion of suitable habitat, certain spatial arrangements of patches, corridors, and stepping stones may sustain the species (Melles et al., 2003). This suggests that a number of habitat characteristics balance or compensate for each other. Patch area may be balanced against total habitat cover, the presence of corridors or stepping stones against patch size, and habitat quality against patch size.

Many corridor types exist and interconnect in a metropolitan area. For example in Shanghai, riparian, coastal, roadside, railway, powerline, hedge, and greenbelt corridors provide diverse benefits, from windbreak to pollution mitigation and provision of nature (Li et al., 2008b). Such a network (if composed of high-quality green corridors), including different scales, should facilitate the movement of many species from habitat patch to patch across the urban area. Remnant grassland corridors may connect urban and rural areas (Cilliers et al., 2008). Of course, significant breaks in the network would disrupt or prevent species movement, just as they would for car or people movement.

Although most urban habitat patches are more or less rectangular, some are long and narrow, effectively grading into corridors. The width of the edge effect thus is important in determining the presence or amount of habitat interior for many specialist species (Christie and Hochuli, 2005; Hamberg et al., 2008). Yet, normally, interior habitat in an urban area is quite limited or absent.

Extensive landscape ecology and conservation biology research has clarified much about the ecological effects of habitat fragmentation (Forman, 1995; Lindenmayer and Fischer, 2006; Collinge, 2009). It is time to focus fragmentation study in urban areas, where species change is rapid, specialist species are disappearing, and species may readily move (Godefroid and Koedam, 2003; Alberti, 2008; Grimm et al., 2008; Cilliers and Siebert, 2011; Dornier and Cheptou, 2012). This may be combined with other societal goals. For example, to cool an urban area, fragmented habitat in the form of a fine mesh of tiny vegetation patches may be better than a single large greenspace or park (see Chapters 5 and 12) (Hough, 2004).

Society could protect certain specialist species in the interior of one or a few large semi-natural patches in the metropolitan area. More promising, however, is to provide a network for effective species movement across an urban area, coupled with large protected habitats at least closely surrounding a metropolitan area (see Chapter 12). The large habitats remain sources of species, especially natives, that readily move across the area mainly using the vegetation network.

Part II Ecological features

Chapter 9

Urban wildlife

Animals must move across land to survive – for water, for food, for minerals. Existence depends upon some kind of movement: you move, or the land kills you where you stand.

Brian Herbert and Kevin J. Anderson, Dune: The Butlerian Jihad, *2002*

And drink pure water from the pump,
I gulp down infusoria
And quarts of raw bacteria,
And hideous rotatorae,
And wriggling polygastricae,
And slimy diatomacae,
And various animalculae
Of middle, high and low degree.

William Juniper, The True Drunkard's Delight, *1933*

Some years ago I tried to call the "National Center for Urban Wildlife" in Washington, D.C. and after futile attempts to find the number, the telephone operator said politely, "Sir, are you really serious?" While cities may be hot spots of human wild life, I was thinking of *urban wildlife*, the non-domesticated terrestrial animals in urban areas (Goode, 1986; Houck and Cody, 2000; USDA Forest Service, 2001; Adams *et al.*, 2006). Wildlife refers to all vertebrates (mammals, birds, reptiles, amphibians) and invertebrates (insects and other groups), except for pets, farm animals, fish, and diverse aquatic animals.

The urban environment, characterized by a concentration of buildings, roads, pollutants, noise, vehicles, and people, differs markedly from natural and agricultural lands. Not surprisingly, urban animals are equally distinctive (Stearns, 1967; Robinson, 1996). In general, urban vertebrates: (1) commonly use human-provided food sources; (2) readily switch diets; (3) may den or nest in artificial structures; (4) have relatively long breeding periods; (5) may occur in rather high density; (6) frequently move from location to location; (7) are *habituated* (accustomed to or behaviorally adjusted to humans); and (8) are generalists (Wheater, 1999; Adams *et al.*, 2006).

Urban birds are of particular interest both to scientists and to the public (Figure 9.1) (Emlen, 1974; Marzluff *et al.*, 2001; Lepczyk and Warren, 2012). Based on East Asia and the USA, most urban birds have non-colonial nesting; occur in groups at least seasonally; feed on the ground or low vegetation; tend to be active and conspicuous; and are non-migratory long-term residents (cited by Wheater, 1999). Many urban invertebrates are small compared with their cousins outside the metro area (Wahlbrink and Zucchi, 1994; Wheater, 1999).

Animals have three overriding "needs," food, water, and cover (Laurie, 1979). Typically *cover* of three types is important, namely, habitat cover for nesting/denning and raising young (especially species with parental care), roosting cover to rest daily, and escape cover to avoid predators, competitors, people and machines. Some wildlife biologists consider the ability of an animal to move from spot to spot to be a fourth basic need,

Figure 9.1. Feeding and appreciating urban wildlife attracts residents. Swans, geese, ducks, corvids, and pigeons. London. R. Forman photo.

which is particularly important where habitat fragmentation and relatively inhospitable impervious surfaces predominate. Urban wildlife survive, or thrive, if they can fit their four basic needs to the spatial arrangement of vegetation, hard surfaces, and diverse stresses.

Although often perceived as depauperate, urban areas may be rich in species. Over 1782 animal species were recorded in a 0.09-ha (0.23-acre) suburban England house plot over 15 years (Owen, 1991), and over 4000 species, mostly insects, were apparently reported from a house plot in Washington, D.C. Of course, some habitats within a city have very few species.

Urban wildlife highlights a range of big questions and issues. Are pests everywhere? Should we enhance urban biodiversity, or protect rare species? Should top predators be eliminated, or encouraged, in urban areas? How about flying mosquitoes, flies, and midges? Do pollinators matter? How about tall structures and lighted glass windows during bird migration? Are street trees important to wildlife? How about habitat fragmentation and park size? Should the widespread garbage-and-dump food sources be maintained to support wildlife? How far do different animals move? Do many species move between city and surroundings? Are green corridors or stepping stones effective for urban wildlife movement? Are urban species resistant or resilient in the face of rapid, climate, or urbanization changes? Are the animals genetically adapted to urban conditions, or simply behaviorally adjusted to them? Are nature's services (ecosystem services) provided by urban wildlife of major or minor importance?

Some of these questions are answered in the pages ahead while others remain scientific challenges. Animal ecology is also introduced in other chapters, e.g., focused on pests (see Chapter 1) (Robinson, 1996; Bolen, 2000; Tello et al., 2005; Adams et al., 2006), domestic animals (Chapter 1), soil animals (Chapter 4), fish and aquatic organisms (Chapter 7), plus animals of roofs and walls (Chapter 10) and airports (Chapter 12). Here we explore urban wildlife in five topic areas: (1) species types; (2) vertical structures, vegetation layers, and animals; (3) spatial habitat patterns and animals; (4) wildlife movement; and (5) changing urban wildlife and adaptation.

Species types

Of the multitude of urban animals, we have chosen to focus on six groups because of their contrasting important ecological roles and evolutionary differences: (1) mammal predators; (2) mammal herbivores; (3) bats; (4) birds; (5) reptiles and amphibians; and (6) invertebrates.

Mammal predators

Two types are usefully differentiated, top predators and mid-size predators and omnivores. Worldwide examples of urban top predators (Adams et al., 2006; Adams and Lindsey, 2011) include coyote (Canis latrans) in cities of Canada, Eastern and Western USA; lynx (Felis lynx) in Sweden; leopard, lion, and hyena in parts of Africa; mountain lion (Felis concolor) in Western USA; and jaguar (Panthera onca) until the 1980s in Rio de Janeiro. Mid-size urban predators and omnivores include red fox (Vulpes vulpes) in Europe, Japan, Australia and North America; monkey species in India and Ribeirao Preto (Sao Paulo, Brazil); wild boar (Sus scrofa) in Germany and Fort Worth (Texas); stone marten (Marten foina) in Europe; raccoon (Procyon lotor) in Russia, Japan, North America and Europe; and striped skunk (Mephitis mephitis), opossum (Didelphis virginiana) and bobcat (Lynx rufus) in North America.

The major mammal predators also live in non-urban areas where they may be more, equally, or less abundant. Many top predators, such as the tiger in India's cities near tiger reserves and mountain lions in the Western USA, primarily live outside urban areas and, especially during food shortages, move into or through suburbs. Similar movement patterns are seen for black bear (Ursus americanus) in some North American cities, grizzly (Ursus arctos) in Jackson (Wyoming, USA), and wolf (Canis lupus) in cities of Russia and Turkey. Such incursions scare people but have little ecological effect in the urban area. On the other hand, mid-size urban predators normally live and reproduce in the urban area, where they are important components of the urban ecosystem.

A study of wildlife in 37 greenspaces (canyons) in San Diego (California) is particularly informative (Soule, 1991). Most of the greenspaces contain managed areas with planted grass and ornamentals, plus semi-natural shrubland areas (chaparral) which support native shrubland mammals and birds. The species diversity of these native animals best correlates with the time since the patch was surrounded and isolated by development (i.e., patch age) (Figure 9.2b). Few native animals survive in greenspaces that were isolated long ago and much disturbed since. On the other hand many native species live in recently isolated

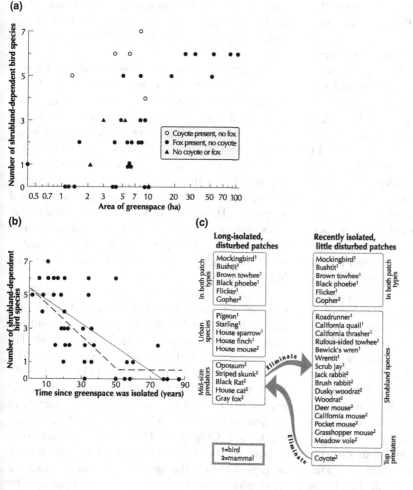

Figure 9.2. Wildlife types relative to patch size, age, and presence of top predator. San Diego (California) greenspace canyons. (a) Species–area curve; note logarithmic horizontal axis. (b) Species richness relative to time since greenspace was isolated. Solid line = linear regression ($r^2 = 0.48$); dashed line suggests a non-linear relationship. (c) Locally breeding species expected to occur in two greenspace-patch types, plus the roles of predators. Adapted from Soule (1991).

and little-disturbed patches. While disturbance level doubtless plays a role, the main effect is hypothesized to be time since isolation. A straight-line (linear regression with $r^2 = 0.48$) drawn through the points suggests that species loss continues until all native species have disappeared about 75 years after isolation. However, a curvilinear response seems to fit the data better, and suggests a steep loss of species in the first 50 years, after which a very few native species persist. Nonetheless, although generally the native birds and mammals can move between greenspaces, it is postulated that they are hesitant to leave the native vegetation of a greenspace when it is completely surrounded by built area.

The species richness (diversity) of native mammals and birds also correlates with the area of a greenspace (Figure 9.2a). However, when the lawn-and-ornamental portions of a greenspace are excluded, the correlation of native vertebrate species number with area of natural vegetation present is stronger.

Native species richness is also higher where the top predator, coyote (*Canis latrans*) is present (Figure 9.2a). Recently isolated and little-disturbed greenspaces contain many native mammals and birds, few non-native birds, and the coyote (Figure 9.2c). In contrast, the long-isolated and disturbed patches contain few native mammals and birds, many non-native birds, many mid-size predators (gray fox, striped skunk, opossum, black rat, house cat), and no coyote. These mid-size predators feed extensively on the eggs, young and adults of the native mammals and birds. The coyote in turn feeds on, or chases away, the mid-sized predators. This means that the native birds and mammals greatly benefit from, and coexist with, the top predator. This pattern illustrates the "meso-predator release hypothesis," whereby without a top predator the mid-size predators are "released" to increase in number (Rogers and Caro, 1998; Crooks and Soule, 1999; Gehrt and Prange, 2007). Therefore, by reducing the mid-size predators that limit the native species, the top predator effectively increases native species richness.

Since the top predator can play such a key ecological role, understanding some of the coyote's characteristics is useful. The coyote is larger than a fox and smaller than a wolf, which will effectively chase both away. Coyotes are extreme generalists, with a wide range of both suitable foods and habitats in city as well as suburb (Gompper, 2002; Wein, 2006). In Los Angeles, its highest density averages 1.6 animals per km^2 (4/mi^2). Coyote home ranges are smaller in residential areas than in undeveloped areas (average 13 versus 17.5 km^2) (Sovada et al., 1995; Grinder and Krausman, 2001). Territories in urban areas usually include a large greenspace, though a cluster of small ones is satisfactory, and coyote activity is centered around such greenspaces (Quinn, 1995). At this density coyotes typically severely limit fox populations (Figure 9.2a), but not domestic cats, dogs, and raccoons that have houses or stormwater drains for escape (Sovada et al., 1995; Gehrt and Prange, 2007). Coyotes mostly feed on herbivores, from mice and rats to rabbits, and coyote removal from golf courses or elsewhere may lead to a large increase in the rodent population. These predators, when in or near their territory, may limit deer populations by feeding on juveniles (fawns) or by sharply altering deer behavior. Coyotes also feed on fruits, and thus disperse seeds in the urban environment. Greenways, railways and river corridors may be especially important for longer-distance coyote movements (Quinn, 1995; Way and Eatough, 2006; Gehrt et al., 2009).

While large top predators play a key ecological role, most cities have none. Instead mid-size predators play a range of roles. Consider three species in Toronto, red fox (*Vulpes vulpes*), raccoon (*Procyon lotor*), and striped skunk (*Mephitis mephitis*), all of which have denser populations in the city than in surrounding rural areas (Rosatte et al., 1991). All three species also have high mortality, especially from vehicle roadkills, dogs, and diseases.

The red fox, found in cities worldwide, has a home range of about 4.1 km^2 (1.6 mi^2) in Toronto (Rosatte et al., 1991). When trapped and released, foxes moved an average straight-line distance of 2.2 km (1.4 mi) in hours or days. However, foxes may also travel tens of kilometers, probably using long green corridors such as along ravines and railways. In Zurich, 15 years after two fox groups colonized separate areas, the population continues to rapidly grow and spread (Wandeler et al., 2003). Like coyotes, foxes are extreme generalists, using a wide range of suitable foods and habitats (Gilbert, 1991).

The striped skunk is a primary carrier of rabies in Toronto and hence of considerable public health interest (Rosatte et al., 1987, 1991). The average home range is 0.51 km^2 (0.2 mi^2). Skunks moved an average 0.91 km (0.57 mi) distance when released at a capture site, and 98% of recaptured skunks were within 1 km of the former capture site. A few skunks moved 2–5 km. Skunks use multiple dens for resting, an average of 14 in a year. In greenspaces, dens are mostly ground burrows and under refuse piles, whereas in residential areas most dens are in ground burrows under houses and associated structures. Toronto skunk densities are high, ranging from 2.5 to 5.7 animals per km^2 (6.5–14.8/mi^2).

Let us look a bit more closely at the still-more-abundant raccoon, which also lives in cities in Asia, Europe, the Caribbean, and across North America, and has relatives in some tropical cities. Toronto has 80–110 raccoons per km^2 (207–285/mi^2) (Rosatte et al., 1987, 1991). The average home range is 0.42 km^2 (0.16 mi^2). When released at point of capture, raccoons moved an average of 0.8 km (0.5 mi) in hours or days, and 97% of recaptured raccoons were within 1 km of the former capture site. Only two of 1723 raccoons studied (with radiotelemetry) moved 11–13 km. A raccoon uses an average of 19 dens through the year. Thus, the loss of a den is no problem for this species. In greenspaces, 89% of the dens are in trees, whereas in residential areas 70% are in houses (e.g., chimneys and attics) or trees (Hadidian et al., 1991).

Raccoons especially move in linear features near water, i.e., along stormwater drainage pipes, ditches, and streams lined with (rip-rap) rocks (Rosatte et al., 1991). The relative ubiquity of stormwater drains along roads and paved surfaces provides abundant convenient escape cover and resting sites. With the urban concentration of people and cars by day, urban raccoons, unlike most non-urban raccoons, are mainly active at night. Residential and commercial garbage, vegetable gardens, pet food dishes, dumpster containers, and dumps provide ample convenient food across the urban area.

Other urban mammal predators include wild pigs or hogs (*Sus scrofa*) in Berlin and some built areas in Florida and Texas (Adams et al., 2006). These animals feed widely in gardens and may use golf courses. In areas with many oaks (*Quercus*) or beech trees (*Fagus*) the pigs root around in the top several centimeters of soil for acorns or beechnuts. Garden vegetables and fruits are favorite foods, and large areas may be dug up overnight.

In summary, urban areas typically support large numbers of mid-size predators that feed on and reduce or eliminate many native vertebrates. The mid-size animals are generalists, also feeding on human-provided food including garbage. Some cities have top predators that feed on and somewhat limit the abundance of these mid-size animals. Consequently, the top predators help sustain a diverse native vertebrate fauna in urban areas.

Figure 9.3. Rats in the urban underground thrive on abundant and dependable food from humans. Stuffed rats in the Paris Sewer Museum, Paris. R. Forman photo.

Mammal herbivores

Worldwide examples of urban mammal herbivores include (Adams *et al.*, 2006; Adams and Lindsey, 2011): white-tailed deer (*Odocoileus virginianus*) in North America, the Caribbean, Europe and New Zealand; beaver (*Castor canadensis*) in North America; gray squirrel (*Sciurus carolinensis*) in the USA and UK; rabbit species (*Lepus, Sylvilagus*); rat species (*Rattus*); and mouse species (*Mus*, etc.).

Some mammal herbivores mainly live in natural areas but are present in urban areas. Moose (*Alces alces*) wander into urban areas, especially in food-shortage times and when juvenile males are looking for a mate. Prairie dogs (*Cynomys*) may persist in colonies surrounded by residential areas and serve as a vector for plague (Collinge *et al.*, 2005; Magle, 2008). Koalas (*Phaseolarctos cinereus*) also will persist for a period feeding on leaves high up in a patch of eucalypt trees (Prevett, 1991). Muskrat (*Ondata zibethicus*) and nutria/coypu (*Myocastor coypus*) may thrive in urban marshes. Beaver (*Castor canadensis*) has moved into some suburbs, creating dams and ponds along streams. The animals cut down yard and park trees, raise water levels, flood driveways, flood basements, and make septic systems malfunction.

Mammal herbivores in urban areas are generalists with a wide range of foods and habitats. Most of the animals forage for food among ground and shrub vegetation. Many will also feed on garbage, which is widely distributed by people. Many species have dens in or under buildings and infrastructure. Some reproduce in burrows or under brush or debris piles, while others are arboreal. Rapid reproduction with relatively large litters of young, probably resulting from the characteristic abundance of food, is common. Yet mortality is high, meaning that the average age of an animal is low. Predators contribute to the mortality, but road kills, disease (often associated with dense populations), and other causes are normally more important. Excluding arboreal and burrowing mammals, most activity occurs at night or dawn-and-dusk, when the human population is mainly indoors or asleep.

Mice and rats are quite familiar to urban residents. The house mouse (*Mus musculus*) and other mouse species, and one to three rat species (*Rattus*), normally have enormous populations in a city (Figure 9.3). House mice are both in and out of buildings, though primarily inside during cold periods, and feed wherever humans leave food (Pennycuik and Dickson, 1989; Lorenz and Barrett, 1990). Rats may be in buildings, especially basements, but also tend to be abundant around docks, storage warehouses, waterways, stormwater drainage systems, and dumps (tips) (Schroder and Hulse, 1979; Boada and Capdevila, 2000; Feng and Hinsworth, 2013). When a regular food source is cut off, such as by closing a dump, rats search for food sources in the surroundings. Both mice and rats are common vectors carrying disease. The 14th-century Black Death, which killed a third of Europe's population, occurred with large numbers of rats, which had lots of fleas, which carried the plague bacteria, which infected people ... and all were packed together within city walls.

Many herbivores use ground burrows, including rabbit species (*Lepus, Sylvilagus*), moles (Talpidae), gophers (Geomidae), and woodchucks (*Marmota monax*). Soil texture is important (Chapter 4), and sandy soils that provide good water drainage are commonly preferred (Bolen, 2000). Rabbits and cottontails that also breed under brush and debris piles are familiar feeding around vegetable gardens and the edges of lawns. Omnivorous moles and gophers commonly create curvy raised lines and/or small mounds in lawns, when they burrow just under the soil surface. Woodchucks in burrow systems may thrive in some residential areas with large house plots, especially near greenspaces and green corridors. Shrubs provide important cover for above-ground wildlife.

Squirrels (*Sciurus, Tamiasciurus*), flying squirrels (*Glaucomys*), and possums (*Trichosurus, Pseudocheirus*) are urban arboreal species that use holes (cavities) in trees, though gray squirrels also build nests of twigs and leaves. Squirrels may mainly feed on nuts, seeds, and buds, which tend to change month by month (Bowers and Breland, 1996; Steele and Koprowski, 2001; Thorington, 2006). However, squirrels also feed on garbage and may become dependent on food put out by people. Squirrels usually store food to get through scarce-food times such as winter. When moving from site to site, squirrels prefer to avoid the ground where predators may lurk, and thus the continuity of trees, wires, and buildings facilitates movement. Greenspaces, such as parks and cemeteries, and residential areas with lines of street trees and property-boundary trees may be squirrel centers.

Deer and kangaroo species are large urban herbivores. Suburban areas with big greenspaces may support very high densities of these species. The large greenspaces normally provide necessary cover for successful reproduction, as well as plant food. Residential areas provide considerable plant food rich in nutrients, because of both the nutrition-rich shrubs and herbaceous species planted and the abundance of fertilizer added. Also, residential areas usually have many shrubs and fences convenient for cover and movement. Although suburban populations of wolf are inadequate to control the herbivore populations, coyotes often reduce deer populations in an area. Suburban dogs will chase deer or kangaroo, but the herbivores readily jump most barriers. The dog, with a relatively limited "home range," (see Figure 1.9) normally gives up the chase. Although road kill may be a major cause of mortality, the herbivores can make babies much faster than drivers can hit herbivores.

Bats

The species of urban bats also vary by region (Kunz and Racey, 1998; Adams *et al.*, 2006; Adams and Lindsey, 2011). Examples of urban bats are little brown bat (*Myotis lucifugus*) and red bat (*Lasciurus borealis*) in North America; vampire bat (*Desmodus rotundus*) and Brazilian free-tailed bat (*Tadarida brasiliensis*) in southern South America; and flying fox (fruit bat) species (*Pteropus poliocephalus, P. scapulatus*) in Australian cities. Bats normally fly and forage across the urban area for food at night. By day they mainly roost in the hollows of large trees or, if such trees are scarce in an urban area, in the roof spaces of buildings.

A horizontal highway bridge across a water body in city center Austin (Texas) serves as a roost for a remarkable 1.5 million free-tailed bats (*T. brasiliensis*). These consume an estimated 10 to 15 tons of insects each night (Keeley and Tuttle, 1999). Citizens have proclaimed the city as "Bat Capital of America," and tourists flock to the evening bat-emergence show.

Bats are social flying mammals that require warmth at night or in cool periods (Parris and Hazell, 2005). Turning bright lights on or off often has an effect on bat movement, because lights attract flying insects, which in turn attract bats (Brigham *et al.*, 1989; Rich and Longcore, 2006). Most species move along linear features in the urban landscape, such as tree lines, forest edges, shrubby property boundaries, and banks of streams and rivers (Gaisler *et al.*, 1998; van Bohemen, 2005).

Many warm-climate cities have large bats, sometimes called fruit bats or flying foxes, that feed on tropical fruits. Also vampire bats may be present feeding on vertebrates in urban areas. Two flying fox species frequently occur in eastern Australian cities, and at large roosts may cover surfaces with bat guano (Parris and Hazell, 2005; Adams and Lindsey, 2011). However, most bats are insectivores, and locate insect prey by echolocation while flying. At least two human diseases are associated with bats, the rabies virus and the histoplasmosis fungus.

A study of bat activity (presumed to correlate with abundance) across Chicago found that activity correlates positively with the proportion of woodland habitat in an area (Gehrt and Chelsvig, 2003). More woodland, more bats. Also bat activity is greater if an industrial or commercial area is adjacent to a greenspace. Bat activity is lower if farmland is adjacent. However, bat activity is greater if both adjacent farmland and a water body, such as a pond, are present.

At a finer scale, bat activity seems to be greater where trees in a greenspace are somewhat separated savanna-like (Gehrt and Chelsvig, 2003). In open areas such as lawn, successional field, and grassland, bat activity decreases at greater distance from the edge of a land use. Lawns, successional fields and grasslands have more bat activity than does cropland.

Although studies are few, it seems that bats are quite sensitive to both the specific habitats present and their arrangement (Fenton, 1997; Gaisler et al., 1998; Everette et al., 2001; Avila-Flores and Fenton, 2005). Some bats seem to thrive in urban areas, foraging around street, yard, and park trees, and having ample roost sites in trees and buildings (Williams et al., 2006). At a broader scale, the bat community might be favored by increasing urbanization (Kurta and Teramino, 1992; Pierson, 1998; Williams et al., 2006).

It seems that urban bats typically use a number of roost locations during the year, and may frequently move among nearby roosts for daytime sleep (Figure 9.4a). In the example diagrammed, animals from a roost mix with other bats on average in 4.8 neighboring roosts (Rhodes et al., 2006). However, a detailed study of a large and highly mobile, insectivorous bat species in Brisbane (Australia) paints a different picture (Rhodes et al., 2006). A hundred radio-tracked bats roosted in 18 trees spread over ca. 200 km², and seldom changed roost trees. Yet one central tree served as a "communal roost" or "hub," and individuals would periodically move to roost in this tree for 1 to 20 days (Figure 9.4b). In this case, most mixing of individuals presumably occurs in the hub rather than among neighboring roosts. With this type of linkage network, the loss of one roost tree has minimal effect on the population, unless the communal roost is lost, which may cause considerable disruption to the population.

A study of big brown bats (*Eptesicus fuscus*) in Fort Collins (Colorado, USA) compared 44 buildings with summer maternity roosts versus the same number of similar buildings selected at random (Neubaum et al., 2007). At this building or microhabitat scale, compared with buildings at random, the roosts were higher above ground, warmer, and served as exit points to large areas. At the urban landscape scale, these building roosts were in areas of lower building density and less vehicle traffic, but with a higher density of streets.

Studies of bats and many other organisms dependent on air in the planetary boundary layer of cities (Chapter 5) have given rise to a scientific research area of *aeroecology* or aerobiology (Lutgens and Tarbuck, 1998; Kunz et al., 2008). In and around urban areas air flows and atmospheric conditions vary widely, especially as they interact with the rough surface of objects in a city (Geiger and Aron, 2003; Erell et al., 2011). Thus, the swifts, swallows, nighthawks, falcons, owls, vultures, flies, mosquitoes, midges, spiders, pollen, spores, seeds, bats, and many more organisms of urban skies are closely tied to these air flows and atmospheric conditions.

Birds

We illustrate the urban bird fauna with three major groups (Lepczyk and Warren, 2012): (1) raptors (birds of prey) and waterbirds; (2) songbirds; and (3) the "big three" city birds (sparrow, starling, pigeon). Urban raptors worldwide include: kestrel (*Falco tinnunculus* and *F. sparvarius*) in Europe and North America; peregrine falcon (*Falco peregrinus*) in Europe, North America, and Australia; and screech owl (*Otus asio*) in North America. Widespread urban waterbirds include: mallard (*Anas platyrhyncos*); coot species (Rallidae); goose species (Anserinae); and heron/egret species (Ardeidae).

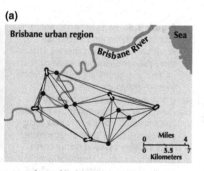

(a)
Number of linkages per roost
Average = 4.8
Maximum = 7
Minimum = 3
Network connectivity very high
Network circuitry (loops) very high

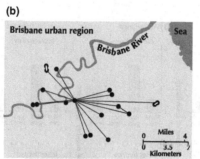

(b)
Number of linkages per roost
Average = 2.3
Maximum = 15
Minimum = 1
Network connectivity very low
Network circuitry (loops) very low

Figure 9.4. Connections among daytime roost sites for an urban bat species. (a) Expected connectivity of roosts somewhat mimicking patterns in the literature (roosts have a similar number of links). (b) Based on radio-tracking of 19 white-striped freetail bats (*Tadarida australis*) in a >200 km² area (77 mi²). Roosts are in large cavities (hollows) in mature *Eucalyptus* trees. The central "communal" roost (or hub) contained 59–291 bats, and "satellite" roosts had 1–21 bats each. 1 km = 0.62 mi. Brisbane, Australia. Adapted from Rhodes et al. (2006).

Widespread urban songbirds include: magpie species (*Pica*); European blackbird (*Turdus merula*) in Europe and Australia; American robin (*Turdus migratorius*); tree swallow (*Passer montanus*) in China; and chimney swift (*Chaetura pelagica*) and mourning dove (*Zenaida macroura*) in North America. The big-three dominants worldwide are house sparrow (*Passer domesticus*), common or feral pigeon or rock dove (*Columba livia*), and European starling (*Sturnus vulgaris*). Corvids (crows, rooks, jays, etc.) and vultures (Cathartidae) are also widespread in urban areas. Tropical cities seem to have a higher diversity of urban birds.

Some proportion of city birds originally came from rock cliff faces where they used crevices and caves (Gilbert, 1991; Larson *et al.*, 2000). The list includes the common or feral pigeon (rock dove), European starling, house sparrow, kestrel (*Falco*), swifts (*Chaetura, Apus*), nighthawk (*Chordeiles*), and gulls (*Laridae*). In the city most of these animals thrive around buildings with ornament and ledges, or in buildings with little maintenance and avian access to attics.

Most urban birds are diurnal, the same as people. Indeed, people and birds have many effects on each other, both positive and negative (Chace and Walsh, 2006; Clucas and Marzluff, 2011; Stracey and Robinson, 2012). Many urban birds use food provided by people, either directly in bird feeders or by feeding in a park, or indirectly by distributing garbage throughout the urban area. Although some species are exceedingly abundant, population size is often limited by few suitable nest sites and low breeding success. Few birds construct open-cup nests, while most species nest in enclosed sites. Predation may or may not be important in limiting urban bird populations.

Urban birds may be highly mobile, continually adjusting to ever changing food sources, nest sites, roost sites, and human disturbances (Marzluff *et al.*, 2001; DeStefano and Webster, 2012). Cities in cooler climes have a higher proportion of birds that migrate, such as swallows (Hirundinidae) and storks (*Ciconia*). Tropical cities have a greater proportion of year-round-resident birds.

Especially for songbirds, the amount of vegetation is a major determinant of avian density, diversity, and distribution. Few birds nest at ground level. The characteristic park with lawn and tall deciduous trees usually contains few breeding species. Holes in tree trunks add avian diversity. Evergreen trees add even more. And shrub cover considerably increases avian diversity and density. Vegetation mostly provides cover for urban birds, though it also provides food particularly for native species.

An Adelaide (Australia) study of four widely planted street trees in the southern hemisphere provides insight into bird use of trees (Young *et al.*, 2007). Nectar- and pollen-feeding "nectarivores," seed-eating "granivores," and insect-eating "insectivores" were recorded in each street-tree type. Nectarivores were most abundant in all four street trees (Figure 9.5). All bird species and all bird feeding types (foraging guilds) differed by tree species. A native tree, red gum, had the most nectarivores (individuals and species). A non-native, plane tree, had the most insectivore birds. Tree use by different avian feeding types did not correlate with environmental conditions measured around the street trees, except for vehicle traffic affecting some nectarivore use. In short, the species of street tree strongly affects the types of birds present.

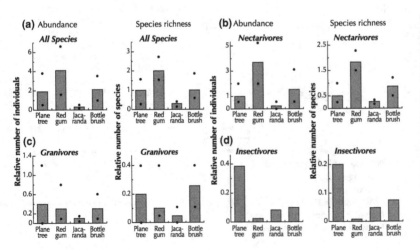

Figure 9.5. Types of bird foraging on street trees in a subtropical city. (a) Results for all songbird species. The three predominant bird feeding groups or "foraging guilds" are (b) nectarivores (nectar- and/or pollen-eating); (c) granivores (seed-eating); and (d) insectivores (insect-eating). Plane tree (sycamore), *Platanus* × *acerifolia*; red gum, *Eucalyptus camauldulensis*; jacaranda, *Jacaranda mimosifolia*; bottle brush, *Callistemon citrinus*. Histograms represent the average (and dots the range) of 5 average values for seasons (autumn, winter, spring, early summer, late summer). Relative numbers on vertical axes are generally *ln*-transformed. Adelaide, Australia. Adapted from Young *et al.* (2007).

A study of birds in 40 urban and suburban woods in a small city (Springfield, Massachusetts, USA) found that the area of woods was the best predictor of avian diversity (Tilghman, 1987). However, bird species richness also correlated with four other interesting variables: density of adjacent buildings (negative correlation); shrub density (positive); distance from trails (positive); and presence of patches of evergreen trees (conifers) (positive). The lower avian diversity next to higher building density emphasizes the importance of adjacency or context.

Suburban residential areas commonly have very high breeding bird densities, even reaching 600–641 breeding pairs per ha (240–256/acre) (excluding house sparrows) (Gilbert, 1991). The suburban area has considerable habitat heterogeneity (see Chapter 11), including numerous diverse shrubs, and provides ample food, cover, water, nest sites, and roost sites. Furthermore residential areas are commonly somewhat close to semi-natural patches and corridors that function as species sources.

A study of breeding bird pairs in 2 km × 2 km areas in suburban and urban Manchester (UK) found that only two of the 25 bird species were denser in the urban portion (Wheater, 1999). Common pigeons were in a 9:1 ratio (urban to suburban) and kestrel (*Falco tinnunculus*) in a 6:4 ratio. Four species were in an equal 5:5 ratio: house sparrow, European starling, mallard, and moorhen (*Gallinula chloropus*). All other breeding species were denser in suburban than urban areas.

To look more closely at urban birds, three categories of birds are usefully recognized in urban areas: (1) raptors and waterbirds; (2) songbirds; and (3) the "big three" city birds.

Raptors (birds of prey) and waterbirds

Several types of hawks in addition to ospreys and vultures fit into this urban category. Some falcons, buteos (large bulky hawks with broad wings), and owls seem to especially thrive in urban areas. For example, territorial pairs of peregrine falcon (*Falco peregrinus*), following a period of bird introductions, are found in more than 60 urban areas of the USA (Cade *et al.*, 1996). Food is normally abundant, but safe nest sites are limited. Most nest sites are on buildings, while others are on bridges, overpasses, and towers. The most frequent prey are common pigeon (rock dove), northern flicker (*Colaptes auratus*), and blue jay (*Cyanocitta cristata*), three birds that commonly fly at tree-canopy level. In London both the peregrine falcon and a kestrel (*Falco tinnunculus*) forage around the tops of buildings. The kestrel in Manchester (UK) feeds mainly on (presumably house) sparrows (87% of the animals eaten), other songbirds, house mice, rats, and pigeons in that order (Gilbert, 1991).

Buteos in North American urban areas are best illustrated by the red-shouldered hawk (*Buteo lineatus*) and red-tailed hawk (*Buteo jamaicensis*). Of 77 nests of the red-shoulders located in urban areas north of San Diego (California), 38% were in non-native trees (eucalyptus species, fan palm, and deodara cedar) and 62% in native trees (western sycamore and coast live oak) (Bloom and McCrary, 1996). Home ranges were 0.45–0.69 km^2 (0.17–0.26 mi^2), a third the size of non-urban red-shouldered hawks. In five cases, when a hawk's territory became more than 50% urbanized, the territory was abandoned.

Red-tailed hawks in urban and suburban Milwaukee (Wisconsin, USA) mainly nest in trees but may nest on artificial structures (Stout *et al.*, 1996). Nesting was successful (fledged at least one young) in 78% of 84 tree nests observed, and in 100% of 15 nests on artificial structures largely in city locations. Red-tails will forage while flying, but mostly use perches to detect prey such as mice and rabbits. Locations of perches in the Boulder (Colorado, USA) area are negatively correlated with developed areas, roadways, and the nearest building (Schmidt and Bock, 2004). Red-tails are often observed at intersections of major highways where ample open land with rodents is present. Other favorite sites for these hawks are cemeteries, institutions, golf courses, and city parks. The distance between active nests is commonly a bit over 2 km, closer in suburbs and further apart in city (Stout *et al.*, 1996).

Many large owls, such as the powerful owl (*Ninox strenua*) in Australia, great fishing owl (*Ketupa blakistoni*) in Japan, and great-horned owl (*Bubo virginianus*) in North America, are characteristic of natural or agricultural areas, and forage or nest at low densities in urban areas. The small common screech owl (*Otus asio*) is a widespread resident in many North American urban areas, where it may have a 4–6-ha (10–15-acre) home range in suburbs, compared with more than 30 ha in the city (Bolen, 2000). Based on a 15-year study, the density of nesting screech owls was greater in a 30-year-old residential suburb than in a 10-year-old suburb of Waco (Texas) (Gehlbach, 1996). No difference in nesting success was present between the two suburbs. However, the suburbs had twice the success rate compared with owls in the surrounding rural area. The suburban populations were more stable over time, and had a reproductive productivity (average 1.8 fledglings per breeding pair) sufficient to maintain the population (Gehlbach, 1996).

The difference in avian feeding between city and outer suburb can be striking, as illustrated by the tawny owl (*Strix aluco*) in London (Gilbert, 1991). Near the city center, 93% of the owl's diet was birds and 7% mammals. In contrast, in the outer suburb, birds were only 10% and mammals 90% of the tawny owl's diet.

Waterbirds, another major component of the urban avifauna, depend on wetlands and/or open water. These birds include waterfowl (e.g., ducks, geese); wading birds (egrets, herons, rails); gulls/terns; shorebirds (sandpipers, plovers); cormorants; pelicans; and more. Some are resident in urban wetlands, ponds, lakes, streams, rivers, and estuaries. Others, such as waterfowl and shorebirds, move through an urban area in huge numbers during annual migrations. Large numbers of resident birds, such as geese or gulls, may become pests. A dense population, for instance in a nesting or roosting wading-bird colony, may lead to disease outbreaks and considerable mortality. In general, however, waterbirds bring pleasure to people, especially egrets/herons on the edge of a marsh, ducks in a park pond, the "honking" of geese, pelicans diving for fish, and shorebirds endlessly skittering along a beach.

Waterbird groups generally sort out in different wetland and open water habitats (Tucker and Evans, 1997; Parsons, 2002). Mudflats and shallow water attract shorebirds (Takekawa *et al.*, 2002). Marshes attract wading birds and waterfowl. Rice fields attract wading birds (and ricebirds) (Fasola and Ruiz, 1996). Shallow ponds of many sorts are duck and heron/egret spots (Wheater, 1999; Huner *et al.*, 2002; Takekawa *et al.*, 2002). Wooded swamps tend to have several waterbird species but may have few individuals (Erwin *et al.*, 1991). Large open water bodies such as lakes and estuary bays attract almost all the waterbird types, especially ducks, geese, gulls, terns, cormorants, and pelicans in the open water, and wading birds along the shoreline.

Habitat characteristics within a water body strongly affect both species richness and composition. Consider the preferred water depths of certain species in North America (Fredrickson and Laubhan, 1994): 0–6 cm (small sandpipers); 2–10 cm (American bittern); 5–17 cm (large sandpipers); 6–15 cm (mallard ducks); 10–25 cm (great blue heron); 15–35 cm (pied-bill grebe, American coot); and 30–400 cm (redhead, ruddy duck, Canada goose, teal species). Water depth, sometimes controlled by gates, is particularly sensitive to changing seasons and dry/wet periods, and generates rapid species responses (Fasola and Ruiz, 1996; Parsons, 2002).

A mosaic of different water depths may be optimal for waterbird diversity. Different salinities in marine areas also provide habitat heterogeneity. Islands add habitat diversity and, if far enough from the shore to escape most terrestrial predators, may be good habitats for roosting or breeding by wading birds and gulls/terns (Parsons, 1995).

A complex of wetlands and small open-water bodies is particularly rich and dense in waterbirds, because it combines large total size with separated and different water-and-vegetation conditions, which frequently change over time. The surroundings or context of a wetland typically has a major effect on waterbirds and their movement patterns (Takekawa *et al.*, 2002). Disturbances and pollutants may come from the surroundings, and waterbirds are sensitive to both (Huner, 2000).

But some species use the surroundings in their daily home-range foraging (Parsons *et al.*, 2001). Certain wading birds feed in crop fields where they may receive heavy doses of toxic chemicals. Gulls may feed on dumps and other sites with urban waste. Alternatively, geese and gulls may roost inland and feed in open water areas. Most of the waterbirds can fly kilometers or tens of kilometers within an urban region. They also function as part of an extensive regional population dispersed in patches over hundreds of kilometers (Parsons *et al.*, 2001).

Songbirds

Songbirds represent a broad category of perching (Passeriformes) birds and other land birds that are mostly relatively small and "sing songs." A few are particularly widespread in cities, such as magpies with long tails and white spots that scare predators, swifts and swallows that seem to endlessly circle and dive in the summer evening sky, woodpigeons and relatives (not the common rock-dove pigeons) often in tree tops, blackbirds of varied types, seed-eating finches, and colorful parakeets in flocks.

The tree swallow (*Passer montanus*) is a predominant bird in Beijing, but the abundance of both summer breeding and winter birds seems to decrease with degree of urbanization (Zhang *et al.*, 2006). Habitat use correlates positively with more (suburban) brick bungalows, coniferous and broad-leaved trees, and even air conditioners. Meanwhile, habitat use by the swallows decreases with more area of tall buildings and paved roads, higher human density, and more vehicle movement.

Most songbirds are highly sensitive to vegetation structure. The vertical layers of foliage in a wooded urban park of a small city (Oxford, Ohio, USA) were compared with those in woods of a surrounding agricultural area (Beissinger and Osborne, 1982). All layers (canopy, subcanopy, understory, shrub, and ground; see Chapter 8) had noticeably less foliage in the urban park (except for the middle understory layer at about 6 m height, which was about the same in urban and rural areas). In the city park the lower vegetation layers were dominated by ornamental plants. Also, at all five foliage-height levels, the horizontal pattern of foliage in the urban park was highly discontinuous.

The Oxford (USA) study also highlighted the birds in different levels by focusing on *bird foraging guilds*, the functional ways of finding food. Insects were the overwhelming food of these (insectivore) birds in both city and farmland areas. In the urban park, the most abundant guild were species of "ground gleaners," which mainly forage for insects and other invertebrates in the soil and herb layer. In contrast, the predominant guilds in the rural woods were "foliage gleaners" and "bark drillers." The former search for insects on leaves and the latter search for insects in and under bark by drilling small holes.

A closer look at the distribution of songbird guilds and vertical vegetation structure compared patterns in city versus suburban residential areas in New England (USA) (DeGraaf and Wentworth, 1981; DeGraaf, 1987). During the breeding season, seed-eating ground foragers predominated at ground level in both city and suburban woods, but were 12 times more abundant in the city woods. The abundance of fruit-eating (frugivore) upper-foliage-and-branch foragers did not differ between city and suburb. However, insect-eaters (insectivores) of five types (bark gleaners, ground gleaners, low-foliage-and-branch gleaners, upper-foliage-and-branch gleaners, and air salliers) were more abundant in the suburban woods. Omnivores were more abundant at ground level (ground foragers) in city woods, whereas little difference was found for three guilds (lower-foliage-and-branch foragers, hover gleaners, and upper-foliage-and-branch foragers) between city and suburban areas. Thus, in summer, the ground layer in city areas supports many more seed-eaters and omnivores than in suburban woods, whereas suburban areas have more insectivores from shrub layer up to tree canopy.

In winter, the ground-layer bird pattern in these urban mainly deciduous New England woods is similar to that in summer (DeGraaf, 1987). However, few birds of any sort forage in the upper foliage and branches. Winter insectivores are scarce except for bark gleaners, which are more abundant in suburban woods. Total summer breeding bird density is >2.5 times greater in city than in suburban areas, while in winter, urban avian density is 1.7 times greater. In contrast, the species richness of summer breeding birds is >2.5 times higher in suburban than city areas, and in winter 1.4 times greater. In short, city areas have higher bird density, while suburban areas have more species.

Residential areas tend to have high habitat heterogeneity (see Chapter 11) including lawn, flower garden, vegetable garden, shrubs, trees (evergreen and/or deciduous), crevices and walls, water, untended successional strips or patches, and brush piles. These microhabitats and heterogeneity are especially valuable for songbird density and diversity. In contrast, city parks often have few shrubs and lower habitat diversity, and most songbirds apparently have lower breeding success.

Within a London park, bird density (excluding house sparrows) was twice as great in a wooded section as in either a flower-garden section or lawn-and-tree section (Gilbert, 1991). All three habitats in the park had higher, though variable, bird densities in winter than in summer. The diversity of summer breeding bird species progressively increased from a park near city center (22 species) to a park 18 km out in the suburbs (45 species), a doubling of species richness.

Within suburbs, bird distribution patterns also vary widely, as illustrated in the Brisbane (Australia) Region (Catterall, 2009). Species richness in a "bare" suburb is half of that in a "well-vegetated-with-trees" suburb. The bare suburb has more non-native bird species, and somewhat fewer seed-eating species. It has many fewer large birds (≥50 grams) and fewer vertebrate-eaters (predators). The bare suburb is much lower in fruit- or nectar-eaters. It has fewer arboreal insectivores. No difference in the number of small-bird (<50 g) species is present between bare and well-vegetated sites. The most frequent birds in bare sites are (in order): house sparrow, common (European) starling, and pale-headed rosella (*Platycercus adscitus*). Meanwhile, the most frequent birds in vegetated suburbs are rainbow lorikeet (*Trichoglossus haematodus*), noisy miner (*Manorina melanocephala*), and pale-headed rosella.

Songbirds typically constitute the bulk of avian diversity and are often of particular conservation importance. Vegetation structure, particularly dense shrub and canopy layers, is of prime importance to songbirds. Safe night-roosting sites for non-migratory birds, which in winter may be exceedingly abundant in the warmth of cities, are important. Evergreen trees, dense shrubs and small trees in urban greenspaces often play key roles as roosting sites.

The "big three" city birds

Cities are the only habitat where the same species tend to predominate from continent to continent, in this case three species: house sparrow (*Passer domesticus*), European starling (*Sturnus vulgaris*), and common pigeon (feral rock dove, *Columba livia*). All three species apparently originated on rocky cliff habitats (Gilbert, 1991). While densest in the city, all three species thrive in urban, suburban and farmland areas, and sometimes starlings nest in natural land. One study found no significant difference in the density of both house sparrow and starling between cities in Quebec (North America) and Rennes (France) in Europe (Clergeau et al., 1998). The abundance of birds, plus their widespread feeding on seeds, human-provided food, and garbage, mean that the big three probably play a major role in cleaning the city. The birds usually feed in flocks, and without these birds many surfaces would likely have much more decaying garbage, "bugs," and microbes. These bird species also produce concentrated spots, plus an endless sprinkle, of bird droppings across the city.

A study of greenspaces in Syracuse (New York, USA) suggests that pigeons, but not the other two species, reduce the species richness of native birds (Johnsen and VanDruff, 1987). The presence of house sparrows and starlings is positively correlated, and hence they often coexist in a greenspace. No other interaction (positive or negative) was significant among the big-three species. Correlations of the presence of the three species with 15 urban variables found different leading variables for each species, and different leading variables for summer versus winter birds.

House sparrows, with their abundance and darting in and about, probably provide pleasure and symbolize nature for people more than does any other urban species. The sparrows especially nest and roost in little holes and crevices. Nests are in rain-protected sites including vegetation, particularly trees, and on artificial structures (Wheater, 1999; Laet and Summers-Smith, 2007). While adults feed insects to their young, adult birds mainly feed on seeds, human food, and garbage. Suitable food seems to be ever-abundant, though the species may presently be in decline in its "native" London.

House sparrows are generally more frequent in weedy patches rather than tidy gardens, and in poor rather than wealthy neighborhoods (Shaw *et al.*, 2008). The sparrows are more abundant in areas with less space committed to paved car parking, and with a higher proportion of native shrubs rather than non-native ornamentals.

European starlings are often abundant in both city and suburb, and nest in somewhat larger holes in trees and built structures. Few people directly feed starlings, though their relatively aggressive nature means that they outcompete most other birds for food sources.

In cooler months, starlings tend to aggregate in large flocks that have nightly roosts in trees or buildings (Gilbert, 1991). The noise and accumulation of bird droppings (guano) annoys people, and the guano may contain fungus diseases (*Crystococcus*, *Histoplasma*) for humans.

The common pigeon (or rock dove, *Columba livia*) colonized London by 1384 (Gilbert, 1991). The pigeon appears to have genetically adapted to urban conditions, and differentiated into at least seven forms during the centuries of urban history (Gilbert, 1991). Pigeons usually breed in groups, typically placing nests on artificial structures in locations protected from rain, cats and hawks (Wheater, 1999). Common pigeons breed nearly all year long in London, but only about a third of the population is breeding at a time. Although these relatively large birds can fly considerable distances, their nesting, roosting, and feeding sites are usually nearby within several hundred meters of each other. Other pigeon and dove species, including woodpigeon (*Columba palumbus*), collared pigeon (*Streptopelia decaocto*) and mourning dove (*Zenaida macroura*), also live in many urban areas and may provide a pleasant "cooing" from tree tops.

A study of common pigeons in Basel (Switzerland) using GPS positioning devices found that the birds were especially abundant in three quite different areas: (1) in streets, squares (plazas), and parks near their nesting site (home site); (2) in agricultural areas surrounding the city; and (3) on docks and railways in harbors (Rose and Nagel, 2006). In Barcelona, many scattered locations contain >300 pigeons/ha (120/acre) (Boada and Capdevila, 2000). In Basel, a third of the common pigeons studied remained within 0.3 km (1000 ft) of the home site (Rose and Nagel, 2006). Very few moved >2 km (1.2 mi) from it, and the maximum distance was 5.3 km. Other (non-GPS) studies of common pigeons report widely differing distances moved, e.g., from 0.3 to 25 km from a nesting site.

While the big-three birds are abundant and conspicuous in cities, the species differ in ecologically significant ways. The house sparrows nest in tiny holes and crevices and prefer foraging in weedy areas. The European starlings are aggressive against many other birds and seasonally form large roosts. The common pigeons prefer relatively bare areas, from urban to rural, and generally remain rather close to their nest sites.

Reptiles and amphibians

Alligators and exotic or venomous snakes in urban areas, whether native or as former pets, often make headlines. Yet most amphibians and reptiles (herpetofauna or simply *herps*) are seldom seen: frogs and toads, salamanders, turtles, and snakes. Lizards, however, are conspicuous in tropical cities. Different herp species populate urban areas in different regions and continents. Amphibians and reptiles are mainly predators, feeding on invertebrates and in some cases vertebrates. Weedy areas rather than manicured greenspaces provide food and cover. Many herps are commonly eaten by meso-predators such as domestic cats, raccoons, and foxes, as well as hawks.

Most amphibians and some reptiles depend on two dissimilar habitats, such as upland vegetation and wet areas, including ponds, streams and wetlands (Parris, 2006; Grant et al., 2011). Thus, habitat loss of either upland or wetland typically reduces herp populations. Habitat degradation, such as chemical contamination, night lighting, cats, urban heat, and noise interfering with calls, also have significant impacts. Habitat fragmentation is particularly serious for these animals because they are relatively immobile, generally only moving short distances. Barriers, such as railways and busy roads, may be nearly impassable and accentuate the habitat fragmentation effect. Small herp populations isolated in small green or "blue" patches are at high risk of local extinction.

Considering these spatial characteristics of the urban environment, it is not surprising that amphibians and reptiles have generally decreased over time.

For example, in Philadelphia over some two centuries and in Washington, D.C. over 57 years, both amphibian and reptile species richness has declined by about 40–50% (Grant et al., 2011). Lizard diversity and snake diversity dropped the most.

Reptiles may be particularly diverse in arid areas, and have been studied along an urban-to-rural gradient in Tucson (Arizona, USA), in a dry area east of Los Angeles, and in the Las Vegas Region (Nevada, USA) (Beatley, 1994; Germaine and Wakeling, 2001). Rare species abound in these arid urban areas, but are subject to habitat loss, degradation, and fragmentation, plus human disturbance. In Southern California, a park surrounded by urbanization had fewer herp species than a similar park with urbanization only partly around it (Morrison et al., 1994). At a finer scale, an institutional campus in Cleveland (Ohio, USA) contained 11 amphibian and 8 reptile species (Dexter, 1955). Ignoring reintroduction efforts, it appears that few if any native amphibians or reptiles are expanding their populations in urban areas, though some non-native species are spreading.

Urban amphibians are especially affected by habitat loss, degradation, and fragmentation (Hamer and McDonnell, 2008). Based on a literature review, seven major factors can be identified that determine good quality habitat for amphibians: (1) ample vegetation through the life cycle; (2) a limited density of predators and competitors; (3) terrestrial habitat suitable for feeding and winter survival, as well as accessible to aquatic habitat; (4) good water quality with limited pollution; (5) a suitably long hydro-period for the aquatic larvae to grow; (6) a paucity of diseases (such as a "Bd" fungus inadvertently transported by people); and (7) minimal direct human disturbance. The authors conclude that amphibian habitat generalists, or those with relatively short dispersal requirements, appear best able to survive in urban and suburban areas.

A study of urban reptiles in 59 remnant greenspaces of Brisbane (Australia) found reptile occurrence to correlate more with habitat structure than with the species composition of vegetation (Garden et al., 2007). The presence of reptiles correlated best with the presence of termite mounds, abundant fallen woody material on the ground, and a moderate cover of weeds. Reptiles decreased with greater soil compaction. The presence of a tortoise (*Testudo hermanni*) in Rome, Italy correlated with only one of eight habitat variables studied, i.e., shrub cover (Rugiero and Luiselli, 2006).

Urban wildlife

Figure 9.6. Fly and bird, conspicuous urban animals. (a) Fly; Zacatecas, Mexico. (b) Kestrel (*Falco*); Puglia, Italy. R. Forman photos.

Amphibians and reptiles that might thrive in urban areas (urbanophiles) manifest somewhat distinctive characteristics (Grant *et al.*, 2011). They have: (1) broad habitat tolerance (e.g., using lawns, buildings, and stormwater drains); (2) a generalist diet; (3) high mobility and dispersal ability; (4) high reproductive output; (5) small body size; and (6) tolerance of humans. Most urban herps lack some or all of these attributes and thus, rather than thriving, only survive, decline, or disappear.

Invertebrates

The species of urban invertebrates also mainly differ by continent and region. For convenience in the face of huge numbers, we refer to species groups such as flies (Figure 9.6), butterflies, beetles, spiders, snails, and so forth. Soil animals were discussed in Chapter 4 (see Figure 4.8) and include many non-arthropod groups such as earthworms, slugs, and snails. Much of the urban invertebrate literature focuses on pest animals, plant pests, public-health disease vectors, and pollutant effects, rather than ecology (Ehler and Frankie, 1978; Fowler, 1983; Bennett and Owens, 1986; Robinson, 1996; Tello *et al.*, 2005). The subject of urban invertebrate ecology is extremely broad, yet in its infancy, so we mainly focus on encapsulated patterns from three recent reviews (Hochuli *et al.*, 2009; McIntyre and Rango, 2009; Kotze *et al.*, 2011).

Urban areas are commonly dominated by a few invertebrate species with high population levels (McIntyre and Rango, 2009). Arthropod species richness (referring to taxa or groups of species) and abundance are usually lower in urban than non-urban areas, and the species composition is typically quite different (McIntyre *et al.*, 2001; Shochat *et al.*, 2004). Some species groups usually disappear or nearly so with increased urbanization, including scorpions, pseudoscorpions, and large ground beetles. However, other groups often increase sharply, such as cockroaches, fruit flies, termites, and Argentine ants.

Consider six major ecological functions or roles of invertebrates: decomposition, herbivory, pollination, seed dispersal, predation/parasitism, and food source for vertebrates. The following effects of urbanization, in this case mainly habitat fragmentation and degradation, on invertebrates and their roles is revealing (Hochuli *et al.*, 2009):

1. *Decomposition* (mainly by the species groups, Collembola, Acarina, Isopoda, Diplopoda, Isoptera, Coleoptera) in urban areas. Characteristic urbanization effects are: loss of invertebrate species, and decrease in nutrient supply and retention, potentially leading to ecosystem "instability."
2. *Herbivory* (Coleoptera, Lepidoptera, Hemiptera, Diptera, Hymenoptera). Changes in host plant quality, increasing the probability of herbivore outbreaks.
3. *Pollination* (Coleoptera, Hymenoptera, Lepidoptera, Diptera). Depauperate pollinator community; decline in pollination levels and seed production; restricted pollen flow; inbreeding.
4. *Seed dispersal* (Hymenoptera, Formicidae). Decline in the diversity of seed dispersers, leading to decreased seed dispersal and plant survival.
5. *Predation and parasitism* (Araneae, Hymenoptera parasitic wasps, Hemiptera, Chilopoda). Reduction of prey abundance, leading to predator/parasite decline; herbivore outbreaks due to release from predator control, leading to plant damage and dieback.

6. *Food source for vertebrates* (most invertebrate taxa or species groups). Decline in insectivorous birds, and perhaps herps, owing to food shortage.

The urban spatial patterns strongly affecting invertebrates are highly diverse. The following patterns, particularly observed in Sydney, may at least suggest trends present in other urban areas (Hochuli *et al.*, 2009):

1. *Hemiptera* (scale insects): trees near roads or concrete support a higher abundance of scale insects.
2. *Lepidoptera larvae* (caterpillars): high mortality further from city where vegetation cover is greater.
3. *Lepidoptera* (butterflies): greater urbanized cover leads to decreased abundance and richness for most species, because of reduction in host plants and nectar sources.
4. *Lepidoptera* (butterflies): increased species diversity with increased habitat area, connectivity, permanent water, vegetation, and flowers.
5. *Lepidoptera* (butterflies): connectivity is a key to butterfly management in urban areas.
6. *Arthropods – ground-dwelling*: small habitat fragments support a fundamentally different fauna than in larger fragments.
7. *Arthropods – ground-dwelling*: arthropod species composition is different in different land-use types.
8. *Non-ant Hymenoptera* (bees, wasps): diversity and abundance are positively related to habitat fragment area, but negatively related to fragment age.
9. *Coleoptera – carabid ground beetles*: higher species richness on early successional sites.
10. *Coleoptera – carabid ground beetles*: abundance and species richness increase from urban to rural.
11. *Hymenoptera – bees*: in general a negative response to urbanization.

These results barely scratch the surface and do not pinpoint general patterns. Still, it seems that arthropods are often sensitive to, and distributed in, some of the same spatial patterns of importance to most urban vertebrates.

A few additional observations broaden the perspective on urban invertebrates. Huge numbers of flying insects are attracted to urban lights ("electric zappers"), where many insects are eaten or otherwise die (Rich and Longcore, 2006; Eisenbeis and Hanel, 2009). Such lights act very much like a vacuum cleaner, sucking insects away from their habitats.

The richness of butterflies, ants, and many other species sometimes correlates with the size of greenspace and the proximity to species source, but sometimes it does not (Giuliano, 2005; Natuhara and Hashimoto, 2009; Kotze *et al.*, 2011). Invertebrate species richness may be particularly high in early succession sites (Gilbert, 1991; Matteson, 2008). Butterflies, grasshoppers, woodlice, and land snails are typically least species-rich on intensively designed and maintained land uses, and richer on neglected urban sites (see Figure 8.12) (Godde *et al.*, 1995). In Britain, generally the longer a tree species has existed in the country the more arboreal insects and insect species it supports (Kennedy and Southwood, 1984).

Invertebrate species thriving in urban areas seem to be food generalists (Kotze *et al.*, 2011). Arthropod larvae [such as of dragonflies (Odonata, Anisoptera) and mayflies (Ephemeridae)] often require a different habitat than that of the adult. Therefore, the proximity of both habitats is important in the urban area. High dispersal ability enhances arthropod success in the city. Many invertebrate species groups have been studied along urban-to-rural gradients (see Chapter 2), with diverse results found. Refuse piles/heaps, roadsides, parks, and early succession sites all tend to be rich in invertebrates, but each harbors a somewhat different fauna. Urban invertebrate ecology remains a rich research frontier.

Vertical structures, vegetation layers, and animals

Urban areas provide a wonderful array of vertical structures, both artificial and natural, used by wildlife. More varied than the structures out in agricultural and natural lands, this urban vertical heterogeneity is used for nest and den sites, foraging for food, movement and escape cover, courtship, and even water (Kunz *et al.*, 2008). Just as on the ground, some vertical objects are beneficial and others hazardous to animals (Figure 9.6b). The hazardous structures are "artificial" (human-made).

We now examine these hazards and habitats for urban animals: (1) towers, buildings, powerlines, bridges; and (2) trees, shrubs, vegetation. Urban ecology studies have primarily emphasized patterns at increasing height above ground, a sort of layers approach. Also interesting, though less known, is how animals fit their diverse daily and seasonal activities with the micro-heterogeneity so evident vertically in urban areas.

Towers, buildings, powerlines, bridges

Hazards and habitats are the two main ecological dimensions of these major human constructions in urban areas.

Hazards

Two decades ago it was estimated that the leading human causes of avian mortality in the USA were, in order (Klem, 1991): hunting; collisions (strikes) with plate glass; and collisions with vehicles. These big-three killed 275 million birds annually, 95% of the total human-caused mortality. Migrating birds are particularly at risk (Hager et al., 2008), and of the collisions with tall structures (towers, stacks, buildings), 98% of the birds hit plate glass. In a New York City study, 82–85% of the "glass collisions or strikes" were fatal. Based on better data and with somewhat more artificial structures, plate glass, and lights present today, estimates now range from 100 million (Adams et al., 2006; Adams and Lindsey, 2011) to 1 billion (Klem et al., 2009) birds killed annually by collisions with tall structures, especially plate glass (see Figure 3.1). Glass strikes are now considered to be the leading cause of human-related bird mortality, greater than hunting and roadkills, and much greater than pollution, poisoning, wind turbines, and domestic cats.

The presence of ground cover, shrubs, and trees in front of a building is hypothesized to increase the rate of bird collisions, and warrants study (Hager et al., 2008; Klem et al., 2009). Also, bird feeders near houses probably increase window collisions of local residential, mainly seed-eating, songbirds. Reducing or changing the nature of plate glass in tall buildings presumably would have the greatest strike-reduction effect. Another significant approach is to turn off the lights in tall buildings at night. Toronto, with reputedly one of the highest bird–building collision rates in the world, has such a program (Gauthreaux and Belser, 2006). In Boston, the mayor and the Massachusetts Audubon Society have collaborated toward this goal, especially during bird migration seasons.

In addition to colliding with tall buildings, birds strike radio, television, communications, cell-phone, electric powerline, wind-turbine, and airport towers (Johnson and Klemens, 2005; Adams et al., 2006; Rich and Longcore, 2006; Kunz et al., 2008; Adams and Lindsey, 2011). Perhaps a quarter of the North American bird species are vulnerable to such collisions.

It appears that red lights rather than white lights, and constant lights rather than flashing lights, are more disruptive of avian flight (Gauthreaux and Belser, 2006). Current research suggests that white flashing lights are best on towers. Floodlights pointing upward to highlight buildings, and the unusual combination of lights around large airports disorient birds, often leading to mortality. In addition, the rapidly rotating ends of wind-turbine blades are lethal to some birds as well as bats (Adams et al., 2006; Adams and Lindsey, 2011).

Of course many hazards exist at a finer scale for wildlife around artificial structures. Predators lurk in diverse locations. Competitors do too. Windstorms destroy nests. Urban heat waves make surfaces too hot. Rainwater pours and drips down through leaks and channels. Snow and ice accumulate. Strong turbulent and vortex air flows alter animal activities. Dust and other pollutants concentrate in certain spaces. Noises and vibrations from inside a building occur. People fix and eliminate holes and broken windows. Humans clean surfaces. The complex of stresses and disturbances for wildlife in the high-microhabitat-diversity vertical world created and maintained by humans remains a little-studied ecological frontier, daily facing us.

Artificial structures as habitats

Storks are generally appreciated for their symbolic role in bringing a baby to the household. The size of a pelican or large heron, storks (*Ciconia*) build large nests atop chimneys and buildings, especially in towns and small cities of Europe and parts of Asia (Figure 9.7). These summit habitats provide good visibility and ample essential space for take-offs and landings. With a tall stork guarding the nest, and with nests often clustered nearby, predators usually stay away.

Raptors, especially certain hawks, are the other large birds nesting high up on built structures including buildings, but in this case the birds are at very low density. Also normally raptor nests are on ledges under an overhang providing protection from rain. In the Milwaukee Region (Wisconsin, USA), red-tailed hawks (*Buteo jamaicensis*) successfully fledged young in nests on electric powerline towers and a tall advertising billboard (Stout et al., 1996). Raptors and ravens (*Corvus corax*) nest on powerline towers, which also serve as perches for foraging (Steenhof et al., 1993).

In addition, the peregrine falcon (*Falco peregrinus*) nests on a variety of tall structures in urban areas,

Vertical structures, vegetation layers, and animals

Figure 9.7. Stork and nest atop a city building. Scores or more of such nests enrich the skyline of Alcala de Henares, Spain. Drawing by and with permission of Barbara L. Forman.

where the bird typically hollows out a shallow depression in accumulated sand or other soil material. Large bridges are often used (Carey, 1998), including at least nine major bridges in the New York City Region, each supporting a pair of peregrines. Such locations provide good visibility, dive space [the birds can dive at 320 km/h (200 mi/h)] for foraging, and some protection from rain, wind, and competitors.

As noted above, bats also nest and roost in bridges (Keeley and Tuttle, 1999; Adams *et al.*, 2006), mainly small or medium-size structures. Building areas in Merida (Yucatan, Mexico) are used in different ways by four bat species (Rydell, 2006). Large fast bats fly above the buildings. Medium-sized fast bats forage between the buildings and above the street lights. Small fast bats forage for insects attracted to street lights. And slow but highly maneuverable broad-winged bats forage for insects just over the low vegetation and ground-level under street lights. Also, mammals around lights on buildings are at risk of being caught by a predator (Rich and Longcore, 2006).

Green roofs support some insects and occasional birds (Chapter 10). The vegetation on green walls, especially vines, supports more insect and bird species, but still is usually low in species diversity (Dunnett and Kingsbury, 2004; Chapter 10). Lacewings, butterflies, moths, and many other invertebrate groups on walls with vegetation are food for migrant birds, resident birds, and bats. Some vines such as ivy (*Hedera helix*) are nectar sources. Wall vegetation of vines and other plants occasionally serves as nest sites for house sparrow, blue jay (*Cyanocitta cristata*), and other birds, and may be important as somewhat protected winter roost sites for a variety of urban songbirds.

Urban wildlife use of the microhabitats associated with built land is illustrated in Barcelona (Boada and Capdevila, 2000). Swallows feed around lights and rest on wires. Mice use building interiors from the basement upward, especially tracking food availability around kitchens. Magpies use railings and ledges of walkways, buildings, and small bridges as perch sites for foraging. Swifts dive into clouds of mosquitoes, midges, and other insects carried upward in wind or rising warm air. Lizards frequent walls, especially near corners or lights, to catch arthropod food. House sparrows slip into small holes. Common pigeons (rock doves) find access to attics, abandoned buildings, protected ledges, and under small-to-medium bridges. Cockroaches and rats move from building to building in drainage pipe networks. The list goes on. This combination of microhabitats is distinctly urban.

In summary, our built structures are both hazard and habitat for wildlife. Lighted glass on buildings seems to be the primary bird killer, but numerous features of our structures cause wildlife mortality. The multitude of building features also provides microhabitat diversity and cover for nesting/denning and feeding, thus supporting a rather distinctive urban fauna.

Trees, shrubs, vegetation

If you were sitting on a park bench or porch just after dark in Melbourne, lights moving around in the tree canopy might not be a surprise. An ecologist is spotlighting urban wildlife – that is, walking with a large spotlight shining a bright beam up into the canopy. Possums and gliders live in holes in the canopy eucalypts and other trees, and when lighted, their eyes shine brightly and species are readily identified. Ring-tailed possum, greater glider, and the abundant brush-tailed possum all live overhead. The vertical structure of urban vegetation (see Chapter 8) holds lots of wildlife surprises.

Some songbirds roost in trees at night, especially high up away from terrestrial predators. During daytime the birds are foraging at different levels in the trees and woods. For instance, as mentioned above, vertical vegetation structure is important for bird species and different foraging guilds (DeGraaf and Wentworth, 1981; Beissinger and Osborne, 1982).

In Nishinomiya City (near Kobe, Japan), birds were sampled in different vegetation layers in woods within

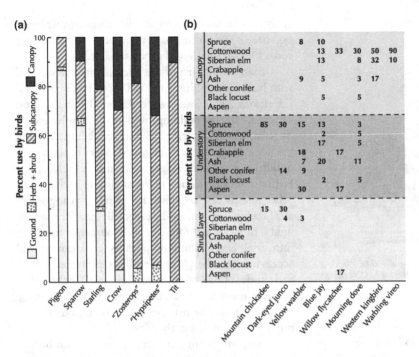

Figure 9.8. Use of urban vegetation layers by different bird species. (a) Nishinomiya City, Japan. Based on vegetation in 20 urban parks, and ≥15 birds per species. From Ichinose (2005). (b) Cheyenne (Wyoming, USA). Based on sampling birds in street trees and house-plot trees along sidewalks at 80 stratified-random sites in a commercial area, plus new, medium-age, and old residential areas. Some columns total <100% due to bird use of other less-common trees. Adapted from Sears and Anderson (1991).

20 urban parks (Ichinose, 2005). For the seven most abundant species, two (common pigeon and sparrow) were observed mostly on the ground layer (Figure 9.8a). Very few birds were in the herb and shrub layer. The other five species were most abundant in the understory layer. Six bird species used the canopy, though only the crow and "hypsipetes" were observed in it ≥30% of the time. Thus, for these common urban birds, the ground layer seems to mainly support two species, and the understory layer five species.

An extensive study of birds in residential and commercial areas of Cheyenne (Wyoming, USA) provides further insight into wildlife usage of trees (Sears and Anderson, 1991). A total of 2438 birds representing 43 species were counted in canopy, understory, and shrub layers of more than 14 tree species. Based on the most common eight bird species and eight tree species, birds preferentially "selected" for spruce (*Picea*) (i.e., used it proportionately more than its abundance would indicate), and selected against the other seven tree species. Overall, birds used the understory more than the canopy, and used the shrub layer very little (Figure 9.8b). The most generalist species appear to be the mourning dove (which used seven trees) and the blue jay and yellow warbler (using five species each). The mountain chickadee only used spruce, and warbling vireo only cottonwood and elm, apparently the two relatively specialist species among these common birds. Clearly, different species focus activity on different vegetation layers, such as the two specialists overwhelmingly in the canopy, and the blue jay primarily in the understory layer. In essence, both tree species and vegetation layer are keys to wildlife usage. Different bird species have preferences for both variables.

The shrub layer is of special importance in urban areas because it is usually scarce and provides cover for ground vertebrates. For example, shrub vegetation is the primary habitat for a tortoise (*Testudo hermanni*) in Rome (Rugiero and Luiselli, 2006). Relatively few urban birds nest in either the ground or shrub layer (Johnson and Klemens, 2005), where eggs and young birds are readily subject to predation. However, many species feed in the shrub and ground layers. Indeed, urban bird species richness has been found to correlate with both shrub cover and shrub height, as well as the height of herbaceous vegetation (Ortega-Alvarez and MacGregor-Fors, 2009).

Tropical cities also have wildlife distributed in vegetation layers, including daytime songbirds, lizards, snakes, monkeys, and more. Yet after the daytime heat passes, dusk-and-dawn and night-time species become active. These are mainly a different set of lizard, frog, snake, and mammal species. Adding these nocturnal species to the diurnal group may double vertebrate biodiversity. "Noises in the night" from animals are familiar in tropical cities, indeed all cities. Tropical frogs are especially vocal, and different species may live on the bark and in upper vegetation layers. Furthermore, as in urban Puerto Rico, different frog species may be active early in the night and others late at night.

Spatial habitat patterns and animals

Focusing on horizontal patterns here, we first introduce (1) species richness and rare species, which are linked to (2) habitat fragments. Then ecological patterns are explored along a gradient: (3) city, suburb, peri-urb/exurb, urban-region ring. Finally, (4) animals of particular habitats are highlighted.

Species richness and rare species

How many species are there in a city? Warsaw has one major river, 1.7 million people, 320 vertebrate species, and at least 3800 species of terrestrial invertebrates in a 517 km^2 (207 mi^2) area (Luniak, 2008). These species are 12% of the total for Poland. Approximately 40 mammals, 247 birds (137 breeding), 5 reptiles, 11 amphibians, and 30 fish constitute the Warsaw vertebrate fauna. Similar numbers are reported for Moscow, Berlin, and Lodz (Poland). Such numbers do not include analyses of the "invisibles," i.e., buried soil animals, tree-canopy invertebrates, and microorganisms (see Figure 1.2).

Habitat heterogeneity, associated with the combination of city center, large parks, other greenspace types, major corridor types, and residential, commercial, and industrial areas, is normally extremely high in a city or metro area. This distinctive habitat heterogeneity prevents internal urban homogenization of the fauna, and helps maintain a high species richness. From city to city the predominant species are often similar, just as for peat bogs in forestland or scattered mountains in a desert region. However, an unending input of species from habitats around cities helps maintain diverse and somewhat distinctive urban faunas.

Comparing the Warsaw species numbers at four spatial scales (Luniak, 2008), (1) region, (2) city, (3) urban parks, and (4) city-center non-park green cover (courtyards, lawns, street trees, etc.), helps place the city scale in perspective. Percentages of the regional invertebrate fauna decrease by scale, from 100% (region) to 75% to 33% to 19% (city-center green cover). Of 10 insect groups plus spiders, earthworms, and terrestrial snails, percentages for almost all groups decreased at about the same rate. It would be interesting to know whether this spatial-scale pattern holds for other cities.

The density of breeding birds across the city of Warsaw is about 300–700 pairs/km^2 (780–1810/mi^2), similar to that in Berlin and Hamburg (Luniak, 2008). In the highly urbanized areas of inner Warsaw, bird density is estimated at 830–1590 pairs/km^2 in the breeding season. In winter the total population for the city averages about 964 birds/km^2. Presumably the "big three" (sparrow, starling, pigeon) predominate. These densities are much higher than in agricultural and natural lands.

Such species inventories lump together species of diverse types and functional roles, such as keystone, native/non-native, abundant/rare, and species dependent on human-supplied food/garbage versus those using more-natural food sources. The top predator, coyote (Figure 9.2), is a *keystone species*, that is, it has a much greater effect on the ecosystem than its limited abundance would suggest. Native and non-native species were discussed in Chapter 8. Abundant and rare species were mentioned above. A few examples relating to non-native animals and rare species in urban areas are useful here.

A Syracuse (New York, USA) study indicates that in summer the presence of common pigeons (rock doves) reduces the diversity (richness) of native bird species (Johnsen and VanDruff, 1987). Richness does not correlate with the abundance of house sparrows or starlings. In winter, native bird diversity decreases with either the presence of common pigeons or starlings, but correlates positively with the presence of house sparrows. Native bird diversity in summer also correlates with several urban-environment variables: building volume (negative correlation, i.e., larger buildings, fewer species); bird feeder density (positive); herbaceous cover-to-pavement ratio (positive); area of coniferous-tree cover (positive); and area of commercial land use (negative). In winter, native species diversity correlates with somewhat different environment variables: area of coniferous cover (positive); abundance of children (negative); age of structures (negative); and area of commercial land use (negative).

In Los Angeles, a rare butterfly (*Euphilotes battoides*) is reported to survive in only two sand-dune locations, an airport site of 122 ha (300 acres) and an industrial site of 0.6 ha (1.5 acres) (Arnold and Goins, 1987). The butterfly has several predators on its scarce dune habitat, where the caterpillar feeds on an uncommon native buckwheat plant (*Eriogonum*). At the small site over

10 years, the host plant species slowly declined, while the rare butterfly suffered a much steeper decline. With 99% of the butterfly's former habitat urbanized, and no known successful dispersal and colonization, the rare butterfly species seems to have a dim future.

In 1987 in Tijuana (Baja California, Mexico), only four pairs of the rare least Bell's vireo (*Vireo bellii pusillus*) were breeding in the 51 ha (135 acres) of riparian habitat suitable for the small insect-eating bird (Beatley, 1994). Since then the human population of Tijuana has grown enormously, and it is quite possible that the bird is now extinct there as a regular breeder. Just to the north in San Diego (USA), 21 nesting pairs of the vireo inhabited the 154 ha of riparian habitat present. San Diego has also grown considerably, but has implemented a habitat conservation plan for the vireo and numerous other rare species. Perhaps the species hangs on there. The vireos in these two adjoining cities are part of a broader-scale population, e.g., (in 1987) of 291 total breeding pairs in 11 areas. All of the sites are affected by urbanization, which continues, or increases, the risk of rare-species extinction.

A portion of the Las Vegas (Nevada, USA) Region has 11 greenspace (brown desert) locations of a rare, slowly reproducing reptile, the desert tortoise (*Gopherus agassizii*) (Beatley, 1994). Only three of the patches have an estimated 100 or more individuals, and six locations only have 1–5 individuals. All 11 "populations" are likely to decrease to extinction as urbanization and other stresses continue.

These rare-species stories can be repeated in urban areas worldwide. The native species barely survive, or disappear, in fragmented remnants of semi-natural vegetation (but see Fuller et al., 2009). All cities probably contain rare species. Paris has 7 nationally protected amphibian species (78% of the city's amphibians), 11 protected mammals (34% of the total), and 119 nationally protected bird species (72% of the city's avifauna) (Beatley, 2012).

Protecting rare species in an urban area may often be a local priority, but not a regional or national priority for conservation. This is because in a metro area the probability of successful protection over human generations seems minuscule. Concentrated human impacts, plus outward urbanization (even without climate change effects added), nearly assure the local extinction of rare native species in urban areas. What to do if the only known population or individual of a species lives in an urban area is perhaps mainly an ethical question.

Habitat fragments

A habitat perspective in urban areas, rather than a species-by-species approach, highlights the ecological effects of habitat fragmentation mainly caused by suburban and exurban spread (Johnson and Klemens, 2005; Haslem and Bennett, 2008). Here we focus on species richness, and secondarily on species composition, since they seem to be the prime concern in urban areas, but many additional ecological dimensions are affected by habitat fragmentation (see equations, Appendix B) (Marzluff and Ewing, 2008).

First consider the *species-area effect* of greenspace patch size affecting species richness (Tilghman, 1987; Adams and Dove, 1989; Fernandez-Juricic, 2000b; Fernandez-Juricic and Jokimaki, 2001; Murgui, 2007; Natuhara, 2008). The diversity of native mammals and birds in San Diego was related to the area of greenspaces [0.4–103 ha (1–258 acres)], as well as to the area (0.1–68 ha) of semi-natural vegetation within the greenspaces (Figure 9.2a) (Soule, 1991). Species richness correlated better with area of semi-natural vegetation than with area of greenspace.

The same pattern was found with avian diversity for wooded areas within parks of 0.2–3.4 ha (0.5–8.5 acres) in Nishinomiya City (Japan) (Ichinose, 2005). In parks, lawn-and-ornamental-plant areas provide more benefit than do the surrounding impervious spaces for the vertebrate species. However, the planted area is still much less ecologically suitable than the semi-natural vegetation.

Bird species richness also increased with area in remnant greenspaces of 1–2 ha, 4–10 ha, and 10–20 ha in Brisbane (Australia) (Catterall, 2009). Furthermore, species composition changed greatly with area. The small greenspaces had more species of large birds than did the medium and large patches. No patch-size effect was present for the number of non-native species or seed-eating species. In Madrid, several groups of birds were denser in the interior than in the edge portion of urban woods, whereas house sparrows and common pigeons (rock doves) were denser in the wooded edges (Fernandez-Juricic, 2001).

Species–area relationships were compared for several groups of organisms in greenspaces of Osaka (Japan) (Natuhara and Hashimoto, 2009). Bird, butterfly, tree, and fern species richness were each correlated with greenspace area (r^2 from 0.88 to 0.64), but ant diversity (in Osaka and Kyoto) was not. Also both bird and butterfly diversity were correlated ($r^2 = 0.60$ and 0.73, respectively) with distance from species source or

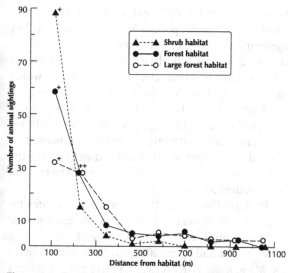

Figure 9.9. Daytime distance of predator from shrub and forest habitats in city. Based on 108 sightings of coyote (*Canis latrans*) in Seattle (USA). Total shrub habitat = 1415 ha (3538 acres) and forest habitat = 634 ha in the 220 km² (85 mi²) city. − and + = significantly less or more sightings than expected on the basis of area alone. Adapted from Quinn (1991).

continuous large forest. Once again, ant diversity was not correlated with distance.

Although urban mammals and birds may nest or den in a greenspace, they normally move out from it, though we still know little about the distance moved by different species. This pattern is illustrated by the top predator, coyote, in Seattle (Washington, USA) (Quinn 1991; Gehrt *et al.*, 2009; Way and Eatough, 2011). Of 108 animals observed, very few moved more than about 275 m from urban forest habitat (Figure 9.9). However, normally large forested greenspaces are the preferred den sites, and animals moved further outward from these (mainly up to about 400 m) into built areas. The animals also mainly moved up to about 400 m from shrub habitat, which provides good cover.

Where a section of an urban area is ≥25% forest, the animals mostly remain within 600 m of forest. If shrub habitat is ≥15% of an area, the animals may often move further from a shrubby area. If neither forest nor shrub cover is abundant, animals commonly roam still further from the main greenspaces. Overall the results highlight the importance of "distance from source" of species.

Another way to visualize the effect of distance between greenspaces is by relating the presence of animals, or species diversity, to the percentage of vegetation within some radius of a site. Thus, the number of native forest-bird species at a site in Seattle decreased from an average of 13 to 4, as the percentage of forest in the surrounding 1 km² decreased from 100% to zero (Marzluff and Ewing, 2008).

Overall, it appears that species richness for vertebrate groups normally increases with area of semi-natural vegetation within greenspaces, and also increases with proximity to species sources or other greenspaces. For invertebrate diversity, it appears that usually neither variable is significant. Nevertheless, positive species–area correlations typically only explain a limited amount of the variability (r^2) in urban studies, and other factors are also important in understanding diversity patterns. Although research is limited and results are variable, the distance from species source or other greenspace may be especially important in urban areas.

A planning study in Syracuse (New York, USA) mapped large, medium, and small urban greenspaces and used the preceding two concepts (species-area and distance-from-source) to estimate the relative number of wildlife species present (Hilliard, 1991). Four groups of greenspaces were hypothesized to vary in species richness from high to low: (1) >12 ha (large greenspace) with direct connection (partially adjacent) to the surrounding non-built area as the key species source; (2) >12 ha with corridor connection to the surroundings; (3) >12 ha with no connection; and (4) <12 ha (small), irrespective of connection. While based on principles, plus evidence from elsewhere, the expectations are useful for planning in the absence of empirical data, and could be validated or sharpened up by ecological study. Additionally, numerous human activities can increase urban biodiversity during landscape design, building design, site work, building process, plantings and vegetation, adding artificial structures, and management/maintenance (van Bohemen, 2005).

Even though urban areas have so many distinctive characteristics, some of the ecological principles learned in natural and agricultural areas should be considered and evaluated for use in cities. Until we have a robust set of urban ecology principles, biodiversity management could cautiously use existing ecological concepts of scale and hierarchy, species composition, habitat fragmentation, habitat degradation, and much more (Savard *et al.*, 2000).

City, suburb, peri-urb/exurb, urban-region ring

In New England (USA) the density of birds in a city is reportedly 2.5 times greater than that in suburbs (DeGraaf, 1986). In contrast, avian diversity is 2.5

times higher in suburbs than in the city. The density or abundance pattern is considered to be largely due to the abundance of human sources of food in the city. The species diversity or richness pattern is correlated with more habitat diversity in the suburbs, and may be linked to habitat connectivity.

The greater density or abundance of organisms in city than suburb is supported by a number of studies, including: birds in Palo Alto (California, USA) (Blair, 1999) and common pigeons (rock doves) in Manchester (UK) (Wheater, 1999). However, the opposite pattern, i.e., greater density in suburb, is also reported for butterflies in Palo Alto (Blair, 1999) and 6 of 25 bird species in London (Wheater, 1999).

Nevertheless, species richness higher in suburb than in city, plus a gradual decrease from suburb to city, seems to be a well-supported widespread pattern. Illustrative studies include: butterflies in Palo Alto (Blair, 1999); lizards in Tucson (Arizona, USA) (Germaine and Wakeling, 2001); and birds in Buenos Aires, London, Palo Alto, and Oxford, Ohio (USA) (Blair, 1999, 2008; Wheater, 1999; Faggi et al., 2008). Considering the different food sources and species present, it is not surprising that bird foraging guilds also differ in city and suburb (DeGraaf and Wentworth, 1991).

Although suburban areas may differ more from each other than do different urban areas, ecological comparisons of suburbs around a city are scarce. In Brisbane (Australia), a ("bare") suburb with few trees differs markedly in many ecological variables from one with considerable tree vegetation: number of bird species; seed-eating species; vertebrate-eating species; arboreal insectivores; and non-native species. Indeed, major differences were present for: number of species of fruit-or-nectar feeders, large birds, and species composition, i.e., the particular species present (Catterall, 2009). In a study of screech owls (*Otus asio*) in 10-year-old and 30-year-old suburbs of Waco (Texas), little difference in life history and population characteristics was present (Gehlbach, 1996). However, owl nest density was 60% higher in the older suburb. Comparison of residential areas of different density in Tucson found lizard diversity highest in low-to-medium house-density areas (Germaine and Wakeling, 2001).

The idea of "rural" in urban–suburban–rural gradient studies has remained a challenge for interpreting results, since non-urban land surrounding a metro area differs markedly, as in cropland, pastureland, successional-regrowth land, deciduous woodland, coniferous forest, and peri-urban area. Most all-built metropolitan areas are surrounded by agricultural land (Forman, 2008). Where semi-natural land adjoins suburban land, it is almost always much altered by urban effects (and probably would not be viewed as "wildland" by most ecologists). In effect, land use type at the rural end of a gradient greatly affects the shapes of curves along city-to-rural gradients. With this caveat, and knowing that it would be surprising if much ecological were the same in urban and rural areas, we can still learn from urban–rural comparisons.

In Basel (Switzerland) two of the prime sites for common pigeons are city-center streets and squares/parks; the third site is agricultural land surrounding the city (Rose and Nagel, 2006). Of 25 breeding birds in the Manchester (UK) region, 14 were most abundant in the rural area (Wheater, 1999). Differences in kestrel (*Falco tinnunculus*) abundance and biology were reported between a city and a rural site (700 km away) in Spain (Tella et al., 1996). Along a gradient of city, suburb, peri-urban, semi-rural, and rural of Buenos Aires, avian diversity was lowest in the city and highest in the semi-rural area (Faggi et al., 2008). Gradient studies also provide ecological comparisons between suburb and rural.

Interestingly, in the Osaka Region (Japan), the city has the highest bird species diversity, which progressively decreases in suburb and rural areas (Natuhara, 2008). This at-first-glance unexpected result seems analogous to an oasis pattern. Trees and diverse food sources are concentrated in the city, while the surrounding rural land is relatively homogeneous cropland supporting few bird species.

The preceding ecological results point to the importance of the urban-region ring surrounding a metro area (Forman, 2008). Lots of important and intriguing spatial patterns in the ring-around-the-city are major determinants of ecological conditions in the city.

Animals of particular habitats

Most habitats characterizing urban areas, such as tree-lined street, park plaza, house plot, and wastewater treatment facility, are scarce or absent elsewhere. Indeed, our zoos are probably the pinnacle of high species diversity. Here exotic non-native species are most conspicuous and appreciated. But the huge amounts of human-provided food and consequent waste attract many other species from city, suburb, and exurb alike.

Indeed, many of the other species thrive at exceptionally high densities.

A study of a small city, Dusseldorf (Germany), identified 75 habitat types present, of which 38 were considered widespread or important (Godde et al., 1995). The diversity of plants, butterflies, land snails, wood lice, and grasshoppers was measured in each habitat type, and ranged from relatively high to zero (see Figure 8.12). Rather than developing an exhaustive list of habitats and their ecological attributes, here we simply list a series of major land uses of cities worldwide, and highlight some key wildlife characteristics.

1. *City center.* House sparrows, common pigeons, starlings, rats, mice, cockroaches, and other insects thrive on an abundance of human food and garbage (Chapter 11). Relatively few species of agricultural and natural lands are present (Gilbert, 1991).
2. *Stormwater drain or basin.* Nocturnal scavengers such as raccoons (*Procyon lotor*), rats and cockroaches rest, feed and move in stormwater structures, which provide escape from vehicles, cats, dogs, and people (Bolen, 2000) (see Chapter 1).
3. *Cemetery.* Grass-only cemeteries are species poor, whereas the presence of diverse shrubs provides cover for vertebrates and some nest sites for native birds. A large tree-covered cemetery is likely to be an overnight feed-and-rest spot for migrating birds in an extensive built urban area (Bolen, 2000; Adams et al., 2006) (Chapter 12).
4. *House plot.* Although a mainly lawn site has few species, the microhabitat heterogeneity of most house plots is relatively high, supporting considerable species richness of many animal groups. Added artificial structures, such as rock wall, bird feeder, or water feature, can raise biodiversity well above natural-area levels (Johnston and Don, 1990) (Chapter 10).
5. *Housing development.* High- and medium-density multi-unit housing normally supports few species other than the common urban scavengers and seed-eaters. On the other hand, high-density single-family-unit housing tends to have a considerable diversity of wildlife, because of the high concentration of front spaces (gardens) and backyards with highly different design and maintenance approaches. Developments with large house-plots may have most of the surrounding natural-area species as well as many generalist species attracted to the more maintained portions (Roth, 1987) (Chapter 11).
6. *Institution.* Often savanna-like mowed grass and trees, sometimes with a water body present, cover the grounds of academic, health, religious, and government building clusters. This pattern supports modest levels of native species and generalist scavengers. Aquatic animals and wildlife attracted to a water body greatly increase species diversity (Dexter, 1955; Swank, 1955) (Chapter 12).
7. *Golf course.* The usual open strips, widely spread people, presence of pond, removal of shrubs, intensive mowing, and abundant use of irrigation water, fertilizer, and pesticides typically produce a relatively sparse but unpredictable wildlife community. With few trees between fairways the area attracts species preferring open areas such as gulls, geese, and hawks. With more semi-natural vegetation, more resident species and migrant birds are present (Moulton and Adams, 1991; Blair, 2008) (Chapter 12).
8. *Airport.* The large open space attracts grassland birds and mammals. Raptors from hawks to vultures and eagles feed on the easily detected rodents and other animals. Because airports often border water bodies or wetlands, water birds including herons/egrets and flocks of gulls, ducks, and geese may be frequent. The large birds and flocks of small birds pose hazards to aircraft, so various control measures are common, from introducing predators to planting certain plants that do not attract birds (Cooper, 1991; Satheesan, 1996; Adams et al., 2006; Bernhardt, 2009; Linnell, 2009) (Chapter 12).
9. *Dump, tip, landfill.* As a major active source of garbage and other refuse, the dump is home to large concentrated populations of local animals, including rats and other mammalian and some avian scavengers. However, some birds, including gulls and corvids (crows, etc.), regularly travel considerable distances to feed (Adams et al., 2006) (Chapter 12).
10. *Island in harbor.* When far enough from the mainland so that foxes and other predators do not reach them, harbor islands may support heron/egret colonies. Coastal birds and mammals use islands to rest and feed (Howell and Pollak, 1991; Parsons, 1995; Parsons et al., 2001) (Chapter 7).

Wildlife movement

We begin with (1) animal biology and movement, and then focus on (2) urban pattern and wildlife movement.

Animal biology and movement

Garbage and other human-provided food play a major role for many wildlife communities, both in city and suburb. Feeding people in a city requires huge amounts, diverse types, and widely distributed food (Figure 9.10), a portion of which becomes "food waste." For instance, the average family in the USA throws away about 0.6 kg (1.3 pounds) of food per day, 14% of the total brought into the home (Adams and Lindsey, 2006). Considering how many people are concentrated in a city, that highlights a lot of garbage or rubbish in home garbage cans, restaurants and hotels, dumpster containers (for accumulation and transport), dump trucks, and dumps (tips, landfills). These are thoroughly scattered across the urban area. The numbers are for a rather consumptive and wasteful population, but which has somewhat efficient waste removal, transport, recycling, and disposal processes. In cities of certain developing nations, people may be less consumptive and wasteful, but the waste removal and disposal processes are often less efficient, so rubbish is more widely spread over the area.

The amount and dependability of the garbage food supply is observable in some dumps and dumpsters. Urban cats are predators and will attack small rats (Adams and Lindsey, 2006). Meanwhile, large rats will consume kittens. But with an abundant regular food supply, it is easier for both species to consume garbage,

Figure 9.10. Open market selling fruits and vegetables. Canterbury, UK. R. Forman photo.

and cats and rats have been observed dining "shoulder to shoulder." Many other animals from insects to foxes, coyotes, raccoons, opossums, possums, corvids, and gulls of course join in the daily feast (Bolen, 2000; Gehrt et al., 2009; Way and Eatough, 2011).

People also directly feed many urban animals, from house sparrows and mallards in the park to varied bird species in the backyard. The density of birds in an area increases with more home bird feeders, but apparently the diversity of birds changes little (Fuller et al., 2008).

Most vertebrate wildlife species have several well-known types of movement. Four types, from shorter to longer distance moved, are of primary importance to fit with urban spatial patterns. Many vertebrate species have a *territory*, a small area around the den or nest that is defended, especially against other individuals of the same species. Overall though, this territoriality is less prevalent in urban areas than in natural and agricultural lands. The *home range* is a larger area covered in day-to-day movements, particularly in foraging for food. *Animal dispersal* refers to a sub-adult animal moving away from its birth home range to mate and establish its own new home range. *Migration* refers to the cyclic movement of certain animals that avoids a stressful environment, such as winter, and provides suitable living conditions throughout the cycle. Urban ecologists usually focus on home range and dispersal.

Territories and home ranges are strongly linked to urban features. Thus, lizards on tree trunks or walls in tropical cities typically display their brightly colored inflatable throat (dewlap), and fiercely defend their territories against other intruding lizards. The size of home ranges for urban wildlife of course varies widely, illustrated by about an 80-m radius (262 ft) for female cats (110 m for males) in Brooklyn (New York), 1100 m radius for raccoons in Toronto, and several kilometers for gulls flying between nest site and rubbish dump (Haspel and Calhoon, 1991; Rosatte et al., 1991). Home range sizes also seem to be highly dependent on amount of food available. Thus, in London, red fox home ranges are 25–40 ha (70–100 acres) in residential suburbs with ample food sources, and >100 ha in industrial and government-building areas with less food available (Gilbert, 1991).

Several other movement patterns are important in urban areas, including escape from predators, courtship, and species range expansion or spread. Escape from predators requires nearby protective cover virtually wherever an animal moves. Raccoons have stormwater drains "everywhere" and readily escape cars, dogs and other predators. Many species apparently most prefer shrub cover, which is often limited in urban areas. Courtship of urban wildlife seems to be little studied though it presumably occurs mostly within a home range.

The geographic range of some species is continually expanding and/or shrinking, often northward or southward, and cities at the border of a species range therefore normally either lose or gain a species. The arrival of an animal species from another city or continent, such as a non-native species, occurs frequently in cities (Elton, 1958; Davis, 2009). Yet the spread of these species within the urban area seems to be rather little studied.

At times a single animal repeatedly causes a problem for people, or a wildlife population becomes abundant enough to be annoying. Such animals are considered to be *pests*, and control measures such as killing or reducing food or habitat are considered (see Chapter 1). Ants and mice entering a kitchen may be trapped and killed, if better approaches are not available.

One control measure related to animal movement involves *translocation* of the animal away to a distant site. If the animal is taken to a suitable habitat, other animals present do not appreciate the newcomer, and the survivorship rate of such translocated animals is usually quite low (e.g., 10% within a year) (Bryant and Ismael, 1991). Some animals quickly leave for home (homing), while others move about in the general vicinity of the release site. For example, white-tailed deer (*Odocoileus virginiana*) were translocated from a Milwaukee (Wisconsin, USA) suburb and released in good deer habitat, apparently several tens of kilometers distant (Bryant and Ismael, 1991). Within a few months the translocated deer settled at sites an average of about 15 km (9 mi) from the release site, whereas the resident suburban deer moved an average of only 1.3 km. Canada geese (*Branta canadensis*) also sometimes become pests in parks and airports, and have been translocated elsewhere. Geese were translocated from Minneapolis-St. Paul (Minnesota, USA) to sites 30–80 km distant and 1100 km distant (Oklahoma) (Cooper, 1987). Interestingly, more birds returned from the distant site (not on the species' migratory pathway from Minneapolis) than from the nearer sites with suitable habitat.

In effect, translocated animals move much greater distances than do resident wildlife. However, the survival rate of translocated animals is normally low.

Therefore, living with and adjusting to urban wildlife is often the preferable option.

Urban pattern and wildlife movement

We start with (1) species sources and adjacencies, and then highlight (2) corridors and stepping stones.

Species sources and adjacencies

Normally, large contiguous natural areas or patches serve as the major sources of native species moving across the urban area. Typically, a large natural or semi-natural patch is either within or adjacent to the all-built metro area. Tiergarten Park in Berlin, Table Mountain Park in Cape Town (South Africa), and Reserva Natural Costanera Sur in Buenos Aires illustrate the large central species source. Table Mountain Park is primarily a remnant of former more-extensive vegetation, while Tiergarten and Costanera Sur are restored or reclaimed vegetation land. Large green patches adjacent to the metro area are illustrated in the Brasilia and Seoul regions (Forman, 2008).

Natural land surrounding a city may be adjacent to the metro area, as for Sapporo (Japan) and Kuala Lumpur (Malaysia), but is usually largely separated from the metro area by agriculture, as for Santiago (Chile) and Portland (Oregon, USA) (Forman, 2008). In a few situations, a wide green wedge or lobe of the surrounding natural land projects well into the city, as in Stockholm and Portland. Also a wide green river corridor may bisect a city, as for Edmonton (Canada). In both the wide green wedge and corridor cases, the species source is therefore close to the core and much of the city (see Figure 12.12c).

Some cities, such as London and Chicago, have no large natural land remaining in their region to serve as a species source. For such cities, the main source of species is agricultural land, either at a heterogeneous fine scale (London) or a less-heterogeneous area of extensive cropland (Chicago). In general, farmland species seem to do little better in urban areas than do natural-land species. Large water bodies, which exhibit similar spatial relationships to cities as natural land, are sources of most aquatic species. The effect of nearby large natural areas on ecological conditions in cities remains an important research frontier (Chapter 12).

These large areas highlight the importance of *adjacencies*, where one habitat or land use is ecologically affected by a different adjoining one (Chapter 2) (Hersperger and Forman, 2003; Dunford and Freemark, 2004; Luck et al., 2004). Bird species from a nature reserve are often present in an adjacent residential neighborhood, as a kind of "spillover effect" (Loss et al., 2009). Such an effect may extend some 0.5 km (1600 ft) from the nature reserve (but see Clergeau et al., 1998). Omnivores and seed-eaters seem more likely to cross a wooded patch edge into a dense residential area than do insectivores and nectar-feeders (Hodgson et al., 2007). Thus, over time the bird fauna of a botanical garden in Java (Indonesia) came to "mirror" the birds of the changing surroundings (Diamond et al., 1987). Adjacency effects may be strong and widespread.

At a broader scale, plant species richness in forest patches of Barcelona is affected by whether agriculture or urbanization is adjacent (see Chapter 2) (Guirado et al., 2006). One study found that bat activity was higher in urban space adjacent to industrial/commercial area, lower if next to farmland, and higher if adjacent to farmland containing a water body (Gehrt and Chelsvig, 2003). However, butterfly species richness and composition in grassland patches near Boulder (Colorado, USA) did not correlate with the amount of development in the surroundings (Collinge et al., 2003). In the same area a small songbird, plumbeous vireo (*Vireo plumbeus*), has its nest "parasitized" more [by cowbird (*Molothrus ater*), a larger bird] if an urban area is adjacent (Chace et al., 2003). Indeed, parasitized nests were closer to the city boundary (average 3.0 km = 1.8 mi) than natural unparasitized nests (4.9 km). The pattern of nesting problems for the songbird was consistent with observations of cowbirds flying from the urban area into the adjacent natural land.

Urban stepping stones for animal movement

In simplest terms *green corridors* are vegetated strips. They function as conduit, filter or barrier, source, sink, and habitat (Forman 1995). *Stepping stones* are small green patches used sequentially in movement. Networks of interconnected green corridors are also important for species movement (see Chapter 12).

To envision fine-scale movement in an urban area, join me looking at how a small lizard species (*Anolis cristatellus*) might spread across Miami (Florida, USA). We are looking for spatial arrangements or features that function as *facilitator patterns* (enhancing movement) or *inhibitor patterns* (blocking or reducing movement). The species was introduced from Puerto Rico some 35 years ago, defends territories mainly on lower trunks of large trees, and has spread over a few square miles of mostly residential area (Losos, 2009; Jason Kolbe and Jonathan Losos, personal communications, 2011).

We see several apparent major inhibitors to movement of the lizards: (1) a large busy boulevard; (2) a commercial strip; (3) a canal about 40 m (125 ft) wide; (4) a recent housing area; and (5) older housing primarily with palms, pines, and/or small trees that are relatively unsuitable for the lizards. Urban patterns that would seem to facilitate or promote spread of the lizard are also conspicuous: (1) lines of large or convoluted-bark street trees; (2) strips of similar-type trees aligned along back boundaries of house plots; (3) a sequence of suitable stepping-stone-like habitat patches, mostly as nearby small parks; (4) a large high-quality-habitat patch as a source of lizards; (5) abundance of medium or large fig trees (*Ficus*) and perhaps live oaks (*Quercus virginiana*); and (6) small bridges mostly lined with vegetation over canals.

Finer-scale conduits, filters, sources, sinks, and habitats also exist, though are harder to detect. For instance, in daytime a subtropical wide street has a hot surface, so the diurnal lizard is likely to cross only where trees on opposite sides cover the street with cool shade. Cats and bicycles can be lethal, so avoiding them is wise. Big hungry green lizards in occasional tree canopies are to be avoided. Longer-distance movement requires getting onto a pile of fallen large palm leaves or other plant waste that are regularly collected by the city and transported to certain park and other locations. In this way the lizard quickly moves a few kilometers. Yet, alas, the destination may or may not be suitable for lizard life.

Dispersed tiny patches of vegetation in a built area seem to be important in enhancing conditions for both nesting/denning and movement by urban vertebrates. For example, avian diversity within a park near Kobe (Japan) correlated with the proportion of woodland within 500 m of the edge of the park (Ichinose, 2005). Analogously, spatial patterns closely surrounding a large woods in Dutch farmland strongly affect species within the woods (Forman, 1995).

For the greater tit (*Parus major*) in Osaka (Japan), a model calculates the probability of occurrence of the bird, based on the area of tree cover within 250 m (675 ft) of a point, plus the number of other habitats within 1 km (Natuhara, 2008; Natuhara and Hashimoto, 2009). Thus, at least 1.8 ha of tree cover within 250 m (9% of the circle), and three other habitats within 1 km, are apparently needed to provide suitable habitat for the species. The model suggests a tradeoff between the two

Wildlife movement

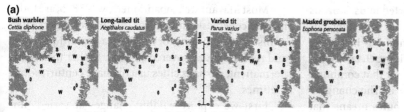

S=Summer breeding habitat patch. W=Winter habitat patch.

Considered as stepping stone:	≥ 25% woody vegetation in 15×15 m space	≥ 50% woody vegetation in 30×30 m space	Predicted movement in city
Bush warbler	200 m	250 m	Can reach some winter patches. Moves readily in shrubs & thickets as fine-scale stepping stones.
Long-tailed tit	–	300 m	Long movement distance, but does not reach winter patches. Small flocks use street trees & street-side wires. Fine-scale patterns used for movement.
Varied tit	100 m	(100 m)	Maximum movement distance inadequate to reach winter patches. Winter habitat in large parks & surrounding forest.
Masked grosbeak	–	250* m	Maximum movement distance sufficient to reach most winter patches. Relatively large birds in wide-ranging winter flocks.

Figure 9.11. Predicted maximum distance moved by songbirds between vegetation stepping-stones in city. Based on a model that compares four resident bird species, different amounts of woody vegetation in grid cells, different potential movement distances from a greenspace patch, and the spatial arrangement of summer and winter habitat. (a) Maps of Kyoto City, Japan (center) and surrounding forest (dark area) for each bird species. S = summer breeding habitat patches (also breeding in the forest); W = winter habitat patches; 0 = species absent. (b) – = not considered to be a suitable stepping stone; * = estimated 250–300 m. 1 m = 3.28 ft. Adapted from Hashimoto (2008).

variables, so if tree cover is greater, fewer other habitats are required.

Another model for four bird species in nearby Kyoto focuses more directly on movement, in this case between summer breeding patches and winter habitats in the city (Figure 9.11) (Hashimoto, 2008). First, woody vegetation is measured in different-sized circles around parks across the urban area. Then, presumably based on empirical observation of birds leaving parks in different directions, suitable stepping stones, and the maximum distance a bird moves between them, are estimated. The stepping stones, distances between them, and potential routes identified are then compared with the actual distribution of summer and winter patches. One species (bush warbler) is predicted to move 200 or 250 m from a point depending on which stepping-stone type is calculated to be present. A second species is expected to only move 100 m, and the other species 250–300 m from a point. The model-predicted distances were sufficient for two of the four species to reach their winter habitats. Although movement patterns were only partially predicted using diffuse-vegetation stepping stones, the results lead to alternative hypotheses for urban movement. The birds may be using finer-scale stepping-stone patterns, or alternatively, broader-scale regional cues, in moving across the city.

Other studies provide added insight into wildlife movement through cities. Species can be expected to reach a well-connected patch more than an isolated patch (Snep et al., 2006; Murgui, 2007; Vergnes et al., 2012). Colonization and extinction processes are related to urban forest cover (see equations, Appendix B) (Marzluff and Ewing, 2008). Termites stream from house to house, while cockroaches and rats run between buildings through connected stormwater drainage pipes. Red foxes in Britain apparently regularly move in and out of urban areas along railway corridors (Kolb, 1985). The broad-scale regional arrangement of land use affects migratory species such as birds, which preferentially fly along natural and farmland areas, even well-vegetated suburbs, rather than highly urbanized areas (Bonter et al., 2009; Loss et al., 2009).

The expansion of "containerized shipping" in recent decades probably alters wildlife, insect, and microbial movements in a big way (Levinson, 2006; Gehrt et al., 2009; Way and Eatough, 2011; Conor O'Shea, personal communication, 2013). Species move long distances in the containers on ships to docks, then on flat-bed trucks, and onward (perhaps via warehouses) to commercial and industrial centers. These destination centers are widely spread across the urban area, and are readily colonized by the transported species. Urban wildlife in turn are attracted to the containers and arriving species.

Bats add to the wildlife movement story. Most insect-eating species apparently move along connective elements in the landscape, such as tree lines, forest edges, lines of buildings, shrubby property boundaries, and banks of streams and rivers (van Bohemen, 2005).

In the city some bat species are reported to use dozens of roosts in trees or artificial structures to rest by day, and frequently move over time among 5–10 roosts, even when mothers have young (Figure 9.4a). A bat thus regularly traces a shifting network that combines roosts, flight paths, and foraging areas. The changing roosts and network of movement routes means that bats can rapidly respond to the ever-changing patches of prey abundance, such as mosquitoes, in the air.

Finally in North America, commercial areas, high-density residential housing, and areas of single-family homes are commonly kept relatively separate, often by zoning. That requires substantial transportation, time, and cost for people. An alternative is the traditional *mixed-use* juxtaposition of land uses of both urban and rural character present in some Asian cities (Chapter 11) (Yokohari et al., 2000). In parts of Kyoto and other Japanese cities small productive rice paddies are partially or completely surrounded by urban housing, plus streets for shopping. With mixed-use patterns including semi-natural vegetation, a relatively high biodiversity of wildlife would coexist with people in urban areas. In this way diverse resources are closer together so barriers to wildlife movement are normally less.

In brief, foraging for food in home ranges accounts for the bulk of urban wildlife movements. Due to the pattern of fragmented vegetation patches and a multitude of tiny vegetated spots, movement through built areas primarily uses stepping stones. Diffuse clusters or arrays of plants and tiny vegetated sites may in effect serve as stepping stones. Species sources and adjacencies seem to be especially important in determining broad patterns of urban wildlife movement.

Changing urban wildlife and adaptation

Visualize a city after the people suddenly left. In 1986, following a massive explosion at the Chernobyl nuclear power plant, the population of the nearby small city of Prypyat (Ukraine) was expelled from a 27 km radius (17 mi) area. Ecological succession immediately began, and species from the surroundings and long-distance arrivals poured into the abandoned city area (Forman, 2008; Adams and Lindsey, 2011). Moose (*Alces alces*), roe deer (*Capreolus capreolus*), wild boar (*Sus scrofa*), fox (*Vulpes vulpes*), river otter (*Lutra canadensis*) and rabbits (*Lepus europaeus*) are now observed in the exclusion zone, but apparently not in most of the land just beyond where people continue living.

Most Mayan cities, abandoned by AD 875, are today mainly covered with tropical rainforest. Desert cities of history and pre-history in Africa, the Middle East, and Asia are today covered by sand. Rather than being permanent, perhaps cities usually have centuries-long lifetimes.

First we explore wildlife changes at varied time scales, from days to centuries. Then we address the question of whether urban animals have altered their behavior or genetically adapted to the urban environment.

Multi-scale changes and succession

Imagine living near a greenspace with thousands of flying foxes. These large fruit bats, as in Melbourne (Australia), sleep hanging from the trees during the day and disappear foraging at night. The million and a half small bats roosting in a Texas bridge also dramatically change locations daily (Keeley and Tuttle, 1999).

Thousands of common pigeons in St. Mark's Square in Venice, or blackbirds roosting in a park, make equally dramatic daily fluctuations, but in this case they feed by day and sleep at night. However, these birds also change location seasonally in equally impressive numbers. In spring the Venice Square is often flooded, so the pigeon population mainly feeds elsewhere until the waters recede. The blackbirds use their park as a winter roost, then move elsewhere for the reproductive period in spring and summer.

However, wildlife changes occur over several time scales. These daily and seasonal fluctuations are cyclic. Normally non-cyclic changes occur over years, decades, centuries, and millennia. Few cities have lasted millennia and relatively little is known about changes in animals in a city over millennia or centuries. So we focus on wildlife changes over years and decades, which are also the typical time scales for ecological succession on an urban site.

Consider censusing birds of a park in Dortmund (Germany) eight times in 43 years (Abs and Bergen, 2008). A total of 33 species was recorded and the censuses varied modestly from 16 to 23 species, with a slight increase over time. Ten bird species remained in all recordings, which means that 23 birds (70% of the total) disappeared, appeared, or both during the four decades. Five species (15%) were only seen once, and two species twice. Five species from the first census disappeared at some point and were not seen again. Six birds were only present for a period in the middle.

Nine species appeared sometime in the middle years and remained to the end. The "turnover rate," i.e., the proportion of bird species from the first census that is present in the last census, was 42% for the whole period. Turnover rates between successive surveys (about 5–6 years apart) ranged widely, from 5% to 24% for individual species.

The population density [expressed as number of bird territories per 10 ha (25 acres)] also changed markedly over the four decades (Abs and Bergen, 2008). Over time the total number of territories varied from 98 to 180. Some of the variation in population density was due to fluctuations in territory number by the ten species always present. The species that disappeared, appeared, or both were almost always at low density. Thus, the number of these temporary species present also contributed to the variation in population density in the park. Note that species appearances and disappearances at this time scale could also be termed colonizations and extinctions (MacArthur and Wilson, 1967; Magle et al., 2010; Marzluff and Ewing, 2008). In effect, the trees in this urban park change very slowly over four decades, while the avian community is highly dynamic.

Shorebirds censused in spring five times in 50 years on tidal mud flats of an Osaka park show some similar patterns of variation (Natuhara, 2008). Species richness varied from 29 to 36, and 20 species remained in all censuses. These are mostly flocking birds in spring migration, and the total number of birds varied enormously, from 226 to 3148, a 14-fold difference. Large flocks of a single species were common, so in each census one species represented 24–46% of the total. Yet different species were most abundant in four of the five censuses. No significant increase or decrease over time for either species richness or abundance was evident. Thus, while overall community measures of richness and abundance often suggest modest change, individual species populations may be appearing and disappearing, as well as fluctuating wildly.

Variation is lower when groups of similar species are lumped. For example, annual winter bird-count data for 7 years in Edmonton (Canada) can be grouped as waterfowl, birds of prey, owls, woodpeckers, finch-like birds, and so forth (Wein, 2006). Total species richness and abundance did not change significantly over the 7-year period. Of the 15 species or species groups identified, only one showed a significant decline or rise (house sparrow declined). Indeed, jays and crows (Corvidae) remained amazingly constant over the 7 years (3075 to 3676 individuals).

Ecological succession in urban areas generally occurs over years and decades. Invertebrates on the soil surface of brick-rubble sites in London, such as many vacant lots and some soil-contaminated brownfields, illustrate certain patterns of animals in succession (Gilbert, 1991) (Figure 9.12). In sites of 0–1, 4–6, and 12–15 years age, species richness increased markedly from the first to second phase (15 to 36 invertebrate species/taxa), but only slightly increased from the second to third phases (36 to 40 species). In contrast, total abundance of animals markedly increased through all stages (ca. 103 to 335 to >785 individuals). Sixty percent of the species in the first stage remained to the end, while 40% of the species disappeared. Fifteen species colonized the youngest sites, 21 new species colonized the 4–6-year sites, and 14 new species colonized the oldest sites, indicating that colonization continues, but fluctuates, during succession. Of species in the first two stages, 71% increased in abundance, and 55% of those in the last two stages increased in abundance.

Changes in species composition also describe succession. At the beginning of this urban rubble succession, many money-spiders parachute in on their silk threads, to be followed by generalist winged beetles and moth/butterfly larvae (Gilbert, 1991) (Figure 9.12). By years 4–6 small accumulations of soil are present. These help attract a few species of woodlouse, carabid (predator) ground beetles, slugs, millipedes, centipedes, and moth/butterfly larvae. By the 12–15-year stage, the soil is more continuous and compact with a root mat. Large soil invertebrates such as earthworms and ants are abundant, along with a diversity of spiders. The predator-and-prey relationships through the successional sequence become more numerous, and the food web more complex. Clearly the soil invertebrates interact in many ways with the changing vegetation, and also vertebrate predators become more diverse and abundant over time.

"Anthropogenic climate change" seems to also mainly produce major effects over decades (even years) (IPCC, 2007). As urban heat gradually increases, warm-weather species may colonize urban areas previously inaccessible due to cold. One listing of climate change effects directly on urban wildlife (ignoring broad issues such as sea-level rise, habitat loss and degradation, spreading development, and so on) highlights the following (Wilby and Perry, 2006): spread of disease and pests; summer drought stress on wetland and woodland wildlife; altered nesting/denning time; invasive and naturalized species favored; less snow and longer frost-free periods. Some species may deal with

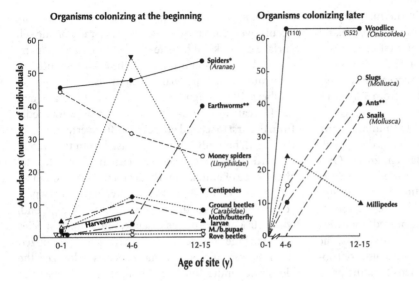

Figure 9.12. Invertebrates in ecological succession on brick rubble sites. Based on collecting and counting organisms in or on the soil surface of a 1-m² plot in each of four sites for each time period. * Excludes money spiders. ** Results for earthworms and ants given as very rare, rare, occasional, and common; thus numbers plotted are only relative for illustration. Note that particular species within an invertebrate group may colonize or disappear at different times. Sheffield, UK. Adapted from Gilbert (1991).

these changes by altering behavior, others by genetically adapting, and still others may disappear.

Finally, at a decades-to-centuries scale, comparisons of the reptiles and amphibians in Philadelphia and Washington, D.C. over time are revealing (Grant et al., 2011). In Philadelphia, records in 1789 and 1999 show reptile diversity dropping from 25 to 11, and amphibians from 16 to 10. In Washington from 1950 to 2007, reptile species richness decreased from 29 to 17, and amphibians from 20 to 13. These are rather consistent 40–50% declines. All individual groups decreased in diversity (lizards, turtles, snakes, frogs/toads, and salamanders), but the sharpest drops were for lizards and snakes.

Thus, over varied time periods, we can recognize a general sequence of species responses to change. Species richness seems to change the least (significantly increase or decrease). Total abundance of animals varies somewhat more. Groups of similar species may typically change still a bit more. Species composition seems to change considerably, even fluctuate. And the abundance or density of individual species populations tends to fluctuate the most. Short-lived animals, mainly small ones, often fluctuate markedly. Long-lived plants (e.g., some trees) may change little in diversity and density. Few long-lived animals such as turtles and eagles thrive in urban areas.

Genetics and adaptation

Change is the norm, constancy the surprise. Although something that remains constant warrants study, here we focus on the way wildlife changes. Almost everything characteristic of urban wildlife differs from that for non-urban animals. Do the urban animals simply change their behavior in the urban environment, or do they genetically adapt to it?

Behavioral adjustment or change (or even "adaptation") refers to an animal altering its activity or response when encountering different environmental conditions. A peregrine falcon (*Falco peregrinus*) changes from nesting on a rock cliff to nesting on a building or large bridge in the city. Such a species is *pre-adapted* for urban conditions, with genes suitable for tolerating or thriving in diverse environments, including the urban (the species has high "phenotypic plasticity"). Behavioral adjustment refers to an individual animal altering its behavior. Physiological change normally accompanies behavioral change (Evans, 2010; Brearley et al., 2012).

Some bats switch roosts from holes in trees to holes in buildings. Blackbirds sing higher-pitched songs in cities than in non-urban areas, presumably a response to low-pitched traffic and other urban noise (Nemeth and Brumm, 2009). In the Eastern USA, chickadees (*Parus*) in winter have adjusted their feeding behavior from seeds of conifer cones in forest to seeds on suburban bird feeders. Mockingbirds (*Mimus polyglottos*) changed from depending on fruits of many woody plants in rural areas to fruits of a new suburban rose species (*Rosa multiflora*). Instead of their natural diverse diet, coyotes (*Canis latrans*) moving into suburbs consume house cats, raccoons, and rats, while cougars (*Felis concolor*) along the edge of housing developments in California add domestic dogs and cats to their diet.

The primary alternative to behavioral adjustment is *genetic adaptation* to the urban environment (Parmesan, 2006; Evans, 2010; Marzluff, 2012). In this case, over generations the population becomes more fit, or better able to live in the new urban conditions. Animals, such as invertebrates, with shorter life cycles or generation times in older cities are more likely to have adapted. Since most cities are only decades or centuries old, urban genetic adaptation is effectively an evolutionary adjustment within a limited number of generations (see Chapter 8) (Ditchkoff et al., 2006). Evolutionary biologists increasingly recognize rapid evolutionary change, including its genetic adaptations, driven by natural selection.

Consider the 100 wild-tufted capuchin (white-faced) monkeys living in a single 158-ha (400-acre) woods in the city of Ribeirao Preto (near Sao Paulo, Brazil) (Amaral et al., 2005). Two relatively separate groups of 60 and 40 animals coexist in this greenspace. Genetic analysis indicates that the genetic diversity of the two groups only differs by 1.9%, indicating that they are of the same population. Also, despite being isolated in a single patch surrounded by urbanization where one would expect considerable inbreeding, the population has high genetic diversity, indicating adaptation to a wide range of environmental conditions. Interesting and surprising patterns appear for wildlife in an urban patch.

To test between the behavioral adjustment and genetic adaptation hypotheses, researchers studied urban and non-urban populations of the common European blackbird (*Turdus merula*) (Partecke et al., 2004; Partecke and Gwinner, 2007). The reproductive gonads develop in urban birds 3 weeks earlier than in non-urban forest birds. Researchers then reared birds from both populations in constant environmental conditions. If gonad development were the same time in both, it would suggest behavioral adjustment, whereas if urban birds developed gonads earlier, probably it is genetically determined. Gonads developed at the same time. The blackbirds were generalists with phenotypic plasticity, and had not genetically adapted for this trait.

At least 500 species of insects and mites of urban areas have developed insecticide resistance (Ehler and Frankie, 1978; Carpenter, 1983; Robinson, 1996). This is genetic adaptation by small many-generation animals. For instance, fleas have developed high resistance, and house flies (*Musca domestica*) and German cockroaches (*Blatella germanica*) very high resistance, to several modern insecticides commonly used in urban environments. Also termites and ants have increased insecticide resistance. These species are widely considered to be pests in urban areas and are extensively treated with insecticides. They are also generalists with wide environmental tolerance. Most invertebrate pests depend on suitable habitats in small isolated urban locations.

The house sparrow (*Passer domesticus*) has spread across North America in a century and a half (Johnston and Selander, 2008). A series of different color and size traits, in widely dispersed populations, together strongly suggest genetic differentiation and adaptation in this time period. Wing length of adult males is significantly greater in Edmonton (North-Central Canada) than in California cities (Figure 9.13). Bill length of adult males is significantly longer in Honolulu, and shorter in California, than it is in Mexican cities. The reflectance of feathers on female breasts at all visible wavelengths is significantly greater for California birds than for Mexico City and Honolulu birds. These geographic differences are similar to those for native species with readily identifiable types or races. The authors suggest that racial differentiation in this sparrow species may occur in only 50 years. The apparent genetic change in sparrow populations across the continent relates not to rural-to-urban differences, but probably to "genetic drift." The urban species was separated in different climatic/environmental conditions where successfully reproducing populations became genetically more distinct over time.

Also, the common pigeon (rock dove, *Columba livia*) differs in size and shape traits from the wild native rock dove (Gilbert, 1991). In the London population seven color varieties have been recognized. These could have arrived from other locations, or may have evolved on site over two millennia from a small Roman city to today's megacity.

The house mouse in Tunisia has two identifiable races or morphs based on chromosome differences (Chatti et al., 1999). One is all over the country, while the second is only in the oldest sections (medinas) of towns and cities, where the first one is absent. Although the environmental conditions differ markedly, population densities differ little. The overlap in spatial distribution of the two mouse races is commonly <500 m.

Amphibians also adapt to urban conditions. Within a large central park in Montreal, four populations of

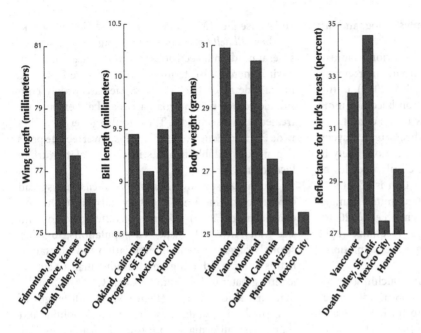

Figure 9.13. Adaptive variations in size and color of house sparrow *Passer domesticus* in widely dispersed cities. Histograms = average. Reflectance is for middle of the visible range (550 nanometers wavelength). Bird size attributes for adult males; reflectance for adult females. 1 mm = 0.0039 in; 1 gram = 0.035 ounce. Adapted from Johnston and Selander (2008).

eastern red-backed salamander (*Plethodon cinereus*) live in forest patches separated by roads, cemeteries, and buildings (Noel *et al.*, 2007). These urban salamanders were compared genetically with four populations out in non-urban continuous forest. Individual salamanders of the city park had lower genetic diversity than those of the non-urban forest. The fragmented urban-park populations were more genetically differentiated into subpopulations, whereas the continuous forest populations were genetically rather homogeneous. Thus, habitat fragmentation within a large urban park is apparently sufficient to cause genetic differences (probably related to genetic drift).

A somewhat similar study of a common urban frog, *Rana temporaria*, found genetic differentiation for different subpopulations in town ponds (averaging 2.3 km apart) to be greater than for subpopulations in much more isolated rural ponds (averaging 41 km apart) (Hitchings and Beebee, 1997). Although the town frog population was apparently not declining, evidence of inbreeding depression (and genetic drift) was present, and the genetic diversity of individual frogs was low. This suggests that, although no obvious major barrier separated town ponds, the frogs did not often cross between them.

Even red foxes (*Vulpes vulpes*) in urban areas may differ genetically from those in surrounding rural areas. Only 15 years before being studied, two separate groups of foxes colonized Zurich (Switzerland) (Wandeler *et al.*, 2003). Despite this origin, the urban boxes were found to have less genetic differentiation into subpopulations than the rural foxes. The species continues to move into the city, so there is considerable incoming gene flow. Also, the city fox population is expanding rapidly.

A review of genetic adaptations and urban wildlife (Evans, 2010) explores evidence for urban–rural and other differences in "species traits" (abundance, body size, communication, physiology, disease risk), and "demographic traits" (timing of reproduction, reproductive success, survival). For population genetic structure, worldwide evidence indicates the following significant effects:

1. Change in genetic diversity of urban populations: 1 moss; 3 vascular plants; 1 insect; 7 amphibians; 2 reptiles; 3 birds; 3 mammals.
2. Divergence of urban and rural populations: 1 moss; 2 vascular plants; 5 amphibians; 1 reptile; 3 birds; 2 mammals.
3. Divergence within urban populations: 1 moss; 2 vascular plants; 3 amphibians; 1 reptile; 2 mammals.
4. No significant differences reported for 2 vascular plants and 5 insects relative to #1; 1 reptile relative to #3.

Overall, the evidence for population genetic differences is still limited though growing, and includes all major wildlife groups.

A further review pinpoints 2 insects, 9 birds, and 1 mammal species that have genetically evolved in response to specific human activities within urban areas (Marzluff, 2012). The evolutionary processes genetically changing these species are selection (11 cases), plasticity (5), and drift (4) (some species have more than one process). The selective forces identified are primarily microclimate (including temperature), pollution, predation, new food, and low-frequency noise.

So, both behavioral adjustment and genetic adaptation are common for urban wildlife. In addition to adaptation through natural selection, other mechanisms of genetic change occur, including genetic drift, "inbreeding depression," and "hybridization." It is too early to know the relative importance or frequency of such changes, and under which conditions and which animals each mechanism would be prevalent. For instance, although the urban heat island effect is primarily at night, studies of possible adaptation by nocturnal species are scarce indeed (Parmesan, 2006; Angilletta *et al.*, 2007). Sometimes we have strong indirect evidence for genetic change, yet the expectation must remain a hypothesis. It also appears that decades of living in urban conditions is sufficient for genetic change in vertebrates, while invertebrates may often adapt in years.

Part III Urban features

Chapter 10 Human structures

The outcome of the cities will depend on the race between the automobile and the elevator, and anyone who bets on the elevator is crazy.

Frank Lloyd Wright, quoted on Public Broadcasting System, May 27, 1974

Rome did not look like the capital of a great empire … no broad avenues and few open spaces … Few streets were wide enough to allow vehicles to pass one another and most of them were unpaved. … to eliminate daytime traffic jams … the night clattered with the cacophony of wooden carts … The rich lived in houses with no outside windows … rooms were grouped around one or more open-air courtyards … Shops lined many of the main streets … Rome was a city of horrible smells. Rubbish and sewage, even occasional human corpses, were tipped into the street. Passersby were so often hit by the contents of chamber pots emptied from the second floor or the roof that laws were passed … City life was made bearable only by the ready availability of water. Four aqueducts … strode across the land, bringing fresh, clean water …

Anthony Everitt, The Life of Rome's First Emperor Augustus, 2006

Once again we explore the core of urban ecology. This time built structures, interacting with organisms and the physical environment, play the leading roles (Bartuska and Young, 1994). Concentrated urban roads and buildings represent a pinnacle of human engineering and architecture, indeed construction.

Hard straight lines, rectangles, grids, and other patterns close to Euclid's geometry predominate. Such patterns, providing many benefits to people packed together, contrast with more natural lands worldwide mainly displaying the soft curves of nature. Yet even the casual observer will notice and feel nature throughout the densest parts of city centers, commercial areas, and high-rise residential areas. Here the hard lines provide contrast to the irregular curves, or vice versa. Indeed, the two forms are well intertwined in urban areas.

Buildings, roads, and their surfaces come and go, while the city carries on. Construction makes a building; degradation, destruction or demolition creates rubble. Ribbon-making machines lay concrete and tarmac/asphalt streets and highways; traffic and weather then pound the ways to pieces, sending the ribbon machines out to unroll yet another smooth surface. Cracks and crevices endlessly appear in and around our structures. Water, soil particles, microbes, spores, seeds, and invertebrates, in an everlasting "rain," quickly find, fill, and widen these cracks.

The persistence or sustainability of structures effectively results from using four "principles of good design," described in Roman times by the architect Vitruvius for buildings. The structure: (1) fits well in the surroundings (i.e., has both positive effects on the surrounding area, and vice versa); (2) is aesthetic or inspirational; (3) is constructed well using good materials; and (4) works well for people. In densely built areas, roads and buildings are overwhelmingly designed by engineers and architects. In less-dense areas, most road construction involves engineers, while many buildings are in essence designed by contractors and residents, without architects. Analogously, landscape architects especially contribute to design in dense areas, and seldom in low-density areas.

Not surprisingly, numerous interactions link organisms, built structures, and the physical environment in densely built areas. Yet important frontiers for understanding remain. For example, how important is vegetation on walls (green walls) in affecting airborne particulate pollution, summer air temperature, and animal habitat around streets? Can green roofs not only improve water and temperature conditions, but also serve as an "archipelago" for habitat, foraging, and migrating species in an extensive built area? How important is the enormous diversity of species in ever-changing house plots (yards, gardens), which in turn reflect the diversity of residents who add, eliminate, and re-arrange their habitats and species? What is the relative importance of railways and highways as ecological barriers, connectors, and habitats around cities? Such tough intriguing questions abound in urban areas.

Most of the ecological dimensions of roads and buildings have been introduced or described in preceding chapters. This chapter pulls threads together to highlight the roads, buildings, and associated structures. Five broad topics are presented in order: (1) railways; (2) roads and associated features; (3) hard surfaces and cracks; (4) house plot, garden, lawn; and (5) buildings.

Railways

Railways and trains began between London and Manchester in 1825–30 (Lay, 1992), so plants and animals have had more than 18 decades to adjust or adapt to the distinctive environmental conditions around tracks. Railways for the transport of people and goods permeate urban areas, and interconnect to form remarkably different networks. In all cases, the central strip of the transport corridor is intensively used and maintained (Carpenter, 1994). Adjoining ditches or pipes accelerate water drainage away, and ditches may contain wetland-related species. The outer edges of most urban railways receive relatively little maintenance attention, and thus typically contain considerable and rather distinctive biodiversity.

Overall these "way" corridors are quite straight and narrow, often forming a trough through higher adjoining land usually of buildings and/or trees (Figure 10.1). Corridor connectivity is essentially complete; no gaps exist. Usually the five corridor functions – conduit, barrier/filter, source, sink, and habitat – are all ecologically important (Forman, 1995; van der Grift and Kuijsters, 1998).

Railway ecology is much less studied than road ecology (Forman, 1995; Spellerberg, 2002; Forman et al., 2003; Trocme et al., 2003; Davenport and Davenport, 2006). Nevertheless, the somewhat distinctive flora along railways, including urban ones, has been of considerable ecological interest (Thellung, 1905; Messenger, 1968; Muhlenbach, 1979; Gilbert, 1991).

In urban areas a rail corridor with two tracks is common (Figure 10.1). The steel rails are attached to ties (sleepers) set in a *rail bed* (cess) of gravel or cinders

Figure 10.1. Commuter train, railway, and gravel bed with essentially no plants. Railway bank (left) with soil erosion mainly related to vegetation cutting under powerline; (right) with little managed trees and shrubs. Outer suburb of Boston, near Henry David Thoreau's 1840s cabin (and periodic walking route home for dinner). R. Forman photo.

(ballast) (Gilbert, 1991; Carpenter, 1994; Wheater, 1999; Cederlunde et al., 2008). The highly porous bed is typically raised above the adjoining surfaces to minimize water accumulation. A cross-section of a typical urban "rail corridor" illustrates considerable habitat heterogeneity: (1) outer fence/wall/bank lined with woody plants; (2) open flattish strip; (3) ditch; (4) slope/embankment of the raised rail bed; (5) nutrient-poor top surface (about 3 m wide per track) of rail bed; and the same types on the other side. The rail-bed slopes facing opposite directions, such as on sunny and shady sides, tend to have different vegetation.

An abundance of distinctive structures along an urban railway, including posts, culverts, bridges, and stations, provides special habitats for diverse species. A wide "rail yard" (marshalling yard) for storing, servicing, and redirecting trains may contain rich biodiversity, especially of non-native species (Muhlenbach, 1979). Although herbiciding generally limits plant cover, a rail yard is best considered as a large greenspace rather than built space.

Trains for freight, commuting, and inter-city travel use the tracks, often relatively frequently. That means repeated loud short-duration noise, vibration that compacts adjoining soils, and brief strong airflows in the direction of train movement. Usually the top surface of the rail bed is frequently and intensively herbicided to minimize plant growth, and the herbicides also affect vegetation on the rail-bed slopes and beyond (Gilbert,

1991; DeSanto and Smith, 1993; Wheater, 1999). Trains also spread pollutants along rail corridors, including hydrocarbons, heavy metals, and coal dust, depending on the fuel used. Some trains spread grain and other seeds along urban tracks, and some emit wastewater from open-pipe toilets. Chemical spills happen.

Hydrology is especially critical for stable railway beds, so rail beds are of coarse material (DeSanto and Smith, 1993). Ditches alongside are designed to drain water out of and away from a rail bed. Wetland plant cover is usually considerably less than that along road ditches, due to the porous substrate, herbiciding, and shower of pollutants.

The outer edges of rail corridors vary considerably, but overall contain relatively high biodiversity since they escape most herbiciding and other maintenance activity (Tikka et al., 2000). Adjoining land owners often dump various solid-waste and yard-waste materials into the corridor edge, a process that adds a diversity of seeds, animals, and plants (ergasiophygophytes, for terminology lovers). But also adjacent properties have a rich variety of plants and animals that colonize or use the rail-corridor edge.

The diverse structures along the railway are effectively different microhabitats. For example, concrete posts may contain crustose lichens, and stations may be special sites for certain ferns (Gilbert, 1991). Rail sidings and unused track areas with shrubs and small trees are often nest sites for birds (Laurie, 1979).

Railway corridors are distinctly polluted (Stengel et al., 2006). Diesel engines, as well as coal-burning ones in parts of China, Eastern Europe, and elsewhere, emit significant amounts of CO, CO_2, NOx, SOx and particulates. On a per-kilometer or per-train (per-km) basis, pollution rates are very high. Yet on a per-person or per-ton (per km) basis, emission rates are very low. Electric engines, particularly in urban areas, mean that the combustion pollutants occur and produce effects near power facilities rather than along the railway.

However, heavy metals come from wear and corrosion. Toxic wood preservatives spread from ties, poles, and other wooden structures. Oils and lubricants are used in many railway operations. Herbicides, which vary in persistence and toxicity, are generally used in high amounts. For example, rail beds represent the largest use of herbicides in Germany (Muhlenbach, 1979; Torstensson et al., 2005; Stengel et al., 2006). Soil microbial activity seems to be especially low. This concentration of pollutants contaminates the soil, as well as local water flowing through it. Thus, a clean-water supply or rare-species habitat near a railway may be at risk.

Unlike highway traffic noise, train noise is louder, briefer, and less frequent. Effectively noise levels in decibels increase exponentially with speed (Pronello, 2003). Apparently at low speeds (e.g., <50 km/h) train noise is mainly produced by the acceleration and deceleration of diesel engines, and by rolling wheels on rails with a rough worn surface. Rolling on rails is the primary noise source at medium speed (ca. 50–275 km/h), and aerodynamic noise at high speed outside cities. Electric trains are much quieter than diesel and steam engines, and are more common in urban areas.

Urban wildlife seem to adjust behaviorally to train noise, though this is little studied. Frequent commuter trains might have noise effects analogous to the chronic traffic noise from busy highways. A 61 m (200 ft) width of forest adjoining a rail corridor is reported to only reduce noise levels from 79 to 73 decibels (dBA) (Marsh, 2005). Very little noise reduction was recorded at low frequencies [<1000 Hz (hertz)], whereas high-frequency noise (4000–8000 Hz) dropped by 18–33%. The leaf litter layer of the forest may be important in decreasing noise transmission.

The long-distance dimension of railways may also affect vegetation and plants. Thus, many railways are connected to ports and to warehouses, both being important sources or entry-points for non-native species (see Figure 3.2). Also, many industrial areas connected by rail contain pollutant-resistant species that may thrive along rail corridors. Trains carrying grain and other seeds often spread some seeds along tracks, and the frequent stopping and starting of trains in urban areas shakes seeds loose from the train cars (wagons/carriages). The Sudgelande Park in eastern Berlin has a rich and unique assemblage of plants largely originating in this way from across the former Soviet Union (Kowarik and Langer, 2005). The uninterrupted open central portion of a rail corridor means that turbulent and vortex airflows readily lift seeds and spores from surfaces. Furthermore, strong streamline airflows, plus the movement of trains, readily transport wind-borne seeds along the rail corridor.

Early successional habitats predominate in rail corridors, basically for train safety. Thus, maintenance activities minimize the presence of plants and soil atop the rail bed, limit woody plants and soil on the rail-bed slopes, and keep woody plants low in the ditch and adjoining open portions of the corridor (Tikka et al., 2000). Trees mainly survive in the outer corridor edges and on abandoned sidings. Thus, high microhabitat diversity describes both the normal cross-section and the lengthwise dimension of a railway corridor (Muhlenbach, 1979; Gilbert, 1991; Cilliers and Bredenkamp, 1998).

Most rail-corridor plants are tolerant of low-nutrient soils (Muhlenbach, 1979). A particularly thorough study of plants of rail corridors and rail-yards in St. Louis (USA) found hundreds of species, of which 393 were spontaneous (adventive, unplanted). Many spontaneous plants were non-native species and considered to be naturalized rather than invasive, since they had apparently reproduced and spread successfully in the railway habitats over many years. Annuals represented 65% of the total railway flora, perennial herbaceous plants 27%, and woody plants 8%. Other studies confirm the relatively high proportion of non-native species in rail corridors (Gilbert, 1991; Hansen and Clevenger, 2005; Williams *et al.*, 2005). Some such species apparently spread into adjoining land, though this is little studied in urban areas.

Wildlife also benefit from urban rail corridors, both as scarce greenspace and for habitat connectivity. The microhabitat heterogeneity is an attractant. For example, the open sun may be used for basking by reptiles, train-killed animals (carrion) attract scavengers, hedgerow wildlife often use the outer corridor edges, the rail-bed slopes attract tunneling animals, and droppings of grain and seed along the track attract sparrows, small mammals, and other species (Edgar van der Grift, personal communication, 2009). Beetles, butterflies, and other invertebrates may thrive, especially if herbicides are limited to the upper surface of the rail bed (Croxton *et al.*, 2005; Saarinen *et al.*, 2005; Small *et al.*, 2006).

Animal mortality due to trains may be of limited importance in urban areas. Certainly urban birds and mammals are hit by trains (van der Grift, 1999; Spellerberg, 2002), but most such species reproduce much faster than trains can kill animals. Mortality may result from a train's high speed and quiet approach, as well as the long interval since the preceding train (Seiler and Helldin, 2006). Removing trees and shrubs may be more effective in reducing train-kills than is the case for roadkills (Jaren *et al.*, 1991).

The "conduit function" for movement of species along an urban rail corridor appears to be relatively important ecologically. Railways slice through a metro area, typically connecting city center with agricultural and natural lands outside (Laurie, 1979), thus being a potential important connector for native species enriching the city. Slight evidence suggests that moose (*Alces*) enter Boston suburbs and wild boar (*Sus*) enter Berlin in this way. Within the metro area, rail corridors are often connected to greenspaces, hence providing valuable connectivity for wildlife movement.

Trains themselves carry and spread seeds, spores, and various animals (Wheater, 1999). The straightness of railways favors the movement of seeds, spores, and certain invertebrates by wind. Grassland and maritime plants have spread along railways (Gilbert, 1991; Tikka *et al.*, 2001). Coyotes (*Canis latrans*) and foxes (*Vulpes vulpes*) have moved along railways, the latter perhaps in dispersal (Trewhella and Harris, 1990; Way and Eatough, 2006). Other mammals and various reptiles/amphibians move along rail corridors (van der Grift, 1999). Many additional species are observed in the urban corridor, but have not been studied for movement patterns (Huijser and Clevenger, 2006). The apparently important ecological conduit function of rail corridors contrasts with the rather limited wildlife movement along urban roads.

The barrier or filter effect on urban wildlife attempting to move across a railway corridor relates both to physical and behavioral factors. The physical structure of an open area, ditches, slopes, rail-bed top, and rails represents concentrated habitat diversity, and doubtless deters many species (van der Grift, 1999; Takehiko *et al.*, 2005). But also, the train noise, light, pollution, and frequency, plus human maintenance activities, behaviorally inhibit some species.

Songbirds in Calgary (Canada) crossed a railway more readily than they did either a road or a river (Tremblay and St. Clair, 2009). However, bees in the Boston Region rarely crossed a railway corridor, even where a patch of flowering shrubs used for feeding was on opposite sides (Bhattacharya *et al.*, 2003). Bees mainly crossed when nectar on one side apparently became scarce. Many non-urban studies have shown that wildlife readily move through culverts, tunnels, and overpasses in crossing railways, and what the key design and ecological factors are for successful crossing (Hunt *et al.*, 1987; DeSanto and Smith, 1993; Yanes *et al.*, 1995; Rodriguez *et al.*, 1996; van der Grift, 1999; Clevenger *et al.*, 2001; Iuell *et al.*, 2003).

In urban areas, the surrounding land usually contrasts markedly with the rail corridor, which in turn is an attractant or a deterrent for wildlife. Also, adjoining land uses change markedly and endlessly along a rail corridor, which is much more homogeneous.

Since rail corridors are relatively straight, adjacent wetlands, ponds, lakes, rivers, and streams are frequently sliced or edged by the railway, with consequent hydrological and pollution effects. The intensity of corridor management also varies spatially and temporally.

The extremely species-diverse Sudgelande Nature Park in Berlin, as mentioned above, effectively represents a former rail-yard. Elsewhere sidings and disused railroads covered with shrubs and small trees support many animals including nesting birds (Laurie, 1979). Where whole rail lines have been abandoned, some have then been converted to walking and/or biking trails (Ryan and Winterich, 1993; Searns, 1995; Poague et al., 2000; Flink et al., 2001). Wildlife seem to readily cross such trails (Poague et al., 2000), though wildlife movement along urban trails heavily used by people and often dogs may be quite limited. Wildlife found along the trails may be mostly resident species, with considerable movement between trail and surrounding lands (Searns, 1995).

Two examples of disused elevated railways, having been converted to walkways with ample plantings, are wonderfully evocative: the Promenade Plantée in Paris and High Line in New York City (Kellert et al., 2008). The plants seem to be mostly horticultural varieties that are rather little-used by most animals. Still, the two green strips in the urban sky would be interesting to evaluate for their local, as well as city-wide, wildlife and biodiversity value. Elevated strips in cities could dovetail recreation and aesthetics into ecological masterpieces.

Roads and associated features

We first introduce the varied types of urban/suburban roads and their ecological features. Second, street and roadside trees plus associated plants and animals are the focus. Third, we consider the objects moving, from vehicles to pedestrians. Finally, a diverse array of objects associated with urban roads is briefly considered. More detailed analyses of the ecological components are presented in the preceding chapters on urban air, plants, wildlife, and so forth.

Highways, roads, streets, networks

Road ecology, elucidating the interactions of roads and vehicles with soil, air, water, plants, and animals, is a small relatively distinct field and still rapidly growing (Figure 10.2) (Forman et al., 2003; Iuell et al., 2003; Trocme et al., 2003; Forman, 2004a; National Research Council, 2005a; Davenport and Davenport, 2006). Non-urban roads have received emphasis, though some studies focus on urban roads.

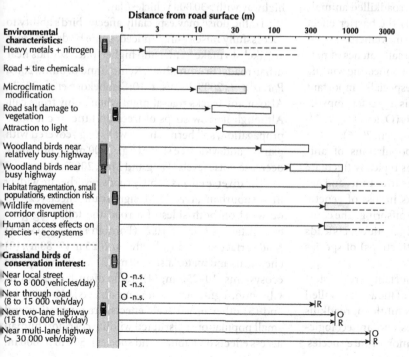

Figure 10.2. Distances of ecological effects from roads with vehicles. Horizontal box or short vertical line indicates approximate range of average and maximum distances from a road that significant ecological effects have been recorded. Upper portion adapted from Forman et al. (2003). Lower portion on grassland birds from Forman et al. (2002), Forman (2004). O = occasional presence of birds; R = regular breeding of birds (5yr period); n.s. = not significant; lm = 3.3 ft.

Urban highways

Multi-lane limited-access highways (e.g., expressway, motorway, autoroute, freeway) mostly radiate outward from a city, though ring highways are prominent around most large cities (Forman, 2008). Portions of these highways may be elevated to avoid disrupting local urban traffic, to permit floodwaters to pass harmlessly on a floodplain, or to protect valuable farmland (e.g., Changsha ring road, Hunan Province, China).

However, most of the length of a multi-lane highway with considerable traffic is a barrier to wildlife crossing and a major source of traffic noise. Such highways also contribute to animal roadkill totals, stormwater flows, stormwater pollutants, greenhouse gas emissions, and unhealthful air pollution. See Chapters 4 to 7 for water and chemical aspects, and Chapters 8 and 9 for vegetation and wildlife dimensions.

Roads

Two-lane highways and other busy roads are prominent features in suburban and exurban/peri-urban areas. The abundance of such roads with relatively continuous traffic, combined with the presence of semi-natural patches and corridors of varied size in these areas, means that busy two-lane roads have diverse ecological effects. Together, four effects of a road reduce wildlife populations (Rajvanshi et al., 2001; Forman et al., 2003; Laurance et al., 2009): (1) habitat loss due to the road; (2) degraded adjoining habitat; (3) roadkilled animals (Seiler and Helldin, 2006); and (4) the barrier effect disrupting connectivity for wildlife movement.

Roads fragment the land into small patches of natural land and farmland, and cut the connecting wildlife corridors. Road width seems to be especially important for small animals, whereas traffic is a greater impediment to mid-size and larger animals (Oxley et al., 1974; Forman et al., 2003; Reijnen and Foppen, 2006).

Small habitats mean small populations of animals and plants. Small populations typically fluctuate demographically in population size over time, and are subject to inbreeding. That results in loss of genetic variation plus more weak or sterile offspring. The combination of these demographic and genetic effects leads to more disappearances (local extinctions) of species from the fragmented habitats.

Overall, roadkills or animal mortality are greatest on two-lane highways. While most of the animals killed can reproduce faster than vehicles hit them, roadkills may reduce local population sizes and are ecologically significant for certain uncommon or rare species. Slowly reproducing predator species such as some reptiles and large mammals, plus some amphibians, seem to be most at risk to urban roadkill.

A rich assortment of mitigation structures is used worldwide to facilitate the movement of animals across roads (Forman et al., 2003; Iuell et al., 2003; Trocme et al., 2003). Impressive *wildlife overpasses* and *underpasses* are often used in Europe and are present in Canada, USA, and elsewhere. Tunnels, wildlife culverts, pipes, overhead poles, rope ladders, and engineered structures are designed for, and used by, varied animal species. Where avoiding an ecological impact appears to be impossible, "mitigation" minimizes the effect. Where mitigation seems impossible, "compensation" to provide the equivalent ecological benefit is appropriate (Cuperus et al., 2002).

Traffic disturbance, especially noise, inhibits certain vertebrate species for considerable distances from a road (Reijnen and Foppen, 1995, 2006; Forman et al., 2002). Most traffic noise is at a low frequency, and mainly results from the tire and road surface interaction, plus truck traffic. Sensitive forest and grassland birds are absent or scarce within a few hundred meters of a road with some 6000 vehicles per day passing a point (Figure 10.2). This effect extends outward for several hundred meters on both sides of a road (e.g., two-lane highway) with about 12 000 vehicles/day. Typically the traffic disturbance effect extends outward >1 km (0.6 mi) for a multi-lane highway with >30 000 vehicles/day.

Traffic noise may especially affect a bird's ability to successfully raise young. Indeed, some birds and frogs are likely to make louder and higher-pitch sounds near urban roads (Brumm, 2004; Nemeth and Brumm, 2009; Parris et al., 2009; Evans, 2010; Slabbekorn et al., 2007). Also, many trucks roar at night when many frogs call. Although narrow strips of trees do little to cut noise propagation, soil berms, noise walls (e.g., covered with plants), and wide tree strips (e.g., >200 m) significantly decrease noise levels (Fang and Ling, 2005).

The diverse effects of busy roads reverberate widely in a suburban area. Most significant effects extend outward on both sides of a road the following general distances (Figure 10.2) (Forman et al., 2003): (1) road surface and roadside effects, 2–25 m (7–83 ft); (2) chemicals and materials, 5–50 m; (3) water and aquatic ecosystems, 10–250 m; (4) traffic disturbance (noise, vibration, light, visible moving objects), 20–1000 m; and (5) other broad-scale effects (fragmentation and small populations; disrupted wildlife corridor; human access effects on habitats and species), 500 to >1000 m.

Suburban roads also cause diverse hydrologic effects, including changing the water-table, blocking flows, redirecting flows, increasing and decreasing wetland sizes, and accentuating floods. Stormwater pollutants from the roads may contaminate groundwater and pollute local water bodies. Road salts for de-icing cause corrosion and pollute local water supplies (US Geological Survey, 2010). But sodium chloride and other salts also degrade soils, water quality, aquatic ecosystems, amphibian populations, trees, and much more (National Research Council, 1970; Shortle and Rich, 1970; Hofstra and Hall, 1971; Hofstra and Smith, 1984; Langton, 1989; Trombulak and Frissell, 2000; Brownlee and Lorna, 2005; Domenico and Hecnar, 2006).

Streets

City and suburb roads permeating built areas include alleys, lanes, streets, avenues, and boulevards (Rowe, 1991; Jacobs, 1993; Watson *et al.*, 2003). The primary difference among them is width. With width comes the number or prevalence of traffic lanes, sidewalks, street trees, and vehicular traffic. Width, the objects along streets, and the height of adjoining buildings (forming a street canyon; see Chapter 5) affect heat, dust, airflows, stormwater pollution, tree growth, animal use, and other ecological dimensions.

A quarter to a third of the surface of most US cities is street. Stormwater is normally channeled along gutters into storm drains. From these it flows through a network of progressively larger pipes typically to a local water body (see Chapter 6). Thus, the street network effectively functions as the "headwaters" of a large downslope pipe system, with rushing stormwater.

Lots of stormwater-related features of streets affect water quantity, water quality, and aquatic ecosystems in local water bodies (US Environmental Protection Agency, 1999). These include the use of street trees, ditches/trenches, swales, vegetation strips, pervious and impervious hard surfaces, curbs (kerbs) and curb cuts, conveyance structures, detention basins, retention basins, wetlands, and biofilters. Overall such features increase evapo-transpiration, water infiltration, absorption, pollutant filtering, and microbial decomposition of pollutants. The street features also decrease water runoff, peak flows, flooding, pollutant runoff, inputs to and pollution of local water bodies, and degradation of aquatic ecosystems and fish populations. Seattle, Portland (Oregon, USA), and various Northern Europe metro areas provide many models of stormwater management that emphasize ecological solutions (Beatley, 2000a; Hill, 2009; Stephanie Hurley, personal communication, 2011).

Street dust is a combination of particles from the atmosphere, adjoining buildings, vehicle wear (including engine, brakes, and tires), and road-surface wear by vehicle use. An analysis of street dust in a small city (Urbana, Illinois, USA) found 35 chemical elements, with the highest concentrations (in order) being: manganese, zinc, barium, nickel, strontium, chromium, and zirconium (Spirn, 1984). The lowest concentrations were for silver, lutetium, lead, and mercury. The movement of vehicles lifts dust off the road surface some meters high, where it is readily inhaled by wildlife and people, and deposited on plants. In Kuala Lumpur, street dust levels were found to be highest around street intersections during vehicle rush-hour periods (Carpenter, 1983).

Summer heat can be fierce along city streets (see Chapter 5). At an intersection in a north-south-east-west street grid, summer heat may be excessive at noon and in the afternoon (Figure 10.3). On the other hand, in a diagonal northeast-southwest-northwest-southeast grid, some shade is present along most sidewalks at those times. These shade and heat patterns provide the basis for optimal street-tree locations (Figure 10.3). Plantings can be arranged to enhance warmth in winter and cooling in summer.

In Vienna, air temperatures in a narrow street, wide avenue, and wide-open square or plaza without trees illustrate patterns of streets and urban heat (Federer, 1971). On a summer afternoon the hottest location was the wide-open square, while the narrow street was coolest. But at night the wide avenue with street trees was the coolest site. In the hotter drier Mediterranean, many design elements of streets, including balcony vegetation, overhanging facades, arcades, street trees, narrow street canyons, and tall adjoining buildings, cool the air somewhat. These features also significantly improve conditions for organisms such as trees and people (Ali-Toudert and Mayer, 2007). The longer a hard surface is exposed to direct sun ("sky view" relative to solar trajectory), as on equator-facing walls, the greater the heat stress is on organisms.

Green walls, as described in a section below, improve street conditions in several ways. These walls may be covered by varied combinations of vines, window-box plants, green roof plants hanging down, balcony plants, outside-stairwell plants, and epiphytes (Beatley, 2000a; Dunnett and Kingsbury, 2004). Biodiversity is higher,

Human structures

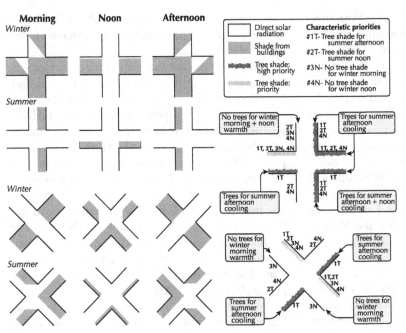

Figure 10.3. Shade and trees in street intersections of regular and diagonal grids. Diagrams assume approximately 15-m-high (50 ft) buildings and 18-m-wide (60 ft) roads. Regular north-south-east-west grids in most cities; diagonal (Cerda or Spanish) grids in Barcelona, Lima (see Figure 6.6a), and other cities (Pallares-Barbera, 2005). On left, gray = street surface shaded by buildings in mid-latitude city. On right, high to low priorities (#1 to #4) are listed by each street side. Adapted from Craul (1999) citing Knowles (1981).

stormwater runoff lower, air temperature lower, and aerial-particle pollution lower. The wall is cooled by shading, insulated, and shielded from ultraviolet radiation by vegetation. The plants also reduce noise, filter air pollutants (especially street dust), and humidify the air. Still, the benefit levels are poorly known in this obvious research frontier.

Traditionally streets have been "full" of domestic animals and animal waste, with associated urban wildlife (Robinson, 1996). Pigs, chickens, and ducks were raised in urban streets. Dogs and cats roamed widely. Until about 1925, horses provided horsepower to move almost all goods and people. With ample seeds in the horse manure layer, house sparrow and starling populations were dense.

Finally, the biodiversity of most street canyons is very low, so even small hotspots of species richness provide a noticeable benefit (see Chapters 8 and 9, plus the roadside section below). Providing microhabitats where species live, plus connectivity with rows of street trees or stepping stones of tiny green cover, enhances streets.

Road networks

Diverse road-related networks vary from city-center grids of varied form to hierarchical suburban networks to the multi-lane-highway networks of urban regions. Network forms are exceedingly diverse (see Figure 6.6)

(Easterling, 1993; Watson et al., 2003; Marshall, 2005b; Pacione, 2005; American Planning Association, 2006). Mapping by road width portrays the hierarchy characteristic of large and small roads in most road networks. However, mapping the flows of traffic emphasizes the functional importance of the road hierarchy in the metro area (Carpenter, 1983; Forman et al., 2002).

Basic or widespread network attributes highlight orientation (relative to north–south), road hierarchy (different road widths or traffic levels), curvy or rectilinear grid, and mesh or grain size (Figure 10.4). Also, special network features include: the relative abundance and arrangement of greenspaces; cul-de-sacs/spur roads (with no outlet); loops (circuits); grid of squares/rectangles/elongated blocks; triangles; diagonals; traffic circles (roundabouts); radial roads; ring roads; dendritic (tree-like) patterns; and discontinuous roads. Each of these attributes has ecological implications on microclimate, air quality, stormwater runoff, stormwater pollutants, vegetation type, habitat diversity, plant species richness, animal species richness, wildlife movement, and aquatic ecosystems (including fish) in nearby water bodies. Both "road density," the total length of road per unit area (see equations, Appendix B), and "network form" strongly affect ecological conditions.

Three simple measures or indices of network pattern provide considerable insight for objects that

Roads and associated features

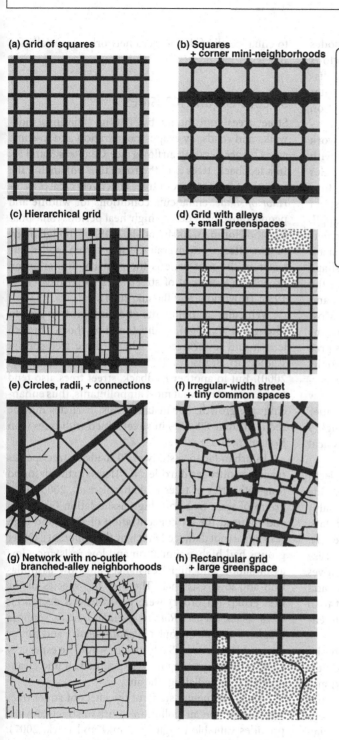

Figure 10.4. Eight alternative forms of street networks. Illustrations from (a) Portland (Oregon, USA), (b) Barcelona, (c) Tokyo, (d) Savannah (Georgia, USA), (e) Paris, (f) Vienna, (g) Cairo, (h) New York. Based on Easterling (1993), Jacobs (1993), Ebrey (1996), Siksna (1997), Panerai et al. (1999), Bianca (2000), Watson et al. (2003), Marshall (2005b), American Planning Association (2006), Maclean and Campoli (2007), Bosselman (2008), and other sources.

move along or alongside roads, including vehicles, pedestrians, and wildlife. Networks composed only of linkages (roads) and nodes (intersections and ends) (Figure 10.4a) are described by connectivity, circuitry, and node connections, using the gamma, alpha, and beta network indices, respectively (see equations, Appendix B) (Forman, 1995). *Connectivity* of a network is the number of linkages (connections, edges) divided by the maximum possible number of linkages. *Circuitry* is the number of loops (circuits, i.e., alternative routes between nodes) divided by the maximum possible number of loops. *Node connection* is the average number of linkages per node. Where nodes differ in size, the "gravity model" (based on Newton's physics) is useful. This model emphasizes that distance between nodes can be expected to have a greater effect than node sizes on the movement or interaction between nodes.

New cities often display regular road-network geometries such as a near-perfect grid, circles with radiating lines, and concentric ring roads. Old cities typically retain geometries related to streams, topography, and solar angle. Consider the early street network of Guangzhou in South China (Ji Kangil, personal communication, 2013). The irregular network apparently contained a hierarchy of street sizes, many spur (no-outlet) streets, varied size-and-shape blocks, and small fine-scale grids embedded in the broad pattern. Wind, seeds, water, animals, and people moved along all streets of the network. Most main streets were perpendicular to active streams lined with tree vegetation. The stream corridors provided cooling, clean area, and a magnet for people, while the street/stream arrangement provided effective stormwater drainage. Pedestrians used narrow shaded alleys, while wagon vehicles passed in wider tree-lined roads. Devoid of monotony, this highly functional network was adapted to both the physical environment and human uses.

Eight forms or types of street networks seem to represent the range present in cities worldwide (Figure 10.4). To ecologically compare these, I evaluated each case for 12 variables related to habitats, plants, animals, soil, air, and water (e.g., see Table 7.1). Negative roles or functions predominated in one case (Figure 10.4a). Positives strongly outweighed negatives in four street network types (d, f, g and h). Water and air variables were considerably more important than habitat/plant/animal variables in ecologically differentiating the street networks.

In considering the form and optimal design, basically road networks are important for four quite-different flows: vehicles; pedestrians/bicyclers; water; and species. These important flows vary both spatially and temporally. For instance, commuter traffic is concentrated at times different than movement of pedestrian shoppers (Berke et al., 2006). Water flows reflect the basic pre-urban topography, as well as rainfall patterns and stormwater-drainage systems (Marsh, 2005). Wildlife movement, both by day and at night when fewer people are around, generally requires some connectivity of vegetation. Creative thinking is needed to enhance all of the flows in networks, especially for species.

Street trees and roadsides

Street trees cool the air by shading buildings, sidewalks and roads, by evapo-transpiration and in some cases by accelerating airflows (see Chapters 5 and 8). In a feedback, trees cool the road, thus enhancing the growth of trees. A row of trees effectively creates a *corridor of shade*, enhancing conditions for wildlife and people movement during high-heat periods (Shashua-Bar and Hoffman, 2004; Marsh, 2005). Although many factors affect trees, overall the stress from inadequate water in the root zone is the primary limitation on growth and survival of street trees (Bradshaw et al., 1995; Trowbridge and Bassuk, 2004).

Vehicular air pollutants often inhibit tree growth (Stengel et al., 2006), as illustrated by shorter branch length and lower seed weight in three tree species studied in Bratislava (Slovakia) (Iqbal and Shafiq, 2000). But in another feedback, street trees also tend to decrease somewhat most air pollutants, thus enhancing tree growth. For instance, dust particles may be reduced some 80–85% in streets lined with trees (van Bohemen, 2005).

In urban Japan, a study of 70 roadside tree species relative to nitrogen dioxide (NO_2) levels in the air found four categories of trees (Takahashi et al., 2005). One group of trees had both high assimilation (absorbed and reduced the air concentration of NO_2) and high resistance (not damaged much by the pollutant). Other groups had high assimilation and low resistance; low assimilation and high resistance; and low assimilation and low resistance. The four top species of the first group (in order) were broad-leaved deciduous trees: black locust (*Robinia pseudo-acacia*); sophora (*Sophora japonica*); poplar (*Populus nigra*); and cherry (*Prunus lannesiana*). Such species would be especially promising for growing in NO_2-polluted air and cleaning (phytoremediating) the air.

An experimental study of the effect of street trees and vehicles on air pollution in urban street canyons provides valuable insight (Gromke and Ruck, 2007). Small tree crowns had little effect on pollutant concentration. For large tree crowns, air pollutant concentration was lowered on the windward side, and elevated on the leeward side. Increased spacing between trees resulted in lower pollution concentration. While stationary vehicles with engines running increased the concentration, moving traffic decreased and tended to homogenize the pollution level along the street. Traffic jams are bad for air quality along streets.

Urban roadsides often have rather continuous strips of chronically disturbed herbaceous vegetation. Disturbance-tolerant plants, weeds, non-native species, small mammals, amphibians, butterflies, and many other species may thrive, mostly because of local habitat conditions (Forman et al., 2003; Huijser and Clevenger, 2006; Harper-Lore et al., 2007). In The Netherlands, 30 butterfly species are found in multi-lane highway roadsides (motorway verges) (van Bohemen, 2005). About half are roadside residents, and 20% are rare (red-listed) species. In general, few animal species move effectively along roads or roadsides, though urban studies are scarce.

The fauna present is highly sensitive to the tree species present in streets and along roads (Fernandez-Juricic, 2000a; Young et al., 2007). Nectar-feeding birds predominate on streets mainly lined by eucalypts (*Eucalyptus*) and jacaranda (*Jacaranda*) in subtropical Brisbane (Australia) (see Figure 9.5). In the small city of Cheyenne (Wyoming, USA), small street trees are little used by any birds (see Figure 9.8b). Several species of mid-height trees are used by many common bird species. Tall canopy-height street trees are also well-used by many bird species, which seem to be more tree-species specific. Thus, two bird species almost only use cottonwood (*Populus*), while another bird essentially avoids the tree. A fourth bird species commonly uses five canopy-level tree species. Uncommon bird species mainly used the mid-level and tall trees. In effect, the species of street trees have a major effect on avian composition and diversity in a neighborhood or city.

Moving objects

Cars and trucks, the primary vehicle users of roads, emit lots of pollutants including noise, particulate matter,

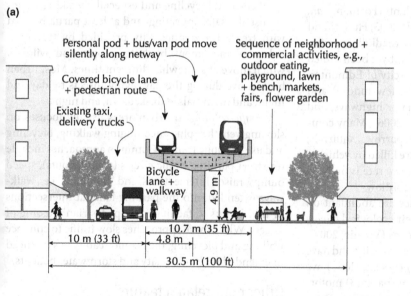

Figure 10.5. Netway-and-pod transportation system in city and suburb. The elevated way is composed of "wings" and "spanners" on pillars about 30 m (100 ft) apart. Lightweight personal pods, public bus/van pods, and freight pods are silently transported, using electric power from a wire buried in the netway surface, and with automated controls. Some pods can leave the no-driving netway system, and be driven at ground level using battery power. (a) city; (b) suburb. Dimensions and characteristics given are illustrative and can be tailored to specific needs, such as only having elevated bus/van pods over selected streets in the city. Adapted from Forman and Sperling (2011).

CO, CO_2, SO_2 and NOx (see Chapter 5) (Forman et al., 2003). The world has over a billion cars (a quarter billion in the USA), and a large portion moves on roads in urban areas (Sperling and Gordon, 2009). In the main cities of Upper New York State (based on 80+ km diameter "commuter sheds"), car commuters are major pollution sources (Hartshorn, 1992). Trucks are the other major transportation pollution source, particularly for particulates and noise.

Plants and animals are transported along roads by vehicles. Thus, seeds of 204 species, half non-native, were vehicle transported along a radial Berlin highway (von der Lippe and Kowarik, 2007). Interestingly, more species moved outward from the city than inward toward the city. In Canberra, seeds germinated from a carwash indicated that 85 plant species came from the city and surrounding cropland, 60 came from the natural-vegetation land of the surrounding region, and 20 species came from afar (probably >150 km) (Wace, 1977; Forman et al., 2003).

Roadkills (animal mortality) by vehicles are also commonly observed by urban residents (Forman et al., 2003; Iuell et al., 2003; Trocme et al., 2003; Huijser and Clevenger, 2006). However, data for roadkills specifically in urban areas apparently has not yet been pulled together to discern patterns. In the city of Edmonton (Canada), the rate of deer (and a few moose) and vehicle collisions is 59/year on major highways, and 44/year on minor highways (Wein, 2006). Many common urban animals such as house sparrows, squirrels, rabbits, house cats, and raccoons are killed by vehicles, but their abundance and reproductive rate is high, so overall the ecological effect is probably low.

Buses, trollies, and motorcycles are abundant on roads in many metro areas, but their ecological effects are yet to be evaluated. In urban areas (Vuchic, 2007): (1) buses are slow, have large people-capacity, and have very large space requirements; (2) cars are slow, have medium capacity, and take medium space; (3) motorcycles are fast, low-capacity, little space-users. These transport mechanisms differ considerably in noise, space consumed, and pollutant emissions. Motorcycles and cars are parked along streets where vegetation often could be planted and grow.

Public transport, bicycling, and walking are three primary ways to commute to work in much of the world. In a study of 20 medium-size cities in the car-dominated USA, the following cities had relatively high and low percentages for these commuters (Adler and Dill, 2004):

High percentages
- *Walking*: Pittsburgh, Portland (Oregon), Seattle, Newark, Milwaukee.
- *Bicycling*: San Jose, Portland, Sacramento, Seattle.
- *Public transit*: Newark, Seattle, Portland, Pittsburgh, Miami, Denver.

Low percentages
- *Walking*: Orlando, Fort Lauderdale, Kansas City, Fort Worth, Indianapolis, San Jose.
- *Bicycling*: Kansas City, Pittsburgh, Cincinnati, San Antonio, Fort Worth, Newark, Indianapolis.
- Public transit: Fort Worth, Kansas City, Indianapolis, Orlando, Norfolk, Fort Lauderdale.

Portland and Seattle are high for all three commuting types, while Kansas City, Indianapolis, and Fort Worth are low in all three. Interestingly, Pittsburgh and Newark are high in two and low in one commuter mode. Nevertheless, all 20 cities have low values in all three modes compared with most European and all Chinese cities.

Well-used bicycling and especially walking routes are usually safe, appealing, and at least partially lined with trees. The connectivity provided by tree lines enhances movement by both people and wildlife, which move at somewhat different times. Most urban birds move during the day, flying insects by day and night, and mammals at dusk, dawn, and night.

Traffic calming in urban areas usually focuses on slowing vehicles, plus encouraging walking, bicycling and use of public transit. Common approaches include on-street parking, traffic circles (roundabouts), speed bumps, raised intersections, and "bulbouts," i.e., walkways extended into the parking lanes at intersections or mid-blocks to facilitate safe pedestrian crossing of streets. While these approaches slow traffic to enhance walking and bicycling, they could readily be enhanced or extended for biodiversity and stormwater benefits.

Other road-related features

Strip or *ribbon development* is a particularly characteristic pattern in the suburban and exurban/peri-urban areas of North American cities, and increasingly cities in some other regions (Davies and Baxter, 1997; Forman et al., 2003). Although development along a road must be an important habitat-fragmenting force and barrier to wildlife movement, studies of the subject are still scarce. Stream and groundwater degradation by strip development is doubtless widespread. Wildlife underpasses and overpasses could be effectively targeted to stretches of roads with strip development. Some cities emerge and expand as a large node around a major road (Garreau, 1991; Hall, 2002). New roads bypassing a city also markedly change land uses and warrant ecological study (Shindhe, 2006). Indeed, the first ring road being built around Rio de Janeiro promises ecological impacts over a vast area (Tangari, 2013). Alternatives to ring highways need evaluation.

Traffic circles or roundabouts usually contain vegetation patches, which, however, are hazardous locations for wildlife. One ecological study of roundabouts found that plant and insect (Hemiptera) diversity are mainly related to patch area, habitat diversity, and management regime (Heiden and Leather, 2004).

Bridges, with their storm drains, culverts, shaded areas, steep banks, and other attributes are widespread and of considerable importance for certain species. Lizards, birds, and insects frequent bridges (Boada and Capdevila, 2000; Adams et al., 2006), while peregrines (*Falco peregrinus*) nest on large bridges (Houck and Cody, 2000). Bats are the champion bridge users for roosting and nesting. More than a million bats use an Austin (Texas) bridge, and 24 species have been recorded as resident on bridges in the USA (Keeley and Tuttle, 1999; Adams et al., 2006).

Carparks, driveways, sidewalks, and alleys each have their distinctive attributes, and thus have different ecological characteristics (see Chapters 5 to 9). Asphalt and concrete cover most urban roads and their associated surfaces. Yet cut stones and cobblestones cover some city streets, while porous pavements are sometimes used for carparks or sidewalks (Ferguson, 1994; Bean et al., 2007). Various fitted block pavers are also used for driveways and some walkways. Water flows and energy fluxes are greatly affected by these different surface types.

Carparks (parking lots for cars) cover a considerable area of most cities (Davis et al., 2010; Ben-Joseph, 2012). A small city in the American Midwest (in Tippecanoe County, Indiana) is 6.6% carpark. The average car occupancy of parking spaces is 28% (maximum 56%), and indeed if all cars registered in the county were in the carparks, 83 000 parking spaces would remain unused. Carparks are major urban heat, stormwater, and pollution sources (see Chapters 5 and 6).

Thus, motor vehicles are major sources of NOx and hydrocarbons, including volatile organic compounds (VOCs) that evaporate even when vehicles are parked (Scott et al., 1999). Tree shade reduces VOC evaporation (emission). Stormwater runoff also transports carpark pollutants, especially in the "first flush" when a rainstorm begins (see Chapter 6) (Greenstein et al., 2004). First-flush carpark pollution, normally rich in hydrocarbons from cars, is considered to be extremely toxic. Heavy metals, apparently mainly deposited from the air in small particles, are readily picked up and carried by stormwater. Grassy swales in parking lots are especially useful in limiting runoff from light rains and the pollutants from most first flushes.

Light-colored or white surfaces reflect more incoming radiation and heat the air less. Porous materials are quieter, but accumulate pollutants such as road salt and heavy metals. Stone surfaces are noisier, but have less stormwater runoff and more water infiltration. Carparks covered with asphalt and hydrocarbons from vehicles heat up more in the absence of tree shade, resulting in more hydrocarbons being emitted as air pollutants. Surfaces with more cracks may have more vascular plants and insects, whereas well-used sidewalks often have trampling-resistant mosses in the cracks. Alleys and driveways may be covered with light-colored gravel so that incoming energy is reflected and runoff is minimized. Roots of trees along these varied structures may expand, causing surface bumps. Meanwhile, the tree canopies provide cool shade, nesting sites, feeding locations, and connectivity for wildlife movement. The opportunities for ecological benefits are rampant around road-related features.

A *netway system with pods* may represent the future of surface transportation (Figure 10.5) (Forman and Sperling, 2011). In essence, roads are replaced by narrow elevated (or sunken) ways, along which sleek lightweight pods silently move. Powered by renewable energy transmitted in buried wires, plus induction technology to run small electric motors, pods move with automated control. Bus, van, and freight pods move quietly over some streets, while car pods or a mixture stream along other routes. No driving. No accidents. No roadkills. No fossil fuel use. No greenhouse gas emissions. No unhealthful air pollutants. Terrain lost to roads and roadsides is recovered. Land is reconnected for people and wildlife. Space is gained for market gardening near communities. New recreational trail networks conveniently develop near towns and cities. In short, the netway system provides both human and ecological benefits thoroughly permeating our land and cities.

Hard surfaces and cracks

Hard relatively impervious surfaces cover our urban areas. But such surfaces are normally inflexible, and lots of mechanisms strain, vibrate, erode, expand, shrink, and damage the surface materials. Consequently, surface cracks or crevices appear nearly everywhere (Wessolek, 2008), so vegetation and animals colonizing the cracks appear as growing linear bits scattered across the city, denser here, sparser there. First we consider (1) the surface types and crack formation, and then (2) crack plants, animals, and succession.

Hard surface types and crack formation

Surface types

Concrete is typically composed of limestone dust, sand/gravel, and water. When combined, these harden to form long-lasting structures such as highways, bridges, buildings, and walls. The surface often seems relatively smooth and homogeneous. Concrete is so inflexible that it is poured in sections, separated by narrow seams or joints to account for expansion and contraction due to temperature changes and other factors.

Tarmac or *asphalt* (black-top) in contrast is a combination of gravel and hot tar (fossil fuel) that hardens upon cooling. Roads, streets, and driveways are often tarmac, which has no expansion joints. At this scale surfaces appear nearly smooth and homogeneous.

The third widespread hard horizontal or vertical surface is composed of rectangular or square *repetitive blocks* in huge numbers. Fired bricks, dried mud or adobe bricks, concrete blocks, cinder blocks, flat cut stones, and cobblestones are widely used in different regions. Often mortar or cement is inserted between the blocks to hold them in place. Without mortar, water readily infiltrates through the surface material between blocks. Over time, mortar tends to dissolve or break, so cracks permitting water seepage may be present or widespread.

Several other hard-surface types may exist in cities. Large flat stones, such as on ancient Roman roads, and coarse boulders, rocks, or stones, as in some classic garden walls, leave lots of space for water flows and plant growth. *Porous pavement*, both of concrete and asphalt/tarmac, has tiny holes for water infiltration (Ferguson, 2005). In light rains this reduces water runoff and flood hazard.

Permeable pavement, in contrast, is constructed with numerous small blocks of geometric shapes fit together to form a surface. One type has blocks containing soil and grass in their centers. Where vehicle traffic is light and vehicles are not constantly parked, this surface becomes grass covered.

Cracks in steel surfaces, such as bridges and skyscrapers, are rare, in part because their appearance may indicate a structural problem that is quickly addressed. Wood surfaces often have natural cracks due to drying, but colonization by plants usually indicates moisture rotting the wood. Finally, various surface coverings such as paint, stucco, and thin tiles may have small cracks, which usually are too small or temporary for much plant colonization. So here we focus on cracks in vertical and horizontal surfaces composed of concrete, asphalt/tarmac, and repetitive blocks with grass (Figure 10.6).

Crack formation

Most horizontal hard surfaces are used by vehicular traffic or walkers. Tarmac/asphalt and concrete cover roads and highways. Tarmac, flat cut stone, and cobblestone cover most streets. Without the weight of truck traffic, carparks and driveways use all types of coverings, including porous pavement and blocks with holes for soil and grass. Walkways predominantly use asphalt/tarmac, concrete, porous pavement, and various smooth block designs, though usually not blocks with holes for soil and grass.

Expansion joints between concrete sections, the boundary between tarmac strips in carparks, and the boundary between a traffic lane and a road shoulder are characteristic locations for linear cracks (Figure 10.6a to g). Roads with surfaces of flat cut stone, cobblestone or blocks, with or without mortar, tend to have a fine-scale rectilinear network of cracks. Vehicular traffic tends to break hard surfaces, especially mortar between blocks, causing cracks and eventually sometimes holes. Heavy truck traffic accelerates the process.

Concrete commonly breaks into large polygonal pieces, particularly triangles plus some 4- and 5-sided polygons (Figure 10.6). Long diagonal cracks, often curvy but seldom convoluted or zig-zag, produce many 30°, 45° and 60° angles. In contrast, tarmac/asphalt cracks are sometimes long but more often short, and commonly somewhat convoluted or zig-zag in appearance. The tarmac/asphalt crack network produces intermixed irregular different-sized, but often small, surface shapes. Oblong patches with somewhat pointed ends may be prominent. In both concrete and tarmac a weak spot (e.g., from settling) in the foundation typically results in a cluster of small surface polygons. Loss of one or more polygon produces a hole (pothole) in the surface that tends to enlarge over time. Holes also often form when corners

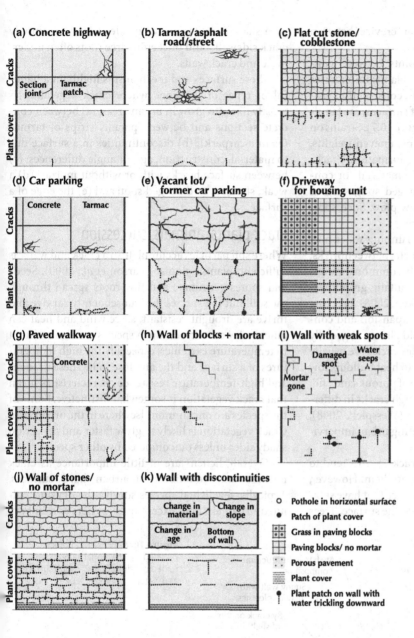

Figure 10.6. Eleven alternative forms of surface cracks. (a) to (g) = horizontal surfaces used by vehicles or walkers; (h) to (k) = vertical surfaces (walls). For each form, upper diagram = typical pattern of surface cracks, and lower diagram = typical plant cover in the cracks.

of concrete sections break, or the edge of a tarmac strip is damaged.

Temporary repair of cracks and small holes is accomplished by pouring hot tar, which covers and kills crack plants. Larger cracked areas and holes are repaired with a patch of tarmac or concrete. Both the tarred strips and the edges of repair patches are prone to further cracking. Periodically the entire surface may be repaved by adding a new surface. For tarmac the surface is readily removed and the material recycled in the new surface.

Walls of buildings, walls separating areas, and walls holding back soil are the primary vertical surfaces. If made of blocks or bricks, long diagonal zig-zag "cracks" appear, typically from the unequal sinking or settling of a foundation (Figure 10.6h to k). Human construction results in some cracks, while nature's processes, especially water-related, produce other cracks. Stone walls without mortar of course have cracks around all stones and may be much covered with vegetation. The most common location and luxuriant vegetation is along the bottom of a wall, where water drains down, soil accumulates, and a deep crack or "crevice" between wall and adjacent surface is typical.

Purposeful cracks, such as joints between concrete sections, may be colonized by plants. Boundaries between different surface materials, construction processes, aged surfaces, and angles of surfaces, are familiar locations for cracks to form (Figure 10.7). Strains on surfaces due to sinking foundations, uneven weights, compression, or tension cause crack formation. Cracks may form at weak spots related to materials or construction, or a subsequently damaged spot, such as caused by an auto accident or snow plow. Such spots commonly lead to holes in a surface.

Frequent seeping, dripping or running water may dissolve surface material, such as mortar between bricks, creating crevices. This is particularly common for walls holding back soil, which normally contains groundwater pushing against the wall (Peterken, 2008). Freezing and thawing of water producing expansion and contraction causes many cracks in cold climes, especially in roads. However, such freeze–thaw cycles are usually less important in urban areas, due to heat buildup, low water-table, and the predominance of porous sandy fill. Considerable water infiltrates through cracks in different horizontal hard-surface types (Wessolek, 2008), which suggests caution in interpreting urban impervious cover percentages.

Rather than causing surface cracks, plants tend to colonize them and widen and deepen them. However, woody plants adjacent to a hard surface that have shallow roots often tend to lift and crack the surface. Trees with shallow roots such as maples (*Acer*) commonly crack sidewalks and thickening tree roots often undermine and crack walls.

These surfaces and mechanisms highlight a handful of types of locations where crack vegetation is mostly likely to grow in urban areas: (a) between concrete sections and between parallel strips of tarmac (as in a carpark); (b) discontinuities in a surface due to material, construction, age, or angle differences; (c) between surface blocks with or without mortar; (d) a weak, stressed, or damaged spot; and (e) the edge of a surface.

Crack plants, animals, succession

Where soil particles accumulate in a crack and water is sufficient, plants colonize (Larson *et al.*, 2000). Seeds and spores germinate and tiny roots spread through the soil. But the species that subsequently survive and thrive are drought-resistant, since wind and heat can quickly dry out a crack. Such species are also resistant to temperature extremes, especially the high temperatures of a surface and the air close to it. These drought and high-temperature resistance characteristics mean that crack vegetation is somewhat distinctive. Some of the species are uncommon elsewhere in the urban area. Crack vegetation is likely to grow faster and denser on shady sides unless outcompeted by other species.

Overall, lichens are of little importance as crack formers and colonizers of urban hard surfaces. Typically, few lichen species survive the dry and polluted urban air, and, except on tile roofs and in very

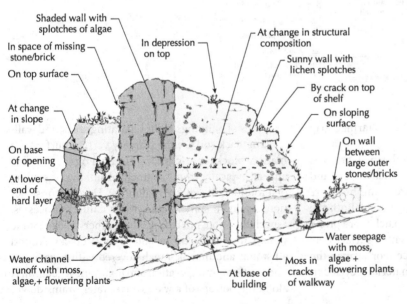

Figure 10.7. Plant colonization providing successional habitats on a ruined building. Cracks, rough surfaces, and water accelerate the processes.

moist climates, hardly any lichens are widely abundant. Algae usually grow over surfaces that are frequently wet. In contrast, a few mosses (e.g., *Bryum argenteum*, *Ceratodon purpureus*) grow quite well in cracks, though probably they are of negligible importance in crack widening. Sometimes called "sidewalk mosses," these are resistant to both surface conditions and moderate trampling by pedestrians.

Vascular plants in cracks, especially flowering plants and sometimes ferns, however, have roots. Successful crack species, such as some grasses, have dense fibrous roots that can penetrate the complex spaces in a crevice. Carrots with tap roots would not be expected. Drought-resistant plants with dense foliage may also be favored, because they "catch" dust and contribute ample leaf litter, both of which accelerate soil accumulation in a crack.

Animals are abundant in many cracks and crevices including those without vegetation, though often these species are well-hidden and little-seen by us. Insects of diverse types live in the crack soil and on plants. Flying insects frequently land on hard surfaces, and spiders build webs, often over cracks, to catch the flying insects. Bees, butterflies and moths, and sometimes hummingbirds and bats, pollinate crack flowers. Lizards forage for insects on hard surfaces and in cracks. Deeper holes and crevices become homes and nest sites for diverse animals, from snakes to colorful beetles. Some birds often forage along surface cracks.

In crack succession, soil particles from airborne dust and the surface material itself initially accumulate in a crack, along with water. Leaves and tiny plant roots die, adding organic matter, a source of energy for bacteria that thrive in these usually high-pH environments. Although microbial decomposition makes some plant nutrients available, ample inorganic nutrients from the accumulation of dust and surface materials suggest that the plants are seldom nutrient limited. The food web of soil animals grows in complexity.

Ecological succession is readily observed on hard surfaces (Figure 10.8) (see Chapter 8). Mosses may be the initial crack colonizers in abundance, or indeed may follow the vascular plants with roots. Irrespective, the rooted plants, overwhelmingly herbaceous (non-woody) species, are especially effective in building soil in a crack. Roots spread and hold soil particles, which accumulate and hold more water. Plant and animal diversity increases.

The next major step follows from successful colonization by shrubs or trees. These woody plants have roots

Figure 10.8. Plant succession on network of cracks of former parking lot. Short white lines indicated parking spaces; long parallel cracks indicate tarmac/asphalt strips originally put down; over time vegetation patches enlarge and interconnecting corridors widen. Sudbury in suburban Boston. R. Forman photo.

that rapidly thicken, contributing to the noticeable enlargement – widening, deepening, lengthening – of crevices. In a positive feedback, a larger crevice favors more woody plant growth, which in turn enlarges the crevice.

The lengthening of cracks tends to form a network over the surface (Figure 10.8). Where shrubs or trees thrive, often with more deeply penetrating roots, a small patch of vegetation is often recognizable. With further cracking by woody plants, the patch becomes a node in the network. Still more woody plants colonize; more cracks form, enlarge, and lengthen; and more vegetation patches appear and enlarge. Meanwhile, the original extensive hard surface becomes composed of progressively more and smaller surface patches. Finally, the vegetation patches begin to coalesce while the hard-surface patches shrink and disappear. At this point woody plant cover, herbaceous plant cover, and soil are all three complete across the former surface area. If that surface area had been a city, natural processes would have erased the city.

But those later successional stages seldom occur. Disturbance intervenes. The hard surface with crack vegetation is herbicided. Or covered with road salt. Or intensely brushed by a street cleaner. Or painted. Or demolished. Or, indeed, tidied up by a maintenance crew. Yet inexorably, ecological succession always follows. Urban hard-surface succession is quite distinctive from that in natural areas.

Vegetation in cracks of many surface types is too sparse to play a significant overall role in a city, though a few important exceptions exist (Figure 10.6). Crack vegetation may often provide considerable ecological benefit in three locations: (1) cracks of vacant lot or former carpark; (2) stone wall without mortar; and (3) home driveway using blocks without mortar and with or without holes for soil and grass. Normally little ecological benefit is provided by cracks in concrete highway, tarmac road/street, active carpark, paved walkway, and wall of blocks and mortar.

The overall roles of crack vegetation remain a research frontier. The drought- and extreme-temperature-resistant vegetation, being somewhat distinctive, raises urban plant diversity. The vegetation is a source of plant and animal species, including pollinators, dispersing across urban areas. The narrow strips serve as stepping stones for movement of species such as bees and butterflies.

Abundant crack vegetation doubtless significantly cools surfaces and adjacent air temperatures at hot times. It also cleans the air somewhat, as particulate matter adheres to leaf and stem surfaces. It reduces water runoff a bit, mainly by friction but also by evapotranspiration. An abundance of crack plants may also somewhat reduce stormwater pollutant runoff.

People express pleasure and delight with crack flowers, pollinating butterflies and hummingbirds, a lizard waiting for an insect, a bird flitting and foraging, even a tiny green plant showing life in a tough urban spot. If urban maintenance budgets and stewardship by neighbors suddenly stopped, residents would be amazed at the profusion of proliferating vegetation across the built land. Society spends funds to hold nature back, but natural processes are powerful and eternal. History well shows us cities that turned to dust. Spreading crack vegetation often played a key role.

House plots, gardens, lawns

To provide an outdoor space for family activities, a small area immediately around a house is commonly delineated. Buildings, walls, fences, and hedges often line the boundaries of a plot (lot) or property. A *private space* is sufficient for sitting outside, children playing, growing vegetables, drying clothes, and much more (Andrzejewski, 2009). Two arrangements of space are particularly widespread. One, the courtyard or patio surrounded by building, especially characterizes tropical Latin America, the Mediterranean Basin, Islamic cities, and the hutongs of China. The other pattern results from placing the building somewhat near the center of the plot, usually producing a *front space* and larger *back space* (typically called "yards" in the USA and "domestic gardens" or gardens in the UK). Detached housing units also normally have narrow side spaces.

Patios and *courtyards* commonly contain one or more trees that provide partial shade. Many plants are usually present, including shrubs, flower beds, and plants hanging or attached to building and tree. Walkways, a water feature, and pets are also common. Patios and courtyards are particularly characteristic of single-level houses. Low-rise buildings of 2–8 levels may have rectangular or square cylinder-like courtyards with windows on each level. Such courtyards in the central portion of buildings may permit smells, pollutants, and exhaust from kitchens, toilets, and other family activities to escape. Also some air movement helps cool the air in buildings. Such low-rise courtyards may have common space for residents, including some plants, at ground level.

The *front- and back-space house plot* is well illustrated in the outskirts of moist tropical cities, where a fence or combination shrub line and fence mark the plot boundaries. Both back space (yard) and front space contain trees and shrubs, including many food-producing plants in layers. A canopy of mango trees, understory of coffee and bananas, shrub layer of manioc and papaya, herbaceous layer of melons and maize, and fruit-producing vines here and there are illustrative. Chickens, dogs, and other animals are frequently present. The house-plot front space is often more-maintained and open beneath the trees, with an attractive diversity of flowers, while the back space may have an outhouse, sheds, storage, and less-tended garden. Such a house plot produces food throughout the year, and provides some financial stability for the family in the face of regional or national economic fluctuations. Biodiversity is relatively high because of the vegetation stratification and productive plants.

In Northern Europe (including Britain), as well as in New Zealand, houses are often attached in a row, creating a continuous strip of front spaces separated from a row of back spaces. Typically a front space is mainly flower garden with a wall or fence or hedge in front, while the back space has extensive flower garden and often a small lawn. In North America, both the smaller front yard and larger back yard are often mainly covered by lawn, with flower gardens in small patches along edges, plus scattered trees and shrubs.

In dry climates with scarce costly freshwater, a property is sometimes "xeroscaped" mainly to reduce water use. Succulent plants, wildflowers, and rock garden plants often predominate, with much of the area covered with pebbles or rocks. Plants may be grouped according to similar water requirements. Such a design dramatically reduces water use compared with lawns and typical flower gardens.

Two other distinctive house-plot designs are less common. For one, an evergreen ground cover is maintained over much of the plot, thus greatly reducing care and management effort. For the second, a somewhat natural meadow of native wildflowers and grasses covers most of the property and is cut once a year. Overall, xeroscaping, evergreen ground cover, and natural meadow are beneficial due to low water use, low use of lawn chemicals, low cost, low maintenance, and the expression of individuality by residents.

Internal structure of house plots

Three spatial aspects of house properties are central to understanding their internal pattern: (1) role of the surroundings; (2) habitat heterogeneity; and (3) spatial arrangement. A fourth dimension, (4) biodiversity, provides insight into the species structure. Maintenance and care of a property also have a large ecological effect (Nassauer, 1988, 1995, 1997). Thus messy and manicured areas differ strikingly in species composition, diversity, soil, water conditions, and wildlife movement.

Role of the surroundings

Because house plots are small, their surroundings have a powerful impact. A wooded plot surrounded by open plots is analogous to an oasis, where wind and heat and open-land species pour in (Forman, 1995). In contrast, if the property is relatively open and surrounded by wooded plots, seeds and animals and cool air pour in. If the property is on a significant slope, water and materials flow in, and out. If it is in a major wildlife corridor, animals charge in and out. If the plot is representative of the surrounding lots, it will be harder to create and maintain distinctiveness against an overall homogenizing force. The proportion of land in natural habitat, say, within 1 km radius of a plot may be a good predictor of, for example, animal diversity on the property.

A study in Montreal finds that the best indicator of the structure of a front space (i.e., the types and locations of plants and flowers) is distance from or proximity to another front space of similar type, which accounts for 20% of the variability (Zmyslony and Gagnon, 2000). The depth, width, and type of front space are secondary predictors. The similarity of front spaces is greater on one side of a street than for the same distance to a front space on the opposite side of the street. Mimicry is alive and well for front-space design. This suggests that conformity and monotony are valued. Yet it is unclear whether a plot that stands out for species richness, interest, and individuality would be mimicked, and thus lead to an entire distinctive street or neighborhood.

A house property may also function as an important part of a vegetation corridor or a sequence of "stepping stones," in which case species continue to move through the plot. However, house plots do not resemble other vegetation types, so this would mainly be the case for species moving along a row of house plots rather than seminatural vegetation (Rudd et al., 2002).

Habitat diversity and spatial arrangement

A house property may be covered with a single type of planting or may be enormously heterogeneous with a richness of habitats and species (Smith et al., 2005; Gaston, 2010). Many types of habitats and many habitats are both keys to high habitat heterogeneity. More habitats also normally means smaller ones. Spatial arrangement, as explored in the following section, also affects heterogeneity.

Consider the five major habitat types in house plots, each serving as a different habitat for species (Figure 10.9): (1) buildings, (2) trees/shrubs, (3) gardens, (4) lawn, and (5) corridors along borders. These of course come in varied forms. House, garage, or shed. Individual trees and shrubs, clumps of them, or combinations. Flower garden or vegetable garden. Intensively managed or little-managed lawns. Strips along front, side or back property lines, and composed of trees, shrubs, fence, or wall.

To these five predominant components of house plots, add other relatively common objects present: driveway, walkway, courtyard/patio, terrace/porch/deck, pond or other water feature, compost pile/heap, and street trees. Finally, an array of specific objects designed to attract wildlife, such as bird feeder, bird bath, bird nest-box, brush pile, bat box, rock wall, and rock pile, may be added. The potential for high house-plot habitat heterogeneity is enormous.

In Sheffield (UK) an estimated 87% of the 175 000 urban house plots have gardens, overwhelmingly flower gardens (Gaston et al., 2005b). The properties contain 350 000 trees >3 m high, an average of two trees per property, though all the trees are on 48% of the properties. Twenty-nine percent of the house plots have a compost heap (51 000 total), 26% have bird nest-boxes (45 000 nest-boxes), 14% have ponds (25 000), and 14% have one or more cats (52 000 total cats). Outdoor spaces in these properties have an average of about 10 plant species per square meter, approximately the same as for "derelict" or vacant land in the city.

Landscape ecology provides the essential theory and framework for understanding a small area such as a house property (see Figure 2.6) (Forman, 1995; Farina 2006). The "patch–corridor–matrix" model applies to the entire surface, which is composed of patches and corridors, with a background matrix frequently present. *Patches* vary from large to small, round to elongated, smooth- to convoluted-margined, and so forth. *Corridors* vary from wide to narrow, straight to curvy, continuous to discontinuous, etc. Wooded, garden, and lawn patches, and corridors as boundary hedges, foundation plantings, and view lines, are widespread in house plots.

Combining and spatially arranging these patch-and-corridor building blocks opens up a huge range of possible patterns. For instance, *adjacencies* refer to conditions influenced by an adjoining land use or habitat. Thus, adjacent pairs of patches or corridors have distinctive conditions, e.g., for a species or for biodiversity (Figure 10.9e and g). *Edge types*, named for the adjacent habitats, such as a shrub-lawn edge or garden-pond edge, generally support different species. *Convergency points* (coverts), where three or more habitats converge, are particularly important for nesting by some seed-eating birds and for foraging by some predators. Trees, especially on the sunny afternoon side of a house, cool the walls (Gartland, 2008). A patch surrounded by three or more different types of habitat is likely to be richer in species than a patch embedded in a single type. For a particular species or process, one type of habitat is often a "source" from which individuals move outward, while another type is a "sink" that absorbs the objects moving.

Convoluted patch edges tend to be much used by wildlife, so S-shaped or serpentine boundaries with coves and lobes are effective designs (Figure 10.9h) (Forman, 1995). Wildlife frequently move along straight boundaries, whereas convoluted boundaries favor animals moving between habitats. A gradual ecotone, rather than abrupt boundary, between habitats supports a large number of species, though normally most are common species. Larger green patches often support some species that are scarce or absent in small patches.

Corridors typically facilitate flows and movements along their length, and form a filter or barrier for movement across. Corridors exhibit edge conditions, and also function as sources and sinks. Corridors are normally the most effective routes for species movement in urban areas and elsewhere. Open sight-line corridors, for example from a back terrace or deck, may be important for some open-habitat species, though probably in a house plot, the size of an open area is more important. Side-boundary shrub corridors or fence-lines or walls are effective movement routes for some species, even in the presence of cats, dogs, and people on both sides. Dense shrub-lines provide good escape cover and sometimes nesting cover, and thus are prime sites for many urban vertebrates.

Ecologically the most important corridor in house plots is the *backline strip* of vegetation (Figure 10.9e and h). If the back property line is more than about 15 m (45–50 ft) behind the house, it tends to escape intensive management or manicuring. Weeds and woody plants colonize and may thrive, forming a strip of semi-natural vegetation. If this vegetation forms on both sides of the property boundary, the backline strip or corridor may be somewhat wide. Such a corridor enhances movement along the line, and also serves as a habitat for nesting and denning species. Frequently house plots along a street are about the same length so backline strips are aligned, forming a long strip. This is likely the most important corridor for wildlife movement from parks and other greenspaces though a residential built area.

Although a continuous corridor is most effective, a discontinuous one, essentially a row of *stepping stones*, is also effective for movement by some species (Figure 10.9c and e). Thus, a sequence of shrub patches often provides adequate connectivity for many wildlife species of house yards. At a broader scale, a row of back space lawns along a block may facilitate movement by some open-habitat species. Or, more importantly, a row of wooded-patch stepping stones in a sequence of house plots is probably frequently used by some wildlife for movement across a built area.

Aggregating habitats on a property with those in adjoining properties adds additional benefit. *Back-corner patches* of vegetation are the key. A back corner is normally the best place for a somewhat large patch of nature containing several vegetation layers, cool moist rich soil, and some less-common specialist

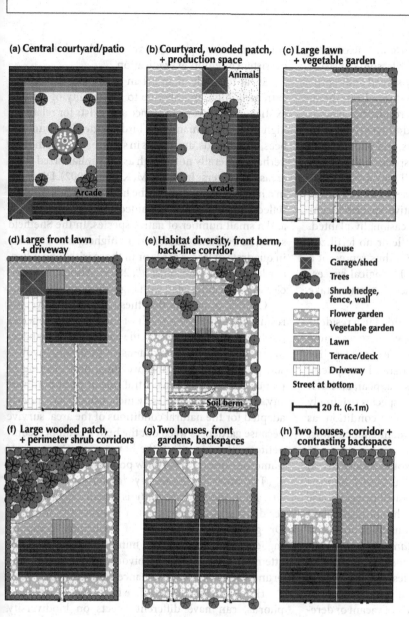

Figure 10.9. Eight alternative forms of house plots. Area of each = 0.03–0.1 ha (1/12 to 1/4 acre). Entrance from street at bottom. Two houses per lot in (g) and (h). 1 foot = 0.3 meter; 1 acre = 0.4 hectare. Based on Goldstein et al. (1981), Kress (1985), Owen (1991), Girling and Helphand (1994), Forman (1995), Gartland (2008), and other sources.

species (Figure 10.9f). But combining this patch with a similar adjacent one on the adjacent plot, or better still, with three others on the plots that converge at the corner, forms a significant patch of natural conditions in a built area.

Eight forms or types of house plots appear to represent the range present in and around cities worldwide (Figure 10.9). To compare these types, I evaluated each with 17 variables related to habitats, plants, animals, soil, air, and water (see, for example, Table 7.1). Negative roles or functions were predominant in two cases (Figure 10.9d and g). Positive roles strongly outweighed the negatives also in two cases (e and f). The presence of a wooded patch with trees in a back corner of the house plot best correlated with an overall positive ecological condition.

Biodiversity and wildlife

Plant diversity can be extremely high on a house property. Many tens of thousands of plant species purchased from nurseries and seed sources are planted in yards. A house plot provides ample space to plant a veritable profusion of species, and such properties seem to exist in nearly every neighborhood. Overwhelmingly these planted species are non-native plants and horticulturally bred cultivars that would not grow well in a nearby natural ecosystem. Native plants would outcompete and kill them. However in house yards, some invasive species (see Chapter 8) successfully spread and reproduce, despite intensive human efforts to eradicate them. Such species become naturalized in residential areas.

Most people like to surround themselves with showy exotic plants, unusual objects, and photographs of stunning far-off scenes. Native species in the local woods too often seem boring (until they are watched carefully). Nevertheless, native house gardens or yards, as an alternative design, with native species essentially throughout the space, are occasionally planted. These gardens generally require little or no fertilizer, pesticide, watering, or mowing, and thus represent a low-cost, low-maintenance space. Ecological succession keeps changing such properties. Let nature take its course.

However, the house plot is embedded in a residential area bursting with non-native species, so over time the low maintenance property looks less and less like a natural ecosystem. Indeed, in a natural ecosystem, species normally increase, decrease, appear, and disappear. So if the resident wants the species originally planted to persist in some semi-stable condition, as most gardeners do, considerable maintenance effort is required.

Still, the profusion of planted exotics (non-natives) and horticultural cultivars is not the only important story. A study of 120 quadrats in 60 private house plots in Sheffield (UK) reports that the abundant plant species, those present in the most quadrats, are native weeds (unwanted plants such as *Taraxacum officinale* and *Epilobium montanum*) (Thompson *et al.*, 2003). The next ten most frequent species are also weeds, both native and non-native. Most of these frequent or widespread species are characteristic of vacant or derelict sites in the city (see Chapter 8). A huge number of planted species, mainly exotics and cultivars, are present in the house plots (Smith *et al.*, 2006). Most planted species are scarce (low population size) and many are "singletons" (represented by a single individual). Each property has a somewhat different set of species so that none of the planted species is found frequently in the samples.

A 25-year study of a Leicester (UK) house plot recorded 95 annual and perennial weed species (unwanted plants) (Tutin, 1973). During the period, there were many species colonizations, as well as some extinctions. Also, some native species became aggressive weeds. Weed species with very different environmental requirements coexisted, and all resisted persistent attempts at eradication.

So, at the scale of an urban area, plant diversity is extremely high, mainly due to the totality of planted exotics and cultivars. Evidence also exists for relatively high diversity of mammals, birds, butterflies, bumblebees, ants, lizards, and plants in suburban areas, though perhaps generally not as high as for similar-sized rural areas with diverse habitats (McKinney, 2002). Diversity in a house plot or yard may be high or low, and mainly reflects an abundance of planted species, some weeds, and a small number of native species. In the Sheffield study, no evidence was found for higher plant diversity in quadrats of larger yards or in those with longer border lengths (Thompson *et al.*, 2003). Also, scarce species were not correlated with high-diversity quadrats.

Overall, it appears that these high plant-diversity results are primarily a result of both the huge species pool of planted species, and the gardening or management effort of residents. Intensive *gardening* involves soil preparation, planting, weeding, and pruning of plants. Weeding plays a central role in plant biodiversity. The planted species, which usually are poorly adapted for the natural conditions of the area, survive because gardeners weed out the better-equipped competitors. Thus, in effect, the house property supports numerous species at very low population sizes.

Two other factors are keys to house-plot biodiversity. First, *plant productivity* is usually high because residents add fertilizers, often in excessive amounts. Or gardeners' techniques of mulching, composting, enriching the soil with humus, adjusting the pH, attempting to control herbivores and diseases, and pruning may be used to enhance plant growth.

High nutrient levels of, e.g., nitrogen or phosphorus, can have different effects on biodiversity. Compared with a low-nutrient habitat, considerably more plant biomass, often nutrient-rich and palatable to herbivores, is produced. That provides a larger base to the food web, and the potential to pack together more species. However, excessive nitrogen or phosphorus commonly leads to one or a few aggressive species dominating, by outcompeting other species. The net effect is a reduction in biodiversity. Overall, a modest nutrient increase may produce high biodiversity while a large nutrient increase reduces the number of species.

Second, biodiversity usually strongly correlates with *habitat heterogeneity*. Increasing the number of microhabitats – house exterior, lawn, shrub patch, rock garden, old fruit tree, vegetable garden, flower bed, hedge, wall, pond, ecotones, and compost heap – progressively increases the number of species (Gilbert, 1991). The particular plant species planted, rejuvenated growth from pruning, and abundant nectar/pollen/seeds/fruits also affect animal diversity. A mosaic of specific habitats can increase or decrease biodiversity depending on their arrangement. Adding an array of structures to attract species, such as bird feeder, bird bath, bird nest-box, bat box, brush pile, and so forth, can be expected to further increase diversity. However, this may depend on other factors, such as the existing level of diversity and the nature of the surroundings.

One study reported little effect on biodiversity by adding such specialized structures to a house plot (Gaston et al., 2005a). No bumblebees colonized the artificial nests that were added. Small ponds added few invertebrates, though amphibians visited the ponds. New logs added were not colonized by saprophytic organisms. Adding nettle clusters did not attract butterflies.

Nevertheless, adding habitats normally does increase species number. As Jenifer Owen's small house plot in Leicester (UK) showed (see Chapter 8), thousands of species may use the space over time (Figure 10.10) (Owen, 1991). A high diversity of microhabitats plus intense management attracted over a third of all the butterfly, ladybird (ladybug), and harvest spider (daddy longlegs) species known in the British Isles. A London house plot had more than 700 beetle species in 47 years (cited by Gilbert, 1991), during which major changes in species composition occurred continuously. No "normal" year or "normal" faunal community existed for the beetles there. Similar results were observed in a 7-year study of macro-fungi in a Bristol (UK) lawn.

Finally, in addition to biodiversity, the density of various groups of organisms may be either higher or lower than in surrounding natural habitats. House plots have the highest breeding bird density of any habitat in Britain, as suggested by 600 and 641 pairs/km^2 (not including house sparrows) in two London area studies (cited by Gilbert, 1991). At a finer scale, some 20–80 slugs (slimy mollusks) per m^2 (2–8/ft^2) are found in a UK garden. Even after removing 10 000 to 17 000 slugs from the garden per year for 4 years, no reduction in slug population was evident (Barnes, 1949). Relative to surrounding natural habitats, normally bird and butterfly densities are higher, whereas the abundance of amphibians and trees is lower, in house plots.

Yards, gardens, lawns

In England and Wales 78% of the houses have private outdoor spaces (yards), while the figure is 56% in Holland and 32% in France (Gilbert, 1991). The 15 million house plots in England and Wales cover 3% of the entire land surface. This abundance of private green spots around houses seems to have been accelerated by the so-called Garden City movement of Howard (1902) and others, who promoted the idea of 12 dwellings per acre (30/ha), e.g., at Letchworth and Hemel Hempstead outside London.

Based on five UK cities, the area covered by house yards ranges from 22% to 27% of an urban area (Loram et al., 2007). In Dunedin (New Zealand), 30% of the urban area is covered by these private spaces, and 42% of residential neighborhoods are house yards (Mathieu et al., 2007b). The UK yards are primarily for gardening, sitting outside, children's play, and clothes drying.

Front and back outdoor spaces

Front spaces or yards cover 26–38% and backyards 62–74% of the outdoor spaces on house properties in five UK cities (Loram et al., 2007). Most combined front- and back-yard spaces are 100–800 m^2 in area, though the commonest yards are 100–200 m^2 (about 30 ft × 30–60 ft) (Gaston et al., 2005b). The common land uses and habitats present are cultivated borders, intensively mowed grass, infrequently mowed grass, vegetable patches, uncultivated areas, ponds, compost heaps (piles), and trees >3 m high. The diversity of land uses (habitat heterogeneity) increases with area of outdoor space (Loram et al., 2007). As the outdoor space decreases, the number and size of sheds and compost piles remains about constant. In smaller yards, shrub hedges and flower beds cover a greater proportion of the area. Lawns essentially disappear in tiny outdoor spaces. Care and maintenance effort in these domestic private spaces is considerable to maintain the land-use diversity present in even a 100–200 m^2 area.

The vegetation and spatial structure of front private spaces in Sheffield seems to be especially correlated with the size and the age of a house plot (Gilbert, 1991). Larger older houses are surrounded by more planted vegetation and habitat diversity. House size and age in turn relate to socioeconomic factors such as income of residents.

A study of trees in the front spaces or yards of 44 areas across a small city (Akron, Ohio, USA), adds insight (Whitney and Adams, 1980). Based on trees,

Human structures

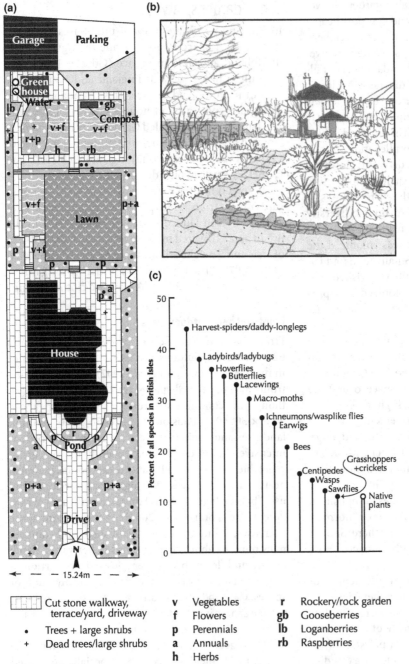

Figure 10.10. House plot designed and managed for both family use and biodiversity. Plot = 0.1 ha (0.25 acre) in housing development (estate) of Leicester, UK. (a) Front space (garden/yard) to south (bottom); back space to north (upper). Front-space trees and large shrubs mapped (30 live, 5 dead) = 7 tree species and 10 shrub species. Side space (12 live, 2 dead). Back-space woody plants (30 live, 5 dead) = 10 tree and 12 shrub species. Total diversity (excluding side-yard woody plants) for 70 woody plants = 14 tree species and 19 shrub species. (b) View from near greenhouse looking southward toward house. (c) Native plants = flowering plants; ladybirds (ladybugs) = Coccinellidae in part; lacewings include allied species. 1 meter = 3.28 feet. Adapted from Owen (1991).

five types of residential areas are recognized: (1) inner city area (e.g., tree of heaven *Ailanthus altissima*, mulberry *Morus alba*, and fruit trees) with fast-growing trees, sometimes spontaneous (self-seeded), and tolerant of urban conditions; (2) maple area (*Acer* spp.) in 30–80-year-old middle-income housing; (3) conifer area (arborvitae *Thuja occidentalis*, blue spruce *Picea pungens*, and other spruces *Picea*) of 40–60-year-old modest houses; (4) mixed suburban area (maples *Acer* spp. and pin oak *Quercus palustris*) in newer <35-year-old housing; and (5) old oak area (including oaks *Quercus* spp. and white ash *Fraxinus americana*) on large wooded plots of mostly native species. The grouping patterns are interpreted to be overwhelmingly due to human, not natural, factors. Fashion at time of planting, availability of different plant species, and house value or resident's income are considered to be primary determinants of tree cover.

Front house-plot spaces without high walls along the street are visible to the public and essentially represent what residents want to show. Thus, sitting areas, children's play areas, clothes drying, vegetable gardens, outhouse toilets, sheds, and so forth are scarce in front spaces, but may be common in the larger more-used back spaces. In addition to entrance walkways and driveways, a profusion of flowers, a golf-green-like lawn, or a distinctive object or sculpture highlighting the resident's individuality is more likely to be conspicuous in front.

Normally front and back spaces are quite different, with habitat heterogeneity and biodiversity greater in back yards. Wildlife is normally more abundant and diverse in back spaces, and often appreciated by residents (Figure 10.11). Indeed, the average house-plot yard, with combined front and back spaces, packs habitats together and displays a richness of native, non-native, and planted ornamental species.

Where properties are wider than houses, side spaces, usually narrow, are present. Mostly a connecting conduit for residents' movement between front and back yards, sometimes side spaces are fenced-off out-of-sight strips used for storage or debris. Side yards are usually of rather little ecological value, though wildlife readily use those with little human use and many accumulated objects.

Flower and vegetable gardens

The main dichotomy in gardens seems to be flower gardens versus vegetable gardens. Flower gardens may include flower beds, ground covers, rock gardens, and so forth. Vegetable gardens also include low fruit-producing plantings, herb gardens, even medicinal gardens.

Flower gardens of course vary greatly but some characteristics are widespread. Overwhelmingly flower garden plants are showy ornamentals. The bulk of the plants are exotic species, hybrids, and cultivars, mostly planted outside their native range and normally poorly adapted to live outside a garden (Gilbert, 1991). Thus, plants with dwarf form, pendulous branches, colored leaves, and white variegated leaves typically compete poorly in surrounding natural vegetation. Such species are usually planted to show a large patch of a single species, medium-size patches of a few species, or a diversity of species packed together.

Another common goal is to show a sequence of blooming flowers over time. To accomplish this in the face of changing weather conditions and a permanent "rain" of weeds (native and non-native wildflowers and other species especially from the surroundings), maintenance and care are intensive. Soils may be enhanced by digging, adding humus, or liming. Water may be intensively and repeatedly added, and fertilizer is commonly added. Pesticides may be added. If the urban garden soil contains some toxic chemicals, soil removal, addition, and/or mitigation is appropriate. Success requires keeping the weeds at bay and the flowers showy over time.

Relatively few insects live in a flower garden due to pesticides, often relatively low plant diversity, and simplified vegetation structure. In addition, many of the cultivars planted are insect-resistant and unpalatable for wildlife. Even the plants weeded out or trimmed are often carted away to a compost pile or discarded, instead of being directly recycled into the soil. These major characteristics combined mean that vertebrate wildlife are remarkably scarce in flower gardens. Few birds, mammals, amphibians, and reptiles prefer foraging, and very few nest or den, in flower gardens. Often pollinators such as bees, butterflies, bats, and birds tend to be relatively few, though specific species and cultivars can be planted to attract such species. For example, butterfly and hummingbird gardens are mainly composed of masses of tubular flowers with ample nectar.

Vegetable gardens have many of the preceding attributes of flower gardens, such as intensive maintenance, soil improvement activities, fertilizer application, pesticide use, heavy repeated watering, change over time, and so forth. However, the differences are interesting. Production of edible plant material for humans is the goal, not showiness. Plants absorb many chemicals from the soil, so toxic substances should be minimal. Pesticides remain attached to leaves and stems or are absorbed by roots, so pesticide use should

Figure 10.11. Private back space of a large house-plot flower garden and lawn. Hornbill at outdoor eating table. Ubud, Bali, Indonesia. R. Forman photo.

be minimal. Adding soil organic matter in the form of humus usually enhances food production and greatly decreases the need for additional fertilizer. With the elimination of pesticides and fertilizers, garden soil often improves. The density of soil animals increases, soil food webs become complex, and larger soil invertebrates increase and perforate the soil, enhancing both drainage and aeration.

Unlike the relatively unpalatable flower garden, vegetable gardens are planted with species delicious and nutritious for us, but also for many animals. Numerous insects suck plant juices or chew holes in leaves. A typical response is to spray pesticides against the insects, which also eliminates beneficial insects.

Mammals also relish many of the vegetable plants, and diverse birds feed on seeds or invertebrates among the garden plants. In Zambia, gardens are sometimes protected from elephant (herbivore) consumption by soaking pieces of cloth in engine motor oil, adding hot-pepper juice, and hanging the cloths around the garden perimeter. An alternative approach, widely familiar to vegetable gardeners, is to add plants that directly effuse chemicals into the air that, in turn, annoy and chase away animals. Zinnias, marigolds, and mints are common herbivore dissuaders.

Lawns

Lawns are estimated to cover perhaps 1% of the entire USA. These include lawns in public spaces and in non-residential areas, as well as on private house properties. One percent of the state of Missouri (USA) is lawn, including 135 000 acres of residential lawn (Robbins and Sharp, 2003). An estimated 23% of urban cover in the USA is lawn.

Across the city of Sheffield (UK) an estimated 60% of the house properties is lawn (Gaston et al., 2005b). Flower gardens are predominant in these properties, so only 41% of the properties, probably mostly the larger ones, contain lawns. In fact lawns cover 75% of these large properties. Since lawns are notoriously low in species number, the bulk of the biodiversity present is in the remaining quarter of these private outdoor spaces.

Lawns have been called "productive vegetation subject to frequent defoliation" (Gilbert, 1991). Mowing is the defoliator. Of the many factors leading to high productivity, two seem most important (Falk, 1980). The species have been genetically selected for high productivity. And mowing stimulates vegetative growth of grasses, the dominants.

First, consider the major characteristics of *lawns*. Normally the soil surface is smoothed, but little is done to improve the soil organic matter and its important organisms. Genetically selected grass seed or grass sod is planted on almost any type of soil. Three things are typically added repeatedly and usually in excessive quantity. Fertilizer, especially containing nitrogen, phosphorus, and potassium, is added. Pesticides, including herbicides against weeds, fungicides against molds, and insecticides against insects, are added. Water is added in small lawns by hand, but usually by sprinklers or inundation for larger lawns. Grass rapidly grows and is mowed to about 2.5–5 cm (1–2 in) height, mostly by rotating blades of a gasoline- or electric-powered vehicle. The grass cuttings are either dispersed over the lawn or removed as "yard waste," for instance, to a compost pile or to be discarded. These processes together represent intensive and costly lawn care, and a closer look is instructive.

Pesticides, fertilizers, and water are major inputs to lawns. One estimate indicates that 250 000 tons of pesticides (including herbicides) and fertilizers are used on lawns each year in the USA (Uhl, 1998; Robbins and Sharp, 2003). Another source estimates that 370 000 tons of pesticides alone are used on lawns annually (Law et al., 2004). Some 70% of the lawns are regularly fertilized. A typical ⅓-acre (0.13 ha) lawn in the USA annually receives 10 pounds (4.5 kg) of pesticide, 20 pounds of fertilizer, and 170 000 gallons (644 000 liters) of added water, that is, in addition to precipitation (Bogo et al., 2002).

Interestingly, most lawn areas are inherently heterogeneous, with variations in solar energy input, shade, adjacent woody plants, soil conditions, slope, and water conditions (Bormann et al., 1993; Uhl, 1998; Thompson et al., 2004). Thus, lawns can be somewhat species rich (Muller, 1990). But the rather homogeneous grass cover and extremely low biodiversity results from purposeful watering, fertilizing, pesticiding, and mowing that together basically smother the habitat heterogeneity.

Weed-free means biodiversity-impoverished. Herbicides reduce or eliminate weeds from the predominant grass cover of lawns, thus reducing plant diversity below its already low level characteristic of a single layer of herbaceous plants. Fungicides kill molds that may inhibit grass growth in wet spots or wet climates, but molds are a minor problem in most lawns. Insecticides kill the insects that eat grass, though such herbivory usually has little effect on the lawn as a whole. However, the negative ecological effect is the array of soil insects killed, greatly reducing soil animal biodiversity and food webs. Loss of the larger soil-perforating insects is particularly important. They create channels in the soil that facilitate water infiltration to roots, water drainage, and aeration of the soil with oxygen. Rachel Carson's *Silent Spring* (Carson, 1962) evolved in part from the loss of birds that fed on pesticide-laden invertebrates of lawns and surroundings.

Pesticides are mainly organic compounds that may decompose rapidly in an organic-matter-rich soil with abundant microorganisms. However, some pesticides persist for a long period, or may partially decompose to release organic products that persist. Pesticides applied to lawns wash from plants into the soil, and are carried onward by infiltrating rainwater and human-added water into the groundwater. From there they may be carried to nearby surface water bodies such as streams and ponds. As long as the pesticide persists, it kills organisms. Thus, pesticide runoff into aquatic ecosystems may significantly reduce macro-invertebrates, a major food of most freshwater fish species, and degrade aquatic habitat quality (Overmyer et al., 2005).

Integrated pest management is an alternative to heavy pesticide use on lawns. This approach emphasizes natural pest-control measures, such as increasing biodiversity to reduce pest population sizes, facilitating ecological succession, and introducing herbivores or predators that feed on the pest. All approaches are usually supplemented by minimal pesticide application. However, few species introductions are appropriate, because they require extensive preceding ecological analysis to evaluate the negative effects on other non-target species and the ecosystem, as well as the potential for an introduced species to become invasive.

Many types of fertilizers are added to the soil to enhance plant growth. These include animal manure, human manure, partially decomposed solid waste, ashes, and combinations of inorganic N, P, and K. Such fertilizers add organic matter, inorganic matter, or both. But too often heavy metals and organic toxins are included (see Figure 2.4) (Maconachie, 2007).

Nitrogen and phosphorus in fertilizer normally can be readily absorbed by the roots and increase lawn-grass production. Without carefully measuring the N and P levels in different portions of the lawn, the amount of nutrients absorbed by the plant roots remains unknown. Residents normally add more fertilizer, often much more, than can be absorbed by the plants (Bormann et al., 1993). The excess then is readily carried by water (again precipitation plus the human-added water) infiltrating into the groundwater, and onward to a nearby water body.

Excess phosphorus reaching freshwater aquatic ecosystems typically stimulates the microscopic floating algae (phytoplankton) to greatly multiply (see Chapter 7). The result is eutrophication: a dense bloom of algae turns the water body green. Algal cells are short-lived and continually die in large numbers, gradually filtering downward in the water where microbes decompose them. Unfortunately, the active microbial decomposition rapidly uses up the oxygen in the water, leading to extensive death of fish and reduction of aquatic biodiversity.

Excess nitrogen from fertilizers has a wider range of negative ecological effects (Bormann et al., 1993; Hubbard Brook Research Foundation, 2003). The lawn grass typically becomes less resistant to drought, less resistant to extreme temperatures, and more susceptible to disease. The soil may become more acid, thus reducing beneficial microbes and soil animals (see Chapter 4). High levels of nitrate from fertilizer can pollute groundwater wells.

Some water bodies, including estuaries, are eutrophicated by nitrogen. Nitrogen fertilizers from suburban lawns of Baltimore County (Maryland, USA) contribute more than half of the total nitrogen input to the watershed (Law et al., 2004). House plots of mid-range value, especially with newly built houses, contribute the most nitrogen. In marine areas near cities, "dead zones" with few organisms may be present. A massive dead zone exists in the Gulf of Mexico beyond New Orleans. However, this is mainly due to nitrogen from excess fertilizer use on agricultural land across the Midwestern USA draining down the Mississippi River to the Gulf.

Available freshwater for human uses worldwide is increasingly scarce and costly, yet in many urban areas watering lawns is a major consumer of water. In the USA a quarter of the non-agricultural water use is for irrigating lawns and gardens (McKinney, 2002). A typical ⅓-acre lawn is reported to receive annually 645 000 liters (170 000 gallons) of added water, as well as precipitation (Bogo et al., 2002).

A major portion of the human-added water is usually wasted. Some of the water sprayed over a lawn directly evaporates in the air to water vapor before hitting the plants, while more of the water is intercepted and evaporated from leaf and stem surfaces without ever reaching the roots. Evapo-transpiration loss is highest with high temperature and windy conditions. As soon as the soil zone containing roots is saturated, all subsequent water added simply puddles, runs off over the soil surface, or in effect infiltrates to the groundwater.

Infrequent heavier watering that soaks the soil deeply is better than frequent lighter watering that only soaks a thin upper layer of soil. The deeper water encourages development of a deeper root system. Such a lawn is therefore more drought-resistant, more extreme-temperature-resistant, and more pest-resistant. In addition, the lawn "requires" less watering, less fertilizer, and less mowing.

Lawn mowing of the typical one-third-acre (0.13 ha) USA lawn reportedly consumes 40 hours a year (Bogo *et al.*, 2002). One hour of mowing with a small two-cycle engine may emit pollutants equivalent to driving some 32 km (20 mi) in an average gasoline-engine car. Gasoline-powered lawn equipment causes almost 5% of the total air pollution each summer in the USA (Woodier, 1998). A small amount of gasoline (and oil) for mowers is spilled or poured by residents into the soil, where it may flow rapidly to groundwater or into a water body, and be highly toxic to aquatic organisms.

The characteristic roar of a mower in a summer neighborhood, which is associated with human hearing loss, doubtless inhibits wildlife, though this seems to be little studied. For example, baby birds in nests or fledglings on the ground that cannot hear the parents' alarm calls about an approaching cat or hawk are unlikely to survive. The parent birds' nest next year will probably be somewhere else.

Frequent mowing usually stimulates growth points (meristems) at the base of most grasses (e.g., *Agrostis capillaris*, *Festuca rubra*) to produce a large number of small tiller-stems that grow outward and upward (Gilbert, 1991). Some species (e.g., *Lolium perenne*) exhibit a rapid vertical regrowth of the mower-damaged leaves.

Lawns cut with the mower blade set for 7.5–10 cm (3–4 in) height appear to have noticeably higher plant diversity than lawns cut at 2.5–5 cm blade height (Broll and Keplin, 1995; Woodier, 1998). Without using herbicides, even cutting at a 2.5 cm height permits a fair number of non-grasses to survive or thrive (Gilbert, 1991). One set, so-called "lawn specialists" such as *Achillea millefolium*, *Hypochoeris radicata*, *Leontodon autumnalis*, and *Senecio jacobaea* in the UK, are virtually absent elsewhere in a house plot. Another set of lawn species, including dandelion *Taraxacum officinalis*, double daisy *Bellis perennis*, buttercup *Ranunculus repens*, and clover *Trifolium repens*, produce attractive low flowers in a lawn, and may grow equally well in cultivated gardens. Indeed, some garden plants (e.g., *Veronica filiformis*, *Chamaemelum nobile*) have successfully invaded lawns.

Yard waste is an odd name for mowed-grass cuttings/clippings and dead leaves, which can be quite useful and valuable in a house property. Secondary components of yard waste are branches, twigs, pulled weeds, and occasionally cut logs. This organic matter may be recycled into soil on the house plot. Alternatively, the resource is often discarded as part of society's solid waste. In the USA, yard waste is the second largest component of solid waste disposal (Sloane, 1996).

Branches may be valuable on a house property to create temporary fences and to maintain brush piles for common wildlife species. Logs and branches may be ground up for wood chips or sawdust to be used on walkways, or as mulch on flower gardens and vegetable gardens, or added to compost piles (heaps).

Mowed-grass clippings and pulled weeds are frequently recycled on-site by using them as mulch. Mulching adds organic matter to the soil and reduces the tendency to add water. Grass clippings can also be added to a *compost* pile or heap where microbes decompose plant material into humus (Brown *et al.*, 2000). Ample oxygen is important for rapid decomposition, so the pile of organic material is periodically turned over or mixed. Adjusting the pH by adding lime or altering the carbon-to-nitrogen ratio by adding dead leaves or other material may be needed for rapid decomposition. The humus produced is then added to gardens to enrich the soil and facilitate the activities of soil invertebrates and microbes, thus enhancing plant growth.

Back-yard outhouse toilets may also use composting to convert human waste to humus (van Bohemen, 2005; McGregor *et al.*, 2006). Solar heat accelerates the process, and plant or synthetic chemicals reduce smell. Food waste, especially of plant origin, is also readily composted, often in bins to keep wildlife from digging and spreading it.

Finally, I know a lawn that has not been seeded, fertilized, pesticided, or watered for 30 years. It is covered with grass species, and commonly used for badminton, bocce, ball-kicking, children's play, and parties. However, along with the predominant grass are numerous small and medium-size patches of flowers, which many people would call weeds. The patches have yellow, orange, white, violet, and purple flowers in a changing sequence through the growing season. Moss patches occur in partially shaded spots. Both native and non-native species enrich the lawn. Low flowers of some species are little-affected by mowing. Species with somewhat higher flowers are often so pretty that, when mowing, some patches are left unmowed to be enjoyed for a few weeks. Butterflies regularly flutter among these flowers.

The lawn is seemingly mowed at the same blade height, but about half as often, as a neighbor's almost golf-green-like yard. The lawn with little management has much higher plant diversity. Both the time and costs required for fertilizing, pesticiding, and watering are avoided. The gasoline used, the roar of the mower, and the mower pollutants emitted are half as much. Patches of different ant species are evident and some years patches of mushrooms appear. Many more birds, e.g., pulling worms and catching bugs, and more bird species use this less intensively managed lawn.

Other house-plot features

In addition to gardens and lawns, numerous mostly smaller components are present in house plots. Key ecological dimensions of these are briefly suggested to illustrate the richness of ecological patterns and processes that in turn are readily used in design by residents.

1. *Woody plants*. Trees, shrubs and vines appear in small and medium-sized patches, separately or intermixed. A larger patch with at least trees and shrubs normally has the widest range of soil and microclimatic conditions, from warm dry to cool moist, on the house plot. Woody plants also form corridors along streets, side plot lines, and, especially important ecologically, back-corner patches and backline strips.

Some woody-plant species attract and are greatly used by wildlife, while others are not. If a planted species retains its fruits and seeds for a long period, as often preferred by a resident, it means they are not very palatable or nutritious for wildlife. Fruits with toxins are avoided, those rich in sugars are eaten fast, and those rich in lipids (high energy per gram) are preferred by migrating birds (Stiles, 1980, 1982).

In the dry plains and mountains of the western USA, birds are particularly attracted to deciduous trees and deciduous shrubs that provide food (Figure 10.12) (Kress, 1985). Although evergreen trees generally are less attractive to birds for feeding, one species (western white pine) is highly attractive.

In the moist northeastern USA, small deciduous trees are most attractive to birds for feeding, though each of the six woody-plant types present contains at least one highly attractive species (Figure 10.13) (DeGraaf and Witman, 1979). In contrast, evergreen trees (especially white pine, *Pinus strobus*) are by far the most important in attracting birds for both cover and for nesting. Again each woody-plant group contains at least one species that is fairly attractive to birds for cover as well as for nesting.

The roles of native versus non-native woody plants, as well as city-tolerant versus other species, differ in detail, but overall are rather similar. These patterns (Figures 10.12 and 10.13) highlight the important roles that woody plant species of several types play in a house plot. Species can be chosen to greatly increase, in this case, the wildlife diversity and usage of a house plot.

2. *Water-related structures*. House plots may include structures with still water, flowing water, seeping water, dripping water, and/or splashing water. Wildlife respond differently to these, so a combination of water conditions may support the greatest diversity of wildlife and plants. Ponds are perhaps most common and provide considerable ecological benefit (Gilbert, 1991). Mini-wetlands may be associated with any of the water conditions. Amphibians such as frogs, toads, and salamanders are especially drawn to water. However, if non-herbivore fish live in the water, amphibian eggs and young may not survive. Many flying and other insects are drawn to water. Birds and mammals frequently drink and sometimes bathe in water structures.

Lots of ecological issues arise when artificial water structures are introduced. Clay or an impervious bottom is needed to hold the water. Some flow is required to oxygenate the water and avoid smelly standing water. Water structures are often added by people in a dry climate, ironically where water is scarce. Flowing water is either recycled by pumping or flows elsewhere. A soil depression may temporarily hold surface runoff from heavy rains and snowmelt. The water-table may be high, so

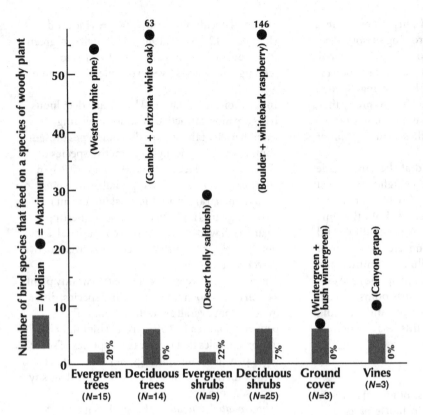

Figure 10.12. Birds feeding on woody plants in a dry region. Based on records of bird species feeding on fruits, seeds, buds, etc. on each of 69 woody plant species in the mountains and deserts of western North America. Species with the most bird usage are listed in the upper portion. The genera, raspberry and winterberry are used where bird use of the different component species is considered by the author to be similar. Percentages at the bottom are the proportion of woody plants with no birds feeding thereon. N = number of species. Based on Kress (1985).

added water creates soggy soil. In effect, outlining a water budget (see Chapter 6), indicating the distribution of water and the amounts of flows in and surrounding a house plot, is valuable before adding a water structure to a house property.

Considerable incoming precipitation means that a drainage system must deal with water runoff from roofs and driveway as well as the yard. Optimally the system channels water directly into the ground or to small stormwater depressions, rain gardens, and the like. These facilitate the infiltration of runoff into the soil, where some may be evapotranspired by plants on-site. If drainage runs off-site, the effects downslope as well as in a nearby water body are important.

A well providing groundwater for drinking needs unpolluted freshwater. Thus, the well is spatially arranged relative to pollution sources, especially the system for disposing of human sewage. In the absence of a sewer system, on-site human sewage is deposited in a hole in the ground covered by an outhouse, or a cesspool, or a septic system (see Chapter 6). Sewage is composed mainly of water, organic matter, bacteria, and mineral nutrients (especially N and P), though heavy metals from pipes are commonly present. Sewage from an outhouse or cesspool is either periodically transported off-site or infiltrates and accumulates on-site, typically causing problems of pathogenic bacteria, smell from anaerobic bacterial decomposition, and eutrophication of a downslope water body. A septic system when working well minimizes or eliminates all three of these problems (Brown *et al.*, 2000), though a surprising proportion of septic systems function poorly.

3. *Garden walls*. Walls can be covered with ornament and niches used by a wide variety of plant and animal species. Garden walls in urban house plots are rather well known ecologically. A survey of 650 walls in Essex (UK) included 278 urban house-plot walls, which had 150 plant species growing on them (Payne, 1978). Common native garden weeds are among the most abundant species. Some plant species primarily live on walls (*Cheiranthus cheiri, Cymbalaria muralis, Antirrhinum majus, Parietaria diffusa*, and in moister areas, ferns *Asplenium* spp.). About 35% of the urban garden walls contain species (e.g., *Antirrhinum, Cheiranthus,*

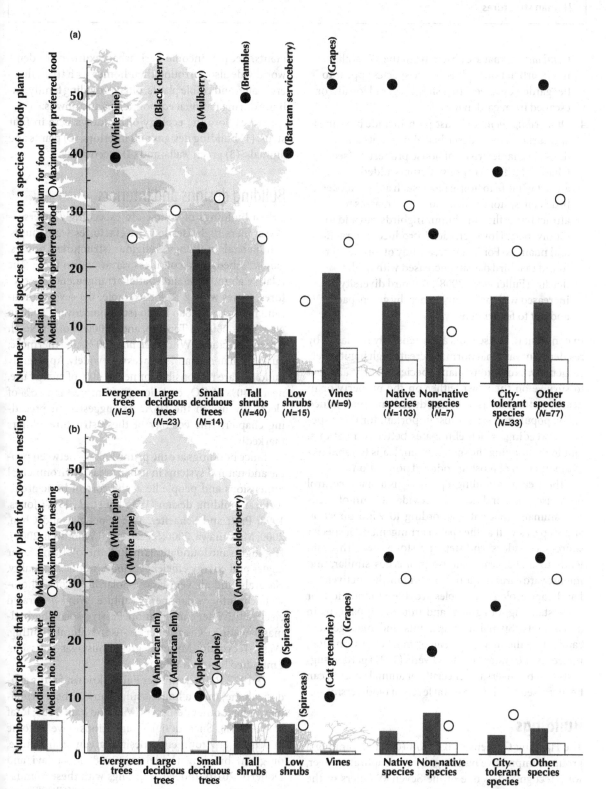

Figure 10.13. Woody plants and bird use in cool moist forest area. Based on records of bird usage on each of 110 woody plant species (useful for attracting birds to residential areas) in the northeastern USA and southeastern Canada. Feeding is mainly on fruits, seeds, or buds; nesting mostly in twigs, branches, cavities, or bark. Shrubs and vines are primarily sources of fruit. Cutoff between large and small deciduous trees is typically 12.2 m (40 ft) height, and tall and low shrubs typically 1.5 m (5 ft). City-tolerant species described as able to withstand city conditions. 1 ft = 0.3 m. Also see Figure 10.12 caption. Based on DeGraaf and Witman (1979).

Cymbalaria) that are absent from the 372 walls not in urban house plots. These species appear to be "garden escapees" that simply spread locally, or escaped from garden to wall.

4. *Other components* of house plots include buildings and surface cracks described above. Cats and dogs are characteristic of house properties (see Chapter 1). The array of structures added to attract wildlife to house plots, such as bird feeders, brush piles, stone walls, and flower masses to attract butterflies and hummingbirds can add to biodiversity. However, each introduces interactions and nuances. For instance, a study of bird feeders found that bird density increased with feeder density (Fuller et al., 2008). But bird diversity increased with proximity to nearby greenspace, and not to feeder density.

In conclusion, house plots are intensively managed by residents in ways that normally create high habitat heterogeneity. Numerous plant species, poorly adapted to survive on their own either on the property or in nearby natural areas, are introduced and maintained at low population sizes. Most important for these species is weeding, which eliminates better competitors. Suddenly stopping the management leads to rapid on-site extinction of most introduced populations.

The area surrounding a property is a major control on house-plot conditions. It provides a "rain of seeds and animals" that enter according to wind direction and slope, as well as the spatial arrangement of species sources, corridors, and stepping-stone rows. This rain tends to make nearby house properties similar, and intense yard care is required to maintain distinctiveness. Landscape ecology principles provide a foundation for understanding, designing, and improving patterns in house plots. Spatial arrangements and management can sustain the recycling of all or most organic matter, nutrients and water on-site. Owen's (1991) garden suggests that biodiversity, especially of animal species, can be increased far above natural levels of biodiversity.

Buildings

The Hanging Gardens of Babylon (600 BC), among a group of impressive buildings by the Euphrates River, were recognized as one of the Seven Wonders of the World. Built by kings on terraces atop stone columns, the gardens apparently had trees rooted in the terraces, water gently flowing along sloping channels, a luxuriance of flowers, and doubtless medicinal and edible plants. Yet low-income peri-urban/exurban residents worldwide also surround their homes with trees, flowers, water, and edible plants, often providing family stability during national economic ups and downs.

We explore the ecology of buildings here in four steps: (1) building designs and interiors; (2) plants and animals; (3) green walls; and (4) green roofs.

Building designs and interiors

Urban buildings of course are exceedingly diverse – skyscrapers, high rises, low rises, factories, warehouses, commercial buildings, religious structures, houses, garages, sheds, and countless variations on these. The relative abundance and spatial arrangement of structures, varies widely, as for example, the several thousand high rises in Sao Paulo (see book cover), 5000 in New York, 2000 in Toronto, and <1000 in other North American cities (Wilcox et al., 2007). Furthermore, building materials and surfaces vary widely. Apparently buildings are responsible for nearly 40% of energy use, nearly 40% of CO_2 emissions, and more than 70% of electricity use in the USA. As suggested in preceding chapters, the ecology of these structures differs markedly.

Since buildings are the prime linkage between people and natural systems in urban areas, environmental dimensions and people lie at the core of basic architectural building design (Woolley et al., 1997; Jones, 1998; Pratt and Schaeffer, 1999; Spengler and Chen, 2000; Matsunawa, 2000; Steele, 2004; Woodwell, 2009). Highlights and foundations include the following. Solar angles and energy efficiency. Good indoor air quality. Natural-wind air ventilation. Solar hot water. Photovoltaic or other solar energy capture. Wind-generated electricity. Water conservation. Passive-solar/thermal-mass heating. Grey-water re-use. Natural lighting. Waste recycling. Minimal greenhouse-gas emission. Embodied life-cycle energy.

In 1927, R. Buckminster Fuller (known for the geodesic dome) built a house using many of these principles, as have other designers (Vale and Vale, 2000; Roaf et al., 2001). Such design foundations have become known as environmental design, green architecture, or green building (Yudelson, 2008; Mostafavi and Doherty, 2010). Indeed, buildings with these foundations missing or weak seem to almost ask "Why?", especially over time. A rating certification-system (LEED) has emerged in the USA highlighting site development, water use, energy efficiency, materials used, indoor air quality, and other environmental attributes (Yudelson, 2008). Overall, this certification system has served as a beneficial stimulus. So far, the ratings are mainly used for commercial and government buildings. Extension of the approach to residences would be a valuable step.

Yet the present certification system, focusing on physical environmental conditions, is only a first step. The next big step is to address the equally important biological dimensions of plants, animals, and microbes. Humans are deeply tied to the biological world. Without adequate vegetation, buildings and cities feel sterilized. The array of opportunities and potential solutions beckons creative thinkers and designers.

"Green architecture" has been characterized with different emphases, such as "minimize environmental impacts," "integrate buildings with local natural ecosystems," or "create positive benefits for the natural environment" (Yeang, 1999). Vernacular architecture tailored to local physical and cultural places may be a prime goal. Design with nature, or with a sense of place that includes nature, seems to best express the idea of green architecture (McHarg, 1969; Kellert, 2005).

Indoor plants play a special role for biofiltration (cleaning of air) and for biophilia, as direct benefits to people. Space exploration scientists have noted the reduction of trace levels of air pollutants when using plants and soil microbes in enclosed spaces (Wolverton et al., 1984). Indoor plants are reported to reduce the concentration of airborne particulate matter (PM) and toxic organic gases (toluene, ethyl-benzene and xylene) (Lohr and Pearson-Mims, 1996; Darlington et al., 2000, 2001). Many environmental hazards, from cooking smoke to toxic gases from furnishings, exist inside buildings (Bartuska and Young, 1994; Nielson and Rogers, 2000; Hardoy et al., 2004; Kellert, 2005; Yudelson, 2008). Nevertheless, biofiltration with more plants, lower pollutant levels, and a low-to-medium ventilation rate can help clean the air.

Biophilia, the inherent human affinity for nature, emphasizes that people evolved with, fundamentally depend on, and are inspired by nature (Ulrich, 1984; Wilson, 1984; Lohr et al., 1996; Kahn and Kellert, 2002; Kellert, 2005; Kellert et al., 2008). Hospital patients mend faster with a view of vegetation rather than a brick wall. Worker attention and productivity are higher with indoor plants present. Prison inmates are more satisfied with a view of farmland or forest. Children's mental development is greater around plants and animals. People have lower stress and more enjoyment with nature nearby. "Bringing buildings to life" indoors is good for people, but also for nature.

Plants and animals

Plants and animals on the exterior walls and roofs of buildings can be diverse and provide important human and ecological benefits. Wall plants can give urban residents a close experience with nature, watching growth, flowering, pollinators, fruits, seeds, herbivores, and nesting animals (Figure 10.14). Cracks, crevices, holes, window boxes, balconies, open stairwells, and roofs are familiar locations to view plants. Mortar decomposes more rapidly than the adjoining bricks or stones, providing roughness suitable for plant germination and growth (Laurie, 1979). The suitability (bioreceptivity) of a material and surface in facilitating use by plants and animals varies widely (Guillitte, 1995; Miller et al., 2010). Modern buildings with smooth surfaces and little ornamentation seem opposed to nature's colonization processes. Of course organisms also can degrade surfaces, extreme cases being a termite nest on a wooden building or a strangler fig (*Ficus*) root growing in a large hole (Robinson, 1996; Miller et al., 2010; Jim and Chen, 2011). Usually spontaneous vegetation on buildings is not a monoculture, but rather a simple natural community, a few species coexisting.

Wall vegetation only 2–10 m above ground level has the greatest benefit in supporting urban wildlife (Figure 10.14). The diversity of nesting birds, small mammals, bees, butterflies, night insects, and other insects is greater at this approximately one- to two-story level than at higher levels.

A film of mainly green algae often thrives on flat or sloping surfaces, which receive ample rainwater, of certain construction materials and exposures (Miller et al., 2010). Other microbes are typically mixed in. Lichens tend to be less tolerant of air pollution including sulfur dioxide, and more tolerant of desiccation. Though less predictable in distribution, lichens may be more frequent on vertical surfaces, high-pH mortar between bricks, and (for certain species) surfaces affected by bird droppings. Mosses also may thrive on somewhat rough moist surfaces, especially on shady sides of buildings (Laurie, 1979).

A study of walls on low-rise and high-rise buildings in old districts of Hong Kong recorded 692 woody plants of 11 species (Jim and Chen, 2010, 2011). Ten species were trees and one a shrub, 74% of which were native species. The plant families represented were Moraceae

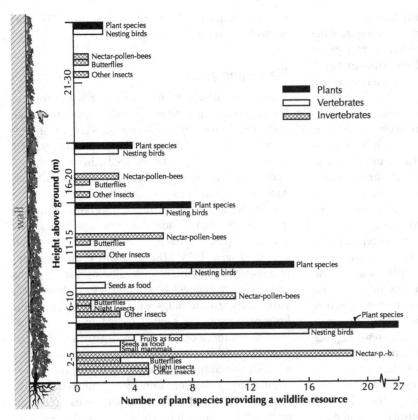

Figure 10.14. Heights of wall plants and their wildlife values. Suitable wall plants for London, a north-temperate city. Highest growing plants = ivy (*Hedera helix*; 30 m), Russian vine (*Polygonum bauldschianicum*; 30 m), vine (*Vitis* spp.; 20 m), wisteria (*Wisteria* spp.; 18 m). Total wall plant species = 5 self-clinging climbers, 14 twining climbers, 8 rambling shrubs. Native plant species not in noticeably higher proportion than non-natives for north-versus-south sides, or for individual wildlife values. 1 m = 3.28 ft. Adapted from Johnston and Newton (1997).

[76%, composed of 7 fig (*Ficus*) species], Apocynaceae, Bombacaceae, and Rutaceae. The building-wall plants seem to have arrived when a fruit-eating bird, bat or other mammal fed on nearby fruits of the species, and then defecated the undamaged seeds on the building. Rainwater probably then washed the seeds into crevices with debris or soil, moist surfaces with cracks, and leaking drainpipes, the three primary locations of the wall woody plants. Strangler fig species such as *Ficus macrocarpa* also grow on buildings in Hawaii, India, Singapore, and Australia (Jim and Chen, 2011). Wall plants in tropical cities without frost can be especially diverse (Dos Reis et al., 2006).

A summary of data from across Europe recognized >1200 vascular plants on walls, though many grew on walls along gardens or property boundaries rather than on buildings (Woodell, 1979; Showell, 1986). The base of a wall commonly has an abundance of plants associated with accumulated soil, mineral nutrients, and organic matter (sometimes from wall plants above) (Woodell, 1979). These species benefit from enhanced moisture, lack of trampling, and minimal maintenance activity.

Worldwide, certain types of species seem frequent or promising for growth on buildings. Planted plants in pots and window boxes often feature *Geranium*, *Impatiens*, and *Petunia*, perhaps because of their desiccation- and pollution-tolerance. Edible, medicinal, and spice plants are also commonly planted by people on buildings. Perhaps the most frequent vines on buildings are ivy (English ivy; *Hedera helix*), creeper (Boston ivy or Virginia creeper; *Parthenocissus tricuspidata*); certain fig species (*Ficus*); and wisteria (*Wisteria*), though many vine species are present. "Epiphytes" (air plants frequently growing on other plants) such as ferns, orchids, and bromeliads, are especially conspicuous on buildings in the tropics. Drought-adapted species from deserts and Mediterranean-type climates may thrive on walls. Plants from rock cliffs are often present. The relatively extreme conditions on building walls may provide an opportunity for certain plants to thrive in the near absence of competitors (Laurie, 1979). Characteristics that tend to tie these diverse plant types together are low maintenance, low water requirement, desiccation resistance, air pollution resistance, paucity of pollinators, and scarcity of herbivores.

Animals, however, are also present on buildings, sometimes in abundance (Bird *et al.*, 1996; Cade *et al.*, 1996; Adams *et al.*, 2006). House sparrows and common pigeons (rock doves) are especially conspicuous on buildings in most urban areas (see Chapter 9). Parakeets may provide pleasure. Rodents, cockroaches, and even monkeys may be pests. Termites and carpenter ants cause structural damage to buildings. Spiders feed on insects. Lizards feed on insects and spiders. Bats roost in buildings, sleeping by day and foraging by night. Raptors (some hawks and owls) nest on or in buildings.

Green walls

Rock cliffs have been extensively studied for energy flux, wind, water, mineral nutrients, crevices, overhangs, microhabitat heterogeneity, surface degradation, numerous plant types, and diverse animals (Gerrath *et al.*, 1995; Ursic *et al.*, 1997; Larson *et al.*, 2000; Kuntz and Larson, 2006). Abandoned or little-used rock quarry faces have similar patterns (Wheater, 1999; Larson *et al.*, 2000). Not surprisingly, cliff ecology has been used to understand urban buildings and walls.

In urban areas, bricks-and-mortar, small wall cracks, large crevices, window boxes, balconies, ornamented facades, outside stairwells, and many other structural features provide habitat heterogeneity supporting diverse species (Dos Reis *et al.*, 2006). However, overall the present plant cover of walls is tiny, effectively appearing as scattered green spots. Walls represent the greatest potential for increasing vegetation in urban areas.

The ecological functions or benefits for society seem impressive. Wall vegetation cools the air, humidifies the air, reduces stormwater runoff, reduces noise, filters air pollutants, and insulates, shades, and reduces UV radiation for buildings. Air cooling is related to wall area covered, not vegetation thickness, and one calculation suggests about a 5°C cooling of air just above street level (Yeang, 1999; Dunnett and Kingsbury, 2004). Wall plants are reported to reduce concentrations of chemicals, e.g., formaldehyde (especially by *Philodendron*), benzene (especially by *Hedera*), and decrease cadmium and lead levels. It seems probable that airborne street dust would also be markedly reduced by wall vegetation. The functions of wall vegetation represent a significant research frontier.

While overhangs, balconies and other structural features help cool street canyons, adding wall vegetation may be an important bonus (Johnston and Newton, 1997; Yeang, 1999; Beatley, 2000a; Shashua-Bar and Hoffman, 2004; Ali-Toudert and Mayer, 2007). Creating these green streets is particularly useful where street trees are absent. Tropical green walls may have an abundance of epiphytes, and typically have a greater cover of moss (and hepatic) species.

Green walls enhance urban plant diversity and provide both habitats and stepping-stone connectivity for movement of animals. For instance, the evergreen ivy (*Hedera*) provides nectar for pollinators, hibernation or overwintering sites for a diversity of insects, and shelter for the winter roosting of small birds. At cool times butterflies absorb heat by spreading their wings on walls or wall vegetation. In Uberlandia (Brazil), 33 butterfly species, representing a quarter of the area's species pool, were recorded on building walls (Ruszczyk and Silva, 1997). Adults were found mainly on exposed warm dry surfaces, while pupae were more often under protected shaded overhangs. Somewhat analogously, spiders were mainly on heat-retaining locations, whereas their webs were mostly around cracks and crevices sheltered from direct sun and rain (Voss *et al.*, 2007).

I have stood in awe, and with great curiosity, before two much-visited impressive green walls (living walls, vertical gardens): in Paris at the Musée du Quai Branly (near the Eiffel Tower and the Musée des Égoutes de Paris); and in Madrid at the Paseo de Prado. Both are walls of four- to five-level buildings, with water trickling down through the plants from the top to a small trough at the bottom (Blanc, 2008). Nutrient fertilizer is apparently added to the water.

Some 15 000 plants of 250 mostly native species were planted on the Madrid wall, which I relished watching over two months during spring (Figure 10.15) (Box, 2011). Sheets of mineral wool or felt are supported by strong metal or plastic frames, and plants are commonly inserted into small pockets in the sheet. Instead of producing a radial arrangement of foliage upward or outward, overall many plants tend to have a narrow downward form, presumably due to gravity and water. Thus, plants, as well as species, can be packed together. Many cliff and waterfall species were planted. From the ground I could usually see about 15 species in flower – reds, yellows, whites, and purples. Fruits were noticed on one species. Habitat heterogeneity is present, as the upper plants receive more wind and light, while lower plants are more shaded and moist.

Ecological succession of course occurs, so presumably numerous plants and species originally planted were outcompeted and died quickly. The more aggressive or dominant species in these unusual conditions

Human structures

Figure 10.15. Green wall with dozens of plant species patches. Gravity and dripping water tend to make many patches extend downward. In counteracting natural ecological succession, the bottoms of several patches have been cut away leaving large holes for plants to grow in. Five-level building, Madrid. R. Forman photo.

spread, a process accelerated by the added nutrients (indeed recycled grey-water could be used). Doubtless the original planted biodiversity plummeted. So, to keep a diverse appearance, maintenance personnel periodically cut back the dominants. Gardening on a wall resets nature's inexorable process of succession.

In brief observations on ten days the paucity of animals surprised me. I never saw a flying insect and only observed three birds (house sparrows, *Passer domesticus*) in the watered vegetation. Even though many flowers were blooming in spring, animal diversity may well be higher in the summer. Green-wall plants could be selected, for instance, to have abundant nectar, pollen, leaves for herbivorous insects, predatory invertebrates, lizards, and birds.

Designing and building a *skyscraper* is the pinnacle of many architects' dreams. Using the solid environmental foundations described at the outset of this section, along with a significant cover of plants for a >40-story (>150 m = 500 ft) building, is rare indeed.

Extensive green walls may be a key component of such an achievement. Creative designs are a delight to ponder (Yeang, 1999; Richards, 2001; Pacione, 2005; van Bohemen, 2005; Newman and Jennings, 2008). Still, four such built skyscrapers or high rises in Colombo (Sri Lanka), Kuala Lumpur, Singapore, and Frankfurt (Germany) stand as pioneers (Yeang, 1999; Gissen, 2003; Boeri and Insulza, 2009/10). Three are in tropical Southeast Asia and one in Northern Europe.

The Colombo building has a long wind-facing façade to reduce solar energy gain; limited height for less elevator transport and energy usage; large overhangs; ground floor open; natural-light staircase; and considerable local materials used, including wood (Robson and Bawa, 2002; Ismail *et al.*, 2011). The Kuala Lumpur design has terraces and balconies; overhangs; large windows; shading devices; double walls; wind scoops and ducts; airflow slots; an open ground floor; "cores" on the sunny hot side; courtyards on every fifth level; and an open rooftop courtyard (Yeang and Richards, 2007; Hart and Littlefield, 2011).

The Singapore structure has outside plants forming a connected spiral or helix from street to summit, and almost all the species grow natively within 1.5 km (Yeang, 1999). Natural ventilation is incorporated, rainwater and recycled grey-water are used, and the considerable surface cover of vegetation cools by evapo-transpiration.

The triangular Frankfurt structure also uses natural ventilation, and has nine "sky gardens," large partially open levels with vegetation (Gissen, 2003). South-facing gardens display Mediterranean plants including citrus, olive, thyme, and lavender. West-facing levels contain North American plants such as maple, rhododendron, and grass, while east-facing sides feature Asian plants, especially magnolia, bamboo, and hydrangea.

Of course, the hard test of future green-wall skyscraper success goes beyond plants and animals for decoration and human enjoyment. The vegetation must also provide benefits in cooling the air, humidifying the air, reducing stormwater runoff, reducing noise, filtering air pollutants, and insulating, shading, and reducing UV radiation for buildings. Then the whole city benefits.

Green roofs

At first glance green roofs all look the same, a cover of plants. Quickly one sees dramatic differences mainly based on two factors (Johnston and Newton, 1997; Beatley, 2000a; US Environmental Protection Agency,

2000b; English Nature, 2003; Peck and Kuhn, 2003; Dunnett and Kingsbury, 2004). First, the basic goal or function of the vegetated roof varies widely, from flower and shrub garden for enjoyment, to growing vegetables and fruits, maintaining a semi-natural meadow with biodiversity, reducing heat loads, or reducing stormwater runoff.

Two green roof types predominate based on soil substrate thickness. An *extensive green roof* has a soil substrate <10 cm (4 in) thick, whereas an *intensive green roof* has deeper soil (Dunnett and Kingsbury, 2004; van Bohemen, 2005; Hien et al., 2007). Soil thickness especially determines how much water, and hence weight, may be present, and what plants will grow and survive. Extensive green roofs with thin soil have low maintenance requirements, are little-visited by people, support relatively few animal species, and may be successful on roof angles up to 25–30 degrees (Figure 10.16). Normally very few plant species survive the desiccation (or freezing) present. In contrast, intensive green roofs with deeper soil hold considerable water and may be irrigated. A relatively wide range of plants grows well, though trees are usually avoided since extreme wind velocity and turbulence occur periodically. People and animals are often common on intensive green roofs.

Vegetated roofs provide several benefits for the building, such as reducing fire hazard, prolonging the life of roof surfaces (by reducing extreme temperatures), insulating for warmth in winter, insulating for cooling the building in summer, and reducing life-cycle cost (Wong et al., 2002; Saiz et al., 2006; Oberndorfer et al., 2007). However, just as for green walls, vegetated roofs provide several outside urban air and water benefits. A green roof contributes to cooling the air in warm periods, cleaning the air of varied pollutants, and reducing stormwater runoff and potential flooding.

In summer the surface temperature of a green roof is almost as cool as a highly reflective white paint surface, and is cooler than any other common roof surface (Hien et al., 2007; Takebayashi and Moriyama, 2007; Gartland, 2008). Rather than reflecting incoming energy, the plant cover cools by evapo-transpiration (giving off latent heat; see Chapter 5). The cool vegetation surface does not heat the air much, so air temperature is lower. Overall, in summer the vegetation surface is cooler than the air, while in winter the surface is warmer than the air. If many urban buildings had green roofs, estimates suggest that the urban air temperature would be 0.5–2°C cooler (Velazquez, 2005; Saiz et al., 2006).

(a)

(b)

Figure 10.16. Green roofs of different type. (a) Atop a five-level hospital, a constructed meadow contains gravel areas, 10-cm-thick soil areas with low plants, 20-cm soil areas with different taller plants, 30-cm-thick soil areas with still-different plant species predominating, plus a few logs as perches for birds. High habitat heterogeneity supports considerable species diversity of beetles and spiders. Basel, Switzerland; courtesy of Stephan Breinnesen. (b) Extensive sloping roof with mostly sedum (*Sedum* spp.) planted and then colonized by taller flowers. Peri-urban Zurich, Switzerland. R. Forman photos.

Air quality is also improved by vegetation (see Chapter 5). Various empirical studies suggest a decrease in airborne particulate matter (PM_{10}) (Getter and Rowe, 2006; Yang et al., 2008), such as 5 m² of plant cover annually removing 1 kg of particulates (van Bohemen, 2005). Decreases in SO_2 and N_2O gases are reported. Also most cadmium, copper, and lead (and some zinc) were removed from rainwater by roof vegetation (Velazquez, 2005; Getter and Rowe, 2006).

Stormwater runoff is noticeably decreased by a green roof (Beatley, 2000a; Scholz-Barth, 2001; Dunnett and Kingsbury, 2004; Nicholaus et al., 2005; Velazquez, 2005; Gartland, 2008). In addition to reducing the amount of water running down drains, the peak flow is generally delayed a few hours. Even green roofs sloping at 25 degrees greatly reduce runoff (Getter et al., 2007). Compared with a conventional black roof, a green roof reduced water runoff from small rains by 90% and from large rainfall events by 50%, in both cases also delaying peak flows (Carter and Rasmussen, 2006). An extensive green roof with only 3 cm of soil (with a drainage layer beneath) also markedly reduced runoff, but only until the soil was saturated, after which there was no stormwater benefit (Bengtsson et al., 2005).

Rainwater may be captured and stored for the irrigation of intensive deep-soil green roofs, such as flower gardens and vegetable gardens (Roaf et al., 2001). Indeed, recycled grey-water may be particularly useful for roof irrigation. The filtration of rainwater chemicals by roof vegetation also helps clean the stormwater runoff (Velazquez, 2005).

Visiting a green roof atop an eight-level (story) hospital in Basel (Switzerland), I unexpectedly experienced a remarkable "meadow in the sky" (Figure 10.16a) (Dunnett and Kingsbury, 2004; Brenneisen and Haenggi, 2006). Indeed, normally meadows in a city are rare and ephemeral, since they quickly become ball fields, lawn parks, or buildings. Dozens, even scores, of green roofs adorn buildings in Basel, as well as Zurich, and two vegetated roofs were close by the hospital meadow.

A luxuriance of native plants and flowers, mostly planted, surprised me. Normally the meadow absorbs all water from rain of up to about 2.5 cm (1 in). One habitat type, a large pebbly patch, had moss and few flowering plants, and enhanced subsurface water runoff.

The other three habitats had sandy-silt soil (very little clay) from nearby areas that contained local seeds. Most common was a matrix of thin soil (about 10 cm thick) covered by moss, sedum (a low succulent plant), and seedlings of many species. Within this low vegetation was an abundance of small mounds or areas about 20 cm high, and some larger mounds 30 cm high. The thicker soils supported grasses, aster, yellow-flowered composites, blue-flowered plants, and many other species. Still, the vegetation on 20 cm of soil looked quite different from that on 30 cm of soil. The four microhabitats differed in water drainage. Habitat heterogeneity had produced a meadow with relatively rich plant species diversity. Naturally, the meadow changes as ecological succession proceeds. About ten seedlings of woody plants appear each year, and over several years grass species have expanded in cover.

Ground beetles and spiders seem to thrive in the meadow, the latter including 90 species, of which 20 are rare or endangered. Pollinating insects are common on the flowers. House sparrows, common pigeons, and wagtails (*Moticilla*) visit the meadow from time to time. Lapwings (*Vanellus vanellus*), a plover-like species, breed in the meadow. They have only successfully raised one young in 4 years, perhaps due to limited food for the young. Heavy logs were added to attract black redstarts (*Phoenicurus ochruros*), so far with limited success. Although not rare here, the redstarts use green roofs in London, where they are rare (Frith and Gedge, 2000). The 30 cm vegetated mounds are apparently important for black redstart breeding.

The soil substrate for most green roofs poses both nutrient and water problems (Craul, 1992, 1999). Fine-particle soils normally hold an abundance of nutrients including N and P. Such high-nutrient or fertilized soils encourage rampant growth by a few species, whereas low-to-medium fertility reduces weed dominance and encourages rich plant diversity. Some vegetated roofs use irrigation, and some contain shaded and unshaded portions. Considerable clay, silt, or soil organic material may hold excessive water, creating weight problems, frost heaving, and potential waterlogging of plant roots. Sandy soils dry out too quickly. Considerable lightweight "vermiculite" or similar horticultural-rooting material in the soil holds water, but readily blows away in the strong winds. Some specialists recommend a 5–8 cm (<10 cm) thick soil for growth of sedum and grasses. Using soil from nearby natural land provides seeds of native species, though as noted in Chapter 8, some non-native species may be better adapted than most native species to the severe environmental rooftop conditions (Kendle and Rose, 2000). As in the meadow example, varying soil depth enhances biodiversity (Dunnett and Kingsbury, 2004; Gedge and Kedas, 2005; Brenneisen and Haenggi, 2006).

A wide variety of plants grows on green roofs, even excluding the rich tropics (Kohler et al., 2002; Snodgrass and Snodgrass, 2006). Thus, 135 species of vascular plant were recorded on 636 green roofs in the UK (English Nature, 2003). The predominant species are *Sedum acre*, 34% of the roofs; *Saxifraga*, 21%; *Poa* (grass), 15%; *Senecio*, 13%; *Stellaria*, 5%; *Cardamine*, 5%; *Acer pseudoplatanus* (a maple), 4%; *Oymbalaria*, 4%; and *Sedum reflexum*, 4%. Deeper-soil intensive green roofs support a much greater diversity of plants, but are more idiosyncratic and different from one another, and ironically less studied ecologically

(Stevens and Harpur, 1997; Dunnett and Kingsbury, 2004).

Extensive thin-soil roofs are mostly covered with sedum (*Sedum* species), though sempervivum, thyme, allium, phlox, and antennaria may also be common (Scholz-Barth, 2001). Sedum is especially suitable in such dry situations because it produces organic compounds for growth in a process (crassulacean acid metabolism) analogous to photosynthesis during daytime, but without opening its stomata (pores) and losing water molecules (Dunnett and Kingsbury, 2004). Plants with wind-dispersed seeds, and growing atop buildings with high winds, may readily colonize other green roofs. Little information is yet available on the species composition and abundance of algae, mosses, and lichens in green roofs (Oberndorfer et al., 2007).

Certain groups of animals are scarce or absent in green roofs. Strong winds, and perhaps an abundance of insect-eating bats, limit butterflies, moths, and most other flying insects (Velazquez, 2005). Frost, if present, eliminates any earthworms, snails, and amphibians that arrive. Little cover is present for animals to escape the eyes of predatory crows and gulls. Ants, bugs, flies, bees, and springtails (collembolans) are often observed in green roofs. However, ground beetles and spiders seem to be common (Brenneisen and Haenggi, 2006; Schrader and Boning, 2006; Oberndorfer et al., 2007). Indeed, the number of rare spider species in Basel green roofs is equivalent to the number in nearby urban ground habitats (abandoned railroad locations) of high nature-conservation value.

Seagulls, common pigeons, house sparrows, and other common birds often briefly visit green roofs. A few songbird species use vegetated roofs for feeding and/or nesting (Yeang, 1999; Frith and Gedge, 2000; Dunnett and Kingsbury, 2004). Some coastal birds find pebble or gravel areas, such as those on green roofs, suitable for nesting (Duncan et al., 2001). Peregrines (*Falco peregrinus*) use fine gravel for nesting but also require cover from precipitation and wind. Grassland birds might be attracted to roof meadows. Adding heavy logs to a green roof provides perches for birds and habitat for some invertebrates. Also, low berry-producing shrubs would attract certain birds.

At ground level, numerous courtyards or patios with trees projecting upward, as in cities of Latin America and the Mediterranean Basin, probably function as stepping stones for the movement of birds and other flying animals across an urban area (Beatley, 2000a; Shashua-Bar and Hoffman, 2004). Perhaps balconies with vegetation provide the stepping stone function at a finer spatial scale (Stevens and Harpur, 1997). Although still a hypothesis, the abundance of green roofs in an urban area, such as in many German and Swiss cities, potentially provides stepping stones for movement of certain species across an urban area (Schrader and Boning, 2006).

Migrating birds over an extensive urban area would potentially stop on a green roof for a day or so en route. However, migrants especially need food, as well as cover for resting, neither of which is present in abundance on green roofs. Furthermore, most songbirds migrate in flocks. A semi-natural greenspace at ground level is best for such migrants, and even ornamental plantings provide cover.

Chicago's ten-level City Hall is covered with a green roof, much being the thin-soil extensive type. Yet a heterogeneous mix of soil depths is also present, including mounds for trees and shrubs over underlying support structures (Daley and City of Chicago, 2002; Dunnett and Kingsbury, 2004). Over 150 plants, mostly native species from the Chicago Region, were planted. The vegetation contributes to slightly reducing local ozone, nitrogen dioxide, and sulfur dioxide levels (Yang et al., 2008). Bird houses added have been used by various species. Many key people visiting the Mayor's tenth-floor office have thus experienced an intriguing hybrid extensive-intensive green roof.

Urban agriculture on rooftops as yet seems to be little-studied ecologically. Rabbits are raised on Havana rooftops (Premat, 2005). Various vegetables and fruits and spices are grown. The constraints on rooftop food production, however, are many: need for irrigation; soil texture limitations; soil desiccation; soil freezing; soil blowing away; waterlogging; damage by high winds; air pollutants; and few pollinators. Still, advantages exist for green-roof farming, including ample sun, relatively few weeds, few herbivores, availability of both water and grey-water; and a paucity of people problems.

Finally, the many uses and benefits of green roofs suggest a bright future and expansion. Environmental benefits to air temperature, air quality, and stormwater runoff in cities are promising. Benefits of cooling, warming, and prolonging roof life accrue to buildings themselves. Urban agriculture, meadows for biodiversity, flower gardens for enjoyment, and extensive greenspaces in the sky enrich the city. Extend green roofs to a wider range of buildings in urban areas, and multiply by the number of buildings present and to be built. How large could the cumulative benefit be? "The sky's the limit."

Part III Urban features

Chapter 11
Residential, commercial, industrial areas

A city that outdistances man's walking powers is a trap.

Arnold Toynbee, quoted in The American Land, *1979*

… features which permit a rich flora and fauna to survive—the squalor, rubbish, old buildings and machinery, derelict huts, rotting dumps, inefficient handling … Bags burst, containers leak, … approved methods of waste disposal are not always followed and the result is a persistent scatter of alien plants …

Oliver L. Gilbert, The Ecology of Urban Habitats, *1991*

In describing your city to a friend on a distant continent, you send stunning photos of a favorite park, a museum, a historic site, the sports stadium, and the skyscraper area. She warmly thanks you for glimpses of these spots, but wonders if you could explain what your metro area is really like. Surprised but intrigued, you quickly get your camera and make a list of places to photograph. Where do we live, work, and shop?

For residential areas, maybe single-family house plots, low-rise apartments/condominiums, high rises, courtyard/patio housing, and even informal squatter settlement are to be photographed. For commercial areas, take photos of the city-center business district, an office center, town center, strip/ribbon development, shopping mall, warehouse trucking center, and certainly neighborhood streets with small shops. For industrial areas, photos of both active manufacturing sites and post-industrial brownfields are important.

These residential, commercial, and industrial types, covering the bulk of your urban area, are introduced in this chapter in an ecological context. Each land-use type listed above has both positive and negative ecological dimensions. Mapping the positives and negatives over an urban area would be extremely interesting and useful for ecological understanding, as well as for wise planning.

Mixed use, in this case where commercial and manufacturing sites are intermixed with residential land, is an alternative to relatively large separate residential, commercial, and industrial areas. The large separate types largely result from common human preferences and shared resources, sometimes enhanced by zoning regulations that try to keep types separate. Mixed use provides richness at a finer scale, such as urban land combining residences, small fields for growing food, small parks, shops, restaurants, local craft or manufacturing shops, entertainment spots, grocery stores, cultural centers, government buildings, walkways, and common-area meeting places (Figure 11.1). A particular value of mixed-use land is a reduction in transportation costs, time, and environmental effects. People mainly work and shop closer to where they live. Mixed use also maintains considerable habitat diversity, and therefore a relative richness of species.

I have an urban-sociologist friend who years ago, figuratively speaking, "radio-tracked" people to determine their daily "human home ranges" in the Boston area. For this, he had families from city center, inner suburb, and outer suburb record where every family member went every day. Children went to schools, husbands and wives to workplaces, and different people went varied directions and distances to meetings, shops, restaurants, parks, entertainment sites, and so forth. The home ranges of daily movements of each person and each family were then mapped. Preliminary data suggested that all family home ranges were highly asymmetric (non-circular), quite large, and not very different from city center to outer suburb. Such studies would be quite useful for planning, especially to reduce human home-range sizes, plus transportation and environmental effects in urban areas. Mixed use is a key way to shrink human home ranges.

The topics in this chapter follow in five groups: (1) city residential areas; (2) suburban residential areas; (3) city-center commercial area; (4) commercial areas

City residential areas

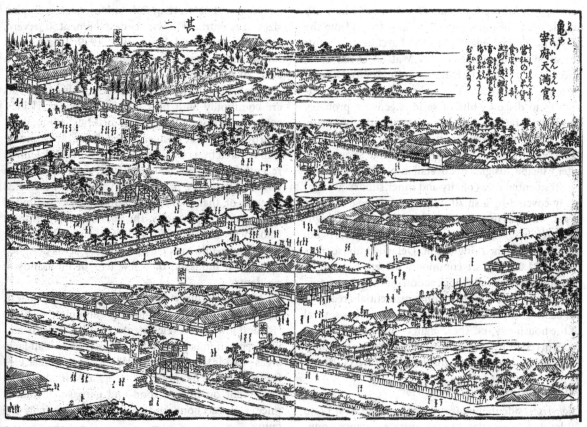

© National Diet Library website. (Partially used and placed two images together)

Figure 11.1. Mixed-use suburban or peri-urban area. Clustered single-unit residential houses (lower right); shops, restaurants (serving delicious fish), craftsman's workshops (local manufacturing), and residential houses (center areas under clouds); food production (rice paddies and vegetable gardens) (lower right); tiny public garden space (center of block above bridge); temple public garden and courtyard community space (left center); tiny horticultural and semi-natural patches (upper right and left); at top, water in gaseous form evokes the sense of liquid water. Walking and boat transportation (for commerce) predominate; few vehicles (not shown); horse (center right). Kamedo in 1818, outer Edo area of today's Tokyo; Tenjin Bridge (bottom); Kamedo Temple (beyond image to left). Courtesy and with permission of the National Diet Library Website, Tokyo; and with appreciation to Taco Iwashima Matthews for assistance.

dispersed across the urban area; and (5) industrial areas. Many of the subjects are introduced in other chapters and other ecological contexts, including house plots (Chapters 8, 9 and 10), suburban developments (Chapters 2 and 8), city center (Chapter 5), and flows and changes (Chapter 3). Thus, the following sections are designed to highlight the subjects, briefly outline key ecological dimensions, and provide entrees into the literature.

City residential areas

For a society espousing ecological or environmental sensitivity, should the goal be to reduce the rate of environmental degradation? Or to have no net degradation? Or to improve existing environmental conditions?

Suppose we took seriously the value of a venerable tree (older than the oldest of us). Or a vernal pool, or an uncommon (but not rare) bird, or even a small patch of native plants. Or at the broader scale, uninterrupted clear fish-rich streams, or connectivity for all wildlife across the land. If we did, no sidewalks and buildings would be built on soil around ancient trees. A street would be closed when the scarce bird begins to nest. Only bridges wider than a stream's floodplain would be built. Stepping on or walking a dog on the patch of plants would be off limits. And so forth.

The present subject of residential areas in city and inner suburb has yet to directly capture the attention of ecologists. Nevertheless, lots of principles and information on residential air, water, plants, animals and more in the preceding chapters are useful here. To supplement this foundation, spatial patterns from the planning, design, engineering, and architecture fields are often useful (Jacobs, 1961; Watson *et al.*, 2003; Sorensen, 2004; Urban Land Institute, 2008). Digging through the extensive, normally anthropocentric, literature to discover bits of serious ecology probably would be a rewarding, as yet unexplored, enterprise. The few rigorous studies of ecological development, sustainability, smart growth, and the like should also offer valuable insights to dovetail into urban ecology.

Residential areas of city and inner suburb are typically covered by a small number of diverse housing types, with lots of variations (Figure 11.2a to e): high-rise; low-rise; informal squatter; courtyard/patio; and linear attached single-unit. Detached house plots (Figure 11.2f to h) are commonly present but will be considered in the following section on suburban housing. Upper- and middle-income residential areas and stable low-income neighbourhoods cut across several of the housing types. These eight contrasting forms or types of residential areas seem to characterize the range of types in urban regions worldwide (Figure 11.2).

Multi-unit high-rise, low-rise

High-rise residential areas with, for instance, 8 (or 12) to 50 levels (stories, floors), concentrate a dense population by stacking numerous housing units atop one another (Figure 11.2a). Some 600 units/ha (240/acre) is characteristic, though the number varies widely. Many features are associated with this concentrated population, including close-by food supermarket, shopping department store, car parking, boulevard or highway, public transport, employment (e.g., light industry), small shops, and restaurants. The ecological dimensions for many of these are explored in sections below.

On the other hand, *low-rise* housing of 3–7 levels (for example with 150 units/ha) supports fewer but still-dense people (Figure 11.2b). Most of the preceding features are present, though perhaps less parking and no department store or supermarket. More park, outdoor-community space, and maybe office space are common.

Insights into high-rise and low-rise residential areas commonly highlight people and housing-unit densities, dimensions of features, and changes over time (Hartshorn, 1992; Lynch and Hack, 1996; Panerai *et al.*, 1999; Watson *et al.*, 2003; Berke *et al.*, 2006; Bosselmann, 2008; Dunham-Jones and Williamson, 2009). "Transit-oriented development," whereby housing and shops are concentrated near urban rail stations, usually emphasizes concentrated low-rise development (Cervero, 1998; Newman and Kenworthy, 1999; Charles and Barton, 2003; Ryan and Throgmorton, 2003; Frumkin *et al.*, 2004; Handy, 2005; Forman, 2008).

High-rise and low-rise clusters or developments are commonly built near existing or potential jobs, sometimes responding to, sometimes stimulating, nearby industry or business. Speculative development is illustrated by a city (Ordos, Inner Mongolia, China) recently built for 300 000 people and covered with high rises and low rises. In the initial years it only attracted a tenth of that number, apparently due to limited employment opportunity. The low rises are widespread (mostly seven levels, which there do not require costly elevators), while each cluster of high rises is designed for several thousand or a few tens of thousands of residents.

Normally impervious surfaces with associated heat, pollutant, and water problems predominate in high-rise and low-rise developments. Almost no semi-natural vegetation is present. Street trees and small lawn areas provide some greenery. Window-box plants, bird feeders, and the like are usually scarce. Animals mainly dependent on human food and garbage, including mice, rats, raccoons, and cockroaches, dominate the fauna.

Informal-squatter and low-income areas

In one generation, from 2005 to 2030, the world's low-income urban population is expected to double from about 1 to 2 billion (UN Population Division, 2007). The ramifications of this seemingly inevitable trajectory are likely to affect almost all aspects of the human and natural worlds. Nothing ecological will escape our explosion of the world's urban poor.

Low-income areas are remarkably diverse. In tidy neighborhoods, stability and pride seem to overcome inadequate funds. "Slums" are variously described as squalid run-down areas to live in a city (*Webster's College Dictionary*, 1991), or urban settlements where more than half the residents live with inadequate housing and lack basic services (UN-Habitat, 2007). Urban scholars and planners have described many spatial, socioeconomic and environmental aspects of slums, from London and Chicago to many cities in developing nations (Hall, 2002; Pacione, 2005; UN-Habitat, 2006). The "Gini coefficient of inequality" has been used to pinpoint the extent of low-income-area patterns in a city (see equations, Appendix B).

City residential areas

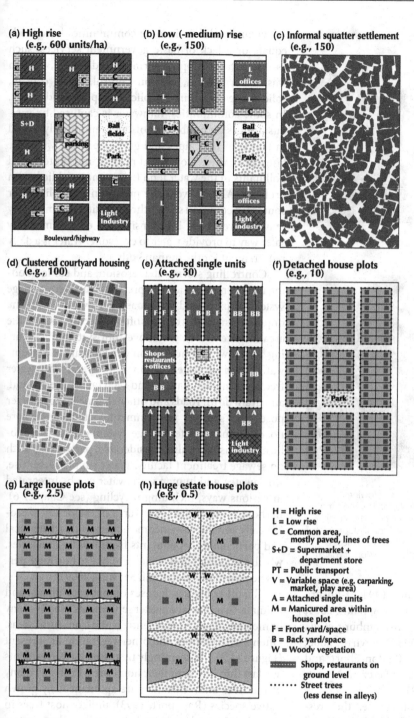

Figure 11.2. Eight forms of urban/suburban residential types. Buildings mainly with (a) 8 (or 12) to 50 levels, (b) 3–7 levels, (c) 1–2 levels, (d) to (h) 1–3 levels. Forms are illustrated in (c) Rio de Janeiro, (d) Fez (Morocco), (e) England, (f) to (h) North America. 1 housing unit/ha = 0.4 units/acre. Based on Rowe (1991), Panerai et al. (1999), Bianca (2000), Habraken (2000), Mendez (2005), American Planning Association (2006), Giusti de Perez and Perez (2008), and other sources.

A rapidly growing type of low-income area is the *informal squatter settlement* (shantytown, favela, asentamiento), particularly in developing nations (Figure 11.2c). This is a residential area of predominantly recent migrants to the city who arrived with little or no funds and have not paid for the land or its occupation (Figure 11.3). A burgeoning literature on informal squatter settlements worldwide highlights their locations in the metro area, environmental "disasters," economic and social dimensions, government and NGO services, employment, and sometimes ingenious solutions for living in a difficult situation

Figure 11.3. Informal squatter settlements as a growing proportion of the world's urban population. (a) Informal housing as the matrix surrounding patches of multi-unit housing. Mumbai. Courtesy and with permission of Niall Kirkwood. (b) Hillslope settlement of 1- to 3-level buildings, scattered trees, and access street on right. Sao Paulo. Courtesy and with permission of William Laurance.

(Hartshorn, 1992; Main and Williams, 1994; El-Bushra and Hijazi, 1995; Gupta and Asher, 1998; Bakir, 2001; Baken, 2003; Hardoy et al., 2004; Brillemburg et al., 2005; Pacione, 2005; Kramer, 2006; Neuwirth, 2006; Swaminathan and Goyal, 2006; Benton-Short and Short, 2008; Forman, 2008; Giusti de Perez and Perez, 2008; Hooper and Ortolano, 2012).

Three widespread urban situations, i.e., the presence of steep slopes, swampy or floodable land, and industrial pollution, create severe environmental conditions for housing (Figure 11.4). In addition, severe disaster events (especially earthquake, landslide, and flood) affect squatter housing markedly more than they affect other housing areas. Daily living is a challenge in an informal squatter settlement.

Virtually all low-income communities have inadequate or a lack of basic government services, such as clean-water supply, stormwater drainage, human wastewater system, erosion control, flood control, solid-waste handling, public transport, and more. In some communities, rainwater is captured by various devices on or above roofs and temporarily stored in tanks, so that pipes and gravity provide water for household uses. However, breeding mosquitoes and pollutants in the water tend to become a public health problem. Alternatively, the provision of public water sources in common spaces throughout the community may be a relatively inexpensive and environmentally safer way to provide water to everyone. Water is a daily requirement for the human body.

Controlling stormwater, erosion, and some flooding usually requires a widespread stormwater drainage system of pipes and often basins. However, with the impervious surfaces of roofs often exceeding 50% of the area (Figure 11.2c), the above-roof system of capturing and slowing runoff becomes a key part of a stormwater drainage system. Human feces and wastewater frequently accumulate in and seep out of the ground, and drain down roads and other channels. Wastewater poses the greatest environmental challenge where no, or an inadequate, sewage system exists. The public health risks are also considerable, particularly with no sewage treatment facility. Water supply, water use, stormwater, and human wastewater can be integrated in various ways, including recycling (see Chapter 6). In short, water-related issues are one major key to ecologically suitable low-income communities. Chemical and particulate air pollutants from nearby industries, which sometimes provide jobs, may also be a huge pervasive problem (Auyero and Swistun, 2009).

The other big key focuses on habitats, plants, and animals. In informal squatter settlements usually many or most residents have recently come from rural areas. Often scattered trees and other vegetation are present (Figure 11.3) and seemingly treasured by the residents as a last thin link with their home area. Probably many of the trees and vegetation are spontaneously grown native species (Rapoport, 1993), unlike most trees in formal neighborhoods and city parks. Thus, native birds, insects, pollinators, and so forth are likely to survive, even thrive, along with the many species dependent on people and their food resources.

The dispersed bits of nature across squatter settlements provide stepping stones for movement of species. Microhabitats are tiny but highly diverse, reflecting

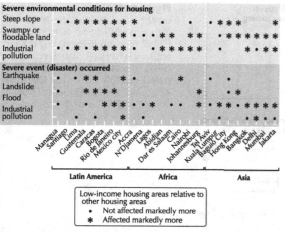

Figure 11.4. Low-income housing areas relative to severe environmental conditions and events. Cities illustrated are mainly among those with the most "severities." Adapted from Main and Williams (1994).

evidence of plantings, whether fruit trees, vegetables or spices, was noted, perhaps suggesting that most people were new or considered their stay to be temporary. Wind eddies twirled dust particles in a dance, mirrored by blue sky at the edge of a metro area. The only visible indication of water was a line of wet soil down the center of a few roads, the telltale sign of wastewater seepage (see Figure 4.10). Small single-level household structures were built of wonderfully diverse materials, mainly from the dump.

Piles of dusty pipes and a bulldozed basin downslope of the squatter settlement suggested that either the community or the government was slowly planning a combined stormwater and wastewater drainage system. I did not feel welcome, and the police in advance emphasized that we must always keep our car windows and doors shut. Nevertheless, as in Rio, this favela was a community of people daily interacting with conspicuous natural processes.

environmental gradients across the settlement, powerful recurring water and earth changes, and the inherent heterogeneity of land and vegetation before the settlement. Then add the heterogeneous distribution of structures, materials, pollutants, heat, stormwater, and wastewater from humans. The movement of animals and seeds, plus the successional processes, must be exceedingly interesting in such a changing micro-mosaic. Overall biodiversity may be rather high in informal squatter settlements. A research frontier lies in wait.

I encapsulated an ecologist's impressions of an informal squatter settlement in the mangrove swamp of a Rio de Janeiro floodplain, especially the remarkable solutions for living by apparently unrelated people without financial resources (Forman, 2008). Since then I visited a "dump favela" in Brasilia, where residents harvest and improve salable discarded resources from a large city dump. Searching through the continuously enriched accumulation of material never ends. Like archaeologists and all explorers, residents essentially ignore most material; some items are of interest (here economically), and occasionally treasures appear. Along the main muddy roads are numerous tiny shops sorting things, cutting away useless parts, and compressing or otherwise preparing materials to sell. A mule pulling a wooden wagon waited by a large scale as men weighed a load of material. The few women seen walked the roads with head high, and scattered children used the little-traffic roads for play.

Trees and bits of vegetation representing early successional habitats were scattered at low density. No

Courtyard/patio and attached single-unit housing

Since housing types in a city basically fit within a street network, it is useful to briefly consider the major forms or types of city blocks present worldwide (Figure 11.5). Eleven city block forms or types were recognized, and compared by qualitatively evaluating each using 13 ecologically related variables (related to habitats, plants, animals, soil, air, water) (e.g., see Table 7.1). Variables with negative roles or functions strongly outweighed positives for five city block forms (Figure 11.5a, b, c, d and e). Positives predominated in two cases (h, i). The total amount of vegetated area was the prime overall predictor of negative and positive environmental conditions for city block types.

The *patio* or *courtyard* residence – characteristic of Latin America, Middle East, North Africa, elsewhere in the Mediterranean Basin, and the hutongs of China – has rooms opening inward to a tiny open space surrounded by a narrow terrace-arcade against sun and rain (Bianca, 2000) (see Figure 10.9a). Typically one or a few trees are present, along with a small central water source and a variety of plants. Basically the plants are planted and cared for, and mainly provide flowers and edible or other useful products. Colonizing plants (or weeds) are usually present in low numbers. In some less-dense areas, the rear portion of a property is open and used for growing vegetables and sometimes farm animals (see Figure 10.9b).

Residential, commercial, industrial areas

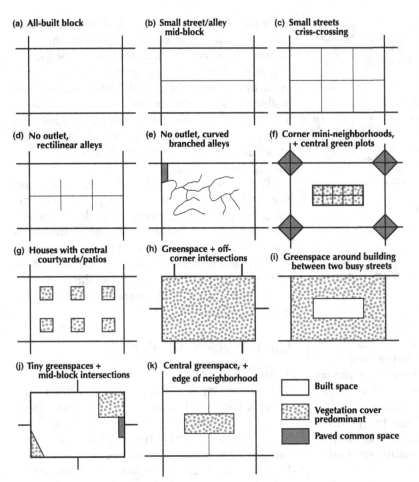

Figure 11.5. Eleven alternative forms of urban blocks. Examples from (e) Cairo, (f) Barcelona, (g) Guatemala City, (h) Hohhot (China), (i) Beijing, (j) Vienna, (k) Auckland and Kansas City (USA). Based on Siksna (1997), Bianca (2000), Dutton (2000), Habraken (2000), Watson et al. (2003), American Planning Association (2006), Bosselman (2008), and other sources.

Life in a patio residence highlights a quiet and private space where residents closely relate to individual plants and animals, and intensely feel nature's overhead power of Sun, Moon, stars, clouds and rain. Family activities, from making things to eating, clothes drying, and moving from room-to-room, are centered in the arcade and patio. As in the preceding case, however, water drainage issues are often a problem.

At a community scale, the trees and sometimes vines protruding upward in patios create a rather regularly distributed abundance of green spots (Figure 11.5g). Although apparently unstudied, this pattern doubtless provides numerous stepping stones, and alternative routes, for the movement of birds, bats, and flying insects such as butterflies and other pollinators (Figure 11.2d). Moreover, the idiosyncratic nature of plantings by residents creates a rich pattern of green spots, effectively high habitat diversity across the courtyard/patio community.

An analogous pattern may occur with patios/courtyards in urban low-rise residential areas. For instance, the Eixample or "Cerda" area of Barcelona generally began with four- or five-level buildings facing streets around a block, and a central common space with plants (Figure 11.5f) (Boada and Capdevila, 2000; Pallares-Barbera, 2005). Often, over time, the common space was subdivided into individual gardens, and progressively shrank as buildings expanded inward. Central courtyards in low-rise urban buildings are often small, deep, covered mainly with impervious surface, embellished with some planted plants, and/or covered overhead with glass. Consequently, their combined ecological value is less than that for one- or two-level residential areas.

I briefly lived in a city-center five-level building backed by a large central courtyard. The 30-m (100-ft) diameter courtyard contained some small trees, shrubs, flowers, water, vines on certain walls, bits of spontaneous successional habitat, and birds roosting and feeding. This large courtyard was probably of particular ecological importance for the neighborhood of low rises (see Figure 2.10).

Lines of *attached single-unit housing*, as in the UK and elsewhere, create distinctive but quite different spatial and ecological patterns (Figure 11.2e) (UCD Urban Institute Ireland, 2008). Each residence has a front space (yard, garden) and a back space (see Chapters 2, 8 and 10). A row of adjoining front spaces lines each side of a street. This effectively produces two parallel green corridors or *front-space strips* separated by a street with traffic. Flower gardens of ornamental and horticultural varieties tend to predominate in front spaces, some of which may also contain a small lawn and/or spontaneous vegetation. Planted street trees, if present, provide shade and other ecological features to the front-space strips. Usually, overall habitat diversity is relatively high along a front-space strip of attached single-unit housing. In addition, the green strips differ on opposite sides of a street.

Yet the back spaces of attached housing are of prime ecological interest. The row of adjacent back spaces for houses on one street adjoins the row of back spaces for houses on the next parallel street. This creates a rather wide green *back-space corridor* down the middle of a block, effectively protected from vehicle-traffic noise and people movement along streets. Furthermore, house-plot *backlines* (see Chapter 10), commonly with a concentration of spontaneous and native species, form the center of the green corridor. In short, usually the double-strip back-space corridors exhibit quite high habitat diversity and biodiversity.

The other important ecological characteristic of attached single-unit housing is the connectivity for species movement provided by the back- and front-space corridors. Although apparently unstudied, presumably the species moving along the front street-side (and sometimes tree-lined) strips differ somewhat from the species using the wider, more diverse, and more natural back-space corridors. Thus in cities, back-space and front-space house-plot corridors tie an attached-single-unit-housing neighborhood together ecologically.

Suburban and peri-urban/exurban residential areas

Apparently "suburbs" originally were settlements that appeared just outside and downslope of the city wall of medieval hilltop towns or cities (Stilgoe, 1988). Today in a general sense, suburbs are the predominately residential areas adjoining cities (see Chapter 1). For instance, several German cities today seem to be ringed by a 10-km (6-mi) wide band of suburbs (Breuste, 2009).

Yet a researcher studying the outer built area of a Swiss city commented that she was studying peri-urban, not suburban, areas. By way of explanation, after decades of watching American movies, Europeans clearly know that *suburbs* are covered by lines of detached houses surrounded by lawns in large lots, and interspersed with commercial strip (ribbon) development and shopping malls. No such pattern existed by her city. Instead, the *peri-urban area* of European cities normally has compact housing developments appearing adjacent to the urban built area.

Rapidly expanding cities in developing nations and elsewhere today usually spread in this European way, with compact development rolling outward from the metro-area border. However, if development is low density (e.g., sprawl) and/or separated from the metro-area border, as typical in North America, we use the term *exurban*. Compact peri-urban development is mainly on farmland, whereas dispersed North American exurban development may occur on both farmland and natural land beyond it. This also means that in most of the world the expansion process is *urbanization*. The North American pattern, also present in scattered areas worldwide, is effectively *suburbanization* or suburban expansion. Urbanization patterns are much studied by urban scholars, planners, and others (Hardoy et al., 2004; Tacoli, 2006; UN-Habitat, 2006; Torres et al., 2007; Biggs et al., 2010). Suburbanization is also well described (Audouin and Loubiere, 1996; Berger, 2004; Stanilov and Scheer, 2004; Caldiron, 2005; Forsyth, 2005; Phelps et al., 2006; Lukez, 2007). Nevertheless, we usually use the familiar term, urbanization, referring to both processes.

Neighborhood ecology, with its promising theoretical and conceptual roots, has been explored in earlier literature (Forman, 1995, 2008), and touched on in preceding Chapters 2, 8 and 10. Also urban studies and planning literature has provided numerous insights into neighborhood spatial patterns, housing, sociology, movement patterns, differences, planning, design, changes, and other factors (Beatley, 2000a; Steiner, 2002; Hardoy et al., 2004; Ozawa, 2004; Klunder, 2005; Pacione, 2005; Erickson, 2006; Newman and Jennings, 2008; Andrzejewski, 2009).

In essence, neighborhood ecology has a functional foundation, similar to that of an urban region based on active linkages between city and ring-around-the-city (see Chapter 2) (Forman, 2008). Active flows and movements of water, air, species, and people tie a neighborhood together, and differentiate it from adjoining neighborhoods. The strength of these flows and movements helps determine the relative distinctness of ecological neighborhoods. Of course, similarities in spatial patterns also help delineate neighborhoods. Planning, based on these linkages, rather than simply spatial pattern, should lead to a more "sustainable" or ecological neighborhood. In such a place, natural systems work much more naturally, and ongoing maintenance, repair, and replacement budgets should be lower.

The distribution of greenspaces and green corridors relative to housing types, road network, and the center of a community is a key to neighborhood ecology. Ten contrasting forms or types of suburban development seem to encapsulate the patterns found in urban regions worldwide (Figure 11.6). I made a qualitative comparison of these types based on 19 ecologically related variables (similar to that of Table 7.1). Overall, negative roles or functions (relative to habitats, plants, animals, soil, air, water) strongly outweighed positives in four cases (Figure 11.6a, b, e and f). Positives predominated in three cases (d, i and j). The presence of a large or medium-sized vegetation patch was the best predictor of overall positive environmental conditions in these suburban development types.

As in the preceding sections, most of the concepts needed to understand the ecology of suburban areas have been outlined in earlier chapters, especially Chapters 2, 8 and 10. Here we pinpoint a few concepts and provide some references as entrees into the literature. Since change produces patterns, we begin with change, and then consider spatial patterns.

Changing patterns

> ... success in the future was surrendering space and privacy. ... solutions were minimalist: good but narrow roads, rooms designed for midgets, jammed subways, tiny restaurants, the whole landscape miniaturized and cemented over. ... the likeliest solution to survival in an overcrowded world ...
>
> Paul Theroux, *Ghost Train to the Eastern Star*, 2008

Typically, outward urbanization occurs on farmland (and former farmland) characterized by farmsteads, local roads, and scattered recent houses of people interacting in the farm community. Wetlands mostly drained or filled, muddy streams, first-order streams transformed to often-dry drainage ditches, narrow hedgerow/roadside/streamside vegetation strips, and scattered patches of semi-natural vegetation (usually on poor agricultural soil) are also characteristic. Often beyond the farmland is continuous natural vegetation (Forman, 2008).

Changing the spatial pattern of an area over time implies creating different conditions, altered flows and movements, and changes in the way an area works. Perhaps the best way to understand the ecology of changing peri-urban or exurban areas is to highlight three key sequential *ecological phases of urbanization*, and their common characteristics.

1. *Altered wildlife pattern phase*. With the onset of urbanization, the routes, movement patterns, population sizes, and distribution patterns of wildlife are significantly altered. This alteration results from (Vail, 1987): the introduction of a few housing developments; widened local roads; increased vehicle traffic; more traffic noise/disturbance; busier town centers; more people and dogs walking in natural areas and field margins; and loss of some farmland hedgerows.

2. *Metapopulation and disrupted water phase*. In the middle phase of outward urbanization, farmland habitat is noticeably decreased and fragmented. Many semi-natural vegetation patches are degraded, reducing species population sizes. Wildlife movement patterns are blocked and disrupted by housing developments and busy roads. "Metapopulations" proliferate, whereby usually small populations of a species are separated in different sites and only occasionally do individuals successfully move between the sites. Consequently, the small populations markedly fluctuate in size, and genetic inbreeding occurs. Both effects increase the probability of a small population disappearing (local extinction). Also, human-related animals [e.g., raccoons (*Procyon*), house mice, domestic pets] spread with the abundance of food sources around houses. Native species biodiversity begins to decrease, and non-native species increase and begin to spread. Water-related characteristics also change in this phase: impervious surface area increases; local flooding often increases; septic systems (if abundant) degrade groundwater quality; many ponds and streams are polluted; stream habitat

Suburban and peri-urban/exurban residential areas

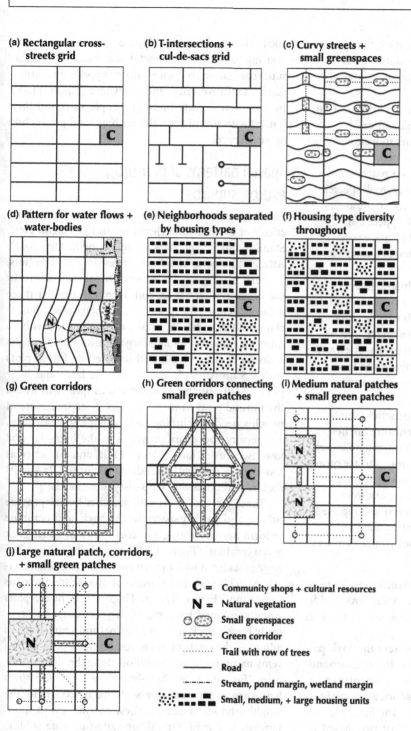

Figure 11.6. Ten alternative forms of suburban developments. Forms integrate road network types, single-unit housing types, patterns for protecting nature, and walking trail networks. To illustrate road networks, a relatively large number of housing units is included. Medium-size houses cover all ten forms except (e) and (f), which also have large houses and small houses. Currently such suburban development is particularly characteristic in North America. Based on Kendig (1980), Sanders (1981), Rowe (1991), Easterling (1993), Girling and Helphand (1994), Bohl (2002), Stanilov and Scheer (2004), Forsyth (2005), Lukez (2007), and other sources.

diversity is degraded and some streams are straightened; scarce wetlands are further reduced; lawns and gardens are irrigated; residential-area chemicals are added to the residual agricultural chemicals in soil; and towns increase their water supplies and associated land protection.

Human changes are conspicuous in this key middle phase: farming noticeably decreases; housing developments begin to coalesce; a shopping center(s) is built outside the town or village center; new local roads are built; and commuter routes to and from the city are widened with more

323

traffic. These concurrent urbanization changes in the exurban/peri-urban area catalyze both an acceleration and a proliferation of ecological changes.

3. *Scattered habitat phase*. With further urbanization, most semi-natural patches shrink and/or disappear; connectivity for species movement among fragmented habitats is limited; metapopulations persist with fewer patches and less movement among them; many small species populations are at risk of extinction in the area; native species biodiversity has dropped; non-native species are abundant and invasive species widespread; and animals attracted to humans and garbage are widespread.

Most stream lengths are typically straightened with loss of stream habitat, and some water courses disappear into pipes; local floods may be common; all water bodies are polluted; some septic systems (if present) are typically eliminated by wastewater pipe-and-sewage-treatment systems; and town water systems commonly have water shortages and/or pollution risks.

Remnant farm fields remain as generally small open patches; semi-natural patches are further shrunk and eliminated by development; built area predominates; extensive impervious surface means increased summer heat and more stormwater runoff and local flooding; municipality/town centers and local roads are often traffic-congested; public transport typically appears; a major highway(s) is usually present, both for commuters and for through inter-city travel.

The three ecological phases of urbanization seem to characterize peri-urban/exurban areas worldwide, though of course variations exist and one can always subdivide the major phases.

Five simple spatial models represent the basic patterns of outward urbanization into the ring-around-the-city (see Chapter 3): bulges model; concentric-rings model; satellite-cities model; transportation-corridors model; and dispersed-sites model. The first three are much better ecologically than the last two, based on analyzing their relative effects on natural systems and human uses of natural systems (Forman, 2008).

The ecological and other dimensions of low-density sprawl (Figure 11.7c), represented by the dispersed-sites model, have been of particular interest (Gillham, 2002; Squires, 2002; Frumkin *et al.*, 2004; Godschalk, 2004; Gutfreund, 2004; Hayden, 2004). Low-density housing especially characterizes North American suburbanization, but increasingly appears in Europe, China, Latin America, and elsewhere (Figure 11.7b). In some low-density housing, people surround their houses with useful activities such as food production or work space.

Spatial patterns of peri-urbs, exurbs, suburbs

House plot sizes are an easy way to evaluate the general effects of urbanization (Jennings and Jarmagin, 2002; Stanilov and Scheer, 2004; Perlman and Milder, 2004). Attached housing on adjoining residential plots (lots, properties) (Figure 11.7a) was introduced above in the urban residential section. Here we consider three *detached-house plot* types (i.e., for separated houses) (Figure 11.2f to h): "normal" (or small) plots with 10 units/ha (1/4-acre plots); large house plots with 2.5 units/ha (1-acre plots); and huge estate house plots with 0.5 units/ha (5-acre plots). The large and huge plots would usually be considered as sprawl, as would the normal plots in many regions. Note that some of people's needs, including food, shopping, workspace, and leisure space, are provided in the high-rise and low-rise illustrations (Figure 11.2a and b), whereas low-density residential residents (Figure 11.2g and h) require additional areas to satisfy those needs.

These "normal" plots with detached houses provide private and garden spaces in the back, and combine to form narrow front green corridors and wider back green corridors (Figure 11.2f), as described above. The large plots form wider green corridors, and often contain a backline strip of somewhat diverse spontaneous vegetation (Figure 11.2g). Huge estate house-plots when combined have still wider corridors, and typically have a back strip of semi-natural vegetation. In addition, the estate plots are sometimes separated by a semi-natural strip of vegetation along the side boundaries (Figure 11.2h). *Side-boundary vegetation strips* help form an extensive fine-scale green network for the neighborhood. In essence, these corridors with spontaneous and semi-natural vegetation provide wildlife habitat, but may be especially important for species movement.

An environmental comparison of the eight urban and suburban residential types, analogous to that for riverside types (see Table 7.1), revealed major differences. The residential types (Figure 11.2a to h) were

Figure 11.7. Different arrangements of houses on properties. (a) Attached residences with front and back spaces along a block (and some low-rise housing). London region. (b) Small houses in different locations within mostly fenced plots. Maun, Botswana. (c) Houses in center of lawn-dominated plots. City of Lake Placid, Florida. R. Forman photos.

qualitatively compared for 22 variables, 11 related to air, water and soil, and 11 related to habitats and species. Overall the results suggest that the high-rise (a), low-rise (b), and clustered courtyard housing (d) have the most negative environmental effects. The huge estate (h) and large house plot (g) residential types [and perhaps informal squatter settlement (c)] seem to be best environmentally per unit area. However, a quite different conclusion results from dividing the environmental positives and negatives by the density of housing units (Figure 11.2). This provides an estimate of environmental effect per housing unit or per person. Per capita, the huge-estate and large-house-plot residential types emerged as the most environmentally negative residential types. Interestingly, informal squatter settlement may produce the best environmental condition per capita, a result warranting research.

Many studies have related biodiversity to suburban residential development (Friesen et al., 1995; Romme, 1997; Theobold et al., 1997; Cohn and Lerner,

2003; Pidgeon *et al.*, 2007). Heterogeneity is especially important ecologically within a residential area where house plots are similar. For instance, cluster housing may be present (Bartuska and Young, 1994; Vince *et al.*, 2005). Also, low-impact development, particularly focused on handling stormwater runoff, creates spatial heterogeneity (see Chapters 6 and 10) (Dietz, 2007).

Planned built communities, resulting from centralized planning and design, also have distinct and widely differing residential patterns. Planned cities, such as Canberra, Brasilia, and Washington D.C., were designed and built on farmland or natural land (Reps, 1997; Forman, 2008). Singapore, in essence, grew naturally into a city and then became intensively planned (Gupta and Pitts, 1992). Various planned towns in the USA and elsewhere have been designed and built mainly on agricultural land, including Reston (Virginia), The Woodlands (Texas), and "new urbanist" examples (Morgan and King, 1987; Duany *et al.*, 2000; Bohl, 2002; Watson *et al.*, 2003; Hough, 2004; Berke *et al.*, 2006; Forsyth and Crewe, 2009). Overall, except notably for The Woodlands, the environmental dimensions of such communities have been conspicuously limited or missing (Forman, 2008).

Based on 26 examples of such planned communities worldwide, seven types have been recognized (Forsyth and Crewe, 2009): social neighborhoods; architectural villages; diverse communities; designed enclaves; ecoburbs; ecocities; and technovilles. Each type is illustrated by three to nine examples. Also, relatively massive developments of about 400–5200 ha (1000 to 13 000 acres), such as Scottsdale Ranch (Arizona) and Pelican Bay (Florida), have been built as "master-planned communities" (Moudon, 1989). These developments seem to first bulldoze almost the entire surface flat, thus effectively obliterating its ecological values and replacing them with the designers' and developers' residential pattern.

This section has highlighted residential areas, and the following sections explore commercial and then industrial areas. However, mixed-use development includes jobs and shopping in addition to housing (Urban Land Institute, 2003). Such areas may have reasonable constraints against inappropriate adjacent land uses, such as a fireworks factory next to a home or school. In the broad perspective, mixed use areas require less transportation and offer significant environmental and social advantages (Figure 11.1) (Urban Land Institute, 2003).

City center

Commercial areas feature office buildings, goods storage, and retail shopping. *City center*, as the core of both the city and the metro area, is the prime location, often representing over half of the total metro-area commercial activity. However, several important and relatively distinct types of commercial areas are spread across

Figure 11.8. City center high-rise area and surrounding land uses. Residential neighborhoods and commercial areas close to city center and airport (top right), which were built on added earthen fill. Wooded hill-slope surroundings provide recreation, biodiversity, flood control, erosion control, and cool air drainage that ventilates the city. Gibraltar. R. Forman photo.

the metro area, from neighborhood streets with retail shopping to office center, commercial strip, and warehouse distribution center.

City center is the heart and heartbeat of a metro area (see Figure 2.8). Sometimes called downtown, high-rise core, or central business district, these centers may be remarkably similar in different geographic regions. (1) We first outline the relatively distinct and unusual characteristics of ecological importance displayed by a city center. (2) Next, key spatial patterns, flows, and changes are pinpointed. Then an array of downtown characteristics is introduced in two categories: (3) buildings, people, and transport; and (4) soil, water, vegetation, animals, and air.

Distinctive and unusual characteristics

Most distinctive and memorable are the high-rise buildings and skyscrapers with elevated lights that contain the city's primary commercial office space and activity (see Figure 3.1). Normally concentrated in an area less than $2.6\,km^2$ ($1\,mi^2$), the vertical surface greatly exceeds the horizontal surface, for instance it is about 10 times more in both Chicago and New York (Hartshorn, 1992; Ursic et al., 1997; Larson et al., 2000). Scattered among these tall buildings normally are older buildings with external ornament, as well as major museums, concert halls, and other cultural buildings. A concentration of government buildings, hotels, and restaurants also characterizes city centers.

In perhaps most cities, commuters pour in and out on weekdays, so evenings and weekends are less congested. Essentially all modes of transportation – train, subway, bus and/or streetcar, car, motorcycle and/or scooter, sometimes bicycle, and walking – are concentrated, providing high accessibility for people. Far more daily trips are accomplished by walking than in other urban areas, and the occasional presence of pedestrianized streets with no vehicles facilitates this efficient mode of movement.

With impervious surface covering the bulk of the area (often >90%), soil surface, water infiltration, and plant evapo-transpiration are minimal (Figure 11.8). Yet underground is a complex concentration of pipe systems, present and former structures, and typically transport systems (see Chapter 4). Water use and sewage production by the concentrated human population is intensive, and commonly almost all stormwater rapidly runs off in a drainage pipe system. Center city air is characterized by strong winds related to tall buildings and street "canyons," a concentration of air pollutants, and warm air, especially in the winter, at night, and on summer days (see Chapter 5).

Vegetation is scarce and mainly limited to bits in window boxes, tubs/planters, trees along some streets, and greenspace in scarce small parks, historic structures, and cemeteries (Rapoport, 1993). Animal diversity is normally quite low, though the density of a few species such as pigeons, rats, and cockroaches is very high, mainly due to abundant food associated with humans.

Spatial patterns, flows, and changes

The spatial patterns and functioning of city center are quite distinctive as well (Franck and Schneekloth, 1994; Kayden, 2000; Watson et al., 2003). Spatially, a dense street network for movements and flows encloses a dense small-patch pattern with low variability in size and shape (see Figure 10.4). Generalist species such as starlings and house sparrows move readily across the city center, as do migrating birds and large-home-range species such as gulls and falcons. But non-city-center species are largely excluded from colonization and even movement in the area.

Manhattan (New York) 200yr ago was covered by 12 long avenues crossing 155 streets, and much later subdivided into today's smaller blocks and building plots. City-center pattern normally changes very slowly due to inertia associated with large buildings and intensively used streets, though economic boom and bust times often produce some change. Floods and earthquakes cause overnight change. Observing an abundance of giant construction cranes on the skyline indicates rather rapid change, while the persistence of vacant lots points to little change.

Also the "border areas" of a city center are frequently a fairly distinct mix, for example, including a major sports stadium, large market with stalls, tourist area, low-income neighborhood (either long-term or squatter settlement), and demolished former-industrial site. In some European cities, a 50–100 m wide greenbelt or greenway, including remnant portions of the city's medieval wall, now partially surrounds the city center.

Buildings, people, and transport

New York has by far the most *skyscrapers* (>40 stories or 500 ft high) in North America (Hartshorn, 1992). Except in earthquake-prone zones, *high-rise buildings* (>8 or 12 stories high) are generally abundant in all medium to large cities. The megacity Sao Paulo has

>5000 high rises (see book cover). Although tight clusters exist there, many high rises are dispersed one or more blocks apart, so that surrounding ground surface exists for playgrounds, tiny park space, and carparks. Skyscrapers overwhelmingly contain commercial office space, a major portion of a city's office space, as for example 30% for Manchester (UK) (Ravetz, 2000). In contrast, high rises often contain both office space and residential space, or only residential space. Tall buildings usually have smooth surfaces and internal air cooling, heating, and circulation systems. However, residential high rises may have penthouse-roof gardens, green roofs, balconies, and window boxes, all containing plants and providing perches used by birds.

High rises and some skyscrapers typically have shops at least at ground level. Although fairly expensive dry goods are the primary merchandise, a range of shop types is present, including specialized goods and tourist shops. Especially important ecologically is the abundance of eateries present, from expensive to ethnic to fast-food restaurants. *Food waste* is considerable. Large restaurants may have food waste transported to pig farms outside the city. Irrespective, pests, from rats and mice to cockroaches and ants, are permanently abundant and widespread at both ground and underground levels.

Hotels and the occasional convention center for tourists and business people also have food waste and pests. Most of the area not covered by streets and tall buildings supports low-rise residential buildings (3–8 levels), commonly with small central courtyards. Some streets lined with low rises also have small shops and eateries at ground level. A high density of people living in center city means numerous kitchens and bags of garbage widely spread across the limited horizontal space. It also means that small grocery stores are abundant, and supermarkets at ground or below-ground levels are present. Food waste and pests are unending challenges throughout most of a city center.

Protected historic buildings, seemingly out of place, are typically scattered over city centers (see Figure 2.8). Most such buildings have ample *ornament*, ledges and crevices that serve as elevated microhabitats for a limited diversity of species such as algae, mosses, lichens, flowering plants, insects, lizards, and roosting or nesting birds (Rapoport, 1993). Government buildings, museums, concert halls, and so forth typically function ecologically much like commercial office buildings.

Most city centers have spread onto marginal sites such as former floodplains, unstable slopes, filled wetlands, tsunami-risk zones, and earthquake-prone fault lines. Dramatic results, such as the tsunami-devastated ancient Alexandria (Egypt), earthquake-toppled skyscrapers in Kobe (Japan), and volcanic lava-flow-covered Pompeii (Italy), can be expected periodically. Yet the nearly unending gradual changes of sinking substrate, rotting foundations and sea-level rise, as in Venice, Dakha, and other cities, are equally devastating to city centers.

Residents overwhelmingly walk to work in cities of some developing nations (e.g., Mumbai, Lagos, and Manila) where highways, buses, and trains between city and surroundings are limited. In essence, if I work in the city I live in the city. The proportion of people walking to work is highest in city centers.

In cities with considerable radial transport movement between surroundings and city as in North America, a crush of commuter traffic arriving and leaving city center each weekday is characteristic. Congestion on workdays is rampant for residents within the city, as in the subways of Tokyo, streets of Shanghai, and sidewalks of New York. However, in city center, non-city residents commuting into the city jam the commuter trains, plus the highways and streets with cars. In the USA the percent of commuters arriving for work is highest in large cities, and the percent arriving for shopping is highest in small cities (Hartshorn, 1992). Also, in large cities, commuters' cars are overwhelmingly parked in outdoor carparks and inside garages, whereas in small cities most cars are parked street-side.

Pedestrianized streets, "sky-bridges" for walkers, underground walkways for shopping and subway access, separated bike lanes, bus lanes, multi-passenger car lanes, raised monorails (e.g., Sydney, Seattle), and other approaches each slightly reduce congestion (Platt, 2004). A novel intense efficient *multi-modal transport system* would be needed to significantly reduce congestion and to provide ample greenspace and common space for the high concentration of people calling city center home.

These and other analyses suggest that five city-center characteristics have the widest range of environmental effects on the physical environment and on organisms/habitats (Figure 11.9 left): (1) medical facilities and university areas; (2) ethnic neighborhoods and low-income areas; (3) hotels and restaurants; (4) low- and high-rise wealthy residential areas; and (5) parks and plazas. Several attributes of the physical environment in city center are significantly affected, especially

City center

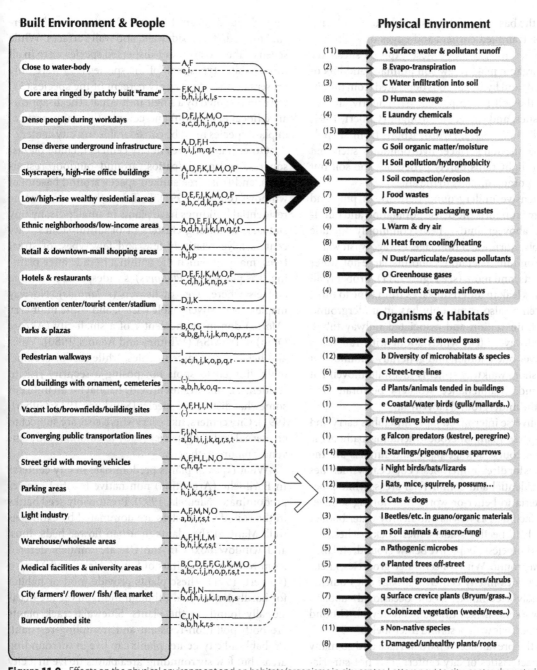

Figure 11.9. Effects on the physical environment and on habitats/organisms in city center. Letters next to city-center characteristics (built environment and people) refer to effects on specific dependent variables on right. Numbers on right indicate the number of city-center characteristics positively or negatively affecting the physical environment or organism/habitat variables. See Gilbert (1991), Hartshorn (1992), Craul (1992), Niemela (1999), Luck *et al.* (2001), Pickett *et al.* (2001), Abbott (2004), Platt (2004), Pacione (2005).

(Figure 11.9 upper right): (a) a polluted nearby water body(s); (b) surface water and pollutant runoff; and (c) paper and plastic packaging waste. These diverse attributes produce both positive and negative effects which permeate the cores of our cities.

Soil, water, vegetation, animals, air

Construction and demolition sites may be the only somewhat large areas of bare soil in city centers. Otherwise, exposed soil is mainly limited to tiny spots, such as at the base of street trees, in planters and window boxes, trampled corners and edges of walkways, and so forth. Virtually all of these exposed sites have little soil organic matter, few soil animals, and lots of heavy metals and hydrocarbons deposited from the air. Normally the soil is compacted, particularly at construction sites, and subject to drought (Gilbert, 1991; McDonnell et al., 1997). Hydrocarbons from air pollution make the soil surface "hydrophobic," such that water from light rains barely penetrates the soil but instead evaporates directly to the air.

An extensive underground network of pipe and tube systems supports the dense population and building space aboveground. These commonly include water supply, stormwater drainage, sewage wastewater, electrical conduits, telephone conduits, large hot-water heating ducts, and more (see Figure 4.11). Probably all pipe systems leak in different places from time to time. The city center also commonly has an underground inter-city train station and associated railway tubes. Often a subway system winds about with rail tubes crossing over one another, and generally a number of relatively short walkway systems facilitate movement underground. Maintenance workers perforate and maintain the underground labyrinth.

These diverse intermixed pipes and tubes carry and widely disperse many microbes including pathogenic bacteria. Of course sewage pipes and tunnels are the prime routes for disease microbes. People carry microbes up into residential and commercial areas, while pipes carry bacteria to better-or-worse functioning sewage-treatment facilities, as well as into nearby water bodies. Mice, rats, bats, beetles, spiders, flies, cockroaches, and sometimes homeless people live below ground, and may move aboveground. With concentrated groceries, kitchens, and eateries in city centers, garbage bags quickly and endlessly accumulate, as the occasional rubbish-removal workers' strikes dramatically reveal. Ground level always crawls with pests and microbes.

Given the prevalence of hard surface and so many other characteristics of city centers, it is not surprising that plant diversity is depauperate. A study of a <4 km^2 area (1.5 mi^2) in Central London, a city known for the abundance of its flower gardens, recorded 157 plant species. This is perhaps but a quarter or a third of the biodiversity in an outer part of the metro area (cited by Gilbert, 1991). Three habitat types were recognized: (1) cultivated sites, including gardens, window boxes, tubs, raised planters, and soil at the base of some trees; (2) uncultivated sites, including building sites, carparks, and vacant lots; and (3) "stonework," referring to road/street, sidewalk, and wall surfaces. While seven of the mostly wind-dispersed species were in all three habitat types, half of the species (77) were only present at one site. Twenty percent of the species (31) were represented by a single individual. The disappearance (local extinction) and colonization (appearance) of plant species must be considerable in a city center.

Other habitat types in city centers include plazas/squares, courtyards/patios, around outdoor restaurants, and frost-free sunken spaces around basement windows (Gilbert, 1991). In Tokyo the fronts of some residential buildings have plants in small pots, by tiny statuary, or among clusters of beautiful stones. City centers have relatively few, and few types, of street trees, many of which die within a decade after planting. The plane tree (*Platanus*) is widespread in Europe and elsewhere. Tree of heaven (*Ailanthus*), Norway maple (*Acer*), silver maple (*Acer*), and white mulberry (*Morus*) dominate the center of a small city (Akron, Ohio) in the USA (Whitney and Adams, 1980). Vines and lichens are scarce in cities, while shrubs are normally limited to spots not posing a security risk to residents. Bombed sites provide habitat for an influx of somewhat distinctive species (Salisbury, 1943; Gilbert, 1991). City centers, like everywhere else, are subject to an eternal "rain" of species ready to colonize and rejuvenate nature in the city.

What types of plants predominate in city centers? In Bariloche (Argentina) non-native (exotic) species predominate (Niemela, 1999). Presumably seed banks in the soil of city centers are dominated by exotics as well. Many of the plants in urban flower beds, tubs, and window boxes in the UK are annuals derived from rocky arid habitats of South Africa and Mexico (Gilbert, 1991). These plants provide food or habitat for hardly any animals, though moth larvae are occasionally found on the plants. Sidewalk-crack plants are both pollution-resistant and trampling-resistant. Probably all city-center plants can live in surrounding areas, but few surrounding-area plants thrive in a city center. Finally, the relatively few plants growing in city centers survive because of, or in spite of, an intensive human maintenance regime.

Certain wildlife thrive in city centers, perhaps largely because of warm winter and night temperatures present and the abundance of food. As mentioned above, grocery stores, restaurants, and home kitchens provide lots of food for wildlife to scavenge (Figure 11.10). People who enjoy seeing wildlife in these "sterile street canyons"

Figure 11.10. Street level and rooftop restaurants on low rises in city center. Rooftops also have scattered swimming pools, small gardens, water tanks, solar-energy collectors, and clothes drying. Sevilla, Spain. R. Forman photo.

put out pet food, or feed feral pigeons and house sparrows from a bench. Compared to wildlife in other urban areas, the relatively few species that thrive in city centers tend to have a higher density, longer breeding season, smaller defended territory, shift in diet, a nest or den in a building, reduced dispersal to elsewhere, lower tendency to migrate, and reduced avoidance of humans (Pickett *et al.*, 2001) (see Chapter 9). The outdoor cat population in city centers is relatively low, whereas indoor dogs are often seen on leash, walking with owners. Rats and other mammals may carry rabies, and pigeons and starlings carry and defecate pathogenic (to humans) fungi (Gilbert, 1991).

In the early 20th century when motor vehicles basically replaced horsepower in cities, the population of house sparrows, which extensively fed on seeds in horse droppings, plummeted (Gilbert, 1991). Yet still the species is a dominant in city centers. Starlings and other birds may form significant roosts in autumn and winter. Breeding bird density may be high, especially for house sparrows and feral pigeons. Bird nests contain a range of invertebrate species, particularly beetles, mites and flies (Gilbert, 1991; Parsons, 1995). The city-center sparrows, starlings, and pigeons may be 80% of the avian population in summer, and 95% in winter (Niemela, 1999; Pickett *et al.*, 2001). City centers in different regions have other birds present at relatively low densities, including magpie, Indian mynah, parakeet, crow/rook, gull, and wood pigeon.

The analogy of tall buildings and rock cliffs, and their characteristic species, has given rise to the *urban cliff hypothesis* (Gilbert, 1991; Ursic *et al.*, 1997; Larson *et al.*, 2000). In effect, many, perhaps half, of the vertebrates associated with the city center originated in caves, cliffs, or talus rock slopes, and thus may thrive among tall buildings. Black rat, house mouse, house cat, pigeon (rock dove), house sparrow, starling, swift, gull, nighthawk, barn owl, kestrel, and peregrine falcon are cited. Also some of the highly urban plants, including species of *Solanum*, *Allium*, *Lactuca*, *Brassicca*, and *Centaurea*, have origins in cliffs and rock slopes.

Peregrine falcons (*Falco peregrinus*), apparently the fastest flying birds known, have been successfully introduced into city centers for over 40 years. Native peregrines mainly nest on ledges of remote rock cliffs, so nest sites in cities are provided high up on buildings (Cade *et al.*, 1996). Today 31% of the Eastern USA peregrines live in cities, and 58% of the Midwestern peregrines do. Of these, 61% of the nests are now on buildings, 30% on bridges and overpasses, and 9% on other tall structures. Pigeons, northern flickers, and blue jays are the main food items along with many other species. The kestrel, a smaller falcon, in Manchester (UK) has a diet of 76% birds, 22% small mammals, 1% insects and 1% earthworms (cited by Gilbert, 1991). Food is far more abundant than these predators can eat.

City center and some industrial sites usually have the worst air pollution (see Chapter 5). The concentration of buildings and transportation, plus poor ground-level air circulation, is supplemented by incoming pollutants from other urban sources and periodic temperature inversions. Together these factors often lead to a pall of pollutants. Between tall buildings are deep street canyons where plants, animals, and people are present (Figure 11.10). Poor air circulation results in pollutant buildup, though moving air in cross streets helps disperse the accumulation. Constantly moving vehicles and people lift particles off streets and sidewalks, accentuating the problem. High winds, including streamlines, turbulence, and vortices occur in spots at street level, but are common well above the ground.

From ground level, tall buildings block distant views and produce considerable shade. East-west streets have different solar radiation patterns than north-south streets. However, the traditional single-level hutong neighborhoods of urban China commonly have wider north–south ways for vehicle movement, while the narrower protected east–west ways (≥2 m wide, with ≥2 m high walls) receive little direct sunlight. Some spots in city centers are never in direct sun, and most plants there grow poorly in permanent shade.

Finally, the urban-heat-island effect (see Chapter 5) is typically most acute in city centers due to the extensive hard surface area, both horizontal and vertical. By day, incoming solar and diffuse radiation is absorbed by these surfaces, which at night emit heat energy, thus heating the air. But city centers are nearly the only place where "anthropogenic heat" from heating buildings during the cool season may be a significant component of the urban heat buildup. Increased warmth at night and in winter benefits some species, but inhibits others. Furthermore, with the scarcity of vegetation and plant evapo-transpiration, city centers tend to have dry air.

Several characteristics of the city-center built environment with widespread ecological effects were pinpointed above (Figure 11.9). The six most-affected organism-and-habitat attributes, whether positive or negative, appear to be: (1) starlings, pigeons, and house sparrows; (2) diversity of microhabitats and species; (3) rats, mice, squirrels, possums, etc.; (4) cats and dogs; (5) night birds, bats, and lizards; and (6) non-native species.

In short, a city's core displays an array of rather distinctive built and human characteristics of ecological importance. These have significant effects on both the physical and biological environments, thus creating a distinctive ecology of city center.

Commercial sites dispersed across the urban area

While city centers represent the commercial core, other commercial activity and associated ecological characteristics are widely dispersed over the metro area. Seven types of commercial areas are generally quite distinct and readily recognized. These are introduced in five groups: (1) neighborhood streets with small shops; (2) marketplace; (3) office center and town center; (4) commercial strip and shopping mall; and (5) warehouse truck-distribution center.

Neighborhood streets with small shops

Numerous small centers of retail shopping serving residential neighborhoods are spread rather evenly across the metro area. "Mom and pop" shops, grocery stores, restaurants, bars/taverns/cantinas/pubs, pharmacies, clothing cleaners, gasoline/petrol stations, and so forth on ground level along a few intersecting streets with tree-lined sidewalks are most characteristic. With strong linkages between shop owner and neighbors, shops tend to remain for long periods. A significant increase in neighborhood median income usually produces a gradual change in goods sold or shops present, while a drop in median income may lead to shops closing and even vacant-lot vegetation in a retail center.

An average *time distance* (travel-distance time) of up to 15 minutes from home to neighborhood retail center is typical (Hartshorn, 1992). Thus, neighborhood retail centers are often located at intervals of about 2–5 km (1.2–3 mi) apart. Neighbors mainly walk, bike, or drive short distances to obtain daily needs and services, though public transport is often present. Car parking is usually street-side, while a truck-delivery area behind the shops may also serve as a carpark. Such retail centers additionally serve as meeting places for the neighborhood.

Walkable neighborhood retail centers greatly reduce the levels of hydrocarbon, heavy metal, particulate, and greenhouse-gas pollutants from vehicles. Long-term gasoline/petrol stations, however, may have leaky tanks and pipes causing serious hydrocarbon pollution of the soil and groundwater. Most of the non-gaseous pollutants on surfaces are washed by precipitation water into the stormwater drainage system and on to local water bodies. However, sites suitable for stormwater basins to reduce flood hazard and treat pollutants typically exist in adjacent residential neighborhoods. Also the presence of small groceries and restaurants indicates a daily production of food waste, and the abundance of pests such as cockroaches, ants, mice, and rats (Feng and Hinsworth, 2013). Pesticides used to control pests wash into the stormwater system.

The overall ecological significance of neighborhood streets with retail shops is that pollutants and pests are widely distributed over the metro area. The pollutants produced are normally at modest levels, but they reach and tend to degrade all metro-area water bodies, such as streams, rivers, ponds, lakes, and estuaries. Analogously, the pest species associated with these retail centers are spread throughout the metro area. It is nearly impossible to escape these pests in urban areas.

Marketplace

Marketplaces come in many forms, including food markets, fish markets, flower markets, farmers' markets, and flea markets. Some are outdoors, including many flea markets mainly selling cheap goods, and others are in buildings such as the Covered Bazaar in Istanbul. Markets are often near tourist centers, squatter settlements, and other concentrated sources of shoppers. Most marketplaces are reached by walking, supplemented by public transit, though some are on commercial strips reached by car. Often many booths or stalls are present for sellers, and crowds of people are channeled between rows of booths. Most markets exhibit an intense daily regime: early morning delivery of goods to sell; a sometimes slow, sometimes frenetic selling period; removal of some or all unsold goods; and finally cleanup of considerable packaging material, food waste, and other wastes.

The short-time concentrations of people attracted to marketplaces produce diverse environmental impacts, from huge parking areas of hard surface, ample food eaten while standing and walking, and considerable human wastewater produced. Ecological impacts also result from the goods sold and wastes from the selling process. Most of the markets have a diversity of plant and animal products for sale.

Numerous insects and microbes from rural areas are carried into the city with the biological products to be sold. Some species simply fly or crawl away from the marketplace. Others get free rides from buyers and are thus dispersed to homes and kitchens in residential areas across the metro area. Often quite a mountain of packaging, food waste, and other plant and animal wastes is accumulated for disposal at the end of each day. Such unending piles are also concentrations of the familiar urban pests and microbes, as well as less familiar ones from the countryside, that disperse into areas around a marketplace. The evening cleanup and disposal process doubtless spreads the past or diverse organisms further.

Office center and town center

A cluster of office buildings, often surrounded by some greenspace, forms the archetypal *office center* (Urban Land Institute, 1998, 2001). Low-rise buildings predominate, though high rises may be present. Sometimes called an office park or office campus, the center is commonly located along a major transportation route in a suburb (Figure 11.11a). Toronto has at least 20 office centers, excluding city center, well dispersed across the metropolitan area (Hartshorn, 1992). One is by a commuter train line, six are by a multi-lane highway, four by both subway and major highway, six by both commuter rail and major highway, one by subway and commuter rail, and two are not by a major transportation route. These office centers have half of the total large-building office space of the metro area, with the other half in the city center (central business district) served by all three major transportation modes.

Office centers mainly serve as financial centers, regional service centers, government centers, and corporate headquarters (Hartshorn, 1992). Information flow usually is the major function or activity. On-site facilities are relatively similar for the various users. Large parking lots or garages provide for suburban employees commuting by car. Luncheon facilities and sometimes shops are provided for employees. Considerable paper and packaging materials, as well as human wastewater, are produced.

Large lawn areas, with dispersed trees, clumps of trees, and mini flower gardens, are especially common in the space surrounding office-center buildings. A pond is sometimes constructed and benches and paths for lunchtime walks may be present. Consistent with the look favored by managers for most office-center users, the space may appear relatively lavish and highly manicured, also with an eye to maintaining security for employees.

Yet some office centers are also designed and managed for ecological goals. Stormwater basins minimize downslope flood hazard and reduce pollution of nearby water bodies. Sewage is treated and grey-water recycled on site, in part using marshes. Green roofs cover the flat-topped buildings. Meadows with a changing array of wildflowers replace most lawns (Joan Nassauer and Olive Thompson, personal communications, 2003, 1990). Semi-natural woods, wetlands, or meadows are arranged to fit with similar habitats in surrounding areas, forming large habitat patches or important corridors for wildlife movement across the surrounding land. Essentially all of these approaches not only ecologically enhance the office center space but also provide benefits to the surrounding community.

The *town center*, whether suburban within the metro area or further out in the urban-region ring, is the commercial hub of its surrounding area (Figure 11.11a) (Urban Land Institute, 2008). Traditionally combined with civic and cultural activities, town centers are mostly walkable for surrounding residents, though car, bus, and sometimes commuter train may also transport many users. Town centers tend to be good examples of mixed-use land patterns, where office employment and services, retail shopping, and restaurants are close to or intermixed with residential areas. This pattern limits fossil-fuel driven transportation. Commonly a town center has buildings a few levels high with, for example, retail shopping on ground level, offices on the second level, and residential apartments on upper levels. Buildings with only offices and only residential units are also usually present.

From a commercial perspective, town centers resemble the neighborhood streets with small shops, supplemented by an office center with little or no greenspace around the buildings. Of 27 large enclosed office centers planned and constructed during 1976–2000 in the UK, 33% were built in existing towns and 26% in rapidly growing "new towns" (41% were located out of town) (Pacione, 2005). These in-town office centers helped channel office and retail activity to the center of residential areas, thus minimizing sprawl and commuter traffic.

Some of the above-mentioned ecological benefits also apply to in-town office centers, including green roofs and fitting with walkways and greenspaces of the town. Sewage is incorporated and treated in the town's sewage system. The town's retail shops greatly benefit economically from shopping by the added office-center personnel, though added restaurant use increases food wastes and pests in the town center.

Commercial strip and shopping mall

Variously called ribbon development, strip development, or commercial corridor, this distinctive pattern depends on convenient car access for shopping.

Commercial strips are mainly on busy highways of two to six lanes, and located on a radial highway outside a city or between two cities (Figure 11.11b). A row of fast-food and other restaurants, gasoline/petrol stations, auto or tire dealers, furniture stores, and home-supply stores typically anchor a commercial strip. A row of parking lots for shoppers is by the highway in front of stores, and parking lots behind the stores mainly serve for truck deliveries and employee parking. Trucks daily deliver retail goods and remove packaging and food wastes. Nearby residential areas may be adjacent to, or separated from, the noise and lights of a commercial strip by a corridor of tree vegetation. A highway constructed to bypass the typical traffic congestion of a commercial strip may turn the strip into a row of cheaper stores and vacant lots, with little management of wastes, water, and pollutants.

From the big-picture perspective, a commercial strip or strip development slices the land and nature into separate areas. A row of stores, parking lots, lights, traffic and concentrated people forms an impassable barrier for most wildlife. For instance, a 50-km (30-mi) highway extending northwestward from Madrid has only two or three locations remaining where natural land is close enough on both sides for most native wildlife to potentially cross. Many animals try, including a wild boar (*Sus scrofa*) smashed by a car following mine.

Years ago in the Taos area (New Mexico, USA) we measured animal tracks crossing commercial corridors extending outward from the small city, and found that most tracks were of domestic dogs and cats. A generalized predator, coyote (*Canis latrans*), also crossed the strip, frequently in culverts, which are usually dry there. Mule deer tracks (*Odocoileus*) crossing were only present in the two largest breaks (>approx. 1 km, or 0.6 mi wide) in the continuous strip development. Breaks and narrows in commercial corridors where large natural areas are present on both sides are prime sites for wildlife underpasses and overpasses to maintain connectivity of the land for nature and people (Forman et al., 2003; Iuell et al., 2003; Trocme et al., 2003).

The extensive hard surface area of parking lots and buildings in connected strips outside the city results in considerable water runoff into local streams, and therefore flood hazard. Chemical pollutants from the store operations, vehicular traffic, road salt, and other sources are also considerable, and pollute the nearby groundwater and streams. Transportation-related leaky tanks and hazardous wastes, if present, are particularly degrading to the area adjoining and

Commercial sites dispersed across the urban area

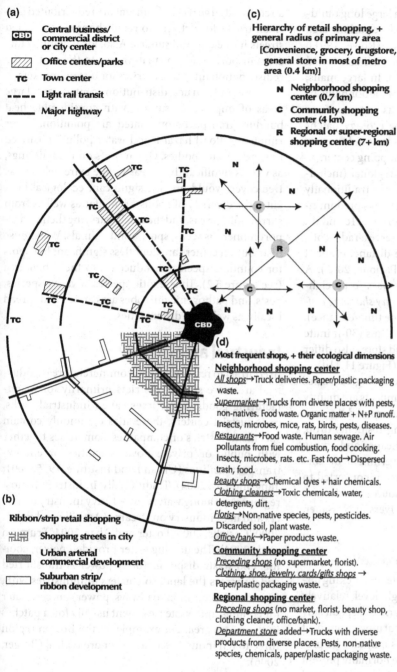

Figure 11.11. Distribution of commercial and shopping areas around city. (a) Major commercial and business centers in city, inner suburbs, and outer suburbs. (b) Strip or ribbon retail-shopping along streets and transportation corridors. (c) Shopping centers hierarchically distributed; lengths of arrows indicate typical distances in providing goods and services; also numerous small grocery, convenience, etc. stores serve local residents throughout the urban area. (d) Prevalent shops and their ecological dimensions in different types of shopping centers. 1 km = 1000 m = 0.62 mile. USA. Based on Hartshorn (1992).

downslope of commercial strips. Green roofs on the flat-topped buildings reduce flood hazard. Abundant trees in parking lots reduce temperatures, air pollution, water runoff, and water pollution. Stormwater basins reduce flood hazard and water pollution. A strip of restaurants means a strip of food waste and a sustained abundance of pests.

The *shopping mall* or shopping center is typically suburban, though shopping centers are also present within and under buildings in the center of large cities (Figure 11.11) (Urban Land Institute, 1999). Like the commercial strip above, access to the mall is mainly by car. Trucks daily deliver goods and remove some wastes, and extensive parking lots serve shoppers, employees, and truck traffic.

335

The shopping mall is commonly a large long building surrounded by an impervious-surface parking area 3-4 or more times the building area (Rowe, 1991). Typically large retail stores at the ends are connected by many small shops (Pacione, 2005). In large malls, small shops are aligned along a central indoor hallway, sometimes with similar types, such as restaurants or clothing shops, clustered. Landscape ecology patch-corridor principles can be used to understand the pattern, movements, and changes of a shopping center.

Neighborhood, community, and regional (including supra-regional) shopping malls are traditionally differentiated, based on size, types of stores predominating, and radius of area served (Figure 11.11c and d). *Neighborhood shopping* malls mainly serve a radius of 1 to 4 km (0.6–2.5 mi), or a driving time-distance of about 5 minutes (Lynch and Hack, 1996; Pacione, 2005). A significant portion of shoppers may walk to a neighborhood mall. Generally the *community shopping center* serves a 4–12 km radius (10-minute time-distance), and the *regional mall* a 12–30+ km radius (30-minute time-distance). The types of shops and stores also differ in this hierarchy of shopping centers (Figure 11.11d).

A shopping mall has many of the ecological issues faced by commercial strips, though impacts are mainly concentrated in a location rather than spread linearly. When income levels of surrounding residents drop, shopping malls are blighted with cheaper and empty stores, and may be recycled to government or corporate use. Occasionally the mall area is converted to greenspace to enhance natural resources and serve surrounding residential areas (Joan Iverson Nassauer, personal communication, 2003).

Warehouse truck-distribution center

The *warehouse truck-distribution center* or trucking terminal is normally an area of long single-level, relatively flat-topped buildings separated by tarmac/asphalt surfaces for the access, parking, and turning around of large trucks (lorries). Railway access may also be present. One or a few such terminals are normally present in suburbs by an entrance/exit to a large city's radial highway. Near the intersection of a radial highway and a ring highway is a prime location. Warehouse distribution centers are also located next to a major airport or shipping port (Figure 11.12).

Extremely diverse, generally non-perishable goods are stored in the warehouses. But the other main purpose of the center is to redistribute goods for transport to different destinations. Thus, goods arriving in ships, aircraft and large trucks from afar are redistributed into small trucks for delivery to retail shops on congested streets in the city and suburb. Also, local manufacturers transport goods in small trucks to the centers for redistribution into large carriers for distant markets.

Warehouse truck-distribution centers are large areas of impervious surface with considerable heat buildup, transportation-related air pollution, water runoff and flood hazard, and water pollution carried to nearby water bodies. Green roofs on the buildings, as well as significantly increasing the extremely sparse tree cover, would provide significant ecological benefits. The diversity of goods, from afar as well as from surrounding areas, and the trucks carrying them, transport numerous seeds, spores, and animals. Warehouse distribution centers are doubtless significant hot spots for countless species introduced into the urban area (see Figure 3.2). Thus, exotic species, invasive species, pests, and pathogenic microbes are carried and spread locally, as well as to distant cities.

Industrial areas

In contrast to residential and commercial areas, industrial areas are sites or districts primarily for manufacturing goods. Sometimes called industrial parks, or industrial centers, these sites commonly contain specific industries or companies, sometimes together with public or private power facilities or energy-transport facilities (Urban Land Institute, 1975, 2001; Yang and Lay, 2004). Traditionally, industries in most regions were along waterways – heavy industry on rivers and light industry on streams (Acebillo and Folch, 2000; Castells, 2000; Foster and Aber, 2004; Frumkin *et al.*, 2004). The flowing water provided power, cooling, and waste disposal. Today power is transported widely across the land, so clusters of industries with, for example, common roadways, power sources, water for cooling, and waste treatment usually form patches in the metro area. For example, in the Boston region three to ten industries in a cluster are typical (Berger, 2006).

Industrial production and post-production at a site

Industrial areas of course vary widely in size. For instance, the average is 120 ha (300 acres) and the minimum about 15 ha in the USA (Lynch and Hack, 1996). One or a few large industries, several medium, or several medium and small companies is typical.

Industrial areas

Figure 11.12. Warehouse and truck distribution center for air cargo. Large flat-topped buildings for storage and redistribution of goods, and with convenient truck access. Eight-lane highway by airport; high water-table indicated by wide water-filled ditch and by pond (upper left). Orlando, Florida. R. Forman photo.

"Boom and bust" production describes the individual industries, as characteristic of market economies. However, some industries have a long boom period, providing some stability to the surrounding urban residential and commercial areas. Even in government-dominated economies, specific companies at an industrial site grow and shrink.

Production inputs, industrial-site features, and outputs

During the *industrial production period* at an industrial site, major inputs and major outputs, in addition to on-site features, determine the ecological conditions present (Figure 11.13). The site itself typically contains a building for management; large outdoor parking lot for employees (e.g., 0.8 to 1.0 parking spots per employee in the USA; Lynch and Hack,1996); tarmac/asphalt space for truck access, parking, and turning around; buildings for production processes; a waste accumulation area; storage areas for raw materials and/or products; and a vegetated area for potential future expansion. Railway access and a pond for waste treatment may also be present. Iron, steel, and concrete structures, specialized for an industry's production process, are often conspicuous.

On-site waste accumulation of course is a central problem (Berger, 2006). Slag heaps, toxic organics, acid drainage, heavy metals, cinders, demolition rubble, and so forth each pose a different challenge for containment, recycling, and cleanup. The toxicity of heavy metal and chromate waste is severe, whereas that of concrete and cinders is slight. Ponds may become toxic. Dumping smothers a soil, typically eliminating the soil's inherent pollutant-decomposition value. Wind blows particles from a heap or mound of waste (Wheater, 1999). Over time an industrial site normally changes markedly, both as accidents and spills occur, and as waste technology and economics evolve (Belanger, 2009).

Some industries offer a ray of hope, as toxic materials are eliminated in manufacturing, and by-products are treated on-site mainly using natural systems (Peter Rogers, personal communication, 2011). Such facilities could become close to self-contained systems.

Inputs to an industrial site feature employees from nearby, raw materials usually from afar, trucks (lorries) or other transport, and power (Figure 11.13). Inputs also include precipitation, many species from nearby, and some species from afar. Wind-blown species arrive in the air or in precipitation, while many types of species arrive in diverse raw materials and on the transport vehicles and trains. Incoming organic materials, such as wood, wool, and skins, typically bring in a rich fauna and microbial population. Such materials may also attract insect and other pests. Some industrial processes use many quite-different raw materials that come from diverse regions with their own faunas and floras.

Outputs from the industrial site are rather different and equally important. Employees, transport, and economic products leave the site. Airflow carries heat and air pollutants, such as particulates and gases, out. Surface stormwater runs off the site and carries water

Residential, commercial, industrial areas

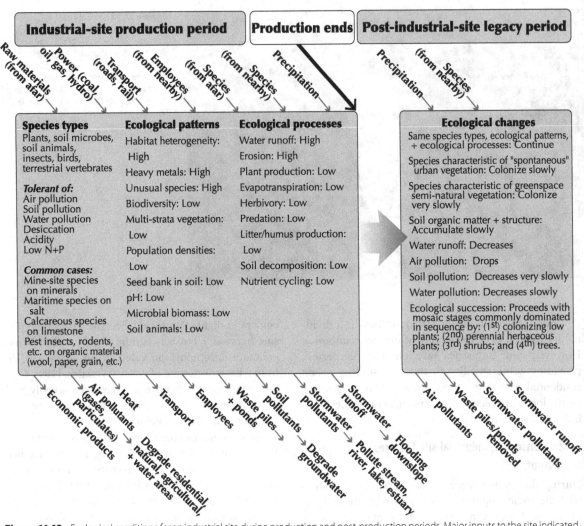

Figure 11.13. Ecological conditions for an industrial site during production and post-production periods. Major inputs to the site indicated at top, outputs at bottom. See Gilbert (1991), Weiss *et al.* (2005).

pollutants. Water also infiltrates into the ground, carrying chemical pollutants to the groundwater, which flows horizontally offsite. Air pollutants bathe the surrounding neighborhoods, food-production areas, and natural areas, especially downwind. The stormwater from extensive hard surfaces races offsite, tending to cause flooding in a downslope built area and a water body. Chemical pollutants contaminate the same water body, as well as the groundwater.

Ecology of the production site

Five distinctive characteristics of industrial sites explain much of the ecological pattern and process (Gilbert, 1991): (1) raw materials; (2) "inhospitable" production processes; (3) waste heat; (4) waste materials; and (5) storage. Numerous species arrive with the raw materials and transport from afar. Production processes usually involve extremely high temperatures or highly toxic chemicals that effectively kill the organisms present. Heat from the production processes disperses to the air, water, soil, and accumulated wastes, which in turn eliminates some species. Yet this heat also favors a set of species from warmer regions. Waste materials from production are mainly unsuitable for most local species, but favor locally uncommon species with specialized requirements, such as maritime species on salty deposits and limestone species on alkaline wastes. The storage of raw materials and of products produced often provides conditions for a few unusual species to thrive.

A simple way to grasp the ecology of an industrial site in active production is to consider the (1) species types, (2) ecological patterns, and (3) ecological processes present (Figure 11.13).

1. *Species types.* Industrial site species are relatively distinctive for being soil-pollution tolerant, air-pollution tolerant, water-pollution tolerant, desiccation-resistant, and successful on soil with little nitrogen and phosphorus. Some species are especially acid-tolerant and others alkaline-tolerant. At a low pH of 4.0–4.5 few species are present, whereas at a high pH of 7.0–8.0 a relatively large number of species may survive. Examples are maritime species on salt accumulation waste, mine-site species on specific mineral accumulations, calcareous/limestone species on alkaline wastes, and pest insects and rodents on organic material such as paper and wool.

   Some industrial site species survive with frequent trampling and weeding, some with infrequent trampling and frequent weeding, and some with little of both (Ohsawa et al., 1988). Lichens are normally scarce. Plant and animal species are a mix of native species from the surroundings, introduced species essentially limited to human-created habitats, and species naturalized in semi-natural habitats (Gilbert, 1991; Attwell, 2000). In the UK, many yellow-flowered members of the mustard family, warm temperate grasses, and feral pigeons, starlings, and house sparrows may be abundant. A rather uncommon bird, black redstart (*Phoenicurus ochruros*), may be present in industrial sites.

2. *Ecological patterns.* Normally, habitat heterogeneity is rather high. Biodiversity is low, while the number of species unusual for the surrounding area is high (Hough, 2004; Kovar, 2004). Vegetation cover and the number of vegetation layers may be high or low. Although one or two species of a group may be rather abundant, most species are present at a low population density (Weiss et al., 2005). Heavy metals tend to be widespread, and the seed bank in the soil is sparse. Soil pH is commonly considerably higher or lower than in surrounding areas. Soil animal density and microbial biomass are low.

3. *Ecological processes.* Plant productivity and the production of litter and humus are normally low. Soil decomposition and nutrient cycling rates are typically low. Water infiltration into the soil is relatively low and evapo-transpiration is low. Surface water runoff is high and soil erosion often high. Both herbivory and predation rates are low.

Post-production inputs, outputs, and ecology

The industrial site and its ecology change dramatically when industries end production. During the *post-production period*, inputs of precipitation and nearby species continue (Figure 11.13). But inputs of raw materials, power, transport, employees, and species from afar essentially stop. Outputs also change. Employees, transport, waste heat, and economic products no longer flow outward from the site.

Continuous change over different time scales is the post-production story. Air pollutants drop quickly, and then gradually decline. Stormwater runoff initially changes little, but later as plants colonize and hard surfaces degrade, runoff drops and eventually levels off. Stormwater pollutants follow essentially the same pattern, except that they continue to spread outward for a much longer period. Soil pollution changes little until decreasing when considerable vegetation cover and soil organic matter accumulate.

Some species types, ecological patterns, and ecological processes described above persist with little change. But species characteristic of "spontaneous" urban vegetation slowly colonize. Species of greenspace semi-natural vegetation colonize very slowly. Plant cover increases more rapidly and later is nearly complete. Soil organic matter and soil structure increase more slowly.

Ecological succession consisting of a series of patchy or mosaic-like stages becomes conspicuous, and theoretically could continue for a century or two. Thus, a stage of colonizing low plants may be followed by a perennial herbaceous-plants stage, followed by a cover of shrubs and tree seedlings, followed by relatively dense small trees, and finally by large trees. The last stage has several layers present, and eventually large trees die, opening up holes in the canopy, such that the forest is a heterogeneous mix of tree sizes and other plants. Reaching the last successional stage is rare in an urban area, because the rate of land-use change is faster.

Alternative approaches for industrial sites

While the preceding insights seem to represent relatively typical conditions for industrial sites, numerous alternative approaches, technologies, and modifications of course exist. Since economic production is involved, all alternatives represent significant trends and have important ecological dimensions.

"Biomimicry" or "biomimetics" represents an especially intriguing long-term approach (Benyus, 2002, 2008; Aizenberg, 2010). In essence, the idea is to manufacture goods and products similarly to the way nature makes natural products, and to create products that mimic the way natural products work or function in nature. Thus, for example, manufacturing processes would avoid the use of high temperatures, and products would be flexible and adaptable. Steel and concrete, for instance, are incompatible with both biomimicry dimensions.

Interdependent industries

Considering that an industrial site or area often contains several unrelated industries, the amount of raw materials and transport, and the total air, soil, and water pollution, are usually huge. The idea of *interdependent industries* (sometimes called industrial ecology, industrial symbiosis, or eco-industrial development) together in an industrial site represents an intriguing idea (Allenby, 2006). In essence, the output of one industry is used as raw material for a second industry, and the output of the second is used as input for the first (or a third) company (Nielsen, 2007; Suh, 2008). This is somewhat like a positive feedback system, though the amounts may or may not control success or failure for a company and the entire system. Interdependent industries not only help one another, but they have smaller external inputs and smaller waste outputs. That means less environmental impact, especially on the surroundings.

The best known several-decade example is Kalundborg (Denmark) with three interdependent industries (Frosch and Gallopoulos, 1989; Beatley, 2000a; Cohen-Rosenthal and Musnikow, 2003). An oil refinery provides surplus gas and cooling water for use by a power station, and the power station provides steam used by the oil refinery. Those two industries are directly interdependent. A third industry, a plasterboard factory, uses surplus gas from the oil refinery, provides condensate used by the power station, which provides steam used by the oil company. This feedback links all three industries.

Other interdependent industry sites include Santa Cruz in Rio de Janeiro, which includes chemical, petrochemical, steel, food producer, paints, and nuclear-electric equipment industries (Veiga and Magrini, 2008). In Kawasaki (Japan), 14 industries, including steel, chemical, cement, and paper companies, exchange seven materials (Van Berkel et al., 2009). A small 15 ha (38 acre) site in Hartberg, Austria accomplishes the goal of interdependence (Liwarska-Bizukojc, 2008).

Stability here is also a key to success (Nielsen, 2007). If one industry chooses short-term profit over long-term stability, the system could fall apart. In market economies, industries typically produce for a while and eventually end their production and disappear. Even when government helps provide stability, at some point government no longer does so. The Kalundborg system has doubtless persisted in part because a central component, the power facility station, has been essentially a public utility or necessity, and appropriately supported by government.

Lichens are classic cases of interdependence or symbiosis. If either the alga or the fungus dies, the lichen, i.e., the entire system, dies. To help provide temporary efficiency and stability, many human systems depend on feedbacks (Todd and Todd, 1994).

In short, interdependent industries are a great idea while they are working (Frosch and Gallopoulos, 1989; Lambert and Boons, 2002; Koenig, 2005; Ashton, 2008). The Kalundborg interdependence results in much less input of water, oil, coal, and raw materials, and much less output of waste, sulfur dioxide, and other pollutants. Furthermore, the varied outputs from the three industries are inputs to a surrounding set of users, including a municipality, fish farm, greenhouses, local farms, cement factory, sulfuric-acid maker, and pharmaceutical plant. The interdependent approach could be improved or expanded using ecosystem and landscape ecology principles (Forman, 1995; Nielsen, 2007). Nevertheless, the inherent instability in such a system, which depends on its weakest link, probably limits the proliferation of interdependent-industries sites.

Stormwater, chemicals, vegetation, ponds, power, location

This set of attributes appears in almost all industrial sites and offers great opportunity for improvement. The number, sizes, shapes, types, and arrangement of such features significantly alter outputs of pollutants. To illustrate with a simple conceptual example, the characteristics of two industries on a site are varied in realistic ways (Figure 11.14a to f). The first image has (a) extensive hard surface; the second (b) stormwater detention ponds; third (c) chemical wastes channeled to the ponds; fourth (d) added wind and solar energy collectors as renewable energy sources; fifth (e) green roofs; and sixth (f) woody vegetation along fencing.

Industrial areas

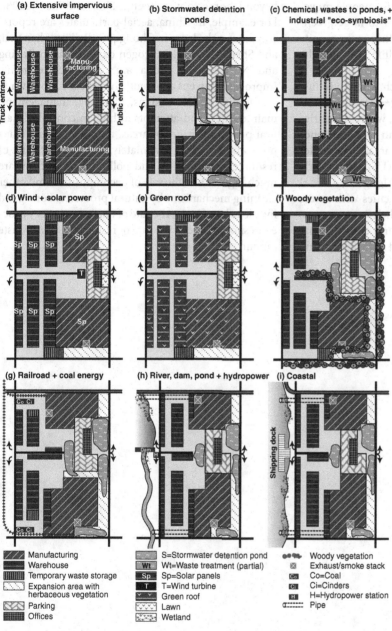

Figure 11.14. Nine alternative forms of industrial sites. All sites have two medium-size industries, a shared office building and parking (entrance on right), paved area for trucks (entrance on left), large manufacturing buildings, warehouses, tall smokestacks, temporary waste-storage locations, areas for future expansion, and highway along boundary at top. Based on Urban Land Institute (1975), Hartshorn (1992), Cohen-Rosenthal and Musnikow (2003), American Planning Association (2006), Wein (2006), and other sources.

Comparing these six options ecologically, based on 17 variables related to air, soil, water, habitat, plants, and animals (e.g., see Table 7.1) reveals qualitative differences. Very few positives and many negatives appear for the first four options, (a) to (d). In contrast, many positive environmental variables and few negatives appear for the vegetation-added options, (e) and especially (f), i.e., woody vegetation along the fencing. Positives and negatives here of course are relative to industrial site types, not other urban land uses.

An additional three alternatives vary characteristics of the power used (coal or hydro) and location (river and coast) relative to the railway (Figure 11.14g to i). Adding rail access and using coal in option (g) has no positive and many negative ecological dimensions. On the other hand, the two options next to water, (h) and (i), have many positives and few negatives.

These sites along railroad, river, and coast are of particular interest because of being located on linear features of the land. Such features are corridors for wildlife movement and can be also for people movement. Adding one or more medium-to-large greenspace in the industrial site would likely make a quite efficient movement route for wildlife through the urban area.

Finally, the *industrial-site* approach, whereby industries are aggregated at designated sites, stands in contrast to the "industrial city" concept, where industries are dispersed over much of the urban area and hence pollutants bathe the whole metro area. Manchester (UK) in the 19th century, Pittsburgh (USA) in the early 20th century, and Baotou (Inner Mongolia, China) today are poster cases of industrial cities with severe environmental conditions. Nevertheless, numerous other cities today in India, China, and elsewhere are covered in pollutants from dispersed industries in and beyond the metro area (Benton-Short and Short, 2008). For example, in China, aerial particulates are reportedly extremely dense in Chongqing, Beijing, Shanghai, and Shenyang, while nitrogen oxide levels in Beijing and Northeastern China are considered extreme. Improvement steps apparently have begun, however.

Despite all the designs, planning, management, maintenance, and attempts at human control of natural processes in urban areas, nature endlessly reappears and grows. Unfortunately, too often the rates of resource consumption and pollution production are much higher than the rate of decomposition and other cleaning mechanisms by natural processes. For a good livable urban future, the rate of nature's recovery processes should exceed the rate of resource use and waste output.

Part III Urban features

Chapter 12 Green spaces, corridors, systems

One might consider an ideal series of parks as you might a great water system, using the metaphor of green water … a weaving, interconnected green mass that changes in size and purpose, but always inter-penetrates forcibly but gently the urban, suburban, and rural scene.
William M. Roth, Conservation Foundation symposium, Washington, D.C., 1971

Rio has edged in between the hills and the sea … But the jungle is still there. You can reach it easily by tram, or through suburban backdoors.
Peter Fleming, Brazilian Adventure, 1933

Is there a better dream than the glorious life of a songbird, gliding over the land from tree to tree, diving into flower gardens, splashing in a pool, and joining hundreds in a chorus with scores of parts? Home may be the luxuriant tropics where life is always sunny. Your ancestors gave you genes for the marathon, when you migrate to foreign land with an explosion of delicious insects or seeds, nice for raising babies. After several nights of flying, an inhospitable metro area stretching to the horizon appears beneath you. But look, a large green patch ahead … and tired wings thankfully carry you there for rest and food (Figure 12.1). While foraging, you meet an unknown relative who grew up in forest beyond the city. The local bird reached your greenspace by moving along a wonderfully wide green corridor, connecting natural land to city center.

Many types of large greenspaces and major green vegetated corridors are introduced in this chapter. The vegetation within them appears in six general forms (Dorney and McClellan, 1984; Kot, 1988; Godde *et al.*, 1995; Rieley and Page, 1995; Breuste, 2009). Examples of each are listed:

1. *Woods*: woodland, forest, natural, semi-natural, remnant, regenerated, all normally containing litter, herbaceous, and shrub layers.
2. *Lawn with trees*: cemetery, golf course, city park, playing field area, and plaza, most containing ornamentals, shrubs, dispersed trees, and flower gardens.
3. *Urban agriculture*: crops/cultivation, paddy, pasture, orchard/vineyard, and community garden.
4. *Wetland*: freshwater or saltwater mudflat, marsh, swamp.
5. *Successional habitat*: meadow, edge of dump, former farm field, former quarry, vacant plot, brownfield.
6. *Combination of the preceding types*: considerable habitat diversity, and often the most common form of greenspace vegetation.

Greenspace patches of course vary from large to small, and green corridors from wide to narrow. At least 75 of these key urban features, varying in spatial scale and from metro area to outer urban-region ring, are readily recognizable (Forman, 2008). In the metro area, examples are large wood-lawn park, zoo, railway corridor, and protected coastline strip, as well as small vacant plot, historical site, tree line, and highway median strip. The exurban or peri-urban area may include large protected semi-natural area, golf course, and greenway, plus small cemetery and pond area. The outer urban-region ring often contains large cropland areas, natural lands, river corridors, and powerline corridors, in addition to small swamp and former quarry.

Considerable research has clarified ecological patterns in urban patches varying from large to small (see Chapters 8 and 9). Yet the ecological effects and functions of urban-corridor width appear to be still little studied. Wide green corridors are usually long, and serve as major connectors across the urban area. Narrow urban corridors tend to be short, and frequently cut up into a row of stepping stones.

Over time, the abundance of greenspace in a metro area decreases as densification and infill occur (see Chapters 3 and 10). Informal squatter settlement is a prime example. From 1950 to 1980, green areas in rapidly growing Mexico City apparently decreased sharply

Figure 12.1. Two large parks important for both city wildlife and migrating songbirds. Parks contain lawn, walkways and small roads, trees scattered and in lines, small semi-natural patches, and water body (lower left). Large urban parks are magnets for migratory birds; large size and some habitat diversity support many resident animals. Tiny bits of green facilitate wildlife movement between the parks and outward across the city. London. R. Forman photo.

from about 42% to 14% of the city. In contrast, with population loss or de-densification, greenspace cover often increases, as in Detroit (USA) in the late 20th century. Disasters such as a hurricane/cyclone in New Orleans (USA) or wars in various cities usually lead to more urban vegetation. Construction sites and associated successional habitats are usually small, appearing plot by plot. Occasionally, extensive construction creates a large greenspace or major corridor.

Greenspace patches, irrespective of size, appear to be more abundant (at higher density) in smaller than larger cities (Forman, 2008). Also, green corridors are more common in smaller cities. Therefore, buildings on average are furthest from an urban park in large cities, and per person the average distance from residence to park is extremely large.

The array of green spaces, corridors, and systems is presented in five groups: (1) urban agriculture; (2) parks; (3) diverse large greenspaces; (4) green corridors and networks; and (5) urban greenspace systems. Railway corridors and rail yards (see Chapter 10), industrial areas (Chapter 11), and informal squatter settlements (Chapter 11 and Epilogue) are excluded.

Urban agriculture

Several hundred million people participate in growing food in gardens and fields of urban areas. Although the process causes lots of human and environmental problems, an extraordinary range of benefits propels the extensive effort forward. From simply the food produced perspective, families may grow 10% or more of their food, while some cities grow 50%, 70%, even more than 90% of the green vegetables and some other products annually consumed by the population (see Figure 12.2a).

From the beginning of cities, residents grew food in urban greenspaces. Turning some parks of these key spaces into lawn-dominated parks interrupted the process, yet urban agriculture persists for good reasons. Food production occurs on remnant large fields in peri-urban/exurban and suburban areas, as well as in temporary vacant lots and little-used interstitial spaces in the city. Despite diverse locations and products, in all cases *urban agriculture* is characterized by growing food and related products as part of an overall urban system. Thus, overwhelmingly the growers, resources used, food produced, environmental effects, and food eaters are of the local urban area (Mougeot, 2005; van Veenhuizen, 2006).

Useful general references on urban agriculture also grow in number (Smit and Nasr, 1992; Lawson, 2005; Mougeot, 2005, 2006; Viljoen *et al.*, 2005; van Veenhuizen, 2006; Nordahl, 2009). Many African cities emphasize urban agriculture (Mougeot, 2005; McGregor *et al.*, 2006; Maconachie, 2007). Other prime urban food-production areas include Germany, East Asia, UK (Moran, 1990; Gilbert, 1991; Ravetz, 2000; Howe, 2002), USA (Hynes, 1996; Lawson, 2005), and Latin America (Losada *et al.*, 1998; Mougeot, 2005; Premat, 2005; Torres *et al.*, 2007; Wright, 2009).

Many key components of the subject were explored in detail in preceding chapters, including soil and chemicals (Chapter 4), air (Chapter 5), water (Chapters 6 and 7), plants (Chapter 8), wildlife (Chapter 9), house plots

Urban agriculture

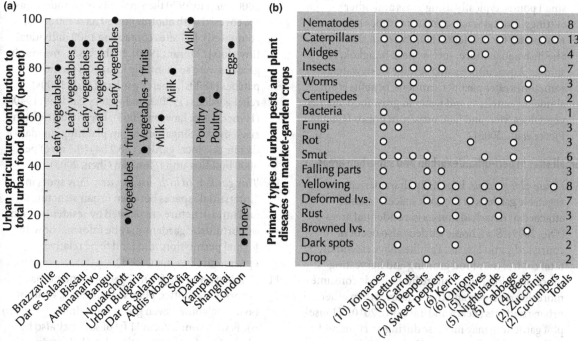

Figure 12.2. Urban agriculture production and pest/disease diversity. (a) All cities are in the Africa Region, except Sofia, Bulgaria, London, Shanghai. (b) Lome (Togo). Ladybugs included with caterpillars; leeches included with centipedes. Insects include crickets, ants, termites, fireworms, white flies, bugs. Adapted from Mougeot (2005).

and buildings (Chapter 10), and residential, commercial, and industrial areas (Chapter 11). We now briefly introduce four dimensions of the subject: (1) locations and types of urban agriculture; (2) growing food; (3) ecological effects; and (4) linkages with neighborhood and urban region.

Locations and types of urban agriculture

The location of a food-growing area helps determine its size. Size and location then largely determine the type of operation and food produced. For convenience, we group urban agriculture types in three categories: (1) large sites; (2) small sites in peri-urban/exurban and suburban areas; and (3) small sites in suburb and city.

Large sites

1. *Livestock on pastureland*. Near cities with high land prices, livestock usually are relatively uncommon and mostly for dairy products (McGregor et al., 2006; van Veenhuizen, 2006).
2. *Fields with national-market crops*. Typically producing grains and beans, such fields are basically linked to rural agricultural economies and communities, not to the urban area. Often these are remnant fields on good soils. Specialized crops are also produced in different areas, including rice in paddies (Song and Gin, 2008), orchards, vineyards, and flower-growing (e.g., in Holland).
3. *Market gardening*. Sometimes called truck farming, these areas typically produce diverse vegetables and fruits aggregated in small fields, intensely using irrigation, fertilizer, and pesticides. Being close by a city, small trucks daily carry fresh produce to markets and restaurants in the urban area, much reducing transportation costs. The agricultural park by Barcelona (Tribo, 1989), huertas by Valencia (Spain), and the Kent and Lee Valley areas by London are characteristic.
4. *Community service agriculture*. CSAs (and various related terms) combine food growing on fields with other goals, such as involving neighbors, volunteers, interns, and inner city children or adults, as well as providing food free to urban people in need, and free or at a discount to neighbors (Donahue, 1999). Educational and recreational dimensions may be included, and often the products produced are quite diverse.
5. *Aquaculture and fish farms*. Fish farms grow mainly herbivorous fish for food in many

small ponds, typically using grass and other cuttings, animal wastes, and sometimes human wastewater for nutrients (Duning et al., 2001). Shellfish aquaculture, such as near Bangkok, is similar. Aquaculture using sewage wastewater, sometimes after partial treatment, benefits from abundant nutrients and water to grow food such as vegetables, rice, and fish, usually in ponds (Costa-Pierce et al., 2005).

Small sites in peri-urban/exurban and suburban areas

6. *House-plot gardens.* Especially characteristic are vegetable gardens in the back spaces/yards of attached or detached houses in residential areas (Ojo, 2009). Such home gardens also occur in courtyard/patio houses. Families have essentially total control over food growing and harvesting processes. Urban agriculture households consume more diverse foods and more calories than other urban residents (Zezza and Tasciotti, 2010). House-plot gardening may increase during wartime and difficult economic periods (Hynes, 1996).

7. *Greenhouse production.* Although sometimes covering large areas, as in Holland, Spain, Italy, and China, most greenhouse production is a small operation. Plant production is normally controlled and often high in greenhouses. Growing and maintenance costs may also be high. Some greenhouses grow flowers, and some contain still-more-expensive hydroponics and recycling facilities (Todd and Todd, 1994). Greenhouses containing imported soil may be built on poor or contaminated soil (Losada et al., 1998).

Small sites in city and suburb

8. *Institutional gardens.* Typically established on the grounds of a university, school, hospital, prison, factory, or government building, institutional gardens often vary considerably over time, as maintenance effort and care change. Usually an educational dimension means a high diversity of products grown.

9. *Community gardens.* A community garden or allotment is typically an aggregation of individually tended tiny plots (Moran, 1990; Gilbert, 1991; UN Development Programme, 1996; Lawson, 2005; Perez-Vazquez et al., 2005). Perhaps as the original urban-agriculture type, allotments were formalized in the UK in the 1800s, and in 1970 the small city of Swindon was 1% covered with allotments (60 ha = 150 acres, composed of 37 sites containing 1300 individual tiny plots) (Moran, 1990). In the USA, community gardens were sequentially known as potato patches (1800s), liberty gardens (early 1900s), relief gardens (1930s), and victory gardens (1940s) (Hynes, 1996; Lawson, 2005). Even in dense high-rise-covered Singapore today, planning standards dictate one local garden (≥0.2 ha = 0.5 acre) per 3000 building units (Jim and Chen, 2003).

10. *Tiny gardens in little-used spaces.* Tiny spots in "interstitial" spaces between urban structures and infrastructure are planted by residents. Such opportunistic gardens may be informal or with formal permission, and tend to be relatively ephemeral (Freeman, 1991).

11. *Buildings growing food.* Vegetables and herbs are grown as potted plants in windows, window boxes, balconies, even green walls. Intensive green roofs with some 30 cm (1 ft) of soil may also be planted for food production. Food plants inside buildings grow in artificial or natural light, even using mirrors (Kellert et al., 2008). Mushroom growing in warm moist soil in the dark typically occurs in building basements or elsewhere under a city (Clement and Thomas, 2001).

Growing food

The setting for growing food in urban areas commonly contains an access road. Other widespread features of urban agriculture are: a scatter of ditches for irrigation and drainage (Geertsema and Sprangers, 2002; Asare Afrane et al., 2004); dispersed trees providing harvestable products and enhancing soil conditions (Maconachie, 2007); lines of woody vegetation along roads, ditches, streams, and hedgerows; shed, fencing, compost pile, and personalized artifacts (Bruno et al., 2006); and farmsteads with farm roads.

Specialized fencing and other features are associated with the growing of different animals for food, such as chickens, rabbits, goats, pigs, and cows (Premat, 2005; McGregor et al., 2006; van Veenhuizen, 2006). Some animals may also be raised on roofs where they are not disturbed by people (Losada et al., 1998). By consuming organic wastes, many animals are effective recyclers.

The plants grown are primarily vegetables, though fruits, herbs, and flowers are also widely grown. Getting started with planting is often a challenge because the urban soil is contaminated with chemicals, including heavy metals (Alam et al., 2003; Kirpichtchikova et al., 2003; Khai et al., 2007; Maconachie, 2007). The basic safe options are to remove the contaminated soil, add clean soil on top, or still better, do both (virtually always the rate of phytoremediation is much too slow). Urban growers often consider organic methods to reduce chemical use and increase biodiversity (Wickramasinghe et al., 2003; Bengtsson et al., 2005).

Pollinators are important for production in many species. In New York City urban gardens, 54 bee species are present, dominated by a native bumblebee and two non-native bees (Matteson, 2008). Weeds are common in urban food-growing spaces, especially if the space adjoins fields and hedgerows (Moran, 1990).

Pest herbivores, our competitors for the food, are attracted to food-producing sites (Figure 12.2b) (Moran, 1990; Gilbert, 1991). Deer, rabbits, rodents, woodchucks, pigeons, sparrows, and lots of insects and other invertebrates may arrive. The herbivores attract predators, so the garden may have foxes, domestic cats, raccoons, snakes, spiders, insect-eating birds, and raptors. Gardeners often plant certain species, such as marigolds and mints, with chemicals that deter herbivores. Deterring people from stealthily removing food produced is often more difficult.

After planting plants, many things are added to enhance growth and food production:

1. *Water* is normally added, particularly in dry climes and dry periods. Partially treated or untreated sewage wastewater is sometimes used for irrigation, and adds nutrients as well (Smit and Nasr, 1992; Keraita et al., 2003).
2. *Chemical fertilizers*, especially containing phosphorus and nitrogen, are commonly added to stimulate growth. The soil pH may be increased with lime or decreased chemically or organically.
3. *Manure* or *humus* provides organic matter as well as mineral nutrients. Composted yard waste is a prime source of humus. The soil organic matter retains water and nutrients, and enhances the soil fauna and microbial components, with added benefits to plant growth. Alternatively, *organic solid waste*, such as household garbage, is sometimes added to provide organic matter and nutrients (Anikwe and Nwobodo, 2002; Hough, 2004; Pasquini, 2006; Maconachie, 2007). However, municipal solid waste, even when sorted for its organic contents, is commonly rich in toxic heavy metals.
4. *Pesticides*, such as insecticides, herbicides, and fungicides, are often added to eliminate pest species that limit food production.

Ecological effects

The air over and downwind of a large urban-agriculture space is likely to be cleaner and cooler than that of nearby urban areas (Figure 12.3). The main source of air pollution would be wind-eroded dust in dry periods. The lower-temperature air results from the near-absence of heat-radiating impervious surfaces, as well as the evapo-transpiration from irrigated soil and a cover of herbaceous plants (Smit, 2006). The cooler air also means that air pollutants in warm air of surrounding areas may be drawn to and deposited on the food plants and soil.

Overall, the soil quality is usually high, a relatively scarce phenomenon in urban areas. Fertilization provides abundant nutrients. Added organic matter normally enhances a whole range of soil properties, including nutrients, moisture, soil animals, soil microbes, root growth, water infiltration, and aeration (see Chapter 4). Pollutants may arrive from the air, but contaminating or toxic levels of pollutants often result from the addition of organic solid waste, sewage wastewater, or dry wastewater sludge (Huang et al., 2006).

Water quantity, water quality, groundwater conditions, and the downslope local water body are all apt to be affected by areas of urban agriculture. Irrigation channels may add water to the food-production site (Figure 12.3). Consequently both infiltration into the soil and evapo-transpiration normally increase. In heavy rainfall times, flooding is increased where the soil is already moist or saturated.

The input of nutrients, especially nitrogen and phosphorus, frequently exceeds the rate of absorption by the plant roots. Excess nutrients are typically carried downward by infiltration and washed downslope by stormwater runoff. Both nutrients and heavy metals reaching groundwater effectively accumulate because (except in limestone areas) groundwater normally flows very slowly. Such groundwater pumped from wells for water supply may be toxic and require water treatment. The abundant nutrients carried downslope eutrophicate the local water body, with reverberating effects often leading to a degraded aquatic ecosystem and scarce fish.

Green spaces, corridors, systems

Figure 12.3. Market gardening agriculture for vegetables and fruits adjacent to city. Intensively cultivated small fields with scattered farm-related buildings and woody-plant corridors along some irrigation/drainage channels. Source of clean cool air for the city, and resting/feeding areas by Mediterranean coast for birds migrating between Europe and Africa. Las Huertas, Valencia, Spain. Photo courtesy and with permission of Arancha Munoz Criado.

Do urban agricultural sites have high or low biodiversity? First, overall, habitat diversity is low, and few habitats normally mean few species. Second, the smoothed food-growing site often has replaced a successional habitat with many species, whether the size of a field, vacant lot, or tiny interstitial spot. Third, heavy pesticide use sharply reduces biodiversity.

Yet three other factors counter these patterns. Water is added, thus benefitting a wide range of non-cultivated species. Soil is enriched by various additions of nutrients and organic matter, thus stimulating growth of numerous plants, and providing invertebrates as food sources for varied predators. Finally, the high production of our nutrient-rich food plants also serves as a significant food source for numerous herbivores. Balancing these six processes suggests that urban agriculture sites generally have a moderately high species richness.

Linkages with neighborhood and urban region

Urban agriculture sites interact with their neighboring area, with the city, and with the broad urban region. At the neighborhood scale, a food-growing site serves as a key species source, from which plants and animals disperse and colonize the surroundings. Wildlife regularly move both directions between site and neighborhood. Pollinators also move from site to surroundings, and vice versa, pollinating flowers and increasing fruit and seed production. Public-health disease vectors and pests, such as mosquitoes, move from ditches and wet areas into the surrounding neighborhood (Brown and Jameton, 2000; Asare Afrane et al., 2004). Pollutants from adjoining busy highways spread into the agricultural site. And farm stands or shops selling the food products attract neighbors to the urban agriculture site (Torres et al., 2007). Indeed, the urban agriculture site is best viewed as a key piece of a "multi-functional landscape mosaic," where pieces fit together and regularly interact (see Chapter 2) (Brandt et al., 2003–2004; Hardoy et al., 2004; McGregor et al., 2006; Tacoli, 2006).

The food-growing site also strongly interacts with the city (Figure 12.3). Fresh food is trucked into city markets and to the periodic farmers' market. Food products go to restaurants, including those serving organically grown food. Soil pollutants absorbed by food-plant roots are passed to consumers in this way. Organic waste from markets and restaurants, in turn, may be sent to peri-urban/exurban farms as valuable fertilizer or pig food. Food-transporting trucks also carry an assortment of "hanger-on" species, including native species, non-native species, and pests. The city serves as a source of such species, yet the urban agriculture site does too. Another interesting research subject awaits study. At the city scale, the food-production site serves as a stepping stone for movement of species across the city.

Viewing an urban agriculture site in the broader context of the urban region highlights several more dimensions of linkage. Birds migrating over an extensive built metro area see a large agricultural patch and stop (Figure 12.3). Although cover for resting is limited, food is often abundant. A different case from the Wuxi area of China, where urban farmland was heavily fertilized with cow manure from a surrounding area, found high levels of heavy metals in both soils and some vegetables (Huang et al., 2006). The cows fed on grass in an area receiving heavy metals from factory emissions. In this way, far-distant factories, distant pastureland, its cows, their cow manure, the urban-agriculture soil, and finally food produced for urban markets were tied together.

A "food-shed" refers to the area surrounding a city where much of a city's food is produced (Lister, 2007). In the USA in 1900, before the spread of motor vehicles and paved roads, <40 km (25 mi) was probably the average distance a piece of food traveled to a plate for eating. Apparently, in 1960 the figure was 425 km (265 mi), and in 2000 on average food traveled 2400 km (1500 mi). Urban agriculture now sharply cuts both transportation distance and cost, also reducing time and food spoilage (Zezza and Tasciotti, 2010).

Outward urban development tends to target farmland and good agricultural soils (Losada et al., 1998; Benfield et al., 1999; Morello et al., 2000; Lopez et al., 2001; Tan et al., 2005; Faggi et al., 2008). In fact, sustaining agriculture near cities in the face of urbanization may be an important societal goal (Johnston et al., 1987; Vail, 1987). Various types of locations have been suggested as particularly desirable for urban agriculture, such as in greenbelts or green wedges (Howard, 1902; Hardoy et al., 2004; Forman, 2008). Also, urban agriculture may flourish near informal squatter settlements full of people recently arrived from rural areas (Freeman, 1991), and thus relatively knowledgeable about farming.

Almost irrespective of location, urban agriculture sites and the intensity of food production tend to be temporary (UN Development Programme, 1996). Near cities and high land prices, dairy farms frequently fold. Community gardens on vacant lots are regularly replaced by buildings. Back-space vegetable gardens in house plots disappear when the next resident moves in. Gardens in tiny interstitial spots are here today, gone tomorrow. Indeed, planting trees for fruit production is sometimes recommended to give the perception of permanence.

Finally, food production in urban agriculture helps provide some "food security" or stability in the face of crises (UN Development Programme, 1996; Koc et al., 1999; Mougeot, 2005; Premat, 2005; Ojo, 2009; Wright, 2009; Zezza and Tasciotti, 2010). Europe has had two major wars in a century, oil markets for transportation remain volatile and seemingly beyond control, and history records many cities degraded by severe prolonged drought. In addition to adaptability from a series of perturbations, flexibility leads to stability. "Land-use flexibility" may be provided by a large area of urban agriculture (Johnston et al., 1987; Jim and Chen, 2003), a large number of separate locations with it, and a large number of types of production. For example, the Berlin and Bamako (Mali) urban regions have many separate agricultural landscapes, but few types (Forman, 2008). On the other hand, the Rome and Barcelona regions have many urban agriculture landscapes and also many types of production (Forman, 2004, 2008). That provides valuable flexibility for getting through future crises.

Parks

From the beginning of cities, urban greenspaces must have provided enjoyment and aesthetic values for residents. Food production was doubtless also important. Since about the 17th century in Europe, urban "parks" with lawns, flower beds, and scattered trees and shrubs have been largely planted and maintained for the enjoyment of people, including leisure and recreation activities (Figure 12.4) (Platt et al., 1994). Such park designs spread worldwide (Ishikawa, 2001; Sorensen, 2004; Havens, 2011). More recently, providing a spot of nature has become an additional important function of parks. Today "city parks" are usually either managed overwhelmingly for the enjoyment of people, or may also have an important nature protection value (Figure 1.11b). Semi-natural vegetation in a park supports both nature and nature-based recreation.

Numerous park roles or benefits are well recognized (Chiesura, 2004; Tyrvainen et al., 2005). Major benefits are: (1) social (recreation, health, educational, cultural, historical values); (2) aesthetic (experience nature, frame/define/screen structures, variations in plant color/texture/density); (3) climatic and physical (cooling, controlling wind, humidity, air cleaning, sound, light, flooding, erosion control); (4) ecological (plants, vegetation, animals, wildlife movement, biodiversity, aquatic ecosystems, fish); and (5) economic

Figure 12.4. City parks mainly for recreation/leisure of residents. (a) Fairground and pond for sailing model boats. Tuileries, Paris. (b) Morning exercises on soil substrate beneath dense tree canopy. Île de la Cité, Paris. (c) Lawn chairs placed in the sun; foreground with soil walkway and bench; group of common pigeons (rock doves) (center left). London. R. Forman photos (bottom photo courtesy of Jessica Newman).

(property values, tourism, and the preceding non-market values).

Park size is typically the most important attribute determining its functions and benefits. This is illustrated with tiny, small, and medium-size city parks and their plants in Rome (Attorre *et al.*, 2003). All three of these Mediterranean-climate parks appear to be well used by people. Apparently all of the trees and shrubs were planted, suggesting rather intensive maintenance that removes colonizing woody plants.

1. *Tiny city square*. The 35 m × 65 m (115 ft × 217 ft) Piazza Cairoli contains one fountain, one monument, two flower beds, nine park benches, a gravel surface, and a surrounding sidewalk. There are 13 deciduous trees (3 species), 6 evergreen trees (1 species), and 13 evergreen shrubs (1 species) present.
2. *Small city park or garden*. The 80 m × 100 m (262 ft × 328 ft) Giardino di Carlo Alberto has connected gravel-walkway areas covering about 40% of the area, plus large flower beds on 40% and a pond covering 20% of the park. Also present are one monument, one fountain area, 12+ park benches, a tiny raised gravel space, and an area of steps. Today there are 61 woody plants representing 25 species. The predominant forms (in order) are conifers, deciduous vines/climbers, deciduous trees, other evergreen trees, palms, and evergreen shrubs. Woody plants in the park have plummeted from 496 (in 1894) to 61 today.
3. *Medium-size city park*. The approximately 320 m × 360 m (1050 ft × 1180 ft) Parco del Pincio is covered about 75% by gravel walkway, 10% asphalt walkway, 5% road surface, and 10% flower beds. One building, 11 fountains, 5+ park benches, and a sidewalk on one side are present. The 1244 woody plants (in order of abundance) are primarily oak (*Quercus ilex*) (approximately 275 trees), laurel (*Laurus nobilis*) (150 shrubs), plane tree (*Platanus* × *acerifolia*) (70 trees), horse chestnut (*Aesculus hippocastanum*) (65 trees), cypress (*Cupressus sempervirens*) (50 trees), pine (*Pinus pinea*) (45 trees), and black locust (*Robinia pseudo-acacia*) (40 trees). Woody plants in the park have dropped from 7049 (in 1870) to 1244 today.

All three parks are mainly covered by gravel for walking park-goers. Gravel cover also facilitates stormwater infiltration, protects against erosion, minimizes heat buildup, and consequently enhances the park's plant growth. All three parks have a conspicuous mix of deciduous trees, evergreen trees, and evergreen shrubs, providing variation and benefits throughout the year. At least in the small and medium parks, woody plant density has dropped enormously over some 13 decades.

Large city parks provide a similar set of resources for people. But in addition, semi-natural areas that escape manicuring by park maintenance personnel are sustained in part due to limited government budgets. Central Park (300 ha = 750 acres) in central New York City contains two semi-natural patches (The Ramble and North Woods), each about 16 ha (40 acres) (Rosenzweig and Blackmar, 1992). Ecologically the North Woods is particularly interesting because much of the tree canopy is composed of non-native species.

Other large parks in the central portions of cities are predominantly used for human enjoyment, but also include semi-natural areas, such as Golden Gate Park (400 ha = 1000 acres in San Francisco), Mount Royal Park (100 ha = 250 acres in Montreal), Monjuic (in Barcelona), Table Mountain Park (in Cape Town, South Africa), a large nature reserve in central Singapore, Tiergarten (215 ha = 530 acres in Berlin), and probably Stanley Park (400 ha = 1000 acres) in Vancouver and Chapultepec Park in Mexico City. The Reserva Natura del Sur in central Buenos Aires established on a former dump site is mainly for nature conservation, and public access is limited.

Plants in ten large urban parks (each ≥400 ha = 1000 acres) near the center and periphery of major cities were studied in the Boston-to-Washington megalopolis (Loeb, 2006). In total, 1391 vascular plant species were recorded, including 490 non-natives. Both natives and non-natives are prominent in the plant mixture dominating all the parks. Perhaps surprisingly, less than 1% of the species are in all ten parks, and <2.5% in nine or more parks. No distinctive or common park flora seems to exist in this region, which has a relatively similar climate and where parks are surrounded by millions of people. Two large parks in the Bronx portion of New York City are more similar in flora to parks in Baltimore and Washington than to nearby parks in other portions of New York City. This suggests that the process of planting and removal of plants by park maintenance personnel has a greater effect on a park's flora than does geographic proximity.

Since convenient human access to a city park is a prime goal, convoluted or elongated shapes are especially effective, as form-and-function theory would indicate (Forman, 1995). In contrast, square parks with a higher interior-to-edge ratio (Figure 12.1) are better for protecting internal resources, such as semi-natural areas. Dense adjacent built areas tend to degrade semi-natural resources, while enhancing human access (Tilghman, 1987). Adjoining built areas generally inhibit mammal movement more than bird movement (Thornton et al., 2011). Adjacent busy roads degrade both semi-natural and human-enjoyment goals, at least in the edge portions of a park (Forman et al., 2003). Even without a semi-natural area, a park with heterogeneous land uses for people may support a moderately high biodiversity.

Trees are especially characteristic and important in parks (Figure 12.4). Flower beds, gravel areas, water features, shrub patches, lines of woody plants, hedges, lawns, and so forth each provide somewhat different conditions and attract different species (Goode, 1986; Gilbert, 1991; Boada and Capdevila, 2000; Houck and Cody, 2000). Tiny fountains and ponds serve as key attractors for a variety of animals in parks. Bright light for night events attract flying insects, plus insect-eating bats. Meanwhile streams, rivers, large ponds, and shorelines are major habitats, indeed tightly packed groups of habitats, in parks.

In effect, this *habitat diversity* or heterogeneity increases species diversity (Tilghman, 1987). High habitat diversity in a park means a large *species pool*, the total number of species present. It also means that the park is a large *species source*, i.e., many species move outward from a park to surrounding areas, including other greenspaces. If the park contains a semi-natural area, both the species pool and species source are still larger and more important for the neighborhood or city.

The simple presence of people frightens away many animals. However, in urban areas perhaps most species are *habituated*, i.e., accustomed and less scared of humans. In parks, people also compact soil by trampling, and damage vegetation in varied ways.

The avian use of four highly different habitats in a medium-size park in Sheffield (UK) is instructive (Gilbert, 1991). Over a year, 643 birds (of 25 species) were recorded in a wooded patch, 322 birds (21 species) on sloping lawn (grass), 253 birds (23 species) in flower gardens, and 854 birds (8 species) on a pond. Thus, in woods and lawn, birds were rather dense and diverse. On grass and gardens, bird density and diversity were highest in winter. On the pond, diversity was low throughout the year, but density was very high in winter (98% of the birds were mallards and black-headed gulls). The park's habitat diversity supported a rather species-rich year-round avifauna, with bird densities highest in winter.

In effect, animals "need" to move in order to find food, nutrients, water, mates and nest/den sites, and

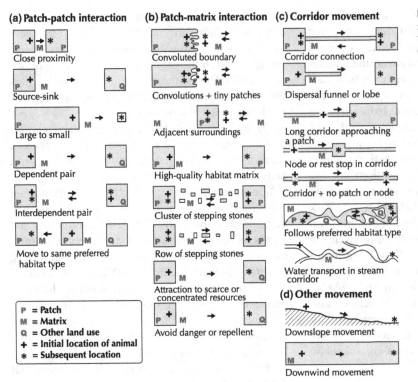

Figure 12.5. Spatial patterns suggesting frequent wildlife routes and a tightly interwoven land mosaic. Letters = different land uses or habitat types.

to escape from predators (Harris and Scheck, 1991). Furthermore, animals may move to exploit sporadic resources, accomplish different stages of a life cycle, colonize new environments, and respond to disturbances and stresses such as climate change. A line of shrubs and/or trees is a linear feature and route, typically enhancing animal movement (Figure 12.5). For us, the same line of woody vegetation provides aesthetics, screens out something on the other side, reduces air flows, and filters aerial dust. However, the linear vegetation typically does not provide sufficient cover for nesting/denning or escape from predators. Instead, a patch or clump of vegetation is optimal (Figures 12.1 and 12.5) (Goldstein et al., 1981). Thus, in park management, providing both vegetation patches and corridors is likely to enhance conditions for wildlife, as well as people. For large urban parks, this is particularly important for migrating birds, which mainly need cover for resting and food energy for tomorrow's trip. Instead, the birds arrive at a typical city park in large numbers after a long flight to find relatively inhospitable lawn, ornamentals, people, and pollution.

Tropical cities have hot air year round (Yu and Hien, 2006; Chang et al., 2007). A study of 61 city parks in Taipei found that large parks have cooler air than in their surrounding built areas. However, approximately 20% of the parks are not cooler. This latter group has ≥50% paved surface and little tree-and-shrub cover. These patterns again emphasize the importance of minimizing hard surface cover, and providing a diversity of land uses or habitats within a park.

Cities in dry climates have both water conservation and high temperature challenges (Pearlmutter et al., 1999; Setha et al., 2005; Brazel et al., 2009). Large parks or grouped parks cool the city air (see Chapter 5) (Shashua-Bar, 2009). Lighter color surfaces absorb less solar radiation, resulting in less urban heat, as evident in white North African cities. Trees, shrubs, and structures shade surfaces, and the plants effectively cool by evapo-transpiration. Street canyons have more shade, but trapped heat rises more slowly, resulting in hot evenings (Pearlmutter et al., 1999). Sprawl residential housing typically has considerable impervious surface and often limited shade. With topographic heterogeneity, cool air drains down valleys into low areas. In arid cities, atmospheric moisture may be high at night due to irrigation of lawns and parks, but much of the moisture quickly evaporates in the morning.

Overall, in dry climes, an abundant cover of scattered low-water-use shrubs and small trees that shade a heterogeneous non-paved surface seems to be a good solution for city parks (Shashua-Bar, 2009). That mimics some desert conditions, which additionally support a rich fauna of animals, especially active at night. Indeed, mimicking nearby natural lands in park design offers varied ecological benefits, and is also likely to reduce maintenance budgets.

Diverse large greenspaces

Of the 75 widespread greenspace types in cities, several often occur as relatively large patches (Forman, 2008). Although essentially all have been introduced from various perspectives in the preceding chapters, it is useful to pinpoint the following seven groups of large greenspaces, representing 13 types, with their characteristic key environmental dimensions:

1. Semi-natural habitat areas and urban woodlands/forests.
2. Cliffs, quarries, mines.
3. Golf courses.
4. Cemeteries.
5. Institutions and municipal facilities.
6. Dumps, brownfields/vacant lots.
7. Airports and military bases.

Other related large urban patches have been considered in different contexts: rail yards (see Chapter 10); construction sites (usually small; Chapter 4); warehouse truck-terminal areas (Chapter 11); informal squatter settlements (Chapter 11); and industrial areas (Chapter 11). Parks, urban agriculture, and major green corridors were considered above in this chapter. Hazardous waste sites are basically non-urban, and demolition sites are usually small.

Each large greenspace has a single specialized function or role, but almost all are multi-functional, multiple use. Conflicting land uses within a greenspace persist as management challenges, and land-use changes appear endless. Consider the large wooded Ajusco area on the southwest edge of Mexico City that contains or provides water supply; air cooling; air cleaning; biodiversity; recreation; weekend houses; informal squatter settlements; drug activities and therefore police, and guerillas (in the past) and therefore the military (Pezzoli, 1998; Forman, 2008). Vegetation in large urban greenspaces remains in flux and usually successional habitats are present, even widespread. Indeed, over time the greenspaces themselves change in area, usually shrinking. Equally important ecologically, their surroundings are constantly changing.

For detailed ecological insights into the following large greenspace types, especially see Chapters 4 (urban soil and chemicals), 5 (urban air), 6 (urban water systems), 7 (urban water bodies), 8 (urban habitats, vegetation, plants), and 9 (urban wildlife).

Semi-natural habitat areas and urban woodlands/forests

Natural and semi-natural habitat areas

Large urban natural-habitat areas are usually better called semi-natural, because of previous and ongoing human impacts (see Figure 1.5). These impacts, combined with the extensive adjoining built area, mean that ecological conditions today are quite different from those of pre-urban times. Such semi-natural areas may be protected as nature reserves or not protected. Although the abundance, e.g., of birds, may be low or high, natural areas are usually species rich (Figure 12.6). The proximity of a large human population means a continued intense pressure on sustaining natural processes and conditions.

Habitat heterogeneity, both natural and human-induced, tends to be rather high. While urban wetlands are often scarce, the presence of wetlands, ponds, streams, and other water bodies greatly augments habitat and species diversity. Woodland or forest is especially characteristic of urban semi-natural areas, though old woodland with big trees, many fallen logs, and a relatively deep soil organic layer is usually absent or scarce. Shrublands, dry habitats, or meadows (Ahern and Boughton, 1994) are sometimes present, and successional habitats often common, in large greenspaces. Large area and considerable habitat diversity mean rich biodiversity (see equations, Appendix B).

In suburban/exurban Wilmington (Delaware, USA), over a century (1890–1990) forest cover increased from 5% to 22% of the area (Matlack, 1997a). Most modern woods are <60 years old, and only 2.5% of the landscape has forest >100 years old, mostly on poor agricultural soils. Most residential development is on well-drained upland areas, which therefore contain little forest. The modern forest is mainly away from roads and on steep valley slopes. Few large woods are present. Almost all forest trees are within 50 m of the margin of a woods, and thus the woods are overwhelmingly edge habitat (Forman, 1995). The average distance from woods to nearest house is 150 m (about 30 houses are

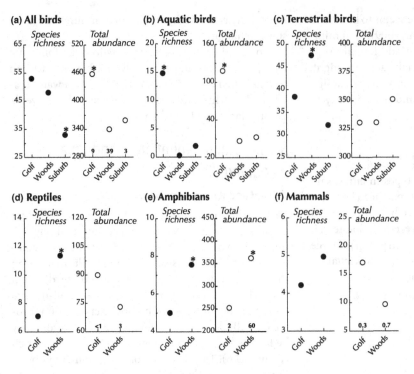

Figure 12.6. Vertebrate diversity and abundance in golf courses, woods, and suburbs. Species richness = number of species; total abundance = an index of relative abundance of individuals; * = significant difference, $p < 0.05$; numbers at bottom = approximate median abundance of regionally threatened species at a site. Based on sampling: 20 golf courses (32–181 ha), 10 woods (nature reserves and forests 40–892 ha), and 10 suburban areas in the Brisbane–Gold Coast Region, Australia. 1 ha = 2.5 acres. Adapted from Hodgkinson et al. (2007).

located within 500 m of a woods), but little evidence of pedestrian damage to the woods is present. An overall rather low plant diversity may mainly result from the woods being young and fragmented.

Natural areas provide many values for people (Spirn, 1984; Breuste et al., 1998; Hough, 2004; Kowarik and Korner, 2005; Adams et al., 2006; Kellert et al., 2008). Large natural areas help clean and cool the urban air, and store ample carbon (Bradley, 1995; Beckett et al., 1998; Rydberg and Falck, 2000; Brack, 2002; Zhu and Zhang, 2008). Some roles of large urban natural areas are suggested by the Tamaki ecological district of Auckland (Galbraith, 2000). Large wooded patches, though much modified over two centuries, seem to be valued by residents for: (1) native people's (Maori) traditions; (2) rich biodiversity; (3) rare plants and animals; and (4) pollinators (small flies, beetles, moths, bees) that pollinate flowers throughout residential and commercial areas.

In Portland (Oregon, USA), Forest Park is a 2000-ha (5000-acre) green wedge on an elevated geological formation projecting inward to the city center area (Houck and Cody, 2000). The park is a large natural area containing infrastructure (pipelines and powerlines), streams, and rich biodiversity. The most striking feature is the presence of large wild mammals in the city, including foxes, coyotes, bobcats, deer, elk, bears and mountain lions (*Vulpes, Canis, Felis, Odocoileus, Cervus, Ursus, Felis*). Eagles and native trout are also present in this area of the city. Habitat heterogeneity and alternative travel routes facilitate wildlife movement (Figure 12.5).

The size of large greenspaces does matter, as illustrated by 12 urban woods in Seoul ranging from 92 to 356 ha (230–890 acres) (Park and Lee, 2000). Breeding bird diversity steadily increases from 14 to 19 species according to large-park size. While there are twice as many resident as summer-visitor breeding birds, the rate of diversity increase along the forest-size gradient is the same. Different bird guilds (see Chapter 9) also increase, though at different rates, along the gradient from large to very-large natural urban greenspaces.

Especially in Mediterranean-type climates (of California, Chile, South Africa, Australia, and the Mediterranean Basin), urban development often spreads into large natural areas composed of fire-prone woodland (Trabaud, 1987; Loehle, 2004; Omni, 2005; Vince et al., 2005; Theobold and Romme, 2007; Keeley et al., 2008). Frequent *wildfires* (uncontrolled burns of areas) lead to an abundance of fire adaptations by plants and animals, such as thick bark, protected buds, burrowing animals, and fire-stimulated sprouting, seed dispersal, and seed germination (Forman, 1979b;

Whelan, 1995; DeBano *et al.*, 1998). Following a fire, such fire-adapted vegetation develops over time to become very similar to the pre-fire vegetation.

Wildfires usually move as a broad ellipse encountering topographic, substrate, and fuel-accumulation patchiness, and subject to changes in wind. Afterward, the burned land is a conspicuous mosaic, often at a fairly fine scale, with skipped patches of green. To reduce building loss, basically the first rule is to keep buildings away from fire-adapted vegetation. Secondly, certain well-known house materials and adjoining woody-plant arrangements can decrease the probability of a house fire. Controlled (prescribed) surface fires are sometimes used in urban areas. However, near built areas, controlled burns tend to be hazardous. This is due to smoke and human respiratory ailments, the risk of fire climbing up into the tree canopy and causing a tree-killing wildfire, and the threat of fire taking off horizontally out of control.

At the landscape scale, special fire danger occurs when about 20–40% of the land is developed (Francisco Rego, personal communication, 2003; Forman, 2004b). With this development level, typically natural vegetation is extensive enough to support big hard-to-stop fires. Also, buildings are frequent enough to likely be in the path of a fire. At higher development levels, typically wildfires are more readily controlled. Land planning and management are a challenge in urban fire-prone lands, often needing to create spatial arrangements that accomplish three goals: (1) limit the loss of trees for wood products; (2) limit the loss of buildings and people; and (3) provide wildfires to sustain uncommon fire-adapted species.

Finally, the complex and delicate natural conditions of a natural greenspace area are highly sensitive to human effects, from recreational activities to altered water-table and baths of air pollutants. Since cities seem to mainly grow and expand, human impacts increase. The future does not seem rosy for individual natural areas. Tying natural and semi-natural habitat areas together into a tightly integrated system, however, offers hope for nature near the people of cities.

Urban woodlands/forests for forestry

Urban forests and woodlands to produce wood products using good silvicultural planning, management, and harvesting practices are scarce in urban areas (Bradley, 1995; Miller, 1997; Kowarik and Korner, 2005; Rydberg and Falck, 2000; Konijnendijk, 2000). Forestry areas normally provide multiple uses or values to society. The prime goal is producing wood products, and usually the secondary goals are protecting and providing clean-water supply, soil erosion protection, and recreational opportunity.

Many German cities have good examples of multiple-use urban forests (Forman, 2008). For 13 such forests, all apparently provided several of the following important values (in order of frequency): (1) production forestry; (2) recreation; (3) water production; (4) mining and reclamation; (5) wildlife habitat (mainly for viewing rather than hunting animals); and (6) climate and noise mitigation. A "group-selection cutting" approach, harvesting small groups of trees, is consistent with or enhances the other values. Berries and mushrooms are also harvested in the urban forests (Konijnendijk *et al.*, 2005).

In somewhat dry climates with savanna-like tree vegetation, as in East Africa, peri-urban residents harvest branches and trees for valuable firewood. Also, in wartime, urban woodlands and forests tend to be valuable resources intensively cut by residents. Where food is short, some residents comb the forest for animal meat to eat.

Occasionally all the trees in an urban area (street trees, yard trees, park trees, and so on) have been called an "urban forest." Such a concept is not used here, since normally hardly any continuous woodland or forest, or even grass-dominated savanna, is present (and harvesting for wood products is normally absent). Tree plantations have occasionally created parks for recreation (e.g., the 875-ha Amsterdam Bos) or to limit bird populations and aircraft-bird strikes (e.g., a monoculture of *Betula pendula* birches by an airport) (Polano, 1992; Berrizbeitia, 1999).

Urban forests producing wood products near cities, like market-gardening areas, provide local resources to urban residents. This is valuable in the face of a global growing population and resource limitation, plus increasing transportation costs. Except for recreational parks, urban forests are the most multi-use large greenspaces available to people. In times of emergency these forests provide stability.

Cliffs, quarries, mines

Rock-outcrop cliffs typically contain crevices, ledges, small caves, and accumulated talus rocks at the base (Larson *et al.*, 2000). This high habitat heterogeneity supports a wide range of vertebrates, invertebrates, and plants. Some algae and lichens (endolithic) even live between minerals in the rock surface.

All cities use sand and gravel for construction and water drainage purposes. Since sand and gravel are heavy, large quarry or mine areas are created nearby (Gilbert, 1991; Wheater, 1999). If ground-level sites are available, especially on a river floodplain, sand and gravel are readily dug out. With a high water-table, such mining sites often create ponds, thus providing habitat for a somewhat distinctive but species-poor flora and fauna (Kelcey, 1984). If sand mining is not available, a "rock quarry" provides rocks that are crushed into gravel. Often older resource-extraction sites have been encircled by development and are no longer active.

The face of a rock quarry typically somewhat resembles a natural rock outcrop. If the quarry is active, very few species thrive, considering the ever-changing surface, heavy machinery, moving trucks, and airborne dust. In an inactive quarry, however, ecological succession proceeds. Various types of plants colonize, soil particles accumulate, plants grow and replace one another, and diverse animals become abundant.

In the Wye Valley by the England/Wales border, natural cliffs have a thin soil, rather dry microclimate, heterogeneous surface, and so forth (Peterken, 2008). The predominant plants are common, though rare species are present. Active quarries have little vegetation. Succession on inactive quarry faces gradually leads to the species composition characteristic of natural cliffs. Dark cave-like spaces tend to have uncommon ferns and bryophytes. Piles or heaps of unused quarry material (without the toxic materials of most industrial mining) are quickly covered by tall herbaceous plants. Small marshes form around quarry ponds. On the infrequent huge quarries, succession leads to species-rich natural communities. However, humans adding a monoculture of quarry-reclamation plantings seems to facilitate non-native species spread, and noticeably slows the species-enrichment process of ecological succession.

Golf courses

About 0.67% of Japan is golf course, a land use rapidly increasing in the UK and USA (Yasuda and Koike, 2004; Hodgkinson et al., 2007). The 16 000 golf courses in the USA average about 55 ha (140 acres) in size, of which 40–70% is typically unplayable space (Kohler et al., 2004; Colding and Folke, 2009). Golf courses usually contain short-grass fairway strips lined with taller-grass "roughs," manicured greens, and elongated tree patches (edge habitat) (Balogh and Walker, 1992; Terman, 1994, 1997; Porter et al., 2005; King et al., 2007). The golf areas are lawn dominated, intensively managed, and have little shrub cover. With very little impervious surface, golf courses normally contribute little heat to the urban air, or surface runoff to stormwater systems.

The earliest Scottish links or golf courses were largely designed by nature; they had considerable reddish-brown mixed grasses, and shrub areas or unusual features were retained for players to go around (Cornish and Whitten, 1988). Today's American golf courses usually result from extensive earth-moving of the surface. This creates mounds, sand traps/pits, ponds, low-diversity grass strips, monoculture greens, a path network for electric carts, and so forth. Considerable maintenance effort and cost are used to maintain patterns against nature's processes. Characteristic widespread problems are: (1) excessive water use; (2) excessive fertilizer use; (3) excessive pesticide use; and (4) habitat loss in the construction phase. In dry climes, 18 strange green patches consume scarce water in a brown land (Merola-Zwartjes and DeLong, 2005).

Typical Midwestern USA golf courses annually add 7 kg/ha (6.2 pounds/acre) of pesticide to reduce pests, plus 41 kg of nitrogen and 4 kg of phosphorus to enhance grass growth (Wheater, 1999; Kohler et al., 2004; King et al., 2007). Many of the pesticides kill non-target animals and plants on site and in nearby water bodies. Much of the nitrogen and phosphorus washes off as surface or subsurface runoff to groundwater and local water bodies, often triggering eutrophic algae blooms and cascading degradation of the aquatic ecosystem.

Phosphorus and nitrogen in stormwater, or in effluent from secondary sewage treatment, can effectively replace the fertilizers and irrigation water added to golf courses. "Constructed wetlands" on a golf course can reduce stormwater runoff and treat (i.e., clean) stormwater pollutants (see Chapter 7) (Kohler et al., 2004). Stormwater basins, ponds, and wetlands often support rich biodiversity (Figure 12.6). Indeed, golf courses could be extremely valuable by absorbing the piped-in stormwater, and treating its pollutants, from surrounding residential and commercial land uses.

In city and suburban areas of Stockholm a quarter of all the ponds present are on golf courses (Colding et al., 2006). Ponds on golf courses attract waterbirds, amphibians, wetland plants and aquatic species (Kohler et al., 2004; White and Main, 2005). The waterbirds are frequently accustomed to people, and feed around the ponds, but usually do not roost or nest there. Amphibians tend to be abundant in vernal pools that dry out seasonally, but the animals are less common in permanent ponds with predatory fish (Boone et al., 2005; Montieth and Paton, 2006).

The grassy rough areas on a golf course are used by some wildlife for both feeding and habitat (Wheater, 1999; Yasuda and Koike, 2004; Hodgkinson et al., 2007). Some grassland species including birds occur on golf courses (Yasuda and Koike, 2004; Cristol and Rodewald, 2005). Large old trees with holes support woodpeckers, owls, arboreal mammals, and other species (Cristol and Rodevald, 2005), and apparently wider wooded strips have more biodiversity than do narrow ones (Yasuda and Koike, 2004). One study suggests that adding natural ground cover adds 1–2 breeding bird species, shrubs add 1–4 species, and a tree layer adds 12–15 bird species (Terman, 1997).

The context or area surrounding golf courses of course also affects this biodiversity. Stockholm contains distinct "green wedges" projecting into the city, and half the golf courses are in these wedges (also 90% of the nature protection areas, plus 25% of the community gardens or allotments) (Colding et al., 2006). Only 3% of the golf courses adjoin a protected natural area. A USA breeding-bird study of six urban land uses (nature reserve, recreation park, golf course, residential area, office center, business district) found avian diversity on the golf course to be about average, whereas bird density was highest (Blair, 1996). In Australia, avian species richness was higher on a golf course than in nearby residential land, while amphibian and reptile diversity was lower than in a nearby woods (Figure 12.6) (Hodgkinson et al., 2007). Another study reports surrounding land use to be a better predictor than on-site golf-course conditions for bird diversity (Porter et al., 2005). Similarly, the amount of forest within a radius of 0.75 km (0.46 mi) was found to be the best predictor for the presence of "species of conservation concern" on a golf course (LeClerc and Cristol, 2005). In short, golf courses are tightly and reciprocally tied ecologically with their surroundings.

Cemeteries

Over time, millions of people have lived and died in a large city and, although today surprisingly few markers indicate their burial, urban cemeteries cover considerable high-value space (Barrett and Barrett, 2001). Typically, burial grounds are on hilltops and other well-drained soil areas near the edge of an urban area. With subsequent urbanization, most cemeteries, as relatively permanent objects in a metro area, are then surrounded by development. In this setting, a sacred burial ground may secondarily serve as a neighborhood amenity or leisure area.

Normally, mowed lawn surrounds a relatively regular distribution of stone markers and a scattering of trees on a cemetery (Gilbert, 1991). Usually few shrubs are present and, unless the lawn is heavily herbicided, a diversity of grasses and other plants thrive. Even though the space may be kept tidy or manicured, some habitat heterogeneity is provided by marker stones, paths, roads, fences, buildings, discarded materials, and water features (Gilbert, 1991; Vezzani, 2007). Also, grave digging, soil disturbance, and variations in herbicide and fertilizer application contribute to habitat diversity.

In addition to chemicals applied on the surface, casket and embalming chemicals (e.g., formaldehyde, a carcinogen) may leak and pollute soil and groundwater (Stowe et al., 2001). "Natural burial," such as with a biodegradable casket and no vault or embalming fluid, promotes rapid return of a corpse to earth.

Lichens and birds have attracted the most biodiversity interest in cemeteries. Lichens often colonize and spread on gravestones, which are dated and usually differ in stone type, angles and carving. Plants and animals in older sections of burial grounds or around older trees tend to be quite different than in other sections. Owls and rare plants may thrive (Adams et al., 2006). Birds from surrounding rural areas use Chicago cemeteries (Lussenhop, 1977; Esteban and Jukka, 2000). A 2.5-year study of birds in a 13-ha (33-acre) cemetery in Dallas (Texas) found common urban birds predominating (Adams et al., 2006): >706 starlings (*Sturnus vulgaris*); 142 house sparrows (*Passer domesticus*); and 47 native species with abundances from 131 to 2 individuals. Recreational use seems to decrease biodiversity (Lussenhop, 1977).

"Old overgrown cemeteries," such as Highgate in London and Mount Auburn in Cambridge (Massachusetts, USA), are rare but of exceptional biodiversity interest (Figure 12.7) (Gilbert, 1991). The 19th-century picturesque 70-ha (174-acre) Mount Auburn Cemetery is a burial ground as well as an arboretum, statuary collection, and wildlife sanctuary (Howard, 1987; Linden-Ward, 1989). Built originally in farmland and now enclosed by residential development, the cemetery lies between two urban water bodies, a river and a large pond. Over 18 decades, the cemetery designs and fashions have changed. Today the place has

Green spaces, corridors, systems

Figure 12.7. Large old overgrown cemetery with high biodiversity. Ivy (*Hedera helix*) on tree trunks and gravestones, abundant shrub- and understory-layer vegetation, and mixed open and shaded areas provide high habitat diversity and extensive cover for wildlife. Highgate Cemetery, London. R. Forman photo courtesy of Jessica Newman.

intensive management and a naturalistic flavor. The savanna-like form has ample lawn, extensive tree cover (4000 trees), 30 000 planted annuals, four ponds, and, most unusual, considerable shrub cover. An extremely high biodiversity is emphasized by 900 tree species and varieties and 130 shrub and groundcover species.

Over 210 bird species have been recorded in the Mount Auburn Cemetery (Howard, 1987). Resident, winter, and summer birds are appreciated by neighborhood birders. However, this large greenspace surrounded by the Boston metro area serves as a "migration magnet." Migratory birds are the highlight, often attracting hundreds of birders on a May migration morning. Most migrants stay for the day (12 hours), during which they need food and rest. Habitat diversity provides a variety of food for the many bird species. Abundant shrubs plus trees provide ample cover for resting and minimizing disturbances by people. Many features of old overgrown cemeteries can be dovetailed into most urban cemeteries to greatly increase both their biodiversity and neighborhood amenity values.

Institutions and municipal facilities

Building clusters in "campus-like settings" (often simply called "institutions") are limited in number, but together are important large greenspaces in a metro area. Universities, colleges, schools, health-care facilities, research centers, government centers, water-supply treatment facilities, and sewage-treatment facilities are characteristic (Urban Land Institute, 1998, 2001). Each type has special features, such as dense residents in universities, daily commuters in research centers, excess nutrients and pathogenic bacteria by sewage-treatment facilities, and water and wetlands around water-supply and sewage-treatment facilities.

The greenspace is commonly covered with lawn, buildings, carparks, paved walkways, a scatter of trees and tree clumps, ornamental plantings, flower beds, and a water feature. Sports fields with compacted soil may be present. Maintenance effort and cost are often considerable to maintain the setting against nature's processes. Fertilizers and pesticides are commonly used, while stormwater is frequently piped off-site. Just as for golf courses presented above, stormwater and its pollutants from surrounding built areas could be piped to the large greenspace and treated in constructed ponds or wetlands (see Chapter 7). Associated high biodiversity would be a bonus.

Somewhat similar to the old-overgrown-cemetery case above, a naturalistic design highlights native plant communities and forms fit to topographic, sun, and water conditions (Jensen, 1990). Local native species could be highlighted, or people could be mentally transported to, e.g., a Japanese garden, lush tropics, or desert environment. The campus may be a mosaic of outdoor rooms, mysterious curves, sweeping vistas, evocative glimpses, symbols of former inhabitants, water features, long tunnel views, and other delights and surprises. People may be encouraged to feel, see, hear, and otherwise experience the place. Institutions and municipal facilities offer rich opportunity for ecologically creative features and flexible solutions, including linkages with other greenspaces.

Dumps and brownfields/vacant lots

Dumps

Solid waste is received and accumulated in a *dump* (landfill, tip, rubbish heap) (Bagchi, 1994; McBean et al., 1995; Wheater, 1999; Boada and Capdevila, 2000). Large dump sites for municipal solid waste (MSW) are normally established on the edge of an urban area, and over time may become surrounded by urbanization (Figure 12.8). In the USA, paper products compose

Figure 12.8. Dump filled with municipal solid waste and capped with clay layer. Eroded soil and chemicals have washed down to the left in several places. Small curved powerline through woods (left), multi-lane highway, detached single-unit housing (lower right), and small warehouse and truck terminal area (upper right). Suburban Boston. R. Forman photo.

35% of the rubbish, and yard trimmings, food scraps, and plastics each compose about 11.5% of the dump content (Benton-Short and Short, 2008). Metabolism by aerobic bacteria in the presence of oxygen and moisture, plus chemical oxidation, readily decomposes organic and other materials, and may warm the dump surface (Cairns, 1995). Indeed, dumps can be, and perhaps will increasingly be, mined for valuable resources (Lee and Jones, 1990).

When a dump is active, trucks and heavy machinery add and spread waste materials on the large mound, which is typically surrounded by a ring of successional habitat (Berger, 2006; Benton-Short and Short, 2008; Belanger, 2009; Marsh, 2010). Local people may sort through the material to find and sell valuable items, as described for a dump "favela" in Chapter 11 (World Resources Institute, 1996; Hardoy et al., 2004). Organic material may be extracted and used for fertilizer (see Figure 2.4) (Maconachie, 2007). Various approaches for recycling diverse valuable materials at dump sites are common (Ravetz, 2000).

Bees, wasps, flies, and other invertebrates readily reside in rubbish (Wheater, 1999). Rats, as well as birds such as gulls and crows, are normally abundant feeding on the concentrated food resources. Coyotes around a Mexico dump are denser and have smaller home ranges than animals further away (Hidalgo-Mihart et al., 2004). The abundance of such animal scavengers on a dependable food source is highlighted when the food input stops, as illustrated by municipal-waste truckers' strikes in Paris, Naples, or New York. Hungry rats quickly invade surrounding neighborhoods.

The typical dump has no clay or other nearly impermeable layer at the bottom. Heavy metals and an array of other chemicals are often present at high levels in a dump (see Chapter 2) (Maconachie, 2007). Precipitation water infiltrates through the waste material, leaching toxic and other substances into the groundwater and local water bodies (Cairns, 1995; Kidder, 2000; Marsh, 2010).

On a closed dump, chemical leaching is minimized by adding an impermeable-layer cover or cap on top, e.g., of clay 1 m (3 ft) thick. Neither water nor oxygen then penetrates into the dump material, so anaerobic bacteria continue the decomposition of organic matter, but at a much slower rate. Methane, a strong greenhouse gas, is produced in this process and, for perhaps a decade, methane production is high. Since methane accumulation can be explosive, pipes project upward to release the gas. Alternatively, the methane is sometimes captured and used as an energy source.

Traditionally trees have not been planted atop the dump cover, because, with high wind velocity on a mound, a tree may blow over leaving a hole where water could penetrate the cover. However, studies suggest that the woody plants present produce shallow roots that do not penetrate the clay layer (Dobson and Moffat, 1993; Robinson and Handel, 1995). As a safety margin, trees may be planted in spots with an extra deep impermeable layer (Stephen Handel, personal communication, 2010). Former large dumps may make good parks with views.

Selected trees and shrubs on closed-dump surfaces can attract fruit-eating birds from neighboring areas (Robinson and Handel, 1993, 2000; Kirmer et al., 2008). These birds bring in and drop seeds, thus accelerating plant colonization. Indeed, vegetation as far as 17 km (10.5 mi) away may affect the species colonization of a dump. Although plant growth is often slow (Rawlinson et al., 2004), this seed input by birds catalyzes ecological succession, which increases biodiversity. Tiny pools and wetlands with associated species often form atop or on the sides of the relatively impervious dump cap. Finally, being in an elevated position, wind-dispersed seeds from dump-top plants are readily spread to surrounding urban areas.

Brownfields and vacant lots

The terms brownfield and vacant lot (plot) have different meanings in the UK and USA, and overall seem to be poor descriptors (Gilbert, 1991; Kirkwood, 2001; Reisch et al., 2003; Hardoy et al., 2004; Lincoln Institute of Land Policy, 2004; Benton-Short and Short, 2008; Hollander et al., 2011). Nevertheless, these urban greenspaces are important, so we use simple descriptive concepts. A *brownfield* is a large area of contaminated soil (substrate), whereas a *vacant lot* is a small unbuilt space, usually between buildings or built plots. Both places may contain areas of bare substrate, but are typically dominated by successional vegetation (herbaceous, often with shrubs and small trees present). Thus, a brownfield is open land and does not refer to non-contaminated meadows or successional habitats A vacant lot may or may not have contaminated soil. Many possible futures, from nature reserve to building, exist for both brownfield and vacant lot.

An estimated 450 000 brownfields are present in the USA, especially in and around metro areas (Kirkwood, 2001; Benton-Short and Short, 2008).With the growing and pervasive use of industrial chemicals, many of which have not previously been encountered by organisms in natural processes, the number of brownfields grows. The process of soil contamination occurs in overlapping phases, often including solid, liquid, and gaseous chemical forms (Brown, 2002). Several characteristics typify a brownfield: resulting from industrial pollution; being an integral part of the urban structure; having adverse effects on urban life; requiring outside mitigation intervention; and offering development or different opportunities that do not encroach on other more-valuable greenspaces.

Following the demolition of structures, brownfield substrates tend to have rapid drainage, high pH, low organic matter, and low nitrogen levels (Wheater, 1999). On the other hand, brownfields on previous industrial sites typically have compacted soil, considerable erosion, and either high alkalinity or high acidity.

Brownfields are typically an early successional mosaic habitat, with numerous plant species, pollinators, and other animals feeding and nesting/denning (Wheater, 1999; Kovar, 2004; Strauss and Biedermann, 2006). Not surprisingly, the brownfield habitat is continually changing, often rapidly, and usually contains patches in different successional stages. Like the golf course, a brownfield essentially does not contribute to surface stormwater runoff or heating the air. However, groundwater and local water bodies are often contaminated by brownfield chemicals. Generally the area is too big to remove all the contaminated soil, so either uncontaminated soil (e.g., 1 m ≈ 3 ft thick) is added on top, or ecological succession or alternative land use follows directly on the contaminated surface. Usually the concentrations of toxic substances decrease very slowly over time with microbial decomposition, oxidation, and other natural processes.

A study of 246 brownfield plots in the Bremen and Berlin areas of Germany found that plant species number was relatively constant (Strauss and Biedermann, 2006). However, species composition varied greatly. Composition correlated best with vegetation structure, landscape context, soil, and age of site.

As indicated in Chapter 4, attempting to speed up the de-contamination process in urban areas using plants ("phytoremediation") is way too slow. This is because the contaminant levels are too high; plants die or grow only slowly; rate of contaminant uptake is too slow; input of pollutants continues; necessary removal and disposal of harvested plant tissue containing pollutants is difficult or too slow; and management, monitoring, and political commitments change. Still, if contamination levels and inputs are low enough, theoretically an adequate rate of bio- or phyto-remediation might be achieved.

Almost any land use may follow brownfield conditions on these large greenspaces (Greenberg et al., 2000, 2001; DeSousa, 2003, 2004; Hollander et al., 2011). In Toronto, using various treatment approaches, 14 brownfields were converted to leisure/recreational parks. In addition to several human benefits, the ecological benefits most frequently cited were creation/expansion of ecological habitat spaces (9 cases); flood control (3 cases); and environmental renewal (e.g., soil and groundwater quality) (3 cases).

Vacant lots are important features of both urban residential and commercial areas (Chapters 2, 3 and 11) (Salisbury, 1961; Davis and Glick, 1978; Vessel and Wong, 1987; Gilbert, 1991; Bastian and Thomas, 1995; Godde et al., 1995; Hodge and Harmer, 1996; Suchman, 2002). Ecological succession, from early colonizing plants and animals to later trees, is commonly conspicuous on vacant lots (Wheater, 1999). Plant species richness increases with the age of a vacant lot (Figure 12.9). As expected from species–area curves, plant diversity also increases with area of a vacant lot.

On 46 urban vacant-lot (plot) sites in the UK, only three tree species were found, mostly colonizing near apparent parent trees (Hodge and Harmer, 1996). Substrates with small stones (gravel, ballast) and little

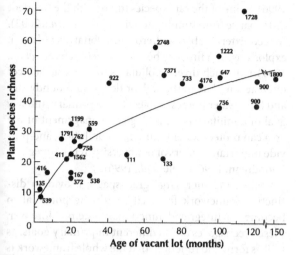

Figure 12.9. Plant diversity in vacant lots varying in age and size. Number next to each point = area in square meters (1 m² = 10.8 ft²). Vacant lots = building plots currently covered mainly by successional habitat (see Chapter 1). Vegetation in lots >30 months old has been cut at least once. Species richness did not correlate with various measures of distance to other lots. Regression line based on log-log model. Chicago. Adapted from Crowe (1979).

grass cover had the most cover of woody plants, suggesting the fastest rate of succession.

Infill, i.e., building on urban vacant lots, highlights a key urban-ecology issue. A green vacant lot provides a spot of habitat, animals, vegetation, evapotranspiration, water infiltration, and air cooling for the immediate neighborhood. Perhaps more importantly, the green plot is part of a system, enhancing species movement and biodiversity across the metro area. Yet by building on a vacant lot within a metro area, infill is often considered ecologically better than constructing the equivalent building in a lower-density exurban/peri-urban area. So, a researchable question remains. At what point does more infill disrupt the urban system of species movement and biodiversity?

Airports and military bases

Airports are the only big places where minimizing or eliminating nature is often a primary goal. Cover it, kill it, prevent its appearance – all to reduce the chance of animals striking a moving aircraft. This applies to both portions of an airport, the "public side" and the "aircraft side."

Most urban regions have one or two major airports, though five or more are present around some cities (Forman, 2008). Airports are typically large open flat areas elongated in the direction of prevailing winds. Commonly located in former farmland, low wetland areas, or by water bodies, this limits the amount of buildings and air-polluting industries nearby (Adams *et al*., 2006; American Planning Association, 2006).

The surroundings of an airport often have considerable protected land for nature, which helps with water quantity and quality issues, but is a source of wildlife moving onto airport land. Certain birds feeding in a nearby dump or in farmland may pose problems for aircraft. A massive planting of feathery birches (*Betula pendula*), essentially without perches for large birds, mainly surrounds the Amsterdam airport, and somewhat seamlessly extends into adjoining neighborhoods.

Aircraft collisions with large herbivores, such as deer and kangaroo, are of concern while the plane is moving along a runway. But bird–aircraft collisions or strikes in the air are the big problem (Satheesan, 1996; Servoss *et al*., 2000; Barras and Seamans, 2002). Heavy-body birds and birds in tight flocks are the prime plane-strike species. The former include waterfowl, gulls, and raptors such as kites and vultures around Indian cities (Cooper, 1991; Satheesan, 1996). Birds in tight flocks include blackbirds and pigeons (Barras and Seamans, 2002). Lots of scare and other deterrent techniques have been used to reduce the plane-strike problem, typically with little or only temporary success (Servoss *et al*., 2000; Avery and Genchi, 2004; Adams *et al*., 2006).

Habitat management approaches seem to be more successful in reducing wildlife around airports. A common goal is to maintain a relatively homogeneous cover of plants that are unattractive to wildlife, and with low fire hazard (Dolbeer *et al*., 2000; Barras and Seamans, 2002; Washburn and Seamans, 2004; Adams *et al*., 2006). Minimizing trees (which provide perches), clumps of evergreens (roosting cover), shrubs (cover and habitat), and tall grass (cover and habitat) is also a common management approach. Yet ecologically, tall species-rich (low-dominance) meadows may support few large birds and few bird flocks. Usually plantings on the public side of an airport are largely attractive ornamentals for us, but are relatively unattractive to wildlife.

Aircraft noise is very loud and a problem for people (Biggs, 1990). With infrequent aircraft landings and take-offs the effect on wildlife may be limited, more like occasional trains passing on a railway corridor (see Chapter 10). With frequent aircraft activity, somewhat like frequent commuter-rail trains, the noise is more analogous to the chronic traffic noise of a busy highway that has a major inhibitory effect on wildlife (Forman et al., 2002, 2003; Reijnen and Foppen, 2006).

Aircraft emit CO_2, NOx, particulate matter, and other pollutants, as do the dense moving vehicles on both the public side and the aircraft side of an airport. Air turbulence generated by moving planes on a runway lifts particulate pollutants into the air, perhaps more strongly than the analogous polluting effect of vehicles moving on a road. Taking off and landing aircraft especially emit pollutants which cover nearby areas. PM2.5 particles cause a human health risk, and presumably damage wildlife. Indeed, pollution from airports of a megacity may reduce air quality over a several-hundred-kilometer-diameter area.

In cool regions, de-icing substances (such as ethylene) are used on both aircraft and runways. Some airports attempt to capture and re-use some of the de-icing liquid. But most of it flows rapidly into, and may severely degrade, groundwater and local water bodies. Other soil and water pollutants emanate from aircraft, airport vehicles, and vehicles on the public side.

Normally, most of an airport area is unpaved, and therefore contributes little to heating the urban air or to surface stormwater runoff. Still, in summer, airports with a high percent cover of black tarmac/asphalt absorb considerable solar radiation, and liberate heat in the late afternoon and evening. This heats the air, and wind may carry the warm air into downwind neighborhoods. But at night typically the wind dies, and then warm air over the heated asphalt/tarmac rises vertically. This draws in air from neighboring land (see Chapter 5). A possible net effect is to provide some ventilation (air cooling and cleaning) for areas surrounding an airport. The amount depends on the surface area and arrangement of tarmac/asphalt and grassy areas.

"Urban-region military bases" typically contain an airport, concentrated living facilities for personnel, and mostly vegetated training areas. Some bases have a port with shipping and some have numerous military training vehicles with associated noise, soil-and-vegetation disturbance, and even fire (Forman et al., 2003). Over time, as development spreads in the surroundings, a military base may close, offering opportunities to "naturalize" the space as well as the city. Recent closure of an army base of >3 km² (740 acres) near the center of Seoul provided the opportunity to create a major green corridor and other park features across a broad area of the city.

Perhaps mainly because these military-base greenspaces have become largely surrounded by development, rare species may be in abundance (Leslie et al., 1996). Some of the rare species thrive with the frequent disturbance from training activities (Smith et al., 2002). Indeed, extensive bare soil, erosion, habitat degradation, explosions, and fire may be common (Milchunas et al., 1999). Extremely diverse pollutants and toxic substances often accumulate in the soil, or flow into groundwater and local water bodies. Subject to the primary military goal of a military base, ecological management of the area can protect water bodies from pollution, and provide important centers of biodiversity in urban regions (Goodman, 1996; Leslie et al., 1996).

Taken together, large greenspaces provide a distinctive framework for a city, with the potential to become an integrated functional system. Moreover, these greenspaces are so different in primary goals, as well as form and function, that the whole framework is extremely rich ecologically. We now turn to the major green corridors and networks that play central roles for urban flows and movements.

Green corridors and networks

Corridors

Green corridors slicing through an urban area are conspicuous. Such strips (1) subdivide the area, (2) block out the other side, (3) are filters for movement across, (4) are thin and dominated by edge species, (5) are likely to be seen or encountered, (6) may be appealing or dangerous, and (7) tend to channel movement along their length. Indeed, these characteristics lead to the five major ecological functions of corridors (see Chapter 3) (Saunders and Hobbs, 1991; Forman, 1995; Forman and Hersperger, 1997; Bennett, 2003): conduit; barrier/filter; source; sink; and habitat. Considering their limited surface area, vegetation corridors (Lyle and Quinn, 1991; Terrasa-Soler, 2006) are major centers of ecological and human activity, strongly determining how an area works.

Few vegetation or ecological corridors simply cross urban land. Even the recent supposedly green "Big Dig" corridor in Boston results from an elevated highway being placed underground. Green urban corridors are usually along streams, powerlines, pipelines, property boundaries, railways, coastlines, and so forth. Many are much used and disturbed by people, and all are used for movement by some wildlife. Small neighborhood corridors such as lines of street trees, front spaces/yards, side-boundary hedges and fences, house-plot backlines, and wider house-plot back spaces/yards

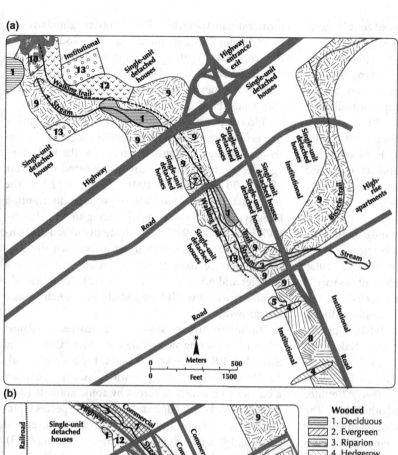

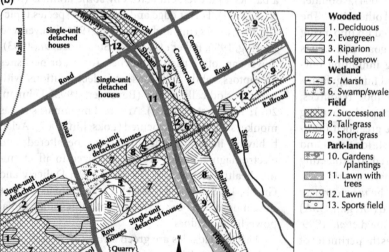

Figure 12.10. Land uses/habitats within and surrounding urban green corridors. Ottawa. Adapted from National Capital Commission (ca. 1981), G. Katz, personal communication, 1998.

are conspicuous and of considerable ecological importance (see Chapter 10) (Rudd et al., 2002).

Major green corridors, typically tens of meters wide, with a lengthwise walking trail and abundant recreational use, are often referred to as *greenways* (Whyte, 1968; Little, 1990; Searns, 1995; Turner, 1995; Ishikawa, 2001; Ahern, 2002; Jongman and Pungetti, 2004; Erickson, 2006; Hellmund and Smith, 2006; Havens, 2011). Like green wedges (see Chapter 2), major radial vegetation-corridors also facilitate the movement of wildlife and walkers/bikers into and out of the city. An exception was the former green strip between East Berlin and West Berlin, apparently militarily maintained in a relatively "sterile" ecological condition. In contrast, today the Korean "de-militarized zone" is non-urban and of considerable biodiversity importance.

In Ottawa, some of the major green corridors or greenways go along streams, railways, and roads, while others do not (Figure 12.10). Urban green corridors are often crossed by roads. Walking and bicycling trails pass along some corridors (Figure 12.10a). Parkland may extend along stretches of green corridors, but it usually occurs as patches at periodic locations. Habitat heterogeneity is often high, illustrated by 10 habitat types and 11 habitat types in the upper and lower maps respectively of Figure 12.10. Such habitat heterogeneity in a corridor supports high species diversity. But, just as for people movement, the heterogeneity must noticeably limit species movement along a green corridor.

Greenbelts, typically as wide protected circular areas around a city or town, normally contain varied land uses and habitats. Greenbelts have been described for an imagined sixteenth century utopia (Moore, 1984), Letchworth (UK) (Howard, 1902), London (Abercrombie, 1945; Elson, 1986), Seoul (Nakagoshi and Rim, 1988; Im, 1992; Jun and Hur, 2001; Bengston and Youn, 2006), Ottawa (Taylor *et al.*, 1995), Tokyo (Yokohari *et al.*, 2000; Sorensen, 2004), Boulder (Colorado, USA) (Forman, 2008), and other cities. The Seoul greenbelt has recently been sliced up into a ring of large remaining greenspaces. The goals and values of greenbelts vary widely, including inhibiting outward urban expansion; recreation and hunting; farmland protection; aesthetics; clean air; food resources; municipal services; military activities; and providing fuel during wartime. For London the greenbelt mimics a geometric ring form. In contrast, for Seoul and Boulder the form was strongly tailored to protecting natural resources, and hence is quite asymmetric.

A related but different concept is the *urban growth boundary*, best known around Portland (Oregon, USA) (Nelson *et al.*, 1995; Porter, 1997; Benfield *et al.*, 1999; Marshall, 2000). This boundary defines the perimeter of the all-built metro area. Outward urbanization is both drastically slowed and usually channeled into environmentally suitable areas. The boundary protects agricultural land, forest resources, and nature. Densification and infill occur inward, though the resulting scatter of urban greenspaces and tiny green spots (Houck and Cody, 2000) may or may not be adequate for effective species movement.

Green corridors along rivers and streams have high habitat diversity, and somewhat enhance water quality. Green strips may follow buried pipelines carrying former stream water, as well as along daylighted streams (see Chapter 7) (Benjamin *et al.*, 2003; Godefroid and Koedam, 2003). Railway corridors tend to be especially important ecologically (see Chapter 10). Former railways sometimes are converted to walking and/or bicycling trails (Flink *et al.*, 2001).

Electric powerline corridors are somewhat similar to *pipeline corridors* for oil, gas, water supply, stormwater, and sewage wastewater, except for the high-voltage electric-transmission lines supported by towers of 15 to 30+ meters (50–100+ ft) (Figure 12.11). The transmission lines lead to lower-voltage distribution lines, typically on wooden poles along street networks (Williams *et al.*, 1997b). The wide disturbed and ofteroded zone beneath transmission lines usually contains a service road commonly used by motorized vehicles and bikes (Luken *et al.*, 1992). These straight green strips are usually herbicided and/or cut to reduce tree growth.

Although urban studies are scarce, predators often forage along powerline corridors (Graves and Schreiber, 1977; Chasko and Gates, 1982; Knight *et al.*, 1995). The strips, especially with flashing lights, are a barrier to movement across by some animals (Rich *et al.*, 1994). Forest edge and shrubland species typically thrive (Niering and Goodwin, 1974; Dreyer and Niering, 1986; Baker *et al.*, 1990; Steenhof *et al.*, 1993). Towers are often used as perches, roosts, or nest sites by raptors, while collisions and electrocutions with wires periodically occur (Bevanger, 1995; Lehman, 2001; Yahner *et al.*, 2003). Arboreal mammals such as monkeys are also electrocuted (Lokschin, 2007). Avian behavior, including migration, may be altered, and electromagnetic fields are hypothesized to affect animal health (Brown and Drewien, 1995; Doherty and Grubb, 1998; Shimada, 2001). Large animals such as deer, moose, and bear may move into cities along radial powerline corridors.

Urban roadsides are green strips but overall seem to be of limited importance as conduits for animal movement (see Chapter 10) (Forman *et al.*, 2003). "Parkways" built around New York City beginning in the 1920s had moderate traffic on curvy highways enclosed by a broad band of woods and lawn, and were impressive vegetation corridors (Rowe, 1991). But today's urban highways have dense traffic and mainly thin grassy roadside strips (Viles and Rosier, 2001). Traffic noise and disturbance, plus mowed roadsides, severely limit habitat use of the strips, and apparently very few animal species move along the urban roadside

Figure 12.11. Corridor with powerline, bicycle/walking path, and canal connecting city and surroundings. Relatively continuous lines of woody and herbaceous vegetation plus aquatic plants and fish indicate high habitat heterogeneity and biodiversity packed together. Northern edge of London. R. Forman photo courtesy of Jessica Newman.

corridors. The same patterns seem to describe roadsides of smaller roads.

A surprising array of values and benefits, including those just presented, are provided by green corridors. The conduit function along a corridor is for the flow of air, water, animals, seeds, and/or people. In addition, green corridors reduce stormwater surface runoff and potential flooding, and increase water infiltration into the soil. Windbreaks reduce streamline windspeed, and alter turbulence and vortex airflows (Brandle et al., 1988; Erell et al., 2011). Lines of street trees create shadow corridors appreciated on hot days. Tree lines filter airborne dust, and visually screen highways and buildings. Soil berms and noise walls, both preferably plant-covered, cut traffic noise. Hedges and plant-covered fences and walls support biodiversity and provide animal movement routes. Habitat heterogeneity both within and along a corridor is often striking and functionally significant (Mason et al., 2007).

Major radial corridors connecting city and countryside, such as along a railway or river, are especially important for air movement, which cools and cleans city air (e.g., breeze from the country; see Chapter 5) (Figure 12.11). Analogously, in hot North African cities, certain streets oriented in the direction of prevailing wind funnel air into the city. In effect evapotranspiration by vegetation in radial green corridors cools the air entering a city. In Tokyo, cool radial green-corridors separate warm corridors of dense development along rail lines (Gartland, 2008). In hilly urban areas, cool air drainage at night flows down valley and stream corridors.

Finally, green corridors are sometimes used to help "shape" development (Walmsley, 1995). In this way the small city of Boulder (Colorado, USA) provides a logical attractive unity to its commercial areas, and helps reduce residential sprawl. Separating small spaces at a relatively fine scale may create an appealing mixed-use mainly walkable urban area.

Overall, three structural or internal characteristics and two external characteristics mainly determine the effectiveness or rate of corridor functions (Henein and Merriam, 1990; Forman, 1995; Bryant, 2006; Mason et al., 2007). For internal structure, (1) the width, (2) the connectivity, and (3) the habitat or land-use quality of the corridor are central. The prime external characteristics are (4) adjoining land uses and (5) attached green patches, if any.

In general, wide urban vegetation-corridors are primarily of habitat value for local wildlife, providing cover and food for nesting/denning, and secondarily of value as a conduit for movement of other animals. For narrow corridors, the lengthwise movement function for local wildlife may often be more important.

Linear features typically cross different substrates, land uses, habitats, and sometimes topographic conditions (Jongman and Pungetti, 2004). Narrow green strips are strongly affected by diverse disturbances from both sides. Corridor vegetation changes over time (Lyle and Quinn, 1991). Therefore, it is not surprising that typically corridor habitat heterogeneity is high. A narrow strip is essentially only suitable for edge species and other urban generalists. Still, with habitat heterogeneity, biodiversity may be relatively high.

Networks

Green networks (ecological networks, ecological infrastructure) at different spatial scales are especially important in providing connectivity for species or people. Fine-scale networks of green corridors without attached vegetation patches are common in certain portions of a metro area. Such networks are mainly discovered when viewing aerial photos and satellite images, or in a descending airplane. At a very fine scale, plants appear along anastomosing cracks and crevices in surfaces, such as former parking lots, stone walkways, and stone walls (see Figures 10.6 and 10.8). Ditch networks for irrigation and/or drainage are often present in peri-urban/exurban areas, and contain strips of changing vegetation (Geertsema and Sprangers, 2002; Blomqvist et al., 2003). More common in residential areas with detached houses are strips of woody vegetation along property boundaries (see Chapter 11). Together the tree lines, shrub lines, fence lines, and plant-covered walls form a distinctive green network of considerable ecological and human importance. At a still-broader perspective, lines of front spaces/yards and back spaces sometimes interconnect to form a network. At the same scale, and often extending into city areas, are networks of street-tree lines.

Broad-scale or regional green networks seem to enhance ecological patterns such as stream biodiversity and the movement and persistence of native plant species (Damschen et al., 2006; Urban et al., 2006). But the regional networks more broadly enhance ecological conditions across a metro area (Jongman et al., 2004; Wu, 2008). While the functions of individual corridors are reasonably well understood, the ecological importance of interconnected corridors as a network remains a little-explored frontier.

The form of green networks also provides ecological insight, just as does the form of other urban network types (see Figures 6.6 and 10.4). Many coastal cities, such as Barcelona and Toronto, have more-or-less parallel streams or piped former-streams draining a slope to the sea or lake, but these do not form a network. Most urban stream networks are *dendritic* (treelike), with a progressive hierarchy of stream sizes and no loops (alternative routes). Usually superimposed on these forms are strongly *rectilinear* patterns (dominated by straight lines and right angles) of roads and buildings. Yet at the urban region scale, a city is effectively a nucleus with corridors radiating outward. Often, for large cities, the radial corridors are interconnected by one or more ring-road corridors (e.g., Beijing has several concentric ring roads).

These dendritic, rectilinear, radials-and-ring, and irregular networks may be evaluated for several characteristics known to be of ecological importance (Forman, 1995; Cook, 2002; Jaeger et al., 2010). Seven attributes are especially valuable for understanding how an urban ecological system works: corridor density; connectivity; circuitry (loops); linkages per intersection; presence of hierarchy; sizes of enclosures; and spatial scale (see equations, Appendix B).

Integrated urban greenspace system

All-built metro areas display a scatter of greenspaces and tiny green spots, plus green corridors and networks at various spatial scales. But do they work together as a system? How could the dispersed elements be converted into an effectively functioning *urban greenspace system*, that is, a group of vegetated patches connected by frequent ecological flows or movements? In this section I explore these questions by considering: (1) groups of greenspaces; (2) functionally linked greenspaces; and (3) keys to an effective greenspace system.

Groups of greenspaces

Using the analogy of urban infrastructure where transportation and utilities provide flows and values, green spaces, especially vegetated corridors and small patches, are sometimes called *green infrastructure* or *ecological infrastructure* (Benedict and McMahon, 2006; Yu et al., 2011). Scattered parks are sometimes referred to together as a system essentially for management purposes, as for instance in Copenhagen, Tokyo, Philadelphia, Kansas City, and numerous other cities (Erickson, 2006). However, the unconnected parks really do not have an inter-linked functional system of flows.

For instance Washington, D.C. was designed for vistas, extensive lawns, and commemorative historic sites. Of the >600 parks, 425 are <0.4 ha (1 acre) in size (Bednar, 2006). Eighty percent of the park area is in parks >20 ha (50 acres), and small green wedges extending outward from parks are prominent. The park system is primarily designed for the movement of people. If nature thrives, that is a bonus.

Boston's Emerald Necklace, a long green corridor connecting some green patches, is often considered as a particularly effective urban structure by providing diverse recreational and natural resources, providing connectivity for movement, and linking well with adjoining neighborhoods (Shea, 1981; Zaitzevsky, 1982; Hough, 2004). The corridor connects large greenspaces that especially support nature, and small ones mainly for social benefits.

A 1929 plan for London recommended walking "parkways" to connect residents to parks and green wedges, and thence to the surrounding greenbelt (Abercrombie, 1945). Similar proposals exist for various cities (Lyle and Quinn, 1991; Ishikawa, 2001; Havens, 2011). To maintain the important scatter of parks as London expanded outward, industrial sites beyond the perimeter were to be identified and converted to parks.

A plan for Syracuse (New York) emphasizing biodiversity identified four types of greenspace (Hilliard,

1991): (1) large green patch adjoining the surrounding rural area; (2) large patch and connected by green corridor to the surroundings; (3) large and isolated in the metro area; and (4) small greenspace located anywhere in the metro area. The plan assumed that surrounding farmland and forest was the primary species source, and predicted that species diversity decreased progressively from greenspace type 1 to type 4. Ecological surveys were needed to provide solidity, but overall the predictions were probably right.

In areas of expanding and coalescing towns and small cities, a plan for the Barcelona Region proposed a "green mesh" network (Forman, 2004b). This was composed of ca. 50–200-m-wide green corridors along municipality borders. The green mesh provided greenspace for habitat and local recreation, a connected walking trail system, a connected system for wildlife movement, and a slight separation of municipalities to highlight and sustain the distinctiveness of each. In much-urbanized Puerto Rico, a proposed green network plan emphasized ecological (ecosystem) services and sustaining biodiversity (Terrasa-Soler, 2006). Different portions of the green network would connect forest areas, protect river corridors, enhance coastal areas, improve urban neighborhoods, ecologically retrofit roads, and conserve lands in the path of development.

Proposed greenway systems for five North American cities are interesting for their distinctive features (Erickson, 2004, 2006):

1. *Chattanooga* (Tennessee):>12 km (7 mi) total length; mostly along streams; several lines radiating outward; very few loops.
2. *Chicago*: 1087 km (674 mi) existing and 1477 km proposed; a dense network; numerous loops; several spurs.
3. *Minneapolis* (Minnesota): hundreds of kilometers existing and proposed; along river and along many streams; connections to scattered lakes; many loops; a few spokes radiating outward.
4. *Portland* (Oregon): 563 km (349 mi) proposed; overall network form is intermediate between those for Chicago and Minneapolis.
5. *Toronto* (Canada): proposed 901 km (559 mi), plus 201 km along the city's lake waterfront; along western and northern ridges beyond the metro area; numerous short relatively parallel strips in stream valleys connecting the ridgetop and waterfront corridors.

The primary objectives in these cities (in order of frequency) are recreation and fitness; conservation and wildlife; linking neighborhoods and parks; water quality protection; economic market values; and non-motorized-vehicle transport. The objectives are expected to change over time.

Waitakere City (New Zealand) chose the goal of "Naturalizing the City" to help address conspicuous air pollution, degraded streams and harbors, degraded soils in plant production and recreation (leisure) use areas, and loss of habitat and wildlife (Wilson, 2000). The first strategic goal focused on developing compact urban "villages" as living and transport nodes to decrease car use. The second strategic goal increased and extended the group of protected greenspaces in order to clean water, clean air, provide wildlife habitat, provide recreation areas, protect biodiversity, connect mountains to the sea, and enclose the urban area.

Functionally linked greenspaces

Years ago, Dutch ecologists, planners, transportation specialists, and others mapped the "National Ecological Network" of The Netherlands (Jongman and Pungetti, 2004; Ministry of Agriculture, Nature and Food Quality, 2004). At the core, this was composed of the nation's large green areas connected by major wildlife and water corridors. Some areas and corridors were incomplete and secondary characteristics were included. Nevertheless, the result was an elegant example of an *emerald network*, where the most-valuable gems (large natural areas) are interconnected by major green corridors. A similar habitat conservation plan drives biodiversity protection for the future in the San Diego Region. An emerald network plan was developed for the Barcelona Region, and was a foundation for successful land protection in a town in suburban Boston (Forman, 2004b, 2008; Forman *et al.*, 2004).

But in urban and many other areas establishing sufficient green corridors would be extremely difficult. Green corridors are generally optimal for both human and wildlife movement. Yet wildlife often moves with varied degrees of effectiveness and in varied routes where no corridors connect large vegetation patches. Thus, the idea of an *ecological network* has evolved, referring to a constellation of green patches connected by species movement routes (with or without green corridors) (Jongman *et al.*, 2004; Ministry of Agriculture, Nature and Food Quality, 2004; Opdam *et al.*, 2006; Vos *et al.*, 2008). Species movement between two patches is likely to be greatest along a continuous

Green spaces, corridors, systems

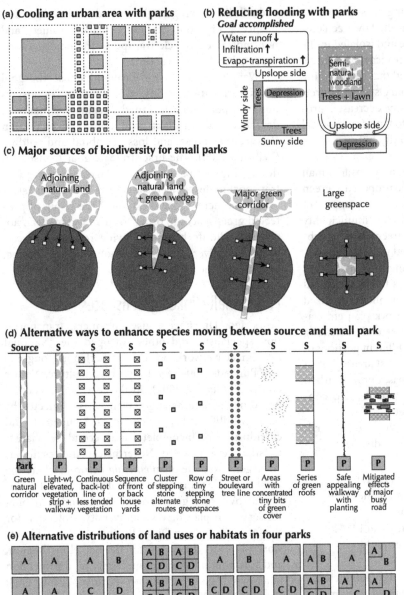

Figure 12.12. Interdependent greenspace system attributes to cool air, reduce flooding, and provide rich biodiversity across a metro area. (a) Illustration roughly based on data in Figures 5.6 and 5.7. (d) With a scarcity of research studies, the 11 alternatives are in hypothesized order from high to low probability of success.

corridor, but also occurs in many non-corridor situations (Figure 12.12d).

A study of 16 common wild animals in urban areas of Finland's 11 largest cities provides insight into the factors affecting wildlife movement among urban greenspaces (Vare and Krisp, 2005). The data suggest that movement between green patches best correlates with low human abundance, low infrastructure density, and the absence of a construction site. In city centers, these conditions were absent, so in effect the greenspaces there are not functionally connected by much wildlife movement. On the other hand, suburban greenspaces seem to be relatively well connected by wildlife movement.

Before considering how to create effective urban greenspace systems, several unrelated insights are useful. (1) A corridor-and-small-patch system often includes special locations where people aggregate, thus inhibiting wildlife movement. (2) Successful flows or interactions decrease with increasing distance moved (often recorded as a "d^{-2} pattern"). (3) Compared with an isolated patch, a patch connected by corridor to another one tends to have a higher rate of species colonization and lower rate of species loss (Figure 12.12d) (Collinge, 1998). (4) A thin line of trees or shrubs is useful for people movement, aesthetics, wind reduction, visual screening, and short-distance wildlife movement. But wider corridors seem better for longer movements, and clumped vegetation better for wildlife habitat (Goldstein et al., 1981). (5) The narrower the green strip, the more important adjoining conditions are in affecting movement along a corridor (Mason et al., 2007). (6) Urban bat movement among diurnal roosts varies from an even distribution of linkages with lots of loops across an area to a centralized roost with radial connections and few loops (see Figure 9.4) (Rhodes et al., 2006). (7) Also, understanding and creating greenspace systems could benefit from the ecological engineering approach of systems analysis (Odum, 1983; Baccini, 2012).

In Washington, D.C., green walkways connect three types of objects: residential neighborhoods, parks, and transportation stops. In dense tropical Singapore, linear green corridors 6–30 m wide with native vegetation are connected to large parks containing rainforest areas (Sodhi et al., 1999). The abundance of park-related birds and rainforest-related birds in these tropical corridors correlates with the type of park or rainforest, and with the amount of vegetation in a corridor. Although avian diversity in a corridor does not correlate with adjoining built-area characteristics, small green woods within 100 m of a corridor seem to greatly increase the number of bird species in it. The planned city of Brasilia has very large parks but few neighborhood parks. Many cities have small intensely managed parks in the central area, and large more-ecological parks near the metroarea edge.

A study of an urban area with 54 greenspaces (637 ha = 1573 acres total area) concluded that biodiversity in a park best correlated with three factors: the total area of parks connected to it; the park size (if >10 ha = 25 acres); and park age (Rudd et al., 2002). Multiple-connected small parks (connected by wildlife movement) were reported to contain more species than the same area in one large park. The degree of connectivity contributes more to biodiversity than does park size. Thus, based on species number, a connected fragmented park system is better than one big park. These urban species are mainly generalist edge species. However, if specialist species are important in urban areas, the large park is more likely to sustain uncommon and rare species.

The study also used several models to evaluate network connectivity, and noted that at least 325 linkages (or corridors) would be needed to effectively connect only half of the 54 greenspaces present (Rudd et al., 2002). That degree of connection would only be accomplished by including and/or enhancing backspace/yard habitats (see Chapter 11), planted boulevards, and utility (infrastructure) rights-of-way.

Keys to an effective greenspace system

In view of the preceding patterns and diverse threads of evidence, I think that an effectively functioning urban greenspace system can be established in almost all cities. Flows tie together patches of the system. Furthermore, this system can accomplish the ambitious objective of sustaining a relatively high biodiversity throughout the entire metro area, not just in major greenspaces. An emerald network of corridor-connected large green patches remains the optimum framework for a metro area. However, this is nearly impossible in the near term for most or all of an urban area.

The three keys to a successful greenspace system, in order of importance, seem to be:

1. Maintain major species-source areas close to the all-built metro area.
2. Maintain an arrangement of urban greenspaces, green corridors, and tiny green spots that are accessible, e.g., within local urban wildlife-movement distances, for almost all species.
3. Design greenspaces and corridors internally to enhance species survival and especially flows between them.

Major species sources effectively provide a "species rain" of organisms into all spaces of the metro area (Figure 12.12c). Most organisms die in unsuitable locations, while others may survive and/or thrive in all greenspaces and tiny green spots. Seeds, spores, spiders, flies, butterflies, bats, and birds often arrive by air, while numerous other species arrive by land, by vehicle, by people, and sometimes by water. Although most populations disappear over time, the species rain maintains a reasonable level of biodiversity. Non-native and native species both arrive in this species rain.

Apparently non-native species mainly arrive in a city on ships, trains, trucks, cars and aircraft from afar (see Figure 3.2). Since most of these carriers come from other cities, often distant, non-native species tend to

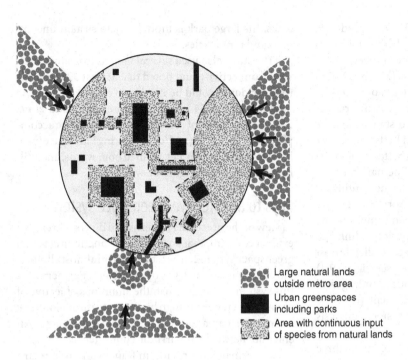

Figure 12.13. Adjacent land and internal greenspaces to provide a "rain" of native species across a metro area. Assumes that species disperse from natural land or a greenspace a distance proportional to its size. Dark shaded area in metro area receives the continual input of species; more strategically located greenspaces needed to provide a species rain over the entire area.

be well-adapted to urban conditions. Trucks serving warehouse storage areas, as well as commercial and industrial sites, effectively spread many of the common types across the urban area.

In contrast, for native species the major sources normally are large natural lands in the urban-region ring (Figure 3.4a) (Forman, 2008). These sources may be adjacent, or in the form of a major corridor, or as a large stepping stone (Figure 12.13). The portion of a metro area directly receiving the rain of species depends on the species-source size and arrangement. Upon arrival in the metro area, the native species progressively spread among patches and corridors according to appropriate species-movement distances. Planning and greenspace restoration/establishment then can fill gaps in the urban pattern to facilitate movement of species across the entire area. Various metro-area and park arrangements greatly facilitate this process (Yu, 1996; Pirnat, 2000). For instance, Dutch studies show that small green patches or corridors close to a large patch increase the species colonization rate, and also decrease the local extinction rate, for large greenspaces (see Figure 3.4c) (Forman, 1995).

Also, various tiny structures or features containing a plant or a few plants, such as widely present in Tokyo, enhance movement between urban greenspaces. An animal navigating through an area makes continuous decisions about attractions, dangers, continuing ahead, turning, or returning. Thus, a multitude of tiny objects (see Figure 1.6) (Hersperger et al., 2012), tiny designed spaces (Krier, 1979; Kostof, 1992), neglected spots (Gilbert, 1991), backyard spaces and front yards (Rudd et al., 2002) both enhance and inhibit movement of animals in an urban area.

Some 23 spatial arrangements of patches, corridors, and matrix increase the rate of movement between, for instance, two green patches (Figure 12.5). These simple spatial patterns are readily amenable to planning, design, and, in many cases, implementation in urban areas.

The third characteristic affecting movement success between parks is their internal design. Habitat diversity supports more species, but the arrangement of habitat and land uses also matters. In Figure 12.12e, the first alternative provides for a specialist species such as a large-home-range animal. The second alternative supports many specialists, while the third option essentially only supports generalist species, which also live in small patches and corridors. The adjacency effect, whereby an adjacent land use or habitat affects a greenspace, especially affects the portion next to the land use (see Figure 2.9). This is important relative to the direction of species movement from another patch. Movement can doubtless be enhanced by maintaining the same habitat on sides of patches facing one another (see Figure 3.4f and 12.12e). Thus, butterfly gardens facing one another (Giuliano, 2005), or duck ponds on sides nearest each other, should facilitate movement of the animals between the greenspace patches.

Could urban heat buildup, as well as biodiversity, be controlled by greenspace arrangement? In Berlin, small parks cool the air about 1°C (1.8°F), medium-size 3°C, and large ones 5°C (see Figure 5.6). The cooling extends outward from the greenspace some 200 m to >1 km, depending on size (see Figure 5.7). Using such information, large, medium and small greenspaces can be arranged to cool the entire urban area, as illustrated in the abstract design of Figure 12.12a.

Stormwater runoff and consequent flooding for the whole area can also be controlled with urban greenspaces. To achieve this, the parks should (1) reduce surface water runoff, (2) increase infiltration into the soil, and (3) increase evapo-transpiration (Figure 12.12b). In addition to arranging greenspaces, their internal design is particularly important. For instance, a strip of trees along the sunny edge, as well as along the windy edge, has a high evapo-transpiration rate. An elongated depression or grassy swale along the upslope edge reduces surface water runoff, and also increases infiltration into the soil. Runoff water from a broad upslope area can be channeled or piped to such depressions. These tree strips and elongated depressions also work well in small narrow greenspaces. A patch of semi-natural vegetation in a greenspace addresses all three stormwater goals, low runoff, high infiltration, and high evapo-transpiration.

What other environmental issues across the entire metro area could be effectively addressed with arrangement, or a functioning system, of urban greenspaces? Consider air quality. The strip of trees along a windy edge filters out airborne particles. Indeed, a strip of shrubs and/or trees on streetside edges of greenspaces probably reduces airborne particulate matter lifted from road surfaces by moving vehicles. Some heat from surrounding built areas flows horizontally (advection) to cooler greenspaces, and carries pollutants in the flows (see Figure 5.3). At higher temperature, hydrocarbons on roads and carparks evaporate to become air pollutants, so cooling the air with an adjacent greenspace (see Figure 5.7) limits the pollution.

Dry hot air may be humidified and cooled using greenspace trees, which evapo-transpire more than shrubs, which in turn evapo-transpire more than herbaceous cover. Using tree lines or shrub lines, streamline airflow can be bent to increase ventilation, or alternatively, to reduce wind. Also greenspace trees can decrease or increase turbulent and vortex airflows created by buildings.

An urban greenspace system of course could also directly benefit people, such as providing leisure and recreation opportunities. For instance, similarly designed parks across an urban area are likely to serve, and in part be cared for by, their local neighborhoods (Forman, 2008). On the other hand, flagship specialized parks normally serve a broader urban area, but largely depend on government for maintenance. These different types and support roles can be fit together into a functioning system. Furthermore, the greenspace system can help delineate and strengthen neighborhoods in urban areas.

In short, it is time to achieve and sustain a range of ecological and human benefits throughout a metro area. The urban greenspace system does just that. The key elements for success seem clear and doable. Wise thinking and action, city-by-city, neighborhood-by-neighborhood, person-by-person, can create the success stories passed on from decade to decade.

Epilogue

One by one they stopped and gasped at the sight that lay beneath them. Millions of stars seemed to have fallen from the sky to land on the Earth far below. It was the lights of the city! Giant buildings crowded into the evening air. Rumblers roared along paved roads. As far as the eye could see the city spread itself over the land. Where were the trees? The birds? The flowers? … The healing tree, the creek, it's all gone.
Garry Fleming, Bollygum, *1995*

There are many themes in nature's symphony, each with its own pace and rhythm. We are forced to choose among these, which we have barely begun to hear and understand.
Daniel B. Botkin, Discordant Harmonies, *1990*

Following our journey through the rich patterns, delights and surprises in urban ecology, let us reflect, even speculate, a bit. I have chosen a handful of themes to explore that seem to be of particular interest and importance now and for the future. Let us consider urban ecology and the (1) distributions of aggregated people; (2) greenspaces and park system; (3) habitat heterogeneity and biodiversity; (4) tight urban water system; (5) ecological flows and urban networks; (6) urban change; and (7) societal goals and applications of urban ecology. In reflecting on these big subjects, my goal is to crack them open enough to catalyze further and useful pondering for scholars and for society.

But first, consider a moment what the preceding text represents. Most likely we have just experienced the first comprehensive scientific portrayal of the ecology of urban areas. We penetrated beyond the familiar human and socioeconomic dimensions describing a city in order to portray equally fundamental ecological patterns and processes, essential to understanding and changing urban areas. Nature in powerful natural systems permeates every spot, every population center. The ecological flows and changes across an urban mosaic provide a compelling view of how cities are structured and how they work. This perspective is essential in wise improvements for our future.

Distributions of aggregated people

Suppose everyone lived in villages connected by roads and surrounded by nature. That means connected villages and fragmented nature. Instead, in this simple spatial model, put people in one city, two towns, and a few villages in one portion of the land. That would mean connected nature (e.g., for stream/river systems, groundwater, biodiversity, wildlife movement) and connected population centers (including city with specialized museum, orchestra, industrial center, rail hub, diverse jobs, diverse housing). But people need to eat, so let us add farmland surrounding the population centers to the model. Now the land has connected nature at some distance from people, connected farmland close to people, and connected population centers including the resources of towns and city. That is good for both nature and people. More variables of course can be added to this simple spatial model to create still better patterns.

Now peer into the city or urban mosaic and consider the typical "home range" of people, that is, the area commonly covered in daily movements. A large home range means that a person's residence, job, shops, school, park, entertainment place, and so forth are far apart. A small average home range invests less energy into transportation/movement and more into a person's neighborhood. Mixed use describes the place. Safe appealing walkways, usually with tree lines, permeate neighborhoods, and parks readily walkable from everyone's residence are densely spread throughout the urban area.

Outward urbanization (ultimately limited by decreasing population growth) in some areas occurs as concentrated development adjoining the existing built metro-area. However, in many regions much urban spread is low-density-development sprawl. This dispersed development may be as large house plots containing large personal spaces, or may be concentrated small plots adjoining dispersed villages and towns. The long-term human problems and widespread ecological problems with dispersed development are well documented. But, assuming that both private spaces and ready access to parks are important, what is the optimum way to urbanize? Concentrate people, but with private spaces and walkable greenspaces that sustain vibrant natural systems across the area. That should be doable.

The entire urban region as a distinct functional unit highlights the interdependence between a city, actually a metro area, and the ring-around-the-city (see Chapter 2). The fundamental form of an urban region points to how it works, almost irrespective of city size and geography or culture. Urbanization expanded from a nucleus, usually surrounded by farmland and outer natural land. Radial transportation corridors with nodes of development subdivide the urban-region ring into sections. In-and-out city-and-ring flows predominate, from people and goods to air, water, and species. Yet multiple built nodes and asymmetries in the urban region create networks that dominate urban ecology flows and highlight how the region works.

At a still-broader scale, "megalopolis ecology" remains a frontier perhaps as well known as bedrock-surface ecology in central Greenland. As essentially the largest urban-ecology object, the megalopolis ties together nearby major cities with associated coalescing development and remnant agricultural and natural land. A megalopolis, such as Boston-to-Washington, Big Dutch Cities Area, Western UK Midlands, Yokohama-to-Tokyo, and Pearl River Delta (China), is almost too big to see. Excesses such as impervious surface, heat, air pollution, water pollution, wastewater, and solid waste are conspicuous. Scarcities, including clean-water supply, recreational greenspace, areas to treat wastes, wetlands, and urban agriculture, increase in importance. But the basic megalopolis patterns of habitat arrangement, species sources, and barriers to and routes of wildlife movement remain little studied. Major human resources are distributed "multi-nodally," and concentrated human movement is multi-directional rather than radial. What is the best pattern for treating and dealing with the massive wastes produced daily? Habitat areas are extensively fragmented, degraded, and shrinking, so how does the emerald network of large connected green areas, and even transit-oriented-development-with-nature (TODN), fit? Perhaps most important is the pattern of big tentacles from the megalopolis permeating the outer agricultural and natural lands of the "greater megalopolis region."

Greenspaces and park system

For most of urban history, urban greenspaces were mainly for intensive local food production. Later many became parks, often with ornamental plantings, fertilized/pesticided/mowed lawn, piped-away stormwater, and extensive walkways plus park buildings. Yet in many cities, greenspaces near potential jobs are prime locations for squatter informal settlements. Small little-used or overlooked spaces, as well as large spaces with severe environmental conditions and hazards, are both suitable for squatter residents. All three uses of greenspaces (intensive food production, lawn/ornamental/walkway, and squatter settlement) serve urban residents. Although some species diversity is present, each case represents an extreme along gradients of degradation of urban-greenspace nature. In the urban greenspaces and park systems designed by society, we typically assume the presence of at least some semi-natural conditions. Virtually all greenspaces can be designed and maintained to limit the severe-degradation uses and sustain some semi-natural conditions.

Could a collection of urban parks be transformed into an integrated park system providing several key functions for the entire urban area? Parks would have to be functionally connected, and success would depend on benefits from an integrated system reaching the whole metro area. For example, a spatial arrangement of greenspaces could cool the summer air across the entire city (see Chapter 12). Furthermore, park designs, both internal and arranged according to sun, wind and stormwater-flow directions, could limit flooding across the urban area. An integrated park system could play additional roles for society, including reducing or increasing air flows, reducing particulate and other air pollutants, treating stormwater pollutants, and enhancing wildlife movement.

For a greenspace system to provide rich biodiversity across the built metro area requires a continual "species rain" from the surroundings (see Chapter 12). Most urban greenspaces are small, ecologically degraded, isolated, dominated by generalist including non-native species, and unlikely to sustain many native specialist-species. The incoming organisms from agricultural and natural lands may not survive long nor reproduce in the new tough milieu, but the endless species rain helps maintain the presence of species (and reduces local extinction) in urban greenspaces.

For instance, assume that species spread outward from a semi-natural area a distance proportional to size of the semi-natural area. We can then readily recognize configurations or arrangements of functionally connected greenspaces in the metro area. Together these configurations highlight the portion of a metro area where rich biodiversity is sustained by species rain from surrounding lands. By strategically adding greenspaces to the system, the entire metro area can be sustained with relatively rich biodiversity.

This biodiversity solution suggests that we must turn outward, focusing on the large natural areas near a metro area. The outward big picture is combined with the major green corridors of the region, the arrangement of greenspaces in and around the urban area, and even bits of green providing functional connectivity between parks. In short, more species will arrive, survive, and thrive in our urban greenspaces, and more will enrich our urban built spaces.

Habitat heterogeneity and biodiversity

Cities or metro areas contain an extraordinary number of different microhabitat types, from tiny flower gardens to rail yards, low-income residential sites, seminatural greenspaces, dumps, and urban water bodies (see Chapter 8). The contrast in species present among habitat types is also striking. In places, the small habitats are packed together, as in many house plots, industrial sites, institutional areas, and medium-size parks. Viewing habitats at different spatial scales provides further insight, as for example the set of microhabitats within a house plot contrasted with, and added to, the set in other house plots along a street or in a neighborhood. In short, although locations such as parking lots and ball fields have few habitats, the metro area as a whole boasts extremely high habitat heterogeneity.

Biodiversity normally strongly correlates with habitat heterogeneity. Three urban features add to the story. First, a richness of native, spontaneous non-native, and planted horticultural/ornamental species is highly intermixed across the metro area. Second, species respond differently along environmental gradients such as air temperature and soil moisture, and the city provides a huge number of environmental gradients (with countless interactions), such as of heavy metals, traffic noise, chemical sprays, trampling, flooding, garbage accumulation, and light at night. Third, diverse network types channel species throughout the urban area, such as cockroaches in storm drains connected to building basements, night mammals along streets, pathogenic bacteria in piped water, and rats through empty little-used or oft-forgotten pipe systems. Again, although conspicuous spots are species-poor, the cumulative effect of high habitat heterogeneity, a large species pool, numerous environmental gradients, diverse network channels for movement, and both habitat and species packing means a cumulative high biodiversity for the metro area.

At the fine scale, a remarkable percentage of the British Isles flora and fauna was recorded in a single 0.1-ha (1/4-acre) house plot (see Chapter 8): 11% for vascular plants; 44% harvest spiders (daddy longlegs); 38% ladybirds (ladybugs); 36% hoverflies; 34% butterflies; 33% lacewings (and allies); and so on. At the broad scale of a metro area [Warsaw with 1.7 million people in 517 km^2 (201 mi^2)], the numbers of species recorded (see Chapter 9) are 3800 terrestrial invertebrates; 320 vertebrates; 40 mammals; 274 birds; 5 reptiles; 11 amphibians; and 30 fish. Comparing these numbers with same-size spaces in nearby agricultural and natural lands, and indeed with the little-known species richness of tropical cities, would be quite interesting.

A finer view of small patterns and processes that escape maps and GIS images is needed to understand urban habitats and biodiversity. Instead of only examining roads, parks, shopping centers, industrial areas, and so forth, look at key "hot spots" for species (e.g., an old mother tree, or rare wet spot), major species' "needs" (food patches, shrub-cover patches, houseplot backlines), and flows of water and wildlife. But also, look sharply at the "invisibles," such as nutrient flows, tree-top organisms, and underground interactions, in addition to familiar visible features (see Chapter 1). Use the human eye, or even move as if using the "eye" of an animal or plant (moving seed). We might then understand habitats and biodiversity of the city.

Tight urban water system

The urban water system is basically a heterogeneous set of "flow-throughs" and receiving bodies, which sharply contrasts with that in agricultural land, and especially natural land. Pipe networks for water supply, stormwater, sewage wastewater, septic wastewater, and stream water, plus groundwater flow, all basically funnel water and pollutants to local water bodies (see Chapter 6). Precipitation provides stormwater and also a water-supply source, either protected and clean, or unprotected and polluted. The stormwater running over impervious and other surfaces picks up pollutants. Meanwhile the water supply picks up and carries household, commercial, industrial and pipe pollutants, as well as human wastewater. Plants pump some water upward in evapo-transpiration, but stormwater is mainly piped to local water bodies. The household water with pollutants and wastewater is commonly piped to sewage wastewater facilities and septic systems that partially clean the water before it drains into water bodies. These mainly separate flows are large, and local water bodies such as streams, river, and estuary receive a heavy dose of both water and pollutants.

Some routes can be beneficially shortened, such as stormwater pipes. Some water can be recycled, as in filtered grey-water from tubs and basins used to flush toilets. Some wastewater and its contents can be used to grow food in aquaculture. However, most short-term tightening of the overall water system decreases interconnections between different flow types. Thus, reduce: stormwater leaks into the wastewater pipe system; stormwater flow through contaminated soil to water bodies; wastewater-pipe leaks into soil and groundwater; sewage flows (CSOs) into stormwater and water bodies during heavy-rain events; and the abundance of malfunctioning septic systems channeling little-treated wastewater toward water body. A tight water system enhances all habitats involved, and saves money. Reduce water use, reduce runoff. Increase infiltration into clean soil, increase evapo-transpiration.

But what can be done with the urban-provided pollutants? Increase stormwater treatment, increase septic-system effectiveness, and increase sewage wastewater treatment efficiency. Stormwater is largely treated (cleaned) by infiltration through uncontaminated soil, for instance, via constructed ponds, wetlands, basins, and biofilters. Sewage wastewater can be treated by a sewage treatment facility (especially tertiary), a pond-and-wetland facility, and theoretically in an aquaculture pond. When functioning well, septic wastewater is cleaned by bacteria and the soil. Still, after cleaning, where have all the pollutants gone?

Some pollutants accumulate in and contaminate the soil, while most pour into local water bodies. Streams, rivers and ponds experience both floods and low-water levels. Aquatic ecosystems and fish in the water bodies depend primarily on aquatic habitat heterogeneity, water quality, and sustained suitable water flows and levels. To maintain these key conditions, tighten the urban water system and treat/clean the water contents. Local water bodies are the assays of the urban system. Make the stream, river, lake, pond and/or estuary both suitable for native fish and appealing to urban residents.

Ecological flows and urban networks

Traditional ecosystem ecology highlights plant productivity and food-chain energy flow, plus mineral-nutrient or biogeochemical cycling. In natural ecosystems, chemicals either cycle within or flow through the system. Thus, nitrogen may cycle from live foliage to dead leaf litter to roots and back to live foliage, or it may flow through an ecosystem, entering in wind and leaving in stream-flow. In natural ecosystems, wind, water, and animals are major transport vectors. Frequently key available nutrients and most human-produced chemicals are limited in amount, while other natural chemicals are exceedingly diverse and mostly present in tiny amounts.

Urban areas seem fundamentally different. Plant productivity is small. The flows of heat energy rather than food-chain energy are primary. Chemical flow-throughs predominate. Little internal cycling occurs. Mineral nutrients are usually abundant or in excess, and typically in a rather high-pH environment. Natural chemicals mainly produced by the plants seem to be little studied, and some (e.g., PAHs) may play important ecological roles. Human-produced chemicals are extremely diverse and abundant (see Chapters 4 and 5), and most originate in the urban system. The flows of human-produced chemicals, involving plants, animals and microbes (and people) in urban areas, should be both interesting and important.

Perhaps equally important and more distinctive is the central role of mostly human-built-and-maintained networks in the urban ecosystem. Railways, powerlines, streets, water-supply pipes, stormwater pipes, sewage wastewater pipes, septic-system pipes are familiar, as are semi-natural streams, groundwater, river, wildlife movement routes, and ecological networks (see Chapter 12). These diverse networks are pervasive, differ markedly in form, connect to large and small nodes, and often interconnect. Other networks are present including truncated food webs, industrial interdependence, and cracks in surfaces (see Chapters 9 to 11). The prime footprints model, spider-like, links key oft-distant resource and waste areas with a metro area (see Chapter 2). In effect, ecological flows in urban areas are centrally driven by people, wind, and water along networks, mainly somewhat rectilinear in form.

The hierarchical street, road and highway network with vehicle traffic is most conspicuous and arguably most ecologically detrimental. Thus, the netway system with pods, designed primarily to recover and reconnect the land and nature in non-urban areas, seems to be a particularly promising change for urban transportation (see Chapter 11). Flexible designs permit alternate-street use, silent bus/van pods, or added bicycle routes, as well as underground use in suburbs. With increased transport efficiency and safety, no fossil fuel use, no greenhouse gas emissions, increased area for market gardening, and increased recreational trail networks close to the city, netways promise many benefits for both humans and nature. Cities with traffic-free streets or zones know that the urban space provided is a boon for people in neighborhoods, convenient local shopping, and more plantings, with associated air, wildlife, and aesthetic benefits.

The urban-waste flows of stormwater, sewage wastewater, and solid waste are especially prominent, and involve water, microbes, mammals, birds, invertebrates, algae, fish, organic matter, and diverse chemicals. The spread of microbes in public health often involves rats, mosquitoes, people, wind, and/or water. Anthropogenic networks are maintained by people; breakdowns occur. Envision a major accident on a ring road, water-main-pipe break, clogged sewage pipe, damaged bridge on a commuter rail line, and urban river pollution blocking migratory fish. Flows are interrupted and often diverted to a different route; indeed network forms change over time. On the other hand, a "string of pearls" path connecting tiny greenspaces and lined with trees has some stability (see Chapter 2), because active users prevent its blockage by informal squatters, flower gardens, or urban agriculture. Every piece counts in a simple connected system. In a complex system, loops and redundancy provide stability. In addition, network lines or linkages may function as barriers as well as conduits, well-illustrated by strip (ribbon) development blocking wildlife movement.

Furthermore, the functional ecological network of flows across the land, and especially the emerald network of connected large green areas, is central to ecological flows in urban areas (see Chapters 2 and 3). Despite the "multi-colored spaghetti" of diverse network lines with massive never-ending flows (and breakdowns), the emerald network seems powerful enough to sustain the flows of biodiversity throughout an urban area.

Urban change

Cities come and go: sometimes appearing, spreading, densifying; sometimes decaying, dying, disappearing. Petra, Angkor Wat, Machu Pichu, and Tikal sank under the sands or rainforests of time. More visible near home are the vacant lots, buildings, and parks that appear and disappear. Since each piece of the urban mosaic plays an ecological role, evaluating and mapping the relative stability of pieces would be informative.

Wars and, in today's world, bombing are targeted to cities. Widespread destruction occurred, for instance, in Tehran (1220), Washington (1814), Atlanta (USA) (1864), Dresden (Germany) (1945), Hiroshima (1945), Manchester (UK) (1996), and Bagdad (2003). In World War II, Berlin's large central Tiergarten Park had virtually all trees removed for fuel, urban agriculture, and military activities. In Tokyo and Seoul, swaths of buildings were transformed into open corridors serving as fire breaks and military zones with anti-aircraft guns. In view of such destructive forces, rather than simply rebuilding in the previous footprint, an adaptable urban design should be able to noticeably and sustainably enhance the ecological conditions of a city.

Nine other "disasters" or big sudden disturbances are particularly serious in urban areas where people and their structures are concentrated: wildfire (e.g., Canberra, 2003; San Diego, California, 2003 and 2007); volcanic eruption (Pompeii, AD 79; Pereira, Colombia, 1985); earthquake (Caracas, 1812; San Francisco, 1906; Kobe, Japan, 1995); tsunami (Alexandria, Egypt, AD 365; Banda Aceh, Indonesia, 2004); flood (Dhaka, Bangladesh, 1998; New Orleans, USA, 2005); hurricane (cyclone/typhoon) (Hong Kong, 1937; Phillipines, 2013); industrial-pollutant release (Bhopal, India, 1984); nuclear-power-plant radiation release (Chernobyl/Prypyat, 1986); disease outbreak (European and Asian cities, 14th century). Just as for war and bombing, urban nature could benefit greatly by implementing a creative urban design for disaster adaptability.

Slow degrading processes, most ecological, also undermine the urban concentration of human structures. Termites chewing, wood decaying, metal rusting, buildings settling, vibrations cracking, groundwater dropping, heat building, salt dissolving, reservoir filling with sediment, and many more processes gradually degrade the urban area. The massive people-movement from Central and Western China to the East Coast over a few decades has transformed – both improved and degraded – population centers across much of the nation. Slow change may be relatively constant but is likely to include noticeable changes in rate, including lulls and spurts. These gradual almost-eternal processes mean that when budgets and maintenance/repair activities are down, ecological change accelerates and more species survive and thrive.

Spatial pattern offers a handle to ecologists for understanding urban change. Spatial processes, including fragmentation, connection, shrinkage, expansion, disappearance, and appearance, operate not only in exurban/peri-urban areas, but at finer scales within the metro area (see Chapter 3). Five models of outward urbanization, such as concentric rings, transportation corridors, and dispersed patches (sprawl), highlight both how we spread and what the ecologically best and worst ways are to urbanize. A more-detailed analysis of alternative ways to expand from a metro-area border highlights the apparent ecologically optimum trajectory. But the approach also provides a changing template to pinpoint at any stage the best and worst locations for the next park, shopping center, or other land-use change. Adding socioeconomic dimensions to these spatial optimization models could highlight a compelling future for the urban billions just ahead.

Stability of a metro area, or portion thereof, may be increased in lots of ways, including strong hierarchy; negative feedbacks; increasing the size (harder to disrupt); maintaining gradual edges (so that responses to disturbance are not all-or-nothing); and loops as optional routes in a network. Providing adaptability, in the sense of a flexible capacity to become somewhat modified in response to disturbances, seems to be a more important goal than stability. An urban mosaic with adaptable changing pieces should provide an ecologically richer trajectory than could a stable system.

Although termites are chewing, bombs going off, and other disturbances brewing, currently global climate change stays in the headlines. Expectations for urban areas differ by region and city, but include higher air temperatures (especially in the surroundings); sea level rise; estuarine encroachment; more precipitation (or less); more extreme-weather events; more and/or higher stream/river flooding; more and/or longer stream/river low flows; and threats to clean-water supply. An array of changes would provide climate change adaptability and ecological benefit, such as recover wetlands in low areas; increase vegetation cover; minimize rebuilding after extreme-weather events; accelerate transit-oriented-development-with-nature away from the coast or river; shorten stormwater-pipe systems; recover and protect vegetation around clean-water supplies; reduce the home-range areas of urban people; establish a protected emerald network; protect suitable habitat for species expected in the future; create and maintain high habitat diversity; replace busy-traffic roads with netways; reduce impervious surface cover; reestablish coastal wetlands; increase soft edges between natural and built areas; make residential development compact; increase plant cover (including with green walls and roofs); find new solutions for sewage wastewater; and plan for environmental surprises. Most are common sense for the reader, as well as the informed public. Most also make sense for dealing with big disturbances or disasters in general. Why wait?

Pondering societal goals and urban ecology applications

Urban ecology is "a study of ...," with the objective of understanding. Applying the theories, principles, models, concepts, examples, evidence, and ideas to solving problems may be creatively accomplished by a range of professions and disciplines. Major portions of the knowledge are useful to engineering, urban planning/design, public health, landscape architecture, water resources, bio/nature conservation, sociology, and economics. With solidification of urban ecology, its knowledge is useful for government in planning, construction, maintenance, and repair. The future is not just what lies ahead; it is something that nature and we create.

An ecologist might highlight goals, such as maximizing native-species biodiversity, or maintaining a relatively natural ecosystem, or establishing small tight water-and-material/chemical flows, throughout the urban area. On the other hand, an urban ecologist probably would not emphasize or recommend maximizing biodiversity, or maintaining only native species, or focusing on rare species protection. In an urban area, success is highly unlikely in all of the cases.

In contrast, using environmental knowledge, a public health official might be more interested in increasing bat populations to control mosquitoes, a water resource manager in how best to treat stormwater to sustain fish populations, an engineer in how best to establish and arrange habitats to reduce flooding, and a landscape architect in how best to arrange which plants to cool air and attract birds. An urban planner/designer has a rich set of ecological principles to use, focused on water, transportation, residential areas, and so forth (see Chapter 2). An additional set of specific spatial patterns, the good, bad, and interesting, can be creatively combined for different situations and different cities (see Appendix A).

Epilogue

Providing ecosystem or nature's services for urban residents might be a promising overall objective. Urban habitats such as street trees, cultivation, and wetlands provide a variety of services, including air filtration, microclimate regulation, and stormwater drainage. Although the services provided are a small portion of the total services required, the general habitat types involved seem to improve residents' quality of life. Yet the measure of success in addressing ecologically dependent goals remains a challenge. A "natural" ecosystem basically does not exist in an urban area and is not a promising goal for a large urban area. We could attempt to achieve some "degree of naturalness," but that is an odd and awkward goal for a largely built area with highly distinctive characteristics. At the other end of the spectrum, the objective could be some measure of human satisfaction, such as quality of life, delight, discovery, surprise, biophilic benefit, or educational, aesthetic, inspirational, health, or political value. A conundrum awaits solution.

Everyone can list societal actions that benefit nature, such as driving less, using less water, growing vegetables, using public transport, and making compact housing developments. Most people can also pinpoint useful goals that directly use urban ecology. Broad-scale goals might include having a city live in balance with resources of its urban region. Or in balance with its present ecological footprint, or even shrinking the footprint. Longer-term goals might be to build structures mimicking nature's time-tested structures, and have them work as nature's processes do. An umbrella goal might integrate several sub-goals, such as widespread tree cover or semi-natural vegetation cooling the air, reducing flooding, cleaning stormwater, and being sources for species dispersing more widely. The usefulness of urban ecology to scholarship and to society will be determined by the perspectives of potential users, and the rate at which ecology becomes important to the life of people in the onrushing urban enterprise on Earth.

As the preceding chapters reveal, I am passionate about understanding the urban mosaic, where I have lived for decades in many states and nations. The pages emphasize that broad perspective and context matter. Cities remain surrounded by extensive farmland and natural land, two places where I have done considerable research and also feel at home. I am an ecologist who scientifically, if not almost *in toto*, grew up in nature. I still go to the most remote places. Now that urban ecology has begun to gel, I ponder an ecology of land and city. Do landscape ecology, road ecology, and urban ecology dovetail enough to spark synergies for scholarship and society?

Appendix A: Positive and negative attributes of an urban region

Ecologically positive and negative attributes of urban regions. Based on analyses of 38 urban regions worldwide (Forman, 2008).

Positive	Negative
City, metro, region	*City, metro, region*
Urban region is an administrative unit	Urban region split between nations
City close to several land types	Urban region split between states/provinces
City on border of two land types	City competes intensively with nearby city
City by protected coastal bay	Metro area reflects former political division
Compact metro area	Elongated metro bisecting a natural landscape
Planned city retained natural attributes	Limited greenspace in metro
Numerous small greenspaces in metro	Creeping development into large natural area
Green network connects greenspaces	*Nature, forest, food*
Distinct border of metro	Little forest remaining in urban region
Urban growth boundary	Few natural areas remain along coast
Scalloped metro border	Limited cropland in urban region
Green wedges present	Only one or two farmland types
One wide long green wedge	Main cropland concentrated near metro border
Greenbelt or ring of large parks	*Water*
Facing hillsides with protected vegetation	Water supply basin partly outside urban region
Towns by farmland–nature boundaries	Water supply covered with cropland
Low total border length of built areas	Major reservoir polluted
Regional planning indicated by metro form	Best aquifer threatened by development
Regional planning evident in urban-region ring	Few rivers surrounded by natural land
Nature, forest, food	Rivers reduced to low flow most years
Many wooded landscapes in urban region	*Transportation, development, industry*
Large forest patches across urban region	Many radial highways reaching region boundary
Large natural patch adjoining metro area	Two-lane ring road likely to be widened
Emerald network well developed	Port located far from city center
Many protected natural areas	Built areas surround most streams/rivers
Many one-day tourist/recreational sites	High total border length of built areas
Native People's lands protected	Towns in urban region threatening to coalesce
Different farmland types present	Dispersed site development predominates

Appendix A

Positive	Negative
Large wooded patches in cropland	Slopes near metro area largely developed
Market-gardening areas close by	Many informal squatter settlements
Agriculture-nature park close by	Heavy industry close to city
Water	Coastal near-shore water polluted
Water supply area mainly woodland/forest	Streams/rivers highly polluted
Natural land around most streams/rivers	Large mine-waste areas in urban region
Main reservoir/lake outside metro area	*Hazards*
Vegetation along lakeshores	Riverside city subject to flooding
Extensive wetlands near metro	Coastal city subject to cyclones, tsunamis
Transportation, development, industry	Area subject to sea-level rise
Commuter rail extends beyond metro	Built areas close to fire-adapted vegetation
Transit-oriented-development along rail line	
Reticulate rail network in urban region	
Highways along borders of land types	
Wildlife under/overpasses for connectivity	
New development only by existing built areas	
Heavy industry mainly by separated port	

Appendix B: Equations

Many of the basic or widely used equations presented additionally exist in other forms, with somewhat different assumptions or variables. Before using a particular equation it is wise to read further about it in the literature. Also see Forman (1995), Turner et al. (2001), Leitao et al. (2006), Butler and Davies (2011), Erell et al. (2011).

Chapter 1
Rank-size rule for city population

$$P_n = P_l \times R_n^{-1}$$

P_n = population of city to be calculated;
P_l = population of the largest city;
R_n = rank of city to be calculated (number in the order from largest to smallest city)

(Hartshorn, 1982; Rowe et al., 2013).

Chapter 2
Patch shape

$$F = \frac{l}{w}$$

F = form;
l = length of long axis;
w = width of patch perpendicular to long axis.

Many measures of patch shape exist, generally indicating the degree of compactness of a patch. Most are based on: lengths of axes; perimeter and area; area; radii; area and length; perimeter and length; perimeter; or a fractal (Forman, 1995; Turner et al., 2001).

Chapter 3
Metapopulation dynamics

$$\frac{dp}{dt} = cp(1 - p - mp)$$

p = the proportion of available locations (patches) with individuals present at any point in time;
c = probability of species colonization of a patch;
m = probability of species extinction in a patch

(Turner et al., 2001).

Network connectivity

$$\text{Gamma index}, \gamma = \frac{L}{L_{max}} = \frac{L}{3(V-2)}$$

L = number of linkages in network;
L_{max} = maximum possible number of linkages;
V = number of nodes (i.e., intersections and linkage ends)

(Forman and Godron, 1986; Forman 1995; Turner et al., 2001).

Network circuitry

$$\text{Alpha index}, \alpha = \frac{\text{number of circuits or loops}}{\text{maximum possible number of circuits or loops}}$$

$$= \frac{L - V + 1}{2V - 5}$$

L = number of linkages;
V = number of nodes

(Forman and Godron, 1986; Forman, 1995).

Node connection

$$\text{Beta index}, \beta = \frac{L}{V}$$

L = number of linkages;
V = number of nodes.

Gravity model

$$I = k\frac{P_1 P_2}{d^2}$$

I = degree of interaction between two centers or nodes;
k = a scaling constant indicating the units of flow;
P_1 = size or population of center 1;

381

P_2 = size or population of center 2;
d = distance between centers

(Turner and Gardner, 1991; Hartshorn, 1992).

Lowry and journey-to-work models
Equations simulate the spatial distribution of population, employment, service, and land use. The models are commonly used based on the hypothesis that residences gravitate toward employment locations, and are independent of environmental variables (Marzluff et al., 2008).

Linear growth

$$\frac{dN}{dt} = r$$

N = population size;
t = time;
r = intrinsic rate of increase for a species

(Smith, 1996; Turner et al., 2001; Cain et al., 2011).

Exponential growth

$$\frac{dN}{dt} = rN$$

N = population size;
t = time;
r = intrinsic rate of increase for a species.

J-shaped curve (Smith, 1996; Turner et al., 2001; Cain et al., 2011).

Logistic or sigmoid growth

$$\frac{dN}{dt} = rN\frac{K-N}{K}$$

N = population size;
t = time;
r = intrinsic rate of increase for a species;
K = carrying capacity (maximum population size sustained by the environment).

S-shaped curve. Verhulst logistic growth (Smith, 1996; Turner et al., 2001; Cain et al., 2011).

Chapter 4

Decomposition of carbon compounds

Organic matter + O_2 = gases (CO_2, NH_3) + water (H_2O) + cations (e.g., K^+, Ca^{2+}, Fe^{2+}) + anions (e.g., NO_3^-, SO_4^{2-}, PO_4^{3-})

By aerobic bacteria and/or fungi (Jenny, 1980).

Network connectivity, network circuitry, node connection
See Chapter 3 equations.

Chapter 5

Shadow length

$$S_l = \frac{h}{\tan(SA)}$$

S_l = shadow length;
h = height of an object;
SA = sun angle

(Lynch and Hack, 1996; Marsh, 2010).

Sky view factor

$$SVF = \frac{\cos\beta_1 + \cos\beta_2}{2}$$

$\beta_1 = \tan^{-1}(H_1/0.5W)$;
$\beta_2 = \tan^{-1}(H_2/0.5W)$;
H_1 and H_2 = the average heights of adjacent structures on, e.g., two sides of a street;
W = width of street

(Craul, 1999; Erell et al., 2011).

Net radiation balance

$$Q^* = (K_{dir} + K_{dif})(1-\alpha) + L\downarrow - L\uparrow$$

Q^* = net radiation balance;
K_{dir} = direct short-wave radiation (solar rays directly from Sun);
K_{dif} = diffuse short-wave radiation (solar radiation reflected from clouds or aerosols in the atmosphere; makes the sky appear bright even when the Sun is hidden);
α = albedo (reflection) of the surface;
$L\downarrow$ = long-wave radiation received from the sky;
$L\uparrow$ = long-wave radiation emitted by the surface

(Erell et al., 2011).

Solar heating

$$SH = S_i(1-A)\sin(SA_g)$$

SH = solar heating (cal/cm²/min or joules/m²/min);
S_i = incoming solar radiation (cal/cm²/min or joules/m²/min);

Equations

A = albedo (or reflection) ($1 - A$ = percentage of energy absorbed);
SA_g = sun angle in degrees (at ground surface)

(Marsh, 2010).

Horizontal windspeed around a barrier

$$\frac{U}{U_h} = f\left(\frac{x}{h}, \frac{z}{h}, \frac{h}{z_0}, \frac{h}{L}, \varphi\right)$$

U = average horizontal windspeed for a long thin windbreak barrier on a large flat surface with wind direction perpendicular to windbreak axis;
U_h = average approach windspeed at top of windbreak;
h = barrier height;
x = perpendicular distance from windbreak;
z = height above the surface;
z_0 = roughness length taken from the uninterrupted wind profile;
L = the Monin–Obukhov stability length (a measure of atmospheric stability);
φ = porosity of the barrier.

The Reynolds number (hU_h/v, where v is the molecular viscosity of air) also affects the average horizontal windspeed, but is unimportant with air mixing over a field (Brandle *et al.*, 1988; Forman, 1995).

Chapter 6
Urban water budget

$$p + I = r + E + F + \Delta A + \Delta S$$

p = precipitation;
I = water supply piped into the urban area;
r = surface and subsurface runoff;
E = evapo-transpiration;
F = water vapor released due to human activities (such as combustion);
ΔA = net advection of moisture in or out of the area;
ΔS = increase or decrease in water storage during the period

(Butler and Davies, 2011; Erell *et al.*, 2011).

Groundwater flow

$$V = -K\frac{h}{L}$$

V = velocity of water flow through a cross-section of a porous medium such as sand;
K = hydraulic conductivity (related to permeability of the medium); negative indicates downward flow;
h/L = the hydraulic gradient (potentiometric head divided by length or distance).

Darcy's law (Jenny, 1980; Baker, 2009; Marsh, 2010).

Stormwater runoff (combined surface and subsurface)

$$SR = f(P, L, S, W, C, V, M)$$

P = precipitation intensity and duration;
L = land surface slope;
S = soil permeability;
W = water-table depth;
C = channel curvilinearity;
V = vegetation width and density in stream/channel corridor;
M = matrix vegetation cover and density surrounding stream/channel corridor

(Forman, 1995).

Surface depression storage

$$d = \frac{k_1}{\sqrt{s}}$$

d = rainwater that becomes trapped in small depressions on a basin/catchment surface (measured in mm of water);
k_1 = coefficient dependent on surface type;
s = surface slope.

Typical values: d = 0.5–2 mm for impervious surfaces, 2.5–7.5 mm for flat roofs, up to 10 mm for gardens (Butler and Davies, 2011).

Peak stormwater discharge

$$Q = A \cdot C \cdot I$$

Q = peak discharge or flow (m³/s);
A = area;
C = coefficient of runoff;
I = intensity of rainfall (cm/h).

Applies to a single rainstorm at the mouth of a basin/catchment (Marsh, 2010).

Chapter 7
Stream density

$$SD = \frac{TL}{A}$$

TL = total length of streams;
A = area.

Analogous equation used for road density, corridor density, edge density, e.g., in km/km² (Forman, 1995; Forman et al., 2003; Marsh, 2010).

Sinuosity ratio

$$SR = \frac{L_s}{L_v}$$

L_s = length of stream or river;
L_v = length of valley.

SR < 1.5 for eroding streams/rivers; SR > 1.5 for meandering sediment-depositing streams/rivers (Gregory and Walling, 1973; Dunne and Leopold, 1978).

Recurrence interval for a particular flow amount

$$T_r = \frac{n+1}{m}$$

T_r = recurrence interval (years);
n = total number of flows;
m = the rank (number in the ordered list of greatest to smallest flows) of a particular flow amount or height

(Marsh, 2010).

Stream flow velocity

$$V = 1.49 \frac{R^{2/3} S^{1/2}}{n}$$

V = velocity (m/s);
R = hydraulic radius, which represents depth, and is equal to the wetted perimeter of the channel divided by its cross-sectional area;
S = slope or gradient of the channel;
n = roughness coefficient for the channel bottom.

Manning equation (Marsh, 2010).

Chapter 8
Species richness of a patch

$S_p = f$[habitat diversity (+), disturbance (−/+), area of patch interior (+)]

+ = positive effect; − = negative effect. Secondary variables, including matrix heterogeneity, patch age, and isolation, may be important in an urban area (Forman, 1995).

Moran index for degree of compactness or clustering

$$I(d) = \frac{n \sum_i \sum_j w_{ij}(z_i - \bar{z})(z_j - \bar{z})}{W_d \sum_i (z_i - \bar{z})^2}$$

n = number of areas or objects;
z_i and z_j = values of the variable at locations i and j respectively;
$\bar{z}$ = the variable mean;
w_{ij} = a weight matrix where a value of 1 indicates a pair of locations i and j that are in the same distance class d, plus a value of 0 for all other cases [commonly used as an indicator of the presence (1) or absence (0) of a connection between two locations];
W_d = the sum of the w_{ij}'s for the dth distance class.

The Moran index is used to indicate the type of compactness/dispersion or spatial autocorrelation of areas or objects in a particular distance class, d. Index values range from +1 (clustered, monocentric, positive autocorrelation) to 0 (a number of smaller clusters, polycentric, no autocorrelation) to −1 (dispersed, decentralized, negative autocorrelation) (Klopatek and Gardner, 1999; Rowe et al., 2013).

Patch shape
See Chapter 2 equations.

Species dominance

$$D = \frac{1}{\sum_i p_i^2}$$

p_i = proportion of a species (i) in a total sample of individuals.

Simpson index (Ricklefs and Miller, 2000; Turner et al., 2001; Rowe et al., 2013).

Chapter 9

Species richness of a patch
See Chapter 8 equations.

Metapopulation dynamics
See Chapter 3 equations.

Chapter 10

Road density
See stream density in Chapter 7 equations.

Network connectivity, network circuitry, node connection
See Chapter 3 equations.

Chapter 11

Gini coefficient of inequality

$$\text{Gini coefficient} = \frac{A}{A+B}$$

This equation compares an observed distribution of a resource in a population with a theoretical distribution. Plotting the cumulative percent of a resource (vertical axis) against the cumulative percent of a population containing it (horizontal axis), a straight line indicates complete equality in the population, while a lower curved line represents an unequal distribution of resources in the population. A = area (on the graph) between the two lines; B = area below the curved line. The Gini coefficient ranges from 0.0 (everyone has the same resource amount) to 1.0 (one person has all the resources). The coefficient can also be a measure of diversity, and may be especially sensitive to the pattern at low resource levels (Frumkin *et al.*, 2004; Rowe *et al.*, 2013; Michael Hooper, personal communication, 2013).

Chapter 12

Corridor density
See stream density in Chapter 7 equations.

Network connectivity, network circuitry, node connection
See Chapter 3 equations.

References

Abbott, C. (2004). Corridors and edges: reshaping downtown Portland. In *The Portland Edge: Challenges and Successes in Growing Communities*, ed. C. P. Ozawa. Washington, D.C.: Island Press, pp. 164–183.

Abercrombie, P. (1945). *Greater London Plan 1944*. London: H. M. Stationery Office.

Able, K. W., Manderson, J. P. and Studholme, A. L. (1998). The distribution of shallow water juvenile fishes in an urban estuary: the effects of manmade structures in the lower Hudson River. *Estuaries*, 21, 731–744.

Able, K. W., Manderson, J. P. and Studholme, A. L. (1999). Habitat quality for shallow water fishes in an urban estuary: the effects of man-made structures on growth. *Marine Ecology-Progress Series*, 187, 227–235.

Abs, M. and Bergen, F. (2008). A long-term survey of the avifauna in an urban park. In *Urban Ecology: An International Perspective on the Interaction Between Humans and Nature*, eds. J. M. Marzluff, E. Schulenberger, W. Endlicher et al. New York: Springer, pp. 373–376.

Acebillo, J. and Folch, R. (2000). *Atles Ambiental de l'area de Barcelona: Balanc de recursos i problemes*. Barcelona: Editorial Ariel.

Adams, C. E. and Leedy, D. L, eds. (1991). *Wildlife Conservation in Metropolitan Environments*. Columbia, Maryland: National Institute for Urban Wildlife.

Adams, C. E. and Lindsey, K. J. (2011). Anthropogenic ecosystems: the influence of people on urban wildlife populations. In *Urban Ecology: Patterns, Processes, and Applications*, eds. J. Niemela, J. H. Breuste, T. Elmqvist et al. New York: Oxford University Press, pp. 116–128.

Adams, C. E., Lindsey, K. J. and Ash, S. J. (2006). *Urban Wildlife Management*. Boca Raton, Florida: CRC Taylor & Francis.

Adams, L. W. and Dove, L. E. (1989). *Wildlife Reserves and Corridors in the Urban Environment*. Columbia, Maryland: National Institute for Urban Wildlife.

Addicott, J. F., Abo, J. M., Antolin, M. F. et al. (1987). Ecological neighborhoods: scaling environmental patterns. *Oikos*, 49, 340–346.

Adler, F. R. and Tanner, C. J. (2013). *Urban Ecosystems: Ecological Principles for the Built Environment*. New York: Cambridge University Press.

Adler, S. and Dill, J. (2004). The evolution of transportation planning in the Portland metropolitan area. In *The Portland Edge: Challenges and Successes in Growing Communities*, ed. C. P. Ozawa. Washington, D.C.: Island Press, pp. 230–256.

Aey, W. (1990). Historical approaches to urban ecology: methods and first results from a case study (Lubeck, West-Germany). In *Plants and Plant Communities in Urban Environments*, eds. H. Sukopp, S. Hejny and I. Kowarik. The Hague, Netherlands: SPB Academic Publishing, pp. 113–129.

Agger, P. and Brandt, P. (1988). Dynamics of small biotopes in Danish agricultural landscapes. *Landscape Ecology*, 1, 227–240.

Ahern, J. (2002). *Greenways as Strategic Landscape Planning: Theory and Application*. Wageningen, Netherlands: Wageningen University.

Ahern, J. (2004). Greenways in the USA: theory, trends and prospects. In *Ecological Networks and Greenways: Concept, Design, Implementation*, eds. R. H. G. Jongman and G. Pungetti. New York: Cambridge University Press, pp. 34–55.

Ahern, J. and Boughton, J. (1994). Wildflower meadows as sustainable landscapes. In *The Ecological City: Preserving and Restoring Urban Biodiversity*, eds. R. H. Platt, R. A. Rowntree and P. C. Muik. Amherst, Massachusetts: University of Massachusetts Press, pp. 172–187.

Ahrens, C. D. (1991). *Meteorology Today: An Introduction to Weather, Climate, and the Environment*. St. Paul, Minnesota: West Publishing Company.

Aizenberg, J. (2010). New nanofabrication strategies: inspired by biomineralization. *MRS Bulletin*, 35, 323–330.

Alam, M. G. M., Snow, E. T. and Tanaka, A. (2003). Arsenic and heavy metal contamination of vegetables grown in Samta Village, Bangladesh. *Science of the Total Environment*, 308, 83–96.

Alberti, M. (2008). *Advances in Urban Ecology: Integrating Humans and Ecological Processes in Urban Ecosystems*. New York: Springer.

Alberti, M. and Marzluff, J. (2004). Ecological resilience in urban ecosystems: linking urban patterns to human and ecological functions. *Urban Ecosystems*, 7, 241–265.

Alberti, M. and Waddell, P. (2000). An integrated urban development and ecological simulation model. *Integrated Assessment*, 1, 215–227.

Alberti, M., Marzluff, J. M., Shulenberger, E. *et al.* (2003). Integrating humans into ecology: opportunities and challenges for studying urban ecosystems. *BioScience*, 53, 1169–1179.

Alcoforado, M. J. and Andrade, H. (2008). Global warming and the urban heat island. In *Urban Ecology: An International Perspective on the Interaction Between Humans and Nature*, eds. J. M. Marzluff, E. Shulenberger, W. Endlicher *et al.* New York: Springer, pp. 249–262.

Aldhous, P. (2012). Protecting New York City from the next big storm. *New Scientist*, 216, 8–9.

Alexander, A., Ishikawa, S., Silverstein, M. *et al.* (1977). *A Pattern Language: Towns, Buildings, Construction*. New York: Oxford University Press.

Ali-Toudert, F. and Mayer H. (2007). Effects of asymmetry, galleries, overhanging facades and vegetation on thermal comfort in urban street canyons. *Solar Energy*, 81, 742–754.

Allan, J. D. (2004). Landscapes and river scapes: the influence of land use on stream ecosystems. *Annual Review of Ecology, Evolution, and Systematics*, 35, 257–284.

Allan, J. D. and Castillo, M. (2007). *Stream Ecology: Structure and Function of Running Water*. New York: Springer.

Allan, T. and Warren, A. (1993). *Deserts, The Encroaching Wilderness: A World Conservation Atlas*. New York: Oxford University Press.

Allen, A. (2006). Understanding environmental change in the context of rural-urban interactions. In *The Peri-Urban Interface: Approaches to Sustainable Natural and Human Resource Use*, eds. D. McGregor, D. Simon and D. Thompson. London: Earthscan, pp. 30–43.

Allen, A., Milenic, D. and Sikora, P. (2003). Shallow gravel aquifers and the urban "heat-island" effect: a source of low enthalpy geothermal energy. *Geothermics*, 32, 569–578.

Allenby, B. (2006). *Progress in Industrial Ecology*. London: Interscience Enterprises Ltd.

Alonso, W. (1964). *Location and Land Use*. Cambridge, Massachusetts: Harvard University Press.

Al-Rashid, H. P. and Sherif, M. M. (2001). Hydrogeological aspects of groundwater drainage of the urban areas of Kuwait City. *Hydrological Processes*, 15, 777–795.

Alston, K. P. and Richardson, D. M. (2006). The roles of habitat features, disturbance, and distance from putative source populations in structuring alien plant invasions at the urban/wildland interface on the Cape Peninsula, South Africa. *Biological Conservation*, 132, 183–198.

Altshuler, A. and Luberoff, D. (2003). *Mega-Projects: The Changing Politics of Urban Public Investment*. Washington, D.C.: Brookings Institution Press.

Aluko, G. A. and Husseneder, C. (2007). Colony dynamics of the Formosan subterranean termite in a frequently disturbed urban landscape. *Journal of Economic Entomology*, 100, 1037–1046.

Amaral, J. M. J., Simoes, A. L. and De Jong, D. (2005). *Genetics and Molecular Research*, 4, 832–838.

Ambrose, J. P. and Bratton, S. P. (1990). Trends in landscape heterogeneity along the borders of Great Smoky Mountains National Park. *Conservation Biology*, 4, 135–143.

American Planning Association. (2006). *Planning and Urban Design Standards*. New York: John Wiley.

Amundson, R., Gong, P. and Guo, Y. (2003). Soil diversity and land use in the United States. *Ecosystems*, 6, 472–482.

Anas, A., Arnott, R. and Small, K. (1998). Urban spatial structure. *Journal of Economic Literature*, 36, 1426–1464.

Anderson, D. L. and Otis, R. J. (2000). Integrated wastewater management in growing urban environments. In *Managing Soils in an Urban Environment*, eds. R. B. Brown, J. H. Huddleston and J. L. Anderson. Madison, Wisconsin: American Society of Agronomy, pp. 199–250.

Andreassen, H. P., Ims, R. A. and Steinset, O. K. (1996). Discontinuous habitat corridors: effects on male root vole movement. *Journal of Applied Ecology*, 33, 555–560.

Andrzejewski, A. V. (2009). Building privacy and community: surveillance in a postwar American suburban development in Madison, Wisconsin. *Landscape Journal*, 28, 40–54.

Angilletta, M. J., Wilson, R. S., Niehaus, A. C. *et al.* (2007). Urban physiology: city ants possess high heat tolerance. *PLoS ONE*, 2(2), e258.

Angold, P. G., Sadler, J. P., Hill, M. O. *et al.* (2006). Biodiversity in urban habitat patches. *Science of the Total Environment*, 360, 196–204.

Anikwe, M. A. N. and Nwobodo, K. C. A. (2002). Long term effects of municipal waste disposal on soil properties and productivity of sites used for urban agriculture in Abakaliki, Nigeria. *Bioresource Technology*, 83, 241–250.

Anthony, A., Atwood, J., August, P. *et al.* (2009). Coastal lagoons and climate change: ecological and social ramifications in U.S. Atlantic and Gulf Coast ecosystems. *Ecology and Society*, 14(1): article 8. [Online]. URL: http://www.ecologyandsociety.org/vol14/iss1/art8/.

Antos, M. J., Ehmke, G. C., Tzaros, C. L. and Weston, M. A. (2007). Unauthorized human use of an urban coastal wetland sanctuary: current and future patterns. *Landscape and Urban Planning*, 80, 173–183.

References

Antrop, M. (2006). Sustainable landscapes: contradiction, fiction or utopia? *Landscape and Urban Planning*, 75, 187–197.

Arnold, C. L. and Gibbons, C. J. (1996). Impervious surface coverage: the emergence of a key environmental indicator. *Journal of the American Planning Association*, 62, 243–258.

Arnold, G. W. (1983). The influence of ditch and hedgerow structure, length of hedgerows, and area of woodland and garden on bird numbers in farmland. *Journal of Applied Ecology*, 20, 731–750.

Arnold, R. A. and Goins, A. E. (1987). Habitat enhancement techniques for the El Segundo blue butterfly: an urban endangered species. In *Integrating Man and Nature in the Metropolitan Environment*, eds. L. W. Adams and D. L. Leedy. Columbia, Maryland: National Institute for Urban Wildlife, pp. 173–181.

Asaeda, T. and Ca, V. T. (1998). A case study on the effects of vegetation on the climate in the urban area. In *Urban Ecology*, eds. J. Breuste, H. Feldmann and O. Uhlmann. New York: Springer, pp. 78–81.

Asaeda, T. and Ca, V. T. (2000). Characteristics of permeable pavement during hot summer weather and impact on the thermal environment. *Building and Environment*, 35, 363–375.

Asakawa, S., Yoshida, K. and Yabe, K. (2004). Perceptions of urban stream corridors within the greenway system of Sapporo, Japan. *Landscape and Urban Planning*, 68, 167–182.

Asare Afrane, Y., Klinkenberg, E. and Drechsel, P. (2004). Does irrigated urban agriculture influence the transmission of malaria in the city of Kamasi, Ghana? *Acta Tropica*, 89, 125–134.

Ashley, R. M., Balmforth, D. J., Saul, A. J. and Blanskby, J. D. (2005). Flooding in the future: predicting climate change, risks and responses in urban areas. *Water Science and Technology*, 52, 265–273.

Ashton, W. (2008). Understanding the organization of industrial ecosystems. *Journal of Industrial Ecology*, 12, 34–51.

Atkinson, I. (1989). Introduced animals and extinctions in conservation for the twenty-first century. In *Conservation for the Twenty-First Century*, eds. D. Western and M. C. Pearl. New York: Oxford University Press, pp. 54–75.

Attorre, F., Bruno, M., Francesconi, F. et al. (2000). Landscape changes of Rome through tree-lined roads. *Landscape and Urban Planning*, 49, 115–128.

Attorre, F., Francesconi, F., Pepponi, L. et al. (2003). Spatio-temporal analyses of parks and gardens of Rome. In *Studies in the History of Gardens and Designed Landscapes*, eds. F. Attorre et al. New York: Taylor and Francis, pp. 293–306.

Attorre, F., Rossetti, A., Sbrega, B. and Bruno, F. (1998). Landscape changes in Rome, Italy. *Coenoses*, 13, 57–64.

Attorre, F., Stanisci, A. and Bruno, F. (1997). The urban woods of Rome (Italy). *Plant Biosystems*, 131, 113–135.

Attwell, K. (2000). Urban land resources and urban planning: case studies from Denmark. *Landscape and Urban Planning*, 52, 145–163.

Audouin, J. and Loubiere, A. (1996). *Les Banlieues*. Paris: Hachette Livre.

August, P., Iverson, L. and Nugranad, J. (2002). Human conversion of terrestrial habitats. In *Applying Landscape Ecology in Biological Conservation*, ed. K. J. Gutzwiller. New York: Springer, pp. 198–224.

Austin, M. P. (1987). Models for the analysis of species response to environmental gradients. *Vegetatio*, 69, 35–45.

Austin, M. P. (1999). A silent clash of paradigms: some inconsistencies in community ecology. *Oikos*, 86, 170–178.

Auyero, J. and Swistun, D. A. (2009). *Flammable: Environmental Suffering in an Argentine Shantytown*. New York: Oxford University Press.

Avery, M. and Genchi, A. (2004). Avian perching deterrents on ultrasonic sensors at airport wind-shear alert systems. *Wildlife Society Bulletin*, 32, 718–725.

Avila-Flores, R. and Fenton, M. B. (2005). Use of spatial features by foraging insectivorous bats in a large urban landscape. *Journal of Mammalogy*, 86, 1193–1204.

Ayres, D. R., Smith, D. L., Zaremba, K. et al. (2004). Spread of exotic cordgrasses and hybrids (*Spartina* sp.) in the tidal marshes of San Francisco Bay, California, USA. *Biological Invasions*, 6, 221–231.

Babbitt, B. (2005). *Cities in the Wilderness: A New Vision of Land Use in America*. Washington, D.C.: Island Press.

Baccini, P. (2012). Designing urban systems: ecological strategies with stocks and flows of energy and material. In *Applied Urban Ecology: A Global Framework*, eds. M. Richter and U. Weiland. New York: Wiley-Blackwell, pp. 55–65.

Bagchi, A. (1994). *Design, Construction, and Monitoring of Landfills*. New York: John Wiley.

Baken, R. (2003). *Plotting, Squatting, Public Purpose, and Politics: Land Market Development, Low Income Housing, and Public Intervention in India*. Burlington, Vermont: Ashgate.

Baker, D., Hinton, A. and Luken, J. (1990). Forest edges associated with power-line corridors and implications for corridor siting. *Landscape and Urban Planning*, 20, 315–324.

Baker, H. G. (1974). The evolution of weeds. *Annual Review of Ecology and Systematics*, 5, 1–24.

Baker, L. A., ed. (2009). *The Water Environment of Cities*. New York: Springer.

Baker, P. J., Bentley, A. J., Ansell, R. J. and Harris, S. (2005). Impact of predation by domestic cats *Felis catus* in an urban area. *Mammal Review*, 35, 302–312.

Bakir, H. A. (2001). Sustainable wastewater management for small communities in the Middle East and North Africa. *Journal of Environmental Management*, 61, 319–328.

Ball, P. (2009). *Nature's Patterns: A Tapestry in Three Parts: Flow*. Oxford: Oxford University Press.

Balling, R. C., Carveny, R. S. and Idso, C. D. (2001). Does the urban CO_2 dome of Phoenix, Arizona contribute to its heat island? *Geophysical Research Letters*, 28, 4599–4601.

Balogh, J. C. and Walker, W. J. (1992). *Golf Course Management and Construction: Environmental Issues*. Chelsea, Michigan: Lewis Publishers.

Barber, L., Murphy, S., Verplanck, P. *et al.* (2006). Chemical loading into surface water along a hydrological, biogeochemical, and land use gradient: a holistic watershed approach. *Environmental Science and Technology*, 40, 475–486.

Barnes, H. F. (1949). The slugs in our garden. *New Biologist*, 6, 29–49.

Barras, S. and Seamans, T. W. (2002). *Habitat Management Approaches for Reducing Wildlife Use of Airfields*. Davis, California: University of Califrnia, pp. 309–315.

Barrat, D. G. (1997). Home range size, habitat utilisation and movement patterns of suburban and farm cats (*Felis catus*). *Ecography*, 20, 271–280.

Barrett, G. W. and Barrett, T. L. (2001). Cemeteries as repositories of natural and cultural diversity. *Conservation Biology*, 15, 1820–1824.

Barrett, M. E. (2005). Performance comparison of structural stormwater best management practices. *Water Environment Research*, 77, 78–86.

Barrett, M. H., Hiscock, H. M., Pedley, S. *et al.* (1999). Major species for identifying urban groundwater recharge sources: a review and case study in Nottingham, UK. *Water Research*, 33, 3083–3097.

Bartuska, T. J. and Young, G. L., eds. (1994). *The Built Environment: A Creative Inquiry into Design and Planning*. Menlo Park, California: Crisp Publications.

Baschak, L. A. and Brown, R. D. (1995). An ecological framework for the planning, design and management of urban river greenways. *Landscape and Urban Planning*, 33, 211–225.

Bastian, L. and Thomas, C. D. (1995). Plant metapopulations and conservation in urban habitat fragments. *Land Contamination and Reclamation*, 3, 70–72.

Bastian, L. and Thomas, C. D. (1999). The distribution of plant species in urban vegetation fragments. *Landscape Ecology*, 14, 493–507.

Batty, M. and Longley, P. (1994). *Fractal Cities: A Geometry of Form and Function*. San Diego, California: Academic Press.

Baxter, C. V., Fausch, K. D. and Saunders, W. C. (2005). Tangled webs: reciprocal flows of invertebrate prey link streams and riparian zones. *Freshwater Biology*, 50, 201–220.

Bean, E. Z., Hunt, W. F. and Bidelspach, D. A. (2007). Field survey of permeable pavement surface infiltration rates. *Journal of Irrigation and Drainage Engineering*, 133, 249–255.

Beard, J. S. (1955). The classification of tropical American vegetation-types. *Ecology*, 36, 89–100.

Beatley, T. (1994). *Habitat Conservation Planning*. Austin, Texas: University of Texas Press.

Beatley, T. (2000a). *Green Urbanism: Learning from European Cities*. Washington, D.C.: Island Press.

Beatley, T. (2000b). Preserving biodiversity: challenge for planners. *Journal of the American Planning Association*, 66, 5–20.

Beatley, T., ed. (2012). *Green Cities of Europe*. Washington, D.C.: Island Press.

Beatley, T., Brower, D. J. and Schwab, A. K. (1994). *An Introduction to Coastal Zone Management*. Washington, D.C.: Island Press.

Beckett, K. P., Freer-Smith, P. H. and Taylor, G. (1998). Urban woodlands: their role in reducing the effects of particulate pollution. *Environmental Pollution*, 99, 347–360.

Bednar, M. (2006). *L'Enfant's Legacy: Public Open Spaces in Washington, D.C.* Washington, D.C.: American Institute of Architects Press.

Beissinger, S. and Osborne, D. (1982). Effects of urbanization on avian community organization. *Condor*, 84, 75–83.

Belanger, P. (2007). Underground landscape: the urbanism and infrastructure of Toronto's downtown pedestrian network. *Tunnelling and Underground Space Technology*, 22, 272–292.

Belanger, P. (2009). Landscape as infrastructure. *Landscape Journal*, 28, 79–95.

Belisle, M. and Desrochers, A. (2002). Gap-crossing decisions by forest birds: an empirical basis for parameterizing spatially-explicit, individual based-models. *Landscape Ecology*, 17, 219–231.

Bell, S., Blom, D., Rautamaki, M. *et al.* (2005). Design of urban forests. In *Urban Forests and Trees*, eds. C. C. Konijnendijk, K. Nilsson, T. B. Randrup and J. Schipperijn. New York: Springer, pp. 149–186.

Ben-Joseph, E. (2012). *Rethinking a Lot: The Design and Culture of Parking*. Cambridge, Massachusetts: MIT Press.

References

Benedict, M. A. and McMahon, E. T. (2006). *Green Infrastructure: Linking Landscape and Communities*. Washington, D.C.: Island Press.

Benedikz, T., Ferrini, F., Garcia-Valdecantos, J. L. and Tello, M.-L. (2005). Plant quality and establishment. In *Urban Forests and Trees*, eds. C. C. Konijnendijk, K. Nilsson, T. B. Randrup and J. Schipperijn. New York: Springer, pp. 231–256.

Benfield, F. K., Raimi, M. D. and Chen, D. D. T. (1999). *Once There Were Greenfields: How Urban Sprawl is Undermining America's Environment, Economy and Social Fabric*. Washington, D.C.: Natural Resources Defense Council.

Benfield, F. K., Terris, J. and Vorsanger, N. (2001). *Solving Sprawl: Models of Smart Growth in Communities Across America*. Washington, D.C.: Natural Resources Defense Council.

Bengston, D. N. and Youn, Y. C. (2006). Urban containment policies and the protection of natural areas: the case of Seoul's greenbelt. *Ecology and Society*, 11 (1): article 3 [Online]. URL: http://www.ecologyandsociety.org/vol11/iss1/art3/.

Bengtsson, L., Grahn, L. and Olsson, J. (2005). Hydrological function of a thin extensive green roof in southern Sweden. *Nordic Hydrology*, 36, 259–268.

Benjamin, T. S., Green, A. and Deshais, K. (2003). New light on the Neponset near Boston, a public/private partnership "daylights" a stretch of river. *Landscape Architecture*, 93(4), 46–54.

Bennett, A. F. (2003). *Linkages in the Landscape: The Role of Corridors and Connectivity in Wildlife Conservation*. Gland, Switzerland and Cambridge, UK: IUCN, The World Conservation Union.

Bennett, A. F. and Radford, J. (2003). Know your ecological thresholds. In *Thinking Bush*. Melbourne, Victoria: State of Victoria, Native Vegetation Research and Development Program.

Bennett, A. F., Radford, J. Q. and Haslem, A. (2006). Properties of land mosaics: implications for nature conservation in agricultural environments. *Biological Conservation*, 133, 250–264.

Bennett, G. W. and Owens, J. M., eds. (1986). *Advances in Urban Pest Management*. New York: Van Nostrand Reinhold.

Bennett, H. H. (1948). Soil conservation in a hungry world. *Geographical Review*, 38, 311–317.

Benson, D. H. and Howell, J. (1990). Sydney's vegetation 1788–1988: utilisation, degradation, and rehabilitation. *Proceedings of the Ecological Society of Australia*, 16, 115–127.

Benton-Short, L. and Short, J. R. (2008). *Cities and Nature*. London: Routledge.

Benyus, J. (2002). *Biomimicry: Innovation Inspired by Nature*. New York: Harper Perennial.

Benyus, J. (2008). A good place to settle: biomimicry, biophilia, and the return. In *Biophilic Design: The Theory, Science, and Practice of Bringing Buildings to Life*, eds. S. R. Kellert, J. H. Heerwagen and M. L. Mador. New York: John Wiley, pp. 27–42.

Berger, A. (2006). *Drosscape: Wasting Land in Urban America*. New York: Princeton Architectural Press.

Berger, M. (2004). *Les periurbains de Paris: De la ville dense a la metropole eclatee?* Paris: CNRS Editions.

Berke, P. R., Godschalk, D. R., Kaiser, E. J. and Rodriguez, D. A. (2006). *Urban Land Use Planning*. Urbana, Illinois: University of Illinois Press.

Berkowitz, A. R., Nilon, C. H. and Holweg, K. S., eds. (2003). *Understanding Urban Ecosystems: A New Frontier for Science and Education*. New York: Springer-Verlag.

Berland, A. (2012). Long-term urbanization effects on tree canopy cover along an urban-rural gradient. *Urban Ecosystems*, 15, 721–738.

Berling-Wolff, S. and Wu, J. (2004). Modeling urban landscape dynamics: a review. *Ecological Research*, 19, 119–129.

Berman, M. and Dunbar, I. (1983). The social behaviour of free-ranging suburban dogs. *Applied Animal Ethology*, 10, 5–17.

Bernhardt, E. S. and Palmer, M. (2007). Restoring streams in an urbanizing world. *Freshwater Biology*, 52, 738–751.

Bernhardt, E. S., Band, L. E., Walsh, C. and Berke, P. (2008). Understanding, managing, and minimizing urban impacts of surface water nitrogen loading. *Annual Review of Environment and Resources*, 33, 61–96.

Bernhardt, G. E. (2009). Management of bayberry in relation to tree-swallow strikes at John F. Kennedy International Airport, New York. *Human-Wildlife Conflicts*, 3, 237–241.

Berrizbeitia, A. (1999). The Amsterdam Bos: the modern public park and the construction of collective experience. In *Recovering Landscapes*, ed. J. Corner. New York: Princeton Architectural Press, pp. 187–203.

Berry, B. J. L. (1990). Urbanization. In *The Earth as Transformed by Human Actions: Global and Regional Changes in the Biosphere Over the Past 300 Years*, eds. B. L. Turner, W. C. Clark, R. W. Kates *et al*. New York: Cambridge University Press, pp. 103–119.

Bertasi, F. (2007). Effects of an artificial protection structure on the sandy shore macrofaunal community: the special case of Lido di Dante (Northern Adriatic Sea). *Hydrobiologia*, 586, 277–290.

Berthier, E., Dupont, S., Mestayer, P. G. and Andrieu, H. (2006). Comparison of two evapotranspiration schemes on a sub-urban site. *Journal of Hydrology*, 328, 635–646.

Bertin, R. I., Manner, M. E., Larrow, B. F. *et al*. (2005). Norway maple (*Acer platanoides*) and other non-native trees in urban woodlands of Central Massachusetts. *Journal of the Torrey Botanical Society*, 132, 225–235.

Bevanger, K. (1995). Estimates and population consequences of Tetraonid mortality caused by collisions with high tension power lines in Norway. *Journal of Applied Ecology*, 32, 745–753.

Bhattacharya, M., Primack, R. B. and Gerwein, J. (2003). Are roads and railroads barriers to bumblebee movement in a temperate suburban conservation area? *Biological Conservation*, 109, 37–45.

Bianca, S. (2000). *Urban Form in the Arab World: Past and Present*. London: Thames & Hudson.

Biggs, A. (1990). *The Impact of Airport Noise: A Case Study of Vancouver International Airport*. Vancouver, British Columbia: University of British Columbia.

Biggs, T. W., Atkinson, E., Powell, R. and Ojeda-Revah, L. (2010). Land cover following rapid urbanization on the US-Mexico border: implications for conceptual models of urban watershed proceses. *Landscape and Urban Planning*, 96, 78–87.

Binford, M. W. and Karty, R. (2006). Riparian greenways and water resources. In *Designing Greenways: Sustainable Landscapes for Nature and People*, eds. P. C. Hellmund and D. A. Smith. Washington, D.C.: Island Press, pp. 108–157.

Binns, T. and Maconachie, R. (2006). Re-evaluating people-environment relationships at the rural-urban interface: how sustainable is the peri-urban zone in Kano, Northern Nigeria? In *The Peri-Urban Interface: Approaches to Sustainable Natural and Human Resource Use*, eds. D. McGregor, D. Simon and D. Thompson. London: Earthscan, pp. 211–228.

Bird, D., Varland, D. and Negro, J., eds. (1996). *Raptors in Human Landscapes: Adaptations to Built and Cultivated Environments*. San Diego, California: Academic Press.

Biro, Z., Lanszki, J., Szemethy, L. et al. (2005). Feeding habits of feral domestic cats (*Felis catus*), wild cats (*Felis silvestris*) and their hybrids: trophic niche overlap among cat groups in Hungary. *Journal of Zoology*, 266, 187–196.

Blair, R. B. (1996). Land use and avian species diversity along an urban gradient. *Ecological Applications*, 6, 506–519.

Blair, R. B. (1999). Birds and butterflies along an urban gradient: surrogate taxa for assessing biodiversity? *Ecological Applications*, 9, 164–170.

Blair, R. B. (2008). Creating a homogeneous avifauna. In *Urban Ecology: An International Perspective on the Interaction Between Humans and Nature*, eds. J. M. Marzluff, E. Schulenberger, W. Endlicher et al. New York: Springer, pp. 405–424.

Blair, R. B. and Launer, A. E. (1997). Butterfly diversity and human land use: species assemblages along an urban gradient. *Biological Conservation*, 80, 113–125.

Blanc, P. (2008). The vertical garden. In *Nature and the City*. New York: Norton.

Blomqvist, M. M., Vos, P., Klinkhamer, P. G. L. and ter Keurs, W. J. (2003). Declining plant species richness of grassland ditch banks – a problem of colonisation or extinction? *Biological Conservation*, 109, 391–406.

Bloom, P. H. and McCrary, M. D. (1996). The urban buteo: red-shouldered hawks in Southern California. In *Raptors in Human Landscapes: Adaptations to Built and Cultivated Environments*, eds. D. M. Bird, D. E. Varland and J. J. Negro. San Diego: Academic Press, pp. 31–39.

Blume, H.-P. (1982). Boden des Verdichtungsraumes Berlin. *Mitteilungen der Deutschen Bodenkundlichen Gesellschaft*, 33, 269–280.

Blume, H.-P. (2009). *Lehrbuch der Bodenkunde*. Berlin: Spektrum Verlag.

Boada, M. and Capdevila, L. (2000). *Barcelona Biodiversitat urbana*. Barcelona: Ajuntament de Barcelona.

Boeri, S. and Insulata, F. (2009/10). The vertical forest and new urban comfort. *Harvard Design Magazine*, 31, 60–65.

Bogo, J., Klinkenberg, J., Bender, K. and Dole, C. H. (2002). Going the extra yard. *Audubon*, 104(2), 64–71.

Bohl, C. C. (2002). *Place Making: Developing Town Centers, Main Streets, and Urban Villages*. Washington, D.C.: Urban Land Institute.

Bolen, E. G. (2000). Wildlife as related to soils. In *Managing Soils in an Urban Environment*, eds. R. B. Brown, J. H. Huddleston and J. L. Anderson. Madison, Wisconsin: American Society of Agronomy, pp. 157–177.

Bolund, P. and Hunhammer, S. (1999). Ecosystem services in urban areas. *Ecological Economics*, 29, 293–301.

Bonter, D. N., Gauthreaux, S. A., Jr. and Donovan, T. M. (2009). Characteristics of important stopover locations for migrating birds: remote sensing with radar in the Great Lakes Basin. *Conservation Biology*, 23, 440–448.

Boone, M. D., Semlitsch, R. D. and Mosby, C. (2005). Suitability of golf course ponds for amphibian metamorphosis when bullfrogs are removed. *Conservation Biology*, 22, 172–179.

Booth, D. B. and Bledsoe, B. P. (2009). Streams and ubanization. In *The Water Environment of Cities*, ed. L. A. Baker. New York: Springer, pp. 93–123.

Booth, D. B. and Jackson, C. R. (1997). Urbanization of aquatic systems: degradation threshold stormwater detection and the limits of mitigation. *Journal of the American Water Resources Association*, 33, 1077–1090.

Booth, D. B. and Leavitt, J. (1999). Field evaluation of permeable pavement systems for improved stormwater management. *Journal of the American Planning Association*, 65, 316–324.

Borchert, J. R. (1993). *Megalopolis: Washington, D.C. to Boston*. New Brunswick, New Jersey: Rutgers University Press.

References

Bormann, F. H., Balmori, D. and Geballe, G. T. (1993). *Redesigning the American Lawn: A Search for Environmental Harmony*. New Haven, Connecticut: Yale University Press.

Bossel, H. (1998). *Earth at a Crossroads: Paths to a Sustainable Future*. Cambridge: Cambridge University Press.

Bosselman, P. (2008). *Land Transformation: Understanding City Design and Form*. Washington, D.C.: Island Press.

Botham, M., Rothery, P., Hulme, P. et al. (2009). Do urban areas act as foci for the spread of alien plant species? An assessment of temporal trends in the UK. *Diversity and Distributions*, 15, 338–345.

Bowen, J. L. and Valiela, I. (2001). The ecological effects of urbanization of coastal watersheds: historical increases in nitrogen loads and eutrophication of Waquoit Bay estuaries. *Canadian Journal of Fisheries and Aquatic Sciences*, 58, 1489–1500.

Bowers, M. A. and Breland, B. (1996). Foraging of gray squirrels on an urban-rural gradient: use of the gradient to assess anthropogenic impact. *Ecological Applications*, 6, 1135–1142.

Box, J. (2011). Building urban biodiversity through financial incentives, regulation, and targets. In *Urban Ecology: Patterns, Processes, and Applications*, eds. J. Niemela, J. H. Breuste, T. Elmqvist et al. New York: Oxford University Press, pp. 309–318.

Boyden, S., Millar, S., Newcombe, K. and O'Neill, B. (1981). *The Ecology of a City and its People: The Case of Hong Kong*. Canberra: Australian National University Press.

Brack, C. (2002). Pollution mitigation and carbon sequestration by an urban forest. *Environmental Pollution*, 116, 195–200.

Bradley, G. A., ed. (1995). *Urban Forest Landscapes: Integrating Multidisciplinary Perspectives*. Seattle, Washington: University of Washington Press.

Bradley, M. P. and Stolt, M. H. (2006). Landscape-level seagrass-sediment relations in a coastal lagoon. *Aquatic Botany*, 84, 121–128.

Bradshaw, A. D. (2003). Natural ecosystems in cities: a model for cities as ecosystems. In *Understanding Urban Ecosystems: A New Frontier for Science and Education*, eds. A. R. Berkowitz, C. H. Nilon and K. S. Hollweg. New York: Springer, pp. 77–94.

Bradshaw, A., Hunt, B. and Walmsley, T. (1995). *Trees in the Urban Landscape: Principles and Practice*. London: E&FN Spon.

Brady, N. C. and Weil, R. R. (2002). *The Nature and Properties of Soils*. Englewood Cliffs, New Jersey: Prentice-Hall.

Braithwaite, L. (1976). *Canals in Town*. London: Adam and Charles Black.

Brande, A., Bockeer, R. and Graf, A. (1990). Changes of flora, vegetation and urban biotopes in Berlin (West). In *Urban Ecology: Plants and Plant Communities in Urban Environments*, eds. H. Sukopp, S. Hejny and I. Kowarik. The Hague, Netherlands: SPB Academic Publishing, pp. 155–165.

Brandle, J. R., Hintz, D. L. and Sturrock, J. W., eds. (1988). *Windbreak Technology*. Amsterdam: Elsevier. [Reprinted from *Agriculture, Ecosystems and Environment* 22–23 (1988)].

Brandt, J., Vejre, H., Mander, U. and Antrop, M. (2003–2004) *Multifunctional Landscapes*. Southampton, UK: WIT.

Brantz, D. and Duempelmann, S. (2011). *Greening the City: Urban Landscapes in the Twentieth Century*. Charlottesville, Virginia: University of Virginia Press.

Braskerud, B. C., Tonderski, K. S., Wedding, B. et al. (2005). Can constructed wetlands reduce the diffuse phosphorus loads to eutrophic water in cold temperate regions? *Journal of Environmental Quality*, 34, 2145–2155.

Brattlebo, B. O. and Booth, D. B. (2003). Long-term stormwater quantity and quality performance of permeable pavement systems. *Water Research*, 37, 4369–4376.

Brauman, K. A., Daily, G. C., Duarte, A. K. and Mooney, H. A. (2007). The nature and value of ecosystem services: an overview highlighting hydrological services. *Annual Review of Environment and Resources*, 32, 67–98.

Braun Blanquet, J. (1964). *Pflanzensoziologie Grundzuge der Vegetationskunde*. Vienna: Springer-Verlag.

Brazel, A. J. and Heisler, G. M. (2009). Climatology of urban long-term ecological research sites: Baltimore Ecosystem Study and Central Arizona-Phoenix. *Geography Compass*, 3, 22–44.

Brazel, A., Selover, N., Voce, R. and Heisler, G. (2000). The tale of two climates – Baltimore and Phoenix LTER sites. *Climate Research*, 15, 123–135.

Brazel, A. J., Chow, W. T. L., Hedquist, B. C. and Prashad, L. (2009). Desert heat island study in winter by mobile transect and remote sensing techniques. *Theoretical and Applied Climatology*, 98, 323–325.

Brearley, G., McAlpine, C., Bell, S. and Bradley, A. (2012). Influence of urban edges on stress in an arboreal mammal: a case study of squirrel gliders in southeast Queensland, Australia. *Landscape Ecology*, 27, 1407–1419.

Breed, A. C., Field, H. E., Epstein, J. H. and Daszak, P. (2006). Infectious disease and mammalian conservation. *Biological Conservation*, 131, 211–220.

Breen, A. and Rigby, D. (1996). *The New Waterfront: A Worldwide Urban Success Story*. New York: McGraw-Hill.

Brenneisen, S. and Haenggi, A. (2006). Green roofs – ecological characterization of a new urban habitat type based on a comparison between the spider fauna found on green roofs and in railway sidings with high conservation interest in Basel (Switzerland). *Mitteilungen der Naturforschenden Gesellschaften beider Basel*, 9, 99–122.

Brenner, M., Schelske, C. L. and Keenan, L. W. (2001). Historical rates of sediment and nutrient accumulation in marshes of the Upper St. Johns River Basin, Florida, USA. *Journal of Paleolimnology*, 26, 241–257.

Brenner, M., Whitmore, T. J., Riedinger-Whitmore, M. A. et al. (2006). Geochemical and biological consequences of groundwater augmentation in lakes of west-central Florida (USA). *Journal of Paleolimnology*, 36, 371–383.

Breuste, J. H. (2004). Decision making, planning and design for the conservation of indigenous vegetation within urban development. *Landscape and Urban Planning*, 68, 439–452.

Breuste, J. H. (2009). Structural analysis of urban landscapes for landscape management in German cities. In *Ecology of Cities and Towns: A Comparative Approach*, eds. M. J. McDonnell, A. K. Hahs and J. H. Breuste. New York: Cambridge University Press, pp. 355–379.

Breuste, J., Feldmann, H. and Uhlmann, O., eds. (1998). *Urban Ecology*. New York: Springer.

Brick, G. (2009). *Subterranean Twin Cities*. Minneapolis, Minnesota: University of Minnesota Press.

Brickell, C., ed. (2003). *The American Horticultural Society Encyclopedia of Gardening*. New York: DK Publishing.

Bridgewater, P. B. (1990). The role of synthetic vegetation in present and future landscapes of Australia. *Proceedings of the Ecological Society of Australia*, 16, 129–134.

Brigham, R. M., Cebek, J. E. and Hickey, B. C. (1989). Intraspecific variation in the echolocation calls of two species of insectivorous bats. *Journal of Mammalogy*, 70, 426–428.

Brillemburg, A., Feireiss, K. and Klumpner, H. (2005). *Informal city: Caracas case*. Munich: Prestel.

Broll, G. and Keplin, B. (1995). Ecological studies on urban lawns. In *Urban Ecology as the Basis of Urban Planning*, eds. H. Sukopp, M. Numata and A. Huber. Amsterdam: SPB Academic Publishing, pp. 71–82.

Brooks, A. (1998). *Channelized Rivers: Perspectives for Environmental Management*. New York: John Wiley.

Browder, J. O. and Godfrey, B. J. (1997). *Rainforest Cities: Urbanization, Development, and Globalization of the Brazilian Amazon*. New York: Columbia University Press.

Browder, J. O., Bohland, J. R. and Scarpaci, J. L. (1995). Patterns of development on the metropolitan fringe: urban fringe expansion in Bangkok, Jakarta, and Santiago. *Journal of the American Planning Association*, 61, 310–327.

Brown, D. B. (2002). The science of brownfields. In *Brownfields: A Comprehensive Guide to Redeveloping Contaminated Property*. Chicago: American Bar Association Publishing, pp. 265–280.

Brown, K. (2000). *Urban Stream Restoration Practices: An Initial Assessment*. Final Report for Office of Wetlands, US Environmental Protection Agency. Ellicott City, Maryland: Center for Watershed Protection.

Brown, K. H. and Jameton, A. L. (2000). Public health implications of urban agriculture. *Journal of Public Health Policy*, 21, 20–39.

Brown, L. R. (1995). *Who Will Feed China? Wake-Up Call for a Small Planet*. New York: W. W. Norton.

Brown, L. R. (2005). *Effects of Urbanization on Stream Ecosystems*. Bethesda, Maryland: American Fisheries Society.

Brown, P. H. (2009). *America's Waterfront Revival: Port Authorities and Urban Redevelopment*. Philadelphia: University of Pennsylvania Press.

Brown, R. B., Huddleston, J. H. and Anderson, J. L., eds. (2000). *Managing Soils in an Urban Environment*. Madison, Wisconsin: American Society of Agronomy.

Brown, W. M. and Drewien, R. C. (1995). Evaluation of two power line markers to reduce crane and waterfowl collision mortality. *Wildlife Society Bulletin*, 23, 217–227.

Brownlee, P. M. and Lorna, J. (2005). Road salts and birds: an assessment of the risk with particular emphasis on winter finch mortality. *Wildlife Society Bulletin*, 33, 835–841.

Brumm, H. (2004). The impact of environmental noise on song amplitude in a territorial bird. *Journal of Animal Ecology*, 73, 434–440.

Brunner, P. H. (2007). Reshaping urban metabolism. *Journal of Industrial Ecology*, 1, 3–19.

Bruno, F., Provantini, R., Francesconi, F. et al. (2006). *Roma Giardino d'Europa: Tipologie de Verde, metodi di analisi e strumenti di gestione*. Rome: Assessorato alle Politiche Ambientali e Agricole di Roma.

Bryant, B. K. and Ismael, W. (1991). Movement and mortality patterns of resident and translocated suburban white-tailed deer. In *Wildlife Conservation in Metropolitan Environments*, eds. L. W. Adams and D. L. Leedy. Columbia, Maryland: National Institute for Urban Wildlife, pp. 53–58.

Bryant, M. M. (2006). Urban landscape conservation and the role of ecological greenways at local and metropolitan scales. *Landscape and Urban Planning*, 76, 23–44.

Buell, L. (2001). *Writing for an Endangered World: Literature, Culture, and Environment in the United States and Beyond*. Cambridge, Massachusetts: Harvard University Press.

References

Buell, L. (2005). *The Future of Environmental Criticism*. Oxford: Blackwell Publishing.

Bulleri, F., Chapman, M. G. and Underwood, A. J. (2004). Patterns of movement of the limpet *Cellana tramoserica* on rocky shores and retaining seawalls. *Marine Ecology Progress Series*, 281, 121–129.

Burchell, R. W., Downs, A., McCann, B. and Mukherji, S. (2005). *Sprawl Costs: Economic Impacts of Unchecked Development*. Washington, D.C.: Island Press.

Burel, F. and Baudry, J. (1999). *Ecologie du paysage: Concepts, methods et applications*. Paris: Technique & Documentation.

Burton, M. L., Samuelson, L. J. and Pan, S. (2005). Riparian woody plant and forest structure along an urban-rural gradient. *Urban Ecosystems*, 8, 93–106.

Busquets, J. (2003). *The Old Town of Barcelona: A Past with a Future*. Barcelona: Ajuntament de Barcelona.

Butler, D. and Davies, J. W. (2011). *Urban Drainage*. London: Spon Press.

Butler, R. B. (1981). *The Ecological House*. New York: Morgan and Morgan.

Byers, D. L. and Mitchell, J. C. (2005). Sprawl and species with limited dispersal abilities. In *Nature in Fragments: The Legacy of Sprawl*, eds. E. A. Johnson and M. W. Klemens. New York: Columbia University Press, pp. 157–180.

Cade, T. J., Martell, M., Redig, P. *et al.* (1996). Peregrine falcons in urban North America. In *Raptors in Human Landscapes*, eds. D. M. Bird, D. E. Varland and J. J. Negro. San Diego: Academic Press, pp. 3–13.

Cadenasso, M. L., Pickett, S. T. A. and Schwarz, K. (2007). Spatial heterogeneity in urban ecosystems: reconceptualizing land cover and a framework for classification. *Frontiers in Ecology and Evolution*, 5, 80–88.

Cadenasso, M. L., Pickett, S. T. A. and Schwarz, K. (2007). Spatial heterogeneity in urban ecosystems: reconceptualizing land cover and a framework for classification. *Frontiers in Ecology and Evolution*, 5, 80–88.

Cadenasso, M. L., Pickett, S. T. A., Weathers, K. C. and Jones, C. G. (2003). A framework for a theory of ecological boundaries. *BioScience*. 53, 750–758.

Cain, M. L., Bowman, W. D., and Hacker, S. D. (2011). *Ecology*. Sunderland, Massachusetts: Sinauer Associates.

Cairns, J. (1995). *Rehabilitating Damaged Ecosystems*. Boca Raton, Florida: CRC Press.

Caldiron, G. (2005). *Banlieue: Vita e rivolta nelle periferie della metropolis*. Rome: Tomacelli.

Callaway, J. C. and Zedler, J. B. (2004). Restoration of urban salt marshes: lessons from Southern California. *Urban Ecosystems*, 7, 107–124.

Calthorpe, P. and Fulton, P. (2001). *The Regional City: Planning for the End of Sprawl*. Washington, D.C.: Island Press.

Campbell, C. and Ogden, M. (1999). *Constructed Wetlands in the Sustainable Landscape*. New York: John Wiley.

Campos, C. B., Esteves, C. F., Ferraz, K. M. P. *et al.* (2007). Diet of free-ranging cats and dogs in a suburban and rural environment, south-eastern Brazil. *Journal of Zoology*, 273, 14–20.

Cano-Hurtado, J. J. and Canto-Perello, J. (1999). Sustainable development of urban underground space for utilities. *Tunnelling and Underground Space Technology*, 14, 335–340.

Cantwell, M. D. and Forman, R. T. T. (1994). Landscape graphs: ecological modeling with graph theory to detect configurations common to diverse landscapes. *Landscape Ecology*, 8, 233–255.

Carey, M. (1998). Peregrine falcons and the Washington State Department of Transportation. In *Proceedings of the International Conference on Wildlife Ecology and Transportation*. Tallahassee, Florida: Florida Department of Transportation, pp. 121–125.

Carinanos, P. and Casares-Porcel, M. (2011). Urban green zones and related pollen allergy: a review. Some guidelines for designing spaces with low allergy impact. *Landscape and Urban Planning*, 101, 205–214.

Carleton, J., Grissard, T. J., Godrej, A. N. *et al.* (2000). Performance of constructed wetlands in treating urban stormwater runoff. *Water Environment Research*, 72, 295–304.

Carolsfeld, J., ed. (2003). *Migratory Fishes of South America: Biology, Fisheries and Conservation Status*. Victoria, British Columbia: World Fisheries Trust.

Carpenter, D., Sinha, S. K., Brennan, K. and Slate, L. O. (2003). Urban stream restoration. *Journal of Hydraulic Engineering*, 129, 491–494.

Carpenter, R. A., ed. (1983). *Natural Systems for Development: What Planners Need to Know*. New York: Macmillan.

Carpenter, T. G. (1994). *The Environmental Impact of Railways*. New York: John Wiley.

Carreiro, M. M. (2008). Using the urban-rural gradient approach to determine the effects of land use on forest remnants. In *Ecology, Planning, and Management of Urban Forests: International Perspectives*, eds. M. M. Carreiro, Y.-C. Song and J. Wu. New York: Springer, pp. 169–186.

Carreiro, M. M., Pouyat, R. V., Tripler, C. E. and Zhu, W.-X. (2009). Carbon and nitrogen cycling in soils in remnant forests along urban-rural gradients: case studies in the New York metropolitan area and Louisville, Kentucky. In *Ecology of Cities and Towns: A Comparative Approach*, eds. M. J. McDonnell, A. K. Hahs and J. H. Breuste. New York: Cambridge University Press, pp. 308–328.

Carreiro, M., Song, Y.-C. and Wu, J., eds. (2008). *Ecology, Planning, and Management of Urban Forests*. New York: Springer.

Carsjens, G. and van der Knaap, W. (2002). Strategic land-use allocation: dealing with spatial relationships and fragmentation of agriculture. *Landscape and Urban Planning*, 58, 171–179.

Carson, R. (1962). *Silent Spring*. New York: Houghton Mifflin.

Carter, T. L. and Rasmussen, T. C. (2006). Hydrologic behavior of vegetated roofs. *Journal of the American Water Resources Association*, 42, 1261–1274.

Castells, M. (2000). *The Rise of the Network Society*. Oxford: Blackwell Publishers.

Catalan, B., Sauri, D. and Serra, P. (2008). Urban sprawl in the Mediterranean? Patterns of growth and change in the Barcelona Metropolitan Region 1993–2000. *Landscape and Urban Planning*, 85, 174–184.

Catterall, C. P. (2009). Responses of faunal assemblages to urbanisation: global research paradigms and an avian case study. In *Ecology of Cities and Towns: A Comparative Approach*, eds. M. J. McDonnell, A. K. Hahs and J. H. Breuste. New York: Cambridge University Press, pp. 129–155.

Catton, W., Jr. and Dunlap, R. E. (1978). Environmental sociology: a new paradigm. *The American Sociologist*, 13, 41–49.

Cederlunda, H., Thierfelderb, T. and Stenstroma, J. (2008). Functional microbial diversity of the railway track bed. *Science of the Total Environment*, 397, 205–214.

Celesti-Grapow, L. and Blasi, C. (1998). A comparison of the urban flora of different phytoclimatic regions in Italy. *Global Ecology and Biogeography*, 7, 367–378.

Cervero, R. (1998). *The Transit Metropolis: A Global Inquiry*. Washington, D.C.: Island Press.

Chace, J. F. and Walsh, J. J. (2006). Urban effects on native avifauna: a review. *Landscape and Urban Planning*, 74, 46–69.

Chace, J. F., Walsh, J. J., Cruz, A. et al. (2003). Spatial and temporal activity patterns of the brood parasitic brown-headed cowbird at an urban/wildland interface. *Landscape and Urban Planning*, 64, 179–190.

Chang, C. R., Li, M. H. and Chang, S. D. (2007). A preliminary study of the local cool-island intensity of Taipei city parks. *Landscape and Urban Planning*, 80, 386–395.

Chapman, M. G. and Underwood, A. J. (2009). Comparative effects of urbanisation in marine and terrestrial habitats. In *Ecology of Cities and Towns: A Comparative Approach*, eds. M. J. McDonnell, A. K. Hahs and J. H. Breuste. New York: Cambridge University Press, pp. 51–70.

Chapman, M. G., Blockley, D., People, J. and Clynick, B. (2009). Effect of urban structures on diversity of marine species. In *Ecology of Cities and Towns: A Comparative Approach*, eds. M. J. McDonnell, A. K. Hahs and J. H. Breuste. New York: Cambridge University Press, pp. 156–176.

Charles, J. A. and Barton, M. (2003). *The Mythical World of Transit-Oriented Development*. Portland, Oregon: Cascade Policy Institute.

Chaskin, R. J. (1995). *Defining Neighborhood: History, Theory and Practice*. Chicago: The Chapin Center for Children, University of Chicago.

Chasko, G. G. and Gates, J. E. (1982). Avian habitat suitability along a transmission-line corridor in an oak-hickory forest region. *Wildlife Monographs*, 82, 1–41.

Chatti, N., Ganem, G., Benzekri, K. et al. (1999). Microgeographical distribution of two chromosomal races of house mice in Tunisia: pattern and origin of habitat partitioning. *Proceedings of the Royal Society of London, Series B-Biological Sciences*, 266, 1561–1569.

Chen, T. C., Wang, S. Y. and Yen, M. C. (2007). Enhancement of afternoon thunderstorm activity by urbanization in a valley: Taipei. *Journal of Applied Meteorology and Climatology*, 46, 1324–1340.

Chen, W. Y. and Jim, C. Y. (2008). Assessment and valuation of the ecosystem services provided by urban forests. In *Ecology, Planning, and Management of Urban Forests: International Perspectives*, eds. M. M. Carreiro, Y.-C. Song and J. Wu. New York: Springer, pp. 53–83.

Cheptou, P.-O., Carrue, O., Rouifed, S. and Cantarel, A. (2008). Rapid evolution of seed dispersal in an urban environment in the weed *Crepis sancta*. *Proceedings of the National Academy of Sciences* (USA), 105, 3796–3799.

Chesterikoff, A., Garban, B., Billen, G. and Poulin, M. (1992). Inorganic nitrogen dynamics in the River Seine downstream from Paris (France). *Biogeochemistry*, 17, 147–164.

Chiesura, A. (2004). The role of urban parks for the sustainable city. *Landscape and Urban Planning*, 68, 129–138.

Childs, J. E. (1986). Size dependence predation on rats (*Rattus norvegicus*) by house cats (*Felis catus*) in an urban setting. *Journal of Mammalogy*, 67, 196–199.

Chin, A. (2006). Urban transformation of river landscapes in a global context. *Geomorphology*, 79, 460–487.

Christaller, W. (1933). *Die Zentralen Orte in Suddeutschland*. Jena, Germany, Gustav Fischer.

Christie, F. J. and Hochuli, D. F. (2005). Elevated levels of herbivory in urban landscapes: are declines in tree health more than an edge effect? *Ecology and Society*, 10(10).

Churcher, P. B. and Lawton, J. H. (1989). Beware of well-fed felines: Britain's five million house cats enjoy both indoor comforts and outdoor hunting. *Natural History*, 7, 40–46.

Cilliers, S. S. and Bredenkamp, G. J. (1998). Vegetation analysis of railway reserves in the Potchefstroom municipal area, North West Province, South Africa. *South African Journal of Botany*, 64, 271–280.

Cilliers, S. S. and Siebert, S. J. (2011). Urban flora and vegetation: patterns and processes. In *Urban Ecology: Patterns, Processes, and Applications*, eds. J. Niemela, J. H. Breuste, T. Elmqvist et al. New York: Oxford University Press, pp. 148–158.

Cilliers, S. S., Williams, N. S. G. and Barnard, F. J. (2008). Patterns of exotic plant invasions in fragmented urban and rural grasslands across continents. *Landscape Ecology*, 23, 1243–1256.

Clark, J. R. (2000). *Coastal Zone Management Handbook*. Boca Raton, Florida: CRC Press.

Clark, W. C. and Dickson, N. M. (2003). Sustainability science: the emerging research program. *Proceedings of the National Academy of Science (USA)*, 100, 8059–8061.

Clement, A. and Thomas, G. (2001). *Atlas du Paris souterrain: La doublure sombre de la Ville lumiere*. Paris: Editions Parigramme.

Clergeau, P., Savard, J.-P. L., Mennechez, G and Falardeau, G. (1998). Bird abundance and diversity along an urban-rural gradient: a comparative study between two cities on different continents. *Condor*, 100, 413–425.

Clevenger, A. P., Chruszcz, B. and Gunson, K. (2001). Drainage culverts as habitat linkages and factors affecting passage by mammals. *Journal of Applied Ecology*, 38, 1340–1349.

Clucas, B. and Marzluff, J. M. (2011). Coupled relationships between humans and other organisms in urban areas. In *Urban Ecology: Patterns, Processes, and Applications*, eds. J. Niemela, J. H. Breuste, T. Elmqvist et al. New York: Oxford University Press, pp. 135–147.

Clynick, B. G. (2008). Characteristics of an urban fish assemblage: distribution of fish associated with coastal marinas. *Marine Environmental Research*, 65, 18–33.

Cohen-Rosenthal, E. and Musnikow, J., eds. (2003). *Eco-Industrial Strategies*. Sheffield, UK: Greenleaf Publishing.

Cohn, J. P. and Lerner, J. A. (2003). *Integrating Land Use Planning and Biodiversity*. Washington, D.C.: Defenders of Wildlife.

Colburn, E. A. (2004). *Vernal Pools: Natural History and Conservation*. Blacksburg, Virginia: McDonald & Woodward Publishing.

Colding, J. and Folke, C. (2009). The role of golf courses in biodiversity conservation and ecosystem management. *Ecosystems*, 12, 191–206.

Colding, J., Lundberg, J. and Folke, C. (2006). Incorporating green-area user groups in urban ecosystem management. *Ambio*, 35, 237–244.

Cole, D. N. (1993). Riparian greenways and water resources. In *Ecology of Greenways: Design and Function of Linear Conservation Areas*, eds. D. S. Smith and P. C. Hellmund. Minneapolis, Minnesota: University of Minnesota Press, pp. 105–122.

Collinge, S. K. (1998). Spatial arrangement of habitat patches and corridors: clues from ecological field experiments. *Landscape and Urban Planning*, 42, 157–168.

Collinge, S. K. (2009). *Ecology of Fragmented Landscapes*. Baltimore, Maryland: Johns Hopkins University Press.

Collinge, S. K. and Forman, R. T. T. (1998). A conceptual model of land conversion processes: predictions and evidence from a microlandscape experiment with grassland insects. *Oikos*, 82, 66–84.

Collinge, S. K., Johnson, W. C., Ray, C. et al. (2005). Landscape structure and plague occurrence in black-tailed prairie dogs on grasslands of the Western USA. *Landscape Ecology*, 20, 941–955.

Collinge, S. K., Prudic, K. L. and Oliver, J. C. (2003). Effects of local habitat characteristics and landscape context on grassland butterfly diversity. *Conservation Biology*, 17, 178–187.

Compagnie National du Rhone. (1996). *La liaison fluviale Saone-Rhin*. Document d'information, Societe Pour la Realisation de la Liaison Fluvial Saone-Rhin, et CETE de Lyon. Lyon, France: Faurite.

Cook, E. A. (2002). Landscape structure indices for assessing urban ecological networks. *Landscape and Urban Planning*, 58, 269–280.

Cooper, J. A. (1987). The effectiveness of translocation control of Minneapolis-St. Paul Canada goose populations. In *Integrating Man and Nature in the Metropolitan Environment*, eds. L. W. Adams and D. L. Leedy. Columbia, Maryland: National Institute for Urban Wildlife, pp. 169–171.

Cooper, J. A. (1991). Canada goose management at the Minneapolis-St. Paul International Airport. In *Wildlife Conservation in Metropolitan Environments*, eds. L. W. Adams and D. L. Leedy. Columbia, Maryland: National Institute for Urban Wildlife, pp. 175–183.

Cornish, S. G. and Whitten, R. (1988). *The Golf Course*. New York: The Rutledge Press.

Corry, R. C. and Nassauer, J. I. (2002). Managing for small patch patterns in human-dominated landscapes: cultural factors and Corn Belt agriculture. In *Integrating Landscape Ecology into Natural Resource Management*, eds. J. Liu and W. W. Taylor. New York: Cambridge University Press, pp. 92–113.

Costa, J. E. and Baker, V. R. (1981). *Surficial Geology: Building with the Earth*. New York: John Wiley.

Costa-Pierce, B., Desbonnet, A., Edwards, P. and Baker, D., eds. (2005). *Urban Aquaculture*. Oxford: CABI International.

Costanza, R., Cumberland, J. C., Daly, H. E. et al. (1997a). *An Atlas to Ecological Economics*. Boca Raton, Florida: St. Lucie Press.

Costanza, R., d'Arge, R., de Groot, R. et al. (1997b). The value of the world's ecosystem services and natural capital. *Nature*, 387, 253–260.

Coulson, R. N. and Tchakerian, M. D. (2010). *Basic Landscape Ecology*. College Station, Texas: Knowledge Engineering Laboratory Partners.

Courchamp, F., Langlais, M. and Sugihara, G. (1999). Cats protecting birds: modeling the mesopredator release effect. *Journal of Animal Ecology*, 68, 282–292.

Craul, P. J. (1992). *Urban Soil in Landscape Design*. New York: John Wiley.

Craul, P. J. (1999). *Urban Soils: Applications and Practices*. New York: John Wiley.

Crawford, B. T. (1950). Some specific relationships between soils and wildlife. *Journal of Wildlife Management*, 14, 115–123.

Creuze des Chatelliers, M., Poinsart, D. and Bravard, J. P. (1994). Geomorphology of alluvial groundwater ecosystems. In *Groundwater Ecology*, eds. J. Gibert, D. L. Danielopol and J. A. Stanford. San Diego, California: Academic Press, pp. 157–185.

Cristol, D. and Rodewald, A. (2005). Introduction: can golf courses play a role in bird conservation? *Wildlife Society Bulletin*, 33, 407–410.

Cronon, W. (1991). *Nature's Metropolis: Chicago and the Great West*. New York: Norton.

Crooks, K. R. and Soule, M. F. (1999). Mesopredator release and avifaunal extinctions in a fragmented system. *Nature*, 400, 563–566.

Crooks, K. R., Suarez, A. V. and Bolger, D. T. (2004). Avian assemblages along a gradient of urbanization in a highly fragmented landscape. *Biological Conservation*, 115, 451–462.

Crowe, T. M. (1979). Lots of weeds: insular phytogeography of vacant urban lots. *Journal of Biogeography*, 6, 169–181.

Croxton, P. J., Hann, J. P., Greatorex-Davies, J. N. and Sparks, T. H. (2005). Linear hotspots? The floral and butterfly diversity of green lanes. *Biological Conservation*, 121, 579–584.

Culley, T. M., Sbita, S. J. and Wick, A. (2007). Population genetic effects of urban habitat fragmentation in the perennial herb *Viola pubescens* (Violaceae) using ISSR markers. *Annals of Botany*, 100, 91–100.

Cullina, W. (2009). *Understanding Perennials: A New Look at an Old Favorite*. Boston: Houghton Mifflin Harcourt.

Cumming, G. S., Barnes, G., Perz, S. et al. (2005). An exploratory framework for the empirical measurement of resilience. *Ecosystems*, 8, 975–987.

Cuperus, R., Kalsbeek, M., Udo de Haes, H. A. and Canters, K. J. (2002). Preparation and implementation of seven ecological compensation plans for Dutch highways. *Environmental Management*, 29, 736–749.

Curiel-Esparza, J., Canto-Perella, J. and Calvo, M. A. (2004). Establishing sustainable strategies in urban underground engineering. *Science and Engineering Ethics*, 10, 523–530.

Curtis, L., Rea, W., Smith-Willis, P. et al. (2006). Adverse health effects of outdoor air pollutants. *Environment International*, 32, 815–830.

Curtis, J. T. (1959). *The Vegetation of Wisconsin: An Ordination of Plant Communities*. Madison, Wisconsin: University of Wisconsin Press.

Curtis, J. T. and McIntosh, R. P. (1951). An upland forest continuum in the prairie-forest border region of Wisconsin. *Ecology*, 72, 476–496.

Cushing, C. E. and Allan, J. D. (2001). *Streams: Their Ecology and Life*. San Diego, California: Academic Press.

Daily, G. C., ed. (1997). *Nature's Services: Societal Dependence on Natural Ecosystems*. Washington, D.C.: Island Press.

Daily, G. and Ellison, K. (2002). *The New Ecology of Nature: A Quest to Make Conservation Profitable*. Washington, D.C.: Island Press.

Dalang, T. and Hersperger, A. M. (2012). Trading connectivity improvement for area loss in patch-based biodiversity reserve networks. *Biological Conservation*, 148, 116–125.

Dale, V. (1997). The relationship between land-use change and climate change. *Ecological Applications*, 7, 753–769.

Daley, R. and City of Chicago. (2002). *A Guide to Rooftop Gardening*. Chicago: Chicago Department of Environment.

Damschen, E. I., Haddad, N. M. and Orrock, J. L. (2006). Corridors increase plant species richness at large scales. *Science*, 313, 1284–1286.

Dandou, A., Santamouris, M., Soulakellis, N. et al. (2008). On the use of cool materials as a heat island mitigation strategy. *Journal of Applied Meteorology and Climatology*, 47, 2846–2856.

Daniels, T. J. and Bekoff, M. (1989). Spatial and temporal resource use by feral and abandoned dogs. *Ethology*, 81, 300–312.

Darlington, A., Chan, M., Malloch, D. et al. (2000). The biofiltration of indoor air: implications for air quality. *Indoor Air*, 10, 39–46.

References

Darlington, A. B., Dat, J. F. and Dixon, M. A. (2001). The biofiltration of indoor air: air flux and temperature influences the removal of toluene, ethylbenzene, and xylene. *Environmental Science and Technology*, 35, 240–246.

Davenport, J. and Davenport, J. L., eds. (2006). *The Ecology of Transportation: Managing Mobility for the Environment*. New York: Springer.

Davies, W. K. D. and Baxter, T. (1997). Commercial intensification: the transformation of a highway-oriented ribbon. *Geoforum*, 28, 237–252.

Davis, A., Hunt, W., Traver, R. and Clar, M. (2009). Bioretention technology overview of current practice and future needs. *Journal of Environmental Engineering*, 135, 109–117.

Davis, A. M. and Glick, T. F. (1978). Urban ecosystems and island biogeography. *Environmental Conservation*, 5, 299–304.

Davis, A. Y., Pijanowski, B. C., Robinson, K. and Engel, B. (2010). The environmental and economic costs of sprawling parking lots in the United States. *Land Use Policy*, 27, 255–261.

Davis, M., Chew, M. K., Hobbs, R. J. *et al.* (2011). Don't judge species on their origins. *Nature*, 474, 153–154.

Davis, M. A. (2009). *Invasion Biology*. Oxford: Oxford University Press.

De Aranzabal, I., Schmitz, M. F., Aguilera, P. and Pineda, F. D. (2008). Modelling of landscape changes derived from the dynamics of socio-ecological systems: a case study in a semiarid Mediterranean landscape. *Ecological Indicators*, 8, 672–685.

De Chiara, J. and Koppelman, L. (1984). *Time Saver Standards for Urban Site Design*. New York: McGraw-Hill.

De Groot, R. (2006). Function analysis and valuation as a tool to assess land use conflicts in planning for sustainable multi-functional landscapes. *Landscape and Urban Planning*, 75, 175–186.

DeAngelis, D. L. and Petersen, J. H. (2001). Importance of the predator's ecological neighborhood in modeling predation on migrating prey. *Oikos*, 94, 315–325.

DeBano, L. F., Neary, D. G. and Folliott, P. F. (1998). *Fire's Effects on Ecosystems*. New York: John Wiley.

Deelstra, T. (1998). Towards ecological sustainable cities: strategies, models and tools. In *Urban Ecology*, eds. J. Breuste, H. Feldmann and O. Uhlmann. New York: Springer, pp. 17–22.

DeGraaf, R. M. (1986). Urban bird habitat relationships: application to landscape design. *Transactions of the North American Wildlife and Natural Resource Conference*, 51, 232–248.

DeGraaf, R. M. (1987). Urban habitat wildlife research – application to landscape design. In *Integrating Man and Nature in the Metropolitan Environment*, eds. L. W. Adams and D. L. Leedy. Columbia, Maryland: National Institute for Urban Wildlife, pp. 107–111.

DeGraaf, R. M. and Wentworth, J. M. (1981). Urban bird communities and habitats in New England. *Transactions of the North American Wildlife and Natural Resources Conference*, 40, 396–413.

DeGraaf, R. M. and Witman, G. M. (1979). *Trees, Shrubs, and Vines for Attracting Birds*. Amherst, Massachusetts: University of Massachusetts Press.

Deinet, S., McRae, L., De Palma, A. *et al.* (2010). *The Living Planet for Global Estuarine Systems*. Gland, Switzerland: World Wildlife Fund and Zoological Society of London.

Del Tredici, P. (2010). *Wild Urban Plants of the Northeast: A Field Guide*. Ithaca, New York: Cornell University Press.

Delcourt, H. R. and Delcourt, P. A. (1988). Quaternary landscape ecology: relevant scales in space and time. *Landscape Ecology*, 2: 23–44.

Dent, B. (2007). *Budapest: A Cultural History*. New York: Oxford University Press.

DeSanto, R. and Smith, D. (1993). Environmental auditing: an introduction to issues of habitat fragmentation relative to transportation corridors with special reference to high-speed rail (HSR). *Environmental Management*, 17, 111–114.

DeSanto, R. S., Glaser, R. A., McMillen, W. P. *et al.* (1976). *Open Space as an Air Resource Management Measure. Vol. II. Design Criteria*. Report EPA-450/3-76-028b. Washington, D.C.: US Environmental Protection Agency.

Desender, K., Small, E., Gaubiomme, E. and Verdyck, P. (2005). Rural-urban gradients and the population genetic structure of woodland ground beetles. *Conservation Genetics*, 6, 51–62.

DeSousa, C. A. (2003). Turning brownfields into green space in the City of Toronto. *Landscape and Urban Planning*, 62, 181–198.

DeSousa, C. A. (2004). The greening of brownfields in America. *Journal of Environmental Planning and Management*, 47, 579–600.

DeStefano, S. and DeGraaf, R. M. (2003). Exploring the ecology of suburban wildlife. *Frontiers in Ecology and the Environment*, 1, 95–101.

DeStefano, S. and Webster, C. M. (2012). Distribution and habitat of Greater Roadrunners in urban and suburban Arizona. In *Urban Bird Ecology and Conservation*, eds. C. A. Lepczyk and P. S. Warren. Berkeley, California: University of California Press, pp. 155–166.

Dexter, R. W. (1955). The vertebrate fauna on the campus of Kent State University. *The Biologist*, 37, 84–88.

Diamond, J. M., Bishop, K. D. and van Balen, S. (1987). Bird survival in an isolated Javan woodland: island or mirror? *Conservation Biology*, 1, 132–143.

Diaz-Chavez, R. A. (2006). Measuring sustainability in peri-urban areas: case study of Mexico City. In *The Peri-Urban Interface: Approaches to Sustainable Natural and Human Resource Use*, eds. D. McGregor, D. Simon and D. Thompson. London: Earthscan, pp. 246–265.

Dietz, M. (2007). Low impact development practices: a review of current research and recommendations for future directions. *Water, Air and Soil Pollution*, 186, 351–363.

Dietz, M. and Clausen, J. (2008). A field evaluation of rain garden flow and pollutant treatment. *Water, Air and Soil Pollution*, 167, 123–138.

Ditchkoff, S. S., Saalfeld, S. T. and Gibson, C. J. (2006). Animal behavior in urban ecosystems: modifications due to human-induced stress. *Urban Ecosystems*, 9, 5–12.

Dobson, M. C. and Moffat, A. J. (1993). *The Potential for Woodland Establishmennt on Landfill Sites*. London: Her Majesty's Stationery Office.

Doherty, P. F. and Grubb, T. C. (1998). Reproductive success of cavity-nesting birds breeding under high-voltage powerlines. *American Midland Naturalist*, 140, 122–128.

Dolbeer, R., Wright, S. and Clearly, E. (2000). Ranking the hazard level of wildlife species to aviation. *Wildlife Society Bulletin*, 28, 372–378.

Domenico, S. and Hecnar, S. J. (2006). Effects of road de-icing salt (NaCl) on larval wood frogs (*Rana sylvatica*). *Environmental Pollution*, 140, 247–256.

Donahue, B. (1999). *Reclaiming the Commons: Community Farms and Forests in a New England Town*. New Haven, Connecticut: Yale University Press.

Dorney, J. R., Guntenspergen, G. R., Keough, J. R. and Stearns, F. (1984). Composition and structure of an urban woody plant community. *Urban Ecology*, 8, 69–90.

Dorney, R. S. and McLellan, P. W. (1984). The urban ecosystem: its spatial structure, its scale relationships, and its subsystem attributes. *Environment*, 16, 9–20.

Dornier, A. and Cheptou, P.-O. (2012). Determinants of extinction in fragmented plant populations: *Crepis sancta* (asteraceae) in urban environments. *Oecologia*, 169, 703–712.

Dos Reis, V. A., Lombardi, J. A. and de Figueiredo, R. A. (2006). Diversity of vascular plants growing on walls of a Brazilian city. *Urban Ecosystems*, 9, 39–43.

Douglas, B. C., Kearney, M. S. and Leatherman, S. P. (2001). *Sea Level Rise: History and Consequences*. San Diego, California: Academic Press.

Douglas, M. (2006). A regional network strategy for reciprocal rural-urban linkages: an agenda for policy research with reference to Indonesia. In *The Earthscan Reader in Rural-Urban Linkages*, ed. C. Tacoli. London: Earthscan, pp. 124–154.

Dreyer, G. D. and Niering, W. A. (1986). Evaluation of two herbicide techniques on electric transmission rights-of-way: development of relatively stable shrublands. *Environmental Management*, 10, 113–118.

du Pisani, P. L. (2006). Direct reclamation of potable water at Windhoek's Goreangab reclamation plant. *Desalinization*, 188, 79–88.

Duany, A., Plater-Zyberk, E. and Alminana, R. (2003). *The New Civic Art: Elements of Town Planning*. New York: Rizzoli International Publications.

Duany, A., Plater-Zyberk, E. and Speck, J. (2000). *Suburban Nation*. New York: North Point Press.

Duany, A., Sorlein, S. and Wright, W. (2008). *Smart Code: Version 9 and Manual*. Ithaca, New York: New Urban News Publication.

Duffy-Anderson, J. T. and Able, K. W. (1999). Effects of municipal piers on the growth of juvenile fishes in the Hudson River estuary: a study across a pier edge. *Marine Biology*, 133, 409–418.

Duncan, A., Duncan, R., Rae, R. *et al.* (2001). Roof and ground nesting Eurasian oystercatchers in Aberdeen. *Journal of the Scottish Ornithologists' Club*, 8, 381–388.

Dunford, W. and Freemark, K. (2004). Matrix matters: effects of surrounding land uses on forest birds near Ottawa, Canada. *Landscape Ecology*, 20, 497–511.

Dunham-Jones, E. and Williamson, J. (2009). *Retrofitting Suburbia: Urban Design Solutions for Redesigning Suburbs*. New York: John Wiley.

Duning, X., Li, X., Chang, Y. and Kun, X. (2001). Rural landscape ecological construction in China: theory and application. In *Landscape Ecology Applied in Land Evaluation Development and Conservation*, eds. D. van der Zee and I. S. Zonneveld. ITC Publication 81. Enschede, Netherlands: International Institute for Aerospace Survey and Earth Sciences, pp. 221–231.

Dunn, C. P. and Heneghan, L. (2011). Composition and diversity of urban vegetation. In *Urban Ecology: Patterns, Processes, and Applications*, eds. J. Niemela, J. H. Breuste, T. Elmqvist *et al*. New York: Oxford University Press, pp. 103–115.

Dunne, T. and Leopold, L. B. (1978). *Water in Environmental Planning*. San Francisco: W. H. Freeman.

Dunnett, N. and Clayden, A. (2007). *Rain Gardens: Managing Water Sustainably in the Garden and Designed Landscape*. Portland, Oregon: Timber Press.

Dunnett, N. and Kingsbury, N. (2004). *Planting Green Roofs and Living Walls*. Portland, Oregon: Timber Press.

Dunning, J. B., Danielson, B. J. and Pulliam, H. R. (1992). Ecological processes that affect populations in complex landscapes. *Oikos*, 65, 169–175.

Dutton, J. A. (2000). *New American Urbanism: Re-forming the Suburban Metropolis*. Milano, Italy: Skira editore.

References

Duvigneaud, P. (1974). L'ecosysteme "Urbs." *Memoires de la Societe Royale Botanique de Belgique*, 6, 5–36.

Easterling, K. (1993). *American Town Plans: A Comparative Time Line*. New York: Princeton Architectural Press.

Ebrey, P. B. (1996). *The Cambridge Illustrated History of China*. New York: Cambridge University Press.

Edmonson, W. T. (1991). *The Uses of Ecology: Lake Washington and Beyond*. Seattle, Washington: University of Washington Press.

Edwards, C. A. (2004). *Earthworm Ecology*. Boca Raton, Florida: CRC Press.

Edwards, P. (2005). Development status of, and prospects for, wastewater-fed agriculture in urban environments. In *Urban Aquaculture*, eds. B. Costa-Pierce, A. Desbonnet, P. Edwards and D. Baker. Wallingford, UK: CABI Publishing, pp. 45–59.

Ehler, L. E. and Frankie, G. (1978). Ecology of insects in urban environments. *Annual Review of Entomology*, 23, 367–387.

Eisenbeis, G. and Hanel, A. (2009). Light pollution and the effect of artificial night lighting on insects. In *Ecology of Cities and Towns: A Comparative Approach*, eds. M. J. McDonnell, A. K. Hahs and J. H. Breuste. New York: Cambridge University Press, pp. 243–263.

El Araby, M. (2002). Urban growth and environmental degradation. *Cities*, 19, 389–400.

El-Bushra, E. and Hijazi, N. (1995). Two million squatters in Khartoum Urban Complex: the dilemma of Sudan's national capital. *GeoJournal*, 35, 505–514.

Elliman, T. (2005). Vascular flora and plant communities of the Boston Harbor islands. *Northeastern Naturalist*, 12(Special issue 3).

Elliott, M. A., Burdon, D. A., Hemingway, K. L. and Apitz, S. E. (2007). Estuarine, coastal and marine ecosystem restoration; confusing management and science: a revision of concepts. *Estuarine, Coastal and Shelf Science*, 74, 349–366.

Ellis, E. C., Wang, H., Xiao, K. et al. (2006). Measuring long-term ecological changes in densely populated landscapes using current and historical high resolution imagery. *Remote Sensing of Environment*, 100, 457–473.

Ellis, S. and Mellor, A. (1995). *Soils and Environment*. New York: Routledge Chapman & Hall.

Elson, M. J. (1986). *Greenbelts: Conflict Mediation in the Urban Fringe*. London: William Heinemann.

Elton, C. S. (1958). *The Ecology of Invasions by Animals and Plants*. London: Methuen & Co.

Emlen, J. T. (1974). An urban bird community in Tucson, Arizona. *Condor*, 76, 184–197.

Emmanuel, R., Rosenlund, H. and Johansson, E. (2007). Urban shading – a design option for the tropics? A study in Colombo, Sri Lanka. *International Journal of Climatology*, 27, 1995–2004.

Endlicher, W., Jendritzky, G., Fischer, J. and Redlich, J. P. (2008). Heat waves, urban climate and human health. In *Urban Ecology: An International Perspective on the Interaction Between Humans and Nature*, eds. J. M. Marzluff, E. Shulenberger, W. Endlicher et al. New York: Springer, pp. 269–278.

English Nature. (2003). *Green Roofs: Their Existing Status and Potential for Conserving Biodiversity in Urban Areas*. Peterborough, UK: English Nature.

Engstrom, R. T. and Mikuminski, G. (1998). Ecological neighborhoods in red-cockaded woodpecker populations. *Auk*, 115, 473–478.

Enyedi, G. and Szirmai, V. (1992). *Budapest: A Central European Capital*. London: Belhaven.

Erell, E., Pearlmutter, D. and Williamson, T. (2011). *Urban Microclimate: Designing the Spaces Between Buildings*. London: Earthscan.

Erickson, D. L. (2004). The relationship of historic city form and contemporary greenway implementation: a comparison of Milwaukee, Wisconsin (USA) and Ottawa, Ontario (Canada). *Landscape and Urban Planning*, 68, 199–221.

Erickson, D. L. (2006). *MetroGreen: Connecting Open Space in North American Cities*. Washington, D.C.: Island Press.

Eriksson, B. K., Sandstro, B. K., Isaeusc, A. et al. (2004). Effects of boating activities on aquatic vegetation in the Stockholm archipelago, Baltic Sea. *Estuarine, Coastal and Shelf Science*, 61, 339–349.

Erwin, R. M., Hatfield, J. S. and Link, W. A. (1991). Social foraging and feeding environment of the Black-crowned Night Heron in an industrialized estuary. *Bird Behavior*, 9, 94–102.

Erz, W. and Klausnitzer, B. (1998). Fauna. In *Stadtokologie*, eds. H. Sukopp and R. Wittig. Stuttgart: Fischer, pp. 266–315.

Evans, C. V., Fanning, D. E. and Short, J. R. (2000). Human-influenced soils. In *Managing Soils in an Urban Environment*, eds. R. B. Brown, J. H. Huddleston and J. L. Anderson. Madison, Wisconsin: American Society of Agronomy, pp. 33–67.

Evans, K. L. (2010). Individual species and urbanisation. In *Urban Ecology*, ed. K. J. Gaston. New York: Cambridge University Press, pp. 53–87.

Evans, K. L., Newson, S. E. and Gaston, K. J. (2009). Habitat influences on urban avian assemblages. *Ibis*, 155, 19–39.

Everette, A. L., O'Shea, T. I., Ellison, L. E. et al. (2001). Bat use of a high-plains urban wildlife refuge. *Wildlife Society Bulletin*, 29, 967–973.

Ewert, A. W., Chavez, D. J. and Magill, A. W., eds. (1993). *Culture, Conflict, and Communication in the Wildland-Urban Interface.* Boulder, Colorado: Westview Press.

Ewing, R., Schmid, T., Killingsworth, R. et al. (2008). Relationship between urban sprawl and physical activity, obesity, and morbidity. In *Urban Ecology: An International Perspective on the Interaction Between Humans and Nature,* eds. J. M. Marzluff, E. Shulenberger, W. Endlicher et al. New York: Springer, pp. 567–582.

Faggi, A. M., Krellenberg, K., Castro, R. et al. (2008). Biodiversity in the Argentinean Rolling Pampa Ecoregion: changes caused by agriculture and urbanisation. In *Urban Ecology: An International Perspective on the Interaction Between Humans and Nature,* eds. J. M. Marzluff, E. Schulenberger, W. Endlicher et al. New York: Springer, pp. 377–389.

Fahrig, L. (2002). Effect of habitat fragmentation on the extinction threshold: a synthesis. *Ecological Applications,* 12, 346–353.

Fahrig, L. and Merriam, G. (1985). Habitat patch connectivity and population survival. *Ecology,* 66, 1762–1768.

Fainstein, S. S. (2000). New directions in planning theory. *Urban Affairs Review,* 35, 451–478.

Falk, J. H. (1980). The primary productivity of lawns in a temperate environment. *Journal of Applied Ecology,* 17, 689–696.

Fang, C.-F. and Ling, D.-L. (2005). Guidance for noise reduction provided by tree belts. *Landscape and Urban Planning,* 71, 29–34.

Fanning, D. S. and Fanning, M. C. B. (1989). *Soil: Morphology, Genesis, and Classification.* New York: John Wiley.

Farina, A. (2006). *Principles and Methods in Landscape Ecology: Towards a Science of Landscape.* New York: Springer.

Fasola, M. and Ruiz, X. (1996). The value of rice fields as substitutes for natural wetlands for waterbirds in the Mediterranean region. *Colonial Waterbirds,* 19, 122–128.

Federer, C. A. (1971). Effects of trees in modifying urban microclimate. In *Trees and Forests in an Urbanizing Environment.* Planning and Resource Development Series 17. Amherst, Massachusetts: Cooperative Extension Service, pp. 23–28.

Feng, A. Y. T. and Hinsworth, C. G. (2013). The secret life of the city rat: a review of the ecology of urban Norway and black rats (*Rattus norvegicus* and *Rattus rattus*). *Urban Ecosystems,* 2013, 1–14.

Fenton, M. B. (1997). Science and the conservation of bats. *Journal of Mammalogy,* 78, 1–14.

Fenton, T. E. and Collins, M. E. (2000). The soil resource and its inventory. In *Managing Soils in an Urban Environment,* eds. R. B. Brown, J. H. Huddleston and J. L. Anderson. Madison, Wisconsin: American Society of Agronomy, pp. 1–32.

Ferguson, B. (1994). *Stormwater Infiltration.* Boca Raton, Florida: CRC Press.

Ferguson, B. K. (1998). *Introduction to Stormwater: Concept, Purpose, Design.* New York: John Wiley.

Ferguson, B. K. (2005). *Porous Pavements.* Boca Raton, Florida: Taylor & Francis.

Ferguson, G. and Woodbury, A. D. (2004). Surface heat flow in an urban environment. *Journal of Geophysical Research: Solid Earth,* 109 (B2).

Fernandez-Juricic, E. (2000a). Avifaunal use of wooded streets in an urban landscape. *Conservation Biology,* 14, 513–521.

Fernandez-Juricic, E. (2000b). Bird community composition patterns in urban parks of Madrid: the role of age, size and isolation. *Ecological Research,* 15, 373–383.

Fernandez-Juricic, E. (2001). Avian spatial segregation at edges and interiors of urban parks in Madrid, Spain. *Biodiversity and Conservation,* 10, 1303–1316.

Fernandez-Juricic, E. and Jokimaki, J. (2001). A habitat island approach to conserving birds in urban landscapes: case studies from southern and northern Europe. *Biodiversity and Conservation,* 10, 2023–2043.

Fetter, C. W. (1999). *Contaminant Hydrogeology.* Upper Saddle River, New Jersey: Prentice-Hall.

Fialkowski, M. and Bitner, A. (2008). Universal rules for fragmentation of land by humans. *Landscape Ecology,* 23, 1013–1022.

Field, R., O'Shea, M. L. and Chin, K. K. (1993). *Integrated Stormwater Management.* Boca Raton, Florida: Lewis Publishers.

Findlay, C. S. and Bourdages, J. (2000). Response time of wetland biodiversity to road construction on adjacent lands. *Conservation Biology,* 14, 86–94.

Finkl, C. W. and Charlier, R. H. (2003). Sustainability of subtropical coastal zones in southeastern Florida: challenges for urbanized coastal environments threatened by development, pollution, water supply, and storm hazards. *Journal of Coastal Research,* 19, 934–943.

Fiorello, C. V., Noss, A. J. and Deem, S. L. (2006). Demography, hunting ecology, and pathogen exposure of domestic dogs in the Isoso of Bolivia. *Conservation Biology,* 20, 762–771.

Fischer, J. and Acreman, M. C. (2004). Wetland nutrient removal: a review of the evidence. *Hydrology and Earth System Science,* 6, 673–685.

Fischer, J., Lindenmayer, D. B. and Manning, A. D. (2006). Biodiversity, ecosystem function, and resilience:

ten guiding principles for commodity production landscapes. *Frontiers in Ecology and the Environment*, 4, 80–86.

Flanders, R. V. (1986). Potential for biological control in urban environments. In *Advances in Urban Pest Management*, eds. G. W. Bennett and J. M. Owens. New York: Van Nostrand Reinhold, pp. 95–127.

Fletcher, T. D., Ladson, A. R., Leonard, A. W. and Walsh, C. J. (2004). *Urban Stormwater and the Ecology of Streams*. Canberra, ACT: Cooperative Research Centre for Freshwater Ecology and Cooperative Research Centre for Catchment Hydrology.

Flink, C. A., Olka, K. and Searns, R. M. (2001). *Trails for the Twenty-First Century: Planning, Design, and Management Manual for Multi-Use Trails*. Washington, D.C.: Island Press.

Forman, R. T. T. (1964). Growth under controlled conditions to explain the hierarchical distributions of a moss, *Tetraphis pellucida*. *Ecological Monographs*, 34, 1–25.

Forman, R. T. T. (1979a). The Pine Barrens of New Jersey: an ecological mosaic. In *Pine Barrens: Ecosystem and Landscape*, ed. R. T. T. Forman. New York: Academic Press, pp. 569–585.

Forman, R. T. T., ed. (1979b). *Pine Barrens: Ecosystem and Landscape*. New York: Academic Press.

Forman, R. T. T. (1990). Ecologically sustainable landscapes: the role of spatial configuration. In *Changing Landscapes: An Ecological Perspective*, eds. I. S. Zonneveld and R. T. T. Forman. New York: Springer-Verlag, pp. 261–278.

Forman, R. T. T. (1995). *Land Mosaics: The Ecology of Landscapes and Regions*. New York: Cambridge University Press.

Forman, R. T. T. (1999). Horizontal processes, roads, suburbs, societal objectives, and landscape ecology. In *Landscape Ecological Analysis: Issues and Applications*, eds. J. M. Klopatek and R. H. Gardner. New York: Springer, pp. 35–53.

Forman, R. T. T. (2004a). Road ecology's promise: what's around the bend? *Environment*, 46, 8–21.

Forman, R. T. T. (2004b). *Mosaico territorial para la region metropolitana de Barcelona*. Barcelona: Editorial Gustavo Gili.

Forman, R. T. T. (2006). Good and bad places for roads: effects of varying road and natural patterns on habitat loss, degradation, and fragmentation. In *Proceedings of the 2005 International Conference on Ecology and Transportation*. Raleigh, North Carolina: C.T.E., North Carolina State University, pp. 164–174.

Forman, R. T. T. (2008). *Urban Regions: Ecology and Planning Beyond the City*. New York: Cambridge University Press.

Forman, R. T. T. (2009). Arrangements of nature and people: using landscape-ecology, coastal-region, and urban-region lenses. In *The Irish Landscape 2009*. Kilkenny, Ireland: The Heritage Council, pp. 28–38.

Forman, R. T. T. (2010a). Urban ecology and the arrangement of nature in urban regions. In *Ecological Urbanism*, eds. M. Mostafavi and G. Doherty. Baden, Switzerland: Lars Muller Publishers, pp. 312–323.

Forman, R. T. T. (2010b). Coastal regions: spatial patterns, flows, and a people-nature solution from the lens of landscape ecology. In *La costa obliqua: Un atlante per la Puglia (The Oblique Coast: An Atlas for Puglia)*, ed. M. Mininni. Rome, Italy: Donzelli editore, pp. 249–265.

Forman, R. T. T. (2012). Infrastructure and nature: reciprocal effects and patterns for our future. In *Infrastructure and Sustainability*, eds. S. Pollalis, A. Georgoulias, S. Ramos and D. Schodek. New York: Routledge, pp. 35–49.

Forman, R. T. T. and Alexander, L. E. (1998). Roads and their major ecological effects. *Annual Review of Ecology and Systematics*, 29, 207–231.

Forman, R. T. T. and Baudry, J. (1984). Hedgerows and hedgerow networks in landscape ecology. *Environmental Management*, 8, 495–510.

Forman, R. T. T. and Collinge, S. K. (1996). The "spatial solution" to conserving biodiversity in landscapes and regions. In *Conservation of Faunal Diversity in Forested Landscapes*, eds. R. M. DeGraaf and R. I. Miller. London: Chapman & Hall, pp. 537–568.

Forman, R. T. T. and Godron, M. (1981). Patches and structural components for a landscape ecology. *BioScience*, 31, 733–740.

Forman, R. T. T. and Godron, M. (1986). *Landscape Ecology*. New York: John Wiley.

Forman, R. T. T. and Hersperger, A. M. (1997). Ecologia del paesaggio e pianificazione: una potente combinazione. *Urbanistica*, 108, 61–66.

Forman, R. T. T. and Mellinger, A. D. (2000). Road networks and forest spatial patterns: comparing cutting-sequence models for forestry and conservation. In *Nature Conservation 5: Conservation in Production Environments: Managing the Matrix*, eds. J. L. Craig, N. Mitchell and D. A. Saunders. Chipping Norton, Australia: Surrey Beatty, pp. 71–80.

Forman, R. T. T. and Sperling, D. (2011). The future of roads: no driving, no emissions, nature reconnected. *Solutions*, 2, 10–23.

Forman, R. T. T., Galli, A. E. and Leck, C. F. (1976). Forest size and avian diversity in New Jersey woodlots with some land-use implications. *Auk*, 93, 356–364.

Forman, R. T. T., Reeve, P., Beyer, H. *et al.* (2004). *Open Space and Recreation Plan 2004: Concord, Massachusetts*.

Concord, Massachusetts: Natural Resources Commission.

Forman, R. T. T., Reineking, B. and Hersperger, A. M. (2002). Road traffic and nearby grassland bird patterns in a suburbanizing landscape. *Environmental Management*, 29, 782–800.

Forman, R. T. T., Sperling, D., Bissonette, J. A. *et al.* (2003). *Road Ecology: Science and Solutions*. Washington, D.C.: Island Press.

Forrest, A. and St. Clair, C. C. (2006). Effects of dog leash laws and habitat type on avian and small mammal communities in urban parks. *Urban Ecosystems*, 9, 51–66.

Forrest, M. and Konijnendijk, C. (2005). A history of urban forests and trees in Europe. In *Urban Forests and Trees*, eds. C. C. Konijnendijk, K. Nilsson, T. B. Randrup and J. Schipperijn. New York: Springer. pp. 23–48.

Forsyth, A. (2005). *Reforming Suburbia: The Planned Communities of Irvine, Columbia, and The Woodlands*. Berkeley, California: University of California Press.

Forsyth, A. and Crewe, K. (2009). A typology of comprehensive designed communities since the Second World War. *Landscape Journal*, 28, 56–78.

Forsyth, A. and Musacchio, L. R. (2005). *Designing Small Parks: A Manual for Addressing Social and Ecological Concerns*. New York: John Wiley.

Foster, D. R. and Aber, J. D., eds. (2004). *Forests in Time: The Environmental Consequences of 1,000 Years of Change in New England*. New Haven, Connecticut: Yale University Press.

Foster, D., Swanson, F., Aber, J. *et al.* (2003). The importance of land-use legacies to ecology and conservation. *BioScience*, 53, 77–88.

Foster, S., Lawrence, A. and Morris, B. (1998). *Groundwater in Urban Development: Assessing Management Needs and Formulating Policy Strategies*. Technical Paper No. 390. Washington, D.C.: World Bank.

Fowler, H. G. (1983). Urban structural pests – carpenter ants (Hymenoptera, Formicidae) displacing subterranean termites (Isoptera, Rhinotermitidae) in public concern. *Environmental Entomology*, 12, 997–1002.

Fox, M. D. (1990). Interactions of native and introduced species in new habitats. *Proceedings of the Ecological Society of Australia*, 16, 141–147.

France, R. L. (2003). *Wetland Design: Principles and Practices for Landscape Architects and Land-Use Planners*. New York: Norton.

Francis, R. A. and Hoggart, S. P. G. (2009). Urban river wall habitat and vegetation: observations from the River Thames through central London. *Urban Ecosystems*, 12, 465–485.

Franck, K. A. and Schneekloth, L. H., eds. (1994). *Ordering Space: Types in Architecture and Design*. New York: Van Nostrand Reinhold.

Franklin, J. F. and Forman, R. T. T. (1987). Creating landscape pattern by forest cutting: ecological consequences and principles. *Landscape Ecology*, 1: 5–18.

Frazer, L. (2005). Paving paradise: the peril of impervious surfaces. *Environmental Health Perspectives*, 113, A456–A462.

Fredrickson, L. H. and Laubhan, M. K. (1994). Managing wetlands for wildlife. In *Research and Management Techniques for Wildlife and Habitat*, ed. T. A. Bookhout. Bethesda, Maryland: The Wildlife Society, pp. 623–647.

Freeman, D. B. (1991). *A City of Farmers: Informal Urban Agriculture in the Open Spaces of Nairobi, Kenya*. Montreal: McGill-Queens University Press.

Frey, H. (1999). *Designing the City: Towards a More Sustainable Urban Form*. London: E & FN Spon.

Freyermuth, G. S. (2010). Edges & nodes/cities & nets: the history and theories of networks and what they tell us about urbanity in the digital age. In *Transcultural Spaces: Challenges of Urbanity, Ecology, and the Environment*, eds. S. L. Brandt, W. Fluck and F. Mehring. Tubingen, Germany: Narr Francke Artempto Verlag.

Friesen, L. E., Eagles, P. F. J. and Mackay, R. J. (1995). Effects of residential development on forest-dwelling neotropical migrant songbirds. *Conservation Biology*, 9, 1408–1414.

Frith, M. and Gedge, D. (2000). The black redstart in urban Britain: a conservation conundrum? *British Wildlife*, 8, 381–388.

Fritsch, T. (1896). *Die Stadt der Zukunft*. Leipzig: Verlag von Theod. Fritsch.

Frosch, R. A. and Gallopoulos, N. E. (1989). Strategies for manufacturing. *Scientific American*, 261, 144–152.

Frumkin, H., Frank, L. and Jackson, R. (2004). *Urban Sprawl and Public Health: Designing, Planning, and Buildings for Healthy Communities*. Washington, D.C.: Island Press.

Fulford, R. (1995). *Accidental City: The Transformation of Toronto*. Toronto: MacFarlane, Walter and Ross.

Fuller, R. A., Tratalos, J. and Gaston, K. J. (2009). How many birds are there in a city of a half million people? *Diversity and Distributions*, 15, 328–337.

Fuller, R., Warren, P., Armsworth, P. *et al.* (2008). Garden bird feeding predicts the structure of urban avian assemblages. *Diversity and Distributions*, 14, 131–137.

Gaisler, J., Zukal, J., Rehak, Z. and Homolka, M. (1998). Habitat use by foraging insectivorous bats. *Journal of Zoology (London)*, 244, 439–445.

Galatas, R. and Barlow, J. (2004). *The Woodlands: The Inside Story of Creating a Better Hometown*. Washington, D.C.: Urban Land Institute.

Galbraith, M. P. (2000). Uruamo Headland (North Shore City, Auckland): an urban remnant conserving ecological goods and services. In *Nature Conservation 5: Nature Conservation in Production Environments: Managing the Matrix*, eds. J. L. Craig, N. Mitchell and D. A. Saunders. Chipping Norton, Australia: Surrey Beatty, pp. 440–447.

Galloway, G. (2011). *A Plea for a Coordinated Water Policy*. Washington, D.C.: National Academy of Engineers.

Galster, J., Pazzaglia, F., Hargreaves, B. *et al.* (2006). Effects of urbanization on watershed hydrology: the scaling of discharge with drainage area. *Geology*, 34, 713–716.

Garaffa, P. I., Filloy, J. and Bellocq, M. I. (2009). Bird community responses along urban-rural gradients: does the size of the urbanized area matter? *Landscape and Urban Planning*, 90, 33–41.

Garden, J. G., McAlpine, C. A., Possingham, H. P. and Jones, D. N. (2007). Habitat structure is more important than vegetation composition for local-level management of native terrestrial reptile and small mammal species living in urban remnants: a case study from Brisbane. *Austral Ecology*, 32, 669–685.

Garreau, J. (1991). *Edge City: Life on the New Frontier*. New York: Anchor Books.

Gartland, L. (2008). *Heat Islands: Understanding and Mitigating Heat in Urban Areas*. London: Earthscan.

Garvin, A. (2002). *The American City – What Works, What Doesn't?* New York: McGraw-Hill.

Gaston, K. J., ed. (2010). *Urban Ecology*. New York: Cambridge University Press.

Gaston, K. J., Smith, R. M., Thompson, K. and Warren, P. H. (2005a). Urban domestic gardens (II): experimental tests of methods for increasing biodiversity. *Biodiversity and Conservation*, 14, 395–413.

Gaston, K. J., Warren, P. H., Thompson, K. and Smith, R. M. (2005b). Urban domestic gardens (IV): the extent of the resource and its associated features. *Biodiversity and Conservation*, 14, 3327–3349.

Gauthreaux, S. A., Jr. and Belser, C. G. (2006). Effects of artificial night lighting on migrating birds. In *Ecological Consequences of Artificial Night Lighting*, eds. C. Rich and T. Longcore. Washington, D.C.: Island Press, pp. 67–93.

Gazal, R., White, M. A., Gilles, R. *et al.* (2008). GLOBE students, teachers, and scientists demonstrate variable differences between urban and rural leaf phenology. *Global Change Biology*, 14, 1568–1580.

Gearheart, R. A. (1992). Use of constructed wetlands to treat domestic wastewater, City of Arcata, California. *Water Science and Technology*, 26, 1625–1637.

Geddes, P. (1914). Cities in evolution. In *Patrick Geddes: Spokesman for Man and the Environment* (1972), ed. M. Stalley. New Brunswick, New Jersey: Rutgers University Press, pp. 111–285.

Geddes, P. (1925). Talks from the outlook tower. In *Patrick Geddes: Spokesman for Man and the Environment* (1972), ed. M. Stalley. New Brunswick, New Jersey: Rutgers University Press, pp. 295–380.

Gedge, D. and Kadas, G. (2005). Green roofs and biodiversity. *Biologist*, 52, 161–169.

Geertsema, W. and Sprangers, J. T. C. M. (2002). Plant distribution patterns related to species characteristics and spatial and temporal habitat heterogeneity in a network of ditch banks. *Plant Ecology*, 162, 91–108.

Gehlbach, F. R. (1996). Eastern screech owls in suburbia: a model of raptor urbanization. In *Raptors in Human Landscapes*, eds. D. Bird, D. Varland and J. Negro. San Diego, California: Academic Press, pp. 69–74.

Gehrt, S. D. and Chelsvig, J. E. (2003). Bat activity in an urban landscape: patterns at the landscape and microhabitat scale. *Ecological Applications*, 13, 939–950.

Gehrt, S. D. and Prange, S. (2007). Interference competition between coyotes and raccoons: a test of the mesopredator release hypothesis. *Behavioral Ecology*, 18, 204–214.

Gehrt, S. D., Anchor, C. and White, L. A. (2009). Home range and landscape use of coyotes in a metropolitan landscape: conflict or coexistence? *Journal of Mammalogy*, 90, 1045–1057.

Geiger, R. (1965). *The Climate Near the Ground*. Cambridge, Massachusetts: Harvard University Press.

Geiger, R. and Aron, R. H. (2003). *The Climate Near the Ground*. Maryland: Rowman & Littlefield Publishers.

Germaine, S. S. and Wakeling, B. F. (2001). Lizard species distribution and habitat occupation along an urban gradient in Tucson, Arizona, USA. *Biological Conservation*, 97, 229–237.

Gerrath, J. F., Gerrath, J. A. and Larson, D. W. (1995). A preliminary account of endolithic algae of limestone cliffs of the Niagara Escarpment. *Canadian Journal of Botany*, 73, 788–793.

Getter, K. L. and Rowe, D. B. (2006). The role of extensive green roofs in sustainable development. *Hortscience*, 41, 1276–1285.

Getter, K. L., Rowe, D. B. and Andresen, J. A. (2007). Quantifying the effect of slope on extensive green roof stormwater retention. *Ecological Engineering*, 31, 225–231.

Ghassemi, F., ed. (2006). *Inter-Basin Water Transfer: Case Studies from Australia, United States, Canada, China and India*. New York: Cambridge University Press.

Gibert, J., Danielopol, D. L. and Stanford, J. A., eds. (1994a). *Groundwater Ecology*. San Diego, California: Academic Press.

Gibert, J., Stanford, J. A., Dole-Olivier, M.-J. and Ward, J. V. (1994b). Basic attributes of groundwater ecosystems and prospects for research. In *Groundwater Ecology*, eds. J.

Gibert, D. L. Danielopol and J. A. Stanford. San Diego, California: Academic Press, pp. 8–40.

Gilbert, O. L. (1970). Urban bryophyte communities in North East England. *Transcript of the British Bryological Society*, 6, 306–316.

Gilbert, O. L. (1991). *The Ecology of Urban Habitats*. London: Chapman & Hall.

Gilbert, O. (1992). *Rooted in Stone: The Natural Flora of Urban Walls*. Peterborough, UK: English Nature.

Giller, P. S. and Malmqvist, B. (1998). *The Biology of Streams and Rivers*. New York: Oxford University Press.

Gillham, O. (2002). *The Limitless City: A Primer on the Urban Sprawl Debate*. Washington, D.C.: Island Press.

Girling, C. L. (2005). *Skinny Streets and Green Neighborhoods*. Washington, D.C.: Island Press.

Girling, C. L. and Helphand, K. I. (1994). *Yard Street Park: The Design of Suburban Open Space*. New York: John Wiley.

Gissen, D. (2003). *Big and Green: Toward Sustainable Architecture in the 21st Century*. New York: Princeton Architectural Press.

Giuliano, W. M. (2005). Lepidoptera–habitat relationships in urban parks. *Urban Ecosystems*, 7, 361–370.

Giusti de Perez, R. C. and Perez, R. A. (2008). *Analyzing Urban Poverty: GIS for the Developing World*. Redlands, California: ESRI Press.

Glasby, T. M. (1999). Differences between subtidal epibiota on pier pilings and rocky reefs at marinas in Sydney, Australia. *Estuarine, Coastal and Shelf Science*, 48, 281–290.

Gleick, P. H. (2002). Water management: soft water paths. *Nature*, 418, 373.

Godde, M., Richarz, N. and Walter, B. (1995). Habitat conservation and development in the city of Dusseldorf, Germany. In *Urban Ecology as the Basis for Planning*, eds. H. Sukopp, M. Numata and A. Huber. The Hague, Netherlands: SPB Academic Publishing, pp. 163–171.

Godefroid, S. and Koedam, N. (2003). Identifying indicator plant species of habitat quality and invasibility as a guide for peri-urban forest management. *Biodiversity and Conservation*, 12, 1699–1713.

Godefroid, S., Monbaliu, D. and Koedam, N. (2007). The role of soil and microclimatic variables in the distribution patterns of urban wasteland flora in Brussels, Belgium. *Landscape and Urban Planning*, 80, 45–55.

Godschalk, D. R. (2004). Land use planning challenges: coping with conflicts in visions of sustainable development and livable communities. *Journal of the American Planning Association*, 70, 5–13.

Goldstein, E. L., Gross, M. and DeGraaf, R. M. (1981). Explorations in bird-land geometry. *Urban Ecology*, 5, 113–124.

Golley, F. B. (2003). Urban ecosystems and the twenty-first century – a global imperative. In *Understanding Urban Ecosystems: A New Frontier for Science and Education*, eds. A. R. Berkowitz, C. H. Nilon and K. S. Hollweg. New York: Springer-Verlag, pp. 401–416.

Golubev, G. and Vasiliev, O. (1978). Interregional water transfers as an interdisciplinary problem. *Water Supply and Management*, 2, 67–77.

Gompper, M. (2002). Top carnivores in the suburbs? Ecological and conservation issues raised by colonization of northeastern North America by coyotes. *BioScience*, 52, 185–190.

Goode, D. (1986). *Wild in London*. London: Michael Joseph.

Goodman, S. W. (1996). Ecosystem management at the Department of Defense. *Ecological Applications*, 6, 706–707.

Goodwin, B. J. and Fahrig, L. (2002). How does landscape structure influence landscape connectivity? *Oikos*, 99, 552–570.

Gordon, J. C., Sampson, R. N. and Berry, J. K. (2005). The challenge of maintaining working forests at the wildland-urban interface. In *Forests at the Wildland-Urban Interface: Conservation and Management*, eds. S. W. Vince, M. L. Duryea, E. A. Macie and L. A. Hermansen. Boca Raton, Florida: CRC Press, pp. 15–23.

Gottman, J. (1961). *Megalopolis: The Urbanized Northeastern Seaboard of the United States*. New York: Twentieth Century Fund.

Gough, K. and Yankson, P. (2006). Conflict and cooperation in environmental management in peri-urban Accra, Ghana. In *The Peri-Urban Interface: Approaches to Sustainable Natural and Human Resource Use*, eds. D. McGregor, D. Simon and D. Thompson. London: Earthscan, pp. 196–201.

Gounot, A. M. (1994). Microbial ecology of groundwaters. In *Groundwater Ecology*, eds. J. Gibert, D. L. Danielopol and J. A. Stanford. San Diego, California: Academic Press, pp. 189–215.

Grant, B. W., Middendorf, G., Colgan, M. J. et al. (2011). Ecology of urban amphibians and reptiles: urbanophiles, urbanophobes, and the urbanoblivious. In *Urban Ecology: Patterns, Processes, and Applications*, eds. J. Niemela, J. H. Breuste, T. Elmqvist et al. New York: Oxford University Press, pp. 167–178.

Graves, J. and Schreiber, K. (1977). Power-line corridors as possible barriers to the movements of small mammals. *American Midland Naturalist*, 97, 504–508.

Gregory, K. J. and Walling, D. E. (1973). *Drainage Basin Form and Process*. New York: Halsted Press.

Greenberg, M., Lowrie, K., Mayer, H. et al. (2001). Brownfield redevelopment as a smart growth option in the United States. *The Environmentalist*, 21, 129–143.

References

Greenberg, M., Lowrie, K., Solitare, L. and Duncan, L. (2000). Brownfields, toads, and the struggle for neighborhood redevelopment. *Urban Affairs Review*, 35, 717–733.

Greenstein, D., Tiefenthaler, L. and Bay, S. (2004). Toxicity of parking lot runoff after application of simulated rainfall. *Archives of Environmental Contamination and Toxicology*, 47, 199–206.

Gregg, J. W., Jones, C. G. and Dawson, T. E. (2003). Urbanization effects on tree growth in the vicinity of New York City. *Nature*, 424, 183–187.

Gresens, S. E., Belt, K. T., Tang, J. A. et al. (2007). Temporal and spatial responses of Chironomidae (Diptera) and other benthic invertebrates to urban stormwater runoff. *Hydrobiologia*, 575, 173–190.

Grey, G. W. and Deneke, F. J. (1992). *Urban Forestry*. Malabar, Florida: Krieger Publishing.

Grime, J. P. (2001). *Plant Strategies, Vegetation Processes, and Ecosystem Properties*. New York: John Wiley.

Grimm, N. B., Baker, L. J. and Hope, D. (2003). An ecosystem approach to understanding cities: familiar foundations and uncharted frontiers. In *Understanding Urban Ecosystems: A New Frontier for Science and Education*, eds. A. R. Berkowitz, C. H. Nilon and K. S. Hollweg. New York: Springer-Verlag, pp. 95–114.

Grimm, N. B., Faeth, S. H., Golubiewski, N. E. et al. (2008). Global change and the ecology of cities. *Science*, 319, 756–760.

Grimm, N. B., Grove, J. M., Pickett, S. T. A. and Redman, C. L. (2000). Integrated approaches to long-term studies of urban ecological systems. *BioScience*, 50, 571–584.

Grimmond, C. S. B., Roth, M., Oke, T. R. et al. (2010). Climate and more sustainable cities: climate information for improved planning and management of cities (producers/capabilities perspective). *Procedia Environmental Sciences*, 1, 247–274.

Grinder, M. I. and Krausman, P. R. (2001). Home range, habitat use, and nocturnal activity of coyotes in an urban environment. *Journal of Wildlife Management*, 65, 887–898.

Groffman, P. M., Bain, D. J. and Band, L. E. (2003). Down by the riverside: urban riparian ecology. *Frontiers in Ecology and the Environment*, 1, 315–321.

Groffman, P. M., Law, N. L., Belt, K. T. et al. (2004). Nitrogen fluxes and retention in urban watershed ecosystems. *Ecosystems*, 7, 393–403.

Gromke, C. and Ruck, B. (2007). Influence of trees on the dispersion of pollutants in an urban street canyon – experimental investigation of the flow and concentration field. *Atmospheric Environment*, 41, 3287–3302.

Grove, A. T. and Rackham, O. (2001). *The Nature of Mediterranean Europe: An Ecological History*. New Haven, Connecticut: Yale University Press.

Grove, J. M. and Burch, W. R., Jr. (1997). A social ecology approach and application of urban ecosystem and landscape analyses: a case study of Baltimore, Maryland. *Urban Ecosystems*, 1, 259–275.

Guan, D.-S., Wang, L.-R. and Li, Z. (1999). Landscape ecological analysis of urban vegetation in Guangzhou, China. *Journal of Environmental Sciences*, 11, 160–166.

Guhathakurta, S. and Gober, P. (2007). The impact of the Phoenix heat island on residential water use. *American Planning Association Journal*, 73, 317–329.

Guillitte, O. (1995). Bioreceptivity: a new concept for building ecology studies. *Science of the Total Environment*, 167, 215–220.

Guirado, M., Pino, J. and Roda, F. (2006). Understorey plant species richness and composition in metropolitan forest archipelagos: effects of forest size, adjacent land use and distance to the edge. *Global Ecology and Biogeography*, 15, 50–62.

Gumprecht, B. (1999). *The Los Angeles River: Its Life, Death, and Possible Rebirth*. Baltimore, Maryland: Johns Hopkins University Press.

Guntenspergen, G. R. and Levenson, J. B. (1997). Understory plant species composition in remnant stands along an urban-to-rural land-use gradient. *Urban Ecosystems*, 1, 155–169.

Guntenspergen, G. R., Baldwin, A. H., Hogan, D. M. et al. (2009). Valuing urban wetlands: modification, preservation, and restoration. In *Ecology of Cities and Towns: A Comparative Approach*, eds. M. J. McDonnnell, A. K. Hahs and J. H. Breuste. New York: Cambridge University Press, pp. 503–520.

Gupta, A. and Asher, M. G. (1998). *Environment and the Developing World: Principles, Policies and Management*. New York: John Wiley.

Gupta, A. and Pitts, J., eds. (1992). *The Singapore Story: Physical Adjustments in a Changing Landscape*. Singapore: Singapore University Press.

Gustafson, E. (1998). Quantifying landscape spatial patterns: What is the state of the art? *Ecosystems*, 1, 143–156.

Gut, P. (1993). *Climate Responsive Building*. St. Gallen, Switzerland: SKAT, Niedermann AG.

Gutfreund, O. (2004). *Twentieth-Century Sprawl: Highways and the Reshaping of the American Landscape*. Oxford: Oxford University Press.

Haase, D. and Schetke, S. (2010). Potential of biodiversity and recreation in shrinking cities: contextualization and operationalization. In *Urban Biodiversity and Design*, eds. N. Muller, P. Werner and J. G. Kelcey. New York: Wiley-Blackwell, pp. 518–538.

Habbel, W., Dias dos Santos, A. C., Pasenau, H. and Armaldo Sales, J. (1998). Air quality in a tropical mega-city – Rio de Janeiro, Brazil. In *Urban Ecology*, eds. J. Breuste, H. Feldmann and O. Uhlmann. New York: Springer, pp. 95–98.

Haber, W. (1990). Using landscape ecology in planning and management. In *Changing Landscapes: An Ecological Perspective*, eds. I. S. Zonneveld and R. T. T. Forman. New York: Springer-Verlag, pp. 217–232.

Habraken, N. J. (2000). *The Structure of the Ordinary: Form and Control in the Built Environment*. Cambridge, Massachusetts: MIT Press.

Hadidian, J., Manski, D. A. and Riley, S. (1991). Daytime resting site selection in an urban raccoon population. In *Wildlife Conservation in Metropolitan Environments*, eds. L. W. Adams and D. L. Leedy. Columbia, Maryland: National Institute for Urban Wildlife, pp. 39–45.

Haeupler, H. (2008). Long-term observations of secondary forests growing on hard-coal mining spoils in the Industrial Ruhr Region. In *Ecology, Planning, and Management of Urban Forests: International Perspectives*, eds. M. M. Carreiro, Y.-C. Song and J. Wu. New York: Springer, pp. 357–368.

Hager, S., Trudell, H., McKay, K. *et al.* (2008). Bird density and mortality at windows. *Wilson Journal of Ornithology*, 120, 550–564.

Hahs, A. K. and McDonnell, M. J. (2007). Composition of the plant community in remnant patches of grassy woodland along an urban-rural gradient in Melbourne, Australia. *Urban Ecosystems*, 10, 355–377.

Hahs, A. K., McDonnell, M. J., McCarthy, M. A. *et al.* (2009). A global synthesis of plant extinction rates in urban areas. *Ecology Letters*, 12, 1165–1173.

Hall, M. J. (1984). *Urban Hydrology*. London: Elsevier Applied Science.

Hall, P. (2002). *Cities of Tomorrow: An Intellectual History of Urban Planning and Design in the Twentieth Century*. Oxford: Blackwell Publishing.

Hallmark, C. T. (2000). Managing soils for construction purposes in urban areas. In *Managing Soils in an Urban Environment*, eds. R. B. Brown, J. H. Huddleston and J. L. Anderson. Madison, Wisconsin: American Society of Agronomy, pp. 251–262.

Hamabata, E. (1980). Changes of herb-layer species composition with urbanization of secondary oak forests of Musashino Plain near Tokyo – studies on the conservation of suburban forest stands I. *Japanese Journal of Ecology*, 30, 347–358.

Hamberg, L., Lehvavirta, S., Malmivaara-Lamsa, M. *et al.* (2008). The effects of habitat edges and trampling on understory vegetation in urban forests in Helsinki, Finland. *Applied Vegetation Science*, 11, 83–86.

Hamer, A. J. and McDonnell, M. J. (2008). Amphibian ecology and conservation in the urbanizing world: a review. *Biological Conservation*, 141, 2432–2449.

Hammer, D. A. (1997). *Creating Freshwater Wetlands*. Boca Raton, Florida: Lewis Publishers.

Handy, S. (2005). Smart growth and the transportation-land use connection: what does the research tell us? *International Regional Science Review*, 28, 146–167.

Hanes, J. E. (1993). From megalopolis to megaroporisu. *Journal of Urban History*, 19, 56–94.

Hansen, A. J. and DeFries, R. (2007). Ecological mechanisms linking protected areas to surrounding lands. *Ecological Applications*, 17, 974–988.

Hansen, M. J. and Clevenger, A. P. (2005). The influence of disturbance and habitat on the presence of non-native plant species along transport corridors. *Biological Conservation*, 125, 249–259.

Hanski, I. (1982). Distributional ecology of anthropochorous plants in villages surrounded by forest. *Annales Botanici Fennici*, 19, 1–15.

Hardoy, J. E., Mitlin, D. and Satterthwaite, D. (2004). *Environmental Problems in an Urbanizing World*. London: Earthscan.

Harms, W. B. (1999). Landscape fragmentation by urbanization in the Netherlands: options and ecological consequences. *Journal of Environmental Sciences*, 11, 141–148.

Harper-Lore, B. L., Johnson, M. and Skinner, M. W. (2007). *Roadside Weed Management*. Publication Number FHWA-HEP-07–017. Washington, D.C.: US Department of Transportation.

Harris, J. A., Birch, P. and Palmer, J. P. (1996a). *Land Restoration and Reclamation: Principles and Practice*. Harlow, Essex, UK: Longman.

Harris, L. D. and Scheck, J. (1991). From implications to applications: the dispersal corridor principle applied to the conservation of biological diversity. In *Nature Conservation 2: The Role of Corridors*, eds. D. A. Saunders and R. J. Hobbs. Chipping Norton, Australia: Surrey Beatty, pp. 189–220.

Harris, L. D., Hoctor, T. S. and Gergel, S. E. (1996b). Landscape processes and their significance to biodiversity conservation. In *Population Dynamics in Ecological Space and Time*, eds O. Rhodes, Jr., R. Chesser and M. Smith. Chicago: University of Chicago Press, pp. 319–347.

Hart, S. and Littlefield, D. (2011). *Ecoarchitecture: The Work of Ken Yeang*. New York: John Wiley.

Hartig, E. K., Gornitz, V., Kolker, A. *et al.* (2002). Anthropogenic and climate-change impacts on salt marshes of Jamaica Bay, New York City. *Wetlands*, 22, 71–89.

Hartshorn, T. A. (1992). *Interpreting the City: An Urban Geography*. New York: John Wiley.

References

Hashimoto, H. (2008). Connectivity analyses of avifauna in urban areas. In *Landscape Ecological Applications in Man-Influenced Areas: Linking Man and Nature Systems*, eds. S.-K. Hong, N. Nakagoshi, B. J. Fu and Y. Morimoto. New York: Springer, pp. 479–488.

Haslem, A. and Bennett, A. F. (2008). Birds in agricultural mosaics: the influence of landscape pattern and countryside heterogeneity. *Ecological Applications*, 18, 185–196.

Haspel, C. and Calhoon, R. E. (1991). Ecology and behavior of free-ranging cats in Brooklyn, New York. In *Wildlife Conservation in Metropolitan Environments*, eds. L. W. Adams and D. L. Leedy. Columbia, Maryland: National Institute for Urban Wildlife, pp. 27–30.

Havens, T. R. H. (2011). *Parkscapes: Green Spaces in Modern Japan*. Honolulu: University of Hawai'i Press.

Hawkins, C. C., Grant, W. E. and Longnecker, M. T. (1999). Effect of subsidized house cats on California birds and rodents. *Transactions of the Western Section of the Wildlife Society*, 35, 29–33.

Hawley, A. H. (1944). Ecology and human ecology. *Social Forces*, 2, 398–405.

Hayden, D. (2004). *A Field Guide to Sprawl*. New York: W. W. Norton.

Hayes, E. B. and Piesman, J. P. (2003). How can we prevent Lyme disease? *New England Journal of Medicine*, 348, 2424–2430.

Head, L. and Muir, P. (2006). Edges of connection: reconceptualising the human role in urban biogeography. *Australian Geographer*, 37, 87–101.

Heiden, A. and Leather, S. R. (2004). Biodiversity on urban roundabouts: Hemiptera, management and the species-area relationship. *Basic and Applied Ecology*, 5, 367–377.

Hellmund, P. C. and Smith, D. A., eds. (2006). *Designing Greenways: Sustainable Landscapes for Nature and People*. Washington, D.C.: Island Press.

Henein, K. M. and Merriam, G. (1990). The elements of connectivity where corridor quality is variable. *Landscape Ecology*, 4, 157–170.

Hermansen, L. A. and Macie, E. A. (2005). An assessment of the Southern wildland-urban interface. In *Forests at the Wildland-Urban Interface: Conservation and Management*, eds. S. W. Vince, M. L. Duryea, E. A. Macie and L. A. Hermansen. Boca Raton, Florida: CRC Press, pp. 25–41.

Hersperger, A. M. (2006). Spatial adjacencies and interactions: neighborhood mosaics for landscape ecological planning. *Landscape and Urban Planning*, 77, 227–239.

Hersperger, A. M. and Forman, R. T. T. (2003). Adjacency arrangement effects on plant diversity and composition in woodland patches. *Oikos*, 101, 279–290.

Hersperger, A. M., Langhamer, D. and Dalang, T. (2012). Inventorying human-made objects: a step towards better understanding land use for multifunctional planning in a periurban Swiss landscape. *Landscape and Urban Planning*, 105, 307–314.

Heschel, M. S. and Paige, K. N. (1995). Inbreeding depression, environmental stress, and population size variation in scarlet gilia (*Ipomopsis aggregata*). *Conservation Biology*, 9, 126–133.

Hickin, N. (1985). *Pest Animals in Buildings*. London: George Godwin.

Hickley, P., Arlinghaus, R., Tyner, R. et al. (2004). Rehabilitation of urban lake fisheries for angling by managing habitat: general overview and case studies from England and Wales. *Ecohydrology and Hydrobiology*, 4, 365–378.

Hidalgo-Mihart, M. G., Cantu-Salazar, L., Lopez-Gonzalez, C. A. et al. (2004). Effect of a landfill on the home range and group size of coyotes (*Canis latrans*) in a tropical deciduous forest. *Journal of Zoology*, 263, 55–63.

Hien, W. N., Yok, T. P. and Yu, C. (2007). Study of thermal performance of extensive rooftop greenery systems in the tropical climate. *Building and Environment*, 42, 25–54.

Hill, K. (2009). Urban design and urban water ecosystems. In *The Water Environment of Cities*, ed. L. A. Baker. New York: Springer, pp. 141–170.

Hiller, D. A. and Meusser, M. (1998). *Urbane Boden*. Berlin: Springer.

Hilliard, E. N. (1991). Hypothetical mammalian habitat potential in Syracuse, New York. In *Wildlife Conservation in Metropolitan Environments*, eds. L. W. Adams and D. L. Leedy. Columbia, Maryland: National Institute for Urban Wildlife, pp. 93–98.

Hitchings, S. P. and Beebee, T. J. C. (1997). Genetic substructuring as a result of barriers to gene flow in urban *Rana temporaria* (common frog) populations: implications for biodiversity conservation. *Heredity*, 79, 117–127.

Hobbs, E. R. (1988). Species richness of urban forest patches and implications for urban landscape diversity. *Landscape Ecology*, 1, 141–152.

Hobbs, R. J. and Suding, K. N., eds. (2009). *New Models for Ecosystem Dynamics and Restoration*. Washington, D.C.: Island Press.

Hochuli, D. F., Christie, F. J. and Lomov, B. (2009). Invertebrate biodiversity in urban landscapes: assessing remnant habitat and its restoration. In *Ecology of Cities and Towns: A Comparative Approach*, eds. M. J. McDonnell, A. K. Hahs and J. H. Breuste. New York: Cambridge University Press, pp. 215–232.

Hodge, S. J. and Harmer, R. (1996). Woody colonization on unmanaged urban and ex-industrial sites. *Forestry*, 69, 245–261.

Hodgkinson, S., Hero, J. and Warnken, J. (2007). The conservation value of suburban golf courses in a rapidly urbanising area of Australia. *Landscape and Urban Planning*, 79, 323–337.

Hodgson, J. G. (1986). Commonness and rarity in plants with special reference to the Sheffield flora. Parts I-IV. *Biological Conservation*, 36, 199–314.

Hodgson, P., French, K. and Major, R. E. (2007). Avian movement across abrupt ecological edges: differential responses to housing density in an urban matrix. *Landscape and Urban Planning*, 79, 266–272.

Hofstra, G. and Hall, R. (1971). Injury on roadside trees: leaf injury on pine and white cedar in relation to foliar levels of sodium and chloride. *Canadian Journal of Botany*, 49, 613–622.

Hofstra, G. and Smith, D. W. (1984). The effects of road deicing salt on the levels of ions in roadside soils in Southern Ontario. *Journal of Environmental Management*, 19, 261–271.

Hogan, D. M. and Welbridge, M. R. (2007). Best management practices for nutrient and sediment retention in urban stormwater runoff. *Journal of Environmental Quality*, 36, 386–395.

Hollander, J. B., Kirkwood, N. G. and Gold, J. L. (2011). *Principles of Brownfield Regeneration: Cleanup, Design and Reuse of Derelict Land*. Washington, D.C.: Island Press.

Holling, C. S. (1996). Surprise for science, resilience for ecosystems, and incentives for people. *Ecological Applications*, 6, 733–735.

Hollister, J. W., August, P. V. and Paul, J. F. (2008). Effects of spatial extent on landscape structure and sediment metal concentration relationships in small estuarine systems of the United States' Mid-Atlantic Coast. *Landscape Ecology*, 23, 91–106.

Holzer, C., Hundt, T., Luke, C. and Hamm, O. G. (2008). *Riverscapes: Designing Urban Embankments*. Basel, Switzerland: Birkhauser Verlag AG.

Hong, S.-K., Nakagoshi, N., Fu, B. and Morimoto, Y., eds. (2008). *Landscape Ecological Applications in Man-Influenced Areas: Linking Man and Nature Systems*. New York: Springer.

Hood, M. J., Clausen, J. C. and Warner, G. S. (2007). Comparison of stormwater lag times for low impact and traditional residential development. *Journal of the American Water Resources Association*, 43, 1036–1046.

Hooper, M. and Ortolano, L. (2012). Motivations for slum dweller social movement participation in urban Africa: a study of mobilization in Kurasini, Dar es Salaam. *Environment and Urbanization*, 24, 99–114.

Hope, D., Gries, C., Zhu, W. et al. (2008). Socioeconomics drive urban plant diversity. In *Urban Ecology: An International Perspective on the Interaction Between Humans and Nature*, eds. J. M. Marzluff, E. Schulenberger, W. Endlicher et al. New York, Springer, pp. 339–347.

Houck, M. C. and Cody, M. J. (2000). *Wild in the City: A Guide to Portland's Natural Areas*. Portland, Oregon: Oregon Historical Society Press.

Hough, M. (1990). *Out of Place: Restoring Identity to the Regional Landscape*. New Haven, Connecticut: Yale University Press.

Hough, M. (1995). *Cities and Natural Process*. New York: Routledge.

Hough, M. (2004). *Cities and Natural Process: A Basis for Sustainability*. New York: Routledge.

Hough, R. L., Breward, N., Young, S. D. et al. (2004). Assessing potential risk of heavy metal exposure from consumption of home-produced vegetables by urban populations. *Environmental Health Perspectives*, 112, 215–221.

Howard, E. (1902). *Garden Cities of Tomorrow*. London: Sonnenschein.

Howard, J. (1987). The garden of earthly remains. *Horticulture*, 65, 46–56.

Howe, J. (2002). Planning for urban food: the experience of two UK cities. *Planning Practice and Research*, 17, 125–144.

Howell, J. A. and Pollak, T. (1991). Wildlife habitat analysis for Alcatraz Island, Golden Gate National Recreation Area, California. In *Wildlife Conservation in Metropolitan Environments*, eds. L. W. Adams and D. L. Leedy. Columbia, Maryland: National Institute for Urban Wildlife, pp. 157–164.

Hsieh, C. and Davis, A. P. (2005). Evaluation and optimization of bioretention media for treatment of urban storm water runoff. *Journal of Environmental Engineering ASCE*, 131, 1521–1531.

Hu, D., Wang, R. and Tang, T. (1995). An analysis of the flora in Tianjin, China. In *Urban Ecology as the Basis of Urban Planning*, eds. H. Sukopp, M. Numata and A. Huber. The Hague, Netherlands: SPB Academic Publishing, pp. 59–69.

Huang, B., Shi, X. Z., Yu, D. S. et al. (2006). Environmental assessment of small-scale vegetable farming systems in peri-urban areas of the Yangtze River Delta Region, China. *Agriculture, Ecosystems and Environment*, 112, 391–402.

Hubbard Brook Research Foundation. (2003). *Nitrogen Pollution: From the Sources to the Sea*. Hanover, New Hampshire, USA.

Huijser, M. P. and Clevenger, A. P. (2006). Habitat and corridor function of rights-of-way. In *The Ecology of Transportation: Managing Mobility for the Environment*, eds. J. Davenport and J. L. Davenport. New York: Springer, pp. 233–254.

References

Huner, J. V. (2000). Crawfish and waterbirds. *American Scientist*, 88, 301–303.

Huner, J. V., Jeske, C. W. and Norling, W. (2002). Managing wetlands for waterbirds in coastal regions of Louisiana and Texas, USA. *Waterbirds*, 25, 67–79.

Hunt, A., Dickens, H. J. and Whelan, R. J. (1987). Movement of mammals through tunnels under railway lines. *Australian Zoologist*, 24, 89–93.

Hunter, M. (2011). Using ecological theory to guide urban planting design: an adaptation strategy for climate change. *Landscape Journal*, 30, 2–11.

Hunter, M. L. (1990). *Wildlife, Forests, and Forestry: Principles of Managing Forests for Biological Diversity*. Englewood Cliffs, New Jersey: Prentice-Hall.

Huntington, H. P. (2000). Using traditional ecological knowledge in science: methods and applications. *Ecological Applications*, 10, 1270–1274.

Hurley, S. E. and Forman, R. T. T. (2011). Stormwater ponds and biofilters for large urban sites: modeled arrangements that achieve the phosphorus reduction target for Boston's Charles River, USA. *Ecological Engineering*, 37, 850–863.

Huste, A., Selmi, S. and Boulinier, T. (2006). Bird communities in suburban patches near Paris: determinants of local richness in a highly fragmented landscape. *Ecoscience*, 13, 249–257.

Hynes, H. (1996). *A Patch of Eden: America's Inner-City Gardeners*. White River Junction, Vermont: Chelsea Green Publishing.

Ichinose, T. (2005). Ecological networks for bird species in the wintering season based on urban woodlands. In *Wild Urban Woodlands: New Perspectives for Urban Forestry*, eds. I. Kowarik and S. Korner. New York: Springer, pp. 181–192.

Im, S.-B. (1992). Skyline conservation and management in rapidly growing cities and regions: successes and failures in Korea. In *International Conference on Landscape Planning and Environmental Conservation Proceedings*. Tokyo: University of Tokyo.

Imanishi, A., Kilagawa, C., Nakamura, S. et al. (2007). Changes in herbaceous plants in an urban habitat garden in Kyoto City, Japan, 9 years after construction. *Landscape and Ecological Engineering*, 3, 67–77.

Imhoff, M. L., Lawrence, W. T., Elvidge, C. D. et al. (1997). Using nighttime DMSP/OLS images of city lights to estimate the impact of urban land use on soil resources in the United States. *Remote Sensing of the Environment*, 59, 105–117.

Imhoff, M. L., Zhang, P., Wolfe, R. E. and Bounoua, L. (2010). Remote sensing of the urban heat island effect across biomes in the continental USA. *Remote Sensing of Environment*, 114, 504–513.

Ingegnoli, V. (2002). *Landscape Ecology: A Widening Foundation*. New York, Springer.

IPCC. (2007). *Climate Change 2007: Impacts, Adaptation and Vulnerability*. Intergovernmental Panel on Climate Change. New York: Cambridge University Press.

Iqbal, M. Z. and Shafiq, M. (2000). Periodical effects of automobile pollution on the growth of some roadside trees. *Ekologia-Bratislava*, 19, 104–110.

Irazabal, C. (2005). *City Making and Urban Governance in the Americas: Curitiba and Portland*. Aldershot, UK: Ashgate Publishing.

Ishikawa, M. (2001). *City and Greenspace* [In Japanese]. Tokyo: Iwanami Syoten Ltd.

Ismail, L. H., Sibley, M. and Wahab, I. A. (2011). Bioclimatic technology in high rise office building design: a comparison study for indoor environmental condition. *Journal of Science and Technology*, 3(2). [Online]. URL: http://penerbit.uthm.edu.my/ojs/index.php/JST/article/viewFile/359/243.

Iuell, B., Bekker, H. (G. J.), Cuperus, R., et al., eds. (2003). *Habitat Fragmentation Due to Transportation Infrastructure: Wildlife and Traffic: A European Handbook for Identifying Conflicts and Designing Solutions*. Brussels: KNNV Publishers, COST 341.

Jackson, J. B. C., Kirby, M. X., Berger, W. H. et al. (2001). Historical overfishing and the recent collapse of coastal ecosystems. *Science*, 293, 629–638.

Jackson, R. M. and Raw, F. (1966). *Life in the Soil*. New York: St. Martin's Press.

Jackson, W. B. (1951). Food habits of Baltimore, Maryland cats in relation to rat populations. *Journal of Mammalogy*, 32, 458–461.

Jacobs, A. B. (1993). *Great Streets*. Cambridge, Massachusetts: MIT Press.

Jacobs, J. (1961). *Death and Life of Great American Cities*. New York: Random House.

Jaeger, J. A. G., Bertiller, R., Schwick, C. et al. (2010). Urban permeation of landscape and sprawl per capita: new measures of urban sprawl. *Ecological Indicators*, 10, 427–441.

Jaeger, J. A. G., Raumer, H., Esswein, H. et al. (2007). Time series of landscape fragmentation caused by transportation infrastructure and urban development: a case study from Baden-Wurttemberg, Germany. *Ecology and Society*, 12(1) article 22. [Online]. URL: http://www.ecologyandsociety.org/vol12/iss1/art22/.

Jaren, V., Andersen, R., Ulleberg, M. et al. (1991). Moose-train collisions: the effects of vegetation removal with a cost-benefit analysis. *Alces*, 27, 93–99.

Jenerette, G. D. and Wu, J. G. (2001). Analysis and simulation of land-use change in the Central Arizona-Phoenix Region, USA. *Landscape Ecology*, 16, 611–626.

Jenerette, G. D., Harlan, S. L., Brazel, A. *et al.* (2007). Regional relationships between surface temperature, vegetation, and human settlement in a rapidly urbanizing ecosystem. *Landscape Ecology*, 22, 353–365.

Jennings, D. B. and Jarmagin, S. T. (2002). Changes in anthropogenic impervious surfaces, precipitation and daily streamflow discharge: a historical perspective in a mid-Atlantic subwatershed. *Landscape Ecology*, 17, 471–489.

Jenny, H. (1980). *The Soil Resource: Origin and Behavior.* New York: Springer-Verlag.

Jensen, J. (1990). *Siftings.* Baltimore, Maryland: Johns Hopkins University Press.

Jim, C. Y. and Chen, S. S. (2003). Comprehensive greenspace planning based on landscape ecology principles in compact Nanjing city, China. *Landscape and Urban Planning*, 65, 95–116.

Jim, C. Y. and Chen, W. Y. (2008). Pattern and divergence of tree communities in Taipei's main urban green spaces. *Landscape and Urban Planning*, 84, 312–323.

Jim, C. Y. and Chen, W. Y. (2010). Habitat effect on vegetation ecology and occurrence on urban masonry walls. *Urban Forestry and Urban Greening*, 9, 169–178.

Jim, C. Y. and Chen, W. Y. (2011). Bioreceptivity of buildings for spontaneous arboreal flora in compact city environment. *Urban Forestry and Urban Greening*, 10, 19–28.

Jim, C. Y. and Liu, H. T. (2001). Species diversity of three major urban forest types in Guangzhou City, China. *Forest Ecology and Management*, 146, 99–114.

Jing, W., Yu, S. L. and Rui, Z. (2006). A water quality based approach for watershed wide BMP strategies. *Journal of American Water Resources Association*, 42, 1193–1204.

Johnsen, A. M. and VanDruff, L. W. (1987). Summer and winter distribution of introduced bird species and native bird species richness within a complex urban environment. In *Integrating Man and Nature in the Metropolitan Environment*, eds. L. W. Adams and D. L. Leedy. Columbia, Maryland: National Institute for Urban Wildlife, pp. 123–127.

Johnson, E. A. and Klemens, M. W. (2005). The impacts of sprawl on biodiversity. In *Nature in Fragments: The Legacy of Sprawl*, eds. E. A. Johnson and M. W. Klemens. New York: Columbia University Press, pp. 18–53.

Johnston, J. and Newton, J. (1997). *Building Green: A Guide to Using Plants on Roofs, Walls, and Pavements.* London: The London Ecology Unit.

Johnston, P. and Don, A. (1990). *Grow Your Own Wildlife: How to Improve Your Local Environment.* Canberra: Greening Australia.

Johnston, R., Thomas, R. and Bryant, C. R. (1987). Agricultural adaptation: the prospects for sustaining agriculture near cities. In *Sustaining Agriculture Near Cities*, ed. W. Lockeretz. Ankeny, Iowa: Soil and Water Conservation Society, pp. 9–21.

Johnston, R. F. and Selander, R. K. (2008). House sparrows: rapid evolution of races in North America. In *Urban Ecology: An International Perspective on the Interaction Between Humans and Nature*, eds. J. M. Marzluff, E. Schulenberger, W. Endlicher *et al.* New York: Springer, pp. 315–320.

Jokimaki, J. (1999). Occurrence of breeding bird species in urban parks: effects of park structure and broad-scale variables. *Urban Ecosystems*, 3, 21–34.

Jokimaki, J. and Suhonen, J. (1993). Biological integrity: a long-neglected aspect of water resource management. *Ecological Applications*, 1, 66–84.

Jones, C. I. (2002). *Introduction to Economic Growth.* New York: Norton.

Jones, D. L. (1998). *Architecture and the Environment: Bioclimatic Building Design.* New York: Overlook Press.

Jones, K. (2008). Strategic planning for urban woodlands in North West England. In *Ecology, Planning, and Management of Urban Forests*, eds. M. M. Carreiro, Y.-C. Song and J. Wu. New York: Springer, pp. 199–218.

Jongman, R. H. G. (2002). Homogenization and fragmentation of the European landscape: ecological consequences and solutions. *Landscape and Urban Planning*, 58, 297–308.

Jongman, R. H. G. and Pungetti, G., eds. (2004). *Ecological Networks and Greenways: Concept, Design, Implementation.* New York: Cambridge University Press.

Jongman, R. H. G., Kulvik, M. and Kristiansen, I. (2004). European ecological networks and greenways. *Landscape and Urban Planning*, 68, 305–319.

Jordan, K. K. and Jones, S. C. (2007). Invertebrate diversity in newly established mulch habitats in a Midwestern urban landscape. *Urban Ecosystems*, 10, 87–95.

Jordan, W. R., Gilpin, M. E. and Aber, J. D., eds. (1987). *Restoration Ecology: A Synthetic Approach to Ecological Research.* New York: Cambridge University Press.

Jump, A. S. and Penuelas, J. (2005). Running to stand still: adaptation and the response of plants to rapid climate change. *Ecology Letters*, 8, 1010–1020.

Jun, M. and Hur, J. (2001). Commuting costs of "leap-frog" new town development in Seoul. *Cities*, 18, 151–158.

Junk, W. J., Bayley, P. B. and Sparks, R. E. (1989). The flood pulse concept in river-floodplain systems. *Fisheries and Aquatic Sciences (Canadian Special Publication)*, 106, 110–127.

Kadlec, R. H. and Knight, R. L. (1996). *Treatment Wetlands.* New York: Lewis Publishers.

Kahn, P. and Kellert, S., eds. (2002). *Children and Nature: Psychological, Sociocultural, and Evolutionary Investigations.* Cambridge, Massachusetts: MIT Press.

References

Kalff, J. (2002). *Limnology.* Upper Saddle River, New Jersey: Prentice-Hall.

Kang, R. S. and Marston, R. A. (2006). Geomorphic effects of rural-to-urban land use conversion on three streams in the central Redbed Plains of Oklahoma. *Geomorphology*, 79, 488–506.

Kaplan, S. and Kaplan, R. (2008). Health, supportive environments, and the reasonable person model. In *Urban Ecology: An International Perspective on the Interaction Between Humans and Nature*, eds. J. M. Marzluff, E. Shulenberger, W. Endlicher *et al.* New York: Springer, pp. 557–565.

Kareiva, P., Tallis, H., Ricketts, T. H. *et al.*, eds. (2011). *Natural Capital: Theory and Practice of Mapping Ecosystem Services.* New York: Oxford University Press.

Karr, J. R. and Chu, E. W. (1999). *Restoring Life in Running Water.* Washington, D.C.: Island Press.

Kates, R. W., Clark, W. C., Corell, R. *et al.* (2001). Sustainability science. *Science*, 292, 641–642.

Katz, B. and Land, R. E., eds. (2003). *Redefining Urban and Suburban America: Evidence from Census 2000.* Washington, D.C.: Brookings Institution Press.

Kayden, J. (2000). *Privately Owned Public Space: The New York City Experience.* New York: John Wiley.

Kaye, J. P., Burke, I. C., Mosier, A. R. and Guerschman, J. P. (2004). Methane and nitrous oxide fluxes from urban soils to the atmosphere. *Ecological Applications*, 14, 975–981.

Kaye, J. P., Groffman, P. M., Grimm, N. B. *et al.* (2006). A distinct urban biogeochemistry? *Trends in Ecology and Evolution*, 21, 192–199.

Kays, B. L. (2000). Environmental site assessment: site characterization methodologies. In *Managing Soils in an Urban Environment*, eds. R. B. Brown, J. H. Huddleston and J. L. Anderson. Madison, Wisconsin: American Society of Agronomy, pp. 69–92.

Kays, R. W. and DeWan, A. A. (2004). Ecological impact of inside/outside house cats around a suburban nature preserve. *Animal Conservation*, 7, 273–283.

Keddy, P. A. (2000). *Wetland Ecology: Principles and Conservation.* New York: Cambridge University Press.

Keeley, B. W. and Tuttle, M. D. (1999). *Bats in American Bridges.* Austin, Texas: Bat Conservation International.

Keeley, J. E., Brennan, T. and Pfaff, A. H. (2008). Fire severity and ecosystem responses following crown fires in California shrublands. *Ecological Applications*, 18, 1530–1546.

Kelcey, J. G. (1984). The design and development of gravel pits for wildlife in Milton Keynes, England. *Landscape Planning*, 11, 19–34.

Kellert, S. R. (2005). *Building for Life: Designing and Understanding the Human-Nature Connection.* Washington, D.C.: Island Press.

Kellert, S. R., Heerwagen, J. H. and Mador, M. L., eds. (2008). *Biophilic Design: The Theory, Science, and Practice of Bringing Buildings to Life.* New York: John Wiley.

Kelly, J. R. (1997). Nitrogen flow and the interaction of Boston Harbor with Massachusetts Bay. *Estuaries*, 20, 365–380.

Kendig, L. (1980). *Performance Zoning.* Chicago: Planners Press.

Kendle, A. D. and Rose, J. E. (2000). The aliens have landed! What are the justifications for 'native only' policies in landscape plantings? *Landscape and Urban Planning*, 47, 19–31.

Kennedy, C. E. J. and Southwood, T. R. E. (1984). The number of species of insects associated with British trees: a re-analysis. *Journal of Animal Ecology*, 53, 455–478.

Kenney, W. A. (2008). Potential leaf area index analyses for the City of Toronto's urban forest. In *Ecology, Planning, and Management of Urban Forests*, eds. M. M. Carreiro, Y.-C. Song and J. Wu. New York: Springer, pp. 336–345.

Kenworthy, J. and Laube, F. (2001). *The Millennium Cities Database for Sustainable Cities.* Brussels: International Union of Public Transport.

Keraita, B., Drechsel, P. and Amoah, P. (2003). Influence of urban wastewater on stream water quality and agriculture in and around Kumasi, Ghana. *Environment and Urbanization*, 15, 171–178.

Kery, M. D., Matthies, D. and Spillman, H.-H. (2000). Reduced fecundity and offspring performance in small populations of the declining grassland plants *Primula veris* and *Gentiana lutea*. *Journal of Ecology*, 88, 17–30.

Khai, N. M., Ha, P. Q. and Oborn, I. (2007). Nutrient flows in small-scale peri-urban vegetable farming systems in Southeast Asia – A case study in Hanoi. *Agriculture, Ecosystems and Environment*, 122, 192–202.

Kidder, G. (2000). Management of organic wastes in urban areas. In *Managing Soils in an Urban Environment*, eds. R. B. Brown, J. H. Huddleston and J. L. Anderson. Madison, Wisconsin: American Society of Agronomy, pp. 93–117.

Kieran, J. (1959). *A Natural History of New York City.* Boston: Houghton Mifflin.

Kim, S. and Rowe, P. G. (2012). Does large-sized cities' urbanisation predominantly degrade environmental resources in China? Relationships between urbanisation and resources in the Changjiang Delta Region. *International Journal of Sustainable Development and World Ecology*, 19, 321–329.

King, K. W., Balogh, J., Hughes, K. and Harmel, R. (2007). Nutrient load generated by storm event runoff from a golf course watershed. *Journal of Environmental Quality*, 36, 1021–1030.

Kinouchi, T., Yagi, H. and Miyamoto, M. (2007). Increase in stream temperature related to anthropogenic heat input from urban wastewater. *Journal of Hydrology*, 335, 78–88.

Kirkby, R., Bradbury, I. and Shen, G. (2006). The small town and urban context in China. In: *The Earthscan Reader in Rural-Urban Linkages*, ed. C. Tacoli. London: Earthscan, pp. 184–197.

Kirkwood, N., ed. (2001). *Manfactured Sites: Rethinking the Post-Industrial Landscape*. London: Spon Press.

Kirmer, A., Tischew, S., Ozinga, W. A. *et al.* (2008). Importance of regional species pools and functional traits in colonization processes: predicting re-colonization after large-scale destruction of ecosystems. *Journal of Applied Ecology*, 45, 1523–1530.

Kirpichtchikova, T., Manceaus, A., Lanson, B. *et al.* (2003). Speciation and mobility of Zn, Cu and Pb in a truck farming soil contaminated by sewage irrigation. *Journal de Physique IV*, 107, 695–698.

Kjerfve, B., ed. (1994). *Coastal Lagoon Processes*. Amsterdam: Elsevier.

Klausnitzer, B. (1993). *Ökologie der Grosstadtfauna. 2e bearbeitete Auflage*. Jena and Stuttgart, Germany: Fischer Verlag.

Klem, D., Jr. (1991). Glass and bird kills: an overview and suggested planning and design methods of preventing a fatal hazard. In *Wildlife Conservation in Metropolitan Environments*, eds. L. W. Adams and D. L. Leedy. Columbia, Maryland: National Institute for Urban Wildlife, pp. 99–103.

Klem, D., Jr., Farmer, C. J. and Delacretaz, N. (2009). Architectural and landscape risk factors associated with bird-glass collisions in an urban environment. *Wilson Journal of Ornithology*, 121, 126–134.

Klopatek, J. M. and Gardner, R. H., eds. (1999). *Landscape Ecological Analysis: Issues and Applications*. New York: Springer.

Klotz, S. (1990). Species/area and species/inhabitants relations in European cities. In *Urban Ecology: Plants and Plant Communities in Urban Environments*, eds. H. Sukopp, S. Hejny and I. Kowarik. The Hague, Netherlands: SPB Academic Publishing, pp. 99–103.

Klunder, G. (2005). *Sustainable Solutions for Dutch Housing: Reducing the Environmental Impacts of New and Existing Houses*. Delft, Netherlands: Delft University Press.

Klysik, K. and Fortuniak, K. (1999). Temporal and spatial characteristics of the urban heat island of Lodz, Poland. *Atmospheric Environment*, 33, 3885–3895.

Knaapen, J. P., Scheffer, M. and Harms, B. (1992). Estimating habitat isolation in landscape planning. *Landscape and Urban Planning*, 23, 1–16.

Knight, R. L., Knight, H. A. and Camp, R. J. (1995). Common ravens and number and type of linear rights-of-way. *Biological Conservation*, 74, 65–67.

Knopf, F. L., Johnson, R. R., Rich, T. *et al.* (1988). Conservation of riparian ecosystems in the United States. *Wilson Bulletin*, 100, 272–284.

Knowles, R. N. (1981). *Sun Rhythm Form*. Cambridge, Massachusetts: MIT Press.

Knowlton, K., Rosenzweig, C., Goldberg, R. *et al.* (2004). Evaluating global climate change impacts on local health across a diverse urban region. *Epidemiology*, 15(4), S100 [Abstract].

Knox, E. G., Bouchard, C. E. and Barrett, J. G. (2000). Erosion and sedimentation in urban areas. In *Managing Soils in an Urban Environment*, eds. R. B. Brown, J. H. Huddleston and J. L. Anderson. Madison, Wisconsin: American Society of Agronomy, pp. 179–197.

Koc, M., MacRae, R., Welsh, J. and Mougeot, L. J. A. (1999). *For Hunger-Proof Cities: Sustainable Urban Food Systems*. Ottawa: International Development Research Centre.

Koenig, A. (2005). Quo vadis EIP? How eco-industrial parks are evolving. *Journal of Industrial Ecology*, 9(3), 12–14.

Kohler, E. A., Poole, V., Reicher, Z. and Turco, R. (2004). Nutrient, metal, and pesticide removal during storm and nonstorm events by a constructed wetland on an urban golf course. *Ecological Engineering*, 23, 285–298.

Kohler, M., Schmidt, M., Grimme, F. W. *et al.* (2002). Green roofs in temperate climates and in the hot-humid tropics – far beyond the aesthetics. *Environmental Management and Health*, 13, 382–391.

Kolb, H. H. (1985). Habitat use by foxes in Edinburgh. *Review of Ecological Applications*, 40, 139–143.

Konijnendijk, C. C. (2000). Adapting forestry to urban demands: role of communication in urban forestry in Europe. *Landscape and Urban Planning*, 52, 89–100.

Konijnendijk, C. C., Nilsson, K., Randrup, T. B. and Schipperijn, J., eds. (2005). *Urban Forests and Trees*. New York: Springer.

Konrad, C. P. and Booth, D. B. (2005). Hydrologic changes in urban streams and their ecological significance. *American Fisheries Society Symposium*, 47, 157–177.

Kopecky, K. (1990). Changes of vegetation and pollen respiratory tract allergies on Prague sample. In *Urban Ecology: Plants and Plant Communities in Urban Environments*, eds. H. Sukopp, S. Hejny and I. Kowarik. The Hague, Netherlands: SPB Academic Publishing, pp. 267–271.

Korhnak, L. V. and Vince, S. W. (2005). Managing hydrological impacts of urbanization. In *Forests at the Wildland-Urban Interface: Conservation and Management*, eds. S. W. Vince, M. L. Duryea, E. A. Macie and L. A. Hermansen. Boca Raton, Florida: CRC Press, pp. 175–200.

Kostof, S. (1992). *The City Assembled: The Elements of Urban Form Through History*. Boston: Little, Brown and Co.

Kot, H. (1988). The effect of suburban landscape structure on communities of breeding birds. *Polish Ecological Studies*, 14, 235–261.

References

Kotze, J., Venn, S., Niemela, J. and Spence, J. (2011). Effects of urbanization on the ecology and evolution of arthropods. In *Urban Ecology: Patterns, Processes, and Applications*, eds. J. Niemela, J. H. Breuste, T. Elmqvist et al. New York: Oxford University Press, pp. 159–166.

Kovar, P. (1995). Is plant community organization level relevant to monitoring landscape heterogeneity? Two case studies of mosaic landscapes in the suburban zones of Prague, Czech Republic. *Landscape and Urban Planning*, 32, 137–151.

Kovar, P., ed. (2004). *Natural Recovery of Human-Made Deposits in Landscape (Biotic Interactions and Ore/Ash-Slag Artificial Ecosystems)*. Prague: Academia.

Kowarik, I. (1990). Some responses of flora and vegetation to urbanization in Central Europe. In *Urban Ecology: Plants and Plant Communities in Urban Environments*, eds. H. Sukopp, S. Hejny and I. Kowarik. The Hague, Netherlands: SPB Academic Publishing, pp. 45–74.

Kowarik, I. and Korner, S., eds. (2005). *Wild Urban Woodlands: New Perspectives for Urban Forestry*. New York: Springer.

Kowarik, I. and Langer, A. (2005). Natur-Park Sudgelande: linking conservation and recreation in an abandoned railyard in Berlin. In *Wild Urban Woodlands: New Perspectives for Urban Forestry*, eds. I. Kowarik and S. Korner. New York: Springer, pp. 287–299.

Kowarik, I. and von der Lippe, M. (2011). Secondary wind dispersal enhances long-distance dispersal of an invasive species in urban road corridors. *NeoBiota*, 9, 49–70.

Kramer, M. (2006). *Dispossessed: Life in Our World's Urban Slums*. Maryknoll, New York: Orbis Books.

Krause, C. W., Lockard, B., Newcomb, T. J. et al. (2004). Predicting influences of urban development on thermal habitat in a warm water stream. *Journal of the American Water Resources Association*, 40, 1645–1658.

Krebs, C. J. (1972). *Ecology: The Experimental Analysis of Distribution and Abundance*. New York: Harper & Row.

Kress, S. W. (1985). *The Audubon Society Guide to Attracting Birds*. New York: Scribners.

Krieger, A., Cobb, D. and Turner, A., eds. (1999). *Mapping Boston*. Cambridge, Massachusetts: MIT Press.

Krier, R. (1979). *Urban Space*. London: Academy Editions.

Kruger, F. (2006). Taking advantage of rural assets as a coping strategy for the urban poor: the case of rural-urban interrelations in Botswana. In *The Earthscan Reader in Rural-Urban Linkages*, ed. C. Tacoli. London: Earthscan, pp. 229–243.

Kubikova, J. (1990). Natural and semi-natural plant communities of the city of Prague, Czechoslovakia. In *Urban Ecology: Plants and Plant Communities in Urban Environments*, eds. H. Sukopp, S. Hejny and I. Kowarik. The Hague, Netherlands: SPB Academic Publishing, pp. 131–139.

Kuhn, A., Ballach, H.-J. and Wittig, R. (1998). Vegetation as a sink for PAH in urban regions. In *Urban Ecology*, eds. J. Breuste, H. Feldmann and O. Uhlmann. New York: Springer, pp. 171–173.

Kuhn, L., Brandl, R., and Klotz, S. (2004). The flora of German cities is naturally species rich. *Evolutionary Ecology Research*, 6, 749–764.

Kuhn, M. (2002). Greenbelt and Green Heart: separating and integrating landscapes in European city regions. *Landscape and Urban Planning*, 57, 1–9.

Kunick, W. (1990). Spontaneous woody vegetation in cities. In *Urban Ecology: Plants and Plant Communities in Urban Environments*, eds. H. Sukopp, S. Hejny and I. Kowarik. The Hague, Netherlands: SPB Academic Publishing, pp. 167–174.

Kuntz, K. L. and Larson, D. W. (2006). Microtopographic control of vascular plant, bryophyte and lichen communities on cliff faces. *Plant Ecology*, 185, 239–253.

Kunz, T. H. and Racey, P. A., eds. (1998). *Bat Biology and Conservation*. Washington, D.C.: Smithsonian Institution Press.

Kunz, T. H., Gauthreaux, S. A., Nristov, N. I. et al. (2008). Aeroecology: probing and modeling the aerosphere. *Integrative and Comparative Biology*, 48, 1–11.

Kurta, A. and Teramino, J. A. (1992). Bat community structure in an urban park. *Ecography*, 15, 257–261.

Kusler, J. A. (1990). *Wetland Creation and Restoration: The Status of the Science*. Washington, D.C.: Island Press.

Kuttler, W. (1993). Stadtklima. In *Stadtokologie*, eds. H. Sukopp and R. Wittig. Stuttgart: G. Fischer Verlag, pp. 113–153.

Kuttler, W. (2008). The urban climate – basic and applied aspects. In *Urban Ecology: An International Perspective on the Interaction Between Humans and Nature*, eds. J. M. Marzluff, E. Schulenberger, W. Endlicher et al. New York: Springer, pp. 233–248.

La Polla, V. N. and Barrett, G. W. (1993). Effects of corridor width and presence on the population dynamics of the meadow vole. *Landscape Ecology*, 8, 25–37.

La Sorte, F. A., McKinney, M. L. and Pysek, P. (2007). Compositional similarity among urban floras with and across continents: biogeographical consequences of human-mediated biotic interchange. *Global Change Biology*, 13, 913–921.

Laakso, L., Koponen, I. K., Monkkonen, P. et al. (2006). Aerosol particles in the developing world: a comparison between New Delhi in India and Beijing in China. *Water, Air and Soil Pollution*, 173, 5–20.

Ladson, A. R. et al. (2006). Improving stream health in urban areas by reducing runoff frequency from impervious surfaces. *Australian Journal of Water Resources*, 10, 23–32.

References

Laenen, A. and Dunnette, D. A. (1997). *River Quality Dynamics and Restoration.* New York: Lewis Publishers.

Laet, J. D. and Summers-Smith, J. D. (2007). The status of the urban house sparrow *Passer domesticus* in north-western Europe: a review. *Journal of Ornithology*, 148, 275–278.

Lai, L.-W. and Chengb, W.-L. (2008). Air quality influenced by urban heat island coupled with synoptic weather patterns. *Science of the Total Environment*, 407, 2723–2724.

Lake, P. S. (2011). *Drought and Aquatic Ecosystems: Effects and Responses.* Chichester, UK: Wiley-Blackwell.

Lake, P. S., Bond, N. and Reich, P. (2007). Linking ecological theory with stream restoration. *Freshwater Biology*, 52, 597–615.

Lambers, H., Chapin III, F. S. and Pons, T. L. (1998). *Plant Physiological Ecology.* New York: Springer.

Lambert, A. J. D. and Boons, F. A. (2002). Eco-industrial parks: stimulating sustainable development in mixed industrial parks. *Technovation*, 22, 471–484.

Landsberg, H. (1981). *Urban Climate.* New York: Academic Press.

Langner, M. and Endlicher, W., eds. (2007). *Shrinking Cities: Effects on Urban Ecology and Challenges for Urban Development.* Frankfurt am Main, Germany: Peter Lang.

Langton, T. E. S., ed. (1989). *Amphibians and Roads.* Shefford, Bedfordshire, UK: ACO Polymer Products Ltd.

Larm, T. (2000). Stormwater quantity and quality in a multiple pond-wetland system: Flemingsbergsviken case study. *Ecological Engineering*, 15, 57–75.

Larson, D. W., Matthes, U. and Kelly, P. E. (2000). *Cliff Ecology.* New York: Cambridge University Press.

Lasserre, P. and Marzolo, A. (2000). *The Venice Lagoon Ecosystem.* Paris: UNESCO and Parthenon Publishing Group.

Laurance, W. F., Goosem, M. and Laurance, S. G. (2009). Impacts of roads and linear clearings on tropical forests. *Trends in Ecology and Evolution*, 24, 659–669.

Laurie, I. C., ed. (1979). *Nature in Cities: The Natural Environment in the Design and Development of Urban Green Space.* New York: John Wiley.

Lavelle, P., Chauvel, A. and Fragoso, C. (1995). Faunal activity in acid soils. In *Plant Soil Interactions at Low pH*, eds. R. A. Date, N. J. Gundon and G. E. Rayment *et al.* Amsterdam: Kluwer, pp. 201–211.

Lavielle, D. and Petterson, T. (2007). Evolution of pollutant removal efficiency in storm water ponds due to changes in pond morphology. *Highway and Urban Environment*, 12, 429–439.

Law, I. B. (2003). Advanced reuse – from Windhoek to Singapore and beyond. *Water*, May 2003, 31–36.

Law, N. L., Band, L. E. and Grove, J. M. (2004). Nitrogen input from residential lawn care practices in suburban watersheds in Baltimore County, MD. *Journal of Environmental Planning and Management*, 47, 737–755.

Lawrence, J. M. and Gresens, S. E. (2004). Foodweb response to nutrient enrichment in rural and urban streams. *Journal of Freshwater Ecology*, 19, 375–385.

Lawson, L. J. (2005). *City Bountiful: A Century of Community Gardening in America.* Berkeley, California: University of California Press.

Lay, M. G. (1992). *Ways of the World: History of the World's Roads and of the Vehicles That Used Them.* New Brunswick, New Jersey: Rutgers University Press.

Le Blanc, F. and Rao, D. N. (1973). Evaluation of the pollution and drought hypotheses in relation to lichens and bryophytes in urban environments. *Bryologist*, 76, 1–19.

Le Gates, R. T. and Stout, F., eds. (1996). *The City Reader.* London: Routledge.

LeClerc, J. E. and Cristol, D. A. (2005). Are golf courses providing habitat for birds of conservation concern in Virginia? *Wildlife Society Bulletin*, 33, 463–470.

Lee, C. S. and Cho, Y. C. (2008a). Restoration planning for the Seoul metropolitan area, Korea. In *Ecology, Planning, and Management of Urban Forests: International Perspectives*, eds. M. M. Carreiro, Y.-C. Song and J. Wu. New York: Springer, pp. 393–419.

Lee, C. S. and Cho, Y. C. (2008b). Selection of pollution-tolerant trees for restoration of degraded forests and evaluation of the experimental restoration practices at the Ulsan Industrial Complex, Korea. In *Ecology, Planning, and Management of Urban Forests*, eds. M. M. Carreiro, Y.-C. Song and J. Wu. New York: Springer, pp. 369–392.

Lee, G. F. and Jones, R. A. (1990). *Use of Landfill Mining in Solid Waste Management.* Chicago: Water Pollution Control Federation.

Lee, H.-S., Shepley, M. and Huang, C.-S. (2009). Evaluation of off-leash dog parks in Texas and Florida: a study of use patterns, user satisfaction, and perception. *Landscape and Urban Planning*, 92, 314–324.

Lee, J. G. and Heaney, J. P. (2003). Estimation of urban imperviousness and its impacts on storm water systems. *Journal of Water Resources Planning and Management*, 29, 419–426.

Lee, S. L., Lee, A. N. and Cho, Y. C. (2008). Restoration planning for the Seoul metropolitan area, Korea. In *Ecology, Planning, and Management of Urban Forests*, eds. M. M. Carreiro, Y.-C. Song and J. Wu. New York: Springer, pp. 393–419.

Lee, S. Y., Dunn, R. J. K., Young, R. A. *et al.*, (2006). Impact of urbanization on coastal wetland structure and function. *Austral Ecology*, 31, 149–163.

References

Lehman, R. N. (2001). Raptor electrocution on power lines: current issues and outlook. *Wildlife Society Bulletin*, 29, 804–813.

Lehner, P. N., McCluggage, C., Mitchell, D. R. and Neil, D. H. (1983). Selected parameters of the Fort Collins, Colorado, dog population, 1979–80. *Applied Animal Ethology*, 10, 19–25.

Leitao, A. B., Miller, J., Ahern, J. and McGarigal, K. (2006). *Measuring Landscapes: A Planner's Handbook*. Washington, D.C.: Island Press.

Leopold, A. (1933). *Game Management*. New York: Scribners.

Lepczyk, C. A. and Warren, P. S., eds. (2012). *Urban Bird Ecology and Conservation*. Berkeley, California: University of California Press.

Lepczyk, C. A., Mertig, A. G. and Liu, J. (2003). Landowners and cat predation across rural-to-urban landscapes. *Biological Conservation*, 115, 191–201.

Leslie, M., Meffe, G. K., Hardesty, J. L. and Adams, D. L. (1996). *Conserving Biodiversity on Military Lands: A Handbook for Natural Resources Managers*. Arlington, Virginia: The Nature Conservancy.

Levin, S. A. (2012). The challenge of sustainability: lessons from an evolutionary perspective. In *Sustainability Science: The Emerging Paradigm and the Urban Environment*, eds. M. P. Weinstein and R. E. Turner. New York: Springer, pp. 431–437.

Levinson, M. (2006). *The Box: How the Shipping Container Made the World Smaller and the World Economy Bigger*. Princeton, New Jersey: Princeton University Press.

Levinton, J. S. and Drew, C. (2006). *Assessment of Population Levels, Biodiversity, and Design of Substrates That Maximize Colonization in the New York Harbor*. New York: Hudson River Foundation and SUNY Research Foundation.

Li, D. W., Shi, Y., He, X. Y. et al. (2008a). Volatile organic compound emissions from urban trees in Shenyang, China. *Botanical Studies*, 49, 67–72.

Li, J., Wang, Y. and Song, Y.-C. (2008b). Landscape corridors in Shanghai and their importance in urban forest planning. In *Ecology, Planning, and Management of Urban Forests*, eds. M. M. Carreiro, Y.-C. Song and J. Wu. New York: Springer, pp. 219–239.

Li, T., Shilling, F., Thorne, J. et al. (2010). Fragmentation of China's landscape by roads and urban areas. *Landscape Ecology*, 25, 839–853.

Liddle, M. (1997). *Recreation Ecology*. London: Chapman & Hall.

Limburg, K. E. and Schmidt, R. E. (1990). Patterns of fish spawning in Hudson River tributaries: response to an urban gradient? *Ecology*, 71, 1238–1245.

Lincoln Institute of Land Policy. (2004). *Recycling the City: The Use and Reuse of Urban Land*. Cambridge, Massachusetts.

Linden-Ward, B. (1989). *Silent City on a Hill: Landscapes of Memory and Boston's Mount Auburn Cemetery*. Columbus, Ohio: Ohio State University Press.

Lindenmayer, D. B. and Fischer, J. (2006). *Habitat Fragmentation and Landscape Change: An Ecological and Conservation Synthesis*. Washington, D.C.: Island Press.

Linnell, M. A. (2009). Using wedelia as ground cover on tropical airports to reduce bird activity. *Human-Wildlife Conflicts*, 3, 226–236.

Lister, N.-M. (2007). Placing food: Toronto's edible landscape. In *Food*, ed. J. Knechtel. Cambridge, Massachusetts: MIT Press.

Little, C. E. (1990). *Greenways for America*. Baltimore, Maryland: Johns Hopkins University Press.

Liu, J., Dietz, T., Carpenter, S. R. et al. (2007). Complexity of coupled human and natural systems. *Science*, 317, 1513–1516.

Liwarska-Bizukojc, E., Bizukojc, M., Marcinkowski, A. and Doniac, A. (2008). The conceptual model of an eco-industrial park based upon ecological relationships. *Journal of Cleaner Production*, 17, 732–741.

Loeb, R. E. (2006). A comparative flora of large urban parks: intraurban and interurban similarity in the megalopolis of the northeastern United States. *Journal of the Torrey Botanical Society*, 133, 601–625.

Loehle, C. (2004). Applying landscape principles to fire hazard reduction. *Forest Ecology Management*, 198, 261–267.

Lohr, V. I. and Pearson-Mims, C. H. (1996). Particulate matter accumulation on horizontal surfaces in interiors: influence of foliage plants. *Atmospheric Environment*, 30, 2565–2568.

Lohr, V. I., Pearson-Mims, C. H. and Goodwin, G. K. (1996). Interior plants may improve worker productivity and reduce stress in a windowless environment. *Journal of Environmental Horticulture*, 14, 97–100.

Loizeaux-Bennett, S. (1999). Stormwater and nonpoint-source runoff: a primer on stormwater management. *Erosion Control*, 6, 56–59.

Lokschin, L. X. (2007). Power lines and howler monkey conservation in Porto Allegre, Rio Grande do Sul, Brazil. *Neotropical Primates*, 14, 76–80.

Lopez, T. D., Aide, T. M. and Thomlinson, J. R. (2001). Urban expansion and the loss of prime agricultural lands in Puerto Rico. *Ambio*, 30, 49–54.

Loram, A., Tratalos, J., Warren, P. H. and Gaston, K. J. (2007). Urban domestic gardens (X): the extent and structure of the resource in five major cities. *Landscape Ecology*, 22, 601–615.

Lorenz, G. C. and Barrett, G. W. (1990). The influence of simulated landscape corridors on house mouse (*Mus musculus*) dispersal. *American Midland Naturalist*, 123, 348–356.

Losada, H., Martinez, H., Vieyra, J. et al. (1998). Urban agriculture in the metropolitan zone of Mexico City: changes over time in urban, suburban, and peri-urban areas. *Environment and Urbanization*, 10, 37–54.

Losch, A. (1954). *The Economics of Location*. New Haven, Connecticut: Yale University Press.

Losos, J. B. (2009). *Lizards in an Evolutionary Tree: Ecology and Adaptive Radiation of Anoles*. Berkeley, California: University of California Press.

Lososova, Z., Chytry, M., Kuhn, I. et al. (2006). Patterns of plant traits in annual vegetation of man-made habitats in central Europe. *Perspectives in Plant Ecology, Evolution, and Systematics*, 8, 69–81.

Loss, S. R., Ruiz, M. O. and Brawn, J. D. (2009). Relationships between avian diversity, neighborhood age, income, and environmental characteristics of the urban landscape. *Biological Conservation*, 142, 2578–2585.

Lucas, M. and Baras, E. (2001). *Migration of Freshwater Fishes*. Malden, Massachusetts: Blackwell Science.

Luck, G. W., Ricketts, T. H., Daily, G. C. and Imhoff, M. (2004). Alleviating spatial conflict between people and biodiversity. *Proceedings of the National Academy of Sciences (USA)*, 101, 182–186.

Luck, M. and Wu, J. (2002). A gradient analysis of urban landscape pattern: a case study from the Phoenix metropolitan region. *Landscape Ecology*, 17, 327–339.

Luck, M. A., Jenerette, G. D., Wu, J. and Grimm, N. B. (2001). The urban funnel model and the spatially heterogeneous ecological footprint. *Ecosystems*, 4, 782–796.

Luken, J. O., Hinton, A. C. and Baker, D. G. (1992). Response of woody plant communities in power-line corridors to frequent anthropogenic disturbance. *Ecological Applications*, 2, 356–362.

Lukez, P. (2007). *Suburban Transformations*. New York: Princeton Architectural Press.

Lundholm, J. T. and Martin, A. (2006). Habitat origins and microhabitat preferences of urban plant species. *Urban Ecosystems*, 9, 139–159.

Luniak, M. (2008). Fauna of the big city – estimating species richness and abundance in Warsaw, Poland. In *Urban Ecology: An International Perspective on the Interaction Between Humans and Nature*, eds. J. M. Marzluff, E. Schulenberger, W. Endlicher et al. New York: Springer, pp. 349–354.

Luo, Z. K., Sun, O. J., Ge, Q. S. et al. (2007). Phenological responses of plants to climate change in an urban environment. *Ecological Research*, 22, 507–514.

Lussenhop, J. (1977). Urban cemeteries as bird refuges. *Condor*, 79, 456–461.

Lutgens, F. H. and Tarbuck, E. J. (1998). *The Atmosphere*. Englewood Cliffs, New Jersey: Prentice Hall.

Lyle, J. and Quinn, R. D. (1991). Ecological corridors in urban Southern California. In *Wildlife Conservation in Metropolitan Environments*, eds. L. W. Adams and D. L. Leedy. Columbia, Maryland: National Institute for Urban Wildlife, pp. 105–116.

Lynch, K. (1981). *A Theory of Good City Form*. Cambridge, Massachusetts: MIT Press.

Lynch, K. and Hack, G. (1996) *Site Planning*. Cambridge, Massachusetts: MIT Press.

Ma, J. (2004). *China's Water Crisis*. Norwalk, Connecticut: EastBridge Publishing.

Ma, S. (1985). Ecological engineering: application of ecosystem principles. *Environmental Conservation*, 12(4), 331–335.

MacArthur, R. H. and Wilson, E. O. (1967). *The Theory of Island Biogeography*. Princeton, New Jersey: Princeton University Press.

McBean, E. A., Rovers, F. A. and Farquhar, G. J. (1995). *Solid Waste Landfill Engineering and Design*. Englewood Cliffs, New Jersey: Prentice Hall.

McCarthy, M. (2009). Using models to compare the ecology of cities. In *Ecology of Cities and Towns: A Comparative Approach*, eds. M. J. McDonnell, A. K. Hahs and J. H. Breuste. New York: Cambridge University Press, pp. 112–125.

McCarthy, R. J., Levine, S. H. and Reed, J. M. (2013). Estimation of effectiveness of three methods of feral cat population control by use of a simulation model. *Journal of the Veterinary Medicine Association*, 243, 502–511.

Maclean, A. and Campoli, J. (2007). *Visualizing Density*. Cambridge, Massachusetts: Lincoln Institute for Land Policy.

McCleave, J. D., Arnold, G. P., Dodson, J. J. and Neill, W. H., eds. (1984). *Mechanisms of Migration in Fishes*. New York: Plenum Press.

McDonald, R. and Marcotullio, P. (2011). Global effects of urbanization on ecosystem services. In *Urban Ecology: Patterns, Processes, and Applications*, eds. J. Niemela, J. H. Breuste, T. Elmqvist et al. New York: Oxford University Press, pp. 193–205.

McDonald, R. I., Kareiva, P. and Forman, R. T. T. (2008). The implications of urban growth for global protected areas and biodiversity conservation. *Biological Conservation*, 141, 1695–1703.

McDonnell, M. J. (2011). The history of urban ecology – an ecologist's perspective. In *Urban Ecology: Patterns, Processes, and Applications*, eds. J. Niemela, J. H. Breuste,

T. Elmqvist et al. Oxford: Oxford University Press, pp. 5–13.

McDonnell, M. J. and Hahs, A. K. (2008). The use of gradient studies in advancing our understanding of the ecology of urbanizing landscapes; current status and future directions. *Landscape Ecology*, 23, 1143–1155.

McDonnell, M. J. and Pickett, S. T. A. (1990). Ecosystem structure and function along urban-rural gradients: an unexploited opportunity for ecology. *Ecology*, 71, 1232–1237.

McDonnell, M. J. and Pickett, S. T. A., eds. (1993). *Humans as Components of Ecosystems: Subtle Human Effects and the Ecology of Populated Areas*. New York: Springer-Verlag.

McDonnell, M. J., Hahs, A. K. and Breuste, J. H., eds. (2009). *Ecology of Cities and Towns: A Comparative Approach*. New York: Cambridge University Press.

McDonnell, M. J., Pickett, S. T. A., Groffman, P. et al. (1997). Ecosystem processes along an urban-to-rural gradient. *Urban Ecosystems*, 1, 21–36. (Reprinted 2008 in *Urban Ecology: An International Perspective on the Interaction Between Humans and Nature*, eds. J. M. Marzluff, E. Schulenberger, W. Endlicher et al. New York: Springer, pp. 299–313).

McDonnell, M. J., Pickett, S. T. A. and Pouyat, R. V. (1993). The application of the ecological gradient paradigm to the study of urban effects. In *Humans as Components of Ecosystems: The Ecology of Subtle Human Effects and Populated Areas*, eds. M. J. McDonnell and S. T. A. Pickett. New York: Springer-Verlag, pp. 175–189.

McElroy, M. B. (2010). *Energy: Perspectives, Problems and Prospects*. New York: Oxford University Press.

McGarigal, K. and Cushman, S. A. (2005). The gradient concept of landscape structure. In *Issues and Perspectives in Landscape Ecology*, eds. J. Wiens and M. Moss. New York: Cambridge University Press, pp. 112–119.

McGarigal, K. and Marks, B. J. (1995). *FRAGSTATS: Spatial Pattern Analysis Program for Quantifying Landscape Structure*. General Technical Report PNW-351. Portland, Oregon: USDA Forest Service.

McGranahan, G. (2006). An overview of the urban environmental burdens at three scales: intra-urban, urban-regional and global. In *The Earthscan Reader in Rural-Urban Linkages*, ed. C. Tacoli. London: Earthscan, pp. 298–319.

McGranahan, G., Balk, D. and Anderson, B. (2007). The rising tide: assessing the risks of climate change and human settlements in low elevation coastal zones. *Environment and Urbanization*, 19, 17–37.

McGregor, D., Simon, D. and Thompson, D., eds. (2006). *The Peri-Urban Interface: Approaches to Sustainable Natural and Human Resource Use*. London: Earthscan.

McHarg, I. (1969). *Design with Nature*. Garden City, New York: Natural History Press.

McIntosh, R. P. (1967). The continuum concept of vegetation. *Botanical Review*, 33, 130–187.

McIntosh, R. P. (1985). *The Background of Ecology: Concept and Theory*. New York: Cambridge University Press.

McIntyre, N. and Hobbs, R. (1999). A framework for conceptualizing human effects on landscapes and its relevance to management and research models. *Conservation Biology*, 13, 1282–1292.

McIntyre, N. E. and Rango, J. J. (2009). Arthropods in urban ecosystems: community patterns as functions of anthropogenic land use. In *Ecology of Cities and Towns: A Comparative Approach*, eds. M. J. McDonnell, A. K. Hahs and J. H. Breuste. New York: Cambridge University Press, pp. 233–242.

McIntyre, N. E., Knowles-Yanez, K. and Hope, D. (2000). Urban ecology as an interdisciplinary field: differences in the use of "urban" between the social and natural sciences. *Urban Ecosystems*, 4, 5–24.

McIntyre, N. E., Rango, J., Fagan, W. F. and Faeth, S. H. (2001). Ground arthropod community structure in a heterogeneous urban environment. *Landscape and Urban Planning*, 52, 257–274.

McKinney, M. L. (2002). Urbanization, biodiversity, and conservation. *BioScience*, 52, 883–890.

McKinney, M. L. (2006). Urbanization as a major cause of biotic homogenization. *Biological Conservation* 127: 247–260.

McKinney, R. E. (2004). *Environmental Pollution Control Microbiology*. New York: M. Dekker.

McMillan, A. M., Bagley, M. J., Jackson, S. A. and Nacci, D. E. (2006). Genetic diversity and structure of an estuarine fish (*Fundulus heteroclitus*) indigenous to sites associated with a highly contaminated urban harbor. *Ecotoxicology*, 15, 539–548.

McNeill, J. R. (2000). *Something New Under the Sun: An Environmental History of the Twentieth-Century World*. New York: Norton.

McPherson, E. G. (1994a). Cooling urban heat islands with sustainable landscapes. In *The Ecological City: Preserving and Restoring Urban Biodiversity*, eds. R. H. Platt, R. A. Rowntree and P. C. Muick. Amherst, Massachusetts: University of Massachusetts Press, pp. 151–171.

McPherson, E. G. (1994b). Benefits and costs of tree planting and care in Chicago. In *Chicago's Urban Forest Ecosystem: Results of the Chicago Urban Forest Climate Project*, eds. E. G. McPherson, D. J. Nowak and R. A. Rowntree. General Technical Report NE-186. Radnor, Pennsylvania: US Department of Agriculture, Forest Service, pp. 115–134.

Maconachie, R. (2007). *Urban Growth and Land Degradation in Developing Cities: Change and Challenges in Kano, Nigeria*. Aldershot, England: Ashgate Publishing.

MacPherson, C., Meslin, F. and Wandeler, A., eds. (2000). *Dogs, Zoonoses and Public Health*. London: CABI Publishing.

McPherson, E. G., Simpson, J. R., Peper, P. J. and Xiao, Q. (1999). Benefit-cost analysis of Modesto's municipal urban forest. *Journal of Arboriculture*, 25: 235–248.

Madronich, S. (2006). Chemical evolution of gaseous air pollutants down-wind of tropical megacities: Mexico City case study. *Atmospheric Environment*, 40, 6012–6018.

Magle, S. B. (2008). Observations on body mass of prairie dogs in urban habitat. *Western North American Naturalist*, 68, 113–118.

Magle, S. B., Reyes, P., Zhu, J. and Crooks, K. R. (2010). Extirpation, colonization, and habitat dynamics of a keystone species along an urban gradient. *Biological Conservation*, 143, 2146–2155.

Main, H. and Williams, S. W. (1994). *Environment and Housing in Third World Cities*. New York: John Wiley.

Maire, R. and Pomel, S. (1994). Karst geomorphology and environment. In *Groundwater Ecology*, eds. J. Gibert, D. L. Danielopol and J. A. Stanford. San Diego, California: Academic Press, pp. 130–155.

Malcom, H. R., Avera, M. E., Bullard, C. M. and Lancaster, C. C. (1986). *Stormwater Management in Urban Collector Streams*. Raleigh, North Carolina: Water Resources Research Institute, North Carolina State University.

Mallin, M. A., Ensign, S. H., Wheeler, T. L. and Mayes, D. B. (2002). Pollutant removal efficiency of three wet detention ponds. *Journal of Environmental Quality*, 31, 654–660.

Mallin, M. A., Williams, K. E., Esham, E. C. and Lowe, R. P. (2000). Effect of human development on bacteriological water quality in coastal watersheds. *Ecological Applications*, 10, 1047–1056.

Mander, U. and Kimmel, K. (2008). Wetlands and riparian buffer zones in landscape functioning. In *Landscape Ecological Applications in Man-Influenced Areas*, eds. S.-K. Hong, N. Nakagoshi, B. Fu and Y. Morimoto. New York: Springer, pp. 329–357.

Mangin, A. (1994). Karst hydrogeology. In *Groundwater Ecology*, eds. J. Gibert, D. L. Danielopol and J. A. Stanford. San Diego, California: Academic Press, pp. 43–67.

Manos, C. G., Pateimandlik, K. J., Ross, B. J. et al. (1991). Prevalence of asbestos in sewage sludges from 51 large and small cities in the United States. *Chemosphere*, 22, 963–973.

Mansuroglu, S., Ortacesme, V. and Karaguzei, O. (2006). Biotope mapping in an urban environment and its implications for urban management in Turkey. *Journal of Environmental Management*, 81, 175–187.

Margat, J. (1994). Groundwater operations and management. In *Groundwater Ecology*, eds. J. Gibert, D. L. Danielopol and J. A. Stanford. San Diego, California: Academic Press, pp. 505–522.

Margolis, L. and Robinson, A. (2007). *Living Systems: Innovative Materials and Technologies for Landscape Architecture*. Basel, Switzerland: Birkhauser.

Marks, B. K. and Duncan, R. S. (2009). Use of forest edges by free-ranging cats and dogs in an urban forest fragment. *Southeastern Naturalist*, 8, 427–436.

Marsh, W. M. (2005). *Landscape Planning: Environmental Applications*. 4th edition. New York: John Wiley.

Marsh, W. M. (2010). *Landscape Planning: Environmental Applications*. 5th edition. New York: John Wiley.

Marshall, A. (2000). *How Cities Work, Sprawl, and the Road Not Taken*. Austin, Texas: University of Texas Press.

Marshall, J. (2005a). Megacity, mega mess. *Nature*, 437, 312.

Marshall, S. (2005b). *Streets and Patterns: The Structure of Urban Geometry*. London: Spon Press.

Martin, D., Bertasi, F., Colangelo, M. A. et al. (2005). Ecological impact of coastal defense structures on sediment and mobile fauna: evaluating and forecasting consequences of unavoidable modifications of native habitats. *Coastal Engineering*, 52, 1027–1051.

Marzluff, J. M. (2005). Island biogeography for an urbanizing world: how extinction and colonization may determine biological diversity in human-dominated landscapes. *Urban Ecosystems*, 8, 155–175.

Marzluff, J. M. (2012). Urban evolutionary ecology. In *Urban Bird Ecology and Conservation*, eds. C. A. Lepczyk and P. S. Warren. Berkeley, California: University of California Press, pp. 167–182.

Marzluff, J. M. and Ewing, K. (2008). Restoration of fragmented landscapes for the conservation of birds: a general framework and specific recommendations for urbanizing landscapes. In *Urban Ecology: An International Perspective on the Interaction Between Humans and Nature*, eds. J. M. Marzluff, E. Schulenberger, W. Endlicher et al. New York: Springer, pp. 739–755.

Marzluff, J. M., Bowman, R. and Donnelly, R., eds. (2001). *Avian Ecology and Conservation in an Urbanizing World*. Norwell, Massachusetts: Kluwer Academic Publishers.

Marzluff, J. M., Schulenberger, E., Endlicher, W. et al., eds. (2008) *Urban Ecology: An International Perspective on the Interaction Between Humans and Nature*. New York: Springer.

Maskell, L. C., Bullock, J. M., Smart, S. M. et al. (2006). The distribution and habitat associations of non-native plant species in urban riparian habitats. *Journal of Vegetation Science*, 17, 499–508.

Mason, J., Moorman, C., Hess, G. and Sinclair, K. (2007). Designing suburban greenways to provide habitat for forest-breeding birds. *Landscape and Urban Planning*, 80, 23–44.

Massa, R., Bani, L., Baietta, M. et al. (2004). An ecological network for the Milan region based on focal species. In *Ecological Netways and Greenways: Concept, Design, Implementation*, eds. R. H. G. Jongman and G. Pungetti. New York: Cambridge University Press, pp. 188–199.

Mathieu, R., Aryal, J. and Chong, A. K. (2007a). Object-based classification of ikonos imagery for mapping large-scale vegetation communities in urban areas. *Sensors*, 7, 2860–2880.

Mathieu, R., Freeman, C. and Aryal, J. (2007b). Mapping private gardens in urban areas using object-oriented techniques and very high-resolution satellite imagery. *Landscape and Urban Planning*, 81: 179–192.

Matlack, G. R. (1993). Sociological edge effects: the spatial distribution of human impact in suburban forest fragments. *Environmental Management*, 17, 829–835.

Matlack, G. R. (1997a). Four centuries of forest clearance and regeneration in the hinterland of a large city. *Journal of Biogeography*, 24, 281–295.

Matlack, G. R. (1997b). Land use and forest habitat distribution in the hinterland of a large city. *Journal of Biogeography*, 24, 297–307.

Matsunawa, K. (2000). The Nikken Sekkei approach to green buildings. In *Sustainable Architecture in Japan*, ed. A. Ray-Jones. New York: John Wiley.

Matter, H. C. and Daniels, T. J. (2000). Dog ecology and population biology. In *Dogs, Zoonoses and Public Health*, eds. C. N. L. MacPherson, F. X. Meslin and A. I. Wandeler. London: CABI Publishing, pp. 17–62.

Matteson, K. C. (2008). Bee richness and abundance in New York City urban gardens. *Annals of the Entomological Society of America*, 101, 140–150.

Matthews, H. and Kazimee, B. (1994). The quest for shelter: squatters and urbanization throughout the world. In *The Built Environment: A Creative Inquiry into Design and Planning*, eds. T. J. Bartuska and G. L. Young. Menlo Park, California: Crisp Publications, pp. 129–136.

Mayer, H. and Provo, J. (2004). The Portland Edge in context. In *The Portland Edge: Challenges and Successes in Growing Communities*, ed. C. P. Ozawa. Washington, D.C.: Island Press, pp. 9–34.

Medley, K. E., McDonnell, M. J. and Pickett, S. T. A. (1995). Forest landscape structure along an urban-to-rural gradient. *Professional Geographer*, 47, 159–168.

Meinig, D. W., ed. (1979). *The Interpretation of Ordinary Landscapes*. New York: Oxford University Press.

Melles, S., Glenn, S. M. and Martin, K. (2003). Urban bird diversity and landscape complexity: species environment associations along a multiscale habitat gradient. *Conservation Ecology*, 7(1) article 5. 5. [Online]. URL: http://www.consecol.org/vol7/iss1/art5/.

Mendez, M. (2005). Latino new urbanism: building on cultural preferences. *Opolis*, 1, 33–48.

Merola-Zwartjes, M. and DeLong, J. P. (2005). Avian species assemblages on New Mexico golf courses: surrogate riparian habitat for birds? *Wildlife Society Bulletin*, 33, 494–506.

Merriam, G., Henein, K. and Stuart-Smith, K. (1991). Landscape dynamics models. In *Quantitative Methods in Landscape Ecology*, eds. M. G. Turner and R. H. Gardner. New York: Springer-Verlag, pp. 399–416.

Mesquita, R. C. G., Delamonica, P. and Laurance, W. F. (1999). Effect of surrounding vegetation on edge-related tree mortality in Amazonian forest fragments. *Biological Conservation*, 91, 129–134.

Messenger, K. G. (1968). A railway flora of Rutland. *Proceedings of the Botanical Society of the British Isles*, 7, 325–344.

Metosi, M. V. (2000). *The Sanitary City: Urban Infrastructure in America from Colonial Times to the Present*. Baltimore, Maryland: Johns Hopkins University Press.

Meurk, C. D., Zvyagna, N., Gardner, R. O. et al. (2009). Environmental, social and spatial determinants of urban arboreal character in Auckland, New Zealand. In *Ecology of Cities and Towns: A Comparative Approach*, eds. M. J. McDonnell, A. K. Hahs and J. H. Breuste. New York: Cambridge University Press, pp. 287–307.

Michopoulos, P., Baloutsos, G., Economou, A. et al. (2005). Biogeochemistry of lead in an urban forest in Athens, Greece. *Biogeochemistry*, 73, 345–357.

Middleton, B. (1999). *Wetland Restoration, Flood Pulsing and Disturbance Dynamics*. New York: John Wiley.

Middleton, B. (2002). *Flood Pulsing in Wetlands: Restoring the Natural Hydrological Balance*. New York: John Wiley.

Mielke, H. W., Gonzales, C. R., Smith, M. K. and Mielke, P. W. (2000). Quantities and associations of lead, zinc, cadmium, manganese, chromium, nickel, vanadium, and copper in fresh Mississippi Delta alluvium and New Orleans alluvial soils. *Science of the Total Environment*, 246, 249–259.

Milchunas, D. G., Schulz, K. A. and Shaw, R. B. (1999). Plant and environmental interactions: plant community responses to disturbance by mechanized military maneuvers. *Journal of Environmental Quality*, 28, 1533–1547.

Millard, A. (2000). The potential role of natural colonization as a design tool for urban forestry – a pilot study. *Landscape and Urban Planning*, 52, 173–179.

Millard, A. (2008). Semi-natural vegetation and its relationship to designated urban green space at the landscape scale in Leeds, UK. *Landscape Ecology*, 23, 1231–1241.

Millennium Ecosystem Assessment. (2005). *Ecosystems and Human Well-Being: Current State and Trends*. Washington, D.C.: Island Press.

Miller, A., Leal, N., Laiz, L. *et al.* (2010). *Primary Bioreceptivity of Limestones Used in Southern European Monuments*. London: The Geological Society of London.

Miller, J. R. and Hobbs, N. T. (2000). Recreational trails, human activity, and nest predation in lowland riparian areas. *Landscape and Urban Planning*, 50, 227–236.

Miller, R. W. (1997). *Urban Forestry: Planning and Managing Urban Greenspaces*. Englewood Cliffs, New Jersey: Prentice Hall.

Mills, G. (2004). *The Urban Canopy Layer Heat Island*. International Association for Urban Climate (2013 website).

Mills, G., Cleugh, H., Emmanuel, R. *et al.* (2010). Climate information for improved planning and management of mega cities (needs perspective). *Procedia Environmental Sciences*, 1, 228–246.

Milne, B. T. (1988). Measuring the fractal geometry of landscapes. *Applied Mathematics and Computation*, 27, 67–79.

Miltner, R. J., White, D. and Yoder, C. (2004). The biotic integrity of streams in urban and suburbanizing landscapes. *Landscape and Urban Planning*, 69, 87–100.

Ministry of Agriculture, Nature and Food Quality. (2004). *Ecological Networks: Experiences in the Netherlands*. Amsterdam.

Mitchell, J. H. (2008). *The Paradise of All These Parts: A Natural History of Boston*. Boston: Beacon Press.

Mitsch, W. J. and Gosselink, J. G. (2007). *Wetlands*. New York: John Wiley.

Mitsch, W. J. and Jorgensen, S. E. (2004). *Ecological Engineering and Ecosystem Restoration*. New York: John Wiley.

Mizell, R. F., III and Hagan, A. (2000). Biological problems and their management in urban soils: integrated pest management of arthropods and diseases. In *Managing Soils in an Urban Environment*, eds. R. B. Brown, J. H. Huddleston and J. L. Anderson. Madison, Wisconsin: American Society of Agronomy, pp. 119–155.

Molnar, J. L., ed. (2010). *The Atlas of Global Conservation: Changes, Challenges, and Opportunities to Make a Difference*. Berkeley, California: University of California Press.

Montieth, K. E. and Paton, P. W. C. (2006). Emigration behavior of spotted salamanders on golf courses in Southern Rhode Island. *Journal of Herpetology*, 40, 195–205.

Moore, J. W., Schindler, D. E., Scheuerell, M. D. *et al.* (2003). Lake eutrophication at the urban fringe, Seattle Region, USA. *Ambio*, 32, 13–18.

Moore, S. (2007). *Alternative Routes to the Sustainable City*. Plymouth, UK: Lexington Books.

Moore, T. (1984). *Utopia* (Translation by J. Sheehan and J. P. Donnelly). Marquette University Press: Milwaukee, Wisconsin.

Moorman, C. E. and DePerno, C. S. (2006). Saving the world one native plant at a time. In *Proceedings 11th Triennial National Wildlife and Fisheries Extension Specialists Conference. Big Sky, Montana*, pp. 53–56.

Moran, D. M. (1990). *The Allotment Movement in Britain*. New York: Peter Lang Publishing.

Moran, J. M. and Morgan, M. D. (1994). *Meteorology: The Atmosphere and the Science of Weather*. New York: Macmillan.

Moran, M. A. (1984). Influence of adjacent land use on understory vegetation of New York forests. *Urban Ecology*, 8, 329–340.

Morello, J., Buzai, G. D., Baxendale, C. A. *et al.* (2000). Urbanization and the consumption of fertile land and other ecological changes: the case of Buenos Aires. *Environment and Urbanization*, 12, 119–131.

Morgan, G. T. and King, J. O. (1987). *The Woodlands: New Community Development, 1964–1983*. College Station, Texas: Texas A&M University Press.

Morin, P. J. (2011). *Community Ecology*. Oxford: Blackwell Science.

Morley, S. A. and Karr, J. R. (2002). Assessing and restoring the health of urban streams in the Puget Sound Basin. *Conservation Biology*, 16, 1498–1509.

Morrison, M. L., Scott, T. A. and Tennant, T. (1994). Wildlife-habitat restoration in an urban park in Southern California. *Restoration Ecology*, 2, 17–30.

Mortberg, U. (2009). Landscape ecological analysis and assessment in an urbanising environment. In *Ecology of Cities and Towns: A Comparative Approach*, eds. M. J. McDonnell, A. K. Hahs and J. H. Breuste. New York: Cambridge University Press, pp. 439–455.

Mostafavi, M. and Doherty, G., eds. (2010). *Ecological Urbanism*, Baden, Switzerland: Lars Muller Publishers.

Moudon, A. V., ed. (1989). *Master-Planned Communities: Shaping Exurbs in the 1990's*. Seattle, Washington: University of Washington, College of Architecture and Urban Planning.

Moudon, A. V. (1997). Urban morphology as an emerging interdisciplinary field. *Urban Morphology*, 1, 3–10.

Mougeot, L. J. A., ed. (2005). *Agropolis: The Social, Political and Environmental Dimensions of Urban Agriculture*. London: Earthscan.

Mougeot, L. J. A. (2006). *Growing Better Cities: Urban Agriculture for Sustainable Development*. Ottawa: International Development Research Centre.

Moulton, C. A. and Adams, L. W. (1991). Effects of urbanization on foraging strategy of woodpeckers. In *Wildlife Conservation in Metropolitan Environments*, eds. L. W. Adams and D. L. Leedy. Columbia, Maryland: National Institute for Urban Wildlife, pp. 67–73.

Mueller-Dombois, D. and Ellenberg, H. (1974). *Aims and Methods of Vegetation Ecology*. New York: John Wiley.

Muhlenbach, J. (1979). Contributions to the synanthropic (adventive) flora of the railroads in St. Louis, Missouri, USA. *Annals of the Missouri Botanical Garden*, 66, 1–108.

Mukherjee, M. (2006). Waste-fed fisheries in peri-urban Kolkata. In *The Peri-Urban Interface: Approaches to Sustainable Natural and Human Resource Use*, eds. D. McGregor, D. Simon and D. Thompson. London: Earthscan, pp. 104–115.

Muller, N. (1990). Lawns in German cities: a phytosociological comparison. In *Urban Ecology: Plants and Plant Communities in Urban Environments*, eds. H. Sukopp, S. Hejny and I. Kowarik. The Hague, Netherlands: SPB Academic Publishing, pp. 209–222.

Muller, N., Werner, P. and Kelcey, J. G., eds. (2010). *Urban Biodiversity and Design*. New York: Wiley-Blackwell.

Mumford, L., ed. (1961). *The City in History: Its Origins, Its Transformations and Its Prospects*. San Diego, California: Jovanovich Publishers.

Mumford, L. (1968). *The Urban Prospect*. New York: Harcourt Brace Jovanovich.

Murcina, L. (1990). Urban vegetation research in European Comecon-countries and Yugoslavia: a review. In *Urban Ecology: Plants and Plant Communities in Urban Environments*, eds. H. Sukopp, S. Hejny and I. Kowarik. The Hague, Netherlands: SPB Academic Publishing, pp. 23–43.

Murgui, E. (2007). Effects of seasonality on the species-area relationship: a case study with birds in urban parks. *Global Ecology and Biogeography*, 16, 319–329.

Murphy, K. J. and Eaton, J. W. (1983). Effects of pleasure-boat traffic on macrophyte growth in canals. *Journal of Applied Ecology*, 20, 713–729.

Murray, D., ed. (1986). *Seed Dispersal*. San Diego, California: Academic Press.

Murthy, R. C., Rao, Y. R. and Inamdar, A. B. (2001). Integrated coastal management of Mumbai metropolitan region. *Ocean and Coastal Management*, 44, 355–369.

Musacchio, L. R. (2009). The scientific basis for the design of landscape sustainability: a conceptual framework for translational landscape research and practice of designed landscapes and the six Es of landscape sustainability. *Landscape Ecology*, 24, 993–1013.

Mussey, G. J. and Potter, D. A. (1997). Phenological correlations between flowering plants and activity of urban landscape pests in Kentucky. *Journal of Economic Entomology*, 90, 1615–1627.

Naiman, R. J. (2009). *Riparia: Ecology, Conservation, and Management of Streamside Communities*. New York: Oxford University Press.

Naiman, R. J., Decamps, H. and Pollock, M. (1993). The role of riparian corridors in maintaining ecological diversity. *Ecological Applications*, 3, 209–212.

Nakagoshi, N. and Moriguchi, T. (1999). Ecosystem and biodiversity planning in Hiroshima City, Japan. *Journal of Environmental Sciences*, 11, 149–154.

Nakagoshi, N. and Rim, Y.-D. (1988). Landscape ecology in the greenbelt area in Korea. In *Connectivity in Landscape Ecology*, ed. K.-F. Schreiber. Munster, Germany: Munstersche Geographische Arbeiten 29, pp. 247–250.

Nassauer, J. I. (1988). The aesthetics of horticulture: neatness as a form of care. *Hortscience*, 23, 973–977.

Nassauer, J. I. (1995). Messy ecosystems, orderly frames. *Landscape Journal*, 14, 161–170.

Nassauer, J. I., ed. (1997). *Placing Nature: Culture and Landscape Ecology*. Washington, D.C.: Island Press.

Nassauer, J. I. and Opdam, P. (2008). Design in science: extending the landscape ecology paradigm. *Landscape Ecology*, 23, 633–644.

National Capital Commission. (ca. 1981). *The Management Plan for the Greenbelt*. Ottawa.

National Research Council. (1970). *Effects of Deicing Salts on Water Quality and Biota: Literature Review and Recommended Research*. NCHRP Report 91. Washington, D.C.

National Research Council. (2005a). *Assessing and Managing the Ecological Impacts of Paved Roads*. Washington, D.C.: National Academies Press.

National Research Council. (2005b). *Valuing Ecosystem Services: Toward Better Environmental Decision-Making*. Washington, D.C.: National Academies Press.

Natoli, E. (1985). Spacing pattern in a colony of urban stray cats (*Felis catus* L.) in the historic center of Rome. *Applied Animal Behavioural Science*, 14, 289–304.

Natuhara, Y. (2008). Evaluation and planning of wildlife habitat in urban landscape. In *Landscape Ecological Applications in Man-Influenced Areas: Linking Man and Nature Systems*, eds. S.-K. Hong, N. Nakagoshi, B. Fu and Y. Morimoto. New York: Springer, pp. 129–147.

Natuhara, Y. and Hashimoto, H. (2009). Spatial pattern and process in urban animal communities. In *Ecology of Cities and Towns: A Comparative Approach*, eds. M. J. McDonnell, A. K. Hahs and J. H. Breuste. New York: Cambridge University Press, pp. 197–214.

Neller, R. J. (1988). A comparison of channel erosion in small urban and rural catchments, Armidale, New South Wales. *Earth Surface Processes*, 13, 1–7.

Nelson, A. C., Duncan, J. B., Mullen, C. J. and Bishop, K. R. (1995). *Growth Management Principles and Practices*. Chicago: Planners Press, American Planning Association.

Nemeth, E. and Brumm, H. (2009). Blackbirds sing higher-pitched songs in cities: adaptation to habitat acoustics or side-effect of urbanization? *Animal Behavior*, 78, 637–641.

Neubaum, D. J., Wilson, K. R. and O'Shea, T. J. (2007). Urban maternity-roost selection by big brown bats in Colorado. *Journal of Wildlife Management*, 71, 728–736.

Neuwirth, R. (2006). *Shadow Cities: A Billion Squatters, A New Urban World*. New York: Routledge.

Newman, P. and Jennings, I. (2008). *Cities as Sustainable Ecosystems: Principles and Practices*. Washington, D.C.: Island Press.

Newman, P. and Kenworthy, J. (1999). *Sustainability and Cities: Overcoming Automobile Dependence*. Washington, D.C.: Island Press.

Newman, P., Beatley, T. and Boyer, H. (2009). *Resilient Cities: Responding to Peak Oil and Climate Change*. Washington, D.C.: Island Press.

Nicholaus, D., van Woert, D., Rowe, B. *et al.* (2005). Green roof stormwater retention: effects of roof surface, slope, and media depth. *Journal of Environmental Quality*, 34, 1036–1044.

Nicholls, R. J., Wong, P. P., Burkett, V. R. *et al.* (2007). Coastal systems and low-lying areas. In *Climate Change 2007: Impacts, Adaptation and Vulnerability*, eds. M. L. Parry, O. F. Canziani, J. P. Palutikof *et al.* Cambridge, UK: Cambridge University Press, pp. 315–356.

Nielsen, S. N. (2007). What has modern ecosystem theory to offer to cleaner production, industrial ecology and society? The views of an ecologist. *Journal of Cleaner Production*, 15, 1639–1653.

Nielson, K. K. and Rogers, V. C. (2000). Determining and managing radon risk. In *Managing Soils in an Urban Environment*, eds. R. B. Brown, J. H. Huddleston and J. L. Anderson. Madison, Wisconsin: American Society of Agronomy, pp. 263–283.

Niemela, J. (1999). Ecology and urban planning. *Biodiversity and Conservation*, 8, 119–131.

Niemela, J., Breuste, J. H., Elmqvist, T. *et al.*, eds. (2011). *Urban Ecology: Patterns, Processes, and Applications*. New York: John Wiley.

Niemela, J., Kotze, D. J. and Yli-Pelkonen, V. (2009). Comparative urban ecology: challenges and possibilities. In *Ecology of Cities and Towns: A Comparative Approach*, eds. M. J. McDonnell, A. K. Hahs and J. H. Breuste. New York: Cambridge University Press. pp. 9–24.

Niering, W. A. and Goodwin, R. H. (1974). Creation of relatively stable shrublands with herbicides: arresting 'succession' on rights-of-way and pasture land. *Ecology*, 55, 784–795.

Nilon, C. and Pais, R. C. (1997). Terrestrial vertebrates in urban ecosystems: developing hypotheses for the Gwynns Falls Watershed in Baltimore, Maryland. *Urban Ecosystems*, 1, 247–257.

Nix, H. A. ed. (1972). *The City as a Life System?* Canberra: Ecological Society of Australia, Vol. 7 Proceedings.

Njorge, J. B., Nakamura, A. and Morimoto, Y. (1999). Thermal based functional evaluation of urban park vegetation. *Journal of Environmental Sciences*, 11, 252–256.

Noe, S. M., Penuelas, J. and Ninemets, U. (2008). Monoterpene emissions from ornamental trees in urban areas: a case study of Barcelona, Spain. *Plant Biology*, 10, 163–169.

Noel, S., Ouellet, M., Galois, P. and Lapointe, F. J. (2007). Impact of urban fragmentation on the genetic structure of the eastern red-backed salamander. *Conservation Genetics*, 8, 599–606.

Nolan, P. A. and Guthrie, N. (1998). River rehabilitation in an urban environment: examples from the Mersey Basin, North West England. *Aquatic Conservation and Freshwater Ecosystems*, 8, 685–700.

Norberg, J. (1999). Linking nature's services to ecosystems: an ecological perspective. *Ecological Economics*, 29, 183–202.

Nordahl, D. (2009). *Public Produce: The New Urban Agriculture*. Washington, D.C.: Island Press.

Norton, D. A. (2000). Sand plain forest fragmentation and residential development, Invercargill City, New Zealand. In *Nature Conservation 5: Nature Conservation in Production Environments: Managing the Matrix*, eds. J. L. Craig, N. Mitchell and D. A. Saunders. Chipping Norton, Australia: Surrey Beatty, pp. 157–165.

Notenboom, J., Plenet, S. and Turquin, M.-J. (1994). Groundwater contamination and its impact on groundwater animals and ecosystems. In *Groundwater Ecology*, eds. J. Gibert, D. L. Danielopol and J. A. Stanford. San Diego, California: Academic Press, pp. 447–504.

Novotny, V. and Olem, H. (1994). *Water Quality: Prevention, Identification, and Management of Diffuse Pollution*. New York: Van Nostrand Reinhold.

Nowak, D. (1994). Air pollution removal by Chicago's urban forest. In *Chicago's Urban Forest Ecosystem: Results of the Chicago Urban Forest Climate Project.* Radnor, Pennsylvania: USDA Forest Service, General Technical Report NE-186.

Nowak, D. and Crane, D. (2002). Carbon storage and sequestration by urban trees in the USA. *Environmental Pollution*, 116, 381–389.

Nowak, D. J., Noble, M. H., Sisinni, S. M. and Dwyer, J. F. (2001). People and trees – assessing the US urban forest resource. *Journal of Forestry*, 99, 37–42.

Numata, M. (1982). Changes in ecosystem structure and function in Tokyo. In *Urban Ecology*, eds. R. Bornkamm, J. A. Lee and M. R. D. Seaward. Oxford: Blackwell Scientific Publications, pp. 139–147.

Numata, M. (1998). The urban ecosystem from the viewpoint of landscape ecology and bioregion. *Modern Trends in Ecology and Environment*, 1998, 91–99.

Nybakken, J. W. (1997). *Marine Biology: An Ecological Approach*. New York: Addison-Wesley Educational Publishers.

Obara, H. (1995). Animals and man in the process of urbanization. In *Urban Ecology as the Basis of Urban Planning*, eds. H. Sukopp, M. Numata and A. Huber. Amsterdam: SPB Academic Publishing, pp. 191–201.

Oberndorfer, E., Lundholm, J., Bass, B. *et al.*, (2007). Green roofs as urban ecosystems: ecological structures, functions, and services. *BioScience*, 57, 823–833.

Ochimaru, T. and Fukuda, K. (2007). Changes in fungal communities in evergreen broad-leaved forests across a gradient of urban to rural areas in Japan. *Canadian Journal of Forest Research*, 37, 247–258.

Odum, E. P. (1971). *Fundamentals of Ecology*. Philadelphia: Saunders.

Odum, E. P. and Barrett, G. W. (2005). *Fundamentals of Ecology*. (5th edition). Belmont, California: Thomson Brooks/Cole.

Odum, H. T. (1983). *Systems Ecology*. New York: John Wiley.

Ohsawa, M., Liang-Jim, D. and Ohtsuka, T. (1988). Urban vegetation: its structure and dynamics. In *Integrated Studies in Urban Ecosystems as the Basis of Urban Planning, III*, ed. H. Obara. Tokyo: Ministry of Education, Culture and Science, pp. 137–142.

Ojo, S. O. (2009). Backyard farming: a panacea for food security in Nigeria. *Journal of Human Ecology*, 28, 127–130.

Oke, T. (1973). City size and the urban heat island. *Atmospheric Environment*, 7, 769–779.

Oke, T. R. (1981). Canyon geometry and the nocturnal urban heat island: comparison of scale model and field observations. *International Journal of Climatology*, 1, 237–254.

Oke, T. R. (1987). *Boundary Layer Climates*. New York: Methuen.

Oke, T. R. (1997a). Urban environments. *The Surface Climates of Canada*, eds. W. G. Bailey, T. R. Oke and W. R. Rouse. Montreal: McGill/Queens University Press, pp. 303–327.

Oke, T. R. (1997b). Urban climates and global change. In *Applied Climatology: Principles and Practice*, eds. A. Perry and R. Thompson. London: Routledge, pp. 273–287.

Oleksyn, J., Kloeppel, B. D., Lukasiewicz, S. *et al.* (2007). Ecophysiology of horse chestnut (*Aesculus hippocastanum* L.) in degraded and restored urban sites. *Polish Journal of Ecology*, 55, 245–260.

Oleson, K. W., Bonan, G. B., Feddema, J. *et al.* (2008). An urban parameterization for a global climate model. Part I: formulation and evaluation for two cities. *Journal of Applied Meteorology and Climatology*, 47, 1038–1060.

Omni, P. N. (2005). *Forest Fires: A Reference Handbook*. Santa Barbara, California: ABC-CLIO.

O'Neill, R. V., DeAngelis, D. L., Waide, J. B. and Allen, T. F. H. (1986). *A Hierarchical Concept of Ecosystems*. Princeton, New Jersey: Princeton University Press.

Opdam, P. (1991). Metapopulation theory and habitat fragmentation: a review of Holarctic breeding bird studies. *Landscape Ecology*, 5, 93–106.

Opdam, P., Foppen, R. and Vos, C. (2002). Bridging the gap between ecology and spatial planning in landscape ecology. *Landscape Ecology*, 16, 767–779.

Opdam, P., Steingrover, E. and van Rooij, S. (2006). Ecological networks: a spatial concept for multi-actor planning of sustainable landscapes. *Landscape and Urban Planning*, 75, 322–332.

Orser, P. N. and Shure, D. J. (1972). Effects of urbanization on the salamander *Desmognathus fuscus fuscus*. *Ecology*, 53, 1148–1154.

Orsi, J. (2004). *Hazardous Metropolis: Flooding and Urban Ecology in Los Angeles*. Berkeley, California: University of California Press.

Ortega-Alvarez, R. and MacGregor-Fors, I. (2009). Living in the big city: effects of urban land use on bird community structure, diversity, and composition. *Landscape and Urban Planning*, 90, 189–195.

Overmyer, J. P., Noblet, R. and Armbrust, K. L. (2005). Impacts of lawn-care pesticides on aquatic ecosystems in relation to property value. *Environmental Pollution*, 137, 263–272.

Owen, J. (1991). *The Ecology of a Garden: The First Fifteen Years*. New York: Cambridge University Press.

Oxford English Dictionary, The. (1998). Oxford University Press.

Oxley, D. J., Fenton, M. B. and Carmody, G. R. (1974). The effects of roads on populations of small mammals. *Journal of Applied Ecology*, 11, 51–59.

Ozawa, C. P., ed. (2004). *The Portland Edge: Challenges and Successes in Growing Communities*. Washington, D.C.: Island Press.

Pacione, M. (2005). *Urban Geography: A Global Perspective*. Oxford: Routledge.

Pallares-Barbera, M. (2005). La percepcio d'optimalitat en el Pla Cerda. El model p-median en el disseny ortogonal de l'Eixample de Barcelona. *Treballs de la Societat Catalunya de Geografia*, 60, 223–253.

Palmer, E. L. and Fowler, H. S. (1975). *Fieldbook of Natural History*. New York: McGraw-Hill.

Palmer, M. A., Bernhardt, E. S., Allan, J. D. et al. (2005). Standards for ecologically successful river restoration. *Journal of Applied Ecology*, 42, 208–217.

Palumbi, S. R. (2001). Humans as the world's greatest evolutionary force. *Science*, 293, 1786–1790.

Panerai, P., Demorgon, M. and Depaule, J.-C. (1999). *Analyse urbaine*. Marseille: Editions Parentheses.

Park, C.-R. and Lee, W.-S. (2000). Relationship between species composition and area in breeding birds of urban woods in Seoul, Korea. *Landscape and Urban Planning*, 51, 29–36.

Park, R. E., Burges, E. W. and McKenzie, R. D., eds. (1925). *The City: Suggestions for Investigation of Human Behavior in the Urban Environment*. Chicago: University of Chicago Press.

Parker, D. E. (2004). Large-scale warming is not urban. *Nature*, 432, 290.

Parlow, E. (2011). Urban climate. In *Urban Ecology: Patterns, Processes, and Applications*, eds. J. Niemela, J. H. Breuste, T. Elmqvist et al. New York: Oxford University Press, pp. 31–44.

Parmesan, C. (2006). Ecological and evolutionary responses to recent climate change. *Annual Review of Ecology, Evolution and Systematics*, 37, 637–669.

Parris, K. M. (2006). Urban amphibian assemblages as metacommunities. *Journal of Animal Ecology*, 75, 757–764.

Parris, K. M. and Hazell, D. L. (2005). Biotic effects of climate change in urban environments: the case of the grey-headed flying-fox (*Pteropus poliocephalus*) in Melbourne, Australia. *Biological Conservation*, 124, 267–276.

Parris, K. M., Velik-Lord, M. and North, J. M. A. (2009). Frogs call at a higher pitch in traffic noise. *Ecology and Society*, 14(1), article 25. [Online]. URL: http://www.ecologyandsociety.org/vol14/iss1/art25/.

Parsons, K. C. and Schuyler, D. (2000). *From Garden City to Greencity: The Legacy of Ebenezer Howard*. Baltimore, Maryland: Johns Hopkins University Press.

Parsons, K. C. (1995). Heron nesting at Pea Patch Island, upper Delaware Bay, USA: abundance and reproductive success. *Colonial Waterbirds*, 18, 69–78.

Parsons, K. C. (2002). Integrated management of waterbird habitats at impounded wetlands in Delaware Bay, U.S.A. *Waterbirds*, 25, 25–41.

Parsons, K. C., Schmidt, S. R. and Matz, A. C. (2001). Regional patterns of wading bird productivity in Northeastern U.S. estuaries. *Waterbirds*, 24, 323–330.

Partecke, J. and Gwinner, E. (2007). Increased sedentariness in European blackbirds following urbanization: a consequence of local adaptation? *Ecology*, 88, 1–10.

Partecke, J., Van't Hof, T. and Gwinner, E. (2004). Differences in the timing of reproduction between urban and forest European blackbirds (*Turdus merula*): result of phenotypic flexibility or genetic differences? *Proceedings of the Royal Society of London, Series B-Biological Sciences*, 271, 1995–2001.

Pasquini, M. W. (2006). The use of town refuse ash in urban agriculture around Jos, Nigeria: health and environmental risks. *Science of the Total Environment*, 354, 43–59.

Paton, P. W. C., Harris, R. J. and Trocki, C. L. (2005). Distribution and abundance of breeding birds in the Boston Harbor. *Northeastern Naturalist*, 12, 144–168.

Patz, J. A., Graczyk, T. K., Geller, N. and Vittor, A. Y. (2000). Effects of environmental change on emerging parasitic diseases. *International Journal for Parasitology*, 30, 1395–1405.

Paul, M. J. and Meyer, J. L. (2001). Streams in the urban landscape. *Annual Review of Ecology and Systematics*, 32, 333–365.

Pauleit, S. and Breuste, J. H. (2011). Land-use and surface-cover as urban ecological indicators. In *Urban Ecology: Patterns, Processes, and Applications*, eds. J. Niemela, J. H. Breuste, T. Elmqvist, et al. New York: Oxford University Press, pp. 19–30.

Pauleit, S., Jones, N., Nyhuus, S. et al. (2005) Urban forest resources in European cities. In *Urban Forests and Trees*, eds. C. C. Konijnendijk, K. Nilsson, T. B. Randrup and J. Schipperijn. New York: Springer, pp. 49–80.

Payne, R. M. (1978). The flora of walls in south-eastern Essex. *Watsonia*, 12, 41–46.

Pearlman, J. (1998). Towards sustainable mega-cities in Latin America and Africa. In *Environmental Strategies for Sustainable Development in Urban Areas: Lessons from Africa and Latin America*. Aldershot, UK: Ashgate Publishing, pp. 109–135.

Pearlmutter, D., Bitan, A. and Berliner, P. (1999). Microclimatic analysis of compact urban canyons in an arid zone. *Atmospheric Environment*, 33, 4143–4150.

Peck, S. and Kuhn, M. (2003). *Design Guidelines for Green Roofs*. Ottawa: Ontario Association of Architects.

Pendall, R. (1999). Do land-use controls cause sprawl? *Environment and Planning B: Planning and Design*, 26, 555–571.

References

Pennycuik, P. R. and Dickson, R. G. (1989). Food restriction mechanisms limiting numbers and home site utilization in populations of house mice, *Mus musculus*. *Oikos*, 55, 159–164.

Perez-Vazquez, A., Anderson, S. and Rogers, A. W. (2005). Assessing benefits from allotments as a component of urban agriculture in England. In *Agropolis: The Social, Political and Environmental Dimensions of Urban Agriculture*, ed. L. J. A. Mougeot. London: Earthscan, pp. 239–266.

Perlman, D. L. and Milder, J. C. (2004). *Practical Ecology for Planners, Developers, and Citizens*. Washington, D.C.: Island Press.

Perman, R., Ma, Y., McGilvray, J. and Common, M. (2003). *Natural Resource and Environmental Economics*. Berkeley, California: University of California Press.

Peterken, G. (2008). *Wye Valley*. London: Collins.

Peterken, G. F. (1996). *Natural Woodland: Ecology and Conservation in North Temperate Regions*. New York: Cambridge University Press.

Peterken, G. F. (2001). Ecological effects of introduced tree species in Britain. *Forest Ecology and Management*, 141, 31–42.

Peterson, T. C. (2003). Assessment of urban versus rural in situ surface temperatures in the contiguous United States: no difference found. *Journal of Climate*, 16, 2941–2959.

Pezzoli, K. (1998). *Human Settlements and Planning for Ecological Sustainability*. Cambridge, Massachusetts: MIT Press.

Pfeiffer, H. (1957). Pflanzliche Gesellschaftsbildung auf dem Trummerschutt ausgebombter stadte. *Vegetatio*, 7, 301–320.

Phelps, N., Parsons, N., Ballas, D. and Dowling, A. (2006). *Post Suburban Europe: Planning and Politics at the Margins of Europe's Capital Cities*. New York: Palgrave MacMillan.

Pickett, S. T. A. and White, P. S. (1985). *The Ecology of Natural Disturbance and Patch Dynamics*. New York: Academic Press.

Pickett, S. T. A., Burch, Jr., W. R., Dalton, S. E. et al. (1997). A conceptual framework for the study of human ecosystems in urban areas. *Urban Ecosystems*, 1, 185–199.

Pickett, S. T. A., Cadenasso, M. L., Grove, J. M. et al. (2001). Urban ecological systems: linking terrestrial ecological, physical, and socioeconomic components of metropolitan areas. *Annual Review of Ecology and Systematics*, 32, 127–157. (Reprinted 2008 in *Urban Ecology: An International Perspective on the Interaction Between Humans and Nature*, eds. J. M. Marzluff, E. Shulenberger, W. Endlicher et al. New York: Springer, pp. 161–179).

Pickett, S. T. A., Cadenasso, M. L., McDonnell, M. J. and Burch, W. (2009). Frameworks for urban ecosystem studies: gradients, patch dynamics and the human ecosystem in the New York metropolitan area and Baltimore, USA. In *Ecology of Cities and Towns: A Comparative Approach*, eds. M. J. McDonnell, A. K. Hahs and J. H. Breuste. New York: Cambridge University Press. pp. 25–50.

Pickett, S. T. A., Cadenasso, M. L. and McGrath, B., eds. (2013). *Resilience in Ecology and Urban Design: Linking Theory and Practice*. New York: Springer.

Pickett, S. T. A., Collins, S. C. and Armesto, J. J. (1987). Models, mechanisms, and pathways of succession. *Botanical Review*, 53, 335–371.

Pidgeon, A. M., Radeloff, V. C., Flather, C. H. et al. (2007). Associations of forest bird species richness with housing and landscape patterns across the USA. *Ecological Applications*, 17, 1989–2010.

Pierson, E. D. (1998). Tall trees, deep holes, and scarred landscapes: conservation biology of North American bats. In *Bat Biology and Conservation*, eds. T. H. Kunz and P. A. Racey. Washington, D.C.: Smithsonian Institution Press, pp. 309–325.

Pignatti, S. (1995). L'ecosistema urbano. In *Contributi del Centro Linceo Interdisciplinare 'Beniamino Segre', N. 90*. Roma: Accademia Nazionale dei Lincei, pp. 137–167.

Pignatti, S. and Federici, F. M. (1989). The synanthropic vegetation from the ecosystemic point of view. In *Spontaneous Vegetation in Settlements*, ed. A. U. Savoia. Camerino, Italy: Braun-Blanquetia, Part I, pp. 29–35.

Pilkey, O. H. and Dixon, K. L. (1996). *The Corps and the Shore*. Washington, D.C.: Island Press.

Pincetl, S., Bunje, P. and Holmes, T. (2012). An expanded urban metabolism method. *Landscape and Urban Planning*, 107, 193–202.

Pinkham, R. (2000). *Daylighting: New Life for Buried Streams*. Snowmass, Colorado: Rocky Mountain Institute.

Pirnat, J. (2000). Conservation and management of forest patches and corridors in suburban landscapes. *Landscape and Urban Planning*, 52, 135–143.

Pitt, R. and Voorhees, J. G., III. (2003). SLAMM, the Source Loading and Management Model. In *Wet Weather Flow in the Urban Watershed: Technology and Management*, eds. R. Field and D. Sullivan. Boca Raton, Florida: Lewis Publishers, pp. 79–102.

Platt, R. H. (2004). Toward ecological cities: adapting to the 21st century metropolis. *Environment*, 46, 12–27.

Platt, R. H., Rowntree, R. A. and Muick, P. C., eds. (1994). *The Ecological City: Preserving and Restoring Urban Biodiversity*. Amherst, Massachusetts: University of Massachusetts Press.

Poague, K. L., Johnson, R. J. and Young, L. J. (2000). Bird use of rural and urban converted railroad rights-of-way in Southeast Nebraska. *Wildlife Society Bulletin*, 28, 852–864.

Polano, S. (1992). The Bos Park, Amsterdam, and urban development in Holland. In *The Architecture of Western Gardens*, eds. G. Teysot and M. Mosser. Cambridge, Massachusetts: MIT Press.

Ponting, C. (2007). *A New Green History of the World: The Environment and the Collapse of Great Civilizations*. New York: Penguin Books.

Pope, S. E., Fahrig, L. and Merriam, H. G. (2000). Landscape complementation and metapopulation effects on leopard frog populations. *Ecology*, 81, 2498–2508.

Porter, D. R. (1997). *Managing Growth in America's Communities*. Washington, D.C.: Island Press.

Porter, D. R. (2002). *Making Smart Growth Work*. Washington, D.C.: Urban Land Institute.

Porter, E. E., Bulluck, J. and Blair, R. B. (2005). Multiple spatial-scale assessment of the conservation value of golf courses for breeding birds in southwestern Ohio. *Wildlife Society Bulletin*, 33, 494–506.

Porter, E. E., Forschner, B. R. and Blair, R. B. (2001). Woody vegetation and canopy fragmentation along a forest-to-urban gradient. *Urban Ecosystems*, 5, 131–151.

Pospisil, P. (1994). The groundwater fauna of a Danube aquifer in the "Lobau" wetland in Vienna, Austria. In *Groundwater Ecology*, eds. J. Gibert, D. L. Danielopol and J. A. Stanford. San Diego, California: Academic Press, pp. 347–366.

Postel, S. and Richter B. (2003). *Rivers for Life: Managing Water for People and Nature*. Washington, D.C.: Island Press.

Potter, R. B. and Salau, A. T. (1990). *Cities and Development in the Third World*. London: Mansell Publishing.

Pouyat, R. V. and McDonnell, M. J. (1991). Heavy metal accumulation in forest soils along an urban-rural gradient in southern New York, USA. *Water, Air and Soil Pollution*, 57–58, 797–807.

Pouyat, R. V., Carreiro, M. M., Groffman, P. M. et al. (2009). Investigative approaches to urban biogeochemical cycles: New York metropolitan area and Baltimore as case studies. In *Ecology of Cities and Towns: A Comparative Approach*, eds. M. J. McDonnell, A. K. Hahs and J. H. Breuste. New York: Cambridge University Press, pp. 329–351.

Pouyat, R. V., McDonnell, M. J., Pickett, S. T. A. et al. (1995). Carbon and nitrogen dynamics in oak stands along an urban-rural gradient. In *Carbon Forms and Functions in Forest Soils*, eds. J. M. Kelly and W. D. McFee. Madison, Wisconsin: Soil Science Society of America, pp. 569–587.

Pouyat, R. V., Parmelee, R. W. and Carreiro, M. M. (1994). Environmental effects of forest soil-invertebrates and fungal densities in oak stands along an urban-rural land use gradient. *Pedobiologia*, 38, 385–399.

Pouyat, R. V., Yesilonis I. D., Russell-Anelli, J. and Neerchal, N. K. (2007). Soil chemical and physical properties that differentiate urban land-use and cover types. *Soil Science Society of America Journal*, 71, 1010–1019.

Prados, M. J., ed. (2009). *Naturbanization: New Identities and Processes for Rural-Natural Areas*. Leiden, Netherlands: Taylor & Francis.

Pratt, D. and Schaeffer, J. (1999). *The Real Goods Solar Living Sourcebook: The Complete Guide to Renewable Energy Technologies and Sustainable Living*. White River Junction, Vermont: Chelsea Green Publishing.

Premat, A. (2005). Moving between the plan and the ground: shifting perspectives on urban agriculture in Havana, Cuba. In *Agropolis: The Social, Political and Environmental Dimensions of Urban Agriculture*, ed. L. J. A. Mougeot. London: Earthscan, pp. 153–185.

Prevett, P. T. (1991). Movement paths of koalas in the urban-rural fringes of Ballarat, Victoria: implications for management. In *Nature Conservation 2: The Role of Corridors*, eds. D. A. Saunders and R. J. Hobbs. Chipping Norton, Australia: Surrey Beatty, pp. 259–272.

Primack, R. B. (1993). *Essentials of Conservation Biology*. Sunderland, Massachusetts: Sinauer Associates.

Pronello, C. (2003). The measurement of train noise: a case study in Northern Italy. *Transportation Research, Part D*, 8, 113–128.

Puliafito, E., Puliafito, C., Quero, J. and Guerreiro, P. (1998). Airborne pollutants from mobile sources for the city of Mendoza, Argentina. In *Urban Ecology*, eds. J. Breuste, H. Feldmann and O. Uhlmann. Berlin: Springer-Verlag, 99–103.

Pulliam, H. R. and Danielson, B. J. (1991). Sources, sinks, and habitat selections: a landscape perspective on population dynamics. *American Naturalist*, 137 (Supplement), 50–66.

Pysek, P. (1993). Factors affecting the flora and vegetation in Central European settlements. *Vegetatio*, 106, 89–100.

Pysek, P. (1995a). Approaches to studying spontaneous settlement flora and vegetation in Central Europe: a review. In *Urban Ecology as the Basis of Urban Planning*, eds. H. Sukopp, M. Numata and A. Huber. The Hague, Netherlands: SPB Academic Publishing, 23–39.

Pysek, P. (1995b). On the terminology used in plant invasion studies. In *Plant Invasions: General Aspects and Special Problems*, eds. P. Pysek, K. Prach, M. Rejmanek and M. Wade. Amsterdam: SPB Academic Publishing, pp. 71–81.

Pysek, P. (1998). Alien and native species in Central European urban floras: a quantitative comparison. *Journal of Biogeography*, 25, 155–163.

Pysek, P. and Pysek, A. (1990). Comparison of the vegetation and flora of the West Bohemian villages and towns. In *Urban Ecology: Plants and Plant Communities in Urban Environments*, eds. H. Sukopp, S. Heijny and I. Kowarik. The Hague, Netherlands: SPB Academic Publishing, pp. 105–112.

Pysek, P., Chocholouskova, Z., Pysek, A. et al. (2004). Trends in species diversity and composition of urban vegetation over three decades. *Journal of Vegetation Science*, 15, 781–788.

Quigley, M. F. (2011). Potemkin gardens: biodiversity in small designed landscapes. In *Urban Ecology: Patterns, Processes, and Applications*, eds. J. Niemela, J. H. Breuste, T. Elmqvist et al. New York: Oxford University Press, pp. 85–92.

Quinn, T. (1991). Distribution and habitat associations of coyotes in Seattle, Washington. In *Wildlife Conservation in Metropolitan Environments*, eds. L. W. Adams and D. L. Leedy. Columbia, Maryland: National Institute for Urban Wildlife, pp. 47–51.

Quinn, T. (1995). Using public sightings to investigate coyote use of urban habitat. *Journal of Wildlife Management*, 59, 238–245.

Rabeni, C. F. and Sowa, S. P. (2002). A landscape approach to managing the biota of streams. In *Integrating Landscape Ecology into Natural Resource Management*, eds. J. Liu and W. W. Taylor. New York: Cambridge University Press, pp. 114–142.

Radeloff, V. C., Stewart, S. I. and Hawbaker, T. J. (2010). Housing growth in and near United States protected areas limits their conservation value. *Proceedings of the National Academy of Sciences* (USA), 107, 940–945.

Radford, J. Q. and Bennett, A. F. (2007). The relative importance of landscape properties for woodland birds in agricultural environments. *Journal of Applied Ecology*, 44, 737–747.

Rajvanshi, A., Mathur, V. B., Teleki, G. C. and Mukherjee, S. K. (2001). *Roads, Sensitive Habitats and Wildlife*. Dehradun, India: Wildlife Institute of India.

Ramalho, C. E. and Hobbs, R. J. (2012). Time for a change: dynamic urban ecology. *Trends in Ecology and Evolution*, 27, 179–188.

Rampanelli, G., Zardi, D. and Rotunno, R. (2004). Mechanisms of up-valley winds. *Journal of the Atmospheric Sciences*, 61, 3097–3111.

Rapoport, E. H. (1993). The process of plant colonization in small settlements and large cities. In *Humans as Components of Ecosystems: The Ecology of Subtle Human Effects and Populated Areas*, eds. M. J. McDonnell and S. T. A. Pickett. New York: Springer-Verlag, pp. 190–207.

Raupp, M. J. (2012). Disasters by design: outbreaks along urban gradients. In *Insect Outbreaks Revisited*, eds. P. Barbosa, D. K. Letourneau and A. A. Agrawal. New York: John Wiley, pp. 313–333.

Ravetz, J. (2000). *City Region 2020: Integrated Planning for a Sustainable Environment*. London: Earthscan.

Rawlinson, H., Dickinson, N., Nolan, P. and Putwain, P. (2004). Woodland establishment on closed old-style landfill sites in N.W. England. *Forest Ecology and Management*, 202, 265–280.

Ray, D. (1998). *Development Economics*. Princeton, New Jersey: Princeton University Press.

Ray, J. C. (2005). *Sprawl and Highly Mobile or Wide-Ranging Species*. New York: Columbia University Press.

Rebele, F. (1994). Urban ecology and special features of urban ecosystems. *Global Ecology and Biogeography Letters*, 4, 173–187.

Redman, C. L. (2011). Social-ecological transformations in urban landscapes: a historical perspective. In *Urban Ecology: Patterns, Processes, and Applications*, eds. J. Niemela, J. H. Breuste, T. Elmqvist et al. New York: Oxford University Press, pp. 206–212.

Reed, D. and Hilderbrand, G. (2012). *Visible/Invisible: Landscape Works of Reed Hilderbrand*. New York: Metropolis Books.

Rees, W. E. (2002). Globalization and sustainability: conflict or convergence? *Bulletin of Science, Technology and Society*, 22, 249–268.

Rees, W. E. and Wackernagel, M. (1996). *Our Ecological Footprint: Reducing Human Impact on the Earth*. Philadelphia: New Society Books.

Rees, W. E. and Wackernagel, M. (2008). Urban ecological footprints: why cities cannot be sustainable – and why they are a key to sustainability. In *Urban Ecology: An International Perspective on the Interaction Between Humans and Nature*, eds. J. M. Marzluff, E. Shulenberger, Endlicher, W. et al. New York: Springer, pp. 537–555.

Register, R. (2006). *Ecocities: Rebuilding Cities in Balance with Nature*. Gabriola Island, British Columbia, Canada: New Society Publishers.

Reifsnyder, W. F. and Lull, H. W. (1965). *Radiant Energy in Relation to Forests*. Technical Bulletin No. 1344. Washington, D.C.: USDA Forest Service, US Government Printing Office.

Reijnen, R. and Foppen, R. (1995). The effects of car traffic on breeding bird populations in woodland. IV. Influence of population size on the reduction of density of woodland breeding birds. *Journal of Applied Ecology*, 32, 481–491.

Reijnen, R. and Foppen, R. (2006). Impact of road traffic on breeding bird populations. In *The Ecology of Transportation: Managing Mobility for the Environment*, eds. J. Davenport and J. L. Davenport. New York: Springer, pp. 255–274.

Reisch, M., Anthony, E. and Bearden, D. M. (2003). *Superfund and the Brownfield Issue*. New York: Novinka Books.

Remmler, F., Hutter, U. and Schottler, U. (1998). Infiltration of storm water runoff with respect to soil- and groundwater protection. In *Urban Ecology*, eds. J. Breuste, H. Feldmann and O. Uhlmann. New York: Springer, pp. 127–132.

Reps, J. W. (1997). *Canberra 1912: Plans and Planners of the Australian Capital Competition*. Melbourne: Melbourne University Press.

Resh, V. H., Brown, A. V., Covich, A. P. et al. (1988). The role of disturbance in stream ecology. *Journal of the North American Benthological Society*, 7, 433–455.

Rhodes, M., Wardell-Johnston, G. W., Rhodes, M. P. and Raymond, B. (2006). Applying network analysis to the conservation of habitat trees in urban environments: a case study from Brisbane, Australia. *Conservation Biology*, 20, 861–870.

Rich, A. C., Dobkin, D. S. and Niles, L. J. (1994). Defining forest fragmentation by corridor width: the influence of narrow forest-dividing corridors on forest-nesting birds in Southern New Jersey. *Conservation Biology*, 8, 1109–1121.

Rich, C. and Longcore, T., eds. (2006). *Ecological Consequences of Artificial Night Lighting*. Washington, D.C.: Island Press.

Richards, I. (2001). *Ecology of the Sky*. Mulgrave, Victoria, Australia: Images Publishing Group.

Richman, T., Worth, J., Dawe, P. et al. (1997). *Start at the Source: Residential Site Planning and Design Guidance Manual for Stormwater Quality Protection*. Los Angeles: Bay Area Stormwater Management Agency Association.

Richter, M. and Weiland, U., eds. (2012). *Applied Urban Ecology: A Global Framework*. New York: Wiley-Blackwell.

Ricklefs, R. E. and Miller, G. L. (2000). *Ecology*. New York: W. H. Freeman.

Rieley, J. O. and Page, S. E. (1995). Survey, mapping and evaluation of green space in the Federal Territory of Kuala Lumpur, Malaysia. In *Urban Ecology as the Basis of Urban Planning*, eds. H. Sukopp, M. Numata and A. Huber. The Hague, Netherlands: SPB Academic Publishing, pp. 173–183.

Riley, A. L. (1998). *Restoring Streams in Cities: A Guide for Planners, Policymakers, and Citizens*. Washington, D.C.: Island Press.

Rink, D. (2009). Wilderness: the nature of urban shrinkage? The debate on urban restructuring and restoration in Eastern Germany. *Nature and Culture*, 4(3), 275–292.

Roaf, S., Fuentes, M. and Thomas, S. (2001). *Ecohouse: A Design Guide*. Oxford, UK: Architectural Press.

Robbins, P. and Sharp, J. T. (2003). Producing and consuming chemicals: the moral economy of the American lawn. *Economic Geography*, 79, 425–451.

Roberts, D. C. (1993). The vegetation ecology of municipal Durban, Natal – floristic classification. *Bothalia*, 23, 271–326.

Roberts, D. G., Ayre, D. J. and Whelan, R. J. (2007). Urban plants as genetic reservoirs or threats to the integrity of bushland plant populations. *Conservation Biology*, 21, 842–852.

Robinson, G. R. and Handel, S. N. (1993). Forest restoration on a closed landfill: rapid addition of new species by bird dispersal. *Conservation Biology*, 7, 271–278.

Robinson, G. R. and Handel, S. N. (1995). Woody plant roots fail to penetrate a clay-lined landfill: management implications. *Environmental Management*, 19, 57–64.

Robinson, G. R. and Handel, S. N. (2000). Directing spatial patterns of recruitment during an experimental urban woodland reclamation. *Ecological Applications*, 10, 174–188.

Robinson, W. H. (1996). *Urban Entomology: Insect and Mite Pests in the Human Environment*. London: Chapman & Hall.

Robson, D. and Bawa, G. (2002). *Geoffrey Bawa: The Complete Works*. London: Thames & Hudson.

Rodriguez, A., Crema, G. and Delibes, M. (1996). Use of non-wildlife passages across a high speed railway by terrestrial vertebrates. *Journal of Applied Ecology*, 33, 1527–1540.

Rogers, C. M. and Caro, M. J. (1998). Song sparrows, top carnivores and nest predation: a test of the mesopredator release hypothesis. *Oecologia*, 116, 227–233.

Rogers, P. P., Jalal, K. F. and Boyd, J. A. (2006). *An Introduction to Sustainable Development*. Cambridge, Massachusetts: Harvard University, Continuing Education Division.

Romme, W. H. (1997). Creating pseudo-rural landscapes in the Mountain West. In *Placing Nature: Culture and Landscape Ecology*, ed. J. I. Nassauer. Washington, D.C.: Island Press, pp. 139–161.

Rosatte, R. C., Kelly-Ward, P. and MacInnes, C. D. (1987). A strategy for controlling rabies in urban skunks and raccoons. In *Integrating Man and Nature in the Metropolitan Environment*, eds. L. W. Adams and D. L. Leedy. Columbia, Maryland: National Institute for Urban Wildlife, pp. 162–167.

Rosatte, R. C., Power, M. J. and MacInnes, C. D. (1991). Ecology of urban skunks, raccoons, and foxes in metropolitan Toronto. In *Wildlife Conservation in Metropolitan Environments*, eds. L. W. Adams and D. L. Leedy. Columbia, Maryland: National Institute for Urban Wildlife, pp. 31–38.

Rose, E. and Nagel, P. (2006). Spatio-temporal use of the urban habitat by feral pigeons (*Columba livia*). *Behavioral Ecology and Sociobiology*, 60, 242–254.

Roseland, M. (2001). The eco-city approach to sustainable development in urban areas. In *How Green Is the City? Sustainability Assessment and the Management of Urban Environments*, eds. D. Devuyst, L. Hens and W. De Lannoy. New York: Columbia University Press, pp. 85–103.

Rosenberg, N. J. (1974). *Microclimate: The Biological Environment*. New York: John Wiley.

Rosenzweig, C., Solecki, W. D., Hammer, S. A. and Mehrotra, S. (2011). *Climate Change and Cities: First Assessment Report of the Urban Climate Change Research Network*. New York: Cambridge University Press.

Rosenzweig, R. and Blackmar, E. (1992). *The Park and the People: A History of Central Park*. Ithaca, New York: Cornell University Press.

Rossin, A., Malizia, A. I. and Denegri, G. M. (2004). The role of the subterranean rodent *Ctenomys talarum* (Rodentia: Octodontidae) in the life cycle of *Taenia taeniaeformis* (Cestoda; Taeniidae) in urban environments. *Veterinary Parasitology*, 122, 27–33.

Roth, M. (2007). Review of urban climate research in (sub) tropical regions. *International Journal of Climate*, 27, 1859–1873.

Roth, R. R. (1987). Assessment of habitat quality for wood thrush in a residential area. In *Integrating Man and Nature in the Metropolitan Environment*, eds. L. W. Adams and D. L. Leedy. Columbia, Maryland: National Institute for Urban Wildlife, pp. 139–149.

Rowe, P. (1991). *Making a Middle Landscape*. Cambridge, Massachusetts: MIT Press.

Rowe, P. G. (2005). *East Asia Modern: Shaping the Contemporary City*. London: Reaktion.

Rowe, P. G. (2010). *A City and Its Stream: An Appraisal of the Cheonggyecheon Restoration Project and Its Environs in Seoul, South Korea*. Cambridge, Massachusetts: Harvard University, Graduate School of Design.

Rowe, P. G., Hunter, K. Jung, S. et al. (2013). *Methodological Notes on the Spatial Analysis of Urban Formation*. Cambridge, Massachusetts: Harvard University Graduate School of Design.

Rowell, D. L. (1994). *Soil Science: Methods and Applications*. Harlow, UK: Longmans.

Rowney, A. C., Stahre, P. and Roesner, L. A., eds. (1999). *Sustaining Urban Water Resources in the 21st Century: Proceedings 7–12 September 1997, Malmo, Sweden*. Reston, Virginia: American Society of Civil Engineers.

Rowntree, R. A. (1986). Ecology of the urban forest – Introduction to Part II. *Urban Ecology*, 9, 229–243.

Roy, A. H., Freeman, M. C., Freeman, B. J. et al. (2005). Investigating hydrologic alteration as a mechanism of fish assemblage shifts in urbanizing streams. *Journal of the North American Benthological Society*, 24, 656–678.

Royal Commission. (1991). *Shoreline Regeneration for the Greater Toronto Bioregion*. Toronto: The Royal Commission on the Future of the Toronto Waterfront.

Rubin, H. D. and Beck, A. M. (1982). Ecological behavior of free-ranging urban pet dogs. *Applied Animal Ethology*, 8, 161–168.

Rudd, H., Vala, J. and Schaefer, V. (2002). Importance of backyard habitat in a comprehensive biodiversity conservation strategy: a connectivity analysis of urban green spaces. *Restoration Ecology*, 10, 368–375.

Rudnicky, J. L. and McDonnell, M. J. (1989). Forty-eight years of canopy change in a hardwood-hemlock forest in New York City. *Bulletin of the Torrey Botanical Club*, 116, 52–64.

Rugiero, L. and Luiselli, L. (2006). Ecological modeling of habitat use and the annual activity patterns in an urban population of the tortoise, *Testudo hermanni*. *Italian Journal of Zoology*, 73, 219–225.

Russell, E. W. (1961). *Soil Conditions and Plant Growth*. London: Longmans.

Ruszczyk, A. and Silva, C. F. (1997). Butterflies select microhabitats on buildings. *Landscape and Urban Planning*, 38, 119–127.

Ryan, K.-L. and Winterich, J., eds. (1993) *Successful Rail-Trails*. Washington, D.C.: Rails-to-Trails Conservancy.

Ryan, S. and Throgmorton, J. (2003). Sustainable transportation and land development on the periphery: a case study of Freiburg, Germany and Chula Vista, CA. *Transportation Research, Part D*, 8, 37–52.

Rydberg, D. and Falck, J. (2000). Urban forestry in Sweden from a silvicultural perspective: a review. *Landscape and Urban Planning*, 47, 1–18.

Rydell, J. (2006). Bats and their insect prey at street lights. In *Ecological Consequences of Artificial Night Lighting*, eds. C. Rich and T. Longcore. Washington, D.C.: Island Press, pp. 43–60.

Saarinen, K., Valtonen, A., Jantunen, J. and Saarnio, S. (2005). Butterflies and diurnal moths along road verges: does road type affect diversity and abundance? *Biological Conservation*, 123, 403–412.

Sachse, U., Starfinger, U. and Kowarik, I. (1990). Synanthropic woody species in the urban area of Berlin (West). In *Urban Ecology: Plants and Plant Communities in Urban Environments*, eds. H. Sukopp, S. Hejny and I. Kowarik. The Hague, Netherlands: SPB Academic Publishing, pp. 233–243.

Saebo, A., Borzan, Z., Ducatillion, C. et al. (2005). The selection of plant materials for street trees, park trees and urban woodland. In *Urban Forests and Trees*, eds.

C. C. Konijnendijk, K. Nilsson, T. B. Randrup and J. Schipperijn. New York: Springer, pp. 257–280.

Sagoff, M. (2011). The quantification and valuation of ecosystem services. *Ecological Economics*, 70, 497–502.

Sahely, H. R., Dudding, S. and Kenned, C. A. (2003). Estimating the urban metabolism of Canadian cities: Greater Toronto Area case study. *Canadian Journal of Civil Engineering*, 30, 468–483.

Sailer-Fliege, U. (1999). Characteristics of post-socialist urban transformation in East Central Europe. *GeoJournal*, 49, 7–16.

Saiz, S., Kennedy, C., Bass, B. and Pressnail, K. (2006). Comparative life cycle assessment of standard and green roofs. *Environmental Science and Technology*, 40, 4312–4316.

Sakamoto, K. (1988). *Remnant Forms of Ulmaceae Woods and Trees in Urban Areas*. Monograph 2, Bulletin of Revegetation Research. Kyoto, Japan: Association of Revegetation Research.

Saley, H., Meredith, D. H., Stelfox, H. and Ealey, D. (2003). *Nature Walks and Sunday Drives 'Round Edmonton*. Edmonton, Alberta: Edmonton Natural History Club.

Salisbury, E. J. (1943). The flora of bombed areas. *Nature*, 151, 462–464.

Salisbury, E. J. (1961). *Weeds and Aliens*. London: Collins.

Sanders, W. (1981). *The Cluster Subdivision: A Cost-Effective Approach*. Planning Advisory Service Report 356. Washington, D.C.: American Planning Association.

Sarlov Herlin, I. L. and Fry, G. L. A. (2000). Dispersal of woody plants in forest edges and hedgerows in a southern Sweden agricultural area: the role of site and landscape structure. *Landscape Ecology*, 15, 229–242.

Sarrata, C., Lemonsub, A., Massona, V. and Guelaliac, D. (2005). Impact of urban heat island on regional atmospheric pollution. *Atmospheric Environment*, 40, 1743–1758.

Satheesan, S. M. (1996). Raptors associated with airports and aircraft. In *Raptors in Human Landscapes*, eds. D. Bird, D. Varland and J. Negro. San Diego, California: Academic Press, pp. 315–323.

Satterthwaite, D. and Tacoli, C. (2006). The role of small and intermediate urban centres in regional and rural development: assumptions and evidence. In *The Earthscan Reader in Rural-Urban Linkages*, ed. C. Tacoli. London: Earthscan, pp. 155–183.

Sauerwein, M. (2011). Urban soils – characterization, pollution, and relevance in urban ecosystems. In *Urban Ecology: Patterns, Processes, and Applications*, eds. J. Niemela, J. H. Breuste, T. Elmqvist et al. New York: John Wiley, pp. 44–58.

Saunders, D. A. and Hobbs, R. J., eds. (1991). *Nature Conservation 2: The Role of Corridors*. Chipping Norton, Australia: Surrey Beatty.

Saunders, D. A., Arnold, G. W., Burbidge, A. A. and Hopkins, A. J. M., eds. (1987). *Nature Conservation: The Role of Remnants of Native Vegetation*. Chipping Norton, Australia: Surrey Beatty.

Savard, J. P. L., Clergeau, P. and Mennechez, G. (2000). Biodiversity concepts and urban ecosystems. *Landscape and Urban Planning*, 48, 131–142.

Schaepe, A. (1990). Grid mapping of bryophytes in Berlin (West). In *Urban Ecology: Plants and Plant Communities in Urban Environments*, eds. H. Sukopp, S. Hejny and I. Kowarik. The Hague, Netherlands: SPB Academic Publishing, pp. 251–254.

Scheer, B. C. (2004). The radial street as a timeline. In *Suburban Form: An International Perpective*, eds K. Stanilov and B. C. Scheer. New York: Routledge, pp. 102–122.

Scheuerell, M. D. and Schindler, D. E. (2004). Changes in the spatial distribution of fishes in lakes along a residential development gradient. *Ecosystems*, 7, 98–106.

Schmid, J. A. (1975). *Urban Vegetation: A Review and Chicago Case Study*. Department of Geography Research Paper 161. Chicago: University of Chicago.

Schmidt, E. and Bock, C. E. (2004). Habitat associations and population trends of two hawks in an urbanizing grassland region in Colorado. *Landscape Ecology*, 20, 469–478.

Schneider, A. and Woodcock, C. E. (2008). Compact, dispersed, fragmented, extensive? A comparison of urban growth in twenty-five global cities using remotely sensed data, pattern metrics and census information. *Urban Studies*, 45, 659–692.

Scholz, M. and Grabowiecki, P. (2007). Review of permeable pavement systems. *Building and Environment*, 42, 3830–3836.

Scholz-Barth, K. (2001). Green roofs: stormwater management from top down. *Environmental Design and Construction*, Jan.-Feb., 63–69.

Schrader, S. and Boning, M. (2006). Soil formation on green roofs and its contribution to urban biodiversity with emphasis on Collembolans. *Pedobiologia*, 50, 347–356.

Schroder, G. D. and Hulse, M. (1979). Survey of rodent populations associated with urban landfill. *American Journal of Public Health*, 69, 713–715.

Schroeder, F.-G. (1969). Zur Klassifizierung der Anthropochoren. *Vegetatio*, 16, 225–238.

Schueler, T. R. (1995). *Site Planning for Urban Stream Protection*. Ellicott City, Maryland: Center for Watershed Protection.

Schueler, T. R. and Holland, H. K., eds. (2000). *The Practice of Watershed Protection*. Ellicott City, Maryland: Center for Watershed Protection.

Schulte, W. and Sukopp, H. (2000). Stadt und Dorfbiotopkartierungen. Erfassung und Analyse okologischer Grundlagen im besiedelten Bereich der Bundesrepublik Deutschland – ein Uberblick (Stand: Marz 2000). *Naturschutz und Landschaftsplanung*, 32, 140–147.

Schulte, W., Sukopp, H. and Werner, P., eds. (1993). Flachendeckende Biotopkartierung im besiedelten Bereich als Grundlage einer am Naturschutz orientierten Planung. *Natur und Landschaft*, 68, 491–526.

Schulze, E.-D., Beck, E. and Muller-Hohenstein, K. (2005). *Plant Ecology*. Berlin: Springer.

Schwartz, H. (2004). *Urban Renewal, Municipal Revitalization: The Case of Curitiba, Brazil*. Alexandria, Virginia: Published by the Author.

Scott, X. L., Simpson, J. R. and McPherson, E. G. (1999). Effects of tree cover on parking lots: microclimate and vehicle emissions. *Journal of Arboriculture*, 25, 129–142.

Searns, R. M. (1995). The evolution of greenways as an adaptive urban landscape form. *Landscape and Urban Planning*, 33, 65–80.

Sears, A. R. and Anderson, S. H. (1991). Correlations between birds and vegetation in Cheyenne, Wyoming. In *Wildlife Conservation in Metropolitan Environments*, eds. L. W. Adams and D. L. Leedy. Columbia, Maryland: National Institute for Urban Wildlife, pp. 75–80.

Seiler, A. and Helldin, J.-O. (2006). Mortality in wildlife due to transportation. In *The Ecology of Transportation: Managing Mobility for the Environment*, eds. J. Davenport and J. L. Davenport. New York: Springer, pp. 165–189.

Sekercioglu, C. (2011). Turkey's globally important biodiversity in crisis. *Biological Conservation*, 144, 2752–2764.

Sekhar, M. C. (1998). Assessment of ambient air quality in an urban ecosystem. In *Urban Ecology*, eds. J. Breuste, H. Feldmann and O. Uhlmann. Berlin: Springer-Verlag, pp. 109–113.

Selman, P. (2006). *Planning at the Landscape Scale*. London: Routledge.

Servoss, W., Engeman, R. M., Fairaizl, S. et al. (2000). Wildlife hazard assessment for Phoenix Sky Harbor International Airport. *International Biodeterioration and Biodegradation*, 45, 111–127.

Setha, M., Low, D. T. and Scheld, S. (2005). *Rethinking Urban Parks: Public Space and Cultural Diversity*. Austin, Texas: University of Texas Press.

Shanahan, P. (2009). Groundwater in the urban environment. In *The Water Environment of Cities*, ed. L. A. Baker. New York: Springer, pp. 29–48.

Shanahan, P. and Jacobs, B. L. (2007). Ground water and cities. In *Cities of the Future: Towards Integrated Sustainable Water and Landscape Management*, eds. V. Novotny and P. Brown. London: IWA Publishing.

Sharma, H. D. and Reddy, K. R. (2004). *Geoenvironmental Engineering*. New York: John Wiley.

Shashua-Bar, L. (2009). The cooling efficiency of urban landscape strategies in a hot dry climate. *Landscape and Urban Planning*, 92, 179–186.

Shashua-Bar, L. and Hoffman, M. E. (2004). Quantitative evaluation of passive cooling of the UCL microclimate in hot regions in summer, a case study: urban streets and courtyards with trees. *Building and Environment*, 39, 1087–1099.

Shaw, L. M., Chamberlain, D. and Evans, M. (2008). The house sparrow *Passer domesticus* in urban areas: reviewing a possible link between post-decline distribution and human socioeconomic status. *Journal of Ornithology*, 149, 293–299.

Shea, N. (2011). Under Paris. *National Geographic*, February 2011, 104–125.

Shea, R. (1981). *Olmsted and the Boston Park System*. Boston: Metropolitan Area Planning Council.

Shimada, T. (2001). Choice of daily flight routes of greater white-fronted geese: effects of power lines. *Waterbirds*, 24, 425–429.

Shindhe, K. C. (2006). The national highway bypass around Hubli-Dharwad and its impact on peri-urban livelihoods. In *The Peri-Urban Interface: Approaches to Sustainable Natural and Human Resource Use*, eds. D. McGregor, D. Simon and D. Thompson. London: Earthscan, pp. 181–195.

Shochat, E., Stefanov, E. L., Whitehouse, M. E. A. and Faeth, S. H. (2004). Spider diversity in the greater Phoenix area: the influence of human modification to habitat structure and modification. *Ecological Applications*, 14, 268–280.

Shochat, E., Warren, P. S., Faeth, S. H. et al. (2006). From patterns to emerging processes in mechanistic urban ecology. *Trends in Ecology and Evolution*, 21, 186–191.

Short, J. R., Fanning, D. S., Foss, J. E. and Patterson, J. C. (1986b). Soils of the Mall in Washington, D.C.: II. Genesis, classification, and mapping. *Soil Science Society of America Journal*, 50, 705–710.

Short, J. R., Fanning, D. S., McIntosh, M. S. et al. (1986a). Soils of the Mall in Washington, D.C.: I. Statistical summary of properties. *Soil Science Society of America Journal*, 50, 699–705.

Shortle, W. C. and Rich, A. E. (1970). Relative sodium chloride tolerance of common roadside trees in southeastern New Hampshire. *Plant Disease Reporter*, 1970, 360–362.

Showell, R. (1986). *Hedges, Walls, and Boundaries*. London: Dryad Press.

Sieghardt, M., Mursch-Radlgruber, E., Paoletti, E. et al. (2005). The abiotic urban environment: impact of urban growing conditions on urban vegetation. In *Urban Forests and Trees*, C. C. Konijnendijk, K. Nilsson, T. B. Randrup and J. Schipperijn. New York: Springer, pp. 281–323.

Sien, C. L. (1992). *Singapore's Urban Coastal Area: Strategies for Management*. Manila, Philippines: International Center for Living Aquatic Resources Management.

Siksna, A. (1997). The effects of block size and form in North American and Australian city centres. *Urban Morphology*, 1, 19–33.

Silliman, B. R., Grosholz, E. D. and Bertness, M. D., eds. (2009). *Human Impacts on Salt Marshes: A Global Perspective*. Berkeley, California: University of California Press.

Sims, V., Evans, K. L., Newson, S. E. et al. (2008). Avian assemblage structure and domestic cat densities in urban environments. *Diversity and Distributions*, 14, 387–399.

Skirgiello, A. (1990). Synanthropization of the Polish mycoflora. In *Urban Ecology: Plants and Plant Communities in Urban Environments*, eds. H. Sukopp, S. Hejny and I. Kowarik. The Hague, Netherlands: SPB Academic Publishing, pp. 255–257.

Sklar, A. (2008). *Brown Acres: An Intimate History of the Los Angeles Sewers*. Santa Monica, California: Angel City Press.

Sklar, F. H. and Costanza, R. (1991). The development of dynamic spatial models for landscape ecology: a review and prognosis. In *Quantitative Methods in Landscape Ecology*, eds. M. G. Turner and R. H. Gardner. New York: Springer-Verlag, pp. 239–288.

Slabbekorn, H., Yeh, P. and Hunt, K. (2007). Sound transmission and song divergence: a comparison of urban and forest acoustics. *Molecular Ecology*, 17, 72–83.

Sloane, D. C. (1996). The American lawn. *American Planning Association Journal*, 62, 126–127.

Small, E., Sadler, J. P. and Telfer, M. (2006). Do landscape factors affect brownfield carabid assemblages? *Science of the Total Environment*, 360, 205–222.

Smit, J. (2006). Farming in the city and climate change: the potential and urgency of applying urban agriculture to reduce the negative impacts of climate change. *Urban Agriculture Magazine* No. 15 (Maiden issue). Leusden, The Netherlands.

Smit, J. and Nasr, J. (1992). Urban agriculture for sustainable cities using wastes and idle land amd water bodies as resources. *Environment and Urbanization*, 4, 141–152.

Smith, M. A., Turner, M. G. and Rusch, D. H. (2002). The effect of military training activity on eastern lupine and the Karner blue butterfly at Fort McCoy, Wisconsin, USA. *Environmental Management*, 29, 102–115.

Smith, R. L. (1996). *Ecology and Field Biology*. New York: HarperCollins.

Smith, R. M., Gaston, K. J., Warren, P. H. and Thompson, K. (2005). Urban domestic gardens (V): relationships between landcover composition, housing and landscape. *Landscape Ecology*, 20, 235–253.

Smith, R. M., Thompson, K., Hodgson, J. G. et al. (2006). Urban domestic gardens IX: composition and richness of the vascular plant flora, and implications for native biodiversity. *Biological Conservation*, 129, 312–322.

Snep, R. P. H., Opdam, P. F. M., Baveco, J. M. et al. (2006). How peri-urban areas can strengthen animal populations within cities: a modeling approach. *Biological Conservation*, 127, 345–355.

Snodgrass, E. C. and Snodgrass, L. L. (2006). *Green Roof Plants: A Resource and Planting Guide*. Portland, Oregon: Timber Press.

Sodhi, N. S., Briffett, C., Kong, L. and Yuen, B. (1999). Bird use of linear areas of a tropical city: implications for park connector design and management. *Landscape and Urban Planning*, 45, 123–130.

Song, I.-J. and Gin, Y.-R. (2008). Managing biodiversity of rice paddy culture in urban landscape. In *Landscape Ecological Applications in Man-Influenced Areas: Linking Man and Nature Systems*, eds. S.-K. Hong, N. Nakagoshi, B. J. Fu and Y. Morimoto. New York: Springer, pp. 193–208.

Song, I.-J., Hong, S.-K., Kim, H.-O. et al. (2005). The pattern of landscape patches and invasion of naturized plants in developed areas of urban Seoul. *Landscape and Urban Planning*, 70, 205–219.

Sorensen, A. (2004). *The Making of Urban Japan*. London: Routledge.

Soule, D. C. (2006). *Urban Sprawl: A Comprehensive Reference Guide*. Westport, Connecticut: Greenwood Press.

Soule, M. E. (1991). Land use planning and wildlife maintenance: guidelines for conserving wildlife in an urban landscape. *Journal of the American Planning Association*, 57, 313–323.

Soule, M. E., Bolger, D. T., Alberts, A. C. et al. (1988). Reconstructed dynamics of rapid extinctions of chaparral-requiring birds in urban habitat islands. *Conservation Biology*, 2, 75–92.

Southworth, M. and Ben-Joseph, E. (1996). *Streets and the Shaping of Towns and Cities*. New York: McGraw-Hill.

Sovada, M. A., Sargeant, A. B. and Grier, J. W. (1995). Differential effects of coyotes and red foxes on duck nest success. *Journal of Wildlife Management*, 59, 61–69.

Spellerberg, I. (2002). *Ecological Effects of Roads*. Plymouth, UK: Science Publishers.

Spengler, J. D. and Chen, Q. (2000). Indoor air quality factors in designing a healthy building. *Annual Review of Energy and the Environment*, 25, 567–600.

References

Sperling, D. and Gordon, D. (2009). *Two Billion Cars: Driving Toward Sustainability*. New York: Oxford University Press.

Spirn, A. W. (1984). *The Granite Garden: Urban Nature and Human Design*. New York: Basic Books.

Squires, G. D., ed. (2002). *Urban Sprawl: Causes, Consequences and Policy Responses*. Washington, D.C.: Urban Land Institute.

Stafford, D. (1999). Secret underground cities – an account of some of Britain's subterranean defence, factory and storage sites in the Second World War. *The Times Literary Supplement* 5004 (February), 33.

Stanilov, K. and Scheer, B. C., eds. (2004). *Suburban Form: An International Perspective*. New York: Routledge.

Stearns, F. (1970). Urban ecology today. *Science*, 170, 1005–1007.

Stearns, F. and Montag, T., eds. (1974). *The Urban Ecosystem: A Holistic Approach*. Stroudsburg, Pennsylvania: Dowden, Hutchinson and Ross.

Steele, J. (2004). *Genzyme Center*. Stuttgart: FMO Publishers.

Steele, M. A. and Koprowski, J. L. (2001). *North American Tree Squirrels*. Washington, D.C.: Smithsonian Institution Press.

Steenbergen, C. M. and Aten, D. (2007). *Sea of Land: The Polder as an Experimental Atlas of Dutch Landscape Architecture*. Netherlands: Stichting Utigeverij Noord.

Steenhof, K., Kochert, M. and Roppe, J. (1993). Nesting by raptors and common ravens on electrical transmission line towers. *Journal of Wildlife Management*, 57, 271–282.

Stein, S. B. (1988). *My Weeds: A Gardener's Botany*. New York: Harper and Row.

Steiner, D. and Nauser, M., eds. (1993). *Human Ecology: Fragments of Anti-Fragmentary Views of the World*. London: Routledge.

Steiner, F. (2000). *The Living Landscape: An Ecological Approach to Landscape Planning*. New York: McGraw-Hill.

Steiner, F. (2002). *Human Ecology*. Washington, D.C.: Island Press.

Stengel, D., O'Reilly, S. and O'Halloran, J. (2006). Contaminants and pollutants. In *The Ecology of Transportation: Managing Mobility for the Environment*, eds. J. Davenport and J. L. Davenport. New York: Springer, pp. 361–389.

Stern, M. and Marsh, W. (1997). The decentered city: edge cities and the expanding metropolis (Editor's introduction). *Landscape and Urban Planning*, 36, 243–246.

Stevens, D. and Harpur, J. (1997). *Roof Gardens, Balconies, and Terraces*. New York: Rizzoli.

Stewart, D. and Streishinsky, T. (1990). Nothing goes to waste in Arcata's teeming marshes. *Smithsonian*, 21(1), 174–179.

Stewart, I. D. (2011). A systematic review and scientific critique of methodology in modern urban heat island literature. *International Journal of Climatology*, 31, 200–217.

Stiles, E. W. (1980). Patterns of fruit presentation and seed dispersal in bird-disseminated woody plants in the eastern deciduous forest. *American Naturalist*, 116, 670–688.

Stiles, E. W. (1982). Fruit flags: two hypotheses. *American Naturalist*, 120, 500–509.

Stilgoe, J. R. (1988). *Borderland: Origins of the American Suburb, 1820–1939*. New Haven, Connecticut: Yale University Press.

Stout, W. E., Anderson, R. K. and Papp, J. M. (1996). Red-tailed hawks nesting on human-made and natural structures in Southeast Wisconsin. In *Raptors in Human Landscapes*, eds. D. Bird, D. Varland and J. Negro. San Diego, California: Academic Press, pp. 77–86.

Stowe, J. P. Jr, Schmidt, E. V. and Green, D. (2001). Toxic burials: the final insult. *Conservation Biology*, 15, 1817–1819.

Stracey, C. M. and Robinson, S. K. (2012). Does nest predation shape urban bird communities? In *Urban Bird Ecology and Conservation*, eds. C. A. Lepczyk and P. S. Warren. Berkeley, California: University of California Press, pp. 49–70.

Strauss, B. and Biedermann, R. (2006). Urban brownfields as temporary habitats: driving forces for the diversity of phytophagous insects. *Ecography*, 29, 928–949.

Strayer, D. L. (1994). Limits to biological distributions in groundwater. In *Groundwater Ecology*, eds. J. Gibert, D. Danielopol and J. A. Stanford. San Diego, California: Academic Press, pp. 287–310.

Suchman, D. R. (2002). *Developing Successful Infill Housing*. Washington, D.C.: Urban Land Institute.

Suh, S. (2008). *Handbook on Input–Output Economics for Industrial Ecology*. New York: Springer.

Sukopp, H. (1990). Urban ecology and its application in Europe. In *Urban Ecology: Plants and Plant Communities in Urban Environments*, eds. H. Sukopp, S. Hejny and I. Kowarik. The Hague, Netherlands: SPB Academic Publishing, pp. 1–22.

Sukopp, H. (1998). Urban ecology: scientific and practical aspects. In *Urban Ecology*, eds. J. Breuste, H. Feldmann and O. Uhlmann. Berlin: Springer-Verlag, pp. 3–16.

Sukopp, H. (2002). On the early history of urban ecology in Europe. *Preslia*, 74, 373–393.

Sukopp, H. (2008). The city as a subject for ecological research. In *Urban Ecology: An International Perspective on the Interaction Between Humans and Nature*, eds. J. M. Marzluff, E. Shulenberger, W. Endlicher *et al.* New York: Springer, pp. 281–298.

Sukopp, H. and Markstein, B. (1989). Changes of the reed beds along the Berlin Havel, 1962–1987. *Aquatic Botany*, 35, 27–39.

Sukopp, H. and Sukopp, U. (1993). Ecological long-term effects of cultigens becoming feral and of naturalization of non-native species. *Experientia*, 49, 210–218.

Sukopp, H. and Trepl, L. (1987). Extinction and naturalization of plant species as related to ecosystem structure and function. In *Ecological Studies*, eds. E.-D. Schulze and H. Zwolfer. Vol. 61. Berlin: Springer-Verlag, pp. 245–276.

Sukopp, H. and Werner, P. (1983). Urban environment and vegetation. In *Man's Impact on Vegetation*, eds. W. Holzner, M. J. A. Werger and I. Ikusima. The Hague, Netherlands: W. Junk Publishers, pp. 247–260.

Sukopp, H., Hejny, S. and Kowarik, I., eds. (1990). *Urban Ecology: Plants and Plant Communities in Urban Environments*. The Hague, Netherlands: SPB Publishing.

Sukopp, H., Numata, M. and Huber, A. (1995). *Urban Ecology as the Basis of Urban Planning*. Amsterdam: SPB Academic Publishing.

Sushinsky, J. R., Rhodes, J. R., Possingham, H. P. *et al.* (2013). How should we grow cities to minimize their biodiversity impacts? *Global Change Biology*, 19, 401–410.

Suzuki, H., Moffatt, S. and Iizuka, R. (2010). *Eco²: Ecological Cities as Economic Cities*. Washington, D.C.: World Bank.

Swaminathan, R. and Goyal, J. (2006). *Mumbai Vision 2015: Agenda for Urban Renewal*. New Delhi: MacMillan India.

Swank, W. G. (1955). Nesting and production of the mourning dove in Texas. *Ecology*, 36, 495–505.

Swanson, F. J., Franklin, J. F. and Sedell, J. R. (1990). Landscape patterns, disturbance, and management in the Pacific Northwest, USA. In *Changing Landscapes: An Ecological Perspective*, eds. I. S. Zonneveld and R. T. T. Forman. New York: Springer-Verlag, pp. 191–213.

Sze, J. and Gambirazzio, G. (2013). Eco-cities without ecology: constructing ideologies, valuing nature. In *Resilience in Ecology and Urban Design: Linking Theory and Practice*, eds. S. T. A. Pickett, M. L. Cadenasso and B. McGrath. New York: Springer, pp. 289–297.

Szold, T. S. and Carbonell, A., eds. (2002). *Smart Growth: Form and Consequences*. Cambridge, Massachusetts: Lincoln Institute of Land Policy.

Tabacchi, E., Correll, D. L., Hauer, R. *et al.* (1998). Development, maintenance and role of riparian vegetation in the river landscape. *Freshwater Biology*, 40, 297–516.

Tacoli, C., ed. (2006). *The Earthscan Reader in Rural-Urban Linkages*. London: Earthscan.

Takahashi, M., Higaki, A., Nohno, M. *et al.* (2005). Differential assimilation of nitrogen dioxide by 70 taxa of roadside trees at an urban pollution level. *Chemosphere*, 61, 633–639.

Takebayashi, H. and Moriyama, M. (2007). Surface heat budget on green roof and high reflection roof for mitigation of urban heat island. *Building and Environment*, 42, 2971–2979.

Takehiko, Y. I., Miura, N., Bhagvasuren, B. *et al.* (2005). Preliminary evidence of a barrier effect of a railroad on the migration of Mongolian gazelles. *Conservation Biology*, 19, 945–948.

Takekawa, J. Y., Warnock, N., Martinelli, G. M. *et al.* (2002). Waterbird use of bayland wetlands in the San Francisco Bay estuary: movements of long-billed dowitchers during the winter. *Waterbirds*, 25, 93–105.

Tallaki, K. (2005). The pest-control system in the market gardens of Lome, Togo. In *Agropolis: The Social, Political and Environmental Dimensions of Urban Agriculture*, ed. L. J. A. Mougeot. London: Earthscan, pp. 51–87.

Tan, M. H., Li, X. B., Xie, H. and Lu, C. H. (2005). Urban land expansion and arable land loss in China – -a case study of Beijing-Tianjin-Hebel region. *Land Use Policy*, 22, 187–196.

Tangari, V. R., Rego, A. Q. and Montezuma, R. C. M. (2012). O Arco Metropolitano do Rio de Janeiro: Integracao e fragmentacao de paisagem a dos sistemas de espacos livre de edificacao. Rio de Janeiro: PROARQ-FAU/UFRJ.

Taniguchi, M., Uemura, T. and Sakura, Y. (2005). Effects of urbanization and groundwater flow on subsurface temperature in three megacities in Japan. *Journal of Geophysics and Engineering*, 2, 320–325.

Tanimoto, S. and Nakagoshi, N. (1999). Landscape ecological characteristic in temporal changes of riverside open space in urbanized area. *Journal of Environmental Sciences*, 11, 155–159.

Tanji, K., Davis, D., Hanson, C. *et al.* (2002). Evaporation ponds as a drain water disposal management option. *Irrigation and Drainage Systems*, 16, 279–295.

Taylor, J., Paine, C. and FitzGibbon, J. (1995). From greenbelt to greenways: four Canadian case studies. *Landscape and Urban Planning*, 33, 47–64.

Taylor, P. L., Fahrig, K., Henein, K. and Merriam, G. (1993). Connectivity is a vital element of landscape structure. *Oikos*, 68, 571–573.

Taylor, S. L., Roberts, S. C., Walsh, C. J. and Hatt, B. E. (2004). Catchment urbanization and increased benthic algal biomass in streams: linking mechanisms to management. *Freshwater Biology*, 49, 835–851.

References

Tella, J. L., Hiraldo, F., Donazar-Sancho, J. A. and Negro, J. J. (1996). Costs and benefits of urban nesting in the lesser kestrel. In *Raptors in Human Landscapes*, eds. D. Bird, D. Varland and J. Negro. San Diego, California: Academic Press, pp. 53–60.

Tello, M.-L., Tomalak, M., Siwecki, R. et al. (2005). Biotic urban growing conditions – threats, pests and diseases. In *Urban Forests and Trees: A Reference Book*, eds. C. C. Konijnendijk, K. Nilsson, T. B. Randrup and J. Schipperijn. New York: Springer, pp. 326–365.

Terman, M. (1994). The promise of natural links. *Golf Course Management*, 62, 52–59.

Terman, M. R. (1997). Natural links: naturalistic golf courses as wildlife habitat. *Landscape and Urban Planning*, 38, 183–197.

Termorshuizon, J. W. and Opdam, P. (2009). Landscape services as a bridge between landscape ecology and sustainability science? *Landscape Ecology*, 24, 1037–1052.

Terrasa-Soler, J. J. (2006). Landscape change and ecological corridors in Puerto Rico: towards a master plan of ecological networks. *Acta Científica*, 20, 57–62.

Terry, N. and Banuelos, G. (1999). *Phytoremediation of Contaminated Soil and Water*. New York: Lewis Publishers.

Thai, K. V., Rahm, D. and Coggburn, J. D. (2007). *Handbook of Globalization and the Environment*. Boca Raton, Florida: CRC Press.

Thaitakoo, D., McGrath, B., Srithanyarat, S. and Palopakan, Y. (2013). Bangkok: the ecology and design of an aqua-city. In *Resilience in Ecology and Urban Design: Linking Theory and Practice*, eds. S. T. A. Pickett, M. L. Cadenasso and B. McGrath. New York: Springer, pp. 289–297.

Thayer, Jr., R. L. (2003). *Life Place: Bioregional Thought and Practice*. Berkeley, California: University of California Press.

Thellung, A. (1905). Einteilung der Ruderal und Adventivflora in genetische Gruppen. *Vierteljahresschrift der Naturforschenden Gesellschaft in Zurich*, 50, 232–305.

Theobold, D. M. (2001). Land-use dynamics beyond the American urban fringe. *Geographical Review*, 91, 544–564.

Theobold, D. M. (2004). Placing ex-urban land-use change in a human modification framework. *Frontiers in Ecology and Environment*, 2, 139–144.

Theobold, D. M. and Romme, W. H. (2007). Expansion of the US wildland-urban interface. *Landscape and Urban Planning*, 83, 340–354.

Theobold, D. M., Miller, J. R. and Hobbs, N. T. (1997). Cumulative effects of development on wildlife habitat. *Landscape and Urban Planning*, 39, 25–39.

Theroux, P. (2008). *Ghost Train to the Eastern Star*. Boston: Houghton Mifflin Harcourt.

Thomas, R. L., Fellowes, M. D. E. and Baker, P. J. (2012). Spatio-temporal variation in predation by urban domestic cats (*Felis catus*) and the acceptability of possible management actions in the UK. *PLOS/ONE*, 7, number 11.

Thompson, J. W. and Solvig, K. (2000). *Sustainable Landscape Construction: A Guide to Green Building Outdoors*. Washington, D.C.: Island Press.

Thompson, K., Austin, K. C., Smith, R. M. et al. (2003). Urban domestic gardens. I. Putting small-scale plant diversity in context. *Journal of Vegetation Science*, 14, 71–78.

Thompson, K., Hodgson, J. G., Smith, R. M. et al. (2004). Urban domestic gardens (III): composition and diversity of lawn floras. *Journal of Vegetation Science*, 15, 373–378.

Thorington, R. W. (2006). *Squirrels: The Animal Answer Guide*. Baltimore, Maryland: Johns Hopkins University Press.

Thornton, D. H., Branch, L. C. and Sunquist, M. E. (2011). The influence of landscape, patch, and within-patch factors on species presence and abundance: a review of focal patch studies. *Landscape Ecology*, 26, 7–18.

Thurlow, C. (1983). *Improving Street Climate Through Urban Design*. New York: American Planning Association, Planning Advisory Service.

Tigas, L. A., van Vurena, D. H. and Sauvajot, R. M. (2002). Behavioral responses of bobcats and coyotes to habitat fragmentation and corridors in an urban environment. *Biological Conservation*, 108, 299–306.

Tikka, P. M., Hogmander, H. and Koski, P. S. (2001). Road and railway verges serve as dispersal corridors for grassland plants. *Landscape Ecology*, 16, 659–666.

Tikka, P. M., Koski, P. S., Kivela, R. A. and Kuitunen, M. T. (2000). Can grassland plant communities be preserved on road and railway verges? *Applied Vegetation Science*, 3, 25–32.

Tilghman, N. G. (1987). Characteristics of urban woodlands affecting breeding bird diversity and abundance. *Landscape and Urban Planning*, 14, 481–495.

Tilley, D. R. and Brown, M. T. (1998). Wetland networks for stormwater management in subtropical urban watersheds. *Ecological Engineering*, 10, 131–158.

Tischendorf, L. and Fahrig, L. (2000). On the usage and measurement of landscape connectivity. *Oikos*, 90, 7–19.

Tivy, J. (1996). *Agricultural Ecology*. London: Longman.

Tjallingii, S. P. (1995). *Ecopolis: Strategies for ecologically sound urban development*. Leiden, Netherlands: Backhuys Publishers.

Todd, D. K. and Mays, L. W. (2005). *Groundwater Hydrology*. New York: John Wiley.

Todd, N. J. and Todd, J. (1994). *From Eco-Cities to Living Machines: Principles of Ecological Design*. Berkeley, California: North Atlantic Books.

Torres, H., Alves, H. and De Oliveira, M. A. (2007). Sao Paulo peri-urban dynamics: some social causes and environmental consequences. *Environment and Urbanization*, 19, 207–223.

Torstensson, L., Borjessonand, E. and Stenstrom, J. (2005). Efficacy and fate of glyphosphate on Swedish railway embankments. *Pest Management Science*, 61, 881–886.

Townsend, C. R., Begon, M. and Harper, J. L. (2008). *Essentials of Ecology*. Oxford: Blackwell Publishing.

Trabaud, L. (1987). *Role of Fire in Ecological Systems*. The Hague, Netherlands: SPB Academic Publishing.

Tremblay, M. A. and St. Clair, C. C. (2009). Factors affecting the permeability of transportation and riparian corridors to the movements of songbirds in an urban landscape. *Journal of Applied Ecology*, 46, 1314–1322.

Trepl, L. (1990). Research on the anthropogenic migration of plants and naturalization: its history and current state of development. In *Urban Ecology: Plants and Plant Communities in Urban Environments*, eds. H. Sukopp, S. Hejny and I. Kowarik. The Hague, Netherlands: SPB Academic Publishing, pp. 75–97.

Trepl, L. (1995). Towards a theory of urban biocoenoses. In *Urban Ecology as the Basis for Urban Planning*, eds. H. Sukopp, M. Numata and A. Huber. The Hague, Netherlands: SPB Academic Publishing, pp. 3–21.

Trewhella, W. J. and Harris, S. (1990). The effect of railway lines on urban fox (*Vulpes vulpes*) numbers and dispersal movements. *Journal of Zoology*, 221, 321–326.

Tribo, G. (1989). *Evolucio de l'estructura agraria del Baix Llobregat (1860–1931)*. Barcelona: CECBLL.

Trocme, M., Cahill, S., de Vries, H. J. G. et al., eds. (2003). *Habitat Fragmentation Due to Transportation Infrastructure: The European Review*. Brussels: European Commission, COST Action 341.

Trombulak, S. C. and Frissell, C. A. (2000). Review of ecological effects of roads on terrestrial and aquatic communities. *Conservation Biology*, 14, 18–30.

Trowbridge, P. J. and Bassuk, N. L. (2004). *Trees in the Urban Landscape*. New York: John Wiley.

Tucker, G. M. and Evans, M. I. (1997). *Habitats for Birds in Europe: A Conservation Strategy for the Wider Environment*. Cambridge, UK: BirdLife International.

Turner, B., ed. (2010). *The Statesman's Yearbook 2010*. New York: Macmillan.

Turner, B. L., Clark, W. C., Kates, R. W. et al., eds. (1990). *The Earth as Transformed by Human Actions: Global and Regional Changes in the Biosphere Over the Past 300 Years*. New York: Cambridge University Press.

Turner, M. G., ed. (1987). *Landscape Heterogeneity and Disturbance*. New York: Springer-Verlag.

Turner, M. G. (1989). Landscape ecology: the effect of pattern on process. *Annual Review of Ecology and Systematics*, 20, 171–197.

Turner, M. G. and Gardner, R. H., eds. (1991). *Quantitative Methods in Landscape Ecology*. New York: Springer-Verlag.

Turner, M. G., Gardner, R. H. and O'Neill, R. V. (2001). *Landscape Ecology in Theory and Practice: Pattern and Process*. New York: Springer.

Turner, T. (1995). Greenways, blueways, skyways and other ways to a better London. *Landscape and Urban Planning*, 33, 269–282.

Tutin, T. G. (1973). Weeds of a Leicester garden. *Watsonia*, 9, 263–267.

Tyrvainen, L. and Miettinen, A. (2000). Property prices and urban forest amenities. *Journal of Environmental Economics and Management*, 39, 205–223.

Tyrvainen, L., Pauleit, S., Seeland, K. and de Vries, S. (2005). Benefits and uses of urban forests and trees. In *Urban Forests and Trees*, eds. C. C. Konijnendijk, K. Nilsson, T. B. Randrup and J. Schipperijn. New York: Springer, pp. 81–114.

Tzoulas, K. and Greening, K. (2011). Urban ecology and human health. In *Urban Ecology: Patterns, Processes, and Applications*, eds. J. Niemela, J. H. Breuste, T. Elmqvist et al. New York: Oxford University Press, pp. 263–271.

Tzoulas, K., Korpela, K., Venn, S. et al. (2007). Promoting ecosystem and human health in urban areas using green infrastructure: a literature review. *Landscape and Urban Planning*, 81, 167–178.

UCD Urban Institute Ireland. (2008). *Green City Guidelines: Advice for the Protection and Enhancement of Biodiversity in Medium to High-Density Urban Developments*. Dublin: UE Urban Institute Ireland.

Uhl, C. (1998). Conservation biology in your own front yard. *Conservation Biology*, 12, 1175–1177.

Ulrich, R. (1984). View through a window may influence recovery from surgery. *Science*, 224, 420–421.

UN Development Programme. (1996). *Urban Agriculture: Food, Jobs, and Sustainable Cities*. New York.

UN-Habitat. (2005). *Water and Sanitation in the World's Cities: Local Action for Global Goals*. London: Earthscan.

UN-Habitat. (2006). *State of the World's Cities, 2006/7*. London: Earthscan.

UN Population Division. (2007). *World Urbanization Prospects: The 2007 Revision*. New York: United Nations.

Urban, D. and Keitt, T. (2001). Landscape connectivity: a graph-theoretic perspective. *Ecology*, 82, 1205–1218.

References

Urban, M. C., Skelly, D. K., Burchsted, D. et al. (2006). Stream communities across a rural-urban landscape gradient. *Diversity and Distributions*, 12, 337–350.

Urban Land Institute. (1975). *Industrial Development Handbook*. Washington, D.C.: Urban Land Institute.

Urban Land Institute. (1998). *Office Development Handbook*. Washington, D.C.: Urban Land Institute.

Urban Land Institute. (1999). *Shopping Center Development Handbook*. Washington, D.C.: Urban Land Institute.

Urban Land Institute. (2001). *Business Park and Industrial Development Handbook*. Washington, D.C.: Urban Land Institute.

Urban Land Institute. (2003). *Mixed-Use Development Handbook*. Washington, D.C.: Urban Land Institute.

Urban Land Institute. (2008). *Creating Great Town Centers and Urban Villages*. Washington, D.C.: Urban Land Institute.

Ursic, K., Kenkel, N. C. and Larson, D. W. (1997). Revegetation dynamics of cliff faces in abandoned limestone quarries. *Journal of Applied Ecology*, 34, 289–303.

US Environmental Protection Agency. (1993). *Guidance Specifying Management Measures for Sources of Nonpoint Source Pollution in Coastal Waters*. Report No. 840-B-92-002. Washington, D.C.: US EPA, Office of Water.

US Environmental Protection Agency. (1999). *Storm Water Technology Fact Sheet: Bioretention*. Washington, D.C.: US EPA, Office of Water.

US Environmental Protection Agency. (2000a). *Low Impact Development (LID): A Literature Review*. Washington, D.C.: US EPA, Office of Water.

US Environmental Protection Agency. (2000b). *Vegetated Roof Cover*. Report EPA-841-B-00-005D. Washington, D.C.: US EPA, Office of Water.

US Environmental Protection Agency. (2001). *Our Built and Natural Environments: A Technical Review of the Interactions Between Land Use, Transportation, and Environmental Quality*. Washington, D.C.: US EPA.

US Environmental Protection Agency. (2007). *Combined Sewer Overflow Management*. Washington, D.C.: US EPA.

US Environmental Protection Agency. (2008a). *Non Point Source Pointers*. Washington, D.C.: US EPA.

US Environmental Protection Agency. (2008b). *Reducing Urban Heat Islands: Compendium of Strategies*. Washington, D.C.: US EPA.

US Geological Survey. (2010). *Mineral Commodity Summary of Salt*. Washington, D.C.: US Geological Survey.

USDA Forest Service. (2001). *The Urban Forestry Manual*. Athens, Georgia, USA: Southern Center for Urban Forestry Research and Information.

Vail, D. (1987). Suburbanization of the countryside and the revitalization of small farms. In *Sustaining Agriculture Near Cities*, ed. W. Lockeretz. Ankeny, Iowa, USA: Soil and Water Conservation Society.

Vale, B. and Vale, R. (2000). *The New Autonomous House: Design and Planning for Sustainability*. New York: Thames & Hudson.

Vale, L. J. and Campanella, T. J., eds. (2005). *The Resilient City: How Modern Cities Recover from Disaster*. New York: Oxford University Press.

Valiela, I. (2006). *Global Coastal Change*. Oxford: Blackwell Publishing.

Van, M. P. and De Pauw, N. (2005). Wastewater-based urban aquaculture systems in Ho Chi Minh City, Vietnam. In *Urban Aquaculture*, eds. B. Costa-Pierce, A. Desbonnet, P. Edwards and D. Baker. Wallingford, UK: CABI Publishing, pp. 77–102.

Van Berkel, R., Fujita, T., Hashimoto, S. and Fuji, M. (2009). Quantitative assessment of urban and industrial symbiosis in Kawasaki, Japan. *Environmental Scientific Technology*, 43, 1271–1281.

van Bohemen, H., ed. (2005). *Ecological Engineering: Bridging between ecology and civil engineering*. Boxtel, Netherlands: Aeneas Technical Publishers.

van der Grift, E. A. (1999). Mammals and railroads: impacts and management implications. *Lutra*, 42, 77–98.

van der Grift, E. A. and Kuijsters, H. M. J. (1998). Mitigation measures to reduce habitat fragmentation by railway lines in the Netherlands. In *Proceedings of the International Conference on Wildlife Ecology and Transportation*, eds. G. L. Evink, P. Garrett, D. Zeigler and J. Berry. Report FL-ER-69-98. Tallahassee, Florida: Florida Department of Transportation, pp. 166–170.

van der Ree, R. and McCarthy, M. A. (2005). Quantifying the effects of urbanization on the persistence of indigenous mammals in Melbourne, Australia. *Animal Conservation*, 8, 309–319.

Van der Ryn, S. (1978). *The Toilet Papers*. Santa Barbara, California: Capra Press.

van der Valk, A. (2002). The Dutch planning experience. *Landscape and Urban Planning*, 58, 201–210.

Van Dolah, R. F., Riekerk, G. H. M., Bergquist, D. C. et al. (2008). Estuarine habitat quality reflects urbanization at large spatial scales in South Carolina's Coastal Zone. *Science of the Total Environment*, 390, 142–154.

van Veenhuizen, R., ed. (2006). *Cities Farming for the Future: Urban Agriculture for Green and Productive Cities*. Ottawa: International Development Research Center Books.

Vannucci, M. (2004). *Mangrove Management and Conservation: Present and Future*. New York: United Nations University Press.

Vare, S. and Krisp, J. (2005). *Ecological Network and Land Use Planning of Urban Areas*. (In Finnish). The Finnish Environment 78. Helsinki: Ministry of the Environment.

Veiga, L. B. E. and Magrini, A. (2008). Eco-industrial park development in Rio de Janeiro, Brazil: a tool for sustainable development. *Journal of Cleaner Production*, 17, 653–661.

Velazquez, L. S. (2005). Organic greenroof architecture: sustainable design for the new millennium. *Environmental Quality Management*, 14, 73–85.

Verboom, J., Schotman, A., Opdam, P. and Metz, J. A. J. (1991). European nuthatch metapopulations in a fragmented agricultural landscape. *Oikos*, 61, 149–156.

Vergnes, A., LeViol, E. and Clergeau, P. (2012). Green corridors in urban landscapes affect the arthropod communities of domestic gardens. *Biological Conservation*, 145, 171–178.

Vessel, M. F. and Wong, H. H. (1987). *Natural History of Vacant Lots*. Berkeley, California: University of California Press.

Vezzani, D. (2007). Review: artificial container-breeding mosquitoes and cemeteries: a perfect match. *Tropical Medicine and International Health*, 12, 299–313.

Viles, R. L. and Rosier, D. J. (2001). How to use roads in the creation of greenways: case studies in three New Zealand landscapes. *Landscape and Urban Planning*, 55, 15–27.

Viljoen, A. T., Bohn, K. and Howe, J. (2005). *Continuous Productive Urban Landscapes: Designing Urban Agriculture for Sustainable Cities*. Oxford/Boston: Architectural Press.

Villard, M.-A. (2012). Land use intensification and the status of focal species in managed forest landscapes of New Brunswick, Canada. In *Land Use Intensification: Effects on Agriculture, Biodiversity and Ecological Processes*, eds. D. Lindenmayer, S. Cunningham and A. Young. Collingwood, Australia: CSIRO Publishing, pp. 85–92.

Vince, S. W., Duryea, M. L., Macie, E. A. and Hermansen, L. A., eds. (2005). *Forests at the Wildland-Urban Interface: Conservation and Management*. CRC Press: Boca Raton, Florida.

Vink, A. (1983). *Landscape Ecology and Land Use*. London: Longman Group.

Vollertsen, J., Astebol, S. O., Coward, J. E. et al. (2007). Monitoring and modeling the performance of a wet pond for treatment of highway runoff in cold climates. *Highway and Urban Environment*, 12, 499–509.

von der Lippe, M. and Kowarik, I. (2007). Long-distance dispersal of plants by vehicles as a driver of plant invasions. *Conservation Biology*, 21, 986–996.

von Stulpnagel, A., Horbert, A. and Sukopp, H. (1990). The importance of vegetation for the urban climate. In *Urban Ecology*, eds. H. Sukopp, S. Hejny and I. Kowarik. The Hague, Netherlands: SPB Academic Publishing, pp. 175–193.

Vos, C. C., Berry, P., Opdam, P. et al. (2008). Adapting landscapes to climate change: examples of climate-proof ecosystem networks and priority adaptation zones. *Journal of Applied Ecology*, 45, 1722–1731.

Vos, C. C., Verboom, J., Opdam, P. F. M. and ter Braak, C. J. (2001). Toward ecologically scaled landscape indices. *American Naturalist*, 157, 24–41.

Voss, S. C., Main, B. Y. and Dadour, I. R. (2007). Habitat preferences of the urban wall spider, *Oecobius navus* (Araneae, Oecobiidae). *Australian Journal of Entomology*, 46, 261–268.

Vuchic, V. R. (2007). *Urban Transit Systems and Technology*. New York: John Wiley.

Vymazal, J., Greenway, M., Tonderski, K. et al. (2006). Constructed wetlands for wastewater treatment. In *Wetland and Natural Resource Management*, ed. J. T. A. Verhoeven. New York: Springer, 69–96.

Wace, N. (1977). Assessment of dispersal of plant species: the car-borne flora in Canberra. *Proceedings of the Ecological Society of Australia*, 10, 167–186.

Waddell, P. (2002). UrbanSim: modeling urban development for land use, transportation, and environmental planning. *American Planning Association Journal*, 68, 297–314.

Wahlbrink, D. and Zucchi, H. (1994). Occurrence and settlement of carabid beetles on an urban railway embankment: a contribution to urban ecology. *Zoologische Jahrbuecher Abteilung fuer Systematik Oekologie und Geographie der Tiere*, 121, 193–201.

Waldheim, C., ed. (2006). *The Landscape Urbanism Reader*. New York: Princeton Architectural Press.

Walesh, S. G. (1989). *Urban Surface Water Management*. New York: John Wiley.

Walmsley, A. (1995). Greenways and the making of urban form. *Landscape and Urban Planning*, 33, 47–64.

Walsh, C. J., Fletcher, T. D. and Ladson, A. R. (2005a). Stream restoration in urban catchments through redesigning stormwater systems: looking to the catchment to save the stream. *Journal of the North American Benthological Society*, 24, 690–705.

Walsh, C. J., Leonard, A. W., Ladson, A. R. et al. (2004). *Urban Stormwater and the Ecology of Streams*. Canberra: Cooperative Research Centre for Freshwater Ecology.

Walsh, C. J., Roy, A. H., Feminella, J. W. et al. (2005b). The urban stream syndrome: current knowledge and the search for a cure. *Journal of the North American Benthological Society*, 24, 706–723.

References

Waltham, N. J. and Connolly, R. M. (2007). Artificial waterway design affects fish assemblages in urban estuaries. *Journal of Fish Biology*, 71, 1613–1629.

Wandeler, P., Funk, S. M., Langiader, C. R. *et al.* (2003). The city-fox phenomenon: genetic consequences of a recent colonization of urban habitat. *Molecular Ecology*, 12, 647–656.

Wang, D. (1999). The impact of the ecological environment and development trend of foreign inter-basin water transfer projects. *Environmental Science News*, 1999(3), 28–32.

Wang, L., Lyons, J. and Kanehl, P. (2001). Impacts of urbanization on stream habitat and fish across multiple spatial scales. *Environmental Management*, 28, 255–266.

Wang, R., Zhao, Q. and Ouyang, Z. (1992). *Ecopolis Planning in China: Principles and Practices for Urban Ecological Regulation*. Beijing: Chinese Academy of Science, Research Center of Eco-Environmental Studies.

Ward-Page, C. A. (2005). Clinoid sponge surveys on the Florida Reef Tract suggest land-based nutrient inputs. *Marine Population Bulletin*, 51, 570–579.

Warner, R. E. (1985). Demography and movement of free-ranging domestic cats in rural Illinois. *Journal of Wildlife Management*, 49, 340–346.

Warren, P. S., Harlan, S. L., Boone, C. *et al.* (2010). Urban ecology and human social organization. In *Urban Ecology*, ed. K. J. Gaston. New York: Cambridge University Press, pp. 172–201.

Washburn, B. and Seamans, T. W. (2004). *Management of Vegetation to Reduce Wildlife Hazards at Airports*. FAA Worldwide Airport Technology Transfer Conference, Atlantic City, New Jersey.

Wasson, K., Lohrer, D., Crawford, M. and Rumrill, S. (2002). *Non-Native Species in Our Nation's Estuaries: A Framework for Invasion Monitoring Program*. Technical Report Series 2002:1. Washington, D.C.: National Estuarine Research Reserve.

Watson, D., Plattus, A. and Shibley, R. G., eds. (2003). *Time-Saver Standards for Urban Design*. New York: McGraw-Hill.

Way, D. S. (1978). *Terrain Analysis: a Guide to Site Selection Using Aerial Photographic Interpretation*. Stroudsburg, Pennsylvania: Dowden, Hutchinson & Ross.

Way, J. G. and Eatough, D. L. (2006). Use of "micro"-corridors by eastern coyotes, *Canis latrans*, in a heavily urbanized area: implications for ecosystem management. *Canadian Field-Naturalist*, 120, 474–476.

Webb, N. R. and Hopkins, P. J. (1984). Invertebrate diversity on fragmented *Calluna* heathland. *Journal of Applied Ecology*, 21, 921–933.

Webb, N. R., Clarke, R. T. and Nicholas, J. T. (1984). Invertebrate diversity on fragmented *Calluna*-heathland: effects of surrounding vegetation. *Journal of Biogeography*, 11, 41–46.

Webster's College Dictionary. (1991). New York: Random House.

Wegner, J. and Merriam, G. (1979). Movements by birds and small mammals between a wood and adjoining farmland habitats. *Journal of Applied Ecology*, 16, 349–358.

Wein, R. W., ed. (2006). *Coyotes Still Sing in My Valley: Conserving Biodiversity in a Northern City*. Edmonton, Canada: Spotted Cow Press.

Weiss, J., Burghardt, W., Gausmann, P. *et al.* (2005). Nature returns to abandoned industrial land: monitoring succession in urban-industrial woodlands in the German industrial Ruhr. In *Wild Urban Woodlands*, eds. I. Kowarik and S. Korner. New York: Springer, pp. 143–162.

Weiss, P. T., Gulliver, J. S. and Erickson, A. J. (2007). Cost and pollutant removal of stormwater treatment practices. *Journal of Water Resources, Planning and Management American Society of Chemical Engineering*, 133, 218–229.

Welch, E. B. and Jacoby, J. M. (2004). *Pollutant Effects in Freshwater: Applied Limnology*. New York: Spon Press.

Welty, C. (2009). The urban water budget. In *The Water Environment of Cities*, ed. L. A, Baker. New York: Springer, pp. 17–28.

Wenger, S. J., Roy, A. H., Jackson, C. R. *et al.* (2009). Twenty-six key research questions in urban stream ecology: an assessment of the state of the science. *Journal of the North American Benthological Society*, 28, 1080–1098.

Wessolek, G. (2008). Sealing of soils. In *Urban Ecology: An International Perspective on the Interactions Between Humans and Nature*, eds. J. M. Marzluff, E. Shulenberger, W. Endlicher *et al.* New York: Springer, pp. 161–179.

Westerhoff, P. and Crittenden, J. (2009). Urban infrastructure and use of mass balance models for water and salt. In *The Water Environment of Cities*, ed. L. A. Baker. New York: Springer, pp. 49–68.

Wheater, C. P. (1999). *Urban Habitats*. New York: Routledge.

Wheeler, A. P., Angermeier, P. L. and Rosenberger, A. E. (2005). Impacts of new highways and subsequent landscape urbanization on stream habitat and biota. *Reviews in Fisheries Science*, 13, 141–164.

Whelan, R. J. (1995). *The Ecology of Fire*. New York: Cambridge University Press.

Whelan, R. J., Roberts, D. G., England, P. R. and Ayre, D. J. (2006). The potential for genetic contamination vs. augmentation by native plants in urban gardens. *Biological Conservation*, 128, 493–500.

White, C. S. and McDonnell, M. J. (1988). Nitrogen cycling processes and soil characteristics in an urban versus rural forest. *Biogeochemistry*, 5, 243–262.

White, L. and Main, M. B. (2005). Waterbird use of created wetlands in golf-course landscapes. *Wildlife Society Bulletin*, 33, 411–421.

White, M. A., Niemani, R. R., Thomston, P. E. and Running, S. W. (2002). Satellite evidence of phenological differences between urbanized and rural areas of the eastern United States deciduous broadleaf forest. *Ecosystems*, 5, 260–273.

Whitehand, J. and Morton, N. (2004). Urban morphology and planning: the case of fringe belts. *Cities*, 10, 275–289.

Whitney, G. G. (1985). A quantitative analysis of the flora and plant communities of a representative Midwestern United-States town. *Urban Ecology*, 9, 143–160.

Whitney, G. G. and Adams, S. D. (1980). Man as maker of new plant communities. *Journal of Applied Ecology*, 17, 431–448.

Whitney, G. G. and Davis, W. C. (1986). From primitive woods to cultivated woodlots: Thoreau and the forest history of Concord, Massachusetts. *Journal of Forest History*, 30, 70–81.

Whittaker, R. H. (1956). Vegetation of the Great Smoky Mountains. *Ecological Monographs*, 26, 1–80.

Whittaker, R. H. (1962). Classification of natural communities. *Botanical Review*, 28, 1–239.

Whittaker, R. H. (1967). Gradient analysis of vegetation. *Biological Review*, 42, 207–264.

Whittaker, R. H. (1975). *Communities and Ecosystems*. New York: Macmillan.

Whyte, W. H. (1968). *The Last Landscape*. Garden City, New York: Doubleday.

Wickramasinghe, L. P., Harris, S., Jones, G. and Vaughn, N. (2003). Bat activity and species richness on organic and conventional farms: impact of agricultural intensification. *Journal of Applied Ecology*, 40, 984–993.

Wiens, J. A. (1989). Spatial scaling in ecology. *Functional Ecology*, 3, 385–397.

Wiens, J. A. (1999). The science and practice of landscape ecology. In *Landscape Ecological Analysis: Issues and Applications*, eds. J. M. Klopatek and R. H. Gardner. New York: Springer, pp. 371–383.

Wilby, R. L. and Perry, G. L. W. (2006). Climate change, biodiversity and the urban environment: a critical review based on London, UK. *Progress in Physical Geography*, 30, 73–98.

Wilcox, A., Palassio, C. and Dovercourt, J., eds. (2007). *Green Topia: Towards a Sustainable Toronto*. Toronto: Coach House Books.

Williams, J. E., Wood, C. A. and Dombeck, M. P., eds. (1997a). *Watershed Restoration: Principles and Practices*. Bethesda, Maryland: American Fisheries Society.

Williams, J. R., Goodrich-Mahoney, J. W., Wisniewski, J. R. and Wisniewski, J. (1997b). *Environmental Concerns in ROW Management*. Oxford: Elsevier Science.

Williams, N. S. G., McDonnell, M. J., Phelan, G. K. et al. (2006). Range extension due to urbanization: increased food resources attract Grey-headed Flying-foxes (*Pteropus poliocephalus*) to Melbourne. *Austral Ecology*, 31, 190–198.

Williams, N. S. G., McDonnell, M. J. and Seager, E. J. (2005). Factors influencing the loss of an endangered ecosystem in an urbanising landscape: a case study of native grasslands from Melbourne, Australia. *Landscape and Urban Planning*, 71, 35–49.

Williams, N. S. G., Schwartz, M. W., Vesk, P. A. et al. (2009). A conceptual framework for predicting the effects of urban environments on floras. *Journal of Ecology*, 96, 8–12.

Williams, P., Whitfield, M., Biggs, J. et al. (2003). Comparative biodiversity of rivers, streams, ditches and ponds in an agricultural landscape in Southern England. *Biological Conservation*, 115, 329–341.

Wilson, D. (2000). Naturalizing the city: a case study of Waitakere City, New Zealand. In *Nature Conservation 5: Nature Conservation in Production Environments: Managing the Matrix*, eds. J. L. Craig, N. Mitchell and D. A. Saunders. Chipping Norton, Australia: Surrey Beatty, pp. 405–410.

Wilson, E. O. (1984). *Biophilia: The Human Bond with Other Species*. Cambridge, Massachusetts: Harvard University Press.

Wilson, E. O. (1994). *Naturalist*. Washington, D.C.: Island Press.

Wilson, M. F. (1992). The ecology of seed dispersal. In *Seeds: The Ecology of Regeneration in Plant Communities*, ed. M. Fenner. Southampton, UK: C.A.B. International, pp. 61–86.

Winter, T. C., Harvey, J. W., Franke, O. L. and Alley, W. M. (1998). *Ground Water and Surface Water: A Single Resource*. US Geological Survey Circular 1139. Denver, Colorado: USDI Geological Survey.

Wittig, R. (2008). Principles for guiding eco-city development. In *Ecology, Planning, and Management of Urban Forests: International Perspectives*, eds. M. M. Carreiro, Y.-C. Song and J. Wu. New York: Springer, pp. 10–28.

Wohl, E. (2004). *Disconnected Rivers: Linking Rivers to Landscapes*. New Haven, Connecticut: Yale University Press.

Wollheim, W. M., Pellerin, B. A., Vorosmarty, C. J. and Hopkinson, C. S. (2005). N retention in urbanizing headwater catchments. *Ecosystems*, 8, 871–884.

Wolman, A. (1965). The metabolism of cities. *Scientific American*, 213, 178–190.

Wolter, C. (2008). Towards a mechanistic understanding of urbanization's impacts on fish. In *Urban Ecology: An International Perspective on the Interaction Between Humans and Nature*, eds. J. M. Marzluff, E. Shulenberger, W. Endlicher *et al*. New York: Springer, pp. 425–436.

Wolverton, B. C., McDonald, R. C. and Watkins, Jr., E. A. (1984). Foliage plants for removing indoor air pollutants from energy efficient homes. *Economic Botany*, 32, 224–229.

Wong, N. H. (2009). *Tropical Urban Heat Islands: Climate, Buildings, and Greenery*. London: Taylor & Francis.

Wong, N. H., Tay, S. F., Wong, R. *et al*. (2002). Life cycle cost analysis of rooftop gardens in Singapore. *Building and Environment*, 38, 499–509.

Woodell, S. (1979). The flora of walls and pavings. In *Nature in Cities: The Natural Environment in the Design and Development of Urban Green Space*, ed. I. C. Laurie. New York: John Wiley, pp. 135–157.

Woodier, O. (1998). Tired of pursuing the perfect lawn? Consider these alternatives. *National Wildlife*, 36, 14–15.

Woods, M., McDonald, R. A. and Harris, S. (2003). Predation of wildlife by domestic cats *Felis catus* in Great Britain. *Mammal Review*, 13, 174–188.

Woodwell, G. M. (2009) *The Nature of a House: Building a World That Works*. Washington, D.C.: Island Press.

Woolley, T., Kimmins, S., Harrison, P. and Harrison, R. (1997). *Green Building Handbook: A Guide to Building Products and Their Impact on the Environment*. London: E & FN Spon.

World Resources Institute. (1996). *World Resources: A Guide to the Global Environment, 1996–97*. New York: Oxford University Press.

Worster, D. (1977). *Nature's Economy: A History of Ecological Ideas*. New York: Cambridge University Press.

Wrenn, D. M., Casazza, J. A. and Smart, J. E. (1983). *Urban Waterfront Development*. Washington, D.C.: Urban Land Institute.

Wright, J. (2009). *Sustainable Agriculture and Food Security in an Era of Oil Scarcity: Lessons from Cuba*. London: Earthscan.

Wu, J. (1999). Hierarchy and scaling: extrapolating information along a scaling ladder. *Canadian Journal of Remote Sensing*, 25, 367–380.

Wu, J. (2004). Effects of changing scale on landscape pattern analysis: scaling relations. *Landscape Ecology*, 19, 125–138.

Wu, J. (2008). Toward a landscape ecology of cities: beyond buildings, trees, and urban forests. In *Ecology, Planning, and Management of Urban Forests*, eds. M. Carreira, Y.-C. Song and J. Wu. New York: Springer, pp. 10–28.

Wu, J. (2010). Urban sustainability: an inevitable goal of landscape research. *Landscape Ecology*, 25, 1–4.

Wu, J. (2012). A landscape approach for sustainability science. In *Sustainability Science: The Emerging Paradigm and the Urban Environment*, eds. M. P. Weinstein and R. E. Turner. New York: Springer, pp 59–77.

Wu, J. and Levin, S. A. (1997). A patch-based spatial modeling approach: conceptual framework and simulation scheme. *Ecological Modeling*, 101, 325–346.

Wu, Z., Huang, C., Wu, W. and Zhang, S. (2008). Urban forest structure in Hefei, China. In *Ecology, Planning, and Management of Urban Forests: International Perspectives*, eds. M. M. Carreiro, Y.-C. Song and J. Wu. New York: Springer, pp. 279–292.

Xu, C., Liu, M. S., Zhang, C. *et al*. (2007). The spatiotemporal dynamics of rapid urban growth in the Nanjing metropolitan region of China. *Landscape Ecology*, 22, 925–937.

Yahner, R. H., Hutnik, R. J. and Liscinsky, S. A. (2003). Long-term trends in bird populations on an electric transmission right-of-way. *Journal of Arboriculture*, 29, 156–164.

Yamaoka, N., Yoshida, H., Tanabe, M. *et al*. (2008). Simulation study of the influence of different urban canyons element on the canyon thermal environment. *Building Simulation*, 1, 118–128.

Yanes, M., Velasco, J. and Suarez, F. (1995). Permeability of roads and railways to vertebrates: the importance of culverts. *Biological Conservation*, 71, 217–221.

Yang, J., Yu, Q. and Gong, P. (2008). Quantifying air pollution removal by green roofs in Chicago. *Atmospheric Environment*, 42, 7266–7273.

Yang, P. P.-J. and Lay, O. B. (2004). Applying ecosystem concepts to the planning of industrial areas: a case study of Singapore's Jurong Island. *Journal of Cleaner Production*, 12, 1011–1023.

Yaro, R. D. and Hiss, T. (1996). *A Region at Risk: The 3rd Regional Plan for the New York-New Jersey-Connecticut Metropolitan Area*. Washington, D.C.: Island Press.

Yasuda, M. and Koike, F. (2004). Do golf courses provide a refuge for flora and fauna in Japanese urban landscapes? *Landscape and Urban Planning*, 75, 58–68.

Yatsukhno, V. and Kozlovskaya, L. (1998). The ecological impact of the Chernobyl catastrophe on sustainable development in Belarus. In *Sustainable Development for Central and Eastern Europe: Spatial Development in the European Context*, ed. U. Graute. New York: Springer, pp. 255–266.

Yeang, K. (1999). *The Green Skyscraper: The Basis for Designing Sustainable Intensive Buildings*. New York: Prestel.

Yeang, K. and Richards, I. (2007). *Ken Yeang: Eco Skyscrapers*. Mulrave, Victoria, Australia: Images Publishing.

Yeh, P. J. and Price, T. D. (2004). Adaptive phenotypic plasticity and the successful colonization of a novel environment. *American Naturalist*, 164, 531–542.

Yokohari, M., Takeuchi, K., Watanabe, T. and Yokota, S. (2000). Beyond greenbelts and zoning: a new planning concept for the environment of Asian mega-cities. *Landscape and Urban Planning*, 47, 159–171.

Yoshikado, H. (1990). Vertical structure of the sea breeze penetrating through a large urban complex. *Journal of Applied Meteorology*, 29, 878–891.

Young, K. M., Daniels, C. B. and Johnston, G. (2007). Species of street tree is important for southern hemisphere bird trophic guilds. *Austral Ecology*, 32, 541–550.

Yu, C. and Hien, W. N. (2006). Thermal benefits of city parks. *Energy and Buildings*, 38, 105–120.

Yu, K. J. (1996). Security patterns and surface model in landscape ecological planning. *Landcape and Urban Planning*, 36, 1–17.

Yu, K., Wang, S. and Li, D. (2011). The negative approach to urban growth planning of Beijing, China. *Journal of Environmental Planning and Management*, 54, 1209–1236.

Yudelson, J. (2008). *The Green Building Revolution*. Washington, D.C.: Island Press.

Zaitzevsky, D. (1982). *Frederick Law Olmsted and the Boston Park System*. Cambridge, Massachusetts: Belknap Press of Harvard University Press.

Zerbe, S., Choi, I.-K. and Kowarik, I. (2004). Characteristics and habitats of non-native plant species in the city of Chonju, southern Korea. *Ecological Research*, 19, 91–98.

Zezza, A. and Tasciotti, L. (2010). Urban agriculture, poverty, and food security: empirical evidence from a sample of developing countries. *Food Policy*, 35, 265–273.

Zhang, S., Zheng, G. and Xu, J. (2006). Habitat use of urban tree sparrows in the process of urbanization: Beijing as a case study. *Biodiversity Science*, 14, 372–381.

Zhao, S. Q., Da, L. J., Tang, Z. Y. et al. (2006). Ecological consequences of rapid urban expansion: Shanghai, China. *Frontiers in Ecology and the Environment*, 4, 341–346.

Zhu, P. and Zhang, Y. (2008). Demand for urban forests in United States cities. *Landscape and Urban Planning*, 84, 293–300.

Zhu, W.-X. and Carreiro, M. M. (1999). Chemoautotrophic nitrification in acidic soils in forests along an urban-rural transect. *Soil Biology and Biochemistry*, 31, 1091–1100.

Ziarnek, M. (2007). Human impact on plant communities in urban area assessed with hemeroby grades. *Polish Journal of Ecology*, 55, 161–167.

Zilliox, L. (1994). Porous media and aquifer systems. In *Groundwater Ecology*, eds. J. Gibert, D. L. Danielopol and J. A. Stanford. San Diego, California: Academic Press, pp. 67–96.

Zipperer, W. C. (2002). Species composition and structure of regenerated and remnant forest patches within an urban landscape. *Urban Ecosystems*, 6, 271–290.

Zipperer, W. C. and Guntenspergen, G. R. (2009). Vegetation composition and structure of forest patches along urban-rural gradients. In *Ecology of Cities and Towns: A Comparative Approach*, eds. M. J. McDonnell, A. K. Hahs and J. H. Breuste. New York: Cambridge University Press, pp. 274–286.

Ziska, L. H., Bunce, J. A. and Goins, E. W. (2004). Characterization of an urban-rural CO_2/temperature gradient and associated changes in initial plant productivity during secondary succession. *Oecologia*, 139, 454–458.

Zmyslony, J. and Gagnon, D. (2000). Path analysis of spatial predictors of front-yard landscape in an anthropogenic environment. *Landscape Ecology*, 15, 357–371.

Zobel, M. van der Maarel, E. and Dupre, C. (1998). Species pool: the concept, its determination and significance for community restoration. *Applied Vegetation Science*, 1, 55–66.

Zollner, P. A. (2000). Comparing the landscape level perceptual abilities of forest sciurids in fragmented agricultural landscapes. *Landscape Ecology*, 15, 523–533.

Zollner, P. A. and Lima, S. L. (1999). Search strategies for landscape-level interpatch movements. *Ecology*, 80, 1019–1030.

Zonneveld, I. S. and Forman, R. T. T., eds. (1990). *Changing Landscapes: An Ecological Perspective*. New York: Springer-Verlag.

Index

All index entries relate to the urban environment unless otherwise noted.
Individual cities are listed under the city name, not the country.

acclimation (acclimatization), 220
actinomycetes, in soil, 109
adaptations, 220, 236, 270, *see also* genetic adaptation; wildlife (urban)
Adelaide (Australia), bird use of trees, 248
adjacency (adjacencies), 51–53, 230
    corridor-centered neighborhood, 56
    house plots, 294
    importance for wildlife, 265–266
    parks, 351
    patch-centered mosaic, 56
    wildlife movement and, 265–266, 370
adjacency arrangement effects, 52, 53
adjacency arrangement model, 56
adjacency effect, 52, 239, 370
advection (net horizontal energy flow) (Q_A), 131, 133, 136
aeration, urban soil, 103
aerobic bacteria, 108, 164, 359
aeroecology (aerobiology), 247
agricultural land, 265, 345
    groundwater pollution, 154
    habitat loss by urbanization, 89–90, 322, 324, 349
    see also cropland
agriculture (urban), 17, 343, 344–349
    ecological effects, 347–348
    food growing, 346–347, 348
    on green roofs, 313
    large sites, types, 345–346
    locations and types, 345–346
    neighborhood and urban region linkages, 348–349
    pest/disease diversity, 345
    small peri-urban/suburban sites, 346
    small sites in suburbs/cities, 346
    temporary nature, 349
air, 125–148
    composition, 143
    cool air drainage, 128, 140
    cooling, greenspace size and, 137–138, 371
    energy and radiation, 129–133
        see also energy; radiation
    global warming and, 147–148
    heat and, *see* heat
    in/out of cities, 126–128
    layers, 128–129
    moisture, 126–127, 352
    "side streak," 142
    temperatures, *see* temperature
    trees' ecological roles, 125, 224

urban boundary layer, 128, 135
urban vs. non-urban, 126–128, 135
urbanization effect, 83–84
air flows (urban), 65, 66, 126, 127–128, 139–143
    air layers and, 128, 129
    around buildings, size/shape effect, 142
    bats and bird affected by, 247
    breezes (local), 139–140
    cool air drainage, 128, 140
    isolated buildings and trees effect, 142–143
    seed dispersal, 220
    skimming, ventilation from, 142–143
    streamline, *see* streamline air flow
    street canyons, 141–142, 327
    turbulent, 65, 127, 140
    vortex, 65, 140, 141
    winds and windbreaks, 140–141
air pollutants/pollution, 25, 126, 143–148
    from aircraft, 362
    airdome development, 128
    cities in arid land, 147
    city centers and, 327, 330, 332
    dominant, in different cities, 145, 146
    gases, 143, 144
    green roofs reducing, 311
    greenspace reducing, 371
    health effects (adverse), 144
    indoor plants reducing, 307
    industrial, 145, 146
    inside buildings/vehicles, 146
    low-income residential areas, 318
    particulates, 25, 100, 143, 144, 145
        land use effect, 146
        urban vs. suburban areas, 145–146
    sources, 143
    spatial distribution, 145–147
    street trees affected by, 284
    temperature effect, 147
    toxic, 144
    traffic causing, 146, 281, 284–285, 287
    types, 143–145
    urban agriculture and, 347
    vegetation reducing, 19
    wall plants reducing, 309
aircraft noise, 361–362
airdome, 128–129, *see also* heat island
airports, 361–362
    air/water pollution, 362
    wildlife habitat, 263, 361
    wildlife movement and collisions, 361

albedo (surface reflection of energy), 131–132, 136, 139
algae, 109, 188, 212, 307
algal blooms, 167, 168, 201
allotment (community garden), 207–208
alpha diversity, 208
alpha index, *see* circuitry (alpha index)
amphibians, 253–254
    adaptation to urban conditions, 271–272
    changes over decades/centuries, 270
    decline with habitat changes/loss, 253
    habitat requirements, 253
Amsterdam, 81, 361
anaerobic bacteria, 108, 164, 167
anaerobic conditions
    urban soil, 103, 104, 106
    wastewater discharge into water bodies, 167
anaerobic decomposition, 116
anastomosing networks, 45
animal(s)
    around dumps, 359
    in cracks in hard surfaces, 291
    dispersal, 67, 264
    for food, in urban areas, 346
    groundwater, 156–157
    in/on buildings, 307, 309
    migration, 67, 264
    pollinators, and seed dispersal, 220
    trees' ecological roles, 224, 225
    in urban soil, 111–113
    urbanization effects, 86
    see also wildlife (urban); *specific animals*
animal biology, 264–265
animal movements, 15–16, 66–68, 70, 264–265
    by boundaries, 69–70
    city centers to residential areas, 330
    control, translocation, 265
    lobes and coves channeling, 70
    narrow wooded corridors, 228
    roadsides, 364–365
    shrubby area, 230
    translocated animals, distance, 265
    tree rows (urban), 227–228, 286
    types, 264
    via vehicles on roads, 286
    see also wildlife (urban), movement
anoxic "dead zones," 167, 186
anthropogenic climate change, 269–270
anthropogenic heat (Q_F), 131, 132, 133, 135, 332

Index

ants, 112
apartments, location, 38
aquaculture, 165–166, 345
aquifers, 151–152, 156
Arcata (California, USA), 165
archaeophytes, 213–214
architecture, 28
Argentina, fertilizer, peri-urban soil contamination, 118
arid land, 135, 147, 253
arthropods, 25, 254
  spatial pattern, 255
  see also insect(s); invertebrates
artifacts, in fill, 117
asbestos, 161
asphalt *see* tarmac (asphalt) surface
Auckland (New Zealand), tree distribution/species, 226
Austin (Texas, USA), bats, 246

back-corner patches, 294–295
backline strips, house plots, 294, 295, 321
back-space corridors, 321,
  see also house plots
bacteria
  aerobic, 108, 164, 359
  anaerobic, 108, 164, 167
  chemo-synthetic, 109
  fecal coliforms, 164, 168, 179
  in urban soil, 108–109
"bank storage," 158
Barcelona (Spain), 230, 257, 266, 320, 367
barrier islands/strips, 201–202
basins, constructed, *see* constructed basins
Basel (Switzerland), pigeons, 252–253, 262
bat(s), 246–247, 268
  activity, and dependencies, 53, 246, 247
  movement routes, 267–268
  roost locations, 247, 257, 268, 270, 369
  species/types, 246
beaches, urban, 202
bees, 220, 278
beetles, 112, 297
behavioral adaptations, 270, 271
behavioral adjustment/change, 270
Berlin, West, bryophyte species, 211
Berlin area (Germany)
  green corridors, 363
  greenspace effect on temperature, 137–138
  land-use changes, 87
  plant density and zones, 214
  rubble and chemicals in, 117
  Sudgelande Railway Park, vegetation, 215
  vegetation, time scales for change, 75
beta diversity, 208
beta index (intersection linkage), 72, 122, 381
Bhopal, India, 145
bicycling, 286
bioaccumulation of heavy metals, 115, 154, 168
biodiversity, 232
  agricultural land (urban), 348
  golf course, 357
  habitat heterogeneity and, 374

house plots, 295–297, 298, 306, 374
human patterns correlated, 234
plants, *see* plant(s) (urban)
primary measure (richness of species), 232
sprawl affecting, 86
suburban residential developments and, 325–326
urban areas, ecological change rates, 84–86
urban ecology solutions, 30
wetlands, 177
wildlife, *see* wildlife (urban)
see also species richness (diversity)
biofilters, 180
biofiltration, indoor plants, 307
biogeochemical cycles, 119
biogeochemical flows, 68–69
biological oxygen demand (BOD), 164, 173, 174
biomimicry/biomimetics, 340
biophilia, 19, 307
"bioreceptivity," 307
bioremediation, 117–118, 181
bioretention basins, 179, 180, 181–182
biotope mapping, 205
bird(s) (urban), 247–253
  agricultural land (urban) and, 349
  air flows and atmospheric conditions, effects, 247
  aircraft collisions, 361
  annual counts/censuses, 268–269
  behavioral adaptations, 270
  big-three city, 248, 252–253
    abundance, and feeding, 252
    effect on native bird diversity, 252
  breeding, density in house plots, 297
  breeding habitat extent required, 240
  on buildings, 309
  cat predation, 22–23
  characteristics, 241
  city centers, 331
  densities
    changes over time, 269
    in cities, 259, 261–262
    in suburbs, 249, 261–262
    tree cover effect, 266–267
  diurnal, food and numbers, 248
  on dumps, 359
  feeding, 264
  "foraging guilds," 248, 251, 257
  genetic changes, 271, 272
  green corridors and, 369
  green roofs and, 312, 313
  groupings and types, 247–248
  habitat selection, 55
  large greenspaces, 354
    cemeteries, 357–358
    golf courses, 354, 356–357
  microhabitats, 257
  migration, 248, 256, 267, 313
  mortality
    floodlights/lights, 256
    glass strikes, tall buildings, 256
    railways and, 278
  movements

  daily fluctuations, 268
  inhibitors, 266
  stepping stones, 266, 267
  turnover rate, 268–269
  origin (rock cliffs), 248
  in parks, 344, 351, 352
  public interest, 241
  seed dispersal, riparian vegetation, 193
  species richness (diversity)
    cities, 215, 261–262
    dogs on leashes and, 20
    habitat fragments, remnant greenspaces, 260
    shrub cover/height relationship, 258
    small wooded patches, 228, 266
    spatial habitat patterns, 259, 260
    suburbs vs. city, 262
    suburbs vs. rural, 262
    Syracuse (USA), 259
    wood area as predictive of, 249
  spillover effect from adjacent land, 265
  tree use in cities, 248, 285
  vegetation layers and, 251, 257–258
  street trees and, 285
  vertical built structures
    as habitats, 256–257
    as hazards, 256
    walls/roof as habitats, 257
  woody plants in house plots, role, 270, 303, 304, 305
  see also raptors; songbirds; waterbirds
bird feeders, 264, 297, 306
Birmingham (UK), tree species, 238, 239
blackbirds (*Turdus merula*), 270, 271
"black-box" models, 46, 68
block surfaces, 288
boat landing areas, 200
"boom and bust" production, 337
Boston Region (USA), 84, 90, 196, 200, 366
boundaries, 44
  convoluted vs. straight, 56–57, 69–70
  flows around, 69–73
  interwoven land mosaic, 51
  movements across, 69–70, 72
    lobes and coves effect, 70
    soft and hard boundaries, 69–70
  parallel movements (animal), 69
  perpendicular movements, 69
  between plants, 231–232, 239, 240
  soil types and, 96
  see also edge(s)
boundary (edge) model, 48, 82, 83
Brasilia, 50–51, 319, 369
Brazil, flooding management, 196
breezes, 66, 139–140, 147
brick surfaces, 288
bridges, 287
Brisbane (Australia), bird diversity, 262
brownfields, 110, 117–118, 359–361
  habitat and succession, 360
  soil contamination, 360
  soil remediation, 181, 360
bryophytes (mosses), 211, 291, 307
buckthorn (*Rhamnus cathartica*), 230
Budapest (Hungary), 80

445

Index

buildings, 306–313
  abundance and spatial arrangement, 306
  air pollution inside, 146
  bird collisions/mortality, 256
  city centers, 327–329
  designs and interiors, 306–307
  green roofs, *see* green roofs
  green walls, *see* green walls
  growing food, 346
  heat due to, 132
  isolated, urban air flow, 142–143
  plants and animals, 307–309, *see also* wall plants
  rating certification-system (LEED), 306
  scale, geomorphic patterns affecting, 37–38
  size/shape effect, air flows around, 142
  urban soil affected by, 100
  wooden foundations, 153
  see also built structures
built area, 7, 137, 202
built environment, 7, 12–14
built objects, 12
built spaces, 7
built structures, 3, 4, 12–14, 125, 275–313
  edge of floodplains, 183
  positive/negative ecological importance, 13–14
  sustainability and design, 275
  urban coastal zone, 197, 198
  urban heat and, 136–137
  vertical, wildlife hazards/habitats, 256–257
  see also buildings
bulges model urbanization, 77, 235, 324
burial grounds, *see* cemeteries
buried structures, 100
buses, 286
business areas, 38
buteos, 249
butterflies, 266, 309

Cairo (Egypt), air pollution, 147
calcium (Ca), 114–115
calcium chloride, 115
calcium-magnesium-acetate (CMA), 115
"campus-like settings," 358
Canada geese, 265
canals, 186
Canberra (Australia), 84, 212
capuchin monkeys, 271
car parking/parks, 287, 332
carbon dioxide, air pollution, 144, 147, 148
carbon monoxide, 144, 146
carbon-to-nitrogen ratio, 110
cars, 30, 285–286
cat(s), 21–23
  abundance estimates, 21, 331
  feral, 19, 21, 22
  food types, 21–22, 264
  free-ranging, types, 21
  predation by, 22, 23
catacombs network, 120, 123
cattle, urban farmland, 349
cellulose, decomposition, 110
cemeteries, 357–358
  ecological succession, 237

lichens, bird and animals, 357–358
  "old overgrown," 357–358
  wildlife habitat, 263
centipedes, 112
central place theory, 46
change(s), 11, 73–76, 376–377
  city centers, 327
  cyclic, 73
  environmental (by century/year), 83–84
  land mosaic, 44
  plant species, 234–238
  time scales, 16
  urban coastal zone, 197
  urbanization and peri-urban area, 322–324
  wildlife, *see* wildlife (urban)
  see also ecological change, rates/trajectories
changes-over-time principles, 16, 73–76
  time scales, 16, 75–76
Charles River (Boston Region), 189–190
Chattanooga (Tennessee, USA), greenway system, 367
chemical(s)
  absorption, tolerant plants, 217
  estuary pollution, 201
  "flushing" from surfaces during storms, 163
  groundwater pollution, 154
  industrial, in wastewater, 161
  urban air, 143, *see also* air pollutants/pollution
  urban river pollution, 189
  urban soil, 91–124
    concentrated chemicals, 94
    contamination, 110–111, 117–118, 119
    flows, 118–119, 375–376
    gases, 116
    from human structures, rubble, artifacts, 116–117
    inorganic, 113–115
    organic, 115–116, 118
    origin and sources, 113, 118–119
    toxicity, 115–116, 119
    types, 113–116
    see also mineral nutrient(s)
  in water bodies after wastewater discharge, 168
chemical flow-through, 119, 120
chemical leaching, 359
chemical oxygen demand (COD), 173, 174, 185
Cheonggyecheon Stream/River (Seoul, South Korea), 195
Chernobyl nuclear power plant, 76, 268
Cheyenne (Wyoming, USA), 258
Chicago (USA), 12, 214, 313, 367
China, 84, 89, 225–226, 349
Chlordane, in urban rivers, 189
chlorofluorocarbons, 145, 147
circuitry (alpha index), 381
  networks, 72, 122, 283
circular gradient, 12
cities, 3, 64
  air in/out of, 126–128
  bird species richness, 215, 262

centers, *see* city centers
coastal, *see* coastal cities
compact vs. spread-out, 80
definitions, 6
development from villages, 36, 212
ecosystem decline, 5
future perspectives, 372–373
gradient pattern, 41
history of nature in, 5–6, 212
largest, population size, 4
location, 36–38
  human needs influencing, 36, 37
  microclimate and, 37
metaphors, 4
"mixed-use patterns," 5
natural forest gradient, 40–41
parks, *see* park(s)
planned built communities, 326
planning, 31, 64, 82–83
population growth and histories, 4–8
population size, *see* population size
satellite, 77, 78, 235, 324
shade from trees, 216–217
shape, 76–77
size
  bioregional limits, 5
  changes over history, 4–6, 236
  greenspace, woodland and trees, 209
  optimum, 5
  radius and number of, 76
  travel time as guide to, 5
small, 63
small food-producing sites in, 346
species richness, vs. in suburbs, 262
transformation patterns, 81
urban mosaic, 50
ventilation, 127–128, 139
waterfront and harbors, 198
wildlife density, 261–262
woodland in, 209
zones, plant species density, 214
city as ecological system model, 49
city centers, 35, 38, 326–332
  air pollution, 332
  birds, 331
  "border areas," 327
  buildings, people and transport, 327–329
  distinctive and unusual characteristics, 327
  habitat types, 330
  plants, habitats/types, 329–332
  shade and urban heat, 332
  soil, water and air, 329–332
  spatial patterns, flows and changes, 326, 327
  spread, disasters and, 328
  wildlife habitat, 263, 327, 330–331, 332
city residential areas, 315–322
  courtyard/patio and single-unit housing, 319–322
  types, 319, 320
  diverse housing types, 316
  low-income areas, 316–319
  multi-unit high-rise or low-rise, 316, 328
  squatter settlements, *see* informal squatter settlement

Index

suburban residential area vs., 324–325
   types/forms, 316, 317
city–suburb–rural gradient, 41, 77, 262
city-to-rural gradient, 10–11, 12
clay, 37, 95, 101–102
clean water supply, *see under* water
climate change, anthropogenic, 269–270
coal-mining spoil, ecological succession, 237
coastal cities, 198, 199
   groundwater flows, 153
   waterfront and harbor, 198, 200
coastal plains, 98
coastal regions, 58–59
coastal saltwater wetlands, 175
coastal wetlands, 175, 176, 177
coastal zones (urban), 197–204
   city waterfront and harbor, 198, 200
   estuary and lagoon area, 198–201
   flows and movements, 59, 197, 202, 203
   natural changes, 197
   natural vs. human differences/influences, 197–198
   outer coastal areas, 201–204
   rocky, 202
   wetlands, 175, 176, 177, 200
Coleoptera, spatial pattern, 255
coliforms, 164, 168, 179
Colombo (Sri Lanka), 310
color surfaces, heat amelioration (cities), 139
combined sewage/stormwater systems, 163
combined sewer overflows (CSOs), 163, 167, 189
commercial areas, 314, 326, 332–336
   city center, 326
   distribution around city, 333, 335
   marketplace, 333
   neighborhood streets with shops, 332–333
   office center and town centers, 333–334
   residential land mixed, *see* mixed-use areas
   warehouse truck distribution center, 336, 337
commercial strips, 334–336
community garden, 207–208, 237, 346
community service agriculture (CSAs), 345
community shopping centers, 336
"commuter shed," 6
commuting, 75, 327, 328
compact-nucleus expansion, 7
composition of ecological community, 14
compost, 109, 110, 118, 297, 302
composting toilets, 169
concentric models, 49
concentric ring model, urbanization, 77, 78, 235, 324
concrete surfaces, 288
   cracks, 288
"cone of influence," wells, 158
connected impervious cover, 171
connectivity (gamma index), 36
   networks, 72, 122, 381
   road network, 283
constructed basins, 179–182
   bioretention/retention, 179, 180, 181–182

detention basins/ponds, 180
   stormwater pollution, 180, 181–182
constructed wetlands, on golf course, 356
construction fill, groundwater pollution, 154
controlled burning, 355
convergency point, 37, 55, 56, 72, 294
cool air drainage, 128, 140
"cool island," urban, 136
cooling
   green roofs, 311
   green walls, 309
   greenspace effect, 137–138, 371
   parks, 352
   strategies in cities, 139
   street trees, 139, 284
coral reefs, 201, 202
corridor(s), 11, 36, 44, 73
   back-space, 321
   characteristics, 44, 362, 365
   continuous vs. discontinuous, 294,
      see also stepping stones
   electric powerline, 364, 365
   functions, 276
   green, 240, 266, 343–371
      benefits/role, 365
      Boston's Emerald Necklace, 84, 366
      characteristics, functions, 362, 365
      land uses, 363, 364, 365
      location and types, 362–363, 364
      radial, 363, 365
      vegetation types, 343
   house plots, 294, 321
      backline strip, 294, 295, 321
   in interwoven land mosaic, flows and, 50, 51
   narrow wooded, 228
   neighborhood, 362
   networks, 45, 72, 365–366
   pipeline, 362, 364
   plant species dispersal, 240
   remnant grassland, 240
   of shade/shadow, 130, 131, 284
   transportation, 77, 276
      railways, *see* railways
      roadsides, 364–365
   types, 51, 240
   urban-to-rural gradient, 42
   wide wooded, 228
   width, 369
   and wildlife movement, *see* wildlife (urban), movement
corridor-and-small-patch system, 368
corridor-centered neighborhood, 56–57
corridor-centered neighborhood, 56
corridors per node (beta index), 72, 122, 381
courtyard(s), 292, 319–322
   plants, 319, 320
coyote (*Canis latrans*), 242, 243, 244, 259, 261
   characteristics and feeding patterns, 244, 270
   around dumps, 359
   railways and, 278
cracks (in hard surfaces), 287–292

animals/invertebrates, 291
concrete surfaces, 288
diagonal, zig-zag, 288
ecological succession, 290–292
formation, 288–290
lengthening, network formation, 291
linear, 288
plants, 290–292
repair, 289
steel or wood surfaces, 288
tarmac surfaces, 288
types/forms, 289
vegetation roles/benefits, 291–292
vegetation sites, 290
in walls, 289–290
created greenspace, definition, 7
cropland (urban), 345
   near streams/rivers, 187, 194
   plant adaptation, 221–222
   see also agricultural land
cultural perspectives, 26, 27
cyclones, 202

daylighting, streams, 186
DDT, in urban rivers, 189
dead zones, 167, 201
decomposition, organic matter, 382
   in dumps, 359
   human-made organic materials, 110
   by invertebrates, 254
   in sanitary sewage pipes, 164
   sewage/wastewater, septic system, 169
   in urban soil, 99, 106, 109, 110–111, 116
de-densification, 80
deer, 246, 265
definitions, urban ecology terms, 6–8
deforestation, 236
de-icing salt, 115, 117, 217–218
delta, 200–201
demographics, 2, 4–5, 63
   low-income population, 316
   megalopolis, 33
   see also entries beginning population
dendritic networks, 45
de-nitrification, 164, 169, 176–177
densification, 7, 76, 79
dependent pairs (dependencies), 53–54
detention basins/ponds, 180
diagonal patterns, 32
direct gradient analyses, 42
disasters, 74, 76, 318, 376
   city centers and, 328
   tree species/density affected by, 227
dispersed development, 372–373
dispersed patches model, urbanization, 77, 78–79, 80, 324
   housing-unit plot sizes, 79
   see also sprawl
dispersion model, 47
dispersion of individual plants, 227
dispersivity, groundwater, 157
dissection (corridor formation), habitat changes, 88–89
"distance from source" of species, 261
distance-decay pattern/model, 46
distribution of aggregated people, 372–373

447

Index

disturbances, 73, 74
  along roadsides, 285
  human, effect on urban soil, 91–92, 96
  resilience/resistance to, 74, 75
  wasteland, plant adaptation, 221–222
ditch networks, 365
diversity, *see* biodiversity; species richness (diversity)
diversity concepts (alpha, beta, gamma), 208
docks, 200
dog(s), 19–21
  diseases, risk for wildlife, 21
  feces/urine, 20, 21, 217
  home ranges, 19, 20
  on leashes, bird diversity and, 20
  packs of, 20–21
  walking, impacts, 20, 21, 228
  wildlife avoidance behaviors, 20
domain of scales, 33
domestic animals, 19–23, *see also* cat(s); dog(s)
dominance of species, 14
donut model, 47
Dortmund (Germany), bird species/turnover, 268–269
downtown, city, *see* city centers
drainage
  house plots, 303–304
  stormwater systems, 120–121
  urban soil, 103
drainage connection, 171
drainage ditches, 172, 173, 180, 181–182
drainfield, for septic systems, 169
dredging, 198
"dump favela," 50–51, 319, 359
dumps (tips), 109, 358–361
  groundwater pollution, 154
  wildlife habitat, 263
dunes, 201
Dusseldorf (Germany), habitat types, 205–206, 263
dust, street, 165, 281, 309

earthworms, 112
East Kolkata (Calcutta, India) wetlands, 177
eco-city, 29
ecological assays, 12–16
ecological change, rates/trajectories, 83–90
  habitat loss/degradation/fragmentation, 88–90
  soil, water and air, 83–84
  urban land uses, 86–88
  vegetation and biodiversity, 84–86
ecological community (natural community), 14
ecological economics, 18
ecological engineering, 68
ecological finiteness, 18
ecological footprint concept, 63
ecological gradients, 42–43
ecological heterogeneity (diversity), 36
ecological infrastructure, 366
ecological mosaics, 60–63, *see also* land mosaic

ecological network, *see* network(s), green/ecological
ecological pattern models, 47–48
ecological processes, urban soil affected by, 99–100
ecological succession, 14, 85, 236–238
  abandoned city building and, 86
  arrested, 237
  brownfields, 360
  cliffs and quarries, 356
  comparisons and positive/negative roles, 14
  in cracks, 290–292
  cyclic, 238
  after disaster (Chernobyl), 268
  on dumps, 359
  examples, 237
  floods and, 193
  forms/types, 14, 15
  green walls, 309–310
  on hard surfaces, 291
  "human-molded vegetation," 86
  industrial sites (post-production), 339
  invertebrates, 269, 270
  plant categories (tolerants, inhibitors and facilitators), 237
  retrogressive, 238
  "sequence of patchy mosaics," 237
  site conditions and plants affecting, 236–237
  stages, 14, 85, 236, 237
  trees (urban), 226–227
  in UK, 85
  urbanization effect, 85–86
  vacant lots, 360
  wildlife, 269
ecology, xii, 3, 8–9, 11, 73–75
  areas involved in field of, 8–9
  environment and, history, 8–9
  history of science/field, 8, 10, 11
  overlapping subjects, 8
  subspecialties, 8
economics, 18–19, 28, *see also* specific types
ecopolis, 29
ecoregions, 58
ecosystem(s), ix, 7, 296, 378
  analysis, 68
  development, 238, 296
  response to disturbances, 74
  stability, 73
ecosystem models, 48
ecosystem services, 30, 225
edge(s), 44, 70
  convoluted, house plots, 294
  estuaries, 199, 200
  filtration effect, 69
  habitat edges, 239, 240
  house plots, 294
  natural habitats, 69
  parks, 69
  straight, in urban areas, 240
  types, 239, 240, 294
  see also boundaries
edge effect, 69, 239–240
edge (boundary) model, 48, 82, 83
"effective mesh size," 89

electric powerline corridors, 364, 365
electric-power generation, 66
elephant hawk-moths (*Deilephila elpenor*), 219
Emerald network, 48, 84, 89, 367, 369
energy, 129–133
  flows, 131, 132, 133
    day–night, 133, 135–136
    urban heat due to, 135–136
  geothermal, 155
  human need for, 17
  solar radiation, 129–131
  storage (Q_s), 131, 133, 136
  surface energy balance of area, 131–133, 136
  surface reflection (albedo), 131–132, 136, 139
  wavelength shifts, 132
  see also heat
energy-driven flow, 66
engineering, 28, 104
engineering soil structure, 102, 103, 104
environment, concept and types, 3, 7
environmental constraints, 49
environmental economics, 18
environmental gradient model, 48
environmental gradients, 42–43
environmentalism, 8
"ephemeral channels," 183, 185
epiphytes, 210, 211, 308
equations/calculations, 381
estuaries, 198–201
  edges, 199, 200
  habitats, 199–200
  pollution, 201
  reconfiguration by flooding, 203
  sediment-covered bottom, 200–201
eutrophication, 179, 201, 301, 356
evaporation, 132, 151
evapo-transpiration, 132–133, 139, 143, 151
  humidification of dry hot air, 371
  latent heat in daytime, 132–133, 136, 143, 347
  reduction, increased stormwater runoff, 171
evergreen ground cover, 293
exponential growth, 382
exurban area, 7, 39, 321
  adjacency effects, 52
  population density, 40
  residential area, 7, 321–326
  small agricultural sites, 346
  spatial patterns, 324–326
  see also peri-urban area

facilitator patterns, 266
farm animals, 19, 345, 346, 349
farmlands, *see* agricultural land
"favela" dump, 50–51, 319, 359
fecal-coliform-bacteria level, 164, 168, 179
feedback loops, 44
feral cats, *see* cat(s), feral
feral dogs, 19
feralization, plants, over centuries, 236
ferns, 211
fertilizers, 40, 114, 217

448

Index

for growing food, 86, 347
for lawns, 300–301
pollution, 40, 41, 173–174, 301
urine recycling and, 169
fill, in soil, 92, 94, 95, 116, 117
Finland, wildlife movement, greenspaces, 368
fire, in large natural areas, 354–355
fire hazardous conditions, 52, 355
"fire-stick" burning, 55
fish
    adaptation to river flooding, 196
    harbors and lagoons, 200
    urban rivers, 190
    urban streams, 186
    water pollution and, 174, 190
fish farms, 345
flash flood, 153, 173, 194
flat areas, large (landform), 98
flooding, 149–150, 153–154, 170, 173, 193
    causes/sources, 194
    combined sewer overflows, 189
    ecological consequences, 194
    flash, 153, 173, 194
    human activities increasing, 194–195
    from land and floodplain, 194–195
    land use affecting runoff and, 194
    Los Angeles River, 187
    risk, measurement, 194
    from sea, 202, 203
    from upriver/upslope land, 194
    by urban rivers, 187, 188, 193–197
        adaptation to, 188, 196
        management, 195–196
        recurrence, 193
    by urban streams, 193–197
floodplain(s)
    irrigation-and-drainage systems, 194–195
    urban rivers, 188, 190, 191, 195
    urban streams, 182–185
floodplain friction, 194
flood-prone areas, 193, 194, 196
flood-pulse model, 188
"floristic similarity," 238
flower gardens, 299–300
flowering plants, 211, 278
flows, 11, 14–16, 65–90, 375–376
    air, see air flows
    around boundaries, 69–73
    biogeochemical, 68–69
    built structures/networks, 142, 375
    changes-over-time, see changes-over-time principles
    chemical, in urban soil, 118–119, 375–376
    city center, 326, 327
    coastal regions, 59, 202, 203
    corridor networks, 72
    curvilinear, 65
    directionality, 57, 70–71, 72
    energy, see energy, flows
    gravity model and, 72–73
    groundwater, see groundwater
    heat, 133
    horizontal, 65, 119
    human-driven processes, 65, 197

    land mosaics and, 50, 51, 70–73
    land-and-sea, 59
    mapping, 55, 72
    mineral nutrient, 68–69, 83, 322
    nature of, 65–69
    network, see network flows
    organisms, 15–16
    patterns, 65–66, 67
    primary routes, 72
    rates/trajectories of ecological change, 83–90
    recurrence interval, 384
    rivers, see river(s) (urban)
    spatial attributes affecting, 70, 71
    system and ecosystem, 68–69
    typical distance, 55
    underground (urban), 122–124
    urban coastal zone, 59, 197, 202, 203
    across urban-region boundary, 63
    urban–rural and rural–urban, 38
    vertical, 65
    water, see water flows
flushing, 20
food
    grown in urban areas, 344, 345–347, 348
        on/in buildings, 346,
        see also agriculture (urban)
    human need, transported, 17, 36
    wastewater aquaculture, 165–166
food security, 349
"food shed," 349
food waste, 264, 328, 330, 332, 333
foraging guilds (bird), 248, 251, 257
forest(s) (urban areas), 353–355
    changes in Rome, 84–85
    city-to-natural forest gradient, 40–41
    ecological succession, 85, 237
    for forestry, and multi-use, 355
    layers/strata, 208, 231
    logging patterns, outward urban expansion, 82, 83
    sprawl development, 79, 80
    unplanned reforestation, land recovery, 89
    urbanization effects, Sydney, 84
    wildlife, 261
    see also tree(s); woodland
forestry, 355
Fort Collins (Colorado, USA), brown bat roosts, 247
fossil fuels, 17, 132, 143
fox, red (*Vulpes vulpes*), 244, 267, 272, 278
fractals, 34
fragmentation (habitats) see habitat(s), fragmentation
fragmentation (land/landscape), 89
Frankfurt (Germany), green wall, 310
"free-spirit" people, 121
freeze–thaw cycles, 290
frogs, 272, see also amphibians
front-space strips, 321, see also under house plots
fruits, 230, 270, 303
function (functioning), land mosaic, 44
fungi, 109, 110, 222
fungicides, lawns, 300

gamma diversity, 208
gamma index, see connectivity (gamma index)
garbage
    cat food supply, 21, 264
    wildlife food source, 264
    see also dumps (tips); waste
garden(s), 2, 292
    community, 207–208, 237, 346
    flower, 299–300
    institutional, 346
    office centers and, 333
    public health issues related to, 25
    tiny, in little-used spaces in cities, 346
    vegetable, 86, 299–300, 346
Garden City movement, 297
garden walls, 304–306
gardening, 17, 296, 299
gases, in urban soil, 116
gasoline-powered lawn equipment, 302
genetic adaptation, wildlife, 271
genetic change, 220, 270–273
genetic diversity, climate-related, 222
genetic drift, 271
genetic variation, plants, 221
genetics, plants, 220–223
geographic information system (GIS) images, 46
geographic regions, 58
geological processes/patterns, 97–99
geomorphic framework, 36–38
geostatistical models, 46
geothermal heat, 155
giant hogweed, 215, 219
Gini coefficient of inequality, 385
glades, 229
global warming, 147–148, 235
golf courses, 354, 356–357
    problems associated, 356
    wildlife habitat, 263, 357
government/politics, urban ecology principles use, 29
gradient patterns, 12, 32
gradients (ecological), 34, 42–43
    in cities, 41
    city–suburb–rural, 41, 77, 262
    city-to-natural forest, 40–41
    city-to-rural, 10–11, 12
    urban–rural, see urban–rural gradient
    vertical, radial, circular, 12
grain size, 32
graph theory, models, 46, 49, 52
gravel, 350, 356
gravity model, 72–73, 284, 381
greater tit (*Parus major*), 266–267
Greater Yellowstone Region, 54
green architecture, 306, 307
green corridors, see corridor(s), green
green cover, 7
green infrastructure, 366
green roofs, 310–313
    agriculture on, 313
    air quality improvement, 311
    animals/invertebrates, 312, 313
    birds, 312, 313
    cooling, 311

449

green roofs (cont.)
  extensive vs. intensive, 311
  goal/function and benefits, 311, 313
  habitats and plants, 312–313
  irrigation, 312
  soils, 312
  stormwater runoff, 311–312
  types, 311
green walls, 309–310
  skyscrapers, 310
  streets, 281–282, 309
  see also wall plants
green wedges, around London, 134
greenbelt(s), 364
greenbelt model, 48
greenhouse effect, 132
greenhouse gas, 147
greenhouse production, 346
greenspace(s), 7, 343–371, 373–374
  adjacent land, "species rain," 369, 370, 373
  changing with urbanization, 84–85, 235
  cities, relative to population size, 209
  distance between, wildlife and, 260, 261
  effective, criteria for and benefits, 369–371
  functionally linked, 367–369, 373
  groups, 366–367
  infill, 80
  large, 353, 354
    airports and military bases, 361–362
    brownfields and vacant lots, 359–361
    cemeteries, 357–358
    cliffs, quarries and mines, 355–356
    dumps, 358–361
    golf courses, 354, 356–357
    institutions/municipal facilities, 358
    natural/semi-natural areas, 353–355
    urban forests, 355
  mammal predators in, 242, 243
  planted vegetation, areas, 209–210
  semi-natural, 7
  size/area, 137, 354
    decrease in metro areas, 343
    species richness (wildlife), 243, 260–261
    species–area effect, 260
    temperatures and, 137–138
  smaller vs. large cities, 344
  stormwater runoff, 170–171
  time scale for urbanization change, 235
  types, 343, 344
  urban heat and, 137–138
  urban wildlife moving out of, 261
  uses, 373
  vegetation on, change during urbanization, 81, 235
greenspace system, integrated, 366–371, 373
  effective, criteria for success, 369–371
  functionally linked greenspaces, 367–369
  groups of greenspaces, 366–367
greenway, 228, 363
greenway systems (USA), 367
Grevillea macleayana, 221
grey-water recycling, 169
groundwater, 151–157
  animals, 156–157
  aquifers, 151–152, 156

cleaning, 154
drinking water source, 151–152, 158
flooding, 149–150, 153–154
flows, 152–154, 383
  coastal cities, 153
habitat conditions, 156–157
heat in, 155–156, 158
industrial uses/usage, 158
microorganisms, 156, 157
pollution, 154–155, see also water pollution
porosity, permeability and dispersivity, 157
pumping, water-table lowering, 152, 153, 196
"recharging," 152
rise in, 153
saltwater intrusion, 153
surface water interaction zone, 157
urban agriculture effect, 347
groundwater emergence zone, 153
groundwater wetlands, 176
growth economics, 18

habitat(s), 7, 11, 14, 205–240
  area/size, plant species and, 232–233, 239
  arrangement, species dispersal, 240
  artificial (built) structures as, 256–257
  breeding, for bird species survival, 240
  changes due to urbanization, 88–90
    agricultural and natural land, 89–90
    loss/degradation/fragments, 88–90, 322, 324, 349
    spatial processes causing, 88–89
  city centers, 330
  classification, 205–206, 207
    grouping-by-mechanism, 206–207
  definition, 7
  diversity (heterogeneity), 36, 207–208, 374
    alpha, beta, and gamma, 208
    cemeteries, 357
    community gardens, 207–208
    fine-scale within land uses, 207–208
    house plots, see habitat(s), in house plots
    parks, 351
    plant diversity see plant(s) (urban)
    rail corridor, 276, 277
    semi-natural areas, 353
    wildlife, see wildlife (urban)
  edge, 70, 240, see also boundaries; edge(s)
  estuaries, 199–200
  fragmentation/fragments, 88–90, 214, 235
    ecological effects, 240
    plants and, 238–240
    reptiles/amphibian decline, 253
    by roads, 280, 286
    spatial measures, 89
    wildlife species richness, 260–261
  groundwater conditions, 156–157
  in house plots, 208, 293–295, 296–297
    specialized structures, effect, 297
    at two scales, 208
    types, 293, 295
  industrial areas, 339

lagoons, 199–200
loss, and degradation, 88–90, 214
  reptiles/amphibian decline, 253
  urbanization causing, 88, 89–90, 322, 324, 349
low-income residential areas, 318–319
management, airports, 361
office centers, 333–334
patch–corridor–matrix model, 238–239
selection, 55
sink or source, 294
small (roads), small populations, 280
spatial patterns, wildlife, see wildlife (urban)
species richness variations, 233
successional, greenspaces and corridors, 343
types, 205–207
underground, urban areas, 121–122
in urban sprawl, 86
habitat arrangement model, 48
habitat selection model, 47
habituation of species, 351
hand test, soil texture type, 101–102
harbor(s), 198, 200
  lakeside, 204
  pollution, 201
harbor islands, 199, 263
hard surfaces, 287–292
  commercial strips, 334
  cracks, see cracks (in hard surfaces)
  ecological succession, 291
  types, 288
  see also impervious surfaces
hawks, 249, 256
headwaters, urban stream, 183–184
healthy ecosystem, definition, ix
heat
  accumulation, greenhouse effect, 132
  airdome development, 128
  airports and, 362
  amelioration in cities, 139
  anthropogenic (Q_F), 131, 132, 133, 135, 332
  causes/sources, 135–137
    energy flow, 135–136
    groundwater heat, 155–156, 158
    surfaces and structures, 136–137
  emission, 132, 133
  flows, 133
  geothermal, 155
  global warming and, 147
  greenspaces and impervious spaces effects, 137–139
  in groundwater, 155–156, 158
  intensity, of heat island, 134, 136
  latent (Q_E), 131, 132–133, 136
  net horizontal advection (Q_A), 131, 133, 136
  pollution, urban rivers, 190
  sensible (Q_H), 131, 132, 133, 136
  storage (Q_S), 131, 133
  streets, 281, 282
  transportation and buildings as sources, 132
  urban air, 126, 131, 133–139, 332

Index

urban streams, 185
see also energy
heat island, 128–129, 133, 135, 148, 155
   characteristics, 134–135
   city centers, 134, 332
   heat intensity, 134, 136
   hot dry and tropical areas, 135
heavy metals
   air pollution, 145
   bioaccumulation, 115, 154, 168
   street dust, 281
   urban river pollution, 189
   urban soil, 40–41, 115, 118
Hemiptera, 255
herbaceous layer, 185–186, 231
herbicides, 276, 277, 300
herbivores
   invertebrate, *see* insect(s)
   mammal, 245–246, 347
      prevention in vegetable gardens, 300, 347
herbivory, 218–220
   distribution in woods, 219
   low levels in urban areas, 218–219
   non-native species, 219
   specific insects and trees, 219
herpetofauna (herps), *see* amphibians; reptiles
hierarchy of spatial scales, 12, 33
high-rise buildings, 316, 326, 327, 328
   skyscrapers, 310, 327, 328
highway bypass, around city, 88
highways, urban, 280
hills, air flow streamlining, 140
hillside development, flash floods and, 194
hill-top, 98
Hiroshima City (Japan), 81, 87
home range, 51, 67
   cats, 22
   dogs, 19, 20
   people, mixed use area and, 314, 372
   wildlife, 55, 264
homes, *see* housing; residential areas
horizontal flows, 65, 119
horizontal natural processes, 34
horizontal patterns, 32, 34, 35
   plant communities, 231
   soil (urban), 92, 95–97
house mouse (*Mus musculus*), 245, 271
house plots, 292–297
   areas surrounding, effects, 306
   biodiversity, 263, 295–297, 298, 306, 374
   designs and management, 292–293, 297, 298
   detached houses, 317, 324
   features, 303–306
   forms/types, 293, 295, 319
   front-/back-space, 292, 297–299, 321
      corridors, species movement, 321
      depth/width and proximity to others, 293
      habitat diversity, biodiversity, 299, 321
      normal vs. large plots, 324
      row of front spaces, 292, 321
      size and uses, 297, 324
   garden walls, 304–306
   internal structure, 293–297
      biodiversity and wildlife, 263, 295–297, 306
      habitat diversity, 208, 293–295
      plant diversity, 295–297
      spatial arrangement, 293–295, 306
      surroundings, role, 293
   private outdoor spaces, 292, 297–299
   side yards, 299, 324
   sizes, 79, 297, 317, 324
   water-related structures, 303–304
   wooded, 293
   woody plants, 303, 304, 305
   yards, gardens and lawns, 297–306,
      see also garden(s); lawn(s)
   see also courtyard(s); patio(s)
house sparrows (*Passer domesticus*), 23, 252, 253, 331
   genetic changes, 271, 272
housing
   attached single-unit, 321
   detached, house plot sizes, 79, 317, 324
   low-density, 324
   multi-unit high-/low-rise, 316, 328
   public health issues related to, 25
   single-family, location, 38
   single-unit, in cities, 79, 319–322
   spatial patterns, 321, 324–326
housing development area, wildlife habitat, 263
human(s)
   inclusion/exclusion in urban ecology, 3
   social patterns, 17–18
human activities, 16–26
   biodiversity correlation, 234
   flooding increased by, 194–195
   natural areas affected by, 355
   over decades, 235
   pests and public health, 23–26
   riverside area (urban), 190–191
   social patterns, economics and needs, 16–19
   soil patterns associated, 92–94
   urban beaches, 202
   urban coastal zone, 198
   urban soil affected by, 100
human needs, 16–17
   city location and, 36, 37
   for village to become a city, 36
human structures, 197, 275–313,
      see also buildings; built structures; underground human structures
human waste, *see* sewage
humidity, 126–127
humus, 95, 106, 302, 347
hurricanes, 202
hydrocarbons, 116, 144, 174, 330
hydrographs, stormwater runoff, 172
hydrologic cycle, *see* water cycle
hydrology, 175, 281
Hymenoptera, spatial pattern, 255, 259–260
hyporheic zone, 156

igneous rocks, 98
impervious surfaces, 136, 170–173
   city centers, 327
   connected (drainage connection), 171
   increase, effect on stormwater discharge rate, 173
   stormwater runoff, 170–173
      amount, 170–171
   urban heat and, 138–139
   water flows and, 170–173
   see also hard surfaces
inbreeding, 239
inbreeding depression, 221, 273
Indian balsam, 219
indirect gradient analyses, 42
indoor plants, 307
industrial areas, 38, 314, 336–342
   alternative approaches, 339–342
      forms/types, 340–342
      interdependent industries, 340
   characteristics, 338
   contaminated soil, 118, 339
   ecological conditions, 337–338
   ecological succession, 339
   groundwater usage, 158
   location, 341–342
   near rivers, flooding and, 195
   plant and animal species, 339, 340–342
   power and water requirements, 336, 340–342
   production/post-production, 336–339
      ecology of sites, 338–339
      post-production inputs/outputs and ecology, 339
      production inputs/outputs, site features, 337–338
   site requirements, 337
   stormwater runoff, 338, 339, 340–342
   wastewater, 161, 338
   woodlands succession, 237
industrial site vs. industrial city, 342
industrial symbiosis, 340
industrial waste, water pollution, 154, 189
infill, and infill threshold, 80, 361
infiltration, water penetration into soil, 103, 104
informal squatter settlement, 316–319, 343, 349
   "dump favela," 50–51, 319, 359
   habitats, plants and animals, 318–319
inhibitor patterns, 266
input–output models, 46, 68, 69
insect(s)
   diversity on non-native trees, 219
   genetic adaptation, 271
   green roofs, 312
   herbivores, 219–220, 254
   on trees, 219–220
insect pests, 24
insecticides, 271, 300
institutional gardens, 346
institutions, 263, 358
integrated pest management (IPM), 23, 301
interdependent elements, 53–54
interdependent industries, concept, 340
"interdigitation"/interface, natural and urban area, 52
"intermittent channels," 183

451

Index

intersection linkage (beta index), 72, 122, 381
"interspersion," natural and urban area, 52
intertidal zone, 202
invertebrates, 254–255
  on/in buildings, 309
  in cracks in hard surfaces, 291
  ecological functions/roles, 254–255
  ecological succession, 269, 270
  gardens, 299
  green roofs and walls, 257, 313
  house plots, 296
  light effect, 255
  spatial patterns (by class), 255, 259–260
  species richness and habitats, 255, 296
  species/types, 254, 269
  underground, 122
  urban soil, 112–113, 269
  urban streams, 184, 186
  see also insect(s)
invisibles, in urban ecology, 1, 2
  importance, 1
irrigation-and-drainage systems, 194–195
island biogeography model, 48, 238
islands
  barrier, 201–202
  harbor, 199, 263
  near lagoon edges, 199, 200
ivy (*Hedera*), 309

Jakarta (Indonesia), wetlands, 177
Japanese knotweed, 215, 219
"jaws-and-chunks" model, 82, 83
journey-to-work model, 382

Kalundborg (Denmark), interdependent industries, 340
kangaroos, 246
Kano (Nigeria), 40–41, 225
Kawasaki (Japan), interdependent industries, 340
kestrel, 249, 262, 331
keystone species, 259
koala, mortality by dogs, 20
"K-selected" species, 216
Kuala Lumpur (Malaysia), green wall, 310
Kyoto (Japan), 237, 267, 268

lagoons, 198–201
  bottoms and sediments, 200
  islands at edge, 199, 200
lakes, 203–204
land cover, 7
land fragmentation, causes, 89
land market value model, 49
land mosaic, 32, 50–51, 372–373
  city center, 50
  corridors in, 50
  dependencies between elements, 54
  flows, 50, 51, 70–73
  interaction strength, 50
  interwoven, 51
    extent determination, 51
  locality-centered, 51
  patch-centered, 56–57
  patches in, 50

persistence/stability, 51, 73
structure–function–change characteristics, 44, 45, 55, 56
urban regions as, 51
land mosaic model, ix, 44, 48
land recovery/restoration, 88–89
land subsidence, 152
land use, 7
  airdome thickness and development, 128–129
  changes during urbanization, 80, 88
  changes over centuries, 236
  habitat diversity (fine-scale), 207–208
  habitat groupings by, 206
  Kano (Nigeria), 40–41
  mapping changes in urban area, 81
  "mixed-use patterns," 5
  patterns, 38–39, 70–73
  planning, principles/categories, 27–28
  time scale of changes, 235
  tree species distribution, 226
  urban soil types and, 96
  during urbanization, 86–88
  water runoff and floods associated, 194
landfill site, 263, 358
landforms, 98–99
landscape architecture, 28
"landscape complementation," 55
landscape ecology, 10, 294, 306
landscape ecology models, 48, 239
"landscape ecology revolution," ix
"landscape metrics," 52, 89
"land-use flexibility," 349
Las Letras neighborhood, Madrid (Spain), 57–58
latent heat (Q_E), 131, 132–133, 136
latrines, 168
lawn(s), 300–303
  in cemeteries, 357
  characteristics, 300
  in greenspaces and corridors, 343
  mowing, 300, 302–303
  pesticides, fertilizers and water, 300–301
    fertilizers, 301
    pesticides, 300–301
    plant biodiversity without, 302–303
    watering, 301–302
"lawn specialists" (species), 302
lead, air pollution, 145
leaf area index, 217
LEED system, 306
legacies of ecological conditions, 75
Leicester (UK), 211, 296
Lepidoptera, spatial pattern, 255
levees, 195
lichens, 211, 290–291, 307, 357
light(s), 255, 256
lignin, decomposition, 109, 110
limestone-karst rock, 156
linear growth, 382
litter (vegetation), 94–95, 105, 106
littoral zone, 153, 178
livestock, 19, 345, 346, 349
lizards (*Anolis cristatellus*), 266
logistic (sigmoid) growth, 382
London (UK)

birds, 249, 252–253
green wedges around, 134
groundwater flow, 152
heat island, 134
Highgate Cemetery, 357–358
long-wave radiation, 130, 131, 132, 133
Los Angeles (USA), 259–260
Los Angeles River, 187, 189
low-density housing, 324
low-impact development, 182
low-income areas, residential, 316, 318, 319
low-rise buildings, 316, 328, 331

macroclimate, 58
macro-fungi, 222
macro-nutrients, in urban soil, 114
Madrid (Spain), 260, 309, 310, 334
  Las Letras neighborhood, 58
magnesium (Mg), 114–115
mammals
  in natural area of Portland (USA), 354
  public health issues, 25
  tree canopy, 257
  vegetable gardens, 300
  see also herbivores; predators; *specific mammals*
Manchester (UK), 40, 144
mangrove swamps, 175, 176, 200, 203, 229
manure, for growing food, 347
maps, as models, 46
Marchetti constant, 5
market, 264
market gardening, 345, 348
marketplace, 333
mass flow, 66
"master-planned communities," 326
matrix, 44
meadows, 293, 312
megacities, 5, 63, 159
mega-floods, 194
megalopolis, 6, 33, 76, 87–88, 373
megalopolis ecology, 373
"mental models," 46
mercury, 189
Merseyside (UK), vegetation change with urbanization, 81
meso-predator release hypothesis, 243
metamorphic rocks, 98
metapopulation, urbanization phase, 322–324
metapopulation dynamics, 71–72, 381
metapopulation models, 48
methane (CH_4), 116, 359
metro area (metropolitan area), 6
  agricultural land around, 60, 62
  convolution of border, 62
  as ecological mosaic, 60–63
  examples, alternative forms, 61–62
  greenspaces, *see* greenspace(s)
  land use change during urbanization, 86–88
  natural land around, wildlife movement, 265
  perimeter-to-area ratio, 60
  planning, 82
  population density, 60

types, negative/positive roles, 61–62
urban regions with, examples, 57, 60
urban–suburban–rural gradient, 262
metro area border models, 48
Mexico City (Mexico), 145, 204, 343, 353
microclimate, 126–129
　air in/out of cities, 126–128
　effects on plant growth, 216
　layers of air and airdome, 128–129
　moisture in air, 126–127
　urban vs. non-urban areas, 128–129
　ventilation of cities, 127–128, 139
microclimatic patterns, 37
microhabitats
　green roofs, 312
　house plots, 208
　informal squatter settlements, 318
　railways, 277, 278
　urban wildlife, 251, 257, 328
micro-nutrients, in urban soil, 114–115
microorganisms (microbes), 24
　bioretention basins, 180
　groundwater, 156, 157
　groundwater pollution effect, 154
　organic matter decomposition,
　　see decomposition
　underground, 121
　urban soil, 108–111
　　distribution, 109–110
　　types, 108–109
　water bodies after wastewater discharge, 168
　wetlands, 176
midges, 174
migration, animals, 67, 264
military bases, 362
millipedes, 112
Milwaukee (USA), tree distribution/species, 226
mineral(s), 36
　in industrial wastewater, 161
　in soil, 91, 92, 95, 99
mineral nutrient(s)
　flows, 68–69, 83, 322
　in urban soil, 99, 114–115
　in wastewater discharge, 167
mineral nutrient cycles, 119
mines, 355–356
Minneapolis (Minnesota, USA), 367
mites, 112
mixed-use areas/land, 268, 314, 315
models, 38, *see also specific models*
moisture
　conditions, vegetation grouping, 206, 207
　in parks, 352
　in urban air, 126–127, 352
　see also rain; water
Mongolian oak (*Quercus mongolica*), 231
Moran index for degree of clustering, 384
mosaic patterns, 12, 31–64
　spatial scales, 32–34
　see also land mosaic; neighborhood mosaics; spatial patterns
mosses, 211, 291, 307
motorcycles, 286

Mount Auburn Cemetery (Cambridge, USA), 357–358
mountain regions, 59–60
mountain top, 98
movement(s), 14–16, 65–69
　across boundaries, *see* boundaries
　animal, *see* animal movements; wildlife (urban)
　coastal zones (urban), 197, 202, 203
　commuting effects, change-over-time, 75
　energy-driven, 66
　fish, in urban rivers, 190
　habitat and land-use preferences affecting, 72
　human-related, coastal zone, 197
　mapping, 72
　materials and objects, underground, 122
　motor-powered, 66
　nature of, 65–69
　patterns, 65–66, 122
　plant, 15–16, 66–68
　primary routes, 72
　rate, 72
　spatial attributes affecting, 70, 71
　species, corridors, 240, 265
　urbanization, *see* urbanization
　see also flows
mowing, lawns, 300, 302–303
mud flats, 200–201
mulch, 302
"multifunctional landscape mosaic," 348
multi-habitat species, 53, 55
multi-modal transport system, 328
multiple nuclei models, 49
municipal solid waste (MSW), 358–359
mushrooms, 222

Nanjing Region (China), outward urbanization, 77
native species, 370, *see also* plant(s) (urban); wildlife (urban)
natural area/land, 7, 89–90, 353–355
　benefits, 354
　habitat loss/degradation, urbanization, 89–90
　native species of wildlife, movement, 265
"natural burial," 357
natural community, plants, 232
natural environment, 3, 7
natural habitats, 7
natural processes, time scales, 234
natural structures, coastal zone, 197
natural systems, 7
　degradation, 2
　wastewater treatment, 165, 173
natural vegetation, protection, priorities, 34
naturalization, species, 213, 235
"naturbanization," 77
nature, definition, 7
nature reserve, "spillover effect" of birds into adjacent area, 265
nature-and-people interaction model, 38, 39
nature's patterns, 34–36
nature's services, 30, 225
negative attributes, urban region, 379

neighborhood, 54
　residential, streets with small shops, 332–333
　sewage treatment facilities, 165
neighborhood corridors, 362
neighborhood ecology, city residential area, 321–322
neighborhood mosaics, 49–58, 87
　adjacencies, 51–53
　central organizing force, 55, 57
　convergency points, 55, 56
　corridor-centered, 56–57
　dependent pairs and surroundings, 53–54
　interactions between, 54–56, 87
　land mosaic, *see* land mosaic
　patch-centered, 56–57
　tightness, strength of interactions, 54–56, 57
　two, tight interactions, 57–58
neighborhood shopping malls, 336
nematodes, 112–113
neophytes, 213–214
nests, parasitized, 266
net radiation balance, 382
The Netherlands, National Ecological Network, 367
netway-and-pod transportation, 285, 287, 375
network(s)
　ditch, 365
　green/ecological, 45, 48, 72, 365–366, 367, 374, 375–376
　broad-scale/regional, 366
　dendritic/rectilinear patterns, 366
　ridge, 45
　road, *see* road networks; street networks
　underground, 124
　　catacombs, 120, 123
　　pedestrian walkways, 122–123
　　quarried limestone tunnels, 123
　　sewage wastewater, 123–124
　urban streams, 366
network flows, underground, 122–124
　city centers, 330
　materials and objects, 122
　transportation, 120, 122
　water, 120–121, 122
network forms, underground, 122
New England (USA), bird density in city, 261–262
New Mexico (USA), commercial strips and wildlife, 334
New York, Long Island (USA), water cycle, 151
Nishinomiya City (Japan), 257–258, 260
nitrogen (N), 114–115
　estuaries, 201
　excess, in wastewater discharge, 167
　fertilizers, 301, 356
　runoff from gardens/lawns, 301
　seawater pollution, 166
　stormwater pollution, 174
　urban agriculture effect on groundwater, 347
　urban streams, 185

453

Index

nitrogen dioxide (NO_2), street tree species and, 284
nitrogen fixation, wetlands, 176–177
nitrogen oxides (NOx), air pollution, 144, 145
node connection, 283, 381
noise, 11
   aircraft, 361–362
   railways, 276, 277
   roads/traffic, 280
non-native species, spread, 67, 68, 369–370, *see also* plant(s) (urban); wildlife (urban)
"non-point sources," pollution, 154
nursery plants, 210
nutrients, *see* mineral nutrient(s)

oak woods/species, 8, 218
"oasis effect," 136
office centers, 333–334
offshore breeze, 140
omnivores, 242
onshore breeze, 140
organic matter, 106
   decomposition, *see* decomposition, organic matter
   sewage/wastewater, groundwater pollution, 155
   urban soil, 102, 104, 106
   carbon-to-nitrogen ratio, 110
organic wastes, 118, 347
organisms, 3, 12–14
   underground, 121–122
   urban soil, *see* soil (urban)
   see also microorganisms; wildlife (urban)
orthogonal grid models, 49
Osaka (Japan), 260–261, 262, 266–267, 269
Ottawa (Canada), green corridors, 363–364
outward expansion, *see* urbanization
owls, 249–250, 262
Oxford (Ohio, USA), songbirds and vegetation, 251
oxidation–reduction balance, 103–104
oxygen
   dissolved, low in wastewater discharge, 167
   levels in groundwater, 157
   root requirement for, 106
ozone, 130, 218
ozone smog, 144, 147

parasitism, by invertebrates, 254
Paris (France), 120, 123, 260, 309
park(s) (city), 344, 349–353, 369
   biodiversity, factors associated, 369, 370–371
   birds, 351, 352
   connected, integrated system, 373
   development, "human-molded vegetation," 86
   distribution, 17
   edges, 69
   functions/benefits, 349, 350
   habitat diversity, 208, 351
   large, 351
   models of urban people and nature, 28
   plants, 351, 352
   scattered, as system, 366
   shape, and adjacent areas, 351
   sizes, 137, 350, 351, 369
   spatial patterns, 351–352
   stormwater runoff control, 371
   temperature and moisture, 352
   trees, 350, 351
   wildlife, 351–352
park system (connected parks), 366, 373–374
"parkways," walking, 366
particulate matter (PM) *see under* air pollutants/pollution
pastureland, 345
patch(es), 11, 34, 44
   characteristics, 44, 50
   dispersed, urbanization model, *see* dispersed patches model
   edge effect, 239–240
   house plots, 294–295
   land mosaic, 50
   occupancy, 238
   shape, 240, 381
   species richness, 384
   urban-to-rural gradient, 42
patch-centered mosaic, 56–57
patch-corridor-matrix model, 44–45, 48, 238–239
   house plots, 294
   uses, 44
pathogens, wastewater discharge, 167, 168
patio(s), 292, 319
   plants, 319–320
pattern perception models, 49
patterns, urban ecology, 11–12
pavements
   permeable, 172, 288
   porous, 172, 288
   see also impervious surfaces
peak flow, 172–173, 194, 383
pedestrian walkways, 122–123
"pedon," 96
people and activities, *see* human activities; human needs
percolation and percolation test, 103, 169
peregrine falcon (*Falco peregrinus*), 249, 256–257, 270, 287, 331
perforation, habitat change due to, 88–89
perimeter-to-area ratio, 60
peri-urban, definition, 7, 39
peri-urban area, 13, 39–40
   adjacency effects, 52
   built objects, 13
   changing patterns, urbanization, 322–324
   description (European cities), 321
   land patterns, 35–36
   residential area, 321–326
   small agricultural sites, 346
   soil, fertilizer, and contamination, 118
   spatial patterns, 35–36, 324–326
peri-urban development, compact, 321
permeability, groundwater, 157
pest(s), 19, 23
   plant/trees, 23, 219
   wildlife (urban), 24, 265
pesticides, 23
gardens, 299
golf courses, 355
growing food and, 347
lawns, 300–301
   runoff into water bodies, 301
in urban rivers, 189
in urban soil, 115–116
pH, soil, elevated, 94
phenotypic plasticity, 270
Philadelphia (USA), 270
Phoenix (USA), 78, 88, 160
phosphorus (P), 114–115
   excess in wastewater discharge, 167
   fertilizers, 301, 356
   reduction, constructed basins effect, 181
   in runoff from gardens/lawns, 301
   stormwater pollution, 174
   urban agriculture effect on groundwater, 347
   urban streams, 185
   wetlands, 176
physical environment, 4, 7, 14
phytoremediation, 181, 217, 360
phytosociology, 205
pigeon (common) (*Columba livia*), 252, 253, 263, 271, 331
   daily fluctuations in Venice, 268
pigeon species, urban areas, 252
pipe systems, *see* sewage system
Pizen (Czech Republic), 235
planned built communities, 326
planning, urban/city and regional, 26, 29, 62–63, 82
   cities, 31, 64, 81, 82–83
   metro area, 82
   towns, 63
   urbanization, 82–83
plant(s) (urban), 205–240
   adaptations, croplands vs. disturbed wasteland, 221–222
   boundaries (between communities), 231–232, 239–240
   on buildings, 307–309
   choice/selection, factors considered, 210
   city centers, 330
   climate-related genetic diversity, 222
   colonization, cracks in hard surfaces, 290–292
   community dominance, 232
   community structure and dynamics, 230–238
      change/dynamics, 234–238
      spatial, *see* spatial structure *below*
      see also ecological succession; time scales
   cropland vs. built-area, 221–222
   density, in city zones, 213, 214
   diseases, 219
   dispersal, corridors, 240
   dispersion, 227
   distribution in urban areas, 213–214
   diversity *see* plant(s) (urban), species richness (diversity)
   emitters of chemicals, 218
   extinction rates, 236
   feralization, over centuries, 236

"floristic similarity," 238
flowering, 211, 278
as generalists, 216
genetic variation, 216
green roofs, 312
groupings, 210
grown for food in urban areas, 346–347
growth inhibition by wastes, 216
habitat fragments, 238–240
herbivory defenses/prevention, 218–219
high plasticity, 222
high uptake/assimilation, 217, 218
indoor, 307
industrial sites, 339
informal squatter settlements, 318–319
invasive, 213, 215, 235
microclimate effects, 216
movements, 15–16, 66–68
native species, 213, 215, 218
   elimination, processes, 214
   gene exchange, 221
   house plots, 296
   time scale for change, 235
natural community, 232
natural hybrids, 236
naturalized, 213, 235
non-native species, 213, 218
   archaeophytes and neophytes, 213–214
   in cities, 213
   herbivory/herbivores, 218, 219
   house plots, 295, 296
   invasion over decades, 235
   mixtures with native species, 214–215
   seed dispersal, 220
   time scale for change, 235
   variation by habitat, 214
normal response curve, 215–216
nursery, 210
origins, 212–213
ornamental, 210
in parks, 351, 352
patios/courtyards, 319–320
pests, 23
phenology, 234, 235
in planted areas, species interactions, 232
preadaptations, 222, 227
productivity, in house plots, 296
railways and, 277–278
resistance and resilience, 216, 217, 222
response to urban stresses, 217–218
roots, *see* root(s)
soil types effect, 216
spatial structure, 230, 231–234
   change over time, 234–236
   horizontal pattern, 231
   vertical stratification, 231–232
species
   green roofs, 312–313
   industrial sites, 339
   planted, in greenspace, 209–210
   spontaneous, in greenspaces, 210
species mixtures (native/non-native), 214–215
species patterns, 230
species pool, 238
species resistance/assimilation variations, 218
species richness (diversity), 232–234, 266
   adjacency arrangement pattern, 56
   adjacency effect, 239
   boundary type and, 239–240
   city centers and, 330
   forest patches in city, 266
   forest size and adjacent land use affecting, 52
   habitat area/size, 232–233, 239
   habitat type and, 233
   Harvard University (Boston), 233–234
   house plots, 295–297
   human design/maintenance efforts and, 232, 233
   planted species, 295, 296
   on vacant lots, 360–361
taxonomic groups, 210–212
tolerance to chemicals, 217
tolerants, inhibitors and facilitators, 237
transport by vehicles on roads, 286
trees' ecological roles for, 224
types, 208–210
in urban soil, 105–108
on walls, *see* wall plants
see also tree(s)
plant biology, 215–223, *see also* herbivory; pollination; seed dispersal
plant ecology, physiological, 215–218
   plant responses to environment, 215–217
   species functions, 217–218
plant genetics and adaptations, 220–223, 236
   factors favoring/limiting change, 221
plantations, in urban areas, 229
"point source," 154
poles and pole arrays (docks), 200
pollination, 220, 221, 234, 254
pollinators, 299, 347, 348
pollutants
   air, *see* air pollutants/pollution
   estuaries, 201
   stormwater, *see* stormwater
   street dust, 281
   urban rivers, 189
   urban soil, 94, 100
pollution
   air, *see* air pollutants/pollution
   cleansing processes of water bodies, 174, 189
   coastal zones of cities, 198
   commercial strips, 334
   environmental economics and, 18
   estuaries, 201
   groundwater, 154–155
   harbors, 201
   industrial areas, 337–338
   neighborhood retail center, 332–333
   ponds, 178–179, 182
   railways, 277
   rivers, *see* river(s) (urban)
   seawater, nitrogen, 166
   stormwater, *see* stormwater
   urban agriculture and, 347
   water bodies, *see* water bodies (urban/local)
   wetlands, 177
polychlorinated biphenyls (PCBs), 189, 201
polycyclic aromatic hydrocarbons (PAHs), 116, 189, 218
pond(s), 178–179
   bioretention, 180
   detention (catch basins), 180
   factors controlling characteristics, 178
   on golf courses, 356–357
   house plots, 303
   pollution, 178–179, 182
   stormwater pollution management, 155
   stormwater runoff, 178
   types, 179
   vegetation, 178
pond-systems, wastewater aquaculture, 166
population density, 36
   exurban, 40
   metro areas, 60
   rural, 40
   suburban, 40
   urban, criteria/definition, 36, 40
   urban regions, 60
population genetics, 272
population size, of cities, 4–5, 76
   optimum and limits, 5, 76
   USA cities, 5
porosity, groundwater, 157
Portland (Oregon, USA), 354, 367
positive attributes, urban region, 379
potholes, 288
prairie dogs (*Cynomys*), 142–143
preadaptation, 222, 227, 270
predators
   escape from, animal movement, 264–265
   invertebrates, 254
   mammal, 242–245
      daytime distance from shrub/forest, 261
      mid-size predators, 242, 243–244
      top predators, 242, 243, 244, 259, 261
prime footprints, 30, 63
private spaces, 292
protozoa, in urban soil, 109
public health issues, 23, 24–26
   air pollution, *see* air pollutants/pollution
   home, plants and sprawl, 25
   low-income residential areas, 318
   urban ecology principles use, 28
   water-related, 24–25
   wildlife-related, 25
public land, 38
public transport, 17, 286
Puerto Rico, 90, 367

Q^* (net all-wave radiation), 131, 135, 382
quarried limestone tunnels, 123
quarries, 355–356

rabbit species (*Lepus, Sylvilagus*), 245–246
raccoons (*Procyon lotor*), 244

Index

radiation, 126, 129–133
　heat in groundwater and, 155
　"sky" (diffuse), 131
　solar, 129–131, 155, 382
　wavelength shift (short- to long-), 132
radioactive isotopes, in air, 143
radius gradient, 12
radon, 116
rail beds, 276, 277
rail yard, 276
railways, 276–279
　barrier/filter effect on wildlife, 278
　corridors, 276, 277, 278, 364
　　land surrounding, 278–279
　disused, walkways, 279
　ditches alongside, 276, 277
　ecological succession, 237, 277
　noise and vibration, 276, 277
　pollution, 277
　spores, seeds and species distribution, 278
　vegetation and plants, 277–278
　wildlife and, 277, 278, 279
rain, 126, 150, 170, see also stormwater
"rain-garden," 180–181
rainwater runoff
　flooding, 194
　green roofs, 311–312
　see also stormwater runoff
rank-size rule for city population, 381
raptors, 247, 249–250, 254
　artificial structures as habitats, 249, 254, 256–257
rare species, 259–260, 362
rats (*Rattus*), 245
rectilinear networks, 45
recycling, 118, 359
　urine, 169
　water, 160, 165, 169
regions (geographic), 58–60
regulatory economics, 18
relative humidity, 126
religion, urban ecology principles use, 29
reptiles, 253–254
　changes over decades/centuries, 270
　characteristics, habitat requirements, 253, 254
　decline with habitat changes/loss, 253, 260
　diversity in arid areas, 253
reservoirs, 158, 159
　sediment in, 158
　stormwater pollution, 203–204
residential areas, 314
　city *see* city residential areas
　commercial area mixed, 268, 314, 315
　green networks, 366
　low-income areas, 316, 318, 319
resilience (to disturbances), 74, 75
resistance (to disturbances), 74
resolution, 32
restaurants, 328, 331
retail business areas, 38, 334
retail center, neighborhood, 332–333
retention basins, 179
Rhododendron punticum, 219
ribbon development, 286

ridge networks, 45
Rio de Janeiro (Brazil), 340
riparian vegetation, 193
　urban rivers, 187–188, 190–193
　urban streams, 183, 185–186
riparian wetlands, 176
riparian zone, disconnected from groundwater, 196
ripple effects, 235
river(s) (urban), 186–193
　algae, herbivores and animals, 188
　benefits, 186, 189
　bottom, 187–188
　channel and water flow, 187–189, 195
　　high flows, *see* flooding
　　low flows, 187, 188–189, 196–197
　channelization, 187, 195
　close human interactions, 190, 191
　fish, 190
　groundwater animals around, 157
　habitat heterogeneity, 190
　migration, 187
　pollution, 189–190
　restoration, 187
　sediment, 187–188, 200–201
　species limited by water flow or pollution, 189
　vegetation, *see* riparian vegetation
　vertical and horizontal dimensions, 187–188, 190
　water quantity/quality patterns, 188
River Thames (London), 187
riverbanks, 187, 190, 191
riverside area, urban, 57, 191
　changes during urbanization, 87
　environmental conditions, 192
　human activities along, 190–191
　natural land uses, 191
　structures, 190–193, 195
　types, positive/negative roles, 191–193
riverside infrastructure, 190
road(s), 279–287
　as conduit for animal movement, 364–365
　extension of effects on sides, 280–281
　moving objects, 285–286
　noise from traffic, effects, 280
　pollution from, 281, 287
　related features, 286–287
　surfaces, 287, 288
　types, 281
　　ring roads, 287
　　streets, *see* street(s)
　　urban highways, 280
　　urban/suburban, 280–281
　widths, 281, 282
　wildlife crossing, mitigation structures, 280
　wildlife population reduction, 280
road ecology, 279
road kill, 245, 246, 280, 286
road networks, 282–284
　patterns/features, 282–284
　types/forms, 283, 284
"road-effect zone," 57
roadsides, 284–285
rock(s), 98, 99

seawalls, 202
rock cliffs, 308, 309, 355–356
rocky coasts, 202
rodents, 25, 245, 359
Roman cities, aqueducts for clean water, 159
Rome (Italy), 84–85, 87, 350
roofs, green, *see* green roofs
root(s), 105, 106–108
　functions, 108
　growth and requirements for, 106–108
　nutrient/mineral absorption, 106, 114
　sizes (large, fine), 108
root hairs, 108
root systems, sizes, 108
"r-selected" species, 216
rubble, 116–117
ruderal vegetation, 207, 221–222

salt marsh, 199, 201, 203, 235
saltwater, 76, 153, 166, 202
saltwater lagoons, 198
San Diego (California), 242–243, 260
sand, 37, 95, 101–102, 158
sand quarry, 356
Sao Paulo (Brazil), 327
saprophytes, 109
satellite cities, 77, 78, 235, 324
scales, 32
　domain of, 33
　see also spatial scales
sea-grass beds, 200
sea-level rise, 203
Seattle (USA), 40–41, 43
seawalls, 202
seawater, 76, 153, 166, 202
sector models, 49
security (defense), 36
sediment
　estuary bottom, 200–201
　lagoons, 200
　reservoirs, 158
　urban rivers/streams, 167–168, 187–188
　in wastewater discharge, 167–168
sedimentary rocks, 98
Sedum, green roofs, 312, 313
seed bank, 236
seed dispersal, 220, 254, 278, 286
self-locomotion, 66
"self-organizing principles," 51
semi-natural greenspace/area, 7, 8, 353–355
　habitat heterogeneity, 353
　parks (city), 351
　resistance to disturbances, 74–75
　woodland, 353–354
septic systems, 168–170, 304
　advantages/disadvantages, 170
　wastewater treatment problems, 169
septic tank, 169
sewage, 118, 123, 151, 160–170, 304
　back-yard outhouse toilets, 302
　city centers and, 327
　components and organisms, 161
　effectiveness of treatment, 164
　primary treatment, 164
　secondary treatment, 164
　septic systems, *see* septic systems

456

Index

sludge, 118, 165, 169
   tertiary treatment, 165, 166
sewage system, 151, 161–166
   combined with stormwater system, 163, 167
   global availability, 160–161
   pipe systems, 161–164
      access to and problems, 162, 163
      decomposition processes in, 164
      separate from stormwater system, 163–164, 166
   septic systems vs., 170
   treatment facilities, 161, 164–165
      neighborhood, 165
   wastewater discharge effect on water bodies, 165, 166–168
sewage wastewater network, 123–124
shade
   city centers, 332
   green walls, 282
   streets, 281, 282
   from trees in cities, 216–217
shade corridors, 130, 131, 284
shadow lengths, 382
Shanghai (China), greenspaces, 85
Sheffield (UK)
   flowering plants, 211
   habitat diversity in parks, 351
   house plots, 293–294, 296, 297, 300
shopping malls, 335–336
shops, 328, 332–333
shorebirds, changes over time, 269
short-wave radiation, 130, 131, 132
shrinkage, habitat change due to, 88–89
shrub layer, 231, 258
shrubs, in urban areas, 229–230
   distribution (Barcelona), 230
   diversity, 229
   house plots, 303
   large wooded patches, 229
   roles, 229, 230
   spontaneous growth, 229, 230
   wildlife and, 258
side-boundary vegetation strips, 324
silt, 37, 95, 98, 101–102
   in reservoirs, 158
silviculture, 355
Singapore, 310, 369
sinuosity ratio, 384
skunks, striped (*Mephitis mephitis*), 244
sky conditions, 126
"sky" radiation (diffuse radiation), 131
sky view factor, 130, 382
skyscrapers, 310, 327, 328
slope of ground, 37–38, 98
sludge, sewage, 118, 165, 169
slugs, 112, 297
slums, 316
snails, 112
social patterns (human), 17–18
societal goals, 377–378
sodium chloride (NaCl), 115, 117, 217–218
soil (urban), 91–124
   aeration, 92
   agricultural "crop," 91
   agriculture effects on, 347

anaerobic conditions, 103, 104, 106
biological/physical components, by size, 107, 108
boundaries between types, 96
bulk density, 102
buried compaction layer, 95
chemicals in, *see* chemical(s)
city center, 329–330
classification of types, 95–96, 101
clay, 95, 101–102
compacted, 91, 93, 94, 330
compaction, 95, 100, 102
   trees resistant to, 105
composition, 91, 92
compressibility, 103
contaminated
   brownfield sites, 360
   detention basins and biofilters, 180
   industrial areas, 118, 339
   Kano (Nigeria), 40–41
core characteristics, 92–94
degradation, 83, 105
designed-and-mixed, 92
development, organic matter role, 106
drainage, 103
under drainfield (septic system), 169
ecosystem development, 105
energy and nutrients from, 106
erosion (by water/wind), 18, 97, 98, 99, 118, 158
fill, 92, 94, 95
food web, 149
functions, 91–92
gases in, 116
for green roofs, 312
horizontal pattern, 92, 95–97
human disturbance effect, 91–92, 96
human structures/artifacts, 93
identification of types, and maps, 96
impervious (hard) surface above, 94
"improved"/"remediated," 92
key natural and human processes affecting, 97–100
   ecological processes, 99–100
   geological processes, 97–99
   human processes, 91–92, 96, 100
loamy, 101, 102
minerals, 91, 92, 95
"natural," 91
organic matter, *see* organic matter
organisms in, 105–113, 114, 121–122
   animals, 111–113
   large invertebrates, 112–113, 114
   microbes, *see* microorganisms
   plants and roots, 105–108
   tiny invertebrates, 112–113, 114, 122
   vertebrates, 111, 122
patterns (associated with human activities), 92–94
permeability and percolation, 103
pH elevation, 94, 114
plant growth, effects on, 216
plasticity and elasticity, 103
pore size distribution, 103
remediation, 181, 360
sandy, 101, 102

shear strength, 103
stormwater infiltration, 93
stratification (A, B, C layers), 94–95
texture, 97, 101–102
texture, properties related to, 102–104
   relative importance (by activity), 104–105
   root growth and, 106
   structural, 102–103
   water-/air-related, 103–104
trees' ecological roles for, 224, 225
types, 91–92, 95–96, 101–102
   Washington, D.C., 96
underground and, *see* underground (urban)
urbanization effect, 83
vegetation and litter, 94
vertical layers (profiles), 92, 94–95
   characteristics, 94, 95
volume change (deformation), 103
water drainage, 92
water infiltration, 103, 104, 151, 185
wetlands, 175, 176
zones/layers (A, B), 91, 92, 94–95, 106
   aerated, groundwater in, 151
zones/layers (C), 94
solar radiation, 129–131, 155, 382
solid waste, 118, 358
solid-waste dumps, *see* dumps (tips)
songbirds, 248, 250–252
   breeding/summer and winter seasons, 251
   crossing railways, 278
   foraging guilds, 251, 257
   habitat affecting density/diversity, 248, 251–252
   migration, parks and, 344
   roosting in trees, 252, 257
   sensitivity to vegetation structure, 251
   in suburbs, 251–252
South Korea, 118, 195, 214, 231
spatial arrangement
   buildings, 306
   in cities, 38–39
   house plots, 293–295, 306
spatial ecological gradients, 34
spatial models
   early models, 46–47, 62
   land-use patterns, 38–39
   patch-corridor-matrix, *see* patch-corridor-matrix model
   random (stochastic), 46
   richness, and model types, 46–49
   simple, 46
   urban-rural gradient, *see* urban-rural gradient, as spatial model
spatial patterns, 12, 26–28, 31–64, 377, 382
   geomorphic framework, 36–38
   greenspaces, 370
   horizontal, *see* horizontal patterns
   house plots, 293–295
   human and nature's, 34–36
   invertebrates, 255
   land-and-sea, 59
   megalopolis, 33
   oblique (diagonal), 32, 59

457

Index

spatial patterns (cont.)
  parks (city), 351–352
  peri-urban (exurban) area, 35–36, 324–326
  plants, see plant(s) (urban)
  scale and, see spatial scales
  universality, 49
  urban regions, metro areas and cities, 58–64, 327
  urbanization, 76–83
  vertical, 31–32
  wildlife habitats, see wildlife (urban)
spatial planning principles, 26–28
spatial processes
  habitat change due to, 88–89
  land recovery/restoration, 88–89
spatial scales, 12, 32–34
  fractals, 34
  hierarchy, 12, 33
  stair-stepped interpretation, 33
  urbanization, 75
spatial sequence models, outward expansion, 82, 83
species, 11
  changes, urbanization effect, 85–86
  "tightness" of interactions, 232
species assemblage, 50, 232
species continuum model, 48
species dominance, 384
species pool, 238, 351
"species rain," 62, 369, 370, 373
species richness (diversity), 14, 89, 232
  birds, see bird(s) (urban)
  fine-scale habitats associated, 208
  habitat type and, 233
  lagoons, 199
  patches, 384
  plants, see plant(s) (urban)
  species–area relationship, 232–233
  suburbs, see suburb
  wildlife, see wildlife (urban)
  see also biodiversity
species source, 265, 351
species–area effect, 260–261
species–area relationship, 232–233, 260–261
spiders, 112
"spillover effect," 265
spores, dispersal by trains, 278
sprawl, 6, 7, 39, 79, 324
  development in forests/woodland, 79, 80
  dispersed patches model, 77, 79
  environmental effects/features, 78–79
  habitats and biodiversity, 86
springtails, 112
squirrels (*Sciurus, Tamiasciurus*), 246
starlings, 252–253, 331
stepping stones, 36, 72, 228, 266
  green roofs, for birds, 313
  house plots as, 293, 294
  plant species dispersal, 240
  wildlife movement and, 266–268
    birds, 266, 267
Stockholm, golf courses, 356–357
storks (*Ciconia*), 256, 257
storms, 185, 188, 194
stormwater, 84, 170–174

discharge rates, 173
drainage systems, 120–121, 170–173
flow patterns, 170–173
groundwater pollution, 154
infiltration into urban soil, 93, 170, 171
pollutants in, sources, 170–174, 180, 198, 281
  clean-water supplies, effect, 174
  USA vs. UK comparisons, 173–174
pollution management, 155
pollution of water bodies, see water bodies
public health issues, 24
  in sanitary sewage system, 162
streets and, 281
structures to contain/absorb, 180–181
use on golf courses, 356
stormwater drain/basin, wildlife habitat, 263
stormwater pipe network, animal movement via, 67
stormwater runoff, 170, 383
  amounts (by surface type), 170–171
  into bioretention basins, 180
  carparks, 287
  control by greenspaces, 371
  estuaries, 201
  first flush, pollution, 173
  flashiness, 173, 184, 185
  green roofs and, 311–312
  greenspaces and woods, 171
  impervious surfaces, 171, 172
  industrial areas, 338, 339, 340–342
  low-income residential areas, 318
  neighborhood retail center, 332
  peak flow, 172–173, 383
  permeable pavements, 172
  ponds, 178
  porous pavements, 172
  small water collectors, residential areas, 172
  swales (drainage ditches), 172, 173, 180, 287
  urban streams, 182, 183–184
stormwater system
  combined with sewage system, 163, 167
  separate from sewage system, 163–164, 166
stream(s), 182–186
  bottom of, 184
  channelization, 184, 195
  daylighting, 186
  definition, and headwaters, 183, 206
  density, 384
  extent in USA, 183
  fish, 186
  flooding, see flooding
  flows, 152, 184, 185
    in storms, 185
    variations and flashiness, 173, 184, 185
    velocity, 184–185, 324, 384
  herbaceous vegetation lining, 185–186
  invertebrates, 184, 186
  networks, 366
  nitrogen/phosphorus levels, 185
  restoration, 186
  temperature, 185
  urbanization effect, 87, 324

wastewater discharge effect, 168
water quantity/quality and biology, 182–183, 185–186
watershed, floodplain and, 182–185
streamline air flow, 65, 127, 140
  isolated buildings effect, 142
  street canyons and, 141
street(s), 281–282
  green walls, 281–282, 309
  heat and shade, 281, 282
  stormwater and, 281
street canyons, 136, 141–142, 281
  air flows in, 141–142, 327
  biodiversity, 282
  tree effects, 143
street dust, 146, 165, 281, 309
street networks, 282–284
"street swales," 180, 287
street trees, see tree(s), street
strip(s), see corridor(s)
strip development, 286
stygobites, 157
stygophiles, 157
subcanopy, 231
subsidence, 152
subsoil, 91
suburb, 38, 39, 321
  bird densities, 249, 261–262
  bird species richness, 262
  definition, 7, 321
  population density, 40
  songbirds, 251–252
  species richness, vs. in cities, 262
suburban area
  changing patterns, 39, 322–324
  development forms/types, 322, 323
  expansion, 321
  gradient from urban, to rural, 41, 77, 262
  land patterns, 35–36
  seed dispersal in, 220
  small agricultural sites, 346
  spatial patterns, 324–326
suburban residential area, 7, 321–326
  inner suburb, 38, 316
    types/forms, 316, 317
  urban areas vs., 324–325
suburbanization, 77, 321
succession, see ecological succession
successional habitats, 14, 15, 85
Sudgelande Nature Park (Berlin), 277, 279
sulfur dioxide, 144
surface depression storage, 383
surface energy balance, 131–133
surfaces
  urban vs. non-urban, effect on heat, 136
  see also hard surfaces; impervious surfaces
"surroundings," importance, 54
sustainability, urban, 29, 322
swales (drainage ditches), 172, 173, 180, 181–182
  street, 180, 287
swamps, 175, 176
Swindon (UK), 346
swine encephalitis, 25
sycamores (*Acer pseudo-platanus*), 219

Index

Sydney (Australia), 81, 84
synanthropic plant species, 52
Syracuse (New York, USA), 259, 261, 366–367
systems ecology, 68
systems models, 46

tarmac (asphalt) surface, 288
   cracks, 288–289
temperature
   green roofs, 311
   groundwater, 155–156
   inversion, 128, 129, 135
   parks, 352
   summer, shade from trees and, 216–217
   urban air, 126, 135
      agriculture effects, 347, 348
      air pollution increase, 147
      increase and global warming, 147–148
      non-urban vs., 136–137
      urban streams, 185
   see also cooling; heat
termites, 67, 112, 267
territory and territoriality, 67, 264
tides, 76
Tiete River (Sao Paulo), 189
Tijuana (Mexico), bird species, 260
time scales, 16
   change-over-time principles, 16, 75–76
   disasters, 76, 234
   natural processes, 234
   outward expansion of urbanization, 235, 322–324
   plant community structure change, 75, 234–236
      centuries and millennia, 236
      decades, 235
      short (hours–seasons), 234
      years, 234–235
   range of (long to instantaneous), 75
   shipping, bridges, cars, transport, 16, 75
   urban changes, 16, 75–76
   wildlife changes, 268–270
      centuries, 270
      daily/seasonal, 268
      decades, 269–270
      years, 268–269
   see also ecological succession
toilets, back-yard outhouse, 302
Tokyo (Japan), 81, 133
topographic features, 97–98
topsoil, 91, 92, 94
Toronto (Canada), 122–123, 204, 217, 244, 333, 367
total nitrogen (TN), 164, 165
total phosphorus (TP), 164, 165
total suspended solids (TSS), 164, 174
town(s), 63–64
   centers, 334
   ecology, 63
   planning, 63
toxic substances, in water bodies, 168
traffic calming, 286
traffic circles, 287
trains, *see* railways
trampling, 239

"transit-oriented development," 316, 373
translocation of animals, to new sites, 265
transpiration, 132, 151, *see also* evapotranspiration
transportation, 16, 17, 373, 375
   in city centers, 327–329
   commuting and time scales for change, 16, 75
   harbors and waterfront area, 198
   heat due to, 132
   human need for, 36
   land use change during urbanization, 87
   motor-powered movement, 66
   seed dispersal by, 220
   time to neighborhood retail center, 332
   underground networks, 120–121, 122
   urban ecology principles use, 28
   see also railways; road(s)
transportation corridors, 77
transportation structures, stormwater runoff, 172
travel time, city size and, 5
tree(s), 125, 223–229
   adjacency effect, 239
   around airports, 361
   assimilation/uptake, 217, 218
   avoidance from dump sites, 359
   benefits in cities, 223, 225
      cooling, 139, 284
      Kano (Nigeria), 225
   bird use in cities, 248, 285
   canopy, 227, 231, 257
   cover, 8, 30
      in USA and Europe, 238
   density, in small wooded patches, 228
   diseases and pests, 219
   distribution and arrangement, 225–229
      arrangement, 227–229
      environmental disasters affecting, 227
      time-related changes (years), 226
   ecological roles, 223–225
      major roles, 224
      minimal roles, 224–225
      minor roles, 224
   ecological succession, 226–227
   in front/back outdoor spaces (house plots), 297–299
   generalists, vs. specialist species, 229
   habitat diversity (Taiwan), 207
   herbivores, 219–220
   house plots, 297–299, 303
   isolated, urban air flow, 142–143
   land use affecting species, 226
   large wooded patches, 228–229, 239
   leaf area index, 217
   minimizing at airports, 361
   narrow wooded corridors, 228
   non-native species, 214, 229, 235
   PAHs assimilation, 218
   in parks, 350, 351
   patterns, in metro area during urbanization, 87
   resistance and resilience, 216
   response to urban stresses, 217–218
   rows of, 227–228
   saplings in shrubby areas, 230

   shade from, 216–217
   single, arranging, 227
   single species in plantations, 229
   in small wooded patches, 228, 229
   soil ecosystem development, 105
   species in Chicago, 214
   species pool, 238
   species resistance/assimilation variations, 218
   street, 284–285, 287
      benefits, 223
      bird species and, 285
      city centers, 330
      cooling by, 139, 284
      mortality, 226
      pollution affecting growth, 284–285
      relative to city population size, 209
      species related to NO_2 levels, 284
   in street canyons, air flow, 143
   urban-to-rural gradient, 226
   vertical stratification, 231
      birds and, 251, 257–258, 285
   volatile organic compound emission, 218
   wide wooded corridors, 228
   see also forest(s) (urban areas); plant(s) (urban); woodland
tree swallow (*Passer montanus*), 250–251
tropical cities, wildlife and vegetation layers, 258–259
trucks, 285–286
tsunami, 202
turbulent air flow, 65, 127, 140
turnover rate, 268–269

UK, house plot biodiversity and coverage, 297, 374
underground (urban), 119–124
   in city centers, 330
   network flows and forms, *see* network flows
   organisms and habitats, 121–122
   rodents, 245
   structures, *see* underground human structures
underground human structures, 93, 116–117, 119–121
   layering/vertical distribution, 121
   networks, 120–121
   types, 119–120
understory layer, 230, 231
urban, definition, 6, 41
urban area(s), xii, 6, 41
   built spaces and greenspaces, 7
   density, changes with urbanization, 79
   ecological succession, 85
   economics, 18
   key concepts, 6–8
   mosaics, 3, 4
   social interactions, 17
   time scale for changes, 75–76
   varying terms in different countries, 6
urban attributes, 11–16
urban backbone, 36, 37
urban boundary layer (UBL), 128, 135
urban canopy layer, 128

459

Index

"urban chemical flows," 68
urban cliff hypothesis, 331
urban ecology, 3
   books on, 10
   breadth of field and uses, 4
   concept, 2–4, 6, 12
   as contrasting concepts, 3
   ecology 'in' vs. 'of' cities, 11
   history of, 4–11
      cities and, 4–8
      current phase, 9–11
      early phases, 9
   key areas to focus on, 11
   major subjects, reflection on and future prospects, 372–378
   principles, for society's solutions, 26–30
   research areas and centers, 10, 11
   for society's solutions, 26–30, 377–378
      for big solutions, 29–30
   use in key disciplines/professions, 28–29
"urban energy flows," 68
urban environment, cultural/regional aspects, 26, 27
"urban forest," concept, 355
urban form, 36, 62
urban form models, 48–49
urban greenspace system, *see* greenspace system, integrated
urban growth boundary, 48, 364
urban hierarchy, 33
urban metabolism, 68
urban objects of ecology study, 3, 12–16
urban regions, 3, 57, 58–64
   agricultural land around metro area, 60, 62
   boundary, flows across, 63
   criteria for mapping/boundaries, 60
   definition, 6, 7, 60
   as ecological mosaic, metro area and, 51, 57, 60–63
   examples, 60
   inner ring, natural land, 62
   planning, *see* planning, urban/city and regional
   population density, 60
   regions surrounding, 60, 62
   spatial features, positive/negative effects, 63
"urban revolution," ix
urban sprawl, *see* sprawl
urban sustainability, 29, 322
urbanization, 2, 4, 6, 76–83, 373
   air changes, 83–84
   China's Changjiang Region, 84
   definition, 7, 76, 321
   ecological change rates, *see* ecological change, rates/trajectories
   ecological phases, 322–324
      altered wildlife pattern phase, 322
      metapopulation, disrupted water phase, 322–324
      scattered habitat phase, 324
   edge model, 82, 83
   habitat loss/degradation and fragmentation, 88–90, 322, 324, 349
   history of concept, 8
   house plot sizes and, 324
   internal changes, 79–81
   "jaws-and-chunks" model, 82, 83
   land use changes, 86–88
      broad scale, 87–88
   natural land, 90
   outward expansion, 7, 76–77, 82, 322, 372–373
      distribution limits for species, 86
      highway bypass, effects, 88
      internal change affecting/affected by, 80–81
      patterns and time scale, 235, 322–324
      spatial sequence models, 82, 83
   outward expansion models, 77–79, 235, 324, 377
      combination of models, 78
      environmental effects, 78
   planning and optimal expansion, 82–83
   soil degradation, 83
   spatial patterns, 76–83
   spatial scales and, 75
   succession and species change, 85–86
   transformation patterns, 81
   vegetation change, 81, 84–86
   water and mineral nutrient flows, 83, 84, 322
urban-region rings, 62, 261–262
urban–rural gradient, 10–11, 40, 42, 43, 49, 262
   challenges and end points, 43
   characteristics measured, 42–43
   city, suburb, peri-urb, farmland and natural land, 38–41
   ecological gradients, 42–43
   as spatial model, 38–43
urban–suburban–rural gradient, 41, 77, 262
urban-to-rural concept, 40, 41
urine, 169
   dog, 20, 21, 217
USA
   air pollution from traffic, 146
   bird mortality due to vertical built structures, 256
   cities and population size, 5
   "food shed" and urban agriculture, 349
   golf courses, 356
   greenway systems, 367
   tree cover, 238
   urban streams, 183

vacant lot ecosystem, 85
vacant lots, 359–361
   building on (infill), 361
   plant diversity, 360–361
valley bottom, 98
vectors, 66, 67
vegetable gardens, 86, 299–300, 346
vegetables, growing, 346–347
vegetation, 205–240
   along rivers/streams, *see* riparian vegetation
   backline strip, 294, 295
   change during urbanization, 81, 84–86
   changes after city abandoned, 268
   in city centers, 327
   classification, 206
      grouping-by-mechanism, 206–207
      by origin, 206, 207
   coastal, time scale of changes, 235
   disturbed along roadsides, 285
   dynamics, *see* ecological succession
   environmental gradients and, 42
   in greenspaces and corridors, 81, 343
   herbaceous, along urban streams/ditches, 185–186
   "human-molded," development sequence, 86
   layers/vertical structure, 231–232
      birds and, 251, 257–258, 285
      wildlife and, 251, 255, 257–259
   planted, in urban greenspace, 209–210
   ponds, 178
   railways and, 277–278
   removal, flooding associated, 194
   riparian, *see* riparian vegetation
   songbirds and, 251
   stability, 234
   succession, *see* ecological succession
   types, 205–207
   urban areas, 84–86
      increase, to ameliorate heat in cities, 139
   on urban soil, 94, 105–108
      ecosystem development, 105
   wetlands, 175
vehicles, 30, 285–286
   types, 285–286
Venice (Italy), 153, 268
ventilation
   airport-surrounding areas, 362
   cities, 127–128, 139
   ground level, burrows, 142–143
vertebrates (urban), 241
   diversity/abundance in large greenspaces, 354
   invertebrates as food source, 255
   pests, 24
   species richness, habitat fragments and, 261
   underground, 122
   in urban soil, 111
vertical gradient, 12
vertical pattern, 31–32
   plants, *see* vegetation, layers
   soil, *see* soil (urban)
Vienna (Austria), 157, 281
villages, city development, human needs, 36, 212, 372–373
vines, 230, 303, 308, 309, 330
violets, genetic variation, 221
vireo, spatial habitat patterns, 260, 266
viruses, in urban soil, 109
visibles, in urban ecology, 1, 2
volatile organic compounds (VOCs), 144, 146, 218, 287
volcano, 76
vortex air flow, 65, 140, 141

Waitakere City (New Zealand), 367
walking, in city centers, 327, 328
walking routes, 286

Index

walkways, 288
wall(s)
   cracks, 289–290
   garden, 304–306
   green, see green walls
wall plants, 304, 307–308, 309
   ecological functions/benefits, 309
   heights and wildlife values, 307, 308
   pollution reduction, 309
   species, source and locations, 307–308, 309
   see also green walls
warehouse truck distribution center, 336, 337
wars, 376
Warsaw (Poland), wildlife, 259
Washington, D.C. (USA), 96, 366, 369
waste
   city centers, 328, 330
   industrial, 154, 189, 338
   plant growth inhibition by, 216
waste site, groundwater pollution, 154
wasteland, 207, 221–222
wastewater, 151, 160–170
   aquaculture for food, 165–166
   commercial, 161
   discharge, effects on water bodies, 165, 166–168
      anaerobic conditions, 167
      estuaries, 201
      microorganisms, 168
      nutrient excess and algal blooms, 167
      sediment excess, 167–168
      toxic substances, 168
   groundwater pollution, 154
   industrial, 161
   in low-income residential areas, 318
   partially treated, in rivers, 189
   public health issues, 24–25
   reused, for potable water, 160, 165
   septic systems, 169
   sewage, see sewage
   treatment
      failure in septic systems, 169
      natural systems, 165, 166
      partial, sludge from, 118
      in water bodies, 167, 181
   treatment facilities, 161, 164–165
   types, 161
wastewater systems, 161–166
   pipe systems, 161–164
water, 204
   absorption by roots, 106–107
   agriculture (urban) effect, 347
   aquifer, uses, 152
   capillary, 106
   city centers and, 327
   clean, supply (urban regions), 18, 150, 151, 158–160
      aqueducts (Roman), 159
      infrastructure to provide, 159
      low-income areas, 318
      stormwater pollutants effect, 174
      supply problems in cities, 160
      transport of supplies, 159
      from urban rivers, 190
      users and usage, 158, 161

      see also water, drinking
   content of urban air, 126–127
   drainage, soil types, 92
   drinking, 154
      contamination, 25
      groundwater, 151–152, 158
      reuse, 160
      sources, 158
      surface-water supply, 158
   erosion of soils, 97, 98, 99, 118, 158
   groundwater, see groundwater
   for growing food, 347
   human need for, 16–17, 36
   infiltration into soil, 103, 104, 151, 185
   inter-basin/inter-regional transfers, 159
   for lawns, 301–302
   potable-water reuse, 160
   quantity/quality
      cities, 160
      urban rivers and, 188
      urban streams and, 182–183, 185–186
   recycling, 160, 165
   runoff, house plots, 303–304
   "surface runoff," 151
   surface-water supply, 158
   trees' ecological roles, 224
   underground networks, 120–121, 121, 122
   urbanization effects, 83, 84
   uses (household), 161
   volume (average) used per household, 160
   waste, see wastewater
water bodies (urban/local), 175–204
   constructed basins, ponds, wetlands, see constructed basins; pond(s); wetland(s)
   "not swimmable or fishable," 168, 189–190
   in parks, 351
   pollution by wastewater discharges, 167
   rivers, see river(s) (urban)
   stormwater pollution, 173–174, 182
      constructed basins, 180, 181–182
      estuaries, 201
      lakes and reservoirs, 203–204
      processes to reduce effects, 174
      quantity/quality effects, 174
      rivers, 189–190
      streams, 182–183
   streams, see stream(s)
   toxic substances in, 168
   urban ecology solutions, 30
   wastewater discharge effects, see under wastewater
   wastewater treatment/clean-up, 167
   waterbirds, 250
   wetlands and ponds, see pond(s); wetland(s)
water budget, 383
water cycle, 150–151
   key characteristics, 151
water flows (urban), 15, 66, 150–151
   groundwater, 152–154
   house plots, 303–304
   non-urban vs. urban flows, 149
   rivers, 187–189

   stormwater, 170–173
      peak flow, 172–173, 194
   stream, to groundwater, 152
   streams, 152, 183, 184
water pipes, 159, 160
   leaks, ecological effects, 159–160
water pollution, 19
   airports, 362
   dog waste and, 21
   groundwater, 154–155
   stormwater, see under water bodies (urban/local)
   stormwater runoff and, 173–174
   strategies for, 375
water resource management, 28
water systems, 149–174, 374–375
   tightening, future prospects, 374–375
   urban vs. non-urban, 149
water tanks, 159
waterbirds, 247, 250
   habitat types, 250
   ponds on golf courses, 356–357
   water depth/salinity effect, 250
water-borne diseases, 168
waterfront, city, 198, 204
water-related changes, urbanization, 322–324
water-related diseases, 168
water-related public health issues, 24–25, 318
water-related structures, in house plots, 303–304
watershed, urban streams, 182–185
water-supply pipes/mains, see water pipes
water-table, 151, 152
   above/below, wastewater leakage from pipes, 162–163
   groundwater pumping reducing, 152, 153, 196
   land subsidence and, 152
   rise in, 153, 196
   urban streams and, 185
water-treatment facilities, 159
weathering of rocks, 98, 99
weeds, 212–213, 296, 299, 347
wells, 158, 304
wetland(s), 18, 149, 175–177
   bioretention, 180
   characteristics, 175–176, 177
   coastal, 175, 176, 177, 200
   constructed, on golf course, 356
   freshwater tidal, 176
   functions, 176–177
   greenspaces and corridors, 343
   groundwater, 176
   habitat heterogeneity, 177
   house plots, 303
   lagoons and estuaries, 199
   microorganisms, 176
   riparian, 176
   surficial, 175
   waterbirds, 250
wetland boundary, 175
wild pigs, 244–245
wildfires, 354–355
"wildland," 52

Index

wildlife (urban), 241–273
  avoidance behaviors to dogs, 20
  changes and adaptations, 157, 268–273
    genetic adaptations, 270–273
    multi-scale changes, succession, 268–270
    population density, 269
    time scales, *see* time scales
  changing to new habitat, 20, 156–157
  city centers, 330–331
  commercial strips, crossing, 334
  cover for, 241–242
  disease risk from dogs, 21
  dog walking impact, 20
  ecological succession, 268–270
  feeding, 264
  garbage as food source, 264
  geographic range of species, 265
  habitat heterogeneity and, 259,
    see also wildlife (urban), spatial habitat patterns
  home range, 264
  house plots, 295–297
  mortality
    railways, 278
    road kill, 245, 246, 280, 286
  movement, 264–268
    airports and, 361
    along corridors, 240, 265, 364, 367, 368, 369
    altered patterns, urbanization phase, 322
    distance, 368
    ecological network, 367
    facilitator/inhibitor patterns, 266
    road networks and, 284
    species sources and adjacencies, 265–266
    stepping stones, 266–268
    into urban areas, factors affecting, 370
    urban patterns, 265–268
    see also animal movements; bird(s) (urban)
  native species, 243–244
  loss, time since isolation in greenspaces, 242–243
  natural areas/land, 265
  needs/requirements, 241–242
  non-native species, 265
  in parks, 351–352
  pests, 265
  population reduction by roads, 280
  protected species, 260
  public health issues associated, 25
  railways and, 277, 278, 279
    noise and vibration effects, 276, 277
  spatial habitat patterns, 259–263
    animals of specific habitats, 262–263
    city, suburb, peri-urb, urban-region ring, 261–262
    greenspace patch size, 260
    habitat fragments, 260–261
    rare species, 259–260
    species richness, 259–260
  species richness, 242, 243, 259–260
    mixed-use land, 268
  species source, and adjacencies, 265–266
  species types, 242–255, *see also individual species and animal groups*
  streets, 282
  suburban, 39
  territories, 264
  traffic noise effects, 280
  urban ecology solutions, 30
  vegetation layers, 251, 255, 257–259
  vertical built structures, 255, 256–257
    as habitats, 256
    as hazards, 256
    see also animal(s)
wildlife underpasses/overpasses, 280, 286
Wilmington (Delaware, USA), 90, 353–354
wind, 140–141
  seed dispersal, 220
  soil erosion, 97, 98, 99, 118
  streamline, *see* streamline air flow in street canyons, 141–142
  ventilation of cities and, 127, 128
windbreaks, 140–141, 227
  distance between, and location, 141, 142
  height and porosity, 140–141, 142
windspeed, 140, 142, 383
wood fiber, 36
woodchucks (*Marmota*), 246
wooded corridors, 343
  narrow, 228
  wide, 228
woodland, 353–355
  area, predictive of bird diversity, 249
  in cities, 209
  conversion to built land, floods and, 194
  ecological succession, 85, 237
  foliage layers (2, 3, or 4), 231
    four-layer, 231
    two-layer, 231
  for forestry, 355
  in greenspaces and corridors, 343
  herbivory distribution, 219
  large wooded patches, 228–229, 239
  layers/strata, 208, 231
  mixed-origin, Berlin, 215
  regenerated, 231
  seed dispersal, 220
  semi-natural areas, 353–354
  small wooded patches, 228, 229, 266
  stormwater runoff, 171
  strips, wide wooded corridors, 228
  see also forest(s) (urban areas); tree(s)
Woodlands, The (Texas), 63
woody plants
  bird feeding, 303, 304
  ecological succession in cracks, 291
  house plots, 303, 304, 305
Wye Valley (UK), cliffs and succession, 356

xeroscaping, 292

yard waste, 118, 300, 302
yards, 292

zones of influence model, 46
zoonotic diseases, 25

Printed in the United States
by Baker & Taylor Publisher Services